1 MONTH OF
FREE
READING

at
www.ForgottenBooks.com

By purchasing this book you are
eligible for one month membership to
ForgottenBooks.com, giving you
unlimited access to our entire
collection of over 1,000,000 titles via
our web site and mobile apps.

To claim your free month visit:

www.forgottenbooks.com/free1002727

ISBN 978-0-331-01414-3
PIBN 11002727

U. S. Department of Agriculture
Forest Service Research Paper SO-83

LABOR TRENDS IN SOUTHERN FOREST INDUSTRIES, 1950 TO 1969

tracts,

Lloyd C. Irland

By raising wages more rapidly than the rest of the Nation, and by expanding employment in the high-wage paper industry, southern forest industries helped improve economic welfare in the rural South. Forest industries in the South provide 2.4 percent of the region's total personal income. They employ 32 percent of the Nation's forest industry workers and 13 percent of the South's manufacturing labor force. From 1950 to 1969, employment fell in lumber and furniture making, while rising strongly in the paper industry.

In recent decades, change has been swift in America's forest industries, and the pace has been most rapid in the South. Rising costs of stumpage and labor since 1950 have promoted mechanization to improve productivity and wood utilization. This trend has altered the industry's status as an employer and contributor to the southern regional economy. This paper describes prominent labor trends in the region's forest industries. It gives recent trends in total employee earnings, employment, and wages. Productivity gains are compared with wage improvements.

OCT , 1973

Per capita personal incomes in the South have long lagged behind those in other regions. In 1971, per capita income in the United States was $4,138. In the South, Virginia had the highest, $3,866; and Mississippi the lowest, $2,766 (fig. 1). Partly as a result of rapid income growth in the South, however, the gap has been closing (Foster 1972, Toal 1972). From 1950 to 1969, per capita real personal incomes in the South rose by 151 percent compared to 120 percent for the Nation as a whole (fig. 2). All States save Mississippi, Arkansas, and Oklahoma exceeded the national rate. Rapid growth in population and manufacturing made Florida, with an increase of 318 percent, a leader in the Nation.

To assess the contribution of the region's forest industries to this progress, employment, earnings, and weekly wages were examined

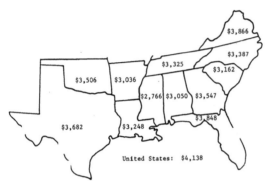

Figure 1.—*Per capita personal incomes in Southern States in 1971 (Source: Bretzfelder, 1972).*

Figure 2.—*Percentage increases in per capita personal incomes, 1950 to 1969.*
* States in which personal incomes grew slower than national average.

by industry groups in the Standard Industrial Classification (U. S. Bureau of the Budget 1957). Industry groups were:

Lumber and wood products	(SIC 24)
Furniture and fixtures	(SIC 25)
Paper and allied products	(SIC 26)

Lumber products and furniture   (SIC 24 + SIC 25)
Forest industries   (SIC 24, 25, 26).

Earnings data were only available for the total of lumber products and furniture.

Kaiser and Dutrow (1971) provide southwide data by SIC 4-digit industries. They report value added and employment trends for 1958 to 1967 in the South relative to the rest of the Nation. Granskog and Guttenberg (1973) estimate employment in logging.

tracts,

For purposes of this paper, *earnings* are defined as the total annual compensation of employees and proprietors; included are wages and salaries, other labor income, and proprietor's income. *Employment* is defined to include production and nonsupervisory workers. Thus, earnings and employment data are not strictly comparable. *Wages* are average weekly earnings rather than hourly earnings of production workers. Since weekly earnings account for differences in length of the work week, wages as reported are close in concept to the annual earnings measure adopted. Since weekly earnings apply only to production workers, however, they are similar in coverage to the employment data.

National importance.—The South is a major timber-producing region. From 1900 to 1929, it led the Nation in the production of lumber. As lumber production waned, new developments in pulping led to creation of a giant southern paper industry. By 1940, southern pine was the leading pulpwood species in the Nation, and in 1970 the South produced 63 percent of the Nation's woodpulp. A favorable timber resource outlook suggests that the region's role will increase in the future.

In 1969, U. S. Forest industries employed 1.8 million workers; 584,500 or 32 percent of these were working in the South (table I). The lumber and wood products sector was heavily concentrated in the South; in 1969, 39 percent of its workers were employed there. The southern share of paper employment is lower—only 23 percent. The low proportion seems striking, since the South consumed 62 percent of the Nation's pulpwood harvest in 1970 (U. S. Bureau of the Census 1972). The reason is that labor-intensive converting industries are located near large cities, most of which are in other regions. Another factor is the concentration of fine paper production in the old papermaking regions of New England and the Mid-Atlantic and Lake States.

Importance to southern economy.—In 1969, the forest products industries employed 13 percent of the South's manufacturing workers. They paid roughly 12.4 percent of the region's manufacturing earnings, an indication that pay levels were roughly even with other regional industries. Of worker earnings paid by forest industries,

.OCT · 1373

· /·:il.

Table I.—*Employment and earnings in southern forest industries relative to United States totals, 1969*

| Industry group | Employment | | Earnings, South |
| | South | United States | South |
	Thousand workers		Percent of U. S.	Percent of U. S.
Lumber and wood products	237.3	607	39	
Lumber, wood products, and furniture	417.5	1,091	38	32
Paper and allied products	167.0	711	23	23
Forest industries	584.5	1,082	32	28
All manufacturing	4,431.2	20,167	22	18

the paper sector accounted for 37 percent, and lumber and furniture accounted for 63 percent.

In terms of total personal income, the contribution of forest industries was small—only 2.4 percent. This value was higher than the national share of 1.8 percent (table II).

Table II.—*Contribution of forest industries to personal income and total manufacturing earnings, 1969*

Region and industry group	Personal income	Manufacturing earnings
	- - Percent - -	
South		
Agriculture	4.1	...
Lumber products and furniture	1.5	7.8
Paper and allied products	.9	4.6
All manufacturing	19.3	100
Total personal income	100	...
United States		
Agriculture	2.8	...
Lumber products and furniture	1.0	4.3
Paper and allied products	.8	3.5
All manufacturing	23.4	100
Total	100	...

Source: Computed from data in Graham et al. 1972.

Earnings

Earnings, the total annual compensation of employees, other labor, and proprietors, are a useful measure of economic welfare. Earnings indicate the contribution of an industry to total personal income in a given time period.

Between 1950 and 1969, total earnings in forest industries grew by 75 percent in the United States and by 103 percent in the South. The rates were lower than those recorded for all manufacturing— 116 percent in the United States and 201 percent in the South (table III).

Table III.—*Increases in personal income and earnings between 1950 and 1969, southern region and United States*

Region and industry group	1950	1969	Increase
	– Million dollars –		*Percent*
South			
Lumber and furniture products	1,334	2,227	67
Paper and allied products	412	1,325	221
Forest industries	1,746	3,552	103
All manufacturing	9,487	28,559	201
Total personal income	58,796	147,596	151
	– Billion dollars –		
United States			
Lumber and furniture products	4.7	6.9	47
Paper and allied products	2.5	5.7	128
Forest industries	7.2	12.6	75
All manufacturing	74.8	161.4	116
Total personal income	313.6	689.6	120

Source: Graham et al. 1972.

Forest industries' contribution to total manufacturing earnings in the South fell from 18.4 to 12.4 percent in the study period. At the same time, the share nationwide fell from 9.6 to 7.8 percent.

These totals obscure important differences among industry groups. In the United States and the South, the lumber products and furniture industry grew slower than all manufacturing, while paper and allied products grew more rapidly. Both industries grew faster in the South than in the rest of the Nation. In terms of total worker earnings, then, paper and allied products contributed to rising incomes while lumber and furniture products lagged behind the rest of manufacturing.

Earnings increases differed by State (fig. 3). State earnings growth in manufacturing was not closely related to that in lumber and paper. Further, earnings gains in manufacturing were unrelated to 1971 per capita incomes and to the increase in per capita incomes from 1950 to 1969. In fact, for 1970 the most industrialized Southern States, such as Tennessee and the Carolinas, had relatively low per capita incomes and average weekly earnings.

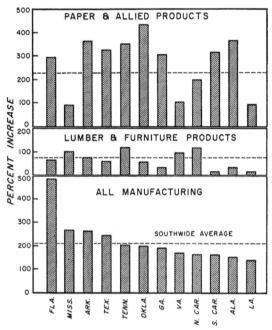

Figure 3.—*Percentage increases in real earnings in Southern States, by industrial sectors, 1950 to 1969 (1967 dollars).*

Employment Patterns

Total earnings are influenced by changes in employment and changes in wage rates. Employment trends, in turn, are influenced by output and productivity changes. The number of jobs an industry provides is a key indicator of its contribution to regional welfare.

United States manufacturing plants increased payrolls by 5 million workers from 1950 to 1969, a gain of 32 percent. The forest industries added only 140,000, for a 9 percent increase. Paper and allied products gained 46 percent and furniture and fixtures 33 percent, while lumber and wood products fell 25 percent. In 1969, the forest industries employed about 1.8 million workers, or 9 percent of the manufacturing labor force, down from 11 percent in 1950 (table IV).

Table IV.—*Employment in forest industries: total, share of all manufacturing, and change, 1950-1969*

Industry	1950		1969		Increase 1950-1969
	Employ-ment	Share of all manu-facturing	Employ-ment	Share of all manu-facturing	
	M workers	Percent	M workers	Percent	Percent
Lumber and wood products	808	5	607	3	−25
Furniture and fixtures	364	2	484	2	+33
Lumber and fur-niture products	1,172	8	1,091	5	− 7
Paper and allied products	485	3	711	4	+46
Forest industries	1,657	11	1,802	9	+ 9
All manufacturing	15,241	100	20,167	100	+32

Source: U. S. Department of Commerce 1971.

In the same period, southern forest industries added 55,000 workers to their payrolls, an increase of 10 percent. Southern manufacturing employment grew even more rapidly, rising by 81 percent. The paper and allied products sector grew more rapidly than all manufacturing, adding 81,000 employees for a 93 percent increase. Lumber and wood products lost 33 percent of its work force from 1950 to 1969. When furniture and fixtures are included, the sector still lost 26,000 workers, or 6 percent of its 1950 labor force (table V).

Table V.—*Patterns of employment change in the South, 1950-1969*

Industry group	1950	1969	Change in employment	Relative change
	-- Thousand employees --			Percent
Paper and allied products	86.3	167.0	+80.7	+93
Lumber and wood products	356.3	237.3	−119.0	−33
Lumber and wood products, and furniture	443.4	417.5	−25.9	−6
Forest industries	529.7	584.5	+54.8	+10
All manufacturing	2,449.3	4,431.2	+1,981.9	+81

In the paper sector, the States of Louisiana, Mississippi, and Virginia grew slowly in employment, while the remaining States outstripped the regional average. Of the 11 States for which data were available, eight grew faster than the industry's national average (fig. 4).

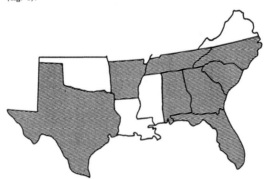

Figure 4.—*States in which employment grew faster than the national average rate in paper and allied products industry.*

The lumber and furniture sector experienced employment gains in only four States: Virginia, North Carolina, Tennessee, and Arkansas. Louisiana and South Carolina had the greatest relative losses. In six States, total employment grew more rapidly than the industry's national rate (fig. 5).

Relative increases in manufacturing employment from 1950 to 1969 were not clearly correlated with 1971 per capita incomes. One reason is that the share of incomes provided by manufacturing is smaller in the South than in the Nation. Another is the tendency for weekly earnings to rise more rapidly from 1950 to 1969 in the least industrialized States. Industrialization is measured by the manufacturing share of the total labor force. In Florida and Louisiana, the least industrialized Southern States in 1970, the average weekly earnings in manufacturing rose by 145 and 154 percent, respectively. In North and South Carolina, both 40 percent industrialized, weekly earnings rose by only 117 and 121 percent. This relationship, however, did not apply to the lumber and paper industries.

U. S. Department of Agriculture
Forest Service Research Paper SO-83

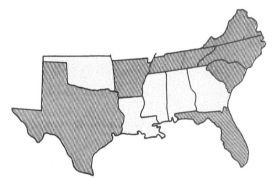

tracts,

Figure 5.—*States in which employment in lumber and wood products and furniture industry grew more than the national average rate.*

Wages

In addition to employment changes, the total earnings paid by an industry are affected by wage rates. Rural labor shortages, competition from high-wage industries, and rising minimum wages have forced wages in forest industries upward since 1950. Industries unable to compete for labor under these conditions will decline. Although employment declines in some industries were acute, the southern forest industries succeeded in raising wages faster than their national counterparts, and expanding their output at the same time. In this study, wages are measured by average weekly earnings.

·OCT · 1973

In the United States, average weekly earnings in manufacturing increased by 122 percent from 1950 to 1969. Earnings in paper and allied products, by contrast, rose 130 percent, while in lumber and wood products the rise was only 115 percent.

·. 7/.:li.

Average weekly earnings in southern forest industries outpaced national averages for the same industries. In the lumber sector, all five States for which data were available (the Carolinas, Florida, Mississippi, Louisiana, and Arkansas) increased wages relative to the national lumber industry average. In paper and allied products, wages in eight States (of nine for which data were available) rose relative to the national industry average. With one exception (North Carolina), paper and allied products wages rose relative to manufacturing in every State for which data could be obtained.

Relation to earnings changes.—Changes in total earnings in an industry result from changes in wage rates and in numbers employed. In southern forest industries from 1950 to 1969, the primary factor affecting earnings shifts was employment. Wage increases by State ranged from 122 to 144 percent in lumber, from 108 to 182 percent in paper, and from 112 to 154 percent in manufacturing. These variations, however, are much smaller than the changes in employment by State. These increases ranged from 24 to 221 percent for manufacturing, from —22 to —48 percent for lumber, and from 4 to 188 percent for paper.

Wage differentials by industry.—The lumber and furniture industries have traditionally paid relatively low wages, and they still do. Wages in the pulp and paper industry are relatively high. In 1969, the average weekly earnings in forest industries as percentages of the average for all manufacturing were:

Lumber and wood products	85 percent
Furniture and fixtures	82 percent
Paper and allied products	108 percent

One reason for low earnings in lumber and furniture is heavy reliance on unskilled labor. An index of skill levels is the proportion of nonproduction workers to total employees. In 1969 nonproduction workers were 26.8 percent of all employees in manufacturing, 22.5 percent in paper, but only 13.2 percent in lumber and 17.0 percent in furniture (U. S. Bureau of Labor Statistics 1971a).

The ranges in average wages paid in different Southern States in 1969 were:

All manufacturing	$94.13 to $128.74
Lumber and wood products	83.02 to 93.39
Paper and allied products	131.66 to 158.40

In no Southern State does lumber and wood products pay wages as high as manufacturing generally, and in no State is the paper and allied products figure below the manufacturing average.

Comparisons between wage levels in Southern States and the national averages for each industry are shown in figures 6, 7, and 8. In lumber and products, weekly earnings in all States were below the national average. Wages ranged from 75 percent of the national level in Tennessee to 85 percent in Louisiana.

In the paper industry, the opposite pattern was evident. In no State were wages below 96 percent of the national paper industry average. In fact, three States—Louisiana, Alabama, and South Carolina—had wages exceeding 110 percent of the national level.

Most Southern States have low manufacturing wages. In Mississippi and Arkansas, wages averaged 73 percent of the national

10

level. In Louisiana, however, wages were 99 percent of the national average. Texas and Oklahoma values were also high. High wages in these States are due primarily to the petroleum and related industries, which have high man-hour productivity.

tracts,

United States average: $110.15

NA = Not available.

Figure 6.—*Average weekly earnings in lumber and wood products industry in Southern States relative to national average for the same industry, percent.*

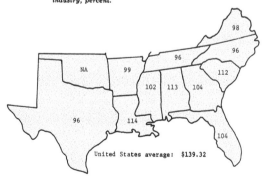

United States average: $139.32

·OCT · 1373

Figure 7.—*Average weekly earnings in paper and allied products industry in Southern States relative to national average for the same industry, percent.*

11

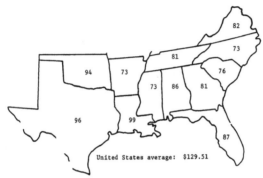

Figure 8.—*Average weekly earnings from all manufacturing in Southern
States relative to the national average, percent.*

Wages and Productivity

A fundamental limit on wage increases is the rate of increase in
labor productivity. Kaiser (1971) measured average annual rates
of increase in output per unit of labor input in forest industries
from 1947 to 1967:

Lumber and wood products	3.2 percent
Paper and allied products	3.6 percent
All manufacturing	3.4 percent

The U. S. Bureau of Labor Statistics (1971a) found that labor productivity rose by 3.9 percent per year in the primary paper industry
during the same period. Productivity trends from 1958 to 1969
within forest industries are shown in table VI. Lumber, logging,
and plywood performed as well as all manufacturing. Primary paper
industries improved productivity significantly more than did manufacturing.

Wages increased slightly faster than productivity in the study
period (table VI). These estimates, however, are only rough indicators because they include only the years at the beginning and
end of the series. In any case, the national average wage increases
stand in the same rank order as the productivity increases. Wages
in lumber rose by 3.9 percent per year, in paper by 4.1 percent, and
in manufacturing by 4.1 percent.

12

Table VI.—*Average annual rates of productivity increase in various forest industries, 1958-1969*

Industry and SIC code	Annual productivity gain
	Percent
Logging camps and contractors (2411)	3.7
Sawmills and planing mills, general (2421)	3.6
Hardwood dimension and flooring (2426)	.5
Millwork (2431)	1.6
Veneer and plywood (2432)	5.3
Wood household furniture (2511)	2.0
Pulpmills (2611)	4.5
Papermills, excluding building paper (2621)	4.5
Paperboard mills (2631)	4.5
Bags, excluding textile bags (2643)	2.2
Folding paper boxes (2651)	3.6
Fiber cans, drums, and related material (2655)	5.8
All manufacturing, weighted average	3.6

Source: U. S. Price Commission 1972.

In lumber, annual wage increases in all States exceeded the national average. This finding is consistent with those of Kaiser and Guttenberg (1970), who showed southern productivity gains exceeding the national average. Annual increases ranged from 4.6 percent per year in South Carolina to 4.1 percent in Arkansas. In paper, wages rose most rapidly in Alabama, 5.3 percent per year, which is a full percentage point above the national rate of 4.3 percent. Wage increases in manufacturing were relatively high in Southern States, ranging from 4.8 percent per year in Louisiana to 3.9 percent in Oklahoma, compared to a national average of 4.1 percent (table VII).

Table VII.—*Annual average rates of increase in average weekly earnings by industry group, 1950-1969*

Region	Lumber and wood	Paper and products	All manufacturing
	– – – – – *Percent* – – – – –		
U. S. Average	3.9	4.3	4.1
South			
Fastest rising State	4.6	5.3	4.8
Slowest rising State	4.1	3.7	3.9

Source: Computed from U. S. Bureau of Labor Statistics data.

Conclusions

Forest industries provided only 2.4 percent of southern personal income in 1969, but their importance was not reflected in that statistic. The South accounts for 32 percent of workers employed in lumber, paper, and furniture manufacturing in the United States. These industries employ 13 percent of the southern manufacturing work force and pay 12.4 percent of total earnings in manufacturing.

The forest sector has contributed substantially to improved incomes. Southern forest industries had a better productivity record from 1950 to 1969 than the national industry, permitting wages to rise faster than the national average. The result has been a closing of the income gap between North and South. The paper industry, which pays high wages and has grown rapidly in the South, has provided a strong boost to employment and incomes in rural areas. The decline of employment in the low-wage lumber industry has reduced the number of low-paying jobs for unskilled workers, but productivity gains have permitted rapid rises in wages. Southern lumber workers gained in wages relative to regional manufacturing, while lumber workers were losing ground in other regions.

As Marshall (1972) has pointed out, 60 percent of the increase in manufacturing employment in recent years in the South has occurred in rural areas. The forest industries have clearly been a factor in this industrial growth.

Differing rates of increase have changed the State wage pattern in the South. Former low-wage States have improved their relative positions. The West Gulf States improved wages rapidly in most sectors, and in 1969 Louisiana firms were paying the region's highest wages in manufacturing, in lumber, and in paper. During the two decades 1950 to 1969, Mississippi improved its relative position and was no longer the low-wage State in paper or in all manufacturing. In 1969, wages were highest for lumber and for manufacturing in the West Gulf States. This advantage is partly due to the softwood plywood industry, which is concentrated in the West Gulf and whose statistics are merged with lumber.

Literature Cited

Bretzfelder, R. B.
　　1972. Total and per capita personal income, 1971. Surv. Curr. Bus. 52(4): 17-21.

Foster, M. I.
　　1972. Is the South still a backward region, and why? Am. Econ. Rev. 62(2): 195-203.

Graham, R. E., Jr., DeGraff, H. L., and Trott, E. A., Jr.
1972. State projections of income, employment, and population.
Surv. Curr. Bus. 52(4): 22-24.

Granskog, J. E., and Guttenberg, S.
1973. Employment trends in southern forest industries. Forest
Prod. J. (In press)

Kaiser, H. F., Jr.
1971. Productivity gains in the forest products industries. Forest
Prod. J. 22(5): 14-16.

Kaiser, H. F., Jr., and Dutrow, G. F.
1971. Structure and changes in the southern forest economy,
1958-1967. USDA Forest Serv. Res. Pap. SO-71, 18 p. South.
Forest Exp. Stn., New Orleans, La.

Kaiser, H. F., and Guttenberg, S.
1970. Gains in labor productivity by the lumber industry. South.
Lumberman 221(2747): 15, 18.

Marshall, F. R.
1972. Some rural economic development problems in the South.
Am. Econ. Rev. 62(2): 204-211.

Toal, W. D.
1972. Manufacturing growth "down South." Mon. Rev. 57: 130-
136. Fed. Reserve Bank Atlanta.

U. S. Bureau of the Budget.
1957. Standard industrial classification manual. 433 p. Executive
Off. Pres., Wash., D. C.

U. S. Bureau of the Census.
1972. Current industrial reports. Pulp, Pap., and Board 1970
Series: M26A(70)-13, 24 p.

U. S. Bureau of Labor Statistics.
1971a. Handbook of labor statistics, 1970. Bur. Labor Stat. Bull.
1666, 400 p.

U. S. Bureau of Labor Statistics.
1971b. Employment and earnings statistics, States and areas, 1939-
1970. Bur. Labor Stat. Bull. 1370-8, 646 p.

U. S. Department of Commerce.
1971. 1971 business statistics. 271 p. Off. Bus. Econ.

U. S. Price Commission.
1972. Rules and regulations—price stabilization. Fed. Regist. 37:
8941-8943.

15

Appendix

Table 1.—*Total earnings from all manufacturing, 1950 and 1969, and percent increase (1967 dollars)*

State and region	1950	1969	Increase
	– – *Million dollars* – –		*Percent*
Virginia	897.8	2,359.2	163
North Carolina	1,506.2	3,943.7	162
South Carolina	738.8	1,926.6	161
Georgia	1,002.5	2,923.7	192
Florida	399.2	2,327.1	483
Tennessee	972.7	2,887.8	197
Alabama	827.0	2,062.9	149
Mississippi	278.8	1,006.9	261
Louisiana	593.8	1,396.4	135
Arkansas	255.8	920.4	260
Oklahoma	321.7	948.0	195
Texas	1,692.3	5,856.4	246
South	9,486.6	28,559.1	201
United States	74,800	161,400	116

Source: Graham et al. 1972.

Table 2.—*Total earnings from paper and allied products industry, 1950 and 1969, and percent increase (1967 dollars)*

State and region	1950	1969	Increase
	– – *Million dollars* – –		*Percent*
Virginia	54.7	108.8	99
North Carolina	41.2	123.8	200
South Carolina	24.1	98.2	307
Georgia	50.5	201.9	300
Florida	36.7	143.1	290
Tennessee	22.7	101.4	347
Alabama	33.1	152.3	360
Mississippi	30.1	56.4	87
Louisiana	72.2	134.4	86
Arkansas	15.0	68.8	359
Oklahoma	1.7	9.0	429
Texas	30.1	127.0	321
South	412.1	1,325.1	221
United States	2,500	5,700	128

Source: Graham et al. 1972.

Table 3.—*Total earnings from lumber products and furniture industry, 1950 and 1969, and percent increase (1967 dollars)*

State and region	1950	1969	Increase
	- - Million dollars - -		Percent
Virginia	135.4	257.7	90
North Carolina	233.8	499.9	114
South Carolina	84.1	92.6	10
Georgia	138.5	179.2	29
Florida	78.6	128.4	63
Tennessee	98.3	209.3	113
Alabama	113.9	146.4	28
Mississippi	100.0	197.6	98
Louisiana	93.7	101.1	8
Arkansas	104.4	177.8	70
Oklahoma	15.9	23.5	48
Texas	136.9	213.8	56
South	1,333.5	2,227.3	67
United States	4,700	6,900	47

Source: Graham et al. 1972.

tracts,

Table 4.—*Employees on manufacturing payrolls in Southern States, 1950 and 1969*

State	1950	1969	Increase, 1950-1969
	- - - Thousands - - -		Percent
Virginia	229.5	370.4	61
North Carolina	418.3	712.7	70
South Carolina	210.4	339.2	61
Georgia	286.5	476.9	67
Florida	102.3	329.1	221
Tennessee	249.9	469.9	88
Alabama	216.1	323.7	50
Mississippi	86.4	181.5	110
Louisiana	145.0	180.6	24
Arkansas	75.7	168.0	122
Texas	363.6	749.2	106
Oklahoma	65.6	130.0	98
South	2,449.3	4,431.2	81

Source: U. S. Bureau of Labor Statistics 1971b.

·OCT 1 1973

TECH. & V

17

Table 5.—*Employment in paper and allied products industry (SIC 26) in Southern States, 1950 and 1969*

State	1950	1969	Increase, 1950-1969
	--- Thousands ---		Percent
Virginia	10.0	14.0	40
North Carolina	8.0	17.2	115
South Carolina	5.7	11.7	105
Georgia	11.3	25.8	128
Florida	7.8	17.9	129
Tennessee	5.1	14.0	174
Alabama	7.4	17.6	138
Mississippi	6.5	7.1	9
Louisiana	15.1	15.7	4
Arkansas	3.4	8.7	156
Texas	6.0	17.3	188
South	86.3	167.0	93

Source: U. S. Bureau of Labor Statistics 1971b.

Table 6.—*Employment in lumber and wood products industry (SIC 24) in Southern States, 1950 and 1969*

State	1950	1969	Change, 1950-1969
	--- Thousands ---		Percent
Virginia	29.4	23.6	−20
North Carolina	42.6	29.2	−31
South Carolina	28.0	14.5	−48
Georgia	45.1	26.3	−42
Florida	21.1	14.0	−33
Tennessee	22.0	17.2	−22
Alabama	40.5	25.5	−37
Mississippi	32.1	23.4	−27
Louisiana	30.0	17.1	−43
Arkansas	30.1	22.3	−26
Texas	32.3	22.2	−31
Oklahoma	3.1	2.0	−35
South	356.3	237.3	−33

Source: U. S. Bureau of Labor Statistics 1971b.

Table 7.—*Employment in lumber and wood products (SIC 24)*
and furniture and fixtures industries (SIC 26) in
Southern States, 1950 and 1969

State	1950	1969	Change, 1950-1969
	--- Thousands ---		Percent
Virginia	44.2	48.7	+10
North Carolina	75.4	97.9	+30
South Carolina	29.8	19.3	−35
Georgia	52.5	36.9	−30
Florida	24.2	23.6	− 2
Tennessee	32.5	42.5	+30
Alabama [1]	40.5	25.5	−37
Mississippi	34.5	27.5	−20
Louisiana	31.6	18.6	−41
Arkansas	34.1	35.8	+ 4
Texas	41.0	39.2	− 4
Oklahoma [1]	3.1	2.0	−35
South	443.4	417.5	− 6

Source: U. S. Bureau of Labor Statistics 1971b.
[1] No furniture data given.

Table 8.—*Average weekly earnings from all manufacturing in*
Southern States and changes, 1950-1969

State	1950	1969	Increase
	-- Dollars --		Percent
Virginia	47.55	106.60	124
North Carolina	43.34	94.13	117
South Carolina	44.51	98.47	121
Georgia	43.52	104.55	140
Florida	46.20	113.30	145
Tennessee	47.84	105.04	119
Alabama	47.32	111.38	135
Mississippi	39.68	95.06	140
Louisiana	50.63	128.74	154
Arkansas	42.33	94.13	122
Texas	57.10	125.03	119
Oklahoma	57.10	121.25	112
United States average	58.32	129.51	122

Source: U. S. Bureau of Labor Statistics 1971b.

19

Table 9.—*Average weekly earnings in paper and allied products industry in Southern States, 1950 and 1969*

State	1950	1969	Increase
	- - Dollars - -		*Percent*
Virginia	53.98	136.28	152
North Carolina	64.21	133.42	108
South Carolina	61.78	156.58	153
Georgia	56.78	145.29	156
Florida	58.18	145.06	149
Tennessee	NA	131.66	NA
Alabama	55.62	157.18	182
Mississippi	52.58	145.25	170
Louisiana	59.36	158.40	167
Arkansas	56.44	138.58	146
Texas	NA	133.25	NA
Oklahoma	NA	NA	NA
United States average	60.53	139.32	130

Source: U. S. Bureau of Labor Statistics 1971b.
NA = Not available.

Table 10.—*Average weekly earnings in lumber and wood products industry in Southern States, 1950 and 1969*

State	1950	1969	Increase
	- - Dollars - -		*Percent*
Virginia	NA	NA	NA
North Carolina	37.05	85.90	132
South Carolina	34.80	85.06	144
Georgia	NA	85.49	NA
Florida	38.68	92.80	140
Tennessee	NA	83.02	NA
Alabama	NA	NA	NA
Mississippi	38.25	87.56	129
Louisiana	38.73	93.39	141
Arkansas	40.17	89.03	122
Texas	NA	88.75	NA
Oklahoma	NA	NA	NA
United States average	51.27	110.15	115

Source: U. S. Bureau of Labor Statistics 1971b.
NA = Not available.

Irland, Lloyd C.

   1972. Labor trends in southern forest industries,
         1950 to 1969. South. Forest Exp. Stn., New
         Orleans, La. 20 p. (USDA Forest Serv. Res.
         Pap. SO-81)

By raising wages more rapidly than the rest of the Nation,
and by expanding employment in the high-wage paper
industry, southern forest industries helped improve economic welfare in the rural South.

U. S. Department of Agriculture
Forest Service Research Paper SO-82

tracts,

Ten-Year Growth
of Planted Slash Pine
After Early Thinnings

Hans G. Enghardt

and

W. F. Mann, Jr.

Southern Forest Experiment Station
Forest Service
U. S. Department of Agriculture

1972

Acknowledgments

Bel Estate-Quatre Parish Company, Lake Charles, La., provided the plantation and assisted in cutting the plots. The USDA Soil Conservation Service marked the D+ plots and provided a soils map of the study area.

U. S. Department of Agriculture
Forest Service Research Paper SO-83

TEN-YEAR GROWTH OF PLANTED
SLASH PINE AFTER EARLY THINNINGS

Hans G. Enghardt and W. F. Mann, Jr.[1]

*Volume growth of slash pine between ages 17 and 27 years
was directly related to residual basal area per acre after thinning. Diameter growth was inversely related to stand density,
and very heavy cutting was required to attain a rate of 3 inches
in 10 years.*

This paper summarizes 10-year volume and diameter growth of
planted slash pine *(Pinus elliottii* Engelm. var. *elliottii)* that was
thinned at age 17 years to a broad array of residual densities. The
site, which has never been in cultivation, is highly productive and
typical of large acreages in the flatwoods region of the West Gulf
Coastal Plain.

Information on growth and yield of slash pine in this region is
limited. It is known, however, that data from slash pine plantations
in the Southeast, whether on old-field or cutover sites, are not applicable here. Differences in soils and climate probably account for
the variations.

The Plantation

The plantation is about 14 miles west of Kinder, Louisiana. The
old-growth longleaf pine was removed from the site in the 1920's,
and when the slash pines were planted the land supported only a
dense stand of native grasses.

The soil is mostly a Beauregard silt loam, slowly permeable.
Since topography is almost flat, surface drainage is also slow. Site
index for slash pine is 91 to 102 feet (age 50 years), a range that is
typical for large areas of central and southwest Louisiana.

The plantation was established in the winter of 1941-42. Seedlings were hand-planted in the grass sod at an approximate spacing

[1] Hans G. Enghardt is Research Forester, Louisiana Forestry Commission, assigned to
the Southern Forest Experiment Station. W. F. Mann, Jr., is Principal Silviculturist
at the Southern Forest Experiment Station, USDA Forest Service, Timber Management Research Project 1102, Alexandria Forestry Center, Pineville, La.

of 6 by 7 feet. No record of early survival is available, but average stocking of 690 trees per acre at age 17 years indicates that it was reasonably good.

In 1957, 1 year before the study was started, a thinning was made to remove an average of 72 small trees per acre whose trunks were cankered by fusiform rust *(Cronartium fusiforme* Hedgc. & Hunt ex Cumm.). Volumes of these trees have been added to cut volumes at age 17 and to mean annual growth. Of the remaining trees, 20 percent had cankers on the trunk.

Description of Study

Eight thinning regimes plus an unthinned check are under test, with four block replications of each treatment. Thinnings, all started at age 17 years, are as follows:

BA 70: Thin to 70 square feet of basal area per acre every 5 years.

BA 85: Thin to 85 square feet every 5 years.

BA 85-10: Thin to 85 square feet every 10 years.

BA 100: Thin to 100 square feet every 5 years.

BA 95/55: Thin to 95 square feet at age 17, and to 55 square feet at age 22 and every 5 years thereafter. Reduction of stocking in two steps was to avoid a heavy release that would leave trees vulnerable to wind and ice damage.

BA 95/115: Thin to 95 square feet at age 17, and to 115 square feet at age 22 and every 5 years thereafter. The first thinning was heaviest, since the aim was to remove badly diseased trees.

Increasing BA: Thin to progressively higher basal areas on a 5-year cycle, starting with 85 square feet and reaching 120 square feet at 52 years.

D+: D+ (X) thinning *(8)*, advocated by the USDA Soil Conservation Service. D is the average diameter (b.h.) of the stand, and in this study X is equal to 4 plus the expected diameter growth in 5 years (in inches). The sum of D + X is expressed in feet for an average spacing between trees after cutting. SCS personnel marked the plots to insure that the system was applied correctly.

Check: No thinning.

Plots are 200 feet square or 0.92 acre in size, but measurements were confined to the central 0.25 acre.

Average stocking at age 17 years was comparable for individual treatments (table 1). Variations from the means of number of

2

Table 1.—*Average stand per acre of trees 3.6 inches d.b.h. and larger before thinning at age 17 years, by treatments*

Thinning treatment	Site index	Trees	Basal area	D.b.h.	Volume, inside bark
	Feet	Number	Sq. ft.	Inches	Cu. ft.
BA 70	97	619	112	6.0	1,746
BA 85	97	622	118	6.0	1,739
BA 85-10	96	650	119	6.0	1,685
BA 100	98	596	116	6.2	1,942
BA 95/55	96	645	120	6.0	1,867
BA 95/115	98	648	122	6.0	1,923
Increasing BA	96	640	120	6.0	1,828
D+	95	639	117	6.0	1,717
Check	95	625	114	5.9	1,707

trees, basal area, and d.b.h. were less than 5 percent. Cubic volumes differed slightly more, ranging from 1,685 to 1,942 cubic feet of peeled wood (i.b.) per acre.

Mean annual growth was rapid during the initial 17 years, a consequence of the good site. It averaged 106 cubic feet (i.b.) per acre and 0.35 inch of diameter for all trees 3.6 inches d.b.h. and larger.

Diameter of all trees 3.6 inches and larger was measured before each thinning. To determine volumes, stand basal areas were multiplied by a volume/basal-area ratio computed from measurements on sample trees. In 1959 and 1964, cut and leave volumes were added to obtain initial volume; in 1969 cut volume was derived by subtracting leave from initial volume.

Sample-tree volumes were obtained by height accumulation (2, 6) in 2-inch taper steps to a top diameter of 4 inches outside bark. Diameter points and heights were determined with a Spiegelrelaskop.[*] To calculate cubic-foot volumes inside and outside bark, bark thickness was measured at breast height, at the midpoint of each merchantable stem, and at the 4-inch top of 100 felled trees distributed over the study plots. Ratios of d.i.b./d.o.b. averaged 0.87, 0.87, and 0.88 in 1959, 1964, and 1969, respectively. Volumes and volume growth reported here are in cubic feet of peeled wood. Approximate values outside bark can be calculated by multiplying inside bark data by 1.306.

Ten well-spaced dominant and codominant trees of good quality were selected on each 0.25-acre plot and numbered as potential crop trees. Measurements on these trees included total height, length

[*] Mention of trade names is for information only, and does not imply endorsement by the U. S. Department of Agriculture.

3

to the base of the live crown, and length of bole clear of branches and branch stubs.

Mortality, which ranged from 5 to 240 cubic feet (i.b.) per acre over the 10-year study period, was included in growth data because it was unrelated to treatment and because losses from suppression were negligible even on the controls. Moreover, losses due to factors like lightning are distorted on small plots. Special measures were taken to control bark beetles, and hence the data offer no information on possible relations between thinning and insect attack.

Site indices are for a base age of 50 years. They were determined from average heights of dominant and codominant trees measured at 27 years and from curves for slash pine in figure 4 of USDA Miscellaneous Publication 50 (10).

Results

Periodic Annual Volume Growth

Periodic annual volume growth from age 17 to 22 years ranged from 195 to 283 cubic feet (i.b.) per acre (table 2). Poorest growth was on plots thinned by the D+ rule, which left the least basal area (67 square feet per acre) after the first thinning. The best growth was on the unthinned check plots, which had 114 square feet of basal area per acre at the start of the period. In general, volume growth during the first 5 years was related directly to residual basal area at age 17 years.

In all but three treatments, residual basal areas at the start

Table 2.—*Periodic annual cubic volume growth* [1] *(inside bark) from age 17 to 27 years*

Thinning treatment	17-22 years	22-27 years	17-27 years
	— — — — — — Cubic feet — — — — — —		
BA 70	201	222	211bc[2]
BA 85	242	197	219bc
BA 85-10	226	289	258a
BA 100	250	266	258a
BA 95/55	243	194	218bc
BA 95/115	236	238	237abc
Increasing BA	233	257	245ab
D+	195	208	201c
Check	283	261	272a

[1] Trees 3.6 inches d.b.h. and larger to a 4-inch top outside bark.
[2] Values followed by the same letter are not significantly different at the 0.05 level.

4

U. S. Department of Agriculture
Forest Service Research Paper SO-83

of the second 5-year period differed from those at the start of the first period. From 22 to 27 years, volume growth on plots with constant residual basal areas averaged about the same as in the preceding period. Rankings changed somewhat, however, and the relationship of volume growth to density was not as strong as initially.

Plots thinned to 85 square feet of basal area per acre on a 10-year cycle had the best growth, averaging 289 cubic feet annually. Growth on the checks ranked third with 261 cubic feet, although these plots had the highest basal area. Least growth was on plots thinned to 55 square feet of basal area per acre (BA 95/55) at age 22 years. Sharp declines on BA 85 and check plots may have been due to sampling errors that inflated volumes at age 22. Such errors would increase calculated volume growth from 17 to 22 years, but decrease it from 22 to 27 years.

Site index, age, number of trees per acre, basal area per acre, and various interactions between these factors were examined by regression analysis (3) for their effect on annual volume growth in both 5-year periods. The best regression contained five of the nine variables tested and accounted for 45 percent of the variation. Because data for only two growth periods were available and some of the treatments had caused large differences in density between periods, the decision was made not to compute prediction equations.

Growth over the 10 years probably offers the best comparison, as it tends to average out weather influences and other unavoidable factors. However, treatments like BA 95/55 are difficult to characterize except with an average residual basal area.

Ten-year annual growth excelled on the unthinned checks, averaging 272 cubic feet per acre. Average basal area during this period increased from 114 to 170 square feet per acre. In contrast, increment was least—201 cubic feet annually—on plots thinned by the D+ rule. On these plots, average basal area per acre left after cutting was 67 square feet at 17 years and 78 square feet at 22 years. Between these two extremes, volume growth increased directly with basal area.

In treatments with a constant residual basal area after each cut, growth increased by about 25 cubic feet for each additional 15 square feet of basal area per acre. This trend is consistent with findings in other studies (7).

Covariance analysis was applied to adjust periodic annual volume growth for differences in site index. Adjusted treatment means differed from actual ones by only a few cubic feet, and rankings were unchanged. Since differences were small, adjusted values are not reported here.

tracts,

OCT • 1373

· 7ि.:H.

5

Mean annual growth to age 27 years ranged from 138 to 167 cubic feet per acre (table 3). These data should not be used to compare thinning treatments, for differences in volumes at the start of the study were larger than differences in growth after treatments. They are given primarily to permit comparison with periodic annual growth and because they may be helpful in certain management decisions.

Table 3.—*Volume cut, mortality, total yield, and mean annual growth per acre to age 27 years*

Thinning treatment	Initial volume— age 17	Cut—age 17 to 27	Mortality— age 17 to 27	Gross yield— age 27	Mean annual growth— to age 27
				– – – – – – – Cubic feet (i. b.) – – – – – – –				
BA 70	1,746	1,889	8	3,860	143
BA 85	1,739	1,662	43	3,933	146
BA 85-10	1,685	1,844	50	4,261	158
BA 100	1,942	1,751	31	4,517	167
BA 95/55	1,867	2,406	36	4,052	150
BA 95/115	1,923	1,237	37	4,293	159
Increasing BA	1,828	1,580	55	4,276	158
D+	1,717	1,304	5	3,731	138
Check	1,707	[1] 49	240	4,427	164

[1] Cut at age 16 years for fence posts.

Cut volumes amounted to 30 to 60 percent of total yield. Generally, they were highest with heavier thinnings, but initial volume at age 17 also influenced the proportion of total growth that was cut.

Mortality was negligible—less than 60 cubic feet—in all treatments except the check, where it totaled 240 cubic feet from age 1 to 27 years. The loss on check plots was caused by annosus root rot (*Fomes annosus* (Fr.) Cke.) on two of the replicates and not by suppression. It is not known why root rot was heaviest on control plots, which were only thinned lightly in 1957.

Saw Log Volumes

Information on saw-log-size trees is given in table 4 to show what might be expected at age 27 years, rather than to compare the performance of treatments. Because of the variation in volumes prior to the first thinning, no significance can be attributed to any of the differences. Moreover, many trees were just below saw log size at age 27, and ingrowth will be substantial during the next 5 to 10 years.

Table 4.—*Number and volume of saw-log-size trees per acre before cutting at age 27 years*

Thinning treatment	Merchantable trees [1]	Saw-log-size trees [2]	Saw log volume	Avg. volume per tree
	– – Number – –		– Board feet [3] –	
BA 70	207	93	6,562	71
BA 85	245	103	6,611	64
BA 85-10	375	95	5,502	58
BA 100	320	97	6,258	65
BA 95/55	162	78	5,385	69
BA 95/115	355	97	6,148	63
Increasing BA	274	91	6,522	72
D+	233	77	5,092	66
Check	491	83	4,850	58

[1] Trees 3.6 inches and larger.
[2] Trees 9.6 inches d.b.h. and larger.
[3] Int. ¼-inch rule, to 8-inch top (o.b.).

Diameter Growth

All merchantable trees.—From 17 to 22 years, diameter growth of trees 3.6 inches d.b.h. and larger varied from a high of 0.26 inch annually on plots thinned to 70 square feet of basal area to a low of 0.20 inch on the unthinned checks (table 5). On the D+ plots,

Table 5.—*Periodic annual diameter growth of all merchantable-size and selected crop trees*

Thinning treatment	Merchantable trees			Crop trees [1]		
	17-22 yrs.	22-27 yrs.	17-27 yrs.	18-22 yrs.	22-27 yrs.	18-27 yrs.
	– – – – – – – – – – Inch – – – – – – – – – –					
BA 70	0.26	0.30	0.28a [2]	0.30	0.30	0.30
BA 85	.24	.26	.25bc	.28	.29	.28
BA 85-10	.22	.21	.21de	.26	.25	.26
BA 100	.22	.22	.22de	.24	.24	.24
BA 95/55	.21	.34	.28ab	.27	.33	.30
BA 95/115	.22	.21	.22de	.28	.25	.26
Increasing BA	.23	.24	.24cd	.28	.25	.26
D+	.24	.27	.25abc	.30	.27	.28
Check	.20	.19	.20e	.26	.22	.23

[1] First measurement made at age 18 years.
[2] Values followed by the same letter are not significantly different at the 0.05 level.

7

which had an initial basal area of 67 square feet, diameter growth averaged 0.24 inch yearly. It was 0.04 inch greater on plots thinned to 70 square feet of basal area than on those thinned to 100 square feet.

In the second 5-year period, growth either remained unchanged or increased in treatments where a constant basal area was maintained on a 5-year cutting cycle. It decreased on checks, BA 85-10 plots, and BA 95/115 plots. On plots cut by the D+ rule, it increased from 0.24 to 0.27 inch. The greatest change occurred on BA 95/55 plots, where the average rose from 0.21 to 0.34 inch.

Averages over the entire 10 years were also inversely related to basal area per acre, ranging from 0.20 to 0.28 inch annually. Thinning back to 70 square feet of basal area stimulated diameter growth 0.06 inch more than light thinning to 100 square feet— a difference of about 27 percent, or an average of 0.03 inch for each 15 square feet of residual basal area.

Covariance analysis was applied to adjust treatment means for differences in site index, but only negligible changes in values were obtained.

Selected sample trees.—Diameter growth of the best, well-spaced crop trees followed the same general trends as for all merchantable trees (table 5). From age 18 to 27 years, averages ranged from 0.23 to 0.30 inch yearly, or 7 to 24 percent more than those for all trees over 3.5 inches d.b.h. Only two of the treatments, BA 70 and BA 95/55, resulted in a growth rate of 3 inches in 10 years.

Average diameter.—Increases of average diameters for all merchantable trees over the 10-year period varied from 2.1 to 3.6 inches, with the largest gains in plots thinned heaviest. Parts of these increases are due to cutting and mortality. On all thinned plots, increment alone accounted for 78 to 92 percent of the increase. On the checks, 95 percent was due to growth and the remainder was from mortality of small trees.

Diameter Distributions

To the land manager, diameter distributions are more informative than average diameters. From the number of trees in each diameter class he can estimate volumes available for various products and appraise the value of the stand.

Table 6 gives the cumulative number of trees per acre in and above each 1-inch d.b.h. class for each treatment. There were more trees 11 inches and larger on thinned than on unthinned plots, and heavy thinnings seemed to have a slight superiority over light thinnings. Most differences in stocking of trees 10 inches d.b.h. and larger were small. In trees 9 inches and larger, BA 95/115 and the unthinned check ranked first and second, and light thinnings excelled heavy thinnings. With each successively lower threshold diameter, unthinned checks attained a greater superiority over other treatments.

Table 6.—*Cumulative number of trees per acre before cutting at age 27, by 1-inch d.b.h. classes*

Thinning treatment	D.b.h.—inches											
	<4	4	5	6	7	8	9	10	11	12	13	14
						Number						
BA 70	222	207	203	201	191	169	141	93	40	12	1	. .
BA 85	253	245	239	234	220	191	143	103	49	13	2	1
BA 85-10	392	375	361	344	316	262	171	95	31	7	. .	. .
BA 100	329	320	306	292	269	230	169	97	32	10	3	1
BA 95/55	171	162	159	158	156	145	123	78	34	11	. .	. .
BA 95/115	364	355	342	324	300	253	179	97	44	14	4	. .
Increasing BA	285	274	270	258	247	210	152	91	39	14	4	1
D+	237	233	233	232	227	198	145	77	39	9	2	. .
Check	507	491	472	424	352	270	173	83	24	6	1	1

Some trees in the upper crown classes were removed by all thinnings at age 17 and 22 years, although cutting was primarily from below. To achieve the desired stocking levels, 179 to 332 trees per acre were cut in the 4-, 5-, and 6-inch d.b.h. classes, with the greatest number in the heavy thinnings. In contrast, there were only 139 trees per acre in these three size classes on the unthinned checks at 27 years. Obviously, harvest costs will not be adversely affected by an excessive number of small trees when unthinned stands are clearcut at age 27.

Discussion

The actual growth and yield values summarized in this paper are probably of greater value to the land manager than are the relationships of growth to stocking. Similar research has already shown that heavy thinning stimulates diameter growth and lowers volume growth *(1, 4, 5, 7)*. In this study, within a range of 70 to 100 square feet of basal area per acre, an increase of 15 square feet resulted in a decline of 11 percent in diameter growth and a gain of equal magnitude in cubic volume growth.

9

Earlier studies have also shown the difficulty of obtaining an average diameter growth rate of 0.30 inch on all merchantable stems. This high rate is achieved only by reducing stocking to about 60 square feet of basal area per acre.

Much more difficult to decipher are the total economics, including logging costs, of thinning versus no thinning on a short rotation primarily for pulpwood. A frequent criticism of unthinned stands is that many small trees have to be handled, with resulting high costs per cord for cutting, limbing, and bucking. This study showed that more small trees may be cut when early thinnings are made than when stands are left unthinned until clearcut at age 27.

At age 27, thinned plots had more trees 10 inches d.b.h. and larger than the checks. A saving in cost of final harvest could therefore be expected, but this advantage might be offset by the higher cost of cutting the small trees in the intermediate thinnings. While logging costs decrease slowly in diameters beyond 8 or 9 inches, they rise rapidly in sizes typically removed in thinnings (table 7). A rough appraisal with Tufts' (9) data indicates there would be little difference in total logging costs between plots thinned heavily at age 17 and 22 years and clearcut at age 27, in comparison to no thinning and clearcutting at age 27.

Table 7.—*Trees per acre removed in thinnings through age 22*

Thinning treatment	D.b.h.—inches									
	<4	4	5	6	7	8	9	10	11	12
	– – – – – – – – – – Number – – – – – – – – – –									
BA 70	39	87	105	86	65	48	17	3		
BA 85	15	86	92	82	56	30	16	1	1	
BA 85-10	11	70	86	67	31	5				
BA 100	32	56	57	66	49	37	10	4	2	
BA 95/55	29	84	98	94	108	65	24	5	0	1
BA 95/115	17	74	83	73	34	15	4	1		
Increasing BA	12	75	101	89	48	29	12	2		
D+	11	83	137	112	56	23	4	3		
Check	0	21	18	5						

Results now available should not influence decisions about plantations from which larger trees are desired. The study was designed to run at least until the trees are 50 years old. The interim findings given here thus apply mostly to pulpwood production. Undoubtedly, thinnings will prove beneficial in rotations where veneer and saw logs are the main product goal.

10

Literature Cited

1. Dell, T. R., and Collicott, L. V.
   1968. Growth in relation to density for slash pine plantations after first thinning. Forest Sci. 14: 7-12.

2. Grosenbaugh, L. R.
   1954. New tree-measurement concepts: height accumulation, giant tree, taper and shape. USDA Forest Serv., South. Forest Exp. Stn. Occas. Pap. 134, 32 p.

3. Grosenbaugh, L. R.
   1967. REX—FORTRAN-4 system for combinatorial screening or conventional analysis of multivariate regressions. USDA Forest Serv. Res. Pap. PSW-44, 47 p. Pac. Southwest Forest and Range Exp. Stn., Berkeley, Calif.

4. Gruschow, G. F., and Evans, T. C.
   1959. The relation of cubic-foot volume growth to stand density in young slash pine stands. Forest Sci. 5: 49-55.

5. Keister, T. D., Crow, A. B., and Burns, P. Y.
   1968. Results of a test of classical thinning methods in a slash pine plantation. J. For. 66: 409-411.

6. Lohrey, R. E., and Dell, T. R.
   1969. Computer programs using height accumulation for tree volumes and plot summaries. J. For. 67: 554-555.

7. Mann, W. F., Jr., and Enghardt, H. G.
   1972. Growth of planted slash pine under several thinning regimes. USDA Forest Serv. Res. Pap. SO-76, 10 p. South. Forest Exp. Stn., New Orleans, La.

8. Mitchell, H. C.
   1943. Regulation of farm woodlands by rule of thumb. J. For. 41: 243-248.

9. Tufts, D.
   1969. Planning a logging operation with analyses of some pulpwood logging systems. In Fourth Timber Harvesting Short Course Proc., 21 p. Am. Pulpwood Assoc. and La. State Univ.

10. U. S. Dep. Agric.
    1929. Volume, yield, and stand tables for second-growth southern pines. USDA Misc. Publ. 50, 202 p.

Wood Machining Abstracts, 1970 and 1971

PETER KOCH

SOUTHERN FOREST EXPERIMENT STATION
FOREST SERVICE
U. S. DEPARTMENT OF AGRICULTURE

1973

Contents

The purpose of this paper is t
significant wood-machining literatur
not been previously digested in Englis
texts and bibliographies.[1] The entr
represent publications of a research
trade-journal articles have been inclu
judged to be of interest. While m
literature is from 1970 and 1971, a
was made to include important paper
been missed in the earlier searches.

Principal sources of information we
journals in wood science and techn
Forestry Abstracts, published by the
wealth Agricultural Bureaux, Farnh
England, and the Bureaux' Card Tit
were invaluable. The search also inc
Abstract Bulletin of the Institute of Paper Chemistry, the USDA *Bibliography of Agriculture,* and publications lists and annual and periodic reports issued by the regional USDA forest experiment stations and the Forest Products Laboratory, Madison, Wisconsin. Patents were largely noted from the listing carried in 1971 by the *Forest Products Journal.*

Abstracts of English-language publications were written by the author after a reading of the original. Abstracts of foreign-language publications were drawn largely from *Forestry Abstracts* and the *Abstract Bulletin,* though with minor editorial changes, deletions, or amplifications. The author recognizes that limits of time and linguistic ability have probably caused omissions and misinterpretations.

ence to the reader,
duplicate the style
review (McMillin
o changes were
field was added:
ntrol of Machines.
nber and Particle-
e fields to permit
ty of hardwoods,
boards, and fiber-

under its major
nce utility, papers
pertinent fields.

Doe. 1966. (Sawing; Bandsawing)

Such a reference indicates that the full citation appears in the Sawing category, under the field Bandsawing. The subject matter heads are summarized in the table of contents. An index of authors appears on page 44.

HIGHLIGHTS OF THE LITERATURE

This review indicated that the total research effort is continuing at a high level. About 66 publications per year appeared in archival journals for the years 1963 through 1965. The reviews of 1966 through the present have references as follows:

---

[1] The chief previous compilations:

Committee on Recent Wood Machining Literature. **Wood machining abstracts, 1957-1958.** For. Prod. Res. Soc., Madison, Wis. 20 p. 1959.

Committee on Recent Wood Machining Literature. **Wood machining abstracts, 1958-1959.** For. Prod. Res. Soc., Madison, Wis. 19 p. [1960.]

Committee on Recent Wood Machining Literature. **Wood machining abstracts, 1959-1961.** For. Prod. Res. Soc., Madison, Wis. 18 p. 1961.

Koch, P. **Wood Machining Processes.** 530 p. N.Y.: Ronald Press Co. 1964.

Koch, P., and McMillin, C. W. **Wood machining review, 1963 through 1965.** For. Prod. J. Part I, 16(9): 76-82, 107-115; Part II, 16(10): 43-48. 1966.

Koch, P. **Wood machining abstracts, 1966 and 1967.** USDA For. Serv. Res. Pap. SO-34, 38 p. South. For. Exp. Stn., New Orleans, La. 1968.

McMillin, C. W. **Wood machining abstracts, 1968 and 1969.** USDA For. Serv. Res. Pap. SO-58, 35 p. South. For. Exp. Stn., New Orleans, La. 1970.

Review and year	References
Koch (1968)	
1966	138
1967	103
McMillin (1970)	
1967	36
1968	187
1969	89
Present review	
Prior to 1969 or undated	35
1969	71
1970	153
1971	156

Thus, references for 1968 through 1971 averaged in excess of 150 per year. Since many titles are inevitably omitted from the second year of a 2-year review (because of delayed receipt), it is likely that 1971 ·- like 1968 -- will prove to have been a record year for reported research on wood machining.

In the current review, 43 fields of activity are identified, yielding a total of 415 papers (table 1) — about one-third more than in either of the previous two reviews. Sawing (8 fields) led with 94 primary and 95 cross-references. Circular sawing was prominent with 23 primary and 31 cross-references. Bandsawing also received much attention, with 13 primary and 23 cross-references. Twenty-seven papers discussed chain sawing, 24 shearing, and 17 sash gang-sawing.

Safety (including noise) led all single fields of activity with 51 primary and 2 cross-references. Of these, 36 were concerned with noise, and 16 with safe operation of chain saws. The prior review (McMillin 1970 (Bibliographies)), carried only 13 primary listings and two cross-references for the entire field of safety; of these only five were concerned with noise. Thus, we see a seven-fold increase in research related to noise of woodworking equipment.

Properties of the cutting edge and cutter (six fields) shows 83 primary and 32 cross-references for a total of 115 citations. In this category, the fields of fitting and sharpening (34 citations) and stability (30 citations) were prominent. Dulling phenomena were reported in 23 papers whereas the prior review showed only 9.

Veneer-cutting research continued active, with 24 primary and 11 cross-references — 39 papers were listed in the prior review. Barking

also received considerable attention, with 27 citations; bark separation from chips was of particular interest, with eight references.

In the category of machinability of species and reconstituted wood (five fields), hardwoods had 31 citations, softwoods 23, particleboard 12, fiberboard 8, and plywood 6.

Jointing, planing, molding, and shaping accounted for 24 references. Ten papers were devoted to orthogonal cutting, 10 to machining with coated abrasives, and 9 to defibrating (including stone grinding). Thirteen primary listings were devoted to research instrumentation and techniques; and the new field of computer, tape, and card control of machines carried four primary listings as did the field of laser cutting (machining with light).

In the category of chipping, flaking, and grinding (five fields), 20 papers discussed conventional chippers, 13 were concerned with chipping headrigs and edgers, 8 with mobile chippers, 6 with flaking, and only 1 with grinding (i.e. pulverizing).

Turning, alone among the 43 fields, has no papers cited in this review.

Despite the obvious risk of offending researchers by omission or misinterpretation, it seems worthwhile to attempt some mention of individual contributions in the various fields.

History and general texts. — Thunell, in articles dated 1971, discussed research trends in woodworking and selection of equipment for sawmills of various capacities. Koch (1971) reviewed significance of the southern pines in the economy of the United States and summarized trends in manufacture.

Properties of wood. — Albers (1971) discussed the crushing strength of a variety of particleboards, veneers, and fiberboards. Futo (1969) elucidated modes of tensile fractures in very small specimens of spruce.

Sharma (1971) reported that silica content in 23 specimens of *Tectona grandis* varied with region of origin from 0.2 to 3.04 percent; it was vitreous and found only in fibers and vessels.

Orthogonal cutting. — Kato et al. (1971) studied the action of vibrating knives in cuts parallel and perpendicular to the grain of a hardwood and a softwood. Hamamoto and Mori

Table 1. — Number of references on wood machining research for 1970 and 1971, classified into 43 fields of activity

Subject	Primary listings	Cross references	Total
History and general texts	3	2	5
Properties of wood	2	1	3
Orthogonal cutting	6	4	10
Peripheral milling parallel to grain	2	2	4
Barking	21	1	22
Sawing[1]			
General	13	13	26
Bandsawing	13	23	36
Sash gangsawing	9	8	17
Circular sawing	23	31	54
Chain sawing	11	16	27
Machining with high-energy jets	1	0	1
Machining with light	4	0	4
Shearing	20	4	24
Total sawing	94	95	189
Jointing, planing, molding, and shaping	7	17	24
Turning	0	0	0
Boring, routing, and carving	8	1	9
Mortising and tenoning	0	3	3
Machining with coated abrasives	9	1	10
Veneer cutting	24	11	35
Chipping, flaking, and grinding			
Chipping	15	5	20
Chipping headrigs and edgers	10	3	13
Mobile chippers	7	1	8
Flaking	1	5	6
Grinding	0	1	1
Total chipping, etc.	33	15	48
Defibrating	8	1	9
Properties of the cutting edge and cutter			
Tool material	10	2	12
Dulling	12	11	23
Fitting and sharpening	24	10	34
Stability	23	7	30
Temperature	7	1	8
Friction	7	1	8
Total cutting edge and cutter	83	32	115
Computer, tape, and card control of machines	4	1	5
Research instrumentation and techniques	13	14	27
Nomenclature	3	0	3
Machinability of species and reconstituted wood			
Hardwoods	13	18	31
Softwoods	2	21	23
Plywood	2	4	6
Particleboards and sandwich boards	8	4	12
Fiberboards, hard and soft	2	6	8
Total machinability	27	53	80
Bibliographies	2	2	4
Reports of research programs and industrial developments	1	10	11
Patents	14	1	15
Safety, including noise[1]	51	2	53
	415	269	684

Noise alone: 36 references. Chain saw safety: 16 references.

(1970) found that surface quality was improved and cutting forces reduced in cuts across end grain (90-90 mode) if the knife was vibrated in the direction of the cut.

McKenzie and Cowling (1971) observed dulling phenomena, and related edge wear to cutting forces (90-90 mode); chip rupture patterns were also explained. Noguchi (1970) and Noguchi et al. (1971) reported cutting data obtained with a pendulum dynamometer.

Woodson and Koch (1970) orthogonally cut southern pine earlywood and latewood in the three principal modes and presented photographs of chip types together with tabulated and graphed data on cutting forces; also given were regression equations for cutting forces, in terms of tool angles and wood properties.

Peripheral milling. — Mori (1970) found that the relationship between cutting force and undeformed chip thickness was affected by the ratio of depth of cut to cutting-circle diameter; there was an optimum ratio.

Stewart (1971) studied chip types resulting from peripheral milling with cutterheads having helical knives. He observed that veneer-like chips resulted. when feed direction was arranged so that wood grain was parallel to the cutting edge.

Solov'ev (1971) developed formulas for calculating cutting forces and their components.

Barking. — Hyler (1969) described types of barking machines used in North America. Kanauchi (1970) reported laboratory data on the effectiveness of various types of barking tools. Shelyapin and Razumovskii (1971) related pulpwood diameter and drum rotation speed to the degree of bark removal from larch. Data on mechanical ring barkers were given by Berlyn (1970) and Birjukov et al. (1970).

The subject commanding most attention was removal of bark from barky chips; papers germane to this subject included: Liiri (1961); Arola (1970); Bumerl (1969); Erickson (1968, 1970, 1971); Lea and Brawn (1970); and Julien et al. (1971).

Sawing. — Causes of inaccurate sawing were reported by Birkeland (1968) and Breznjak and Herak (1970). Sawmill layout and developments were discussed by the Home Merchants Association of Scotland (1970), Mügge (1970), Neville (1970), and Thunell (1971).

Riberolles (1968) and Dargan (1969) de-scribed novel methods of sawing wooden tubes and round dowels.

Reynolds (1969) explained the SHOLO (SHOrt LOg) process for manufacturing pallet parts from low-grade hardwood logs.

Clark (1968, 1969) described the operations of mills using high-strain bandsaws and thin-kerf circular saws. Cumming (1969) and Porter (1971) further discussed high-strain bandsaws; and Sugihara (1970) defined stresses in bandsaws.

Thin-kerf circular sawing was discussed by Clark (1968), Kintz (1969), Salemme (1969), and Schliewe (1969).

Feeds and speeds when ripping with circular saws was the subject of work by Kubasak (1968), Breznjak and Moen (1969), McKenzie (1970), Stefaniak (1970), Belikov (1971), Koshunyaev et al. (1971), Nakamura (1971), Pashkov and Bodalev (1971), and Radu et al. (1971).

Crosscutting of roundwood was studied by Kubasak (1968), Degermendzhi (1969), Kondrat'ev (1969), and Vasil'ev (1970, 1971). Bramhall and McLauchlan (1970) described a method of circular sawing microsections as thin as 100 mμ.

Macomber (1969) provided detailed procedures for smooth-ripping 2 by 4's from wider lumber in such a manner that the edges need no further planing, and the corners are rounded.

Yu (1966) was able, under certain conditions, to friction-saw softwood with a 14-inch, 14-gauge, toothless steel disk rotated at 4,620 r.p.m.

The chain saw will likely remain for 20 years a mainstay in Canadian logging, according to the *Pulp and Paper Magazine of Canada* (Anonymous 1969). The chain saw is, however, the subject of much research aimed at reducing vibration and noise (Fushimi and Shigaki 1971; Szepesi 1970; *see also* Safety, including noise). Subjects of much study were chain saw guidebars (Gul'ko and Matjuskin 1970; Tudoran et al. 1971) and teeth (Janco 1967; Pahlitzsch and Peters 1968; Szepesi 1969; Kuhnke 1970). Reynolds et al. (1970) analyzed chain saw cutting forces, and Shackelberg (1971) studied cutting time.

Franz (1970) found that corrugated board slit with high-energy liquid jets had smooth edges free of damage and debris.

McMillin and Harry (1971) reported successful laser cutting of various southern pine products. With an air-jet assisted carbon-dioxide laser of 240-watt output, maximum feed speed decreased with increasing workpiece thickness in both wet and dry samples, but was unrelated to grain direction. Feed speeds averaged 99.1 and 14.6 inches per minute for lumber 0.25 and 1.00 inch thick, respectively. Doxey (1970), March (1970), and Ready (1970) also reported on industrial applications of lasers.

Arola (1971) sheared roundwood bolts with a single blade cutting against an anvil and found shear force positively correlated with blade thickness and wood density; wood temperature was negatively correlated with shear force; low-friction coatings reduced shear force, but did not reduce damage from splits.

Johnston and St-Laurent (1970, 1971) found that V- and U-shaped shear blades yielded promise of reduced splitting damage as did Shilke et al. (1971), they observed that cutting force and wood damage increased with decrease in temperature, increase in moisture content, and increase in specific gravity.

In another study, Johnston and St-Laurent (1971) found that conveyorized boards could be end-trimmed with shears; sheared ends, while smooth, frequently had splits one-half-inch deep.

In a study of slow-speed shearing of southern pine logs, Koch (1971) found that shear forces were positively correlated with sharpness angle, wood density, and log diameter; shear-induced splits were most severe (averaging 1.4 inches deep) in low-density logs sheared with a blade having a large sharpness angle.

Kubler (1960) reported on Russian experiments in which vibrated knives substantially reduced force required to cut wood across the grain. Hamamoto and Mori (1970) confirmed these observations.

McIntosh and Kerbes (1969) found that shear-caused splits resulted in lumber losses varying from 1 percent per tree in warm weather to 5 or 7 percent in very cold weather. Tuovinen (1970) also studied lumber degrade caused by shearing saw logs; trimbacks shortened lumber 8 to 20 cm. depending on species and temperature.

White et al. (1969) found that operators of mobile hydraulic tree shears spent 11 percent of their time positioning the shears, 22 percent shearing, and 67 percent moving from tree to tree.

Jointing, planing, molding, and shaping. — Goodchild (1967) noted that power requirement increases with feed speed, depth of cut, and peripheral velocity of the cutting edges; it also increases as rake angle is reduced and as the width of heel caused by jointing becomes greater.

Stewart (1970) found that planing hard maple panels across the grain yielded flakes suitable for high-strength particleboard. In another paper dated 1970, he observed that depths of cut up to 1/16-inch did not significantly influence the amount of chipped grain when planing dry *Platanus occidentalis;* chipped grain was minimized by reducing rake angle to 20° or less.

Raknes (1969), Roth (1970), and Tsuzuki et al. (1971) reported on methods of finger jointing and end-splicing. Kryazhev (1971) discussed the advantages of down-milling when cutting end-grain wood.

Boring, routing, and carving. — From observations of thrust, tool life, and hole quality, Edamatsu and Nagahara (1970) made recommendations for boring tropical woods. Noak (1970) discussed economic considerations for dowel borers.

Radu (1969) evaluated five drill types and recommended an improved twist drill with smooth flutes; in 1970 he presented methods of calculating torque and thrust when boring beech, oak, spruce, and particleboard.

Data on the routing process were reported by Pahlitzsch and Sandvoss (1969), Theien (1970), Kondo et al. (1971), and Mori et al. (1971).

Machining with coated abrasives. — Pahlitzsch (1970) provided a review of the state of the art. Noda and Umetsu (1971) studied particle sharpness, fracture, and loss during abrasive machining of *Fagus crenata.* Meyer (1970, 1971) developed a technique for evaluating the progression of dulling of an abrasive belt.

Potrebic (1969) related the surface smoothness of beech wood and the grade of sandpaper used to machine it. And Hirst (1971) described wood finishes that most successfully hid machining marks; marks were less perceptible on flat

5

and satin surfaces than on gloss. Black flat paints hid marks better than white or clear coatings. Abrasive score marks were nearly invisible under three coats of paint.

Sternal (1969) noted that achievement of sawing accuracy must precede widespread adoption of abrasive planers. Stewart (1970) found that abrasive planing required about six times more power than knife planing; power was positively and linearly correlated with width of cut, depth of cut, and feed rate. Power to sand across the grain was 20 to 25 percent less than that needed to sand parallel to the grain. With increased feed rate, quality of abrasive-planed surfaces remained constant.

Veneer cutting. — Grosshenig (1971) summarized veneer production from log preparation through clipping of the veneers. Baldwin (1971) described a program found effective in increasing veneer yields in Pacific Northwest softwood mills.

Feil and Godin (1970) described a number of common defects in veneer and outlined causes and corrective measures for each. Meriluoto (1971) noted that if splitting is to be avoided, frozen birch bolts should be thawed in cold water (0 to 50° C.) before being heated for processing. Sevast'yanov (1971) published formulas and nomograms easing computation of temperatures in a cooling veneer bolt. Nikolaev (1970, 1971) reported on the stability of veneer bolts during peeling.

Basic studies of veneer formation were reported by Hayashi et al. (1970), Mori and Kubota (1970), Palka (1970), Plesko (1970), Lyubchenko (1971), and Sugiyama and Mori (1971).

Porter et al. (1969) successfully tested a roller nosebar lubricated by high-pressure water; such a nosebar need not be power-driven.

Hayashi and Tochigi (1969), and Tochigi and Hayashi (1971) reported that Japanese red birch was successfully peeled with an air-jet nosebar; the air jet emanated from a slot in the position normally occupied by a metal nosebar.

Veneer cutting of various species was reported as follows: Numerous Indian woods by Jain et al. (1967, 1968, 1969, 1970); redwood by Cumming and Collett (1970); spruce by Feihl and Godin (1970); larch by Cernenko and Zolotov (1970); southern pine by Woodson and Koch (1970); and West Coast softwoods by Baldwin (1971).

Chipping, flaking, and grinding. — Neusser and Krames (1970) defined the most important measurements required to characterize pulp chips. Optimum chip thickness, a critical dimension in kraft pulping, was determined to be 3 mm. by Borlew and Miller (1970). Men'shikova et al. (1971) found that roller-pressing of resinous pine chips affected rate of impregnation with benzine.

McKenzie (1970) illustrated and discussed cutting action, characteristics, and advantages of six types of chippers and described methods of choosing a suitable size of chipper and estimating its power requirement. Hyler (1970) reviewed chipping and chip handling methods for sawmills. Chiviksin (1971) and Koperin et al. (1971) described the economics and techniques of chipping low-grade wood.

Knife angles of disk chippers were studied by Rushnov (1969) and Kalashnikov (1971).

Altman (1969) described a disk chipper for converting lumber trim ends to pulp chips; and Isomaki (1970) evaluated the performance of 16 types of chippers designed for this purpose.

McLean (1969) described the configuration and performance of a machine for making veneer-like pulp chips from log ends trimmed from peeler bolts (lily pads).

Uusvaara (1971) found that birch veneer chips from Finnish plywood plants were thinner and shorter than conventional pulp chips.

Chipping headrigs continued to receive much attention. Aas (1969) recommended their application in Scandinavian mills. Koch (1969, 1970) and Anderson and Kaiser (1970) cited their advantages for southern pine mills in the United States. Dobie (1970) concluded that chipping headrigs would be 11 to 153 percent more productive if logs were sorted by diameter class prior to conversion.

Installations of various types of chipping headrigs were described by Neville and Bennie (1969), Anonymous (1970), Kotesovec (1970), Sullivan (1970), and Galloway and Thomas (1971).

Mobile chippers were evaluated by Cadenhead (1970), USDA Forest Service (1970), Hensel (1971), Laubgan (1971), Plokhov (1971), Sterle (1971), and Swan (1971).

Pahlitzsch and Sommer (1971) cut pine flakes 0.3 mm. thick with a drum-type chipper and

6

concluded that, with wood of 30-percent moisture content, optimum knife angles were as follows: clearance 6°, sharpness 49°, and rake 35°.

Defibrating. — Perry (1970) discussed pulp grinding technology and results of a cooperative program involving manufacturers of grinding machines, pulp stones, and groundwood pulp. Norton Company (1968) described stones for grinders in sizes that may absorb 14,000 horsepower and produce more than 200 tons of pulp per day.

Atack and May (1970), by photographic techniques, showed that particles and fiber bundles align radially between the counter-rotating disks of a disk refiner. May (1970) shed new light on misalignments that occur during operation of disk refiners.

Properties of the cutting edge and cutter. — Hard alloy cutting tips were studied by Wuttkowski (1968), Borikov and Pozdeev (1970, 1971), Quezada (1970), Stefaniak (1970), Dunaev (1971), Mihut (1971), Plastov (1971), and Tallack (1971).

Dem'janovskij (1970) and Ohsako and Domoto (1971) related composition of knife steel to cutting performance.

Chardin and Froidure (n. d.), in a 3-volume work devoted to tropical species, diagrammed cutting-edge wear as affected by wood moisture content and density, chip thickness, duration of cutting, and cutting speed.

Additional data on wear of cutting edges were provided by Cowling and McKenzie (1969), Deger et al. (1969), Timonen (1969, 1970), Barz (1970), Ivanovskij et al. (1970), Pahlitzsch and Sandvoss (1970), St -Laurent (1970), Stefaniak (1970, 1971), Vlasov and Litvinov (1970), Barz and Breier (1971), Bridges (1971), McKenzie and Cowling (1971), Ohsako et al. (1971), and Plastov (1971).

Welding, tensioning, straining, and stability of bandsaws were studied by Feoktistov (1960), Dobrynin (1962, 1970), Berlin (1969), Trubnikov (1969), Abeels (1970), Ayoma (1970, 1971), Foschi and Porter (1970), Lyth (1970), Saito and Mori (1970), Thunell (1970), McKenzie (1971), and Yur'ev and Veselkov (1971).

Stability of sash gangsaws received the attention of Derjagin (1969), Simienski (1969), Wassipaul (1970), and Veselkov (1971).

Data on circular-saw stability were provided by Pahlitzsch and Rowinski (1966), Kostrikov (1969), Salemme (1969), Schliewe (1969), Stahiev (1969), Barz (1970), Mote (1970), Stahiev (1970), Friebe (1970), Mote (1970), Stahiev (1970), Strzelecki (1970), and Pahlitzsch and Friebe (1971).

Temperature data, a subject closely related to stability, were given for circular saws by Grube et al. (1969) and by Okushima and Sugihara (1969). Sanev (1969), and Sanev and Pljusnin (1969, 1970), discussed temperature variations in bandsaws.

Saw swaging and sharpening techniques were described by Kostrikov (1969), Anonymous (1970), Barz (1970), Borovikov and Pozdeev (1970), Dogaru and Capraru (1970), Beck (1971), and Borovikov and Pozdeev (1970).

Pahlitzsch and Peters (1968) recommended optimum grinding angles for chain saws; they found that chipper-type chains have substantial advantages over scratch-type chains.

Dunaev (1970, 1971) recommended procedures for grinding planer knives. Godin (1969) explained the technique of grinding veneer knives; and Gusev and Rushnov (1971) proposed a method for sharpening swept-back knives in helical-disk chippers.

Schmutzler (1969) found that centrifugal forces tending to dislodge cutter knives are greater than the cutting forces; he analyzed the clamping force required to retain knives in a cutterhead.

The friction of wood against tool steel was studied by Yu (1966), Cuprin (1969), Ettelt (1969), Grubov (1970), Lemoine et al. (1970), McMillin et al. (1970), and Pool (1971). The friction coefficient is about 0.2; it can be reduced by Teflon coatings to perhaps half this value.

Computer, tape, and card control of machines. — Hallock and Bulgrin (1970) and Bulgrin (1971) reviewed the state of research on application of computers to make ripping and cross-cutting decisions during conversion of hardwood boards into dimension lumber.

Mundler (1971) described systems for thickness control of lumber cut on log bandsaws and resaws.

Rugge (1971) reviewed the state of the art and concluded that tape- or card-controlled machinery can produce parts economically.

Research instrumentation and techniques. — Barz and Breier (1969) described an apparatus evaluating cutting edge wear. Lin and Cumming (1970) devised a method of printing grid patterns on green wood; such grids are useful in studies of chip formation. Malcolm and Koster (1970) explained the use of photoelastic plastic to examine the location and magnitude of stresses in tooth assemblies of inserted-tooth saws. Peters and Mergen (1971) and Porter et al. (1971) described methods for measuring surface roughness.

Nomenclature. — Abeels (1970) suggested standardized nomenclature for bandsaws; and Wuster (1970) described the problems of standardizing all kinds of sawblades. Ford-Robertson (1971) published volume 1 in a proposed multilingual compilation of forestry terms.

Machinability of species and reconstituted wood. — Data on machining of softwoods are given in a number of references:

Species	Cutting process	Reference
Siberian larch	Slicing	Cerenco and Zolotov, 1970
Redwood	Peeling	Cumming and Collett, 1970
White spruce	Peeling	Feihl and Godin, 1970
Southern pine	Shearing	Koch, 1971
Lodgepole pine and spruce	Shearing	McIntosh and Kerbes, 1969
Southern pine	Laser cutting	McMillin and Harry, 1971
Southern pine	Orthogonal cutting	Woodson and Koch, 1970
Scots pine	Sanding	Zajac, 1970

Bosquet and Flann (1970) described yield of hard maple lumber when logs were converted by three sawing patterns and five edging practices. Howland (1971) found that only 27 percent of the cubic volume of *Eucalyptus saligna/grandis* logs was recoverable as dry lumber. Jaìn et al. (1967, 1968, 1969, 1970) continued studies of veneer cutting of Indian

timbers. Kukachka (1970) and Chardin and Froidure (n. d.) presented data on a wide range of tropical woods. Stewart, in papers dated 1970 and 1971, gave basic data on orthogonal cutting of white ash against the grain and on planing of hard maple and sycamore.

Hayashi et al. (1970) found that the cutting resistance of plywood assembled with thermosetting resins was greater than that of plywood glued with thermoplastics.

Machinability of particleboard was evaluated by Fukui and Kimura (1968), Neusser and Schall (1970), Pahlitzsch and Sandvoss (1970), Radu (1970), Theien (1970), Bridges (1971), Plastov (1971), Shen (1971), and Stefaniak (1970, 1971).

Fiberboard cutting was studied by Fukui and Kimura (1968), Franz (1970), McKenzie (1970), McKenzie and Cowling (1970), Neusser and Schall (1970), Pahlitzsch and Sandvoss (1970), and McMillin and Harry (1971).

Bibliographies. — McMillin (1970), in continuation of a series inaugurated in 1957, abstracted the 1968-1969 literature on wood machining. His compilation listed some 300 articles and commented on notable developments.

Thunell (1971) listed 388 citations of world literature pertaining to bandsaws.

Smith (1971) listed 27 citations in a review of the literature on noise of woodworking machines.

Safety, including noise. — Sakurai (1969) described the effect of chain saw vibrations on the blood pressure and pulse rate of operators. Fujii (1970) found that 90 percent of chain saw vibrations are transmitted to the back of the operator's hand, but only 10 percent to his shoulder. Sakurai et al. (1968) and Gebert (1971) determined which types of vibration isolators were most effective. Harstela (1971) found that incidence of "white fingers," caused by chain saw vibrations, was faster than recovery, which required 2 to 15 minutes or more. Keighley (1970) noted that incidence of "white fingers" could be reduced if the chain saw was rested as much as possible against tree or thigh, and gripped lightly with gloved hands; he recommended that hands and arms be warmed before using a chain saw, and that rests be taken frequently.

Bulboaca (1971) summarized causes of bandsaw accidents, described a rapid braking system, and suggested preventive measures.

Calin (1971) described modifications to veneer slicers to prevent accidents, and Wiklund (1971) discussed kickback of power saws.

Safety practices applicable to a range of woodworking equipment in England are described by the Department of Employment and Productivity (1970).

The large number of references on noise caused by woodworking machines attests to current interest in the subject. These references are summarized in the following tabulation.

Machine type and subject	Reference
Woodworking plants in general	
General information and review	Klimkiewicz (1969)
Literature review	Smith (1971)
Noise level in English woodworking shops	Zeller and Paul (1969)
Noise level in Jugoslav factories	Prijatelj (1970)
Noise level in Soviet sawmills	Birjukov and Koz'jakov (1970)
Walsh-Healey Act	Lamb (1971)
Instrumentation for noise survey	Bohrer (1971)
Noise control and hearing conservation programs	Lamb (1971); Landry (1971); Tröger (1971)
Noise control by equipment design	Allen (1970); Gatley (1971)
Noise control by encapsulation	Schmutzler (1970)
Reduction of chain-conveyor noise	Kudrijasov (1970)
Chain saws	
Noise evaluation	Montaner (1970); Fushimi (1971); Myles et al. (1971)
Physiological effects	Gellert (1971)
Noise reduction	Degtjarev (1970)
Chain mortisers	
Noise evaluation	Montaner (1970)
Circular saws	
Noise evaluation	Montaner (1970)
Causes of noise	Pahlitzsch and Friebe (1971)
Noise reduction	Cizevskij and Ceremnyh (1970); McKenzie (1970); Malcolm (1971); Strzelecki (1971); Werner (1971)
Disk chippers	
Noise evaluation	Starzinskij (1970, 1971)
Drum barkers	
Noise evaluation	Tanaka and Sakurai (1970)
Planers	
Noise evaluation	Pahlitzsch and Liegmann (1956)
Variables affecting sound pressure	Kitayama and Sugihara (1969); Tanaka et al. (1970, 1971)
Cutterhead modification	Potanin (1970)
Sound insulation of 4-side planer	Bobin (1970)
Spindle shapers	
Noise evaluation	Tröger (1969, 1970)

9

HISTORY AND GENERAL TEXTS

Koch 1969,1970. (Sawing; Chipping headrigs and edgers)

Koch, P. 1971. TECHNOLOGICAL DEVELOPMENTS IN THE SOUTHERN PINE INDUSTRY. For. Farmer 30(7): 16-20.
Reviews the significance of the southern pines in the U.S. economy and summarizes trends in manufacture.

Thunell, B 1971. RESEARCH TRENDS IN MECHANICAL WOODWORKING. Wood Sci and Technol. 5: 63-72.
Cutting theory, cutting energy, tool wear, kerfless cutting, kerf

chips, chipping headrigs, and systems engineering are briefly discussed. The bibliography contains 96 citations.

Thunell. 1971. (Sawing, General)

Thunell, B. 1971. SELECTING MACHINERY FOR VARIOUS ANNUAL OUTPUTS OF SAWN TIMBER. Schweiz. Z. Forstwes. 3: 91-104.
Volume and composition of output define equipment requirements Final selection is determined by economic considerations.

PROPERTIES OF WOOD

Albers, K. 1971. BEHAVIOUR OF WOOD-BASED MATERIALS UNDER TRANSVERSE COMPRESSION. Holz Roh- und Werkst. 29: 94-96.
Allowable crushing strength cannot be deduced from stress at failure, because failure is preceded by irreversible crushing. Allowable transverse crushing stress should therefore be related to the initial portion of the stress-strain curve that depicts elastic behavior. Crushing strengths are given for a variety of particleboards, veneers, and fiberboards

Futo, L. P. 1969. QUALITATIVE AND QUANTITATIVE EVALUATION OF THE MICROTENSILE STRENGTH OF WOOD. Holz Roh- und Werkst. 27: 192-201.
Describes apparatus for tensile testing of small samples down to

individual fibers, and for microscopic observation of the developing fractures. Results reported include the variation in strength across the individual growth ring of spruce, effect of thickness on strength of micro-specimens of earlywood and latewood cut in different planes, and microscopic mode of fracture.

Kukachka. 1970. (Machinability of species and reconstituted wood; Hardwoods)

Sharma, M. 1971. A NOTE ON SILICA CONTENT IN TEAK. J. Indian Acad. Wood Sci. 2(1): 25-26.
Silica content in 23 samples of Tectona grandis varied with region of origin from 0.2 to 3 04 percent. Silica was of the vitreous type and was found only in fibers and vessels.

ORTHOGONAL CUTTING

Hamamoto, K, and Mori, M. 1971. FUNDAMENTAL STUDIES ON LOW FREQUENCY CUTTING OF WOOD. I. MECHANICS OF VIBRATORY CUTTING PARALLEL TO FEEDING DIRECTION. J. Jap. Wood Res. Soc. 17(5): 137-146.
Wood was cut orthogonally in all three modes with a knife vibrated parallel to the feed direction (visualize a carpenter tapping a chisel with a mallet). Vibration amplitudes of 0.05 to 800 μm improved surface quality when cutting across end grain in the 90-90 mode. Mean cutting forces in this mode were negatively correlated with the product of amplitude and frequency (10 to 60 Hz); i.e., as amplitude and frequency of vibration increased, cutting force decreased.

Hayashi et al. 1970. (Machinability of species and reconstituted wood; Plywood)

Hayashi et al. 1971. (Veneer cutting)

Kato, K., Tsuzuki, K., and Asano, I. 1971. STUDIES ON VIBRATION CUTTING OF WOOD. I. J. Jap. Wood Res. Soc. 17(2): 57-65.
Air-dry Chamaecyparis obtusa Endl and Fagus crenata Blume were orthogonally cut parallel to the grain (90-0 mode) and perpendicular to the grain (90-90) with knives vibrated in directions parallel to the cutting edge (i.e., laterally) or perpendicular to the edge (i e. in the feed direction). Frequency was about 20 kHz and double amplitude was varied from 15 to 30 μm. Clearance angle was held constant at 5° and sharpness angle was varied from 21.5 to 50°. When the vibrating knife was cutting parallel to the grain, Franz Type II chips occurred with smaller knife sharpness angles (greater rake angles) and deeper cuts than in conventional cutting. As amplitude of vibration was increased from zero to the maximum, chip type changed from Franz Type I to Franz Type II. When cutting was parallel to the grain, parallel cutting force on the vibrated knife was 10 to 50 percent of that exerted by a conventional knife. When cutting

was perpendicular to the grain, parallel force on the vibrated knives was 5 to 10 percent of that exerted by a conventional knife. Also, normal forces were less for vibrated than for conventional knives; normal force was least with knives vibrated laterally. Horizontal cutting force was the same whether knives were vibrated laterally or in the feed direction.

Kubler. 1960. (Sawing; Shearing)

McKenzie, W. M., and Cowling, R. L. 1971. A FACTORIAL EXPERIMENT IN TRANSVERSE-PLANE (90/90) CUTTING OF WOOD. I. CUTTING FORCE AND EDGE WEAR. Wood Sci. 3: 204-213.
With data from slow orthogonal 90/90 cutting of wood species having a wide range of specific gravity and moisture content, the effects on average cutting force of rake angle, edge radius, chip thickness, and wood properties were incorporated into semiempirical regression equations and nomograms. A positive linear correlation between maximum and average force and the square root of the edge radius was confirmed. However, the effects of wear were not mainly due to increase in extreme edge radius. Because even the sharpest edges did not approach ideal, and also because edge radius changed during the experiment, the results could not be used to test theoretical models. Blunting produced an edge profile that was not circular but more like a Gothic arch. The radius of the extreme edge tended toward a value of approximately 1 μm. whether the initial radius was greater or smaller than this. Recession rate of the edge increased with increasing rake angle (decreasing sharpness angle) and decreased with increasing initial edge radius.

McKenzie, W. M., and Cowling, R.L. 1971. A FACTORIAL EXPERIMENT IN TRANSVERSE-PLANE (90/90) CUTTING OF WOOD. II. CHIP FORMATION. Wood Sci. 4: 55-61.
As part of an experiment in slow orthogonal 90/90 cutting of wood, chip formation was photographed to record the effects of

wood properties, tool rake angle (15-45°), edge bluntness (radius 0.2-5 μm.), and chip thickness (0.1-0.8 mm.). Moisture content and species were very important determinants of chip rupture pattern, and it is suggested that their effects are related to changes in the ratio between modulus of rupture and shear strength. Depth of damage below the cut surface was predominantly affected by chip thickness (positive correlation). Rake angle and edge radius had minor effects.

Noguchi, M. 1970. WOOD CUTTING WITH A PENDULUM DYNAMOMETER. Mem. Coll. Agric., Kyoto Univ. 96, p. 9-46. Japan.
Rake angle (0° to 40°) had strong negative correlation with cutting energy. Cutting across the grain in the 90-90 mode required more energy than 90-0 or 0-90 cutting. Depth of cut, wood specific gravity and moisture content (0 to 60 percent) were positively correlated with energy required – but not always linearly. Cutting energy, as measured by the pendulum dynamometer, is as useful as cutting force in evaluating wood machinability.

Noguchi, M., Imamura, K., and Sugihara, H. 1971. MEASUREMENTS OF ENERGY CONSUMPTION IN CUTTING WOOD WITH A SWAGE-SET SAW-TOOTH BY MEANS OF A PENDULUM DYNAMOMETER (VIII). Holztechnologie 12(2): 104-108.
Machinability of Southeast Asian timbers (red lauan, kapur, selangan batu, and manggasinoro) and Japanese beech was investigated with a single swage-set saw tooth mounted on a

pendulum dynamometer. Factors evaluated included cutting direction, amount of swage, rake angle, and wood moisture content.

Stewart. 1971. (Machinability of species and reconstituted wood; Hardwoods (hard maple))

Woodson, G. E., and Koch, P. 1970. TOOL FORCES AND CHIP FORMATION IN ORTHOGONAL CUTTING OF LOBLOLLY PINE. USDA For. Serv. Res. Pap. SO-52, 25 p. South. For. Exp. Stn., New Orleans, La.
When cutting (at a velocity of 2 inches per minute) was in the planing or 90-0 direction, shallow cuts, intermediate to high wood moisture content, and rake angles of 5 and 15° favored formation of the Franz Type II chip and accompanying good surfaces.
In the 0-90 direction, a knife with 70° rake angle cut the best veneer; wood cut saturated yielded the highest proportion of continuous veneer, although saturated earlywood developed some compression tearing
When cutting was across the grain in the 90-90 direction, McKenzie Type I chips were formed and the best surfaces were achieved when a knife having 45° rake angle was used to cut saturated wood; earlywood was more difficult to surface smoothly than latewood.
For each cutting direction, regression equations were developed to state average cutting forces (normal and parallel) in terms of rake angle, depth of cut, specific gravity, and moisture content.

PERIPHERAL MILLING PARALLEL TO GRAIN

Les'kiv, V. D. 1969. DETERMINING THE SPECIFIC CUTTING FORCE WHEN WORKING WOOD WITH ROTARY CUTTERS. Lesn. Zh. 12(3): 84-89. Arkhangelsk.
A study of the cross-sectional areas of chips and changes in cutting force.

Mori, M. 1970. AN ANALYSIS OF CUTTING WORK IN PERIPHERAL MILLING OF WOOD. II. THE CUTTING FORCE, POWER, AND ENERGY REQUIREMENTS IN UP-MILLING PARALLEL TO WOOD GRAIN. J. Jap. Wood Res. Soc. 16: 1-9.
Continuation of a study of a knife during formation of a single chip. The relationship between cutting force and undeformed chip thickness was affected not only by feed per knife but also by the ratio of depth of cut to cutting-circle diameter. Equations indicated that specific cutting power is positively correlated with number of knives in the cutterhead, and with all factors fixed except cutting-circle diameter, there is an optimum diameter at which power demand is at a minimum. Other equations showed that specific cutting energy is negatively correlated with feed per knife, and that specific cutting energy is minimized at an optimum ratio of depth of cut to cutting-circle diameter.

Mori. 1971. (Boring, routing, and carving)

Simonov, A. S. 1969. THE RESISTANCE OF WOOD TO CUTTING IN ROTOR-TYPE MACHINES. Lesn. Zh. 12(6). 73-77. Arkhangelsk.

Solov'ev, A A. 1971. ERRORS IN DETERMINING THE MEAN CUTTING FORCE FROM THE MEAN VALUES OF ITS COMPONENTS. Derevoobrab. Prom-St'. 20(4): 8-10. Moscow.
Develops formulae for calculating mean values of the cutting force and its components in machining with rotating cutters.

Stewart, H. A. 1971. CHIPS PRODUCED WITH A HELICAL CUTTER. For. Prod. J. 21(5): 44-45.
Most peripheral milling (planing) heads carry straight knives; lumber is customarily fed longitudinally in such a manner that the grain is perpendicular to the rotational axis of the cutterhead. In the research here reported, the cutterhead had helical knives (30 and 45° helix angles), and grain direction in relation to cutterhead axis was varied from 45° to 90°. Two basic chip types were observed, one type was similar to a conventional planer shaving, and the other resembled thin veneer chips. The veneer-like chips were produced when feed direction was arranged so that wood grain was parallel to the cutting edge.

BARKING

Arola, R.A. 1970. TREMENDOUS CHALLENGE REMAINS – EFFECTIVE DEBARKING OF CHIPS. Pulp and Pap. 44(1): 79-83.
Illustrated review of all attempts to separate bark from chips. No economical method has been found. The author thinks that two or more complementary methods will be required, and that combinations of methods will vary with species.

Bizjukov, M. S., Ganzura, L. P., and Pankov, J. P. 1970. REDUCING THE IMPACT WHEN A LOG MEETS THE BARKING BLADES IN BARKING MACHINES. Derevoobrab. Prom-St'. 19(5): 14-16. Moscow.

Berlyn, R. W. 1970. SLITTING FROZEN BARK TO IMPROVE THE PERFORMANCE OF A RING BARKER. Pulp and Pap. Res. Inst. Can., Woodland Pap. 21, 16 p.
Slits 1/4-inch wide were cut lengthwise through the bark of frozen (0° F.) spruce logs, which were then sent through a ring barker. As number of slits was increased from none to eight, removal of inner bark increased substantially and variation diminished. The effect of the slits was most pronounced on logs that were difficult to bark. A tendency of the scraping tool to gouge into wood near the slits led to substantial losses. The slits also contributed to dynamic instability of the tool and resulting loss of barking efficiency.

11

Berlyn, R.W. 1970. THE EFFECT OF SOME VARIABLES ON THE PERFORMANCE OF A RING BARKER. Pulp and Pap. Res. Inst. Can., Woodland Pap. 15, 22 p.
A 30-inch Beloit Hi-Speed Ring Debarker with twin rings was used on tree-length black spruce, red spruce, balsam-fir, and jack pine harvested over a 7-month period preceding the study. When feed speed was high and loads on the scraper arms were low, bark removal became less complete as log diameter, bark thickness, and bark-to-wood bond strength increased. For diameters in the 4- to 10-inch range, quality of barking was improved by slowing the feed or increasing the load on the arms; either measure diminished the effect of the other A mathematical model was postulated in which the percent of bark removed over a 4-foot length was expressed as a function of bond strength and log diameter; the equation accounted for 81 percent of the observed variation.

Bumerl, M. 1969. A NEW METHOD OF WOOD TREATMENT AND DEBARKING. Lesn. 15(11): 1019-1030.
In a method covered by Czechoslovakia patent 130,192 (September 21, 1968), electricity is used to reduce bark-wood cohesion by raising the temperature of the cambium layer − thus softening it. Most tests and measurements were conducted on Norway spruce with 220 volts AC and a frequency of 50 Hz. For best results, logs should have a moisture content in excess of 70 percent.

Butora, A. 1970. BARKING STEMWOOD WITH THE 'RINDAB'. Ber. Eidg. Anst. Forstl. Versuchswes. 30,22 p. Birmensdorf, Zurich.

Butora, A. 1967. STUDIES ON BARKING LONG LOGS. Ved. Pr. Vysk. Ust. Lesn. Hosp. Zvolen 8, p. 225-251. Czech.

Erickson, J. R. 1968. BARK SEPARATION DURING CHIPPING WITH A PARALLEL KNIFE CHIPPER. USDA For. Serv. Res Note NC-54, 3 p. North Cent. For. Exp. Stn., St. Paul, Minn.
Frozen and unfrozen winter-cut bolts of Acer saccharum, Acer saccharinum, Tsuga canadensis, Populus tremuloides, and Picea glauca were chipped with knives cutting parallel to the grain Chips from frozen bolts contained a higher percentage of bark-free wood, bark, and fines (and a lower percentage of bark bonded to wood) than did chips from unfrozen bolts.

Erickson, J. R. 1970. BARK REMOVAL AFTER CHIPPING − A PROGRESS REPORT. Pulp and Pap. Mag. Can. 71(5): 78-79.
Research is handicapped by lack of basic data on the physical and mechanical properties of bark and its adhesion to wood. Much bark is separated from wood during chipping. Additional methods of breaking the wood-bark bond include chemical, thermal, and pressure processes. Principal methods of bark segregation include soaking, air aspiration, air flotation, Vac-sink, compression between nip rolls, and segregation based on differences in di-electric properties of bark and wood. Methods used will depend partly on the amount of bark that can be tolerated by chip users.

Erickson, J. R. 1971. BARK-CHIP SEGREGATION: A KEY TO WHOLE-TREE UTILIZATION. For. Prod. J. 21(9): 110-113.
Discusses three methods of bark-chip segregation: compression with nip rolls, air flotation, and use of immiscible liquids. The compression method is most promising. Air flotation removes some types of bark that the compression method does not. Aspen bark and wood have been separated in the laboratory through the use of two immiscible liquids of different specific gravities. Combinations of these or other methods may be necessary when pulp chips are made from entire trees or barky residues.

Hyler, J. E. 1969. MACHINES FOR BARK REMOVAL. South. Lumberman 219(2720): 31-34; (2721): 29-34; (2722): 31-36; (2723): 21-25.

Describes and illustrates various types of barkers available in North America, their modes of operation, capabilities, and ancillary equipment.

Julien, L. M., Edgar, J. C., and Conder, T. M. 1971. A DENSITY-GRADIENT TECHNIQUE FOR OBTAINING WOOD AND BARK CHIP DENSITY. USDA For. Serv. Res. Note NC-112, 4 p. North Cent. For. Exp. Stn., St. Paul, Minn.
Describes a density-gradient column to measure wood and bark chip densities for segregation purposes. Tests with the column indicated that if aspen chips with moisture content of 95 percent or higher (dry basis) are placed in a liquid with density of 0.90, the wood should float and the bark should sink. If chip moisture content is less than 90 percent and density of the liquid is 1.00 the wood should float and the bark should sink.

Kanauchi, T. 1970. MECHANICAL CHARACTERISTICS OF BARKING TOOLS AND MACHINES. I. FUNDAMENTAL EXPERIMENTS ON TYPES OF TOOLS AND BARKING VELOCITIES. II. EFFECT OF BARKING VELOCITY ON EFFICIENCY OF BARKER MANUFACTURED FOR TRIAL. III. EFFECTS OF LOG DIAMETER ON EFFICIENCY OF BARKER MANUFACTURED FOR TRIAL. J. Jap. Wood Res. Soc. 16: 65-69. 219-229.
Reports degree of bark removal and energy consumption when barking birch, larch, sen, alder, and basswood with rotating cutters and chains of various designs.

Lea, N. S., and Brawn, J. S. 1970. A NEW WOOD RECLAMATION PROCESS. TAPPI 53. 622-624.
Describes a new flotation process for recovering pulpable wood from chip-screen rejects. Illustrates the prototype installation in Everett, Wash., which removes all except 0.06 percent of bark and supplies 1.5 percent of the total wood for the digesters in a sulfite mill. The prototype handles 50 tons (ovendry basis) input per day with output of about 26 tons (ovendry) of bark-free wood.

Liiri, O. 1960. INVESTIGATIONS ON THE DEBARKING OF BIRCH CHIPS BY THE SOAKING METHOD. Pap. ja Puu 43: 711-715. Fini.
Bark-free birch chips vacuum-impregnated with water, were heavier than water and could be removed from the bottom of the vat. Bark loosened from the chips floated on the surface and could be skimmed off for use as fuel. A diagram of the system is provided

Lonn, K. A., and Plotnikov, V. L. 1970. PNEUMATIC BARKING OF WOOD. Lesn. Zh. 13(2): 33-36. Arkhangelsk.

Onyakov, E. F. 1971. CALCULATING THE CUTTER INSTALLATION ANGLE IN ROTOR-TYPE BARKING MACHINES. Lesn. Zh. 14(3): 52-57. Arkhangelsk.
A theoretical analysis of the angle at which multibladed barking cutters should be fitted. The results are also applicable to branch-trimming

Pavelescu, I. M. 1970. THE STAGE OF MECHANIZATION OF WOOD BARKING FOR INDUSTRIAL VENEERS. Ind. Lemnului 21(5): 161-165. Bucharest.
Compares the performance of several different barking machines, mainly local inventions, at Rumanian plywood factories. Concludes that the imported Hepke barker (swinging cutter-head type) is economically and technically superior.

Schelshorn, H. 1970. TRIALS WITH A FINNISH BARKING MACHINE OPERATING WITH CUTTER HEADS. Holz Zentralbl. 96(109): 1581-1582.

Shelyapin, F. N., and Razumovskii, V. G. 1971. INVESTIGATION OF THE BARKING OF LARCH WOOD IN DRUM BARKERS. Bum. Prom-Sť. (8), (9-10) Moscow.
Graphs show how pulpwood diameter and drum rotation speed were related to duration and quality of barking. The study was

12

made in Siberia on floated and unfloated larch pulpwood in both
summer and winter.

Tanaka and Sakurai. 1970. (Safety)

Vlasov, G. D., and Pokryshkin, O. V. 1971. EFFECT OF THE
DEGREE OF BARK THAWING ON THE POWER PARAM-

ETERS IN BARKING. Derevoobrab. Prom-St'. 20(7): 6-7.
Moscow.
*Barking force and power requirements were measured during the
barking of spruce logs at various air temperatures down to -40°
C., and after total or partial immersion for 1 to 1-1/2 hours in a
log pond at water temperatures of 5 to 25° C.*

SAWING

GENERAL

Anderson and Kaiser. 1970. (Chipping, flaking, and grinding;
Chipping headrigs and edgers)

Anonymous. 1970. (Properties of the cutting edge and cutter;
Fitting and sharpening)

Birkeland, R. 1968. RESULTS OF MEASUREMENTS OF THE
SAWING ACCURACY OF CIRCULAR SAW MILLS AND
BANDSAW MILLS. Nor. Skogind. 22(6): 192-211.
*In six mills cutting frozen and unfrozen logs with circular saws,
accuracy diminished with increasing depth of cut. Carriages
accounted for some of the inaccuracies, but most were related to
the wood or to the sawblade. In five bandmills, depth of cut did
not appear to influence sawing accuracy; most inaccuracies were
traceable to the carriage. In both types of mill, only a small part
of the inaccuracies was attributable to incorrect methods of
feeding logs into the saws.*

Bousquet and Flann. 1970. (Machinability of species and reconstituted wood; Hardwoods)

Breznjak, M., and Herak, V. 1970. SAWING QUALITY WITH
MODERN HEADSAWS. Drvna Ind. 21(1/2): 2-13. Zagreb,
Yugosl.
*Thickness variation and surface quality was studied on softwood
and hardwood lumber sawn by vertical frame saws and log
bandsaws in Yugoslav mills. Thickness variation was somewhat
greater with the bandsaws, but surface quality was better.*

Bulgrin. 1971. (Computer, tape, and card control of machines)

Dargan, E. E. 1969. BY-PRODUCT PRODUCTION AT
ROUNDWOOD CORPORATION. *In* Wood utilization, p. 27-31.
Proc., Annu. Meet. Mid-South Sect., For. Prod. Res. Soc.,
Lufkin, Tex.
*In a new process for manufacturing hardwood turning stock, logs
are chucked between centers and dowels are then holesawn
endwise with a long, tubular, core-drill-like saw. The first cut is
taken near the periphery of the log, and successive cuts are made
as the log is rotated and indexed, each bore overlaps into the
kerf space of the previous bore. The process is applicable to logs
8 to 40 inches in diameter and 36 to 52 inches in length. Dowels
up to 3.5 inches in diameter can be cut at rates up to 12 per
minute. In larger logs, utilization should be about 20 percent
higher than that achievable by conventional methods. Faster
drying times, better workability, fewer rejects, easier storage,
reduced transportation costs, and greater recovery per log are
advantages cited for the new technique. The Terrell Machine Co.
of Charlotte, N.C., has acquired all patent rights to the
Roundwood dowel machines.*

Dobrynin. 1962. (Properties of the cutting edge and cutter;
Stability)

Gunnerman. 1971. (Patents)

Hallock and Bulgrin. 1970. (Computer, tape, and card control
of machines)

Home Timber Merchants Association of Scotland and Forest
Products Research Laboratory. 1970. SYMPOSIUM ON SAW-
MILLING. [Proc.] 81 p. For. Prod. Res. Lab., Dep. Trade Ind.,
Princes Risborough, Eng.
*Thirteen papers on the general subject of sawmilling and log
supplies. Papers directly related to machining describe recent
sawmill developments, the Linck chipping headrig, and several
finger jointers.*

Koch. 1971. (Patents)

Mügge, H. F. 1970. THE MODERN SAWMILL. II. PRODUC-
TION LAYOUT AND CONVEYING SYSTEMS. Holz Roh- und
Werkst. 28: 453-470.
*Liberally illustrated discussion of sawmill design. Unimpeded
flow of logs and lumber is achieved by closely correlating
operator functions, operator control, machine output, and
conveyor layout.*

Neville, C. W. 1970. RECENT DEVELOPMENTS IN SAW-
MILLING. Timber Trade J. 275(4919). 50-52.
*Reviews unconventional methods of cutting wood (plasma arc,
carbon-dioxide gas laser, water jet), and developments in slicing
techni-.e. Discusses cant resaws, thin-kerf bandsaws, computerized headrig control, and climb-cutting circular saws that
produce pulp chips. Chipping headrigs are mentioned only
briefly.*

Noguchi et al. 1971. (Orthogonal cutting)

Porter et al. 1971. (Research instrumentation and techniques)

Reynolds, H. W. 1969. THE SHOLO (SHOrt LOg) MILL.
South. Lumberman 219(2778): 182-185.
*Using computer simulation to determine the values that could be
obtained by sawing low-grade oak and hickory logs in many
different ways, the USDA Forest Service's Forest Products
Marketing Laboratory at Princeton, West Virginia, has developed
the SHOLO (SHOrt LOg) process for manufacturing parts for
40- by 48-inch standard warehouse pallets. The steps in the
process are debarking tree-length logs, bucking into lengths of
44 or 52 inches, live-sawing into 1- or 2-inch boards, ripping into
needed widths, planing to finished thickness, and end-trimming.
Selective bucking of tree-lengths yielded 57 percent acceptable
short logs and 43 percent random-length round residue suitable
for chipping. Sixteen percent of the original long-log volume was
converted into pallet parts. Cost for a mill of this type (including
yard, mill, and tools) was estimated at $550,000. With a 16-man
crew, such a mill could utilize approximately 70 cords of
tree-length logs in one 8-hour shift. Annual sales were estimated
at $640,000.
For detailed reports, see USDA For. Serv. Res. Pap. NE-180 and
NE-187, Northeastern For. Exp. Stn., Upper Darby, Pa.*

Reynolds and Gatchell. 1969. (Research instrumentation and
techniques)

Riberolles, J. P. de. 1968. RUHO: A NEW CONVERSION
TECHNIQUE: SAWING IN THE ROUND. Rev. Bois Appl.
23(1/2): 8-11. Paris.
Describes a method of sawing half-tubes (visualize tubes of

13

various diameters split longitudinally) from bolts of wood previously split in a plane through the pith.

St.-Laurent. 1970. (Properties of the cutting edge and cutter; Dulling)

Stol'cer, N. G. 1970. SAW MACHANISMS AS OBJECT OF REGULATION. Lesn. Zh. 13(3): 115-119. Arkhangelsk.

Thunell, B. 1971. CHOICE OF SAWMILL MACHINERY FOR DIFFERENT ANNUAL PRODUCTION RATES. Schweiz. Z. Forstwes. 122(3): 91-104.
Discusses recent developments and trends in Swedish sawmills. Emphasizes the need for analyzing intended production in terms of lumber volume, size, and quality.

Thunell, B. 1971. PRODUCTION OF SAWN TIMBER IN INTEGRATED MACHINE UNITS. Sven. Papperstidn. 74: 475-479.
Briefly describes and illustrates multiple bandsaws and sash gangsaws, and Chip-N-Saw, Beaver, Soderhamn, Bruzaholm, and Linck chipping headrigs. Illustrates the Strojimport multiple ripper. Tabulates advantages and disadvantages of chipping headrigs.

Timonen. 1969. (Properties of the cutting edge and cutter; Dulling)

Vetsheva, V. F. 1971. INDICES OF THE UTILIZATION OF LARGE LOGS IN SAWING TO PRODUCE A SINGLE CANT OF VARIED THICKNESS. Derevoobrab. Prom-St'. 20(3): 12-13. Moscow.

Vetsheva, V. F. 1971. INDICES OF THE UTILIZATION OF LARGE LOGS IN SAWING TO PRODUCE TWO CANTS OF VARIED THICKNESS. Derevoobrab. Prom-St'. 20(4): 13-15. Moscow.

Wuster. 1970. (Nomenclature)

BANDSAWING

Abeels. 1970. (Nomenclature)

Abeels. 1970. (Properties of the cutting edge and cutter; Stability)

Alexandru, S., and Tudor, E. 1970. DRIVE SYSTEMS FOR THE LOG CARRIAGES OF BANDSAWS. Ind. Lemnului 21(5): 182-189. Bucharest.

Aoyama. 1970. (Properties of the cutting edge and cutter; Fitting and sharpening)

Aoyama. 1971. (Properties of the cutting edge and cutter; Fitting and sharpening)

Berlin. 1969. (Properties of the cutting edge and cutter; Stability)

Bulboaca. 1971. (Safety, including noise)

Clark, F. 1968. REDUCING FIBRE LOSS WITH HIGH-STRAINED BAND SAWS AND REDUCED-KERF CIRCULAR SAWS. B.C. Lumber Manuf. Symp., 11 p. Vancouver.
Describes operation of three western Canadian softwood mills using high-strain bandsaws with 1/8-inch kerf, and circular saws with 1/4-inch kerf. Bandsaw strain (pounds) equals blade thickness in thousandths of an inch x blade :width in inches x 23; this is 2.3 x the normal strain
Power meters were placed on the circular saws to permit

operating each saw at its maximum feed capacity. Horsepower required per swaged saw may be calculated as follows: for saws with a kerf of 1/4-inch multiply the gullet area (per tooth) in square inches by the number of teeth cutting per minute and by the constant 0.003. Horsepower for kerfs narrower than 1/4-inch is proportionately less in a linear relationship*

Clark, F. 1969. HIGH STRAIN BAND MILLS. *In* Proc. For. Prod. Res. Soc., North. Calif. Sect., p. 2-5. Calpella, Calif.
Describes high-strain bandsaws of the type used in a western Canadian softwood mill that recovers 68 percent lumber, 12 percent sawdust, and 20 percent pulp chips from incoming logs.

Cowling, R. L. 1971. A SURVEY OF WIDE BAND SAWS. Aust. Timber J. 37(8): 27-35.
Reports a survey of 148 companies in Australia, Papua, and New Guinea using bandsaws wider than 3 inches. Thirty-eight percent of the companies responded and indicated no need for research on wide bandsaws. The response did indicate, however, a shortage of skilled saw filers.

Cumming, J. D. 1969. STRESSES IN BAND SAWS. *In* Proc. For. Prod. Res. Soc. North. Calif. Sect., p. 21-23. Calpella, Calif.
By formulas relates stress in bandsaws to velocity of blade, cutting power, bending around band wheels, straining, and tensioning. Concludes that strain levels should be calculated from working stresses and dynamic endurance ratings of the saw steel.

Dobrynin. 1970. (Properties of the cutting edge and cutter; Stability)

Dogaru and Capraru. 1970. (Properties of the cutting edge and cutter; Fitting and sharpening)

Feoktistov. 1960. (Properties of the cutting edge and cutter; Stability)

Foschi and Porter. 1970. (Properties of the cutting edge and cutter; Stability)

Lyth. 1970. (Properties of the cutting edge and cutter; Fitting and sharpening)

McKenzie. 1971. (Properties of the cutting edge and cutter; Fitting and sharpening)

Mundler. 1971. (Computer, tape, and card control of machines)

Pochekutov. 1969. (Properties of the cutting edge and cutter; Stability)

Porter, A. W. 1971. SOME ENGINEERING CONSIDERATIONS OF HIGH-STRAIN BAND SAWS. For. Prod. J. 21(4): 25-32.
Tests of saw steel gave average values of 207,000 p.s.i. for ultimate tensile strength and 179,000 p.s.i. for the 0.2 percent yield stress. Sixty-three samples of bandsaw welds gave average values of 137,000 p.s.i. for ultimate tensile strength and 107,000 p.s.i. for the 0.2 percent yield stress. Theoretical calculations and experimental measurements indicated that the sum of the tensile, bending, and centrifugal stresses acting on normal and high-strain bandmills is approximately 40,000 to 50,000 p.s.i. Graphs illustrate improvement of lateral saw stability resulting from changes in effectiveness of strain and in blade geometry. The critical buckling load on the edge of a plate is very nearly proportional to the tensile force acting on the plate. An equation for predicting the critical edge buckling load in terms of blade parameters and tensile stress is presented. Increasing the strain level from 14,900 to either 23,250 or 28,800 pounds on a high-strain bandmill was found to reduce the thickness variations in sawn lumber by approximately 1/32 of an inch.

14

Porter and Olson. 1971. (Properties of the cutting edge and cutter; Fitting and sharpening)

Quezada. 1970. (Properties of the cutting edge and cutter; Tool material)

Saito and Mori. 1970. (Properties of the cutting edge and cutter; Stability)

Sanev. 1969. (Properties of the cutting edge and cutter, Temperature)

Sanev, V. L. 1971. THE SLIPPING OF BANDSAWS ON THEIR PULLEYS. Lesn. Zh. 14(2): 58-65. Arkhangelsk.

Sanev and Pljusnin. 1969. (Properties of the cutting edge and cutter; Temperature)

Sanev and Pljusnin. 1970. (Properties of the cutting edge and cutter; Temperature)

Sugihara, H. 1970. FORCES ACTING ON A BAND SAW. Can. Dep. Fish. and For. Transl. OOFF-130, 32 p. (Wood Ind. 8(5): 38-45. Tokyo 1953.)
Mathematically defines stresses acting on a bandsaw when strained but static, when running but not cutting, and when cutting. Numerical examples are based on a 48-inch, 19-gage saw, 6 inches in width and running at 11,000 f.p.m. Discusses factors limiting stability and strength of band saws.

Thunell. 1970. (Properties of the cutting edge and cutter; Stability)

Thunell. 1971. (Bibliographies)

Trubnikov and Leikhtling. 1969. (Properties of the cutting edge and cutter; Fitting and sharpening)

Yur'ev and Veselkov. 1971. (Properties of the cutting edge and cutter; Fitting and sharpening)

SASH GANGSAWING

Batin, N. A., and Januskevic, A. A. 1971. CALCULATION OF THE ARRANGEMENT OF BLADES IN A FRAME SAW FOR LOG BREAKDOWN BY THE 'SECTOR' METHOD. Derevoobrab. Prom-St'. 20(2): 11-12. Moscow.
Presents a nomogram, and gives a specific example of this method of breakdown.

Borovikov, E. M. 1970. EFFECT OF FEED RATE ON CUTTING FORCES DURING THE SAWING OF WOOD BY FRAME SAWS. Lesn. Zh. 13(4): 65-71. Arkhangelsk.

Borovikov, E. M., and Pozdeev, A. S. 1970. EFFECT OF FEED RIGIDITY ON THE CHANGE IN THE CUTTING AND THRUST FORCES IN FRAME SAWS. Derevoobrab. Prom-St'. 19(3): 12-14. Moscow.

Borovikov and Pozdeev. 1970. (Properties of the cutting edge and cutter; Fitting and sharpening)

Borovikov and Posdeev. 1970. (Properties of the cutting edge and cutter; Tool material)

Borovikov and Pozdeev. 1971. (Properties of the cutting edge and cutter; Tool material)

Curtu, I., and Fleischer, H. 1971. THE KINEMATICS OF A HORIZONTAL FRAME SAW. Ind. Lemnului 22(8): 303-307. Bucharest.
An analytical and graphical study of this type of saw – commonly with only one horizontal blade – in the form used for breaking down large logs. The characteristics vary with the height of setting of the frame. A dynamic study is to follow.

Derjagin. 1969. (Properties of the cutting edge and cutter; Stability)

Fonkin, V. F. 1971. IMPROVEMENTS IN FRAME-SAW MACHINES AND IN FEEDING AND TAKE-OFF DEVICES. Holztechnologie 12(3): 159-164.
Possibilities of improving the construction of frame saws are indicated from analysis of the cutting process. Improvements in advancing mechanisms and in feed and take-off devices are discussed, and prospects of automation are considered.

Jur'ev, J. L., and Gernet, G. M. 1969. GRIPPING FORCE ON THE LOG BY THE DOGS OF A FRAME-SAW CARRIAGE. Lesn. Zh. 12(5): 79-80. Arkhangelsk.

Krutikov, N. S. 1970. DETERMINATION OF OPTIMUM SPEED FOR FEEDING FLITCHES INTO A FRAME SAW. N. Z. For. Serv., Wellington. 9 p. (Transl. from Lesn. Zh. 11(2): 112-115. Arkhangelsk.)
Develops a formula for optimum feed speed of a roll case designed to deliver cants to a sash-gangsaw.

Potyarkin, L. P. 1971. CRITERIA OF THE IRREGULARITY OF ROTATION OF A FRAME-SAW CRANKSHAFT. Lesn. Zh. 14(1): 84-89. Arkhangelsk.

Potjarkin, L. P. 1971. INVESTIGATION OF THE IRREGULARITY OF ROTATION OF THE CRANKSHAFT AND THE OPERATING CONDITIONS OF THE MAIN BELT DRIVE ON TWO-STORY FRAME SAWS. Derevoobrab. Prom-St'. 20(1): 11-13. Moscow.
Results of factory studies.

Sieminski. 1969. (Properties of the cutting edge and cutter; Fitting and Sharpening)

Veselkov. 1971. (Properties of the cutting edge and cutter; Fitting and sharpening)

Wassipaul. 1970. (Properties of the cutting edge and cutter; Stability)

CIRCULAR SAWING

Anonymous. 1969. BRITISH RIGHTS TO "SAWPLANES." Timber Trade J. 271(4857): 52.
Spear and Jackson, Ltd., have acquired exclusive British rights for the "Sawplane" reverse-taper splitting saw. The saw leaves a highly planed finish, thus eliminating further processing and reducing sawmilling time and costs. It is made in high-speed steel for softwoods and is tipped with tungsten carbide for hardwoods.

Barz. 1970. (Properties of the cutting edge and cutter; Fitting and sharpening)

Belikov, V. M. 1971. CALCULATION OF FEED SPEED FOR CIRCULAR SAWS IN RIP SAWING. Derevoobrab. Prom-St'. 20(4): 12-13. Moscow.
Theoretical analysis.

Bramhall, B., and McLauchlan, T. A. 1970. THE PREPARATION OF MICROSECTIONS BY SAWING. Wood and Fiber 2: 67-69.

A precision saw makes possible the cutting of smooth micro-sections as thin as 100 mμ in the radial and tangential directions, and 900 mμ in the transverse direction, variations are less than 10 mμ. The method reduces problems of tensile strength loss due to flexing and wood permeability change due to damage of specimen. Disadvantages include 1.25 mm. loss in saw kerf and slightly woolly appearance of specimens as viewed under a microscope. Cost of saw and grinding was approximately $15.

Breznjak, M., and Moen, K. 1969. SAWING WITH SWAGE SET CIRCULAR SAWBLADES WITH HIGH BITES PER TOOTH. Nor. Skogind. 23: 325-332.
Sawing Norwegian spruce logs with large feeds per tooth (0.6, 1.2, 1.8, 2.4, and 3.0 mm.) seemed mechanically feasible and did not cause serious operational problems other than reduction in surface quality of the lumber (deep torn grain, woolly appearance, and torn board edges) and increased variation in board thickness. Kerf appeared to be less, and specific cutting energy decreased, as feed per tooth increased.

Briggs, J. L. 1970. A REVOLUTIONARY NO. 1 BENCH. Aust. Timber J. 36(2): 44, 47.
Describes the Briggs automatic No. 1 sawbench designed to supersede the No. 1 bench now used in hardwood mills in Australia. It is claimed that, with the new bench, one sawyer can readily flitch, cant, and slab wood from any type of headrig, obtaining the same recovery as a sawyer and three assistants using a standard bench. The machine may also be used as a headrig for cutting logs not exceeding 6 feet in girth.

Cizevskij and Ceremnyh. 1970. (Safety, including noise)

Clark. 1968. (Sawing; Bandsawing)

Curtu and Serbu. 1970. (Properties of the cutting edge and cutter; Stability)

Degermendzhi, G. A. 1969. CALCULATION OF THE OPTIMUM MOTOR POWER OF A SAW FOR CROSS CUTTING ROUNDWOOD. Lesoinzh. delo i Mekh. Teknol. Drev., p. 66-69. Krasnoyarsk.
Mathematical analysis.

Dunaev. 1971. (Properties of the cutting edge and cutter; Tool material)

Foley Manufacturing Company. [n.d.] (Properties of the cutting edge and cutter; Fitting and sharpening)

Friebe. 1970. (Properties of the cutting edge and cutter; Stability)

Fukui and Kimura. 1968. (Machinability of species and reconstituted wood; Particleboards and sandwich boards)

Grube et al. 1969. (Properties of the cutting edge and cutter; Temperature)

Howland. 1971. (Machinability of species and reconstituted wood; Hardwoods)

Kintz, A. H. 1969. THE ROLE OF THE TOP ARBOR EDGER IN PRECISION THIN SAWING. In Proc. For. Prod. Res. Soc. North. Calif. Sect., p. 6-10. Calpella, Calif.
Describes a top-arbor, climb-cutting, multiple saw edger designed to rip softwood cants up to 6 inches thick at feed speeds to 150 feet per minute. Kerf is 1/8-inch. The machine has guides and collars designed for use with circular saws having no tension, a rake angle of 7° is practical. A six-saw machine requires 150 horsepower; each additional saw requires another 15 horsepower.

Kondrat'ev, V. I. 1969. THE JAMMING OF SAWS DURING THE CROSSCUTTING OF LONG LOGS. Lesoinzh. delo i Mekh. Teknol. Drev., p. 70-72. Krasnoyarsk.
A theoretical analysis of the cause of blade jamming in slashers and trimmers. Installation of copying devices tracing all the unevennesses of the log surface would prevent the trouble.

Koshunyaev, B. I., Pizhurin, A. A., and Fergin, V. R. 1971. QUALITY OF WORK IN RIPSAWING WOOD WITH CIRCULAR SAWS. Lesn. Zh. 14(1): 151-154. Arkhangelsk.
Develops regression equations and nomograms for determining (1) the depth of unevennesses on the sawn surface, and (2) deviations from nominal dimensions, in relation to several process variables.

Kostrikov. 1969. (Properties of the cutting edge and cutter; Fitting and sharpening)

Kotesovec. 1970. (Chipping, flaking, and grinding; Chipping headrigs and edgers)

Kubasak, E. 1968. COMPARING THE CUTTING CAPACITY OF CROSSCUTTING SAWS. Ved. Pr. Vysk. Ust. Lesn. Hosp. Zvolen 11, p. 195-208. Czeck.
Presents technical and performance data for several commercial or prototype Czechoslovak circular saws for crosscutting small or large softwood logs, and, for comparison, a Dolmar crosscutting chain saw.

McKenzie, W. M. 1970. FEED PER CUT, A V.I.P. IN WORKING WOOD-BASED MATERIALS. Aust. Timber J. 26(7): 133, 135.
Feed per cut, which determines chip thickness, has dominant effects on power requirement, surface and edge quality, blade stability, heat generation, and dulling during sawing. Increase in feed per cut decreased overheating, motor overload, and dulling of blades. Sawdust became more chip-like, and blade stability increased.

McKenzie. 1970. (Safety, including noise)

Macomber, D. 1969. SMOOTH SPLITTING. Stetson-Ross Mach. Co., Seattle, Wash. 5 p.
Detailed description of equipment and technique for multiple smooth ripping of 2-inch-thick lumber on a planer equipped with double profilers. Ripped planks have planed and rounded edges. The 4- to 12-knife cutter assemblies turn at 3,600 revolutions per minute and cut a 1/4- to 5/16-inch kerf. Feed speed is approximately 50 feet per knife, e.g., 500 f.p.m. for 10-knife heads.

Malcolm. 1971. (Safety, including noise)

Malcolm and Koster. 1970. (Research instrumentation and techniques)

Mote. 1970. (Properties of the cutting edge and cutter; Stability)

Nakamura, G. 1971. ON NUMBER OF TEETH AND FEEDING DIRECTION IN CIRCULAR SAWING. J. Jap. Wood Res. Soc. 17: 203-208.
Lumber was fed at 4.5 to 17.6 meters per minute into carbide-tipped circular ripsaws having 4 to 48 teeth, rotational speed of the saws varied from 2,350 to 5,850 r.p.m. At a given feed speed, and also with feed per tooth held constant, tangential cutting force and feed force both decreased with decreasing number of teeth. Relation between tangential cutting force T (kg.) and feed force F (kg.) was expressed as: $T = 0.041 F^{0.81}$ for swage-set teeth. Tangential cutting forces in climb sawing were about equal to those for counter sawing.

16

Neville and Bennie. 1969. (Chipping headrigs and edgers)

Ohsako et al. 1971. (Properties of the cutting edge and cutter; Dulling)

Okushima and Sugihara. 1969. (Properties of the cutting edge and cutter; Temperature)

Pahlitzsch and Friebe. 1971. (Properties of the cutting edge and cutter; Stability)

Pahlitzsch and Friebe. 1971. (Safety, including noise)

Pahlitzsch and Rowinski. 1966. (Properties of the cutting edge and cutter; Stability)

Pashkov, V. K., and Bodalev, V. G. 1971. NOMOGRAM FOR SELECTING WORKING SCHEDULES FOR CIRCULAR SAWS. Lesn. Zh. 14(1): 57-60. Arkhangelsk.

Poole. 1971. (Properties of the cutting edge and cutter; Friction)

Radu, A., Dogaru, V., and Vaida, A. 1971. POSSIBILITIES OF INCREASING THE SPEED OF CUTTING IN SAWING SOLID WOOD WITH CIRCULAR SAWS. Ind. Lemnului 22(7): 255-260. Bucharest.
Analyzes influence of peripheral cutting speed on the sawing process. Advantages of high speed (e.g. 70 meters per second) include reduced vibration and cutting force, and the possibility of using thinner saws with corresponding thinner kerf.

Salemme, F. J. 1969. THIN CARBIDE SAWS IN THE SAW-MILLS. *In* Proc. For. Prod. Res. Soc., North. Calif. Sect., p. 10-13. Calpella, Calif.
Reviews efforts of California cedar products in reducing the kerf of circular saws sufficiently to rip 3-inch incense cedar squares into 11 pencil slats. Also reports experience in reducing the saw kerf on a Schurman double-arbor edger from 0.175 inch to about 0.105 inch when cutting 3- and 4-inch squares from planks. Concludes that saw collars should be as large in diameter, as thick, and as flat as possible and that the saw blade should be of highest quality steel (e.g. Crucible #55) with the smallest possible eye, not tensioned or hammered, and of about 48-50 Rockwell C hardness for carbide tipping. Rim speed of 15,500 feet per minute is recommended, with direct drive. Teeth are ground with 28° rake angle, 55° sharpness angle, and 7° clearance angle.

Schliewe, R. 1969. THIN KERF SAWING CONSIDERA-TIONS. *In* Proc. For. Prod. Res. Soc., North. Calif. Sect., p. 14-18. Calpella, Calif.
Describes research of the Weyerhaeuser Co. to develop 1/8-inch kerf, climb-cutting saws for edgers with overhead arbors. Saws should require minimum filing skill and yield softwood lumber within plus or minus 1/64-inch of target thickness, and surfaces should be smooth enough for finishing with abrasive planers.

Shen. 1971. (Machinability of species and reconstituted wood; Particleboards and sandwich boards)

Shuin, V. E. 1971. EFFECT OF CUTTING SPEED ON THE PROCESS OF RIPSAWING WOOD WITH CIRCULAR SAWS. Lesn. Zh. 14(2): 162-164. Arkhangelsk.
Tangential and radial components of cutting force were deter-mined for saws rotating at high speeds (cutting speeds 20-104 m/s). The force/speed curves are discussed, with special reference to the lodging of fine sawdust in the gullet, and a formula is presented for calculating the optimum cutting speed for a given saw diameter.

Stahiev. 1969. (Properties of the cutting edge and cutter; Stability)

Stahiev. 1970. (Properties of the cutting edge and cutter; Fitting and sharpening)

Stahiev. 1970. (Properties of the cutting edge and cutter; Stability)

Stefaniak. 1970. (Properties of the cutting edge and cutter; Dulling)

Stefaniak, W. 1970. RAPID DETERMINATION OF SELECTED PARAMETERS OF THE TOOL AND THE WORK-ING SCHEDULE IN SAWING WITH CIRCULAR SAWS AND IN MOLDING. Przem. Drzewny 21(8): 7-11. Warsaw.
Presents a set of nomograms for determining cutting speed, critical rotation speeds, minimum saw-blade diameter, and feed speeds.

Strzelecki. 1970. (Properties of the cutting edge and cutter; Stability)

Strzelecki. 1971. (Safety, including noise)

Tallack. 1971. (Properties of the cutting edge and cutter; Tool material)

Thrasher. 1971. (Patents)

Vasil'ev, G. M. 1970. SPEEDING UP THE CROSSCUTTING OF WOOD WITH CIRCULAR SAWS. Derevoobrab. Prom-St'. 19(3): 3-5. Moscow.
Analyzes the technical factors that militate against speeding up the work cycle in crosscutting logs, and shows how cycle time can be reduced.

Vasil'ev, G. M. 1971. INCREASING THE PERFORMANCE OF CIRCULAR SAWS FOR CROSSCUTTING WOOD. Lesn. Zh. 14(3): 87-90. Arkhangelsk.

Werner. 1967. (Safety, including noise)

Yu, K. C. 1966. THE FRICTION SAWING OF WOOD. M.S. Thesis. Univ. B. C., Vancouver. 87 p.
Douglas-fir and western red cedar lumber of various moisture contents was friction-sawn with a 14-inch-diameter, 14-gauge steel disk with a smooth edge. Rotational speed was 4,620 r.p.m. Frictional forces, power consumption, and cutting temperatures increased with feed speeds. Moisture content of the wood had no noticeable effect on the sawing action. Kerfs were narrow and sawn surfaces were straight, smooth, and polished. At low feed speed (4 inches per minute) the calculated cutting edge temperatures were well below the ignition point of the wood. At high feed speeds (18 inches per minute) power demand increased and calculated edge temperatures were high. Excessive power consumption and high edge temperatures were believed to be related to difficulty in ejecting waste from the kerf. Reasonably high feed speeds were used in cutting plywood and veneer; surfaces were clean and smooth. When thicker lumber was cut by this method, feed speed was confined to a very low level and power consumption was far higher than that required for ordinary sawing. Disk modifications are suggested to improve cutting action and waste disposal.

CHAIN SAWING

Altman. 1971. (Safety, including noise)

Anonymous. 1969. POWER CHAIN SAWS: THE FACTS AND THE FUTURE. Pulp and Pap. Mag. Can. 70(18) 55-57.

The chain saw has been continuously improved since it was first designed 100 years ago. It is a mainstay in Canadian logging and will remain so for at least another 20 years. Approximately 85,000 chain saws are sold annually in Canada at an average price of $200. Life expectancy of a saw in continuous industrial use is about 1 year.

Degtjarev. 1970. (Safety, including noise)

Fujii. 1970. (Safety, including noise)

Fushimi. 1971. (Safety, including noise)

Fushimi, T., and Shigaki, H. 1971. ON THE VIBRATION OF THE CHAIN SAW IN CROSS CUTTING TIMBER. Ehime Univ. For. Bull. 8, p. 33-42. Japan.

Gebert. 1971. (Safety, including noise)

Gellert. 1971. (Safety, including noise)

Gul'ko. 1971. (Properties of the cutting edge and cutter; Stability)

Gul'ko, L. I., and Matjuskin, V. Z. 1970. INCREASING THE RELIABILITY OF THE GUIDE BARS OF CHAIN SAWS. Lesn. Prom. (1), (27-28). Moscow.

Harstela. 1971. (Safety, including noise)

Huzl et al. 1969. (Safety, including noise)

Janco, J. 1967. THE PRH CHIPPER-TYPE CHAIN IN CROSS-CUTTING SOFTWOODS. Ved. Pr. Vysk. Ust. Lesn. Hosp. Zvolen 8, p. 15-53. Czech.
Determination of optimum parameters for Czechoslovak PRH chains, and their comparison with scratch-type chains.

Keighley. 1970. (Safety, including noise)

Kennemore. 1971. (Patents)

Kuhnke, A. 1970. CHAIN-SAW CUTTING TEETH HAVING GOOD HEAT CONDUCTION. Holz Zentralbl. 96(19): 282.
Describes a patented tooth with a narrow, rounded rib on the curved part of the cutter. The rib reduces friction and produces a cooling air stream below it.

Myles et al. 1971. (Safety, including noise)

Pahlitzsch and Peters. 1968. (Properties of the cutting edge and cutter; Fitting and sharpening)

Polishchuk, A., Fedin, V., Belovzorov, L., and Simonov, B. 1971. A HYDRAULIC JACK FOR FELLING LARGE TREES. Lesn. Prom. (6), (13-4). Moscow.
Describes, with brief specifications, the DGM-16 hydraulic jack (driven from the engine of a chain saw), the method of using it, and results of trials in fir-beech forests in the Caucasus.

Reynolds, D. D., Soedel, W., and Eckelman, C. 1970. CUT-TING CHARACTERISTICS AND POWER REQUIREMENTS OF CHAIN SAWS. For. Prod. J. 20(10): 28-34.
An apparatus was developed for measuring horizontal and vertical cutting forces on chain saw teeth, engine torque output, sprocket speed, and feed rate. These factors, together with net cutting horsepower, power loss, engine horsepower, chain speed, feed rate, depth of cut, and man input power, were all correlated in mathematical models of the cutting process. Power loss had a

positive linear correlation with chain speed; it resulted primarily from friction at the sprocket and between the chain and chain bar. Before a new chain will yield uniform force data, it must be deburred by a series of preliminary cuts. Horizontal and vertical cutting forces were positively and linearly correlated with depth of cut per tooth, and an increase in horizontal cutting force was accompanied by an increase in vertical force.

Sakurai. 1968. (Safety, including noise)

Sakurai et al. 1969. (Safety, including noise)

Shackelberg, S. von. 1971. CUTTING TIME OF POWER CHAIN SAWS AS A FUNCTION OF THE STEM DIAMETER. N.Z. For. Serv., 5 p. Wellington. (Transl. from Allg. Forst- und Jagdztg. 138(12): 273-275. 1967.)
When spruce was crosscut with a chain saw of 4.5 horsepower, time per cut increased linearly with increasing basal area. Production of cut timber volume per unit time was positively correlated with stem diameter, the effect was considerable up to a diameter of 15 cm., moderate with diameters between 15 and 25 cm., and small with diameters over 25 cm. Cutting time was 40 percent of that with a double-handed crosscut saw for stems 20 cm. in diameter; for stems 45 cm. in diameter chain saws required only 20 percent of the time needed with a double-handed crosscut saw.

Szepesi, L. 1969. PRACTICAL EXPERIENCES FROM AN INTERNATIONAL COMPARATIVE TEST OF PLANE TEETH CHAINS. Erdeszettud. Kut. 65(1): 215-218. Budapest.
East German, Polish, Russian, Czech, and Hungarian saw chains with pitches of 10 to 15 mm. were evaluated along with chains from the United States, Sweden, West Germany, Japan, Austria, and other countries.

Szepesi, L. 1970. RESULTS AND PROBLEMS IN REDUCING POWER SAW VIBRATION. Erdeszettud. Kut. 66:197-210. Budapest.
General discussion of factors influencing chain saw vibration, measurement of vibration, and problems related to vibration.

Tudoran, P., Giacomelli, I., and Ionescu, M. 1971. THE CHOICE OF MATERIAL AND THERMAL TREATMENT FOR THE GUIDE BAR OF A CHAIN SAW. Ind. Lemnului 22(4): 126-131. Bucharest.
A study, as part of the development of the Retezat chain saw in Rumania, of the design and material (analyzed metallurgically) of imported bars, with practical recommendations.

MACHINING WITH HIGH-ENERGY JETS

Franz, N. C. 1970. HIGH-ENERGY LIQUID JET SLITTING OF CORRUGATED BOARD. TAPPI 53: 1111-1114.
Corrugated board slit with high-energy liquid jets has smooth edges free of damage and debris. The concept is simple and ideal for setups requiring high-speed, numerically controlled cuts of great accuracy. Laboratory experiments have demonstrated slitting speeds that are well above the board-forming capabilities of present-day corrugating machines. Mechanical feasibility of jet cutting appears to be fully established, and pumping equipment, fittings, nozzles, and controls are commercially available. Data pertaining to operating parameters are needed, and a small-scale prototype slitting device is suggested for the purpose.

MACHINING WITH LIGHT

Doxey, B. C. 1970. LINE-FOLLOWING LASER SYSTEM FOR CUTTING DIE BOARD. Boxboard Containers 78(1): 50-55.
The carbon dioxide laser developed by British Oxygen Company, Ltd., is applied in cutting plywood bases for steel die rules used

in the carton container industry. The workpiece is guided past the stationary laser by an automatic line-following system. The 200-watt laser emits a beam 15 mm. in diameter at a wavelength of 10.6 μm. A typical 15-up die took 27 hours to mark out and cut by conventional methods, but only 5-1/4 hours by the laser process (of which 2-3/4 hours was actual cutting time at 9 inches per minute and 2-1/2 hours was preparation time). The machine consumes about $0.60 worth of gas and electricity per hour.

McMillin, C. W., and Harry, J. E. 1971. LASER MACHINING OF SOUTHERN PINE. For. Prod. J. 21(10): 35-37.
When cutting with an air-jet-assisted carbon-dioxide laser of 240 watts' output power, maximum feed speed at the point of full penetration of the beam decreased with increasing workpiece thickness in both wet and dry samples the trend was curvilinear. Feed speeds averaged 99.1 and 14.6 inches per minute for samples 0.25 and 1.00 inch thick. Somewhat lower speeds were required for wet than for dry wood. In wet wood, maximum speed was unrelated to specific gravity. In dry wood, slightly slower speeds were required when wood density was high than when it was low. The laser cut along and across the grain with equal speed. Scanning electron micrographs showed that the laser-cut surfaces, while blackened, were far smoother than sawn surfaces. There was little damage to wood structure, but some carbon deposits were evident on cell walls and in lumen cavities.

March, B. W. 1970. LASER PROFILING MACHINE GIVES HIGH QUALITY CUTS IN PLYWOOD. FIRA Bull., Furn. Ind. Res. Assoc. 8(30): 34-35. Stevenage, Eng.
*The machine was developed by the British Oxygen Company, Ltd. Its heart is a 200-watt carbon-dioxide laser whose beam is 15 mm. in diameter and has a wavelength of 10.6 microns. The focused beam is surrounded by a nozzle that ejects a stream of high-velocity gas along the beam axis. Kerf in 5-cm.-thick plywood is about 0.75 mm.
The machine will accommodate plywood to a maximum size of 180 cm. by 115 cm. Motion of the beam is controlled by a two-axis system deriving information from a full-scale drawing, this optical-line-following system permits cuts to be made with an accuracy of ± 0.13 mm.*

Ready, J. F. 1970. USING THE LASER IN PRODUCTION: THE DESIGN IMPLICATIONS. Mech. Eng. 92(9): 18-24.
Use of lasers for specialized production will increase. The carbon-dioxide laser and the neodymium-doped yttrium aluminum garnet (Nd·YAG) laser are the most appropriate of those now available. Cost of operating a 100-watt pulsed carbon-dioxide laser is about $1.00 per hour for gas supply, electricity, and routine maintenance. Continuous, 250-watt, carbon-dioxide lasers to cut organic materials or ceramics will soon be ready for application. Nd·YAG lasers in several-hundred-watt capacities will melt and vaporize metals; carbon-dioxide lasers are less effective in cutting metals.

SHEARING

Arola, R. A. 1971. CROSSCUT SHEARING OF ROUND-WOOD BOLTS. USDA For. Serv.·Res. Pap. NC-68, 20 p. North Cent.·For. Exp. Stn., St. Paul, Minn.
Forces required to shear 10-inch-diameter bolts were determined for a single blade cutting against an anvil. Force for hard maple rose ,57 percent when blade thickness increased from 1/8- to 1/4-inch, and 46 percent more with an increase to 1/2-inch. With basswood the same increases in blade thickness increased force 73 and 41 percent. Very thin blades tended to buckle laterally even when prestressed. Guillotine blades, with the edge obliquely inclined to the direction of cut, required less force than blades with no obliquity when cutting unfrozen aspen, white spruce, and hard maple; splitting damage in spruce and aspen, however, increased with obliquity. Tapering of blades (-¡° to +3°) did not greatly affect force to shear unfrozen aspen, spruce, or hard

maple. A positive taper (blade thicker at root than near cutting edge) increased the splitting damage in aspen and spruce
*Low-friction coatings applied to blade surfaces reduced force requirements but not splitting damage. Dulling did not greatly increase forces required to shear unfrozen aspen, spruce, and yellow birch; damage to sheared ends, however, was generally greater with dull blades. Variation of shearing speed (2 to 12 inches per second) did not affect force required.
Wood temperature (0 to 60° F.) was negatively correlated with shear force and splitting damage, i.e., at 0° F. damage was worst and shear force greatest. Wood specific gravity was the best single indicator of shear force required; wood of highest specific gravity required the greatest force.*

Bryan, R. W. 1970. FELLING ATTACHMENT LOGS SMALL TREES AND UNDERSTOCKED STANDS. For. Ind. 97(7): 52-54.
Describes new Fleco feller-buncher attachment for a crawler loader; the system is designed to harvest small timber in understocked stands under adverse conditions In southern pine stands averaging approximately 7 inches in diameter and with a harvest potential of one cord per acre, trees can be cut at a rate of about 100 per hour. The shear has a capacity of approximately 15 inches and cuts stumps to a 4-inch height.

Deberdeev, A. A. 1969. THE EFFECT OF TRANSVERSE DEFLECTIVE OSCILLATIONS OF THE STEM ON THE IMPACT OF THE TREE. Lesn. Zh. 12(1): 72-92. Arkhangelsk.
Analyzes forces resulting from felling trees directly onto harvesting machines. A series of formulae are developed for finding the impact force, duration of impact, and impact impulse, the formulae take into account transverse deflective oscillations of the stem.

Fulghum. 1971. (Patents)

Gorodeckij, M. 1969. ROTOR-TYPE BRANCH REMOVER. Lesn. Prom. (12), (17-28). Moscow.
Describes a machine similar to a rotor-type mechanical barker, but used to shear branches from stems.

Graf, E. 1971. PROBLEMS IN THE MECHANIZATION OF BRANCH TRIMMING IN WOOD HARVESTING. Soz. Forstwes. 21(6): 178-180, 192.

Hamamoto and Mori. 1970. (Orthogonal cutting)

Hensel, J. S. 1970. KOEHRING SHORT-WOOD HARVESTER. Am. Pulpwood Assoc. Tech. Release 70-R-56, 5 p.
The harvester combines manual, semiautomatic, and automatic processes to allow its single operator to fell (with shears), delimb, buck, and forward up to 40 cords of 8-foot pulpwood in a single shift. It will process trees from 3 to 16 inches in diameter. Designed for year-round, three-shift operation, it has been found by one Canadian paper company to produce an average of 2.60 cords/hour at a direct cost of $9.80/cord. Cost of the harvester is approximately $133,000

Johnston, J. S., and St-Laurent, A. 1970. V-SHAPED TREE SHEAR BLADES SPELL END TO BUTT SPLITTING. Can. For. Ind. 90(8): 39-45.
V- or U-shaped shearing blades appeared promising in a study on wood specimens 2 inches square. In cold weather most fiber and splitting damage occurred on only one face of the cut, which in tree-felling operations could be the stump face. In the 2-inch-square specimens, splitting damage was least on the convex surface of the sheared wood. The blades were stiffer than straight blades, but the V-shaped blades tended to assume a U-shape after use.

Johnston, J. S., and St-Laurent, A. 1971. CROSSCUT SHEAR-ING OF FROZEN TREES. Pulp and Pap. Mag. Can. 72(5): 97-108.

19

Trees were felled with a double-blade shear having 3/8-inch or 1/2-inch-thick knives (45° sharpness angle) that closed at 0.8 to 1.5 inches per second. Wood temperature was as low as -15° F. Four coniferous species common in Canada were included in all tests; for two deciduous species (birch and aspen) only the 1/2-inch blades proved practical. Cutting force and wood damage increased with decrease in temperature, increase in moisture content, and increase in specific gravity of wood. When softwood trees were felled at low temperature (down to -15° F.) a 15° F. decrease in temperature or a 25-percent increase in moisture content increased maximum cutting force by approximately 10 percent. Brief discussions are included of wood temperature in relation to air temperature, effect of machine rigidity, and methods of assessing wood damage.

Johnston, J. S., and St-Laurent, A. 1971. HOW SHEARS COULD REPLACE SAWS FOR LUMBER TRIMMING. Can. For. Ind. 91(5): 35-39.
Shears are fast, simple, easily sharpened, and can readily trim conveyorized boards at angles other than 90°; moreover, shears make no sawdust. A single blade cutting against an anvil yielded better surfaces than opposed knives when cutting green white spruce (Picea glauca (Moench) Voss) and balsam-fir (Abies balsamea (L.) Mill) boards of 1x4, 2x4, and 1x6 dimensions. Sheared surfaces of the spruce boards were superior to those of the balsam-fir. Knots were readily sheared without loss in surface quality, but decayed boards yielded rough surfaces. When a knife 3/16-inch thick was used to cut through knots in 2-inch-thick green spruce, maximum force ranged up to 2,960 pounds per inch of board width. Sheared ends, while smooth, frequently had splits one-half-inch deep, quality would probably be acceptable for studs.

Kempe. 1971. (Patents)

Kessler and Davis. 1971. (Patents)

Koch, P. 1971. FORCE AND WORK TO SHEAR GREEN SOUTHERN PINE LOGS AT SLOW SPEED. For. Prod. J. 21(3): 21-26.
When logs of three diameter classes and two specific gravity classes were sheared with a 3/8-inch-thick knife travelling at 2 inches per minute, shearing force and work averaged greatest for dense 13.6-inch logs cut with a knife having a 45° sharpness angle (73,517 pounds, 49,838 foot-pounds). Force and work averaged least for 5.1-inch bolts of low density when cut with a knife having 22-1/2° sharpness angle (9,975 pounds, 2,885 foot-pounds). Values for 9.7-inch bolts were intermediate. Shear force reached a maximum about three-fourths of the way through the log, it then dropped rapidly as the knife travelled the remaining distance. Momentary peaks of force commonly occurred near the three-quarter point. The greatest observed force to shear was 92,000 pounds required for a 13.6-inch log of 0.51 specific gravity (ovendry weight and green volume) when cut with a knife having 45° sharpness angle. When sheared logs were viewed in radial section, each annual ring showed a check at the earlywood-latewood boundary. Checks were least severe in small logs sheared with the 22-1/2° knife, where they averaged 0.8 inch deep, they were most severe in large logs of low density sheared with the 45° knife, where they averaged 1.4 inches deep. Each sheared log generally also had one to several rather lengthy checks that formed just prior to emergence of the knife. Regression expressions were developed to predict force and work to shear as well as average and maximum check depth – all in terms of sharpness angle, wood specific gravity, and log diameter.

Kubler, H. 1960. CUTTING TIMBER WITH VIBRATING KNIVES. Holz Zentralbl. 86(115): 1605-1606.
Reports Russian experiments on cutting wood with knives vibrating in the direction of feed (visualize a carpenter tapping a

chisel with a mallet). Cuts across the grain were more successfully accomplished than those made parallel to the grain. Vibration of the knife substantially reduced cutting forces required.

Nesterenko, V. G. 1969. POWER SEVERANCE OF BRANCHES BY RIGID AND FLEXIBLE HINGED CUTTERS. Lesn. Zh. 12(4): 34-39. Arkhangelsk.

McIntosh, J. A., and Kerbes, E. L. 1969. LUMBER LOSSES IN TREE SHEAR FELLING. B. C. Lumberman 53(10): 43-46.
Frozen and unfrozen lodgepole pine and white spruce trees were felled with a hydraulic shear having a single 1.5-inch-thick blade closing against an anvil. Because the shear caused splits, trim losses in boards cut from butt logs were greater than if the trees had been felled with chain saws. When trees less than 14 inches d.b.h. were sheared at air temperatures of 45°F. the losses were approximately 1 percent of the total lumber output per tree, this loss was more than compensated by the very low stumps, which increased utilized tree volume by as much as 5 percent. Operators reported acceptable results down to 15° F., but at -25° F. the trim loss was about 5 percent for lodgepole and 7 percent for white spruce.

Manuchin, G. F., and Kamenetzkii, S. A. 1971. SOLVING THE PROBLEM OF MECHANICAL BRANCH TRIMMING. Soz. Forstwes. 21(4): 116-119.

Obrosov, M. Y. 1971. CUTTING FORCE IN RELATION TO DEPTH OF PENETRATION OF THE CUTTER INTO THE WOOD DURING BRANCH TRIMMING. Lesn. Zh. 14(2): 156-159. Arkhangelsk.
A theoretical analysis in which an equation is derived to express the relation between cutting force and depth of cutter penetration. Curves constructed by use of this equation agree well with experimental data.

Raileanu, V. 1970. A MODERN WOODWORKING PROCESS: DIE-STAMPING. Ind. Lemnului 21(5): 173-176. Bucharest.

Ramey. 1971. (Patents)

Schilke, A., Liebold, E., Kittner, H., and Töpfer, H. 1971. IN-VESTIGATIONS ON THE CHIPLESS CROSS CUTTING OF SPRUCE LOGS. Holztechnologie 12(2): 94-97.
Describes the bucking of logs with shears. Shears produced surfaces comparable to sawn surfaces. They were fast and safe, and made no sawdust (and therefore resulted in greater log yield). Shearing force was positively correlated with wood density and log diameter; below the freezing point, force was negatively correlated with wood temperature. Thin knives with sharp cutting edges required least force to shear.

Tuovinen, A. 1970. WOOD DAMAGE RESULTING FROM CROSSCUTTING BY SHEAR BLADES (PIKA 50 PROCESSOR). Tied., Metsäteho Rep. 300, 12 p. Finl.
Thirty-two pine and spruce logs (14 to 25 cm. in top diameter inside bark) were sheared in January and February during extremely cold weather, 40 more were sheared at the end of April in warmer weather. The shear was of the guillotine type, 12 mm. thick, and with a sharpness angle of 30°. The logs were sawn into lumber, and the lumber kiln-dried and trimmed to remove shear-caused split ends. For logs sheared in April, loss of dry lumber per log was 1.27 litres for spruce and 1.21 litres for pine, comparable figures for logs sheared in winter were 3.73 and 2.70 litres. Trimbacks shortened lumber 8.9 cm. for spruce sheared in summer and 19.7 cm. for winter-sheared wood. Corresponding figures for pine were 7.7 cm. and 16.4 cm. Losses were of such magnitude that shear-blade bucking of tree stems into saw logs requires careful analysis before adoption.

Vasil'ev, G. M. 1971. FOLLOWING TRUNK IRREGULARI-
TIES IN TRIMMING BRANCHES FROM TREES IN FIXED
MECHANISMS. Can. Dep. Fish. and For. Transl. OOFF-181, 16
p. (Transl. from Lesn. Zh. 5(2): 143-145. 1962. Arkhangelsk.)
*Discusses types of curvature and irregularities found in trunks of
trees, explores the demands these irregularities impose upon
trimming mechanisms, and describes several types of mechanisms
designed to trim irregular trunks.*

White, M. C., Foil, R. R., and McDermid, R. W. 1969. TREE
VOLUME AFFECTS PRODUCTIVITY OF HYDRAULIC
TREE-SHEARS IN SOUTHERN LOGGING. LSU For. Notes
85, 2 p. La. State Univ. Sch. For., Baton Rouge.
*Time-study data on 14 tree shears operating in southern pine
stands in six States indicated that, of total working time, 11
percent was spent positioning the shear, 22 percent shearing, and
67 percent moving from tree to tree. Average volume per tree
cut accounted for 74 percent of variation in tree-shear productivity, and operator skill accounted for another 13 percent.*

JOINTING, PLANING, MOLDING, AND SHAPING

Bobin. 1970. (Safety, including noise)

Dem'janovskij. 1970. (Properties of the cutting edge and cutter;
Tool material)

Dunaev. 1970. (Properties of the cutting edge and cutter;
Fitting and sharpening)

Dunaev. 1971. (Properties of the cutting edge and cutter;
Fitting and sharpening)

Goodchild, R. 1967. FACTORS AFFECTING THE POWER
USED IN PLANING. Woodworking Ind. 24(7): 28-29.
*Power requirement increases with feed speed, depth of cut, and
peripheral velocity of the cutting edges. It also increases as rake
angle is reduced and as the width of heel caused by jointing
becomes greater.*

Ivanovskij et al. 1970. (Properties of the cutting edge and
cutter; Dulling)

Kitayama and Sugihara. 1969. (Safety, including noise)

Kryazhev, N. A. 1971. THE ADVANTAGE OF FEED IN THE
DIRECTION OF ROTATION IN THE MOULDING OF END-
GRAIN WOOD. Derevoobrab. Prom-St'. 20(4). 10-11. Moscow

Macomber. 1969. (Sawing; Circular sawing)

Pahlitzsch and Liegmann. 1956. (Safety, including noise)

Pahlitzsch and Sandvoss. 1969. (Boring, routing, and carving)

Potanin. 1970. (Safety, including noise)

Raknes, E. 1969. FINGER JOINTING STRUCTURAL TIM-
BERS WITH A HIGH MOISTURE CONTENT. Nor. Skogind.
23(6): 184-190.
*In 3- by 8-inch Norway spruce (Picea abies) finger-jointed at 14-
to 22-percent moisture content, wood at the higher moisture
contents yielded joints 5 to 7 percent weaker than that at lower
moisture contents. When both pieces were at high moisture the
joints were weaker than when one piece had low and the other
high moisture content. However, the weakening effect – which
was probably due to drying stresses – did not appear to be
sufficient to require reduction of allowable working stresses.
Confirmatory tests will follow.*

Roth, A. 1970. A NEW TYPE OF FINGER JOINT. Pap. ja Puu
52(1): 25-28. Finl.
*Describes a joint in which the fingers run diagonally across the
board rather than parallel to the face or the edge. Bending*

*strengths of 2- by 4-inch and 2- by 6-inch pine assembled with
the diagonal joint approached that of solid pine, whereas the
conventional joint had about two-thirds of this strength;
moreover, standard deviation of bending strength of the diagonal
joint was less than that of a conventional joint. Tensile strength
of the diagonal joint was not greater than that of a conventional
joint, however.*

Schmutzler. 1969. (Properties of the cutting edge and cutter;
Stability)

Stefaniak. 1970. (Sawing; Circular sawing)

Stewart, H. A. 1970. CROSS-GRAIN KNIFE PLANING HARD
MAPLE PRODUCES HIGH-QUALITY SURFACES AND
FLAKES. For. Prod. J. 20(10): 39-42.
*Hard maple panels of short, clear cuttings were planed at 36
combinations of four rake angles (10, 20, 30, and 45 degrees),
three depths of cut (1/32, 1/16, and 1/8 inch), and three feed
rates (10, 20, and 30 knife marks per inch). Maximum surface
roughness was less than that for planing across the grain and
parallel to the grain, but average roughness was about the same for both
methods. Shavings from cross-grain planing resembled high-
quality flakes for particleboard.*

Stewart, H. A. 1970. EFFECT OF CUTTING ANGLE AND
DEPTH OF CUT ON THE OCCURRENCE OF CHIPPED
GRAIN ON SYCAMORE. USDA For. Serv. Res. Note NC-92, 4
p. North Cent. For. Exp. Sta., St. Paul, Minn.
*When Plantanus occidentalis was planed at 7-percent moisture
content, depths of cut up to 1/16-inch did not significantly
influence the occurrence of chipped grain, this generalization
was true for all rake angles tested (10, 15, 20, and 25°). Chipped
grain was most frequent with knives having rake angles of 25°,
and was minimized by reducing rake to 20° or less. Rake angle
was inversely correlated with power demand and dulling rate of
cutters.*

Stewart. 1970. (Machining with coated abrasives)

Stewart. 1971. (Machinability of species and reconstituted
wood; Hardwoods (white ash))

Tanaka et al. 1970. (Safety, including noise)

Tanaka et al. 1971. (Safety, including noise)

Tröger. 1969. (Safety, including noise)

Tsuzuki, K., Asano, I., Kato, K., and Nishiwaki, M. 1971.
STUDIES ON THE PLANE SCARF JOINTING OF PLYWOOD.
I. CUTTING AND JOINTING OF PLANE SCARF. Wood Ind.
26(12): 14-20. Tokyo.

TURNING

No publications found.

BORING, ROUTING, AND CARVING

Chivaksin, L. E. 1969. INVESTIGATION OF THE POWER REQUIREMENT AND FORCE PARAMETERS OF THE PROCESS OF REMOVING ROTTEN WOOD BY LONGITUDINAL MILLING. Lesn. Zh. 12(4): 79-85. Arkhangelsk.

Edamatsu, N., and Nagahara, Y. 1970. MACHINE BORING TESTS ON TROPICAL WOOD. Wood Ind. 25(4): 19-23. Tokyo. *Specimens of 11 tropical woods were evaluated in terms of plunge rate and thrust on the bit at spindle speeds of 610, 1450, and 3420 r. p. m. Tool life and quality of the bored hole were also assessed. Recommendations are made for boring tropical woods.*

Kondo, T., Fukui, H., and Kimura, S. 1971. FUNDAMENTAL MACHINING PERFORMANCE OF ROUTER AND CHIP PRODUCING MECHANISM IN ROUTING. II. WEIGHT DISTRIBUTION OF CHIP. Wood Ind. 26(9): 26-29. Tokyo.

Mori, M. 1971. AN ANALYSIS OF CUTTING WORK IN PERIPHERAL MILLING OF WOOD. III. VARIATION OF CUTTING FORCE IN INSIDE CUTTING OF WOOD WITH ROUTER-BIT. IV. THE POWER REQUIREMENT IN INSIDE CUTTING OF WOOD WITH ROUTER-BIT. J. Jap. Wood Res. Soc. 17. 437-448. *Equations predicting cutting force, work done, and net cutting power.*

Noack, E. 1970. HIGH-PERFORMANCE DOWEL BORERS. I. COMPARATIVE INVESTIGATIONS ON CUTTER HEADS AND DOWEL BORERS. II. ECONOMIC CALCULATIONS ON DOWEL BORERS. Mod. Holzverarb. 5(6/7): 410-416, 458-461.

Pahlitzsch, G., and Sandvoss, E. 1969. MILLING (ROUTING AND SHAPING) OF WOOD AND WOOD MATERIALS. Holzbearbeitung (5): 23-30.

Radu, A. 1969. COMPARATIVE STUDIES ON THE MOST IMPORTANT TYPES OF DRILLS USED IN THE WOOD-WORKING INDUSTRY. Holztechnologie 10(1): 33-37. (Also FIRA Transl. 125, Furn. Ind. Res. Assoc., Stevenage, Eng.) *Evaluates five major drill types in terms of manufacture, productive capacity, hole quality, and maintenance. The author recommends that the bit he developed (improved twist drill with smooth flutes) be adopted as standard in the wood industry.*

Radu, A. 1970. FORMULAE FOR CALCULATIONS RELATING TO WOOD BORING. Inst. Polit. Bull. 12, Ser. B., p. 301-314. Brasov. *Presents exponential formulae, and graphs with logarithmic scales on both axes, as a convenient way of calculating the torque and axial force in terms of tool diameter and other factors. Data are for beech, oak, spruce, and particleboard.*

Theien. 1970. (Machinability of species and reconstituted wood. Particleboards and sandwich boards)

MORTISING AND TENONING

Kryazhev. 1971. (Jointing, planing, molding, and shaping)

Raknes. 1969. (Jointing, planing, molding, and shaping)

Roth. 1970. (Jointing, planing, molding, and shaping)

MACHINING WITH COATED ABRASIVES

Hirst, K. 1971. PERCEPTIBILITY OF MACHINING MARKS ON PAINTED TIMBER SURFACES. CSIRO For. Prod. Newsl. 380, p. 2-4 Australia. *Samples of alpine ash, myrtle beech, and Monterey pine were planed at various feed speeds, or sanded with belts of various grit sizes, and then painted. Machining marks were generally less perceptible on flat and satin surfaces than on gloss. Black flat paints hid defects better than white or clear coatings, with gloss coatings, however, paint color had little effect on perceptibility of marks. A second coat of paint reduced the visibility of marks on sanded but not on planed surfaces.*
With gloss coatings, 32 cutter marks per inch were barely perceived, with flat or satin coatings, however, 20 marks per inch could not be noticed. Under most finishes, 4 to 8 cutter marks per inch were readily visible.
When sanding was done along the grain, score marks of 120-grit paper were nearly invisible under three coats of paint. Score marks made by sanding across the grain were much more difficult to hide.

Meyer, H. R. 1970. MEASUREMENT PROCESS FOR THE EVALUATION OF ABRASIVE PARAMETERS OF ABRASIVE BELTS FOR THE POLISHING OF WOOD AND WOOD MATERIALS. Ann. C. I. R. P. 18: 279-287. Great Britain. *Analysis of grinding ability and the effects of grit distribution, grain shape, and distribution of grain heights.*

Meyer, H. R. 1971. A CONTINUOUSLY OPERATING MEASUREMENT PROCESS FOR EVALUATING ABRASIVE BELTS IN TERMS OF PARAMETERS. Ann. C. I. R. P. 19: 667-672. Great Britain.
By application of a previously observed relationship between the distribution and orientation of grains and the grinding ability of belts, a continuous device for measuring grain spacing was developed. A foil of 20 μm. thickness is pressed into the belt surface; resulting perforations in the foil are evaluated continuously, thus permitting qualitative determination of the number of grains and average grain shape. The device should be usable by industry.

Noda, S., and Umetsu, J. 1971. STUDIES ON SHAPE CHANGES OF ABRASIVE GRAINS IN BELT SANDING OF WOOD. II. Wood Ind. 26(4): 21-26. Tokyo. *Fracture characteristics and wear of abrasive grains, loading of the belt face, loss of abrasive weight from the belt, abrasive sharpness, and belt life were studied on buna (Fagus crenata).*

Pahlitzsch, G. 1970. INTERNATIONAL STATE OF RESEARCH IN THE FIELD OF WOOD SANDING. Holz Roh- und Werkst. 28: 329-343. *Describes the interaction of abrasive belt, sanding machine, and workpiece. The belt progresses through three stages of wear initial sharpness, working sharpness, and bluntness. The curve relating stage of wear to abraded volume of wood per unit time per unit of contact area has a downward trend through all three stages – curvilinear in the first and last stages, but linear during the period of working sharpness. Belt pressure and speed are*

22

positively, and nearly linearly, correlated with volume of wood abraded per unit time. Cutting force increases almost in proportion to increase in belt pressure, but is virtually unrelated to belt speed. Quality of surface finish is almost independent of belt pressure and speed. Because of blunting, specific cutting force increases linearly with increased sanding time. Service life of a belt is defined by the lower permissible (i. e., economic) limit of abraded volume per unit of time.

Potrebic, M. 1969. RELATION BETWEEN THE ROUGHNESS OF BEECH WOOD AND THE GRADE OF SANDPAPER USED TO SAND IT. Tre og Mobler (6): 13-19.
Develops techniques for measuring four parameters that permit numerical expression of surface smoothness.

Sternal, L. S. 1969. ABRASIVE PLANING – A MUST AFTER PRECISION SAWING. *In* Proc. For. Prod. Res. Soc., North. Calif. Sect., p. 18-20. Calpella, Calif.
Achievement of sawing accuracy must precede widespread adoption of abrasive planers.

Stewart, H. A. 1970. ABRASIVE VS. KNIFE PLANING. For. Prod. J. 20(7): 43-47.
Power requirement was positively and linearly correlated with width of cut, depth of cut, and feed rate for both knife planing and abrasive planing. Abrasive planing, however, required approximately six times more power when cutting was parallel to the grain and 20 to 25 percent less power across the grain than parallel to the grain. While knife-planed surfaces were generally of higher quality than abrasive-planed surfaces, the machining defects extended farther below the surface, and required more subsequent processing to eliminate, than did scratches from abrasive planing. With increased feed rate, knife-planing surface quality decreased and abrasive-planing surface quality remained constant.

Zajac, I. M. 1969. CHOICE OF OPTIMUM CUTTING AND FEED SPEEDS IN SANDING SOFTWOODS WITH ABRASIVE DISKS. Lesn. Zh. 12(6). 87-91. Arkhangelsk.

Zajac. 1970. (Machinability of species and reconstituted wood, Softwoods)

VENEER CUTTING

Baldwin, R. 1971. HIGH-YIELD MILL MANAGEMENT. I. HOW TO INCREASE VENEER RECOVERY IN PLYWOOD AND VENEER PLANTS. II. PRACTICAL TECHNIQUES ACHIEVE INCREASED VENEER RECOVERY. III. THREE MORE AIDS TO ACHIEVING INCREASED VENEER RECOVERY. For. Ind. 98(1): 45-49; (2): 53-57; (3): 56-60.
Information is based on results obtained in Pacific Northwest mills, and on interviews with mill operators. Includes tables of volume recovered from veneer blocks, a formula for accurately setting lathes, a process control system for green clippers, a discussion of shrinkage control at the dryer, a graph showing percentage of block volume in cores of various sizes, a discussion of back-up rolls, a system for waste control, and a review of the factors that control panel thickness.

Calin. 1971. (Safety, including noise)

Cerenko and Zolotov. 1970. (Machinability of species and reconstituted wood; Softwoods)

Cumming, J. D., and Collett, B. M. 1970. DETERMINING LATHE SETTINGS FOR OPTIMUM VENEER QUALITY. For. Prod. J. 20(11): 20-27.
Young redwood was rotary-peeled at room temperature on a modified milling machine at 30 inches per minute. A roller nosebar ½-inch in diameter was used to control quality of the veneer. Surface smoothness and uniformity of veneer thickness were better indicators of quality than check depth and frequency. Best veneer resulted when roller bar gap was approximately equal to roller bar exit gap; compression (difference between veneer thickness and gap) optimums varied from 0.030 inch for 1/8-inch veneer to 0.055 inch for ¼-inch veneer. Gap was defined as the horizontal distance from the vertical plane of the knife top to the face of the roller bar; exit gap was defined as the minimum distance from the back of the knife (that surface in contact with the veneer) to the roller bar.

Dyskin, I. M. 1971. THE CALCULATION OF LOG TAPER IN THE PRODUCTION OF SLICED VENEER. Lesn. Zh. 14(1): 75-81. Arkhangelsk.

Dyskin, I. M., Plahov, V. N., and Tajc, O. G. 1970. USING THE TAPERING PART OF LOGS IN THE PRODUCTION OF SLICED VENEER. Derevoobrab. Prom-St'. 19(4): 12-14. Moscow.

Feihl, O., and Godin, V. 1970. PEELING DEFECTS IN VENEER – THEIR CAUSES AND CONTROL. Can. Dep. Fish. and For. Publ. 1280, 18 p.

Describes a number of common defects in veneer and outlines causes and corrective measures for each. Includes a list of publications prepared by the Ottawa Forest Products Laboratory on veneer cutting.

Feihl, O., and Godin, V. 1970. ROTARY CUTTING OF VENEER FROM WHITE SPRUCE. Can. Dep. Fish. and For. Publ. 1281, 12 p.
White spruce (Picea glauca (Moench) Voss) logs from the Ottawa Valley and New Brunswick were rotary-peeled successfully without heating, the knife was given a 0.015-inch-long micro-bevel (30° sharpness angle) to resist damage caused by knots. Heating the logs did not appreciably improve veneer quality. On a lathe with fixed nosebar (14° nosebar angle), veneer tended to jam in the cut and chucks often chewed log ends and caused spinout; with a 5/8-inch motor-driven roller bar, these problems were lessened. Veneer yield was 64 percent of bolt volume. In this laboratory test bolts were 4 feet long, averaged 12.6 inches in diameter, and were peeled to a 4.5-inch core.

Godin. 1969. (Properties of the cutting edge and cutter; Fitting and sharpening)

Grosshennig, E. 1971. MODERN VENEER PRODUCTION. I. LOG PREPARATION AND PEELING. II. TRANSPORTATION, DRYING, AND CLIPPING. III. ECCENTRIC PEELING AND SLICING OF VENEERS. Holz Roh- und Werkst 29: 129-142, 169-178, 209-216.
Part I: Surveys developments in log transport systems, barkers, feeding and centering devices, and in the peeling operation. Describes methods of improving veneer quality and peeling velocity and operation of the lathe.
Part II Summarizes designs of veneer transport systems, including veneer trays, reeling and unreeling devices, dryers, clippers, sorters, stackers, and jointers
Part III: Illustrates equipment for manufacture of very thin face veneers (0.2 mm. thick), and describes technique of producing thick veneers (to 25 mm.) for laminated products.
Part I has been translated (1971) into English by the New Zealand Forest Service, Wellington (22 p.).

Hayashi, D., and Tochigi, T. 1969. VENEER CUTTING WITH JET AIR PRESSURE. I. J. Jap. Wood Res. Soc. 15(7): 308.
Curling of Japanese red birch veneer was reduced by replacing a fixed, rigid nosebar with a high-pressure air jet Veneer peeled with the air-jet nosebar had fewer but deeper lathe checks.

Hayashi, D., Tochigi, T., and Inoue, H. 1970. MACHINEABIL-

23

ITY OF THIN THICKNESS VENEERS. Wood Ind. 25(12): 25-28. Tokyo.
A study, illustrated with graphs, on the cutting of thin veneers from walnut (Juglans spp.), Fraxinus mandshurica and Tectona grandis.

Hayashi, D., Tochigi, T., and Inoue, H. 1971. OBSERVATION ON RUPTURE FORMATION OF CELL WALLS AT THE CELLULAR LEVEL IN THE WOOD CUTTING PROCESS. I. EFFECTS OF INCLINATION ANGLE OF RAYS UPON CUTTING RESISTANCE AND RATIO OF U-SHAPED CELLS. II. EFFECTS OF INCLINATION ANGLE OF RAYS UPON PREVIOUS CHECKS, LATHE CHECKS AND CHARACTERISTICS OF CUTTING SURFACE. Wood Ind. 26(8): 358-363. Tokyo.
Veneer in thicknesses from 0.1 to 1.0 mm. was cut at 10 mm./second from airdry Japanese red birch (Betula maximowicziana Regel). Clearance angle was held constant at 45 minutes while sharpness angle was varied from 23 to 60°. Inclination of rays (to the direction of cut) was varied in 30-degree increments from 0 through 150°. The maximum number of U-shaped cells visible at the veneer surface (as opposed to cells separating at the intercellular layer) was found when the rays made an angle of 30° with (not against) the direction of cut. this observation was true when sharpness angle was 23 or 30° and veneer thickness was 0.6 or 1.0 mm. At these sharpness angles, cutting resistance was little affected by ray inclination. Depth of lathe checks appeared not related to inclination angle of rays, but at 23 and 30° sharpness angles, check length was maximum at ray inclination angle of 30°. In general, lathe checks appeared less damaging when the knife travelled with the rays (inclination angle less than 90°) than when it travelled against the rays (inclination angle 90-180°).

Hayashi, D., Tochigi, T., and Yamazaki, M. 1970. STUDIES ON VENEER CUTTING AT THE CELLULAR LEVEL - RUPTURE FORMATION OF CELL WALLS. J. Jap. Wood Res. Soc. 16(2). 70-75.
Observations of the types of cellular failures whereby veneer is formed, i.e., trans-wall, between-wall, and cellular tearing.

Jain et al. 1967, 1968, 1969, 1970. (Machinability of species and reconstituted wood; Hardwoods)

Koch. 1971. (Patents)

Lane. 1971. (Research instrumentation and techniques)

Lyubchenko, V. I. 1971. ANALYSIS OF THE CONDITIONS FOR OBTAINING A HIGH-QUALITY SHAVING (VENEER) IN TRANSVERSE CUTTING OF WOOD. Lesn. Zh. 14(3): 80-84. Arkhangelsk.
An analysis, mainly of check formation as affected by geometry of the pressure bar and its position relative to the knife of veneer peelers and slicers.

Mensbruge, G. de la. 1971. THE WOOD INDUSTRIES OF THE IVORY COAST. III. VENEER PEELING AND SLICING, AND PLYWOOD MANUFACTURE. Bois For. Trop. (137), (53-69).
This third part of the account of Ivory Coast timber industries deals with the veneering, slicing, and plywood industry, Though the industry is new, there are five veneer plants, two of which are equipped to make plywood. Timber species depends on the grades of plywood required. Much bombax is used. Production meets local needs, except where certain special grades are concerned. Some of it is also exported, mostly as veneer. Local secondary industries use some of the veneer for making flush doors, lath for packing cases, and particularly matches.

Meriluoto, J. 1971. THAWING BIRCH BOLTS. Pap. ja Puu 53(9): 493-497. Finl.
If splitting is to be avoided, frozen birch veneer bolts should be thawed in cold water (0 to 50° C.) before being heated for processing.

Mori, M., and Kubota, T. 1970. VENEER CUTTING WITH ROLLER-BARS. I. EFFECTS OF ROLLER DIAMETER AND BAR RESTRAINT. II. EFFECTS OF ROLLER DIAMETER AND BAR LEAD. Wood Ind. 25(10/12): 22-26; 19-24. Tokyo.
*Part I. Veneer 10 mm. thick was sliced from flat-grain saturated Shorea spp. at ambient temperature. Variable factors were roller-bar diameter (10 to 52 mm.) and bar lead (0.7 to 8.5 mm.), held constant were bar gap (8 mm.) and knife angles (sharpness angle 19° and clearance angle 0°30').
Bar force component perpendicular to (but not parallel to) feed direction increased with bar diameter. With some exceptions, the perpendicular component of combined knife and bar force decreased with a decrease in bar lead, the horizontal component, however, increased. The thickness of veneer produced after attainment of equilibrium cutting conditions was less than nominal, the difference was greatest with roller bars of large diameter set with small lead. Bar restraint was the major factor affecting actual veneer thickness.
Part II. Depth of lathe checks became shallower with decrease in bar lead and increase in roller diameter. When bar restraint was set to 1.5 mm., check depth varied from 5.5 to 6 mm. Curvature in the veneer was most pronounced with roller bars of small diameter set with small lead. Roller-bars yielded smoother veneers than double-bevelled solid bars. With roller-bars, surface smoothness was not affected by bar diameter or lead. Guide bars of two designs controlled the flow of veneer, removed curvature from the veneers, and prevented aggravation of lathe checks formed during cutting.*

Nikolaev, A. F. 1971. EFFECT OF MODULUS OF ELASTICITY OF THE VENEER BOLT ON STABILITY OF THE PEELING PROCESS. Lesn. Zh. 14(1): 108-112. Arkhangelsk.
A theoretical study, supported by experiments on veneer bolt cores 70 to 92 mm. in diameter.

Palka, L. C. 1970. PREDICTION OF VENEER LATHE SETTINGS. Wood Sci. 3: 65-82.
The theoretical prediction of optimum lathe settings as a function of veneer thickness, roller-bar diameter or nosebar profile, knife bevel angle, wood species, wood moisture content, and cutting temperature was investigated. A simple physical model of veneer formation is proposed. A segment of the continuous veneer sheet, bounded by two adjacent lathe checks, is identified as the repeating unit in the model. A working formula derived from this model showed some agreement with past experimental results from Douglas-fir. It also revealed some relationships not previously known, the practical implications of which are discussed.

Plesko, I. 1970. INFLUENCE OF NOSEBAR SHAPE ON THE QUALITY OF THIN SLICED VENEER. Drev. Vysk. (2), (75-82). Bratislava, Czech.

Porter, A. W., and Sanders, J. L. 1970. A HYDROSTATIC ROLLER BAR FOR VENEER LATHES AND THICK SLICING STUDIES. For. Prod. J. 20(10): 42-49.
A new roller bar design is based on the theory of hydrostatic lubrication. The bar is forced away from the roller-bar housing by high-pressure water. Alignment is maintained on the principle of self-compensation, i. e., four fine-control needle valves permit equalizing pressures to be developed in four pockets, symmetrically located about the bar. The main advantages of the design are significantly less wear of the roller, elimination of the drive mechanism for rotating the bar, and perfect synchronization with veneer velocity.

Porter, A. W., Walser, D. C., Kusec, D. J., and Sanders, J. L. 1969. DESIGN AND TESTING OF A HYDROSTATIC

24

ROLLER BAR FOR COMMERCIAL VENEER LATHES. Can.
Dep. Fish. and For. For. Prod. Lab. Inf. Rep. VP-X-48, 15 p.
*A new type of roller bar has been successfully tested on a
60-inch lathe. High-pressure water is used as a lubricant to force
the bar away from the stationary housing. The normal drive
system for rotating the roller bar is not required. Since the log
being peeled is the means of rotating the bar, veneer speed is
perfectly synchronized with bar speed. When veneer 0.125 inch
thick was peeled with a constant 15-percent compression,
quality decreased as bar diameter increased from 0.75 to 2
inches.*

Sevast'yanov, K. F. 1971. CALCULATION OF THE TEMPER-
ATURE FIELD IN A VENEER LOG DURING COOLING
AFTER HEATING. Derevoobrab. Prom-St'. 20(7): 8-10.
Moscow.
Formulae and two nomograms, with examples of their use.

Simonov, A. S. 1969. FORMATION OF THE PRODUCT
(SLICED VENEER) WHEN CUTTING WOOD ON ROTOR-
TYPE MACHINES. Lesn. Zh. 12(3): 72-77. Arkhangelsk.

Sugiyama, S., and Mori, M. 1971. FUNDAMENTAL STUDIES
ON MECHANISM OF VENEER CUTTING. I. ON BEHAVIOR
OF DEFLECTION CAUSED IN WOOD BY PRESSURE BAR

COMPRESSION. II. AN ANALYSIS ON DEVELOPMENT OF
DYNAMIC EQUILIBRIUM CONDITION IN VENEER CUT-
TING. J. Jap. Wood Res. Soc. 17(3): 96-110.
*Pressure bars were moved – under pressure – across specimens
of Shorea sp.p. When conditions were static, contours of wood
deflection (recordable from grid lines printed on the wood)
caused by single- and double-bevelled solid bars were asymmetrical
about the pressure point, whereas those for roller bars were
symmetrical. Under dynamic conditions, however, both types of
bar caused asymmetric deflections. Formulae are presented to
predict bar indentation and wood deflection.
The literature (R. B. Hoadley) contains reports on a transitional
period during which successively cut veneers vary in thickness.
The present research develops equations that explain this
variation.*

Swillinger and Hobbs. 1971. (Patents)

Tochigi, T., and Hayashi, D. 1971. VENEER CUTTING WITH
JET AIR PRESSURE. II. J. Jap. Wood Res. Soc. 17: 326-334.
*Continuation of an earlier study in which the same authors
showed that curl in veneer cut from Betula maximowicziana
Regel was reduced if a jet of air was directed against the sheet of
veneer as it formed. The air jet emanated from a slot in the
location ordinarily occupied by a nosebar.*

CHIPPING, FLAKING, AND GRINDING

CHIPPING

Altman, J. A. 1969. THE TRIMBLOCK CHIPPER. Am. Pulpwood Assoc. Tech. Release 69-R-47, 3 p.
*Earl Ballew of Stuart Manufacturing Company, Morton, Miss.,
has designed a disk chipper, primarily for the trim ends of
lumber. The machine will also chip pole trim ends and small
roundwood pulpsticks. It is gravity-fed through an inclined
spout; the 61-inch disk is distinguished by its arrangement of 22
small knives in a single spiral extending from the outer edge to
the center. Each small knife, sharp on two edges, cuts a swath
through the wood and exhausts chips through an opening in the
disk for bottom discharge; each following knife cuts a slightly
overlapping swath, in such a manner that uniform chips are
produced with low specific cutting energy. Feed rate is 20
ft./min. and power can be as little as 30 horsepower for lumber
trim ends. For wastewood and roundwood up to 10 inches in
diameter about 40 horsepower would be needed to drive the disk
at 300 to 400 r.p.m. Feed speed of 40 ft./min. can be attained if
44 knives are arranged in a double spiral, in which case 75
horsepower might be required for installations accepting roundwood up to 18 inches in diameter.*

Borlew, P. B., and Miller, R. L. 1970. CHIP THICKNESS: A
CRITICAL DIMENSION IN KRAFT PULPING. TAPPI
53:2107-2111.
*Reviews literature dealing with penetration and diffusion of
alkaline liquors into wood chips and with observations that
eventually led to the concept of chip thickness as a critical
dimension in kraft cooking. A value of 3 mm. is given as an
optimum thickness for delignification, yield, screenings, Roe
number, and physical properties of kraft pulps. Commercially
available devices for routine mechanical determination of chip
thickness are described. Discusses chippers designed to control
both length and thickness of chips.*

Chiviksin, L. E. 1971. THE ECONOMICS OF CONVERTING
LOW-GRADE WOOD INTO INDUSTRIAL CHIPS. Lesn. Zh. 14
(1): 117-121. Arkhangelsk.

Gorohovskij, K. F. 1969. INVESTIGATION OF THE PROC-
ESS OF COMMINUTING BRANCHES WITH ROTARY
CUTTER-BLOCKS. Lesn. Zh. 12(2): 38-41. Arkhangelsk.

Gusev and Rushnov. 1971. (Properties of the cutting edge and
cutter; Fitting and sharpening)

Hyler, J. E. 1970. CHIPPING AND CHIP HANDLING METH-
ODS FOR SAWMILLS. I.-IV. South. Lumberman 221(2743):
19-22; (2744): 27-36; (2745): 19-27; (2746): 25-28.
*A four-part discussion, with special attention to types of
chipping headrigs and problems incurred by the chipping
industry.*

Isomaki, O. 1970. STUDIES ON THE PREPARATION OF
CHIPS FROM END TRIMMINGS AND OVER-SIZED CHIPS.
Can. Dep. Fish. and For. Trans. OOFF-88, 11 p. (Transl. from
Pap. ja Puu 51(4): 369-372. 1969.)
*Of 16 types of chippers evaluated, 13 were designed to chip end
trimmings from boards, and three to rework oversize chips (i.e.,
those over 32 mm. in length). Crusher-type chippers with built-in
screens produced most sawdust (chips under 6 mm. in length)
but few oversize chips. Cutting chippers with built-in screens
gave similar results. Without a screen, machines of both types
delivered many oversized chips. Disk or drum knife chippers
without a feeding device generally made the best chips from end
trimmings, but it was desirable to screen the output and rework
the oversize chips. Chip thickness desired (for sulfate pulp) was
3.5 mm.; length specifications called for 95 percent of acceptable chips to be 6 to 32 mm.*

Kalashnikov, Y. 1971. SHARPNESS ANGLE OF CHIPPER
KNIVES AND CHIP QUALITY. Lesn. Prom. (8), (28-29).
Moscow.
*The effect of knife sharpness angle on chip quality was
investigated in a test bench and under production conditions
with a helical-disk chipper. Results were best with a sharpness
angle of 35° 20' to 31°30', but at temperatures below - 25° C. the
angle should be reduced about 1°.*

Koperin, F., Korobov, V., and Matjunin, V. 1971. A LINE
FOR PRODUCING INDUSTRIAL CHIPS. Lesn. Prom. (2),
(12-15). Moscow.
Describes highly mechanized layout for making pulp chips from

low-grade wood. Recommends barking schedules and provides data on quality of barking and wood losses for aspen, birch, and spruce debarked at various air temperatures in a drum machine.

McKenzie, W. M. 1970. CHOOSING YOUR CHIPPER – CHIPPING FOR PULP PRODUCTION. Aust. Timber J. 36(4): 21-31.
Illustrates and discusses the basic cutting action, characteristics, advantages, and disadvantages of disk, CCL, spiral, drum, and shredder-type chippers, and chipping headrigs. Describes methods of choosing a suitable size of chipper, and estimating its power requirement.

McLean, G. D. 1969. GLOBAL LILY PAD CHIPPER DISCUSSED BY REPRESENTATIVE. Am. Pulpwood Assoc. Tech. Pap. 68-20, p. 8-9.
The "Global Lily Pad Chipper," with 68 drum-mounted knives cutting in the 0-90 mode, produces 3/4-inch-long, veneer-like chips from log ends trimmed from peeler bolts (lily pads). Unscreened chips were 93 to 97.5 percent acceptable; fines totaled only 1.4 percent. The chipper drum is driven at 50 to 100 r.p.m. by a 50-horsepower motor through a gear reducer and roller chain. Costs of sharpening and replacing knives were $0.50 per 200 cubic feet of chips.

Men'shikova, L. V., Kalugina, Z. S., Sidorovskaya, I. P., and Korotov, S. A. 1971. EFFECT OF ROLLER-PRESSING CHIPS OF RESINOUS PINE WOOD ON THE PROCESS OF IMPREGNATION WITH BENZINE. Gidroliz. Lesokhim Prom-St'. 24(4): 3-4. Moscow.

Meyer. 1969. (Research instrumentation and techniques)

Neusser, H., and Krames, U. 1970. DETERMINATION OF SOME IMPORTANT WOOD CHIP INDEXES. USDA For. Serv. Transl. FPL-709, 13 p. (Holzforsch. und Holzverwert. 21(4): 77-80.)
Four classes of measurement are required to characterize a mass of chips (1) chip dimensions of length, width, and thickness (and, dependent on these, chip surface and volume); (2) chip surface characteristics (smooth, fibrous, rough, even, curved, curled) and chip shape (cubical, rod-like, needle-shaped, or scale-like). (3) individual chip weight; and (4) density of chip mixtures. Methods are given for measuring these parameters and their distribution curves.

Pahlitzsch, G., and Sommer, I. 1971. PRODUCTION OF WOOD CHIPS WITH A CYLINDER-TYPE CHIPPING MACHINE. II: EFFECT OF KNIFE GEOMETRY ON KNIFE BLUNTING, CUTTING FORCE, FEED FORCE, AND CHIP QUALITY. USDA For. Serv. Transl. FS-584, 26 p. (Holz Rohund Werkst. 24(3): 109-117. 1966.)
Pine at 30-percent moisture content was chipped so that "chipping was always carried out in the direction of the fiber." (The reviewer interprets this statement to mean that the cylinder chipper was fed to cut more or less in the 0-90 mode, i.e., with cutting edge parallel to the grain and with knife motion perpendicular to the grain.) Chip (or flake) thickness was 0.3 mm. The knife geometry that minimized power consumption was clearance angle 3°, sharpness angle 42°, and rake 45°. For best chip quality and least knife wear, however, the researchers found that clearance angle should not be less than 6°, the sharpness angle should be not less than 50 to 52°, and the rake angle should be 32 to 34°. Considering all factors of power demand, chip quality, and knife wear, the knife angles judged best were clearance 6°, sharpness 49°, and rake 35°.

Rushnov, N. P. 1969. EFFECT OF THE SWEEP-BACK ANGLE OF CHIPPER KNIVES ON CHIP QUALITY. Bum. Prom-St'. (8), (21-22). Moscow.
Describes studies on the effect of knife sweep-back angle ω (0, 10, 20, 30, and 40°) on chip quality. Optimum ω is 0-10°, but

disk design is difficult with $\omega \rangle 0°$, and hence the most practical arrangement of the knives is $\omega = 0°$, i.e. radially arranged.

Starzinsky. 1970. (Safety, including noise)

Tröger. 1970. (Safety, including noise)

Uusvaara, O. 1971. ON THE PROPERTIES OF CHIPS PREPARED FROM PLYWOOD PLANT WASTE. Inst. For. Fenn. Folia For. 107, 17 p. Helsinki.
Veneer chips (with some plywood chips included) from birch plywood plants in Finland had a moisture content of 54 percent of dry weight after transport to the pulp mill. Compaction occurred during transport; weight of a cubic meter of chips was 265 kg. before and 272 kg. after transport. Veneer chips were thinner and shorter (82 percent were 6 to 32 mm. in length) than conventional pulp chips. Average bark content was 0.67 percent.

CHIPPING HEADRIGS AND EDGERS

Aas, P. S. 1969. CHIPPING SAW MACHINES – IN SCANDINAVIAN SAWMILLS OF THE FUTURE. Nor. Skogind. 23(10): 263-266.
A general article reviewing Koch's work on the development of chipping headrigs and Dobie's work on their productivity, and supporting their use in Scandinavian sawmills.

Anderson, W. C., and Kaiser, H. F., Jr. 1970. ECONOMIC IMPLICATIONS OF CHIPPING HEADRIGS FOR MILLING SOUTHERN PINE. For. Prod. J. 20(3): 42-46.
To analyze how installation of chipping headrigs affects sawmill operating methods and net revenue, a sawmill was synthesized by modifying a conventional mill. Data for the modification were collected from mills having such headrigs. Linear programming indicated that, where small logs predominate, a sawmill built around a chipping headrig and without a conventional side would be most advantageous.

Anonymous. 1970. COMPACT STUDMILL PROCESSES PEELER CORES IN TWO SIZES. For. Ind. 97(3): 102-103.
A compact new studmill operated by 2 men produces about 20,000 board feet of precision end-trimmed 2 x 4's daily from peeler cores of two sizes: 6-5/8 inches and 5-3/8 inches. Central to the operation is an ADCO West Stud Cant machine with four chipping heads and two saws. Studs are processed at the rate of 60 lineal feet per minute.

Dobie, J. 1970. ADVANTAGES OF LOG SORTING FOR CHIPPER HEADRIGS. For. Prod. J. 20(1): 19-24.
Simulated log processing on chipping headrigs indicates that advantages are to be gained from sorting logs by diameter classes. The objective of sorting is to reduce the frequency of chipperhead adjustments. For 1,500 logs of each of three diameter distributions tested, production gains from sorting rose from 11 percent to 153 percent, in direct proportion to increases in feed speed. There was little advantage in lengthening the deck for sorted logs from 30 to 60 or 90 feet. Sorting by 2-inch diameter classes was less advantageous than sorting by 1-inch classes, but was still worthwhile. Achievement of full potential from sorting would probably require modifications to the auxiliary systems, e.g., barkers, resaws, and edgers.

Galloway, L. R., and Thomas, P. R. 1971. CHIP QUALITY FROM A "CHIP-N-SAW" CHIPPING HEADRIG. Pap. Trade J. 155(50): 55.
Chips produced by several Chip-N-Saw headrigs have very acceptable dimensions for both the kraft and acid sulfite processes, and pulp from them processed normally in both paper pulp and dissolving processes. Chip lengths (along the grain) averaged 0.75 inch and thicknesses 0.137 inch. Excellent kraft

26

and sulfite pulping characteristics were found in pilot plant studies. In kraft processing, Chip-N-Saw chips cooked faster to at least comparable yields and with less rejects than chips from disk chippers. Chip-N-Saw pulp had more long fibers and gave higher tear values. On the negative side, Chip-N-Saw output had a higher percentage of fines than chips from the disk machine, but this objection can probably be corrected by modifications in the design of the chipping headrig.

Hickman. 1971. (Patents)

Koch, P. 1969. SHARING THE TREE. South. Lumberman 219(2718): 18.
Chipping headrigs have made possible a system of wood procurement that efficiently channels tree segments into products of highest value. Lumbermen, veneer manufacturers, and pulpmills can cooperate in the system to resolve their individual log-supply problems.

Koch, P. 1970. NEW PROCUREMENT APPROACH IN-CREASES PINE UTILIZATION. For. Ind. 97(3).46.
The chipping headrig is the key to a system under which sawmillers do all logging of small trees, regardless of land ownership, and then sell chips to pulpmills and high-quality bolts to plywood plants.

Kotesovec, V. 1970. EXPERIMENTS ON CUTTING WOOD WITH THE 10 VTR-2 TEN-SPINDLE MULTIPLE-BLADE CIRCULAR SAW. Holztechnologie 11(4): 225-233.
Describes the construction, operation, and performance of the machine. Defines operating conditions for making chips suitable for particleboard and for pulp or fiberboard. The machine produces a good quality of sawn timber.

Morton. 1971. (Patents)

Neville, C. W., and Bennie, A. 1969. PROGRESS WITH THE F.P.R.L. CHIPPING SAW SYSTEM. Timber Trade J. 271(4862): 59, 61-62.
Describes a double-arbor rip saw (i.e., one ripping simultaneously from top and bottom) carrying special saws that produce strand-like material (instead of sawdust) for use in particle boards.

Sullivan, M. D. 1970. SMALL LOG SIDE USES QUAD, CHIP–CANT MILLS. For. Ind. 97(8): 82-83.
The Fruit Growers Supply Co., at Hilt, Calif., has constructed a facility for processing small logs. The center of the system is a Beloit chipper/canter with two chipping heads designed to handle logs with a maximum diameter of 18 inches and a minimum top diameter of 5 inches. The new facility is adjacent to the firm's large sawmill, and total mill production has improved over 10 percent since installation in November of 1969. Further gains are expected when the method of log segregation has been perfected and a stable supply of small timber can be maintained.

Thunell. 1971. (Sawing; General)

MOBILE CHIPPERS

Cadenhead, E. S. 1970. ROADSIDE BARKING AND CHIP-PING. Pulp and Pap. Mag. Can. 71(9): 83-85.
Describes a roadside barking and chipping trial carried out in small Canadian timber (88 percent balsam fir and 12 percent spruce) in which trees averaged slightly over 5 cubic feet in volume. Trials lasted 3 months, including some winter months; equipment was a Morbark Chiparvestor Model 1450 with connected mechanical debarker. The unit was reliable with mechanical availability of 87.5 percent. Logs were skidded about 900 feet direct from stump to chipper, and chips were

transported in two vans 40 to 60 miles to a kraft linerboard mill. About 2-1/2 hours of operating time was required to fill a van. Maximum log that could be chipped was 14 inches in diameter. Roadside chipping had several advantages: higher man-day productivity between stump and chip storage pile (3.28 cunits per man-day); improved utilization due to processing in tree-length form; elimination of bark and debris around the mill; delivery of clean wood fiber without the mud, rock, ice, or snow often found in roundwood; and reduction of roundwood inventories in the woods. There were also some disadvantages, the cost of power at roadside was higher than at the mill, and chips had a higher bark content than those produced from small wood in the mill woodroom. Since the bark stayed in the woods, the mill was deprived of a fuel source, but this loss was offset by elimination of handling costs and by reduction in boiler maintenance.

Côte. 1971. (Patents)

Hensel, J. S. 1971. MORBARK SATELLITE SAUCHIP SYSTEM. Am. Pulpwood Assoc. Tech. Release 71-R-40, 9 p.
Morbark Industries, Winn, Michigan, have developed a system whereby tree-length wood is processed for lumber and chips. Equipment consists of a heel-boom loader, log cutoff saw, mechanical debarker, circular sawmill, chipper, chip screen, blower, and conveyors. Trees are first cut to length and then debarked. Logs suitable for lumber are sawn, and all smaller roundwood, together with slabs and edgings, is converted into pulp chips. Cost of the system (installed) is estimated at about $289,000, including land, buildings, and equipment. With a five-man crew and a weekly output of 1,250 tons of chips and lumber, the mill can show a profit of $1.50 to $2.00 per ton if stumpage is priced at $1 to 1.72 per green ton.

Laubgan, A. A. 1971. COMMINUTION OF SMALL WOOD FROM TENDING OPERATIONS IN PROTECTIVE STANDS. Lesn. Khoz. (1), (64-68). Moscow.
Reviews the design and performance of various Soviet and foreign disk and drum chippers, both stationary and mobile types, particularly with regard to the chipping of brushwood.

Plokhov, V. 1971. A FLOATING FACTORY FOR CONVER-SION OF WASTE WOOD. Lesn. Prom. (6), (17-18). Moscow.
Describes proposed design, for use in the rivers of the Archangel region, of a floating factory consisting of two pontoons fitted with equipment for barking and chipping waste wood and residues.

Sterle, J. R. 1971. MORBARK METRO CHIPARVESTORS. Am. Plywood Assoc. Tech. Release 71-R-19, 5 p.
Describes, with specifications, two portable chippers manufactured by Morbark Industires of Winn, Michigan. Both models are diesel powered and designed for on-site chipping of trees, including stem, bark, and branches, they are trailer-mounted for towing on or off the highway. The larger unit has a three-knife disk chipper 75 inches in diameter and will accept full-length trees up to 20 inches in diameter. The smaller unit has a 48-inch chipper, with 2 or 3 knives, and will accommodate 10-inch trees. Both are operated by one man. Trees and brush are placed on the infeed conveyor by a Prentice hydraulic loader and fed into the chipper through compression rollers. Chips are discharged through a spout into waiting trucks. Production depends on number and size of trees available, but averages about 1 ton per minute for the larger machine and 1/2-ton per minute for the smaller.

Swan, D. A. 1971. THE ASPLUNDH TRAILER CHIPPER. Am. Pulpwood Assoc. Tech. Release 71-R-46, 3 p.
The machine is designed with a 10- by 12-inch or a 10- by 16-inch throat to chip roadside brush and blow the chips 25 to 30 feet back onto the roadside. Two or three men are required

27

for its operation. one to feed the machine, one to cut the brush, and sometimes a third man to assist in cutting and feeding. A pickup truck easily pulls the trailer-mounted chipper. The 12-inch model is powered by a Ford 172 or 300 gasoline engine, and the 16-inch model by a Ford 300 or 330 engine. It is reported that both models will chip brush about as fast as they can be fed.

USDA Forest Service. 1970. RESULTS OF FIELD TRAILS OF THE TREE EATER, A TREE AND BRUSH MASTICATOR. USDA For. Serv. Equip. Dev. and Test Rep. 7120, 26 p. San Dimas, Calif.
The Tree-Eater is a flail-type drum hog powered by a 325-horsepower diesel engine. it can be mounted on the front of a tractor with 18-inch tracks. The drum, which turns at 1,800 r.p.m., is 30 inches in diameter, 72 inches in length, and has four rows of cutting arms (70 in all) mounted on four 72-inch hinge pins. Cost (with spares) of the 25,500-pound unit is about $38,500, not including the tractor. It is manufactured by Tree Eater Corporation, Gurdon, Ark. During cutting and masticating operations, the height of cutters above ground can be varied from 0 to 8 inches, to aid in pushing trees forward as they are cut, a hydraulically operated bar is located above the drum cutter. During tests by the USDA Forest Service the machine demonstrated its ability to fell trees and masticate all but the main stems (10 inches in diameter and over) of the larger and tougher species. *In mastication of pine slash and undesirable shrubs, and in strip thinning, juniper control, and cactus control, ground could be covered at 0.4 to 1.5 acres per hour. Brush, including chaparral, and trees up to 6-inch stump diameter were almost completely shredded and reduced to a mulch.*

FLAKING

Demidov, Ju. M. 1970. THE SHAPE AND DIMENSIONS OF CHIPS TO BE COMMINUTED IN A CENTRIFUGAL CHIPPER. Derevoobrad. Prom-St'. 19(12): 9-11. Moscow.

Hyler. 1970. (Chipping)

Kershaw. 1971. (Patents)

Pahlitzsch and Sommer. 1971. (Chipping, flaking, and grinding)

Stewart. 1970. (Jointing, planing, molding, and shaping)

Stewart. 1971. (Peripheral milling parallel to grain)

GRINDING

Efremova and Mitusov. 1971. (Safety, including noise)

DEFIBRATING

Atack, D., and May, W. D. 1970. HIGH SPEED PHOTOG-RAPHY OF PARTICLE MOTION IN A DISC REFINER. *In* Proc., 9th Int. Congr. on High-Speed Photogr., p. 526-527. Soc. Motion Pict. and Telev. Eng., N.Y.
Motion pictures, with each exposure timed to occur in front of a transparent window cut in one disk, showed that particles and fiber bundles align radially between counter-rotating disks spaced about 0.05 inch apart. The radially aligned particles rolled (in the sense of a needle bearing) at high speed, and the lateral projections on the particles, whose presence was originally thought to exclude rolling, proved to be splinters peeling off the particle. These projecting splinters were flung laterally into the space between the disks and then folded into the particle as the rolling action forced them against the disk surfaces Details of the photographic technique are given.

Belova, T. A., and Laskeev, P. H. 1971. PRODUCING MECHANICAL PULP FROM SCOTS PINE. Bum. Prom-St'. (5): 15. Moscow.

Corson, S. R. 1971. APPLICATION OF SIZE REDUCTION THEORIES TO DISC REFINER PULP PRODUCTION. N.Z. J. For. Sci. 1(1): 125-127. Rotorua.
General comminution theories may be used to describe the breakdown of wood chips in a disk refiner.

Lyon. 1971. (Patents)

Martin, A. C. 1969. A STUDY OF GRINDER VARIABLES AS PRODUCTION IS INCREASED. Pulp and Pap. Mag. Can. 70(22): 81-88.
Production of a continuous chain grinder operating on 4-foot sticks of spruce and balsam was increased from 18.4 to 29.8 cunits per day – a gain of 61.5 percent – in experiments conducted at the Abitibi Paper Company's mill at Ste. Anne de Beaupre', Quebec. In large part, the technical bases for the trials were Klemm's equation relating production to surface speed of the stone and to average tracheid diameter, Atack and Heffell's method of grit count, and Gavelin's formula relating production to stone sharpness. When 80-inch-diameter stones turning at 257 r.p.m. were substituted for 72-inch stones turning at 225 r.p.m. output rose 26.5 percent. *The remaining 35-percent gain was obtained by changing burr-patterns so as to increase the number of active grits in the stone. At the 29.8-cunit level, about 31.1 airdry tons of pulp were produced per day at a specific grinding energy* of *73.9 horsepower days per airdry ton, or 80.3 horsepower days per cunit ground. Analyses of heat balances showed that only 26.2 percent of the energy input at the stone was transferred to the pulp in the grinding zone, 49 5 percent was dissipated in heating ground stock in suspension in the pit and in heating shower water; 24.3 percent was exhausted from the grinder with the pulp-water mixture.*

May, W. D. 1970. THE MEASUREMENT OF DISC MIS-ALIGNMENT IN REFINERS. Pulp and Pap. Mag. Can. 71(15): 47-56.
A device is described that measures out-of-tram and other misalignments in a double-disk refiner while the machine is in operation. (Out-of-tram refers to the condition in which the two shafts of a double-disc refiner are tilted with respect to each other.) It was shown that the tram changes with changes in temperature distributions over the refiner base and that oscillatory changes in plate clearance due to unbalance are damped out by passage of refiner stock. Pulp properties were therefore unaffected by the degree of unbalance. The instrument revealed a new effect, that of dynamic out-of-tram, whereby a disc of the refiner was forced out of tram by the passage of stock. The direction of tilt was found to rotate at up to three c.p.s. The amplitude of tilt was much greater than that normally associated with out-of-tram due to misalignment of the shafts, and indicated that severe bending of the shafts was taking place. The amount of dynamic out-of-tram was sufficiently large to cause the plates to touch under normal operating conditions; it was concluded that dynamic out-of-tram is a normal mode of operation of this refiner.

Norton Company. 1968. NORTON PULPSTONES. 42 p. Worcester, Mass.
Describes stones designed to produce mechanical pulp in grinders that will accommodate cylinder-shaped pulpstones up to 100 inches in diameter and 104 inches in face width. Such large stones may absorb 14,000 horsepower and produce more than

200 tons of pulp per day. Abrasives are of fused alumina or silicon carbide with particle sizes from 0.136 to 0.0034 inch, and are bonded with vitrified or glass-type material of specified hardness (J to T in order of increasing hardness). Gives instructions for handling, storing, mounting, and operating pulpstones. Tabulates weight, volume, and inertial data for a variety of stones.

Perry, J. H. 1965. THE NEW AGE OF GROUNDWOOD. 10 p. Norton Co., Worcester, Mass.
Briefly discusses pulp grinding technology and grinder configuration. Includes a description of the Anker technique of pitless grinding.

Perry, J. H. 1970. FACTORS ESSENTIAL TO MAXIMUM GRINDER PRODUCTION. Norton Int., Inc., Worcester, Mass. 12 p.

Describes some results from research project "Camel," a cooperative program involving manufacturers of grinding machines, pulpstones, and groundwood pulp. Discusses recent developments of pitless grinding, high-speed grinding (surface speeds of the stone were as high as 50 meters per second), increased power (to 10,000 horsepower on the stone), controlled stone grinding area (detailed description of burring technique provided), and improved control of temperature and shower water. Clear white water at pressures from 5 to 7 atmospheres should be directed via fixed-flow showers to clean the stone and by adjustable showers to control the grinding zone temperature. Water should be at $100°$ to $140°$ F. Recommended burrs create sharp grooves and flat-topped lands in the stone to yield effective grinding pressures (on freshly cut lands only) of 101 to 242 pounds per square inch. A 10,000-horsepower grinder can produce 130 to 150 short tons of pulp per 24 hours.

PROPERTIES OF THE CUTTING EDGE AND CUTTER

TOOL MATERIAL

Borovikov, E. M., and Pozdeev, A. S. 1970. EFFECT OF HIGH-FREQUENCY HEATING DURING BRAZING AND TEMPERING ON THE MICROSTRUCTURE AND STRENGTH OF HARD-ALLOY (TIPPED) FRAMESAW TEETH. Lesn. Zh. 13(2): 98-103. Arkhangelsk.
Data show that heating of the steel during brazing to harden the tips of the teeth changes its structure and reduces its strength, but that subsequent tempering almost doubles the mechanical strength of the teeth

Borovikov, E. M., and Pozdeev, A. S. 1970. THE EFFECT OF THE MAKE OF METAL/CERAMIC HARD ALLOY ON THE STRENGTH OF FRAMESAW TEETH. Lesn. Zh. 13(6): 77-81. Arkhangelsk.
Describes tests on teeth tipped with five Soviet makes of hard alloy, three of which proved to be significantly stronger than the others.

Borovikov, E. M., and Pozdeev, A. S. 1971. EFFECT OF THE POSITION OF HARD-ALLOY TIPS ON THE STRENGTH OF FRAMESAW TEETH. Lesn. Zh. 14(3): 66-68. Arkhangelsk.

Dem'janovskij, K. I. 1970. INVESTIGATIONS ON THE STRENGTH OF THE CUTTING PART OF WOOD-CUTTING TOOLS. Lesn. Zh. 13(3): 69-73. Arkhangelsk.
Bending tests were made on cutters of several Soviet types of steel; loading was at various distances from the edge of the cutters. The nature of the breakages is discussed.

Dunaev, V. D. 1971. THE EFFICIENCY OF HARD-ALLOY (-TIPPED) (CIRCULAR-) SAW BLADES ON EDGERS. Derevoobrab. Prom-St'. 20(1): 17-18. Moscow.
In operational trials with tipped blades on standard equipment, Scots pine boards 16 and 22 mm. thick were edged in both summer and winter. Results were promising, although output was reduced in winter by the increase in feed resistance and rate of blunting.

Mihut, I. 1971. THE DEVELOPMENT OF KNIVES FOR WOOD PROCESSING, THE PROCESS OF KNIFE WEAR, AND METHODS OF DECREASING IT. Ind. Lemnului 22(8). 297-303. Bucharest.
Reviews the materials used. Special reference to the development and testing of carbide-tipped blades, and various theories of the geometry of the cutting profile and its erosion.

Norton Company. 1968. (Chipping, flaking, and grinding; Defibrating)

Ohsako, Y., and Domoto, K. 1971. INVESTIGATION OF THE CHANGE ON THE TOOL SURFACE CAUSED BY CORROSION OF TOOL MATERIALS AND OF THE CONSEQUENT CHANGE IN CUTTING ENERGY. Kyoto Univ. For. Bull. 42, p. 235-244. Japan.
Describes surface effects and changes in cutting energy resulting when knives of low-carbon steel, alloy steel, and high-speed steel were corroded with a 5-percent salt solution. Hard knives resisted corrosion.

Porter. 1971. (Sawing; Bandsawing)

Quezada, A. F. 1970. STELLITING OF BANDSAWS. Chile Inst. For. Nota Tec. 15, 12 p.
Describes the technique and the results of experience in a Chilean sawmill cutting Nothofagus dombeyi and Weinmannia trichosperma Stelliting improved output by 14 percent.

Tallack, R. F. 1971. SOLID TIPPING OF CIRCULAR SAW TEETH WITH COBALT-BASED ALLOYS. Timber Trades J. 276(4921): 42-46. London.
Describes in detail a method devised and used at the College of Technology, Belfast, Northern Ireland, for solid-tipping the teeth of circular saws (rip and cross-cut) with alloys of cobalt, chromium, and tungsten.

Wuttkowski, H. 1968. ECONOMY IN USING CARBIDE-TIPPED TOOLS. Drevo 23(1): 30-32. (FIRA Transl. Furn. Ind. Res. Assoc. 103, Stevenage.)
A cost study.

DULLING

Barz, E. 1970. RESULTS ABOUT EFFECTS OF WOOD AND WOOD MATERIALS ON THE DEVELOPMENT OF WEAR. In Mechanical Conversion of Wood Meet. Proc., p. 46-52. Sect. 41, IUFRO. Braunschw.

Barz and Breier. 1969. (Research instrumentation and techniques)

Barz, E., and Breier, H. 1971. ACCELERATED TESTING OF WEAR EFFECTS AND CHIPPING PROPERTIES OF WOOD AND WOOD-BASED MATERIALS. Holz Roh- und Werkst. 29: 142-149.
A single-edged tool of alloy steel, employed in a machine designed to accomplish accelerated wear of the cutting edge, yielded comparative data on a variety of softwoods, hardwoods, and particleboards.

Bridges. 1971. (Machinability of species and reconstituted wood; Particleboards and sandwich boards)

Chardin. A., and Froidure, J. [n.d] THE WEAR OF SAW TEETH. Introduction, Vol. I, INFLUENCE OF WOOD MOISTURE CONTENT AND CHIP THICKNESS. Vol II, INFLUENCE OF CUTTING SPEED. Cent Tech. For. Trop., Nogent-sur-Marne, France, 752 p.
Cent. Tech. For. Trop., Nogent-sur-Marne, France, 752 p.
The introduction (82 p) establishes the importance of study of saw tooth wear in the milling of tropical woods. It describes equipment to measure weight of metal worn from teeth, technique of making tooth replicas for evaluation of edge profiles, and methods for photographing teeth before and after wear. Illustrates standard method of data presentation for publication of results on edge profiles, and progression of tooth wear
Volumes I (378 p) and II (292 p) diagram cutter wear data on numerous tropical species as affected by wood moisture content and density, duration of cutting, cutting speed, and chip thickness

Cowling and McKenzie. 1969 (Properties of the cutting edge and cutter, Fitting and sharpening)

Deger, M , Fischer, R , and Tröger, J. 1969. EXTENSION OF LIFE OF WOODWORKING TOOLS BY MODIFIED BORON TREATMENT. Holztechnologie 10(4) 229-232.
Boron-treating the back of tools may extend time between sharpenings. The method was developed by the Technical University of Dresden

Hayashi et al. 1970. (Machinability of species and reconstituted wood; Plywood)

Ivanovsky, E. G., Vasilevskaja, P. V., and Lautner, E. M 1970. INVESTIGATION OF THE PROCESS OF CUTTER DULLING DURING THE PLANING OF WOOD Lesn Zh. 13(5): 65-67. Arkhangelsk

McKenzie, W. M , and Cowling, R. L. 1970. THE EARLY STAGE OF EDGE WEAR IN CUTTING WOOD. *In* Proc , IUFRO Sect 41 Meet. on Mechanical Conversion of Wood, p. 39-45. Braunschweig
Eucalypt hardboard laminated with PVA adhesive and averaging 1 05 specific gravity and 8-percent moisture content was cut with steel knives containing 12-percent chromium Knives made sharper than usual by stropping techniques that produced a "gothic arch" profile at the cutting edge appeared to withstand initial cutting impact better, be sharper in the work-sharp state, and stay sharper longer than edges sharpened in the usual manner.

McKenzie and Cowling. 1971. (Orthogonal cutting)

Neusser and Schall. 1970. (Machinability of species and re-constituted wood, Fiberboards, hard and soft)

Ohsako, Y., Toyama, Y , and Sugimoto, R. 1971. INVESTIGATION ON TOOL LIFE AND WEAR CHARACTERISTICS OF THE CIRCULAR SAW. Kyoto Univ. For. Bull. 42, p. 245-258. Japan.
Saw life between sharpenings is primarily limited by dulling from wear, but also by buckling or cracking of the blade, and by tooth damage (from mishandling the saw). In tests, wear on most clearance (back) surfaces of cutting edges exceeded that on rake surfaces, possibly because compressed material on cut surfaces expanded after teeth passed and therefore rubbed against clearance surfaces Each saw tested had a characteristic amount of set remaining after it became dulled to a degree requiring resharpening. The amount of set was one factor controlling the degree to which cut surfaces were burned or burnished Tool edges became rougher as they wore. Power to drive saws was

proportional to wear, so that power demand is one indicator of time to resharpen.

Pahlitzsch and Sandvoss. 1970. (Machinability of species and reconstituted wood; Fiberboards, hard and soft)

Pahlitzsch and Sandvoss. 1970. (Machinability of species and reconstituted wood; Particleboards and sandwich boards)

Plastov. 1971. (Machinability of species and reconstituted wood; Particleboards and sandwich boards)

St-Laurent, A., 1970. EFFECT OF SAWTOOTH EDGE DEFECTS ON CUTTING FORCES AND SAWING ACCURACY. For. Prod. J. 20(5): 33-40.
Loads on swaged ripsaw teeth with edges not perpendicular to cutting direction differ from loads on normal swaged teeth whose cutting edges are perpendicular When edges were ground 12-1/2 degrees from perpendicular, lateral forces on each tooth were substantial (about 10 pounds for a 1/4-inch-wide inserted-bit tooth cutting 0.005-inch deep in a variety of woods). Damaged teeth also caused lateral loads, lumber cut on a 12-inch bandsaw with a succession of identically damaged teeth had thickness variations of 1/8-inch.

Stefaniak, W 1970. EFFECT OF CUTTING SPEED ON THE BLUNTING OF HARD-ALLOY-TIPPED TEETH OF CIRCULAR SAWS DURING THE SAWING OF PARTICLE BOARDS Folia For. Pol. (Drzewn.) 9, p. 57-78.
Blunting increased curvilinearly with cutting speed in the range from 33 to 124 m./sec. Amplitude of blade oscillations also influenced rate of blunting.

Stefaniak. 1971. (Machinability of species and reconstituted wood; Particleboards and sandwich boards)

Tanaka et al. 1970. (Safety, including noise)

Timonen, S. M. 1969. WEAR AND DULLING OF SAW TEETH AS A FUNCTION OF THE CUTTING PATH. Lesn. Zh. 12(3) 67-72. Arkhangelsk.

Timonen, S. M. 1970. EFFECT OF CUTTING SPEED ON THE WEAR AND DULLING OF SAW TEETH. Lesn. Zh. 13(3): 85-88. Arkhangelsk.

Vlasov, V. A., and Litvinov, A. B. 1970. EVALUATION OF THE DURABILITY OF A WOOD-CUTTING TOOL. Derevoobrab. Prom-St'. 19(3): 11-12. Moscow.

FITTING AND SHARPENING

Anonymous. 1970. A MACHINE FOR HOT SWAGING AND HARDENING OF SAW TEETH. Holz Zentralbl. 96(114): 1663.
Briefly describes the Bansomat (made by F. Banholzer, Heilbronn-Sontheim) in which the saw-teeth are heated to 800° C. at the moment of swaging. The method increases wear resistance and reduces maintenance time.

Aoyama, T. 1970. TENSIONING OF BAND SAW BLADE BY ROLLS. I. CALCULATION OF CROWN. II. CALCULATION OF TENSION. J. Jap. Wood Res. Soc. 16: 370-381.

Aoyama, T. 1971. TENSIONING OF BAND SAW BLADE BY ROLLS. III. EFFECT OF SAW BLADE THICKNESS. J. Jap. Wood Res. Soc. 17(5): 188-202.
Changes in blade crown and thickness profile are related to blade thickness, roll pressure, and location of application.

Barz, E. 1970. THE DEVELOPMENT OF CRACKS IN CIRCULAR SAW BLADES. Holz Zentralbl. 96(98): 1427-1428.

30

Discusses cracks developing as a result of faults in manufacture, use and wear (e.g., overheating caused by flutter), and faulty maintenance, with notes on the effects of type of sharpening and grain direction, and of metal hardness and temperature. Gives practical recommendations

Beck, E. 1971. HOT OR COLD SWAGE-SETTING OF FRAME- AND BANDSAWS? Holz Zentralbl. 97(105): 1495.
Discusses metallurgical aspects and technology of both methods. Favors the established cold system.

Borovikov, E. M., and Pozdeev, A. S. 1970. EFFECT OF THE BACK BEVEL OF FRAMESAW TEETH ON CUTTING FORCE AND THRUST FORCE. Derevoobrab. Prom-St'. 19(11). 19-20. Moscow.

Böttcher, A. 1966. CORRECT PREPARATION OF CUTTING TOOLS. Mod. Holzverarb. 1(10): 646-649.

Clark. 1969. (Bandsawing)

Cowling, R. L., and McKenzie, W. M. 1969. LAPPING OF STEEL TOOLS FOR MAXIMUM SHARPNESS. Wood Sci. 2: 20-25.
Describes methods for sharpening edges used in research on the effects of bluntness It was possible to produce a stable edge of about 0.2 μm. radius on tools of steel with 2 percent carbon and 12 percent chromium. The influence materials, techniques, and limitations on initial sharpness are discussed.

Cumming. 1969. (Sawing; Bandsawing)

Dogaru, V., and Capraru, M. 1970. REDUCING THE CONSUMPTION OF BANDSAW BLADES FOR SAWING WOOD. Bull. Inst. Politeh. Brasov (Ser. B. Econ. For.) 12, p. 217-225.
A study of factors tending to decrease tension at the base of teeth, showing a relation between service life and regularity of the surface in this zone.

Dunaev, V. D. 1970. FINAL GRINDING OF PLANER KNIVES WITH FINE-GRAINED GRINDING DISKS. Derevoobrab. Prom-St'. 19(4): 24-25. Moscow.
Describes a series of investigations, and recommends procedures and equipment.

Dunaev. V. D. 1971. 'HYDRO-ABRASIVE' FINAL GRINDING OF PLANER KNIVES. Derevoobrab. Prom-St'. 20(4): 16-17. Moscow.

Foley Manufacturing Company. [n.d.] MONEY SAVING FACTS ON CARBIDE SAW AND TOOL SHARPENING. 27 p. Minneapolis, Minn.
A short manual on grinding practices for circular saws tipped with cemented carbide cutting edges.

Fukui and Kimura. 1968. (Machinability of species and reconstituted wood, Particleboards and sandwich boards)

Godin, V. 1969. THE GRINDING OF VENEER KNIVES. Can. For. Serv. Publ. 1236F, 24 p.
Illustrated explanation of techniques for grinding veneer knives.

Gusev, O. N., and Rushnov, N. P. 1971. CORRECTING THE SHARPNESS ANGLES OF THE KNIVES IN DISK CHIPPERS. Derevoobrab. Prom-St'. 20(7). 13-14. Moscow.
Method of introducing corrections for sharpening swept-back knives in helical-disk chippers.

Kalashnikov. 1971. (Chipping)

Kintz. 1969. (Sawing; Circular sawing)

Kostrikov, P. V. 1969. CONDITIONS FOR STABLE OPERATION OF SWAGE SAWS. Derevoobrab. Prom-St'. 18(9) 14-16. Moscow.
Studies were made of swage saws (500 mm. diameter, angle of taper 0° 47') at four angles of incidence equal to 0, and equal to, less than, and greater than the angle of taper. The saws worked properly if the angle of incidence was equal to the angle of taper Good results also were obtained with angle of incidence = 0, provided that the teeth on the taper side of the saw were spring-set 0.1 to 0.3 mm. more than those on the flat side.

Lyth, C. 1970. THE WELDING AND METALLURGY OF WIDE BANDSAW BLADES. Aust Timber J. 26(3): 37-43.
Explains metallurgical principles in welded joints of bandsaw blades and summarizes a method to prevent premature joint failure. (1) preheat areas to be welded to a very dull red/black color (this should prevent cracking during welding by reducing the cooling rate and the thermal stresses); (2) weld with a slightly reducing flame; (3) allow to cool to black heat, (4) reheat to cherry red and forge quickly; (5) normalize by reheating to dull cherry and allow to air-cool, the whole of the weld and the heat-affected zone should be heated, (6) harden by heating the whole of the affected area to a dull cherry red and quench out with an oil soaked rag; (7) temper at 350° C. (dark blue temper color); (8) finish weld off flush by longitudinal filing and emery dressing.

McKenzie, W. M. 1971. SAWBLADE TENSIONING – WHAT IS IT ALL ABOUT? CSIRO For. Prod. Newsl. 383, p. 2-4.
Tensioning maintains stiffness of thin blades during sawing Blades lose their stiffness because of stresses imposed by heating (usually from cutting friction), saw guides, and sawing force. Additionally, circular saws are stressed by centrifugal force, and bandsaws are stressed by the tightening (straining) apparatus and inertial forces arising from motion of the blade around top and bottom wheels. Blades are conventionally tensioned by rolling, hammering, and heating in such a manner that the portion near the teeth will not buckle from compressive stresses when sawing This paper describes tensioning procedures in simple terms.

Noguchi et al. 1971. (Orthogonal cutting)

Pahlitzsch, G., and Peters, H. 1968. INVESTIGATIONS ON CUTTING WITH CHAIN SAWS. II: THE INFLUENCE OF CUTTING ANGLES AND THE IMPORTANCE OF CHIP-THICKNESS-LIMITER ON CHAIN SAWING. Holz Roh-und Werkst. 26 382-388.
Recommends optimum grinding angles for chains of two types. Chipper-type chains have substantial advantages over scratch-type.

Porter, A. W., and Olson, S. L. 1971. DEVISE ROLLER-TYPE SHAPER FOR SIDE-DRESSING BANDSAW TEETH. Can. For. Ind. 91(9). 51-53.
A new British shaper for side-dressing swaged bandsaw teeth has been evaluated on saws currently used in western North America. Instead of the conventional hardened side dies, the new device employs two sets of rollers, ground to the desired tooth profile The original roller design had certain limitations when applied to thick saws with long tooth spacing and deep gullets. A Canadian revision of the British design has increased stability, is faster to set up and operate, and is as precise as any commercially available shaper, like the British machine, however, it rounds over the tooth corners. Best opportunity for automating the swaging-shaping process probably is to combine the two operations by one machine, with an eccentric die for swaging and hardened side dies for shaping saw teeth.

St-Laurent. 1970. (Properties of the cutting edge and cutter; Dulling)

Salemme. 1969. (Sawing; Circular sawing)

31

Schliewe. 1969. (Sawing; Circular sawing)

Siemuński, R. 1969. EFFECT ON FRAME-SAW BLADE RIGIDITY OF ROLLER-TENSIONING WITH ROLLERS OF DIFFERENT SHAPES. Przem. Drzewny 20(11): 1-4. Warsaw.
Graphs the effect on blade rigidity of the size and shape of the rollers used for tensioning. Results were best with rollers leaving flat marks 3 mm. wide on the blade surface.

Stahiev, J. M. 1970. THE DEGREE OF HAMMERING AND ROLLING OF FLAT CIRCULAR SAWS. Lesn. Zh. 13(4): 157-161. Arkhangelsk.

Trubnikov, I. I., and Leikhtling, R. A. 1969. EFFECT OF SURFACE HAMMERING ON THE FATIGUE STRENGTH OF BANDSAW BLADES. Lesoinzh. delo i Mekh. Teknol. Drev., p. 169-171. Krasnoyarsk.
Presents a diagram of limiting stresses, and describes its use for evaluating the effect of hardening on the fatigue strength of bandsaw blades.

Veselkov, V. I. 1971. INVESTIGATION OF THE RESIDUAL STRESSES IN FRAMESAW BLADES. Lesn. Zh. 14(2): 69-73. Arkhangelsk.
Describes method and apparatus employed to investigate stresses in blades of band resaws before and after the teeth are cut.

Yamanishi. 1970. (Research instrumentation and techniques)

Yur'ev, Y. I., and Veselkov, V. I. 1971. CALCULATIONS OF SCHEDULES FOR THE ROLLING OF BANDSAW BLADES. Lesn. Zh. 14(1): 69-75. Arkhangelsk.
Develops expressions for calculating the residual stresses in the rolled zones of the blade, and compares the results with measurements made with an induction profilograph.

STABILITY

Abeels, P. 1970. CONTRIBUTION TO THE STUDY OF BANDSAWS BEHAVIOUR DURING SAWING. *In* Mechanical Conversion of Wood Meet. Proc., p. 53-66. Sect. 41, IUFRO. Braunschw.
Relates bandsaw stability and tooth forces to steel hardness, sharpening technique, saw speed, form and dimensions of swaged tooth portion, and wood species.

Berlin, E. P. 1969. EXPERIMENTAL INVESTIGATION OF THE EFFECT OF AEROSTATIC GUIDES ON THE TRANS-VERSE STABILITY OF A BANDSAW BLADE. Lesn. Zh. 12(2). 85-87. Arkhangelsk.

Birkeland. 1968. (Sawing, General)

Curtu, I., and Serbu, A. 1970. THE CRITICAL VELOCITIES OF ROTATION IN CIRCULAR SAWS. Ind. Lemnului 21(2): 49-55. Bucharest.
Presents equations and nomograms for speeds at which the saw blade or arbor is unstable (irrespective of any imposed cutting forces), for systems of 1 to 3 degrees freedom, in relation to blade radius and other parameters.

Derjagin, R. V. 1969. THE DYNAMIC · STABILITY OF FRAMESAWS. Lesn. Zh. 12(5): 89-94. Arkhangelsk.
A theoretical study showing that, under certain conditions, a framesaw may lose dynamic stability, the most dangerous time being at the start of sawing.

Dobrynin, E. D. 1962. STRESSES AND STRAINS IN OPER-ATING SAW BLADE. Lesn. Zh. 5(5): 123-127. Arkhangelsk.

Dobrynin, E. D. 1970. STRESSES IN BANDSAW BLADES

RESULTING FROM THE FORCES OF RESISTANCE TO CUTTING AND COMPRESSION. Lesn. Zh. 13(2): 83-88. Arkhangelsk.
Presents measurements on a test rig and calculations that define magnitude of deformations.

Feoktistov, A. E. 1960. THE STABILITY OF (BAND) SAW BLADES UNDER FEED PRESSURE. Lesn. Zh. 3(3): 95-106. Arkhangelsk.

Foschi, R. O., and Porter, A. W. 1970. LATERAL AND EDGE STABILITY OF HIGH-STRAIN BAND SAWS. Can. Dep. Fish. and For. For. Prod. Lab. Inform. Rep. VP-X-68, 13 p. Vancouver.
Recently, North American machinery manufacturers have introduced bandsaws capable of carrying two or three times the conventional strain, i.e., the force with which upper and lower band wheels are kept apart. Lateral stability is determined primarily by the strain level and the free length of a blade. Edge stability is governed primarily by the blade thickness but is also affected by changes in width, free length, and tensioning force.

Friebe, E. 1970. STIFFNESS AND VIBRATION BEHAVIOUR OF CIRCULAR SAW BLADES. Holz Roh- und Werkst. 28: 349-357.
Circular sawblades, not prestressed, were investigated as to vibration behaviour and stiffness. Static stiffness and natural frequencies are closely interrelated and can be derived from the theory of plates by Kirchhoff. Static stiffness can be determined by means of nomogram. In a circular blade under point load in the root of a tooth it increases in proportion to Young's modulus and with the cube of thickness of the sawblade; it decreases with the square of blade diameter. The constant of proportionality — static stiffness of non-prestressed sawblades — rises parabolically with increasing rate of rotation. The natural frequencies of a disc increase in proportion to sawblade thickness and decrease with the square of diameter. The constant of proportionality depends on sawblade diameter, clamping ratio, number of nodes, and prestressing.

Gul'ko, L. 1971. EFFECT OF ACTIVE RESISTANCE ON THE STABILITY OF A SAW AGAINST VIBRATION. Lesn. Prom. (12): 25-27. Moscow.
Develops formulae to express various relations of the resistances (inertial, elastic, and active) to the vibration of the chain and bar of stationary crosscutting saws. Discusses the effects of bar length and weight, and the stability of various bar designs.

Hallnor, G. 1971. PRECISION IN WOODWORKING MACHINES. Timber Trades J. Annu. Spec. Issue: S/13-S/17.
Describes the development by SKF, Gothenburg, Sweden, of improved spindles to carry tungsten-carbide cutters on planers, routers, and slitting saws.

Malcolm and Koster. 1970. (Research instrumentation and techniques)

Mote, C. D., Jr. 1970. FORMULATION OF DISCRETE ELE-MENT MODELS FOR STRESS AND VIBRATION ANALYSIS OF PLATES. Univ. Calif. For. Prod Lab. Serv. Rep. 35.01.77, 68 p.
Summarizes the theoretical basis for computer codes to analyze stress and vibration in circular saws, which, because of their unusual geometry and operating environment, require special computational techniques. Finite-element numerical procedures possessing sufficient generality for nearly all saw problems are developed for such analyses. Applications are not discussed.

Okushima and Sugihara. 1969. (Properties of the cutting edge and cutter; Temperature)

32

Pahlitzsch, G., and Friebe, E. 1971. ON THE BEHAVIOR OF CIRCULAR SAW BLADES DURING CUTTING. I. EFFECT OF THE CUTTING CONDITIONS ON VIBRATION BEHAVIOR AND LOADING OF CIRCULAR SAW BLADES. II. EFFECT OF THE CUTTING CONDITIONS ON THE QUALITY OF SAWN WOOD SURFACES. Holz Roh- und Werkst. 29: 149-157, 265-269.
Because the blade was stiffened by centrifugal forces, higher saw speeds resulted in lower amplitudes of vibration; surface quality was therefore better. Total cutting force increased in proportion to blade thickness and wood specific gravity but was unaffected by saw speed or protrusion.
At usual feeds per tooth, surface roughness increased slightly with blade thickness — the result of increased amplitude of vibration. Maximum roughness increased with blade protrusion or with depth of cut.

Pahlitzsch and Friebe. 1971. (Safety, including noise)

Pahlitzsch, G., and Rowinski, B. 1966. CRITICAL (NUMBER OF) REVOLUTIONS OF CIRCULAR SAW BLADES. Mod. Holzverarb. (11): 702-704.

Pochekutov, S. P. 1969. INVESTIGATION OF THE ACCURACY OF LOG SAWING WITH A STENNER BANDSAW. Lesoinzh. delo i Mekh. Teknol. Drev., p. 243-247. Krasnoyarsk.
Describes theoretical analysis and investigations in a Siberian sawmill, and discusses ways of improving sawing accuracy.

Porter. 1971. (Sawing; Bandsawing)

Saito, Y., and Mori, M. 1970. ON THE BUCKLING OF A BANDSAW BLADE. Parts 1 and 2. Can. Dep. Fish. and For. Lib. Transl. OOFF-TR 125, 25 p. (Wood Ind. (Tokyo) 8 (7 and 8). 1953.)
A 42-inch bandsaw carrying 2- to 4-inch-wide tensioned blades of 21 to 26 gauge (BWG) with punched teeth (no spring-set) was used to evaluate factors causing buckling of teeth in thin saws. Tooth shape is the major factor. Buckling strength of teeth loaded in the running direction of the saw is about double that of teeth loaded in the direction of lumber travel. Equations are given for computing loads causing buckling. Such loads are greatest if saws are tensioned by rolling or hammering, if "strain" (force exerted to separate wheels) is high, and if distance between wheels is short.

Schmutzler, W. 1969. THE CLAMPING OF KNIVES IN CUTTER HEADS AND ITS CALCULATION. Holz Roh- und Werkst. 27(4): 153-157.
Analyzes the stresses tending to dislodge cutters during rotation, and shows that for the cutterhead examined a moment of 1,400-2,200 kg.-mm. on the fixing turnscrews is necessary for a maximum working speed of 6,000 rev./min. If only one of the two screws in the slot is tightened, a moment of 2,900 is needed but cannot normally be achieved, and the cutter would break loose. The component of maximum cutting power at the cutting edge was computed from the drive power of the machine tested, and was smaller than the centrifugal power component.

Stahiev, J. M. 1969. EFFECT OF THE DIAMETER OF THE COLLARS ON THE INITIAL FORM OF THE LOSS OF STABILITY OF CIRCULAR-SAW BLADES. Lesn. Zh. 12(5): 67-71. Arkhangelsk.

Stahiev, J. M. 1970. RESONANCE VIBRATIONS OF FLAT CIRCULAR SAWS. Lesn. Zh. 13(5): 80-84. Arkhangelsk.

Strzelecki, A. 1970. DAMPING THE VIBRATIONS OF A ROTATING CIRCULAR SAW IN A MAGNETIC FIELD. Folia For. Pol. (Drzewn.) (9): 29-56. Warsaw.
Damping reduced vibration 0 to 56 percent when the saw was running idle and as much as 12-1/2 percent when it was cutting.

Sugihara. 1970. (Sawing; Bandsawing)

Thrasher. 1971. (Patents)

Thunell, B. 1970. THE STABILITY OF THE BANDSAW BLADE. Holz Roh- und Werkst. 28: 343-348.
Discusses stresses in stationary, idling, and cutting blades.

Wassipaul, F. 1970. DETERMINATION OF THE TENSIONING FORCE OF FRAMESAW BLADES. Holzforsch. und Holzverwert. 22(5): 85-91.
There is a relationship between the tensioning force on a gangsaw blade and the frequency with which the blade vibrates. Knowledge of this relationship permits computation of the tensioning force through the use of measurements obtained with a frequency meter. The force should be as great as possible, but must not overstress sawblades or clamping mechanisms.

TEMPERATURE

Grube, A. E., Sanev, V. I., and Pashkov, V. K. 1969. TEMPERATURE CHANGE ALONG THE RADIUS OF CIRCULAR SAWS, AND THE EFFECT OF THE TEMPERATURE GRADIENT ON THE TRANSVERSE OSCILLATIONS OF SAWS. Lesn. Zh. 12(4): 60-66. Arkhangelsk.

Leikhtling, R. A., and Trubnikov, I. I. 1969. COMPUTATIONAL AND GRAPHICAL DETERMINATION BY THE METHOD OF CONTACT PAIRS, OF THE HEATING TEMPERATURE OF A WOOD-CUTTING INSTRUMENT. Lesoinzh. delo i Mekh. Teknol. Drev., p. 232-237. Krasnoyarsk.
Describes the procedure.

Okushima, S., and Sugihara, H. 1969. TEMPERATURE DISTRIBUTION OF CIRCULAR SAW BLADE MEASUREMENT WITH INFRARED RADIOMETRIC MICROSCOPE. J. Jap. Wood Res. Soc. 15(1): 11-19.
Nonuniform distribution of heat in circular saw blades causes deflection and inaccurate cutting. In this study, an infrared radiometric microscope was used to measure radial temperature distribution in a rotating steel disk heated on its peripheral surface, and in a circular saw blade in operation. The internal stress distribution was then calculated from the observed temperature gradients.

Sanev, V. I. 1969. HEATING OF BANDSAW BLADES WHILE SAWING WOOD. Lesn. Zh. 12(3): 64-67. Arkhangelsk.

Sanev, V. I., and Pljusnin, V. N. 1969. THEORETICAL INVESTIGATION OF THE TEMPERATURE DISTRIBUTION ACROSS THE WIDTH OF THE BLADE OF A BANDSAW EDGER. Lesn. Zh. 12(5): 61-67. Arkhangelsk.
Develops two formulae for calculating temperature, illustrates their use, and summarizes general features of the temperature distribution.

Sanev, V. I., and Pljusnin, V. N. 1970. TEMPERATURE GRADIENT ALONG THE LENGTH OF BANDSAW BLADES WHILE SAWING WOOD. Lesn. Zh. 13(2): 59-64. Arkhangelsk.
Reports laboratory investigations complementing previous measurements of temperatures across the blade. A gradient of 1.3 to 4.1° C. along the blade was recorded.

FRICTION

Cuprin, V. I. 1969. THE COEFFICIENT OF FRICTION IN CUTTING PINE AND OAK WOOD. Lesn. Zh. 12(6): 91-93. Arkhangelsk.

Ettelt, B. 1969. WHAT ARE THE ADVANTAGES OF PLASTIC-SURFACED TOOLS? Investigations on saws surfaced with Teflon S. Mod. Holzverarb. 4(12): 798-803.

33

Grubov, S. 1970. EFFECT OF DURATION OF CONTACT ON THE STATIC COEFFICIENT OF RESISTANCE (TO MOVEMENT). Lesn. Prom. (3): 28-29. Moscow.

Lemoine, T. J., McMillin, C. W., and Manwiller, F. G. 1970. WOOD VARIABLES AFFECTING THE FRICTION COEFFICIENT OF SPRUCE PINE ON STEEL. Wood Sci. 2: 144-148.
Wood of Pinus glabra Walt. was factorially segregated by moisture content (0, 10, and 18 percent), specific gravity (less than 0.45 and more than 0.45), and extractive content (unextracted and extractive-free), and the kinetic coefficient of friction on steel (having surface roughness of 9 microinches RMS) determined for tangential earlywood, tangential latewood, radial, and transverse surfaces. The friction coefficient was at maximum for the unextracted transverse surface of high moisture content pulled parallel to the rings (avg. 0.263). It was at minimum for the extractive-free transverse surface of high moisture content pulled perpendicular to the rings (avg. 0.165). Among unextracted samples, it was least for the tangential latewood surface of dense wood having intermediate moisture content and pulled perpendicular to the grain.

McMillin, C. W., Lemoine, T. J., and Manwiller, F. G. 1970. FRICTION COEFFICIENT OF OVEN-DRY SPRUCE PINE ON STEEL, AS RELATED TO TEMPERATURE AND WOOD PROPERTIES. Wood and Fiber 2: 6-11.
The coefficient of friction ranged from 0.10 to 0.25 and was negatively correlated with temperature for tangential earlywood, tangential latewood, and radial surfaces pulled parallel and perpendicular to the grain and for transverse surfaces pulled parallel and perpendicular to the annual rings. For transverse surfaces pulled parallel to the annual rings, tangential latewood surfaces pulled parallel or perpendicular to the grain, and tangential earlywood surfaces pulled parallel to the grain, the coefficient

also increased with increasing extractive content for a given temperature. No significant relationships were detected between the coefficient and wood specific gravity after the effects of temperature and extractive content had been accounted for.

McMillin, C. W., Lemoine, T. J., and Manwiller, F. G. 1970. FRICTION COEFFICIENT OF SPRUCE PINE ON STEEL – A NOTE ON LUBRICANTS. Wood Sci. 3: 100-101.
Generally, the introduction of water and ethanol increased the friction coefficient for ovendry samples but decreased it for saturated samples. Octanoic acid decreased the coefficient when samples were wet. Coefficients ranged from 0.14 to 0.78.

Poole, C. I. 1971. FLOUROCARBON RESIN FINISHES FOR WOOD CUTTING TOOLS. For. Prod. J. 21(3): 40-42.
Teflon-S nonstick, self-lubricating finishes are combinations of tetrafluoroethylene and binders. They are applied as air-sprayed, dipped, or rolled coatings and cured at temperatures from 450 to 700° F. No unusual metal preparation is required. This chemically inert finish has a water absorption rate of less than 4 percent, a static coefficient of friction of 0.08 to 0.15, and a service temperature range of $-50°$ to $500°$ F. Tests indicate that circular saws coated with the material can run 50 percent longer between sharpenings than can uncoated blades. The increase is attributed to reduction of heating by elimination of resin or pitch buildup on the sides of the blade. Industrial trials indicated a reduction of 25 percent in the time required to cut through identical cross sections, but tests of power savings have been inconclusive. Costs for custom application of this finish range from $0.50 to $1.50 per square foot, depending upon size, shape, and weight. Coatings applied as part of the manufacturing process add 10 to 25 percent to the cost of the product.

Yu. 1966. (Sawing, Circular sawing)

COMPUTER, TAPE, AND CARD CONTROL OF MACHINES

Bulgrin, E. H. 1971. THE COMPUTER CAN MAKE THE SAWING DECISIONS. North. Logger 20(1): 16-17, 30-32.
Reviews present state of research at the USDA Forest Products Laboratory, Madison, Wis., on the use of computers to make ripping and crosscutting decisions during conversion of hardwood boards into dimension lumber. Research objectives are listed, as are citations to reports already published. (See also Can. For. Ind. 91(2) 46-49, 1971.)

Hallock, H., and Bulgrin, E. H. 1970. TOMORROW: COMPUTER-MADE SAWING DECISIONS? For. Prod. J. 20(9): 52-57.
Much progress has been made to provide mechanisms by which computers can determine – and initiate – cutting patterns in hardwood mills. A computer program has been developed that can determine the optimum cut-up pattern to yield predetermined cuttings from a mathematically described board. Also

completed is a successful program for accurately and almost instantly grading hardwood lumber by rules of the National Hardwood Lumber Association. In defect sensing, ultrasonics holds promise for locating knot-type defects.

Mundler, H. U. 1971. ELECTRONIC CONTROLLING IN THE WOODWORKING INDUSTRY. I. CONTROL OF DIMENSIONS OF LOG BANDSAWS AND RIPSAWS. Holz Roh- und Werkst. 29: 89-93.
Describes systems for electronic control of lumber thickness.

Reynolds and Gatchell. 1969. (Research instrumentation and techniques)

Rügge, K. 1971. PROGRAMME CONTROLLED WOODWORKING MACHINERY. Holz Roh- und Werkst. 29: 5-10.
Reviews state of the art. Tape- or card-controlled machinery can produce accurate parts economically.

RESEARCH INSTRUMENTATION AND TECHNIQUES

Atack and May. 1970. (Defibrating)

Barz, E., and Breier, H. 1969. SHORT-METHOD FOR TESTING THE WEAR EFFECT AND THE CHIPPING PROPERTIES OF WOOD BASE MATERIALS. Holz Roh- und Werkst. 27: 148-152.
Describes an apparatus with a replaceable cutter for rapid determination of the relations between moment, cutting and feed forces, and wear of various types of blades. The relations, determined on different timbers, are shown in graphs and are used to define optimum cutting conditions with conventional edges.

Changes in angles at the cutting edge had a greater effect on forces and quality of cut than did changes elsewhere in the tooth.

Barz and Breier. 1971. (Properties of the cutting edge and cutter; Dulling)

Bohrer. 1971. (Safety, including noise)

Bramhall and McLauchlan. 1970. (Sawing; Circular sawing)

34

Chardin and Froidure. [n.d.] (Properties of the cutting edge and cutter; Dulling)

Cowling and McKenzie. 1969. (Properties of the cutting edge and cutter; Fitting and sharpening)

Fuchs, F. R. 1971. ELECTRONIC MEASUREMENT OF VENEER SURFACE AREAS. Holz Zentralbl. 97(34): 477-478.
Techniques and economics of measuring parcels of veneers photoelectrically.

Lane, P. H. 1971. IDENTIFYING VENEER IN RECOVERY STUDIES. For. Prod. J. 21(6): 32-33.
Describes a color coding system useful in studies to determine veneer recovery from individual peeler blocks. Nozzles for spraying paint or dye are mounted directly at the outfeed side of lathe to provide a continuous stripe of one or more colors on the green veneer.

Lin, C. C., and Cumming, J. D. 1970. PRINTING GRID PATTERNS ON GREEN WOOD. For. Prod. J. 20(11): 37.
After the surface had been smoothed and coated with epoxy, Kodak Photosensitive Resist was applied and a grid pattern was photographically printed. Such grids are useful in studies of chip formation.

Malcolm, F. B., and Koster, A. L. 1970. LOCATING MAXIMUM STRESSES IN TOOTH ASSEMBLIES OF INSERTED-TOOTH SAWS. For. Prod. J. 20(10): 34-38.
Techniques were investigated for examining the location and magnitude of stresses. The most satisfactory material was photoelastic plastic, which has birefringent or double-refraction characteristics under polarized light. The technique for using this material is explained. Stress values as high as 57 percent of the strength of the steel were created merely by the action of inserting the tooth and holder in the socket.

May. 1970. (Defibrating)

Meyer, B. 1969. THE PROBLEM OF DETERMINATION OF THE SPECIFIC EXTERNAL SURFACE OF WOOD AND WOOD PARTICLES. Holztechnologie 10(3): 168-172
Discusses eight methods for determining the effective specific external surface, which is an important measurement in some technical processes. Concludes that existing methods are not accurate enough.

Meyer. 1970. (Machining with coated abrasives)

Mote. 1970. (Properties of the cutting edge and cutter; Stability)

Neusser and Krames. 1970. (Chipping)

Noguchi. 1970. (Orthogonal cutting)

Orlicz, T., and Bajkowski, J. 1970. INDICATOR INSTRUMENTS FOR RAPID MEASUREMENT OF THE DIMENSIONAL ACCURACY OF WOODEN ITEMS. Folia For. Pol. (Drzewn.) (9): 79-96. Warsaw.
Describes gages developed at the woodworking department of the Agricultural University in Warsaw for quick, accurate measurement of deviations in length, width, thickness, angles, smoothness of surfaces, and straightness of edges of dimension stock.

Pahlitzsch, G., and Dziobek, K. 1970. THE EVALUATION OF PROCESSED WOOD SURFACES. USDA For. Serv. Transl. FPL-706, 18 p. (Holztechnologie 6(3): 153-160.)
Surface characteristics of wood in use are dependent on the interaction of machining process, wood properties, and use conditions. Functional demands on wood surfaces, therefore, dictate processes by which surfaces should be prepared. Aspects of measuring surface roughness with a profile sensing instrument are discussed. Surface characteristics other than profile roughness are listed but not discussed.

Peters, C. C., and Cumming, J. D. 1970. MEASURING WOOD SURFACE SMOOTHNESS: A REVIEW. For. Prod. J. 20(12): 40-43.
An extensive review of literature on measurement of the surface smoothness of wood showed that three methods predominate – visual, light-sectioning, and stylus tracing. No system has been completely satisfactory, and no standard method exists. A method should be developed that will be rugged and easy to use, and will give reproducible results. It should be capable of measuring profiles up to several inches in length along a board or panel, and have a depth and sensitivity range suitable for rough-sawn as well as finely finished surfaces.

Peters, C., and Mergen, A. 1971. MEASURING WOOD SURFACE SMOOTHNESS: A PROPOSED METHOD. For. Prod. J. 21(7): 28-30.
A precise method for measuring wood smoothness would be useful. Because stylus tracing seemed the most suitable of the methods tried, a stylus tracing head was designed specifically for wood surfaces. Tests showed it to have sufficient range and sensitivity to show wood anatomy as well as 1/4-inch-deep roughness. The complete system, including a head assembly, balancing network, amplifier, calibrating networks, and recorder, was capable of performing on microtomed, sanded, planed, and sawn specimens.

Porter, A. W., Kusec, D. J., and Sanders, J. L. 1971. AIRFLOW METHOD MEASURES LUMBER SURFACE ROUGHNESS. Can. For. Ind. 91(7): 42-45.
A simple, inexpensive device assesses the smoothness of sawn timber by measuring the amount of air required to maintain a constant pressure differential between the atmosphere and an inner-pocket reservoir placed over the wood surface. As surface roughness increases, more air is required to maintain the differential.

Potrebic. 1969. (Machining with coated abrasives)

Reynolds, H. W., and Gatchell, C. J. 1969. SAWMILL SIMULATION: CONCEPTS AND COMPUTER USE. USDA For. Serv. Res. Note NE-100, 5 p. Northeast. For. Exp. Stn., Upper Darby, Pa.
Product specifications were fed into a computer so that the yield of products from a sample of logs could be determined for a variety of simulated sawing methods. Since different sawing patterns were tested on the same sample, variation among log samples was eliminated; conclusions therefore, were precise in a statistical sense.

Reynolds et al. 1970. (Sawing; Chain sawing)

Shen. 1971. (Machinability of species and reconstituted wood, Particleboards and sandwich boards)

Yamanishi, K. 1970. STUDIES ON POLISHING PROCESS OF WOOD CUTTING TESTING KNIVES. Wood Ind. 25(6): 16-19. Tokyo.
Shadowgraph and microscopic methods were used to assess the results obtained with simple sharpening apparatus developed to improve on hand methods.

35

NOMENCLATURE

Abeels, P. 1970. SUGGESTIONS FOR STANDARDIZATION OF WOODWORKING TOOLS TERMINOLOGY. *In* Mechanical Conversion of Wood Meet. Proc., p. 67-72. Sect. 41, IUFRO. Braunschw.
Illustrated proposal for standardized nomenclature useful in description of bandsaws, including their sharpening and fitting.

Ford-Robertson, F. C. 1971. TERMINOLOGY OF FOREST SCIENCE, TECHNOLOGY, PRACTICE AND PRODUCTS. 349 p. Wash., D.C.: Soc. Am. For.
This English-language version forms volume 1 in a proposed multilingual compilation of forestry terms. The definitions include many terms applicable to harvesting and processing timber, some of which are illustrated in an appendix.

Wuster, E. 1970. PROBLEMS OF THE INTERNATIONAL STANDARDIZATION OF WOOD SAW BLADES. Holz Roh- und Werkst. 28: 357-362.
The standardization of saw blades began in 1928. Today about 220 pertinent national standards exist in 15 countries, but there is no international recommendation yet. In April 1970, five Draft Proposals were discussed at a meeting of the International Organization for Standardization. Two main reasons explain the lack of basic research that would aid standardization: saw makers and consumers lack theoretical knowledge in the entire field of saw engineering; and wood technologists devote too little time to the subject.

MACHINABILITY OF SPECIES AND RECONSTITUTED WOOD

HARDWOODS

Barz and Breier. 1971. (Properties of the cutting edge and cutter; Dulling)

Bousquet, D. W., and Flann, I. B. 1970. LIVE-SAWING UPS LUMBER RECOVERY. Can. For. Ind. 90(12): 37-39.
Describes effects of three sawing patterns and five edging practices on the yield of lumber and furniture dimension stock from 21 hard maple butt logs 12 feet in length and 14 to 17 inches in diameter. "Live" or "through-and-through" sawing, followed by grade-ripping, yielded the greatest total lumber value. If lumber is merely an intermediate product in the manufacture of dimension stock by a company which operates both sawmill and rough mill, however, the use of unedged boards from live-sawn logs will maximize both total surface yield and total value.

Bulgrin. 1971. (Computer, tape, and card control of machines)

Chardin and Froidure. [n.d.] (Properties of the cutting edge and cutter; Dulling)

Cuprin. 1969. (Properties of the cutting edge and cutter; Friction)

Edamatsu and Nagahara. 1970. (Boring, routing, and carving)

Hallock and Bulgrin. 1970. (Computer, tape, and card control of machines)

Hayashi and Tochigi. 1969. (Veneer cutting)

Hayashi et al. 1970. (Veneer cutting)

Hayashi et al. 1971. (Veneer cutting)

Howland, P. 1971. SAWING EUCALYPTUS SALIGNA/GRANDIS. Malawi For. Res. Inst. Res. Rec. 47, 14 p.
Ten eucalyptus trees of the hybrid saligna/grandis, approximately 2 feet d.b.h., and 160 feet in height, were bucked into 13-foot logs and made into lumber with a circular saw cutting a 1/4-inch kerf. The logs were turned frequently to reduce growth stress evenly, produce wide flat boards, and box out defective log centers. Volume recovery after oversize sawing averaged approximately 39 percent, this was reduced to 33 percent after seasoning and to 27 percent after the rejection of unserviceable material. About 19 percent of the final recovery was of prime quality, 35 percent of moderate quality, and 46 percent of poor quality. Large differences were noted between trees. Log

positions in the tree had little effect on recovery, with the sixth log being about 70 percent as valuable as the butt log. Mill time required was about four times that for cutting an equivalent amount and size of softwood by accepted patterns.

Iacob, M. 1971. CONVERSION OF BEECH WOOD WITH BANDSAWS. Ind. Lemnului 22(6): 207-219. Bucharest
One of the few Rumanian mills equipped with band headsaws and band resaws is at Pitesti. In both yield and quality, the lumber is superior to that cut with frame saws. Discusses problems due to cant distortion on the carriage.

Ionescu, N. H. 1970. THE TECHNOLOGY OF CONVERSION OF BEECH LOGS INTO SAWN TIMBER AND HALF-FINISHED AND PREFABRICATED PRODUCTS. Ind. Lemnului 21(3): 81-88. Bucharest.

Jain, N. C., Gupta, R. C., Tandon, R. C., and Bagga, J. K. 1968. PEELING CHARACTERISTICS OF INDIAN TIMBERS. Part 5. DIPTEROCARPUS TURBINATUS (GURJAN). Holzforsch. und Holzverwert. 21(2). 35-37.
Optimum steaming time for logs was 16 hours at 55° C. Optimum knife angle was 90.5°. Cutting speed in the range examined (24-50 m./min.) did not affect peeling characteristics.

Jain, N. C., Gupta, R. C., Tandon, R. C., and Bagga, J. K. 1969. PEELING CHARACTERISTICS OF INDIAN TIMBERS. Part 6. SHOREA ROBUSTA (SAL). Holzforsch. und Holzverwert. 21(6). 140-143.
Shorea Robusta (sal) yielded comparatively smooth veneers when rotary-peeled at a temperature of 55° C. and cutting speeds of 46 to 52 meters per minute; a knife angle of 90° gave smoothest veneers, but cutting resistance was appreciably higher than for an angle of 91.2°.

Jain, N. C., Gupta, R. C., Tandon, R. C., and Bagga, J. K. 1970. PEELING CHARACTERISTICS OF INDIAN TIMBERS. Part 7. CULLENIA EXCELSA (KARANI); Part 8. SHOREA ASSAMICA (MAKAI). Holzforsch. und Holzverwert. 22(1): 4-9.
Optimum conditions for Karani and Makai were steaming of logs for 16 hours at 83° and 64° C., knife angle 90° and 89.5°, and cutting speed 39.6 and 33.5 m./min.

Jain, N. C., Gupta, R. C., Srivastava, V. K., and Bagga, J. K. 1970. PEELING CHARACTERISTICS OF INDIAN TIMBERS. Part 9. PINUS ROXBURGHII (CHIR). Holzforsch. und Holzverwert. 22(5): 91.

Jain, N. C., Gupta, R. C., Tandon, R. C., and Dev, I. 1967. PEELING CHARACTERISTICS OF INDIAN TIMBERS. Part 2. CEDRELA TOONA. Holzforsch. und Holzverwert. 19(5): 77-79.

Kukachka, B. F. 1970. PROPERTIES OF IMPORTED TROPI-
CAL WOODS. USDA For. Serv. Res. Pap. FPL-125, 67 p. For.
Prod. Lab., Madison, Wis.
*Descriptions of more than 100 tropical genera and generic
groups of wood, with emphasis on properties that affect
utilization. Tables present drying schedules, total shrinkage from
green to ovendry condition, and strength values for both green
and dry wood.*

Liiri. 1961. (Barking)

Meriluoto. 1971. (Veneer cutting)

Mori and Kubota. 1970. (Veneer cutting)

Noda and Umetsu. 1971. (Machining with coated abrasives)

Noguchi et al. 1971. (Orthogonal cutting)

Radu. 1970. (Boring, routing, and carving)

Reynolds. 1969. (Sawing; General)

Sharma. 1971. (Properties of wood)

Stewart. 1970. (Jointing, planing, molding, and shaping)

Stewart, H. A. 1971. CHIP FORMATION WHEN ORTHOGO-
NALLY CUTTING WOOD AGAINST THE GRAIN. Wood Sci.
3: 193-203.
*White ash (Fraxinus americana L) conditioned to 7-percent
moisture content was cut with the knife edge perpendicular to
the grain, and with cutting direction varied by 5-degree
increments from parallel to grain (0°) to 30° angularity against
the grain. Chipped grain occurred at grain angles up to 20°, but
was prevented with rake angles equal to or less than 20°,
optimum combinations of rake angle and depth of cut were
found for specific grain angles Extremely shallow cuts produced
poorer surfaces than deep cuts. An approximate analysis,
expanded from the Franz analysis for determination of chip
type, related wood mechanical properties and tool force components to chip type for orthogonal cutting against the grain.*

Stewart, H. A. 1971. RAKE ANGLE FOR PLANING HARD
MAPLE DETERMINED BEST BY DEPTH OF CHIPPED
GRAIN. USDA For. Serv. Res. Note NC-116, 4 p. North Cent.
For. Exp. Stn., St. Paul, Minn.
*Optimum rake angle for hard maple (and perhaps other
hardwoods) should be determined from depth rather than
frequency of machining defects In wood at 8-percent moisture
content, rake angles of 20 and 25 degrees resulted in the
shallowest chipped grain (1/64 to 1/32 inch, depending on slope
of grain) at 30 knife marks per inch.*

SOFTWOODS

Baldwin. 1971. (Veneer cutting)

Barz and Breier. 1971. (Properties of the cutting edge and
cutter; Dulling)

Belova and Laskeev. 1971. (Defibrating)

Cernenko, S. A., and Zolotov, G. P. 1970. OPTIMUM TEM-
PERATURE FOR HEATING LARCH FLITCHES. Derevoobrab.
Prom-St'. 19(2): 6. Moscow.
*In trials in Siberia, flitches of Larix sibirica were sliced at
temperatures ranging from 60 to 84°C. Those sliced at 75 to 78°
C. yielded the best veneer.*

Cumming and Collett. 1970. (Veneer cutting)

Cuprin. 1969. (Properties of the cutting edge and cutter;
Friction)

Feihl and Godin. 1970. (Veneer cutting)

Futo. 1969. (Properties of wood)

Koch. 1971. (Sawing, Shearing)

Lemoine et al. 1970. (Properties of the cutting edge and cutter;
Friction)

McIntosh and Kerbes. 1969. (Sawing; Shearing)

McMillin et al. 1970. (Properties of the cutting edge and cutter;
Friction)

McMillin and Harry. 1971. (Sawing; Machining with light)

Quezada. 1970. (Properties of the cutting edge and cutter; Tool
material)

Radu. 1970. (Boring, routing, and carving)

Schulke et al. 1971. (Sawing; Shearing)

Shelyapin and Razumovskii 1971. (Barking)

Tuovinen. 1970. (Sawing, Shearing)

Vlasov and Pokryshakin. 1971. (Barking)

Woodson and Koch. 1970. (Orthogonal cutting)

Yu. 1966. (Sawing, Circular sawing)

Zajac, I. M. 1970. REGIMES FOR SANDING SCOTS PINE
WOOD WITH ABRASIVE DISKS. Lesn Zh. 13(2): 79-83
Arkhangelsk.

PLYWOOD

Doxey. 1970. (Sawing; Machining with light)

Hayashi, D., Tochigi, T., and Yamashita, A. 1970. EFFECTS OF
GLUE LINE ON CUTTING RESISTANCE. I. GLUE LINE
CUTTING RESISTANCE DURING THE ORTHOGONAL CUT-
TING PROCESS IN PLYWOOD. II. SOME ADHESIVES GLUE
LINE CUTTING RESISTANCE DURING THE ORTHOGONAL
CUTTING PROCESS. Wood Ind 25(3): 21-26. Tokyo.
*Plywood glued with a urea melamine resin offered substantially
more resistance to orthogonal cutting than did matched veneers
laminated without glue The cutting resistance of plywood
assembled with thermosetting resins was greater than that of
plywood glued with thermoplastic adhesives or thermosetting
resin emulsions.*

McMillin and Harry. 1971. (Sawing; Machining with light)

March. 1970. (Machining with light)

Tsuzuki et al. 1971. (Jointing, planing, molding, and
shaping)

PARTICLEBOARDS AND SANDWICH BOARDS

Albers. 1971. (Properties of wood)

Barz and Breier. 1971. (Properties of the cutting edge and
cutter; Dulling)

37

Bridges, R. R. 1971. A QUANTITATIVE STUDY OF SOME FACTORS AFFECTING THE ABRASIVENESS OF PARTICLEBOARD. For. Prod. J. 21(11). 39-41.
In three-layered southern pine particleboards from a single wood source, the correlation between board density and degree of abrasiveness was positive and linear; i.e., the densest boards dulled knives most quickly. Wood from the same source was then made into boards with varying proportions of resin solids, abrasiveness varied little within the range of 5- to 8-percent resin solids but almost doubled when solids content was raised to 9-11 percent. When shavings from a number of sources were made into boards of fixed density and resin content, abrasiveness varied widely and was linearly and positively correlated with board silica (grit) content.

Fukui, H., and Kimura, S. 1968. FACE BEVEL ANGLE (θ_f) and rake angle (β) of carbide-tipped saws in sawing wood particle board and fiber board. Wood Ind. 23(3): 21-24. Tokyo.
Carbide-tipped circular saws, 254 mm in diameter and with eight teeth, were ground with rake angles of $-25°$ to $30°$ and with alternate face bevels of from $0°$ to $30°$. The saws were rotated at 3,000 r.p.m. and fed at 3.4 m./min. while cutting particle board 15 or 20.5 mm. thick, hardboard 5 mm. thick, and fiber insulation board 19 mm. thick. When face bevel was $0°$, cutting power on all boards (except the fiber insulation board) decreased substantially as rake angle was increased. With insulation board, however, power requirement increased nearly 50 percent as rake angle was varied from $-25°$ to $+30°$. In hardboard and insulation board, power diminished as face bevel angle was increased from $0°$ to $30°$; this was true at all rake angles tested. Results were similar for the particleboards sawn at rake angles of $+8°$ and less, with $30°$ rake angle, power increased with increasing face angles in the range $10°$ to $30°$.

Neusser und Schall. 1970. (Machinability of species and reconstituted wood, Fiberboards, hard and soft)

Pahlitzsch, G., and Sandvoss, E. 1968. EXTRACTION RESISTANCE OF SCREW ELEMENTS IN PARTICLEBOARDS AND NATURAL WOOD. Mod. Holzverarb. 2, 8 p.

Pahlitzsch, G., and Sandvoss, E. 1970. INFLUENCE OF CUTTING CONDITIONS ON TOOL WEAR AND TEAR DURING THE MILLING OF PARTICLEBOARDS. Holzbearbeitung (5). 25-30.

Pahlitzsch, G., and Sandvoss, E. 1970. INFLUENCE OF THE TOOL DIAMETER ON THE STRESS, AND WEAR AND TEAR OF THE CUTTING EDGE DURING THE MILLING OF PARTICLEBOARDS. Holzbearbeitung (6): 21-26.

Plastov, E. F. 1971. EFFECT OF THE DULLING OF HARD-ALLOY CUTTERS ON QUALITY AND FORCES IN ROTARY CUTTING OF PARTICLEBOARD EDGES. Derevoobrab. Prom-St'. 20(6). 9-10. Moscow.
A note on the dulling of cutters working on veneered particleboard. the defects that occur on the board edges, and the specific work of cutting in relation to dulling.

Radu. 1970. (Boring, routing, and carving)

Shen, K. C. 1971. EVALUATING PARTICLEBOARD PROPERTIES BY MEASURING SAW-CUTTING FORCE. For. Prod. J. 21(10). 46-52.
A cutting-force pattern in a particleboard can be conveniently obtained by measuring the torque developed by a circular saw when slope-sawing through the board thickness. This pattern will be closely correlated with the physical and mechanical properties of the particleboard. The highest cutting force is developed in high-density high-strength surface layers, and the minimum

cutting force is usually found in the center plane of the board. The shape of the pattern reflects the board stratification (layer construction), while the magnitude indicates the mechanical properties. This new technique may have potential for nondestructive testing of board quality on the production line.

Stefaniak. 1970. (Properties of the cutting edge and cutter; Dulling)

Stefaniak, W. 1971. INFLUENCE OF CUTTING SPEED ON BLUNTING OF CARBIDE-TIPPED SAWBLADES IN CUTTING OF PARTICLEBOARD. Holztechnologie 12(1): 9-15.
Degree of blunting was proportional to peripheral velocity (i.e., cutting speed). At speeds of 50 m./sec. and greater, dulling proceeded rapidly because of the larger amplitudes of saw vibration.

Theien, C. M. 1970. ROUTING AND SHAPING OF PARTICLEBOARD. For. Prod. J. 20(6): 30-32.
Because numerous material and operational variables affect the machining of particleboard, power feed is recommended. With it, clearance on the cutting edge can be kept minimal, with resulting increase in tool life. Cutting depths up to 1-1/2 inches at feed speeds of 300 to 400 in./min. can be achieved with a 3/4-inch carbide-tipped two-flute router. Where cutters less than 3/4-inch in diameter must be used, a shank diameter of 3/4-inch is recommended to reduce tool breakage caused by tool deflection or vibration. Also recommended are air-jets to cool the cutter and aid in chip removal, a carbide drill point on the cutter if plunge cutting is necessary, and dullness or horsepower indicators to facilitate decisions on tool replacement.

FIBERBOARDS, HARD AND SOFT

Albers. 1971. (Properties of wood)

Franz. 1970. (Sawing; Machining with high-energy jets)

Fukui and Kimura. 1968. (Machinability of species and reconstituted wood; Particleboards and sandwich boards)

McKenzie. 1970. (Sawing; Circular sawing)

McKenzie and Cowling. 1970. (Properties of the cutting edge and cutter; Dulling)

McMillin and Harry. 1971. (Sawing; Machining with light)

Neusser, H., and Schall, W. 1970. TOOL WEAR IN MACHINING OF CHIPBOARD AND SEMI-HARD FIBREBOARD. Holzforsch. und Holzverwert. 22(6): 110-116.
Tool wear is positively correlated with density and mineral (sand) content of boards. Size of mineral particles and their adhesion to boards also affect rate of wear. If mineral content is less than 100 mg. per kg. of board, wear is significantly slower than in boards of higher mineral content.

Pahlitzsch, G., and Sandvoss, E. 1970. ANALYSIS OF EDGE BLUNTNESS IN THE MILLING OF HARDBOARDS. In Mechanical Conversion of Wood Meet. Proc., p. 5-38. Sect. 41, IUFRO, Braunschw.
When hardboard was down-milled with a one-knife peripheral milling cutterhead, parallel and normal cutting forces increased with feed speed and with chip thickness. The knife initially dulled through development of a 50 μm.-radius at the cutting edge, later this radius developed into a flat chamfer extending from the cutting edge into the clearance surface. Coefficient of friction between wood and knife surfaces was 0.2, and was independent of clearance and rake angles.

BIBLIOGRAPHIES

McMillin, C. W. 1970. WOOD MACHINING ABSTRACTS, 1968 AND 1969. USDA For. Serv. Res. Pap. SO-58, 35 p. South. For. Exp. Stn., New Orleans, La.
Abstracts of some 300 references. An introduction mentions notable papers in various categories of endeavor and briefly surveys trends in wood utilization research.

Smith. 1971. (Safety, including noise)

Thunell. 1971. (History and general texts)

Thunell, B. 1971. LIST OF PUBLICATIONS ON BANDSAWS. Swed. For. Prod. Res. Lab., 27 p. Stockholm.
A bibliography of world literature, prepared by the Working Group on Mechanical Conversion within the IUFRO, Section 41, Forest Products. Contains 388 citations.

REPORTS OF RESEARCH PROGRAMS AND INDUSTRIAL DEVELOPMENTS

Anderson and Kaiser. 1970. (Chipping, flaking, and grinding; Chipping headrigs and edgers)

Baldwin. 1971. (Veneer cutting)

Erickson. 1970. (Barking)

Grosshennig. 1971. (Veneer cutting)

Home Timber Merchants Associations of Scotland and Forest Products Research Laboratory. 1970. (Sawing; General)

Koch. 1969, 1970. (Chipping, flaking, and grinding; Chipping headrigs and edgers)

Koch. 1971. (History and general texts)

Neville. 1970. (Sawing; General)

Pahlitzsch, G. [n.d.] TEACHING AND RESEARCH AT THE INSTITUTE FOR TOOL MACHINES AND MANUFACTURING TECHNIQUES OF THE TECHNICAL UNIVERSITY CAROLA-WILHELMINA AT BRAUNSCHWEIG. Short Reports of the University Group for Manufacturing Techniques of the Technical Universities and Universities of the German Federal Republic. Verlag W. Girardet, 2 p. Essen.

Perry. 1970. (Chipping, flaking, and grinding; Grinding)

Thunell. 1971. (History and general texts)

Thunell. 1971. (Sawing; General)

PATENTS

Bumerl. 1969. (Barking)

Côte, R. 1971. CHIP CONVERTER. (U.S. Pat. No. 3,576,203) U.S. Pat. Off., Wash., D.C.
An apparatus for chipping an upright standing tree, from top to bottom.

Fulghum, O. T. 1971. TIMBER SHEAR CONSTRUCTIONS. (U.S. Pat. No. 3,627,002) U.S. Pat. Off., Wash., D.C.
A tree shear. Patent assigned to Fulghum Enterprises, Inc., Wadley, Ga.

Gunnerman, R. J. 1971. SAW SPLITTER AND TIPPER ASSEMBLY. (U.S. Pat. No. 3,583,450) U.S. Pat. Off., Wash., D.C.
Cut boards are tipped away from a saw blade and an adjacent splitter by spring members that apply a maximum tipping force to the upper portion of the cut boards. Tipping is further assured by an abutment which the cut board strikes as it falls.

Hickman, B. D. 1971. WOOD CHIP CUTTING APPARATUS. (U.S. Pat. No. 3,570,567) U.S. Pat. Off., Wash., D.C.
An end-milling configuration of a cutterhead for chipping headrigs. The cutterhead features a bent knife to cut wood across the grain and leave a flat surface parallel to direction of log movement.

Kempe, C. 1971. FELLING HEAD RECIPROCATING BLADE TYPE. (U.S. Pat. No. 3,625,268) U.S. Pat. Off., Wash., D.C.
A mobile feller skidder having two hydraulically actuated shear blades.

Kennemore, P. H. 1971. CHIP EJECTOR CHAIN TOOTH. (U.S. Pat. No. 3,625,266) U.S. Pat. Off., Wash., D.C.
A saw chain designed to eject chips efficiently.

Kershaw, R. G. 1971. WOOD COMMINUTING APPARATUS. (U.S. Pat. No. 3,572,594) U.S. Pat. Off., Wash., D.C.
The apparatus has adjacent disks, rotatable in opposite directions, in a housing with cutting blades at one side of the first disk positioned to contact a length of wood fed sidewise. Openings adjacent to blades on this disk convey cut wood to the adjacent side of the second disk, which rotates faster than the first disk and in opposite direction. Cutters on one side of the second disk comminute wood, which then passes through openings in the disk and is discharged from the housing by the impellers.

Kessler, K. Q., and Davis, E. C., Jr. 1971. TREE SHEAR. (U.S. Pat. No. 3,627,003) U.S. Pat. Off., Wash., D.C.
Patent for the shear assigned to Deere and Co., Moline, Ill.

Koch, P. 1971. PROCESS OF MAKING LAMINATED WOOD PRODUCT UTILIZING MODULUS OF ELASTICITY MEASUREMENT. (U.S. Pat. No. 3,580,760) U.S. Pat. Off., Wash., D.C.
Process of making very strong and stiff beams from woods-run timber. Laminae are graded by modulus of elasticity, and the most limber are then assembled into the center of the beam and the progressively stiffer ones to the outside.

Lyon, C. R. 1971. ATTRITION MILL. (U.S. Pat. No. 3,572,595) U.S. Pat. Off., Wash., D.C.
An apparatus to break up pulp sheets and continuously deliver the resultant chips into an attrition mill. The mill is comprised of one stationary and one opposed rotating disk. Chips enter through a central aperture in the stationary disk and leave the periphery of the mill in fibrous form.

Morton, D. L. 1971. MACHINE FOR CUTTING PEELER CORES OR LOGS INTO STUDS AND CHIPS. (U.S. Pat. No. 3,627,005) U. S. Pat. Off., Wash., D.C.

39

A chipping headrig in which veneer cores or small logs move longitudinally past a series of cutterheads and saws that produce parallel upper and lower surfaces. At least one horizontal saw kerf is made after the cores or logs have been shaped.

Ramey, K. J. 1971. ROTATABLE TREE SHEAR DEVICE. (U. S. Pat. No. 3,627,004) U. S. Pat. Off., Wash., D. C.
A shear for attachment to a vehicle and permitting control of felling direction to either side.

Swillinger, M., and Hobbs, C. F. 1971. METHOD AND MEANS FOR PRODUCING WOOD VENEER. (U. S. Pat. No. 3,627,006) U. S. Pat. Off., Wash., D. C.

Method of supporting a flitch so that warp is removed as flitch thickness approaches zero during slicing.

Thrasher, E. W. 1971. APPARATUS FOR SAWING WOOD TO PRODUCE SMOOTH ROTARY PLANED SURFACES. (U. S. Pat. No. 3,568,738) U. S. Pat. Off., Wash., D. C.
A machine and a method in which the circular sawblade is guided by flexible and resiliently supported saw guides. Also provided is a splitter that maintains the sawed wood in spaced relation and out of substantial contact with the back edge of the saw; the splitter is shifted transversely in response to transverse deflections of the saw.

SAFETY, INCLUDING NOISE

Allen, C. H. 1970. GUIDELINES FOR DESIGNING QUIETER EQUIPMENT. Mech. Eng. 92(1): 29-34.
Noise control is complicated by the need to consider both airborne sound and solidborne vibration. These are two essentially different phenomena responding to different control techniques, yet energy is readily transformed from airborne sound to solidborne vibration and vice versa. Practical suggestions, diagrammatically illustrated, are presented for reducing airborne noise at its source — i. e., from solid vibrating parts and from air in motion — and for minimizing solidborne vibrations emanating from moving parts and from sound impact. Suggestions are presented for noise isolation of airborne sound by separation of source and receiver, acoustical absorbing treatment, and sound barriers. Recommendations are also given for isolation of noise-causing vibrations in solid structures by separation of source and isolated point, vibration brakes, blocks, and damping materials.

Altman, I. A. 1971. "BODY GUARD" FOR YOUR CHAIN-SAW. Am. Pulpwood Assoc. Tech. Release 71-R-14, 2 p.
Describes aluminum guard designed for simple installation on chainsaws (either bow or straight types) used for limbing or topping. Guard shields chain at the points closest to power source.

Burjukov, M., and Koz'jakov, A. 1970. INVESTIGATION OF NOISE IN MILLS IN THE BRATSK FOREST-INDUSTRY COMPLEX. Lesn. Prom (7): 13-15. Moscow.
Shows in graphs and a table the levels and spectra of noise of various machines and ancillary equipment in sawmills of this complex. Noise significantly exceeds the Soviet health standards.

Bobin, E. V. 1970. SOUND INSULATION OF A 4-SIDED PLANER. Derevoobrab. Prom-St'. 19(12): 11-12. Moscow.

Bohrer, D. A. 1971. INSTRUMENTATION FOR 'WALSH-HEALEY" NOISE SURVEYS. For. Prod. J. 21(9): 88-89.
Equipment and training needed to comply with the Department of Labor's occupational noise standards are relatively inexpensive and easy to obtain. Basic equipment includes a sound-level meter and a sound-level calibrator. Noise exposure meters and impact noise analyzers may be necessary under certain circumstances to evaluate special areas and problems. If a serious problem does exist, more elaborate equipment may be necessary to establish a satisfactory hearing conservation program and to control the noise at its source.

Bulboaca, I. 1971. WORKER PROTECTION DEVICE ON THE BAND SAW. Ind. Lemnului 22(8). 308-310. Bucharest.
Gives details of a rapid braking system developed in Rumania. Summarizes causes of accidents and suggests preventive measures.

Calin, S. 1971. DEVICES FOR PREVENTING ACCIDENTS

ON VENEER SLICING MACHINES. Ind. Lemnului 22(8): 308-309. Bucharest.
Discusses modifications developed in Rumania means of trapping the carriage if it should break free, guards between the operator and the front of the carriage, main control panel with view of the machines, and louder warning signals.

Cizevskij, M. P., and Ceremnyh, N. N. 1970. REDUCING THE WORKING NOISE OF THE CA-2 (CIRCULAR SAW). Derevoobrab. Prom-St'. 19(9): 24-25.Moscow.

Degtjarev, V. I. 1970. REDUCING THE NOISE OF PETROL-DRIVEN CHAIN SAWS. Lesn. Prom (3): 21-22. Moscow.

Department of Employment and Productivity. 1970. SAFETY IN THE USE OF WOODWORKING MACHINES. Health & Safety at Work Booklet 41, 140 p. Her Majesty's Stationery Office, London.
Describes safety practices in England (generally useful elsewhere) applicable to circular saws, bandsaws, planers, routers, and mortisers. Contains suggestions on plant layout and training.

Efremova, T. K., and Mitusov, V. A. 1971. THE CRITICAL SPEED OF AN AIR CURRENT BLOWING AWAY DEPOSITS OF WOOD DUST. Derevoobrab. Prom-St'. 20(6): 11-12. Moscow.
Experiments in wood-flour factories showed that air currents of about 1 m./sec. can blow away deposits of dust, thus creating an explosion hazard.

Fuju, Y. 1970. ON THE TRANSMISSION OF THE VIBRATION OF CHAIN-SAW MACHINES TO THE HUMAN BODY. Koyoto Univ. For. Bull. (41): 137-143. Japan.
Percentages of vibration transmitted from the chain saw to various parts of the operator's body were about as follows back of hand, 90; wrist, 40; elbow, 30, and shoulder, 10 percent. The frequency of the chain saw engine piston (about 86 c. p. s.) and about half of this frequency (42 c. p. s.), were the predominate frequencies transmitted to the body of the operator. Higher saw frequencies (about 170 and 270 c. p. s.) were not transmitted.

Fushimi, T. 1971. THE NOISE OF THE CHAIN SAW. A SIMPLE METHOD OF EVALUATION OF CHAIN-SAW NOISE BY MEANS OF SOUND-LEVEL METER READINGS; AND THE CHAIN-SAW-NOISE-CONTROL CRITERION IN BUCKING OPERATION. Ehime Univ. For. Bull. (8). 17-31. Japan.

Fushimi and Shigaki. 1971. (Chain sawing)

Gatley, W. S. 1971. INDUSTRIAL NOISE CONTROL — PAST, PRESENT, AND FUTURE. Mech. Eng. 93(4): 29-37.
Reviews fundamental aspects of noise control. Discusses terminology, instrumentation for measurement and analysis of sound, annoyance criteria, damage-risk criteria for hearing loss as

40

specified in the Walsh-Healey Act, and examples of noise control in industry.

Gebert, P. 1971. ANTI-VIBRATION DEVICES ON MODERN CHAIN SAWS. Erdö 20(4): 169-171. Budapest.
The antivibration devices of the Stihl-050-AVL, the Husqvarna-180-S, and the Homelite-XL-923 chain saws were the most effective of those tested, but none gave complete protection to the worker.

Gellert, E. 1971. THE EFFECT OF NOISE AND VIBRATION ON FOREST WORKERS, AND PROTECTIVE MEASURES. Soz. Forstwirtsch. 21(7): 205-207, 210.
Physiological effects and resulting disturbances to health, practical hygiene, and types of equipment responsible, with special reference to power saws.

Harstela, P. 1971. ON THE EFFECT OF MOTOR SAW VIBRATION ON THE HANDS OF FOREST WORKER. Folia For. 118, 14 p. Inst. For. Fenn., Helsinki.
Skin temperature in the worker's hands, sensation threshold, and manipulating capacity were lowered by chain saw vibrations that led to "white fingers," but study of these indicators did not throw light on the causative mechanism. Incidence of the effect was faster than recovery, which required 2 to 15 minutes or more.

Harstela, P. 1971. THE EFFECT OF THE SEQUENCE OF WORK ON THE PREPARATION OF APPROXIMATELY 3-M., ROUGH-LIMBED SPRUCE PULPWOOD. Folia For. 105, 23 p. Inst. For. Fenn., Helsinki.
Reports an effort to determine the effect of work sequence during pulpwood production on the time a worker is exposed to chain saw vibrations and the time it takes him to recover.

Huzl, F., Stolarik, R., Mainerova, J., and others. 1969. RESULTS OF A STUDY OF THE HEALTH OF A GROUP OF FOREST WORKERS EMPLOYED IN FELLING WITH POWER SAWS. Lesn. Pr. 48(1): 21-24. Prague

Keighley, G. D. 1970. EFFECTS OF CHAINSAW VIBRATION. Timber Trade J. Suppl., p. 39, 41.
Occurrence of Reynaud's Phenomenon, commonly called "white fingers," has increased among chain saw operators. A condition caused by prolonged and excessive vibration of the hands and arms, it may result in pain and insensitivity as well as possible change in bone structure. While manufacturers have made considerable progress in reducing the vibration level of chain saws, users can further reduce effects by (1) resting the saw as much as possible on the tree (or occasionally on the thigh) and holding it as lightly as possible, (2) wearing chain saw gloves, (3) warming up the hands and arms before using a chain saw, and (4) stopping frequently to rest.

Kitayama, S., and Sugihara, H. 1969. ON NOISE ANALYSIS OF SINGLE SURFACE PLANER. J. Jap. Wood Res. Soc. 15(3): 87-92.
Sound pressure rose with an increase in width of wood planed or with an increase in cutterhead speed. Feed speed, depth of cut, thickness of wood, and wood species were not significant. In the planer studied, the primary noise was generated by rotation of the cutterhead; sound level peaks appeared at frequencies from 350 to 400 cycles per second at cutterhead speeds of 4,990 to 6,660 r. p. m.

Klimkiewicz, W. 1969. PROBLEMS WITH THE NOISE OF WOODWORKING MACHINES. Przem. Drzewny 20(10): 20-25. Warsaw.
General information on noise perception and measurement. Summarizes some German and Russian literature on the noise of woodworking machines and its control.

Kudŋasov, V. 1970. REDUCING THE NOISE OF THE DRIVES OF CHAIN CONVEYORS. Lesn. Prom. (5): 14. Moscow.

Lamb, F. M. 1971. INDUSTRIAL NOISE AND NOISE EX-POSURE. For. Prod. J. 21(9): 84-87.
The Walsh-Healey Public Contracts Act of May 1969 specifies that all agencies wishing to sell the Federal Government goods valued at more than $10,000 (or services valued at more than $2,500) must initiate a program to protect their employees from excessive noise; excessive is defined as above 90 dBA. Many woodworking tools generate noise above this level. Sound intensity, sound pressure, and sound power are defined, and exposure limitations called for by the Act are discussed.

Lamb, F. M. 1971. INDUSTRIAL NOISE CONTROL GUIDE-LINES. For. Prod. J. 21(11): 12-16.
Noise exposure standards have been established and industry must develop control procedures and hearing-conservation programs. These programs should include: (1) noise survey of plant by reliable techniques; (2) medical (audiometric) test of all employees, with periodic follow-up checks, (3) engineering control of noise at its sources, (4) administrative control to limit workers' exposure in accordance with regulations.

Landry, Y. 1971. NOISE AND ITS IMPLICATIONS. Pulp and Pap. Mag. Can. 72(2): 109-111.
Briefly discusses the nature of deafness and the measurement and conservation of hearing. Advocates inclusion of noise-control measures in every timber company's safety program.

McKenzie, W. M. 1970. WHISTLING OF CIRCULAR SAWS. Aust. Timber J. 36(6): 55.
Whistling of circular saws can be lessened by reducing the number of teeth, preferably so that the ratio of tooth pitch to tooth width is greater than 8. Alternatively, symmetrically placed radial slots in the periphery of a saw plate are effective if slot depth (i. e., length) is at least 1/6th the blade radius. As a third alternative, effective on small saws, single air jets can be directed against the rotating saw to provide a frictionless force on the blade; whistling of larger blades was suppressed by a "pad" of air issuing through small (1/32-inch) holes in metal disks 2 inches in diameter set close to the blade – one on each side near the rim.

Malcolm, F. B. 1971. SYMMETRICALLY PLACED EXPAN-SION SLOTS SOLVE PROBLEM OF 'SCREAMING' SAWS. For. Ind. 98(6): 76-77.
A simple, effective means of reducing the high-frequency scream of circular saws has been developed by two German scientists and tested at the USDA Forest Products Laboratory at Madison, Wis. The method calls for cutting four or five symmetrically placed expansion slots in the rim. The slots, which must be at least 1/6 the radius of the blade in length, were shown to reduce sound by 14-16 dbs. Slots tested had a length-to-width ratio of 4 to 1; they were ground as extensions of the faces of raker teeth.

Montaner, M. 1970. NOISE MEASUREMENT AND CON-TROL. Acta Manilana 5A, p. 31-48.
Gives measurements and analyses of noise from a chain mortiser, chainsaw, circular wood saw, and metal-cutting saw. Makes suggestions for reducing noise.

Myles, D. V., Hirvonen, R., Embleton, T. F. W., and Toole, F. E. 1971. AN ACOUSTICAL STUDY OF MACHINERY ON LOGGING OPERATIONS IN EASTERN CANADA. Can. Dep. Fish. and For. Inf. Rep. FMR-X-30, 41 p.
Reports on noise produced by typical logging operations and by various machines. The influence of forest conditions on attenuation of such sounds was also studied. It was found that chain saw and skidder operators, with few exceptions, were working in an environment so noisy that extended exposure

41

might damage hearing. Under some conditions logging noise carried up to 1-1/2 miles, but at that distance was barely identifiable above the ambient level. It is recommended that logging operations be kept at least a mile distant from areas where the noise might cause concern.

Pahlitzsch, G., and Friebe, E. 1971. CAUSES OF DISCRETE FREQUENCIES IN THE IDLE-RUNNING NOISE SPECTRUM OF CIRCULAR SAWBLADES. Holz Roh- und Werkst, 29: 31-37.
Whistling of circular sawblades is a result of vibrations caused by flow of air around the teeth — initially down the rake faces, then in a vortex on the backs, and finally in a separation of vortex and tooth backs. Whistling occurs when the frequency of vortex separation coincides with the natural frequency of the blade. The frequency of vortex separation increases in proportion to the flow velocity (initial), and decreases with blade thickness. Initial flow velocity is approximately 25 percent of the peripheral speed of the saw. Frequency of whistling is not related to number of teeth.

Pahlitzsch, G., and Liegmann, E. 1956. NOISE INVESTIGATION ON A SURFACE PLANING MACHINE. Holz Roh- und Werkst. 14: 121-135. (Available in English transl. FIRA Furn. Ind. Res. Assoc. Transl. 78, Stevanage, Eng.)

Potanin, V. A. 1970. REDUCING THE NOISE OF LONGITUDINAL ROTARY CUTTERS. Derevoobrab. Prom-St'. 19(5): 11-12. Moscow.
Describes trials of a cutterhead (cylinder) for planers, comprised of many disks so placed that the cutting edges are staggered in a helical pattern.

Pryatelj, T. 1970. NOISE – A MODERN EVIL. Les, Ljubljana 22(7/8): 123-128. Jugosl.
Measurement in three Jugoslav woodworking factories employing some 1,400 people showed that the permitted noise level is exceeded at 78 percent of the work-places. The harmful effects of noise on man are discussed.

Reeves, E. H. 1971. DESIGNING OF LOGGING EQUIPMENT FOR OPERATING EFFICIENCY AND COMFORT. Pulp and Pap. Mag. Can. 72(1): 149-153.
To achieve the maximum in efficiency and safety from man and machine, designers of logging machinery must consider not only dimensions and capacities of the average man, but also the wide natural deviations from these averages.

Sakurai, T., Takahashi, A., and Fujita, S. 1968. ON THE MECHANICAL VIBRATION OF CHAIN SAW. II. THE ISOLATING EFFECTS OF RUBBER MOUNTS ROUND THE HANDLE. Shimane Univ. Fac. Agric. Bull. (2): 76-80. Japan.
Rubber insulators substantially reduced vibration at chain saw handles.

Sakurai, T., Tanaka, C., and Takahashi, A. 1969. ON THE MECHANICAL VIBRATION OF CHAIN SAWS. III. PHYSIOLOGICAL RESPONSES UPON THE HANDLER. Shimane Univ. Fac. Agric. Bull. (3): 40-45. Japan.
Describes effect of chain saw vibrations on blood pressure and pulse rate of operators.

Schmutzler, W. 1970. NOISE PREVENTION IN WOODWORKING MACHINES BY ENCAPSULATION. Holzindustrie 23(6): 170-174.
Illustrated account of principles, techniques, and materials.

Smith, J. H. 1971. NOISE IN THE WOODWORKING INDUSTRY – A REVIEW OF THE LITERATURE. For. Prod. J. 21(9): 82-83.
A review of literature selected (27 citations) as pertinent to

control of noise in the wood industry. Although the subject has been of increasing concern during the last 20 years, little has been published in English-language journals. Obviously, increased research accompanied by an increased flow of technical information will be necessary if industry is to comply with health and safety standards.*

Starzinskij, V, N. 1970. INVESTIGATION OF THE NOISE OF DISK-TYPE CHIPPERS. Derevoobrab. Prom-St'. 19(6): 11-13. Moscow.

Starzinskij, V. N. 1971. CALCULATION OF THE NOISE OF DISK-TYPE CHIPPERS. Derevoobrab. Prom-St'. 20(1): 13-15. Moscow.

Strzelecki, A. 1971. NOISE REDUCTION AS A RESULT OF DAMPING THE VIBRATIONS OF A ROTATING CIRCULAR SAW IN A MAGNETIC FIELD. Folia For. Pol. (Drzewn.) (10): 35-53. Warsaw.
Noise was decreased by 15 decibels in some cases but was unaffected in others. Maximum reduction occurred where undamped vibration was of highest frequency.

Szepesi. 1970. (Sawing; Chain sawing)

Tanaka, C., Sakurai, T., and others. 1970. INVESTIGATIONS OF VIBRATIONS AND NOISE FROM A DRUM BARKER. Wood Ind. 25(10): 27-30. Tokyo.

Tanaka, C., Sakurai, T., and Horie, T. 1970. VIBRATION AND NOISE ON HAND FEED PLANER. II. EFFECT OF KNIFE WEAR ON PLANING NOISE. Shimane Univ. Fac. Agric. Bull. (4): 93-96. Japan.
Noise increased as planer knives dulled.

Tanaka, C., Sakurai, T., and Horie, T. 1971. VIBRATION AND NOISE OF HAND FEED PLANER. I. VIBRATION AND NOISE IN NO-LOAD RUNNING. J. Jap. Wood Res. Soc. 17: 111-116.
Noise level was influenced by knife projection, width of gap in the table (bed-plate), and degree of unbalance of the cutterhead. The primary noise source is air periodically compressed by the knives passing an obstacle (e. g. the bed-plate); frequency is equal to number of knives multiplied by revolutions per second of the cutterhead.

Tröger, J. 1969. THE MECHANISM OF NOISE PRODUCTION (BY A SPINDLE SHAPER) IN CHIPPING. I. and II. INVESTIGATIONS ON THE WORK NOISE. Holztechnologie 10: 181-184, 265-269.
Part I. Describes investigations made on a spindle shaper, in which several chip-production conditions were varied: speed, bite, feed speed, chipping width, volume chipped per time unit, and direction of rotation relative to material. Tool design and nature of the material worked were also varied. Results are graphed. Part II. Eddies of air generated by obstacles near the rotating knives cause siren-like noise. Discusses possibilities for reducing noise levels through changes in tool construction and housing.

Tröger, J. 1970. THE MECHANISM OF NOISE PRODUCTION IN CHIPPING. THEORETICAL INVESTIGATIONS. I and II. Holztechnologie 11: 41-47, 75-80.
Part I is an analysis of noise frequency based on a mathematical-physical model. Part II discusses test results.

Tröger, I. 1971. FUNDAMENTALS OF NOISE ABATEMENT IN THE WOOD INDUSTRY. I and II. Holzindustrie 24(11,12): 331-333, 372-376.

Werner, H. 1966. CIRCULAR SAW BLADES WITH FEW

TEETH AS PROTECTION AGAINST NOISE. FIRA Transl. 65, 8 p. Stevenage, Engl.1967. (Holzindustrie (12): 344-346. 1966.)
Blades with few teeth are less noisy than those with many teeth. The author strongly recommends that circular saws have variable-speed mandrels, which he feels would facilitate use of saws with few teeth.

Wiklund, M. 1971. KICKBACK OF POWER SAWS. Tek. ForsknStift. Skogsarb. 2, 3 p.

Zeller, W., and Paul, H. H. 1969. NOISE IN WOODWORKING SHOPS. FIRA Transl. 58, 26 p. Stevenage, Engl. (Lärmbekämpfung 1964, 2 and 3, p 37-45.)
Of the machines evaluated (planers, circular saws, and bandsaws), panel planers were the most noisy. Noise levels were evaluated in eight shops; based on 90 percent of maximum observed values, levels were as follows, according to frequency measured:

Cycles per second	Decibels
63	83
125	92
250	100
500	97
1,000	94
2,000	92
4,000	90
8,000	78

The damping of sound on its path between carpenter shop and the buildings to be protected was related by formula to height of shop and distance from the shop. Specific recommendations for construction of walls, windows, doors, and roofs in buildings to be protected were based on the formula and on knowledge of the sound level at its source.

Index of Authors

Koch, Peter

    1973. Wood machining abstracts, 1970 and 1971. South. For. Exp. Stn., New Orleans, La. 46 p. (USDA For. Serv. Res. Pap. SO-83)

Lists 415 references, most with abstracts. An introduction mentions notable papers in various fields of endeavor and surveys trends in mechanical processing of wood.

REGENERATING LOBLOLLY PINE BY DIRECT SEEDING, NATURAL SEEDING, AND PLANTING

Thomas E. Campbell and William F. Mann, Jr. [1]

Fourteen studies comparing pine regeneration methods in Louisiana and Texas showed that planting was most reliable, and that direct seeding was preferable to natural seeding.

Additional keywords: Site preparation, early growth, *Pinus taeda.*

This paper summarizes 14 pilot-scale studies comparing artificial and natural methods of reforesting recently cutover sites with loblolly pine *(Pinus taeda* L.). Studies were started in 1960, when direct seeding was a new technique, and continued through 1968. They pro-

[1] Respectively, Silviculturist and Chief Silviculturist, Alexandria Forestry Center, which is maintained at Pineville, Louisiana, by the Southern Forest Experiment Station, USDA Forest Service.

vided evaluations under a variety of climatic, site, and soil conditions.

The original plan was to compare only artificial and natural seeding. As the research progressed, however, the interest of some cooperating landowners changed from regeneration by the seed-tree method to planting of nursery seedlings. Consequently, six installations compared natural and direct seeding, six compared planting and direct seeding, and two compared all three methods.

Study Areas and Methods

Installations were in Louisiana and east Texas (fig. 1). Soils varied from deep sands with fast internal drainage to heavy silt and clay

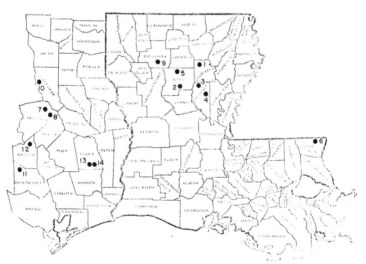

Figure 1.—*Locations of test areas.*

Review and year	References
Koch (1968)	
1966	138
1967	103
McMillin (1970)	
1967	36
1968	187
1969	89
Present review	
Prior to 1969 or undated	35
1969	71
1970	153
1971	156

Thus, references for 1968 through 1971 averaged in excess of 150 per year. Since many titles are inevitably omitted from the second year of a 2-year review (because of delayed receipt), it is likely that 1971 -- like 1968 -- will prove to have been a record year for reported research on wood machining.

In the current review, 43 fields of activity are identified, yielding a total of 415 papers (table 1) — about one-third more than in either of the previous two reviews. Sawing (8 fields) led with 94 primary and 95 cross-references. Circular sawing was prominent with 23 primary and 31 cross-references. Bandsawing also received much attention, with 13 primary and 23 cross-references. Twenty-seven papers discussed chain sawing, 24 shearing, and 17 sash gangsawing.

Safety (including noise) led all single fields of activity with 51 primary and 2 cross-references. Of these, 36 were concerned with noise, and 16 with safe operation of chain saws. The prior review (McMillin 1970 (Bibliographies)), carried only 13 primary listings and two cross-references for the entire field of safety; of these only five were concerned with noise. Thus, we see a seven-fold increase in research related to noise of woodworking equipment.

Properties of the cutting edge and cutter (six fields) shows 83 primary and 32 cross-references for a total of 115 citations. In this category, the fields of fitting and sharpening (34 citations) and stability (30 citations) were prominent. Dulling phenomena were reported in 23 papers whereas the prior review showed only 9.

Veneer-cutting research continued active, with 24 primary and 11 cross-references — 39 papers were listed in the prior review. Barking

also received considerable attention, with 27 citations; bark separation from chips was of particular interest, with eight references.

In the category of machinability of species and reconstituted wood (five fields), hardwoods had 31 citations, softwoods 23, particleboard 12, fiberboard 8, and plywood 6.

Jointing, planing, molding, and shaping accounted for 24 references. Ten papers were devoted to orthogonal cutting, 10 to machining with coated abrasives, and 9 to defibrating (including stone grinding). Thirteen primary listings were devoted to research instrumentation and techniques; and the new field of computer, tape, and card control of machines carried four primary listings as did the field of laser cutting (machining with light).

In the category of chipping, flaking, and grinding (five fields), 20 papers discussed conventional chippers, 13 were concerned with chipping headrigs and edgers, 8 with mobile chippers, 6 with flaking, and only 1 with grinding (i.e. pulverizing).

Turning, alone among the 43 fields, has no papers cited in this review.

Despite the obvious risk of offending researchers by omission or misinterpretation, it seems worthwhile to attempt some mention of individual contributions in the various fields.

History and general texts. — Thunell, in articles dated 1971, discussed research trends in woodworking and selection of equipment for sawmills of various capacities. Koch (1971) reviewed significance of the southern pines in the economy of the United States and summarized trends in manufacture.

Properties of wood. — Albers (1971) discussed the crushing strength of a variety of particleboards, veneers, and fiberboards. Futo (1969) elucidated modes of tensile fractures in very small specimens of spruce.

Sharma (1971) reported that silica content in 23 specimens of *Tectona grandis* varied with region of origin from 0.2 to 3.04 percent; it was vitreous and found only in fibers and vessels.

Orthogonal cutting. — Kato et al. (1971) studied the action of vibrating knives in cuts parallel and perpendicular to the grain of a hardwood and a softwood. Hamamoto and Mori

Table 1. — Number of references on wood machining research for 1970 and 1971, classified into 43 fields of activity

Subject	Primary listings	Cross references	Total
History and general texts	3	2	5
Properties of wood	2	1	3
Orthogonal cutting	6	4	10
Peripheral milling parallel to grain	2	2	4
Barking	21	1	22
Sawing[1]			
General	13	13	26
Bandsawing	13	23	36
Sash gangsawing	9	8	17
Circular sawing	23	31	54
Chain sawing	11	16	27
Machining with high-energy jets	1	0	1
Machining with light	4	0	4
Shearing	20	4	24
Total sawing	94	95	189
Jointing, planing, molding, and shaping	7	17	24
Turning	0	0	0
Boring, routing, and carving	8	1	9
Mortising and tenoning	0	3	3
Machining with coated abrasives	9	1	10
Veneer cutting	24	11	35
Chipping, flaking, and grinding			
Chipping	15	5	20
Chipping headrigs and edgers	10	3	13
Mobile chippers	7	1	8
Flaking	1	5	6
Grinding	0	1	1
Total chipping, etc.	33	15	48
Defibrating	8	1	9
Properties of the cutting edge and cutter			
Tool material	10	2	12
Dulling	12	11	23
Fitting and sharpening	24	10	34
Stability	23	7	30
Temperature	7	1	8
Friction	7	1	8
Total cutting edge and cutter	83	32	115
Computer, tape, and card control of machines	4	1	5
Research instrumentation and techniques	13	14	27
Nomenclature	3	0	3
Machinability of species and reconstituted wood			
Hardwoods	13	18	31
Softwoods	2	21	23
Plywood	2	4	6
Particleboards and sandwich boards	8	4	12
Fiberboards, hard and soft	2	6	8
Total machinability	27	53	80
Bibliographies	2	2	4
Reports of research programs and industrial developments	1	10	11
Patents	14	1	15
Safety, including noise[1]	51	2	53
	415	269	684

[1] Noise alone: 36 references. Chain saw safety: 16 references.

3

(1970) found that surface quality was improved and cutting forces reduced in cuts across end grain (90-90 mode) if the knife was vibrated in the direction of the cut.

McKenzie and Cowling (1971) observed dulling phenomena, and related edge wear to cutting forces (90-90 mode); chip rupture patterns were also explained. Noguchi (1970) and Noguchi et al. (1971) reported cutting data obtained with a pendulum dynamometer.

Woodson and Koch (1970) orthogonally cut southern pine earlywood and latewood in the three principal modes and presented photographs of chip types together with tabulated and graphed data on cutting forces; also given were regression equations for cutting forces, in terms of tool angles and wood properties.

Peripheral milling. — Mori (1970) found that the relationship between cutting force and undeformed chip thickness was affected by the ratio of depth of cut to cutting-circle diameter; there was an optimum ratio.

Stewart (1971) studied chip types resulting from peripheral milling with cutterheads having helical knives. He observed that veneer-like chips resulted when feed direction was arranged so that wood grain was parallel to the cutting edge.

Solov'ev (1971) developed formulas for calculating cutting forces and their components.

Barking. — Hyler (1969) described types of barking machines used in North America. Kanauchi (1970) reported laboratory data on the effectiveness of various types of barking tools. Shelyapin and Razumovskii (1971) related pulpwood diameter and drum rotation speed to the degree of bark removal from larch. Data on mechanical ring barkers were given by Berlyn (1970) and Birjukov et al. (1970).

The subject commanding most attention was removal of bark from barky chips; papers germane to this subject included: Liiri (1961); Arola (1970); Bumerl (1969); Erickson (1968, 1970, 1971); Lea and Brawn (1970); and Julien et al. (1971).

Sawing. — Causes of inaccurate sawing were reported by Birkeland (1968) and Breznjak and Herak (1970). Sawmill layout and developments were discussed by the Home Merchants Association of Scotland (1970), Mügge (1970), Neville (1970), and Thunell (1971).

Riberolles (1968) and Dargan (1969) de-

scribed novel methods of sawing wooden tubes and round dowels.

Reynolds (1969) explained the SHOLO (SHOrt LOg) process for manufacturing pallet parts from low-grade hardwood logs.

Clark (1968, 1969) described the operations of mills using high-strain bandsaws and thin-kerf circular saws. Cumming (1969) and Porter (1971) further discussed high-strain bandsaws; and Sugihara (1970) defined stresses in bandsaws.

Thin-kerf circular sawing was discussed by Clark (1968), Kintz (1969), Salemme (1969), and Schliewe (1969).

Feeds and speeds when ripping with circular saws was the subject of work by Kubasak (1968), Breznjak and Moen (1969), McKenzie (1970), Stefaniak (1970), Belikov (1971), Koshunyaev et al. (1971), Nakamura (1971), Pashkov and Bodalev (1971), and Radu et al. (1971).

Crosscutting of roundwood was studied by Kubasak (1968), Degermendzhi (1969), Kondrat'ev (1969), and Vasil'ev (1970, 1971). Bramhall and McLauchlan (1970) described a method of circular sawing microsections as thin as 100 mµ.

Macomber (1969) provided detailed procedures for smooth-ripping 2 by 4's from wider lumber in such a manner that the edges need no further planing, and the corners are rounded.

Yu (1966) was able, under certain conditions, to friction-saw softwood with a 14-inch, 14-gauge, toothless steel disk rotated at 4,620 r.p.m.

The chain saw will likely remain for 20 years a mainstay in Canadian logging, according to the *Pulp and Paper Magazine of Canada* (Anonymous 1969). The chain saw is, however, the subject of much research aimed at reducing vibration and noise (Fushimi and Shigaki 1971; Szepesi 1970; *see also* Safety, including noise). Subjects of much study were chain saw guidebars (Gul'ko and Matjuskin 1970; Tudoran et al. 1971) and teeth (Janco 1967; Pahlitzsch and Peters 1968; Szepesi 1969; Kuhnke 1970). Reynolds et al. (1970) analyzed chain saw cutting forces, and Shackelberg (1971) studied cutting time.

Franz (1970) found that corrugated board slit with high-energy liquid jets had smooth edges free of damage and debris.

4

McMillin and Harry (1971) reported successful laser cutting of various southern pine products. With an air-jet assisted carbon-dioxide laser of 240-watt output, maximum feed speed decreased with increasing workpiece thickness in both wet and dry samples, but was unrelated to grain direction. Feed speeds averaged 99.1 and 14.6 inches per minute for lumber 0.25 and 1.00 inch thick, respectively. Doxey (1970), March (1970), and Ready (1970) also reported on industrial applications of lasers.

Arola (1971) sheared roundwood bolts with a single blade cutting against an anvil and found shear force positively correlated with blade thickness and wood density; wood temperature was negatively correlated with shear force; low-friction coatings reduced shear force, but did not reduce damage from splits.

Johnston and St-Laurent (1970, 1971) found that V- and U-shaped shear blades yielded promise of reduced splitting damage as did Shilke et al. (1971), they observed that cutting force and wood damage increased with decrease in temperature, increase in moisture content, and increase in specific gravity.

In another study, Johnston and St-Laurent (1971) found that conveyorized boards could be end-trimmed with shears; sheared ends, while smooth, frequently had splits one-half-inch deep.

In a study of slow-speed shearing of southern pine logs, Koch (1971) found that shear forces were positively correlated with sharpness angle, wood density, and log diameter; shear-induced splits were most severe (averaging 1.4 inches deep) in low-density logs sheared with a blade having a large sharpness angle.

Kubler (1960) reported on Russian experiments in which vibrated knives substantially reduced force required to cut wood across the grain. Hamamoto and Mori (1970) confirmed these observations.

McIntosh and Kerbes (1969) found that shear-caused splits resulted in lumber losses varying from 1 percent per tree in warm weather to 5 or 7 percent in very cold weather. Tuovinen (1970) also studied lumber degrade caused by shearing saw logs; trimbacks shortened lumber 8 to 20 cm. depending on species and temperature.

White et al. (1969) found that operators of mobile hydraulic tree shears spent 11 percent of their time positioning the shears, 22 percent shearing, and 67 percent moving from tree to tree.

Jointing, planing, molding, and shaping. — Goodchild (1967) noted that power requirement increases with feed speed, depth of cut, and peripheral velocity of the cutting edges; it also increases as rake angle is reduced and as the width of heel caused by jointing becomes greater.

Stewart (1970) found that planing hard maple panels across the grain yielded flakes suitable for high-strength particleboard. In another paper dated 1970, he observed that depths of cut up to 1/16-inch did not significantly influence the amount of chipped grain when planing dry *Platanus occidentalis;* chipped grain was minimized by reducing rake angle to 20° or less.

Raknes (1969), Roth (1970), and Tsuzuki et al. (1971) reported on methods of finger jointing and end-splicing. Kryazhev (1971) discussed the advantages of down-milling when cutting end-grain wood.

Boring, routing, and carving. — From observations of thrust, tool life, and hole quality, Edamatsu · and Nagahara (1970) made recommendations for boring tropical woods. Noak (1970) discussed economic considerations for dowel borers.

Radu (1969) evaluated five drill types and recommended an improved twist drill with smooth flutes; in 1970 he presented methods of calculating torque and thrust when boring beech, oak, spruce, and particleboard.

Data on the routing process were reported by Pahlitzsch and Sandvoss (1969), Theien (1970), Kondo et al. (1971), and Mori et al. (1971).

Machining with coated abrasives. — Pahlitzsch (1970) provided a review of the state of the art. Noda and Umetsu (1971) studied particle sharpness, fracture, and loss during abrasive machining of *Fagus crenata.* Meyer (1970, 1971) developed a technique for evaluating the progression of dulling of an abrasive belt.

Potrebic (1969) related the surface smoothness of beech wood and the grade of sandpaper used to machine it. And Hirst (1971) described wood finishes that most successfully hid machining marks; marks were less perceptible on flat

ms. While several of the driest sites were
own to be poor seeding chances, they were
:luded to test a wide array of soils. Topo-
aphy ranged from flat to rolling, but none
the areas were steep enough to have an ad-
rse effect on the regeneration methods.
acts for both seeding methods were 14 acres
larger; planted stands were a minimum of 5
res.

Most of the study areas were selected imme-
itely before the existing stand was harvest-
, but some were chosen after the cut had
en made. Stands were mixed loblolly and
ortleaf *(P. echinata* Mill.) pines with nu-
rous hardwoods in the understory (fig. 2).
ocking varied from poor to excellent. Dom-
ants ranged from 13 to 36 inches in d.b.h.
erchantable hardwoods were present in one
the understocked pine stands.

Areas scheduled for direct seeding and plant-
; were usually clearcut. On tracts to be re-
nerated naturally, 5 to 12 loblolly pine seed
es per acre were retained. Site preparation
ried, depending on the judgment of the land-
ner; burning, chopping, and the two in com-
ation were most common (fig. 3). Methods
d for each installation are described in
following sections.

With three exceptions to be noted later,
artificially sown loblolly pine seeds were strat-
ified, coated with currently recommended bird,
rodent, and insect repellents, and then broadcast with hand-operated grass seeders at the
rate of 1 pound per acre. Sowing was during
February in 13 trials and during November in
the other. Natural seedfall was sampled in
four of the eight seed-tree areas.

Seedling stocking was estimated in May or
June of the first year and at the end of the first
three growing seasons. Stocking estimates
were made on circular milacres for seeded
areas, and on 1/20-acre circular plots for planted areas. Total height of each seedling on all
sample plots was measured at age 3 years.

Regeneration By Direct and Natural Seeding

Results of each installation are presented
individually. Table 1 shows number of seedlings per acre, stocking percents, and seedling
heights as recorded in the final inventory of
all stands.

Test 1.—Caldwell Parish, La.; 1960.—The
study area and procedures were recorded as
follows: Rolling topography, fine sandy loam

Figure 2.—*Typical loblolly-shortleaf pine stand selected
for harvest and regen-
eration.*

Figure 3.—*Chopping followed by burning 6 weeks later provided an excellent seedbed for planting and seeding.*

soil, and a highly productive site. Fully stocked stand of loblolly and shortleaf pines up to 36 inches d.b.h. with dense understory brush; cut in early 1959. Burned in late 1959. Seven seed trees per acre produced 60,500 sound seeds in 1959. Direct-seeded in February 1960.

Initial stocking from direct seeding, measured in May of the first year, averaged 4,700 seedlings per acre. At the end of the third year

2,600 were left and 78 percent of the milacres had one or more seedlings (fig. 4). Average height of the largest tree on each milacre was 48 inches, as compared to 40 inches for all seedlings.

Natural seeding produced 19,700 seedlings per acre initially, but 7,500 died during the next 3 years. Additional seedfall in the third year increased total stocking to 17,800 per acre, with 100 percent of the milacre stocked.

Table 1.—*Number of seedlings per acre and size of pines at age 3 years for each test*

Test No.	Seedlings			Proportion of sample plots stocked		Mean total height				
	Direct seeding	Natural seeding	Planting	Direct seeding	Natural seeding	Direct seeding		Natural seeding		Planting, all trees
						All trees	Tallest per milacre	All trees	Tallest per milacre	
	– – – Number – – –			– Percent –		– – – – – – – Inches – – – – – – –				
DIRECT SEEDING AND NATURAL SEEDING										
1	2,600	17,800	...	78	100	40	48	33	50	...
2	1,900	1,800	...	68	56	41	49	13	15	
3	1,500	2,600	...	54	63	51	65	48	62	...
4	600	3,300	...	26	69	61	68	55	62	...
5	4,200	1,600	...	81	66	38	54	30	39	
6	1,000	1,200	...	50	60	25	29	23	30	...
DIRECT SEEDING, NATURAL SEEDING, AND PLANTING										
7	1,100	650	400	53	43	36	48	32	44	70
8 [1]	700	50	548	37	5	4	5	8	9	16
DIRECT SEEDING AND PLANTING										
9	2,400	...	414	56	...	20	25	...	...	42
10	1,000	...	392	40	...	26	32	...	...	38
11 [1]	350	...	496	31	...	7	9	...	.	20
12 [1]	550	...	537	27.	...	5	7	..	.	16
13	2,600	...	496	81	...	21	29	..	.	57
14	3,600	..	446	89	...	29	39	..	.	60

[1] Measurements were made at the end of the first year, when direct or natural seeding was classified as unsatisfactory and the trial was terminated.

Figure 4.—*Three-year-old stand of direct-seeded loblolly pine.*

an height of the largest seedling per milacre
ᵢ 50 inches, while all trees averaged 33
hes. Because of the large number of young-
seedlings, the total stand averaged 7 inches
ᵢ in height than on the direct-seeded area.

ᵉst 2.—Winn Parish, La.; 1961.—Level,
ᵢrly drained silt loam soil; good site. Mod-
ᵢtely stocked with loblolly-shortleaf pine
to 24 inches d.b.h.; some merchantable
ᵢdwoods; heavy hardwood underbrush. Cut
late 1959. Clearcut portion burned in July
0 and direct-seeded in February 1961.
ᵢd-tree area unburned, but hardwoods 4
hes d.b.h. and larger girdled.

nitial stocking on the direct-seeded area
raged 3,400 seedlings per acre and 85 per-
t milacre stocking. Mortality reduced the
ᵢd to 1,900 and 68 percent stocking at age
ears. Heights averaged 41 inches for all
ᵢs and 49 inches for the tallest tree per mil-
ᵢ.

eed trees failed to produce the first year,
second- and third-year seed crops yielded
nal stand of 1,800 seedlings per acre on 56
ᵢent of the milacres. Heights averaged 13
ᵢes for all stems and 15 inches for the tall-
ᵢn each milacre. Despite a lack of site prep-
ᵢion, most seedlings appeared free to grow.

ᵉst 3.—La Salle Parish, La.; 1962.—Rolling
ᵢain; well-drained fine sandy loam soil; site

index about 90. Fully stocked loblolly-shortleaf stand averaged about 18 inches d.b.h.;
moderate understory of small brush. Cut in
1960. Burned in 1961 and direct-seeded in
February 1962. Five trees per acre yielded
30,000 sound seeds in 1961 and 41,000 in 1962.

The 3-year-old direct-seeded stand had 1,500
seedlings per acre—half the initial catch—on
54 percent of the milacre plots. The tallest
seedling per stocked milacre averaged 65
inches, and all stems averaged 51 inches.

The natural stand had 1,400 3-year-old pines
per acre on 47 percent of the milacres; second-
and third-year seedfall boosted the total stand
to 2,600 with 63 percent milacre stocking.
Average heights of all seedlings and of the
largest stem on each milacre were 3 inches less
than for direct-seeded pines at age 3, probably
an effect of the younger average age.

Test 4.—La Salle Parish, La.; 1962.—Site,
cover, and treatment were the same as in test
3, but seedfall in the first year averaged only
4,500 sound seeds per acre.

Direct seeding yielded 1,000 seedlings per
acre initially. Early losses reduced the 3-year
stand to 600 seedlings and 26 percent milacre
stocking. Much of the mortality was due to
cattle trampling, and to overtopping brush that
had not been killed by the preparatory burn.
Although 600 seedings per acre ware adequate,

4

their distribution was unacceptable. Growth in this installation was better than in all others, averaging 61 inches for all seedlings at age 3 and 68 inches for the tallest per milacre.

Initial stocking in the natural stand averaged 1,500 seedlings per acre on 34 percent of the sample plots. Although 400 died through the third year, additional seedfall in the second year increased the final stand to 3,300 per acre and 69 percent milacre stocking. Growth of the natural pines was also excellent, with heights averaging 6 inches less than in the direct-seeded stand. The difference in size was due to the large number of younger seedlings in the natural stand.

*Test 5.—Winn Parish, La.; 1963.—*Level terrain, imperfect surface and internal drainage, sandy loam topsoil with impervious clayey subsoil, and low site index. Fully stocked stand of mature loblolly-shortleaf, averaging about 23 inches d.b.h., had dense hardwood underbrush up to 4 inches d.b.h. Cut in late 1962. Burned in December 1962, but many stems survived the fire. Hardwoods over 1 inch d.b.h. injected in April 1963. Direct-seeded by company personnel in February 1963 with 1.5 pounds of seed per acre.

Initially, the direct-seeded stand averaged 6,600 seedlings per acre and 89 percent milacre stocking. After 3 years seedlings per acre averaged 4,200, and 81 percent of the milacres were stocked. The high sowing rate was thus an error, but the decision was a logical one because of the adverse site. The stand at age 3 averaged 38 inches in height, and the tallest pines per milacre were 54 inches.

The area reproduced by natural seeding had 3,900 seedlings per acre after the first year, with 94 percent stocking. At age 3, the stand had 1,300 seedlings per acre plus 300 four- and five-year-olds that had survived the site-preparation burn. Height of natural seedlings averaged only 39 inches for the tallest per stocked milacre; this average included the older, larger trees.

*Test 6.—Washington Parish, La.; 1965.—*Rolling topography, fine sandy loam overlying red silty clay subsoil; moderate site productivity. Young, second-growth loblolly pine averaged 12 inches d.b.h. in a fully stocked stand; dense, small brush in the understory. Cut in summer of 1964. Burned in early Oc-

tober 1964 and direct-seeded in February 1965. six trees per acre produced 8,600 sound seeds.

The direct-seeded stand averaged 1,700 per acre and 62 percent stocking initially, and 1,000 per acre with 50 percent stocking at age 3 years. Heights averaged 25 inches for all seedlings and 29 for the tallest trees.

At age 1 year, the natural stand had 900 seedlings per acre and 40 percent milacre stocking. In the second and third years, additional seedlings increased the stand to 1,200 per acre and 60 percent milacre stocking. Tree heights were about the same as in the direct-seeded stand.

Regeneration By Direct Seeding, Natural Seeding, and Planting

*Test 7.—Houston County, Texas; 1966.—*Gently rolling terrain, sandy loam surface soil underlain by red plastic clay, high site index. Fully stocked with loblolly-shortleaf pine averaging 20 inches d.b.h. and moderately dense understory brush up to 3 inches d.b.h. Cut to seed trees in 1965, leaving three per acre on sites to be direct-seeded and planted and seven per acre on the seed-tree area. Site prepared by one pass of single-drum rolling chopper in September 1965; planted and seeded in February 1966.

As the seed trees bore no cones, initial stocking of 5,300 seedlings per acre was entirely from direct seeding. By the third year, mortality had reduced the stand to 1,100 per acre with 53 percent stocking. The largest trees were 48 inches tall, while the whole stand averaged 36 inches.

First-year seedfall in the natural stand was low, and initial stocking was only 250 seedlings per acre. There was a light seed crop in the second year, and by age 3 the total stand amounted to 650 seedlings per acre with 43 percent of the milacres stocked. However, many of the younger seedlings were overtopped by invading brush, and the number of pines free to grow was below acceptable levels. After 3 years, seedling heights averaged 4 inches less than on the direct-seeded area.

A total of 834 seedlings per acre were planted. Survival was 66 and 48 percent at the end of the first and third years. Survivors averaged 70 inches in height at age 3.

5

est 8.—*Houston County, Texas; 1966.*—
el terrain, sandy loam surface with a red
subsoil; an excellent site. Well-stocked,
ure loblolly-shortleaf pine averaging 22
es d.b.h. with a light understory of small
:h. Cut to heavy seed-tree stand in 1963
single-chopped in August 1964. As no re-
luction followed, half the area (50 acres)
clearcut and half was cut back to seven
trees per acre in June 1965; the entire
was burned in August 1965. Forty acres
e seeded and 10 acres planted in early
ch 1966.

irect seeding resulted in 3,200 seedlings
acre initially, but heavy losses in the first
· reduced the stand to 700 per acre with
ercent milacre stocking. The land man-
· judged the stand unacceptable and plant-
he site.

ecause natural seeding produced only 50
ling seedlings per acre, this plot was also
ted.

fter the first year the planted stand had
seedlings per acre, averaging 16 inches tall.

:generation By Direct Seeding and Planting

:st 9.—*Bienville Parish, La.; 1964.*—This
was actually the sixth to be installed and
first to include planting. The area and
ment were described as follows: gently
ig topography; sandy clay loam surface
red sandy clay subsoil with excellent
iage; site index about 100 feet (age 50).
' stocked loblolly-shortleaf pine stand
iged about 24 inches d.b.h., with a dense
rstory of small brush up to 2 inches d.b.h.
·cut in the summer of 1963. Forty acres
ed and direct-seeded in February 1964;
s and hardwoods on 40 acres bulldozed
windrows and area planted in February
800 trees per acre.

e early stand from direct seeding averaged
seedlings per acre. Half the trees sur-
through the third year; milacre stocking
6 percent. Height of tallest seedlings on
ed milacres averaged 25 inches; all stems
ged 20 inches.

;htly more than half the planted seedlings
/ed after 3 years, and they averaged 42

inches in height. The stand was satisfactory
to the land manager.

Test 10.—Cherokee County, Texas; 1966.—
This was the first of three installations on
drouthy sites known to be poor seeding
chances. Techniques deviated from recommended procedures in that unstratified seed
was sown in the fall with only leaf cover for
protection. It was thought that fall sowing
might boost initial establishment.

Terrain gently rolling; fine, sandy surface
soil up to 36 inches deep dries out rapidly but
site is good nevertheless. Fully stocked stand
of loblolly-shortleaf pines, with dense underbrush, cut to eight seed trees per acre in 1962.
Area failed to regenerate, and in 1966 all sawtimber was cut, leaving some small seed-pro-
ducing trees. Twenty-five acres prescribe-
burned in November 1966 in advance of seeding, but kill of hardwoods was erratic. Sown
with 1¼ pounds of unstratified seed per acre
in mid-November 1966, so that seeds would
be covered by fresh leaf fall. Twenty acres
with patchy reproduction left unburned for
planting; seedlings planted in openings at rate
of 1,000 per acre in December.

The direct-seeded area had 6,700 seedlings
per acre and 98 percent milacre stocking initially. It is not known how many of these originated naturally. First-summer mortality was
severe, reducing the stand to 1,800 pines per
acre on 62 percent of the sample plots. Natural
seeding in the second year established 1,900
new trees per acre, of which 50 percent died
in their first year. The stand was further
reinforced by seedfall the next year. At the
start of the third growing season there were
1,600 two-year-old, 950 one-year-old, and 950
juvenile pines per acre. Total milacre stocking
was 83 percent. The third summer, 1969, was
extremely dry and hot, and losses reduced the
stand to 900 three-year-old and 100 two-year-
old seedlings per acre; 40 percent of the milacres were stocked. Hardwood sprouts about
5 feet tall completely dominated the site and
overtopped the pines, which averaged 26 inches
in height (fig. 5).

Planted seedlings averaged 392 per acre and
38 inches tall at age 3 years. Most were free to
grow, for hardwood sprouts on this unburned
area were less numerous and less vigorous than
on the one burned and direct-seeded.

6

Figure 5.—*After a poor burn, hardwoods sprouted vigorously and overtopped young pines.*

After the fourth growing season, a cursory examination revealed that nearly half the seeded pines had died from intense hardwood competition, no new ones had seeded in, and the brush was getting progressively worse. Most planted trees were still thrifty and relatively free of hardwoods.

Test 11.—Montgomery County, Texas; 1967. —Gently rolling terrain, fine sandy loam A horizon ranging from 30 to 40 inches deep, a moderately productive site. Well-stocked loblolly pine stand that averaged 16 inches d.b.h. was clearcut in 1966. A single-drum chopper was pulled over area once, but dense brush and large pine tops prevented adequate site preparation. Fourteen acres were direct-seeded and 76 acres planted with nursery stock in February.

Direct seeding produced an unacceptable yearling stand averaging 350 trees per acre and 31 percent milacre stocking. Failure was due to high summer mortality on the sandy soil, which dried out rapidly after every rain.

Of 1,086 seedlings planted per acre, 496 remained after the first year; they averaged 20 inches in height.

Test 12.—Walker County, Texas; 1967.—The area was clearcut in the early 1960's, but precise history of the stand is unknown. Terrain rolling to hilly, with a loamy fine sand surface soil about 30 inches deep. Small hardwood brush chopped in early 1966 and burned late in the year, providing an excellent seedbed. In February 1967, 400 acres sown by helicopter and 160 acres planted by machine.

Seeds germinated well, but early mortality was heavy. At the end of the first year there were 550 seedlings per acre and 27 percent milacre stocking. Measurements were discontinued because of the low stocking.

Planting was successful. Only 10 percent of the seedlings died during the first year, leaving 537 stems per acre that averaged 16 inches tall.

Test 13.—Tyler County, Texas; 1968.—Flat terrain; loamy sand surface soil from 12 to 24 inches deep underlain by an impervious clay loam. Early history of area unknown. Dense brush averaging 1 inch in diameter and 8 feet tall dominated the site after pine was clearcut in 1966. Area chopped in July 1967 and large hardwoods killed by injection. A burn in September left an excellent seedbed. In February 1968 loblolly pine was direct-seeded on 40 acres, and 560 acres were machine-planted with 650 slash pine *(P. elliottii elliottii* Engelm.) seedlings per acre (slash was planted through an oversight).

Initially, the direct-seeded stand had 5,100 seedlings per acre distributed on 89 percent of the milacres. After 3 years there were still 2,600 seedlings per acre and 81 percent milacre stocking. Early growth was disappointing, probably because of the deep, sandy, topsoil. Average heights were 21 inches for all trees and 29 inches for the tallest on each milacre.

Mortality of planted pines was 24 percent, leaving 496 seedlings per acre at the end of the third year. Here, too, growth was slow, with 3-year-old slash pines averaging 57 inches tall.

Test 14.—Tyler County, Texas; 1968.—The final installation was contiguous to and in conjunction with test 13. Soils, terrain, and cover on the two areas were similar. Site prepared by a single pass with a heavy wildland disk in summer of 1968, uprooting most of the brush but leaving the site rough for planting (fig. 6); hardwood sprouting was a minor problem. Loblolly pine was direct-seeded on 29 acres, and slash pine planted on five acres in February 1968.

There were 6,400 direct-seeded pines per acre at the initial inventory, and all sample milacres had one or more seedlings. Mortality through age 3 was 44 percent, leaving a stand of 3,600 seedlings per acre and 89 percent milacre stocking. Average height was 29 inches;

Figure 6.—*A heavy wildland disk left the site in excellent condition for seeding, but rough for planting by machine.*

tallest trees on each milacre averaged 39 inches. Growth was apparently stimulated somewhat by disking, since dominant trees were about 10 inches taller than on the chopped and burned site.

After 3 years, planted trees averaged 446 per acre and 60 inches in height (fig. 7). Cattle had congregated on the planted area and killed some trees by trampling.

Discussion

Criteria for judging success or failures of pine regeneration are difficult to define. They vary by such factors as species, site, and land-owner objectives. However, 50 percent milacre stocking at the end of the first year and 40 percent after 3 years are generally considered the lower thresholds for acceptable stands. Upper limits of satisfactory stocking are based on number of trees per acre rather than distribution of seedlings, and here the arbitrary figure of 2,500 seedlings per acre will be taken for discussion purposes. Stocking in excess of this figure may require precommercial thinning. The same criteria for success or failure should apply to both artificial and natural seeding.

Four of the 14 direct-seeding tests had less than 40-percent milacre stocking after 3 years (three of these were ruled unsatisfactory after

Figure 7.—*Three-year-old slash pine planted on a disked site. Disking retarded hardwood incursion.*

the first year). Test 4 failed largely because of excessive cattle trampling. Since recommendations for direct seeding have long specified protection from concentrations of livestock, this must be judged as a human error rather than a failure of the technique. Failures also occurred in tests 11 and 12. Both sites have deep, sandy soils that dry out fast on the surface, and moisture for sustained germination is inadequate. Both sites would normally be rejected as seeding chances, but were included to test a broad array of soils.

Third-year stocking exceeded 2,500 seedlings per acre in four direct-seeding tests. In two trials, however, the stands averaged only 100 more than this upper limit. The highest stocking was 4,200 seedlings per acre in test 5, where the sowing rate was 1½ pounds per acre. The other overstocked stand (test 14) was sown on a disked bed where conditions were excellent for germination. With prescription sowing, in which the rate is adjusted to soil, cover, and climate, most of the excessively stocked stands could have been avoided. Such mistakes are eliminated as landowners gain experience with the technique.

Considering the mistakes that were made, only one direct-seeding failure (test 8) can be attributed to the method.

In eight trials of natural seeding, stocking was inadequate twice. Test 7 was ruled a failure at age 3 because stocking was marginal and most seedlings were overtopped by brush. The other failure, in test 8, resulted from a poor seed crop the first year. Excessive stocking was obtained in three tests, but one of these had only 2,600 seedlings per acre. Two of the trials were overstocked from continued reinforcement of a satisfactory initial stand by subsequent seed crops.

The greatest shortcoming of natural seeding was failure to obtain adequate stocking in the first year. In four of the eight tests, two to three seed crops were needed. Delays give brush an advantage over the young pines, and also retard production by 1 or 2 years. Even with delayed establishment, however, natural regeneration was more successful than is generally recognized in the West Gulf Coastal Plain.

Planting success is also elusive to define. Cooperators in these tests agreed, however, that 400 or more trees per acre at age 3 years were satisfactory. Planting was successful in every trial, even on the driest sites. The range in stocking was from 392 to 548 seedlings per acre. Planted pines consistently outgrew those from seeding, with differences averaging about 1 year's height growth at age 3.

As shown by these tests, direct seeding is preferable to natural regeneration. Reforestation from seed depends mainly upon the right amount of seed, at the right time, on an acceptable site, and on a well-prepared seedbed. Direct seeding can provide all of these prerequisites, while seed trees can assure only two of them.

Prescription seeding allows land managers to adjust sowing rates to site requirements. Sowing can be timed, within limits, to minimize the period in which seeds and seedlings are exposed to predators and bad weather, to assure ample soil moisture for rapid germination, and even to allow additional time to complete site preparation. Dependence upon seed trees forfeits such control.

Planting nursery seedlings is the most reliable technique, and most landowners prefer it when feasible. But planting is often too slow to complete a job in the allotted time, it is expensive, and it is impossible to use under some site conditions.

Results from this study show that loblolly pine can be successfully regenerated by all three methods. But it is unwise to make snap decisions on which is best. Among other things, the landowner must consider probable weather conditions, site conditions and acreage to be reforested, availability of labor, and economic factors including the premium on speed in getting a new stand established.

Long-Term Contracts
For

U. S. Department of Agriculture
Forest Service Research Paper SO-85

nber

Polymorphic Site Index Curves
For Shortleaf Pine
In the Ouachita Mountains

David L. Graney and Harold E. Burkhart

Southern Forest Experiment Station
Forest Service
U. S. Department of Agriculture

1973

Polymorphic Site Index Curves For Shortleaf Pine In The Ouachita Mountains

David L. Graney and Harold E. Burkhart [1]

Polymorphic site index curves for shortleaf pine at ages 25 and 50 were derived from stem analysis data collected in the Ouachita Mountain Province. Curve shape was significantly related to site quality. The polymorphic curves are preferable to standard anamorphic curves for stand ages less than 30 to 40 years, especially for site indices greater than 60. The polymorphic curves produce unbiased estimates for all ages and site qualities and they reduce estimation errors.

Additional keywords: *Pinus echinata*, productivity, growth patterns, stem analysis.

In the 5-million-acre Ouachita Mountain Province of west-central Arkansas and southeastern Oklahoma, shortleaf pine *(Pinus echinata* Mill.), the most important timber species, grows at widely varying rates on the different soils and topography on which it is found. Forest managers in the Province commonly use site index to estimate productive potential for shortleaf pine because the method is simple, easy to apply, and accurate when site index curves are reliable. In recent years, however, there has been increasing evidence that conventional harmonized site index curves for broad regions can be improved upon.

This paper presents improved curves for shortleaf pine in even-aged stands in the Ouachita Mountains. The curves given are polymorphic. That is, the shapes of the curves vary with site quality. Tree height at age 50 was the base for calculations, but curves for tree height at age 25 are also presented.

Until now, the most reliable references for computing site index of shortleaf pine in the Ouachita Mountains were conventional harmonized curves for the whole South *(12)* and adjusted curves for the Piedmont Plateau *(6)*. Since direct measurement of site index is the only method presently available for measuring shortleaf pine site quality in the Ouachita Mountains, the improved curves should be valuable to land managers.

[1] The authors are Associate Silviculturist stationed in Fayetteville, Arkansas, and Assistant Professor of Forest Biometrics, Virginia Polytechnic Institute and State University, Blacksburg, Virginia.

METHODS

The stems of two dominant or codominant trees on each of 216 plots were analyzed. The plots were located to represent the normal range of site conditions in natural even-aged shortleaf pine stands in the Ouachita Mountain Province (table 1). Stand age ranged from 45 to 89; thus, site index was either determined directly or required only short extrapolations.

Table 1.—*Shortleaf pine plot distribution*

Total age (years)	Site index class					
	33-44	45-54	55-64	65-74	75-84	Total
	– – – – – – – Number of plots – – – – – – –					
45-49	1	5	4	6	5	21
50-59	2	28	34	24	10	98
60-69	3	20	27	16	3	69
70-79	...	7	11	3	...	21
80-89	2	1	2	2	...	7
Total	8	61	78	51	18	216

Stems were sectioned at 0.5 foot, 4.5 feet, and at intervals of 4 to 12 feet up to the growing tip of the trees. Total height was recorded for each tree, and annual rings were counted at each of the sectioned points.

Ring counts were converted to tree age at each section point. Heights and ages of individual trees were then plotted and inspected for indications of suppression or top breakage. Growth values of individual trees on each plot were next combined to form one average curve per plot. Each average plot curve was adjusted for possible bias resulting from section points falling below terminal buds *(5)* and then placed in a 10-foot site index class based on tree height at age 50 years.[1] Possible bias due to correlations between stand age and site quality was compensated for by the proportional correction method of Curtis *(8)*. These adjusted plot curves were used for all subsequent computations.

The individual adjusted plot curves were then combined to form one average height-age curve for each 10-foot site index class. These curves indicated the general pattern of height growth of each site index class and were compared with all mathematically derived curves.

---

[1] Heights were extrapolated to age 50 for 21 plots with ages of 45 to 49 years.

ANALYSIS OF DATA

Several linear growth equations and one nonlinear equation were tested to determine which most nearly described the observed height growth patterns of the individual site index classes. Although the linear equations produced high coefficients of multiple determination (R^2), the curves derived from these equations did not adequately describe growth patterns. Curves derived from the nonlinear model described average height growth patterns for all site index classes.

The nonlinear model was that of Richards (10); it has been successfully applied to height growth patterns of several species (1, 2, 5, 7). The form of the equation was:

$$H = b_1 [1 - \exp(-b_2 \text{age})]^{b_3} \qquad (I)$$

where H is height at a given age and b_1, b_2, and b_3 describe ultimate tree height, rate of tree height growth, and initial pattern of height growth, respectively.

The shapes of the average height-age curves varied somewhat between high and low site classes. Each site index class, therefore, was analyzed separately, and a height-age curve was calculated for the class from its equation. These curves also indicated slightly different shapes between high and low sites. The relationships between the various coefficients (b_1, b_2, b_3) and site index were then analyzed graphically and by regression. First, the coefficients for each site class equation were plotted versus site index. The graphs for b_1 and b_2 indicated linear relationships with site index. Regression analysis indicated that both b_1 and b_2 were significantly correlated with site index. As an expression of ultimate tree height, b_1 was expected to be significantly related to site index. The significant correlation between site index and b_2, which expresses rate of height growth, indicated a polymorphic relationship between levels of site index.

The relationship between curve shape and site index indicates that shortleaf pine height growth, rather than being a function of age only (as in equation I), could be expressed in terms of age and site index. This possibility was investigated by substituting for b_1 and b_2 in equation I thus producing the five parameter model:

$$H = [a_0 + a_1 (SI)] [1 - \exp[-(a_2 + a_3 (SI)) \text{Age}]]^{a_4} \qquad (II)$$

which expresses height as a function of site index (observed height at age 50) and age. Fitting this model to the combined data for all plots produced the equation:

$$H = [20.975 + 1.2113 (SI)] [1 - \exp[-(0.012362 + 0.00013639 (SI)) \text{Age}]]^{1.0018} \qquad (III)$$

$$R^2 = 0.991$$

Standard error of estimate $= 2.14$.

3

This model has the advantage of requiring only a single equation for generation of polymorphic height-age curves. Since the model was derived from the entire data set, the curves produced should be more representative of average growth patterns than would the individual site curves derived from the data representing discrete site index classes. In the 40- and 80-foot site classes, for example, there is evidence of a skewed distribution of plot site index. Only eight plots were observed in the 40-foot class and 18 plots in the 80-foot class. Mean site indices for these classes were 42.7 and 77.2. Height-age curves derived for the discrete site index classes based on this skewed data distribution might be representative of the high or low portions of the class rather than the entire class.

Polymorphic site index curves (fig. 1)[2] were generated from equation III. The shapes of these curves agree closely with those of the average site class curves (fig. 2) for all classes and ages except 40 feet at older ages. In that class, height growth shown by equation III leveled off more with advanced age than in the average curve for site class 40.

Although a 50-year base for site index curves is most common in the study area, there is increasing interest in site curves for younger base ages. Shortleaf pine site index curves for a base age of 25 years were also generated with equation III (fig. 3). From the equation, the site indices at age 50 required to produce even 20- through 60-foot heights at age 25 were calculated. These values were substituted in equation III, which was solved for heights at ages 10 through 50.

COMPARISONS WITH CONVENTIONAL CURVES

In the Ouachita Mountain Province, Miscellaneous Publication 50 (12) and Coile and Schumacher (6) are the most common sources for shortleaf pine site index curves. The Miscellaneous Publication 50 curves were constructed from average height-age data collected from 186 stands throughout the South according to the guide curve procedure described by Bruce (3). The Coile and Schumacher curves were prepared for the Piedmont Plateau by adjusting the Miscellaneous Publication 50 curves to correct for observed overestimation of site index in young stands.

In recent years the ability of such regional harmonized anamorphic curves to accurately represent the growth patterns of individual stands has become increasingly suspect. Curtis (8) indicated that one principal source of error in the guide curve could result from correlation of site quality and age of the sample stands. A second

[2] When preparing figure 1, it was necessary to make a slight adjustment in the predicted heights so that the curves passed through site index at age 50.

4

source of error results from the assumption of a constant proportional relationship between the growth curves for all sites and stand conditions. The assumption of proportionality between curves has been proven invalid for many species *(1, 2, 4, 5, 9, 11)*.

' a single equation
Since the model
s produced should
ns than would the
resenting discrete
uses, for example,
t site index. Only
nd 18 plots in the
were 42.7 and 77.2
ndex classes based
presentative of the
entire class.

closely
sses and
t shown
t the av

ves is mo
t in site
ves for a

o p
t. These values
for heights at ag

AL CURVES
neous Publication

harmonized anami
atterns of individu
s (8) indicated thi
e could result fr
le stands. A secu
djustment in the predi
h

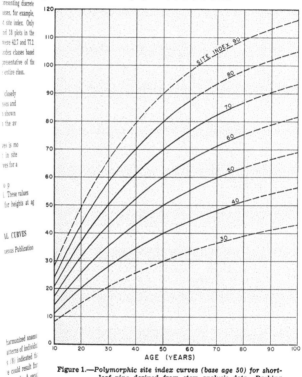

Figure 1.—*Polymorphic site index curves (base age 50) for short-leaf pine derived from stem analysis data. Dashing indicates extrapolations beyond the range of original data.*

5

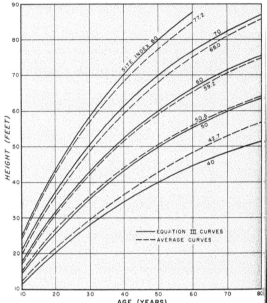

Figure 2.—*Comparison of Equation III curves and average curves for individual site index classes.*

The curves of Coile and Schumacher generally agree with the equation III curves for ages greater than 50 years, but differ markedly at younger ages at which they would give consistently lower estimates of site index (fig. 4). For example, a tree at age 20 that is 37 feet tall would have a site index of 60 according to the Coile and Schumacher curves and a site index of 70 using the equation III curves. The differences between the curves is the most pronounced for stands of age 30 or less.

Agreement between the Miscellaneous Publication 50 and equation III curves was also fairly close for all sites and ages above index age. For younger age classes the two sets of curves are very close

6

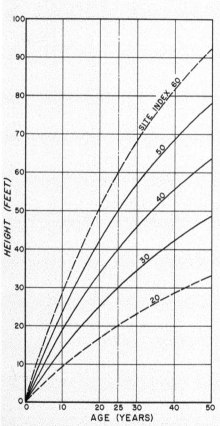

Figure 3.—*Polymorphic site index curves (base age 25) for shortleaf pine. Dashing indicates extrapolations beyond the original data ranges.*

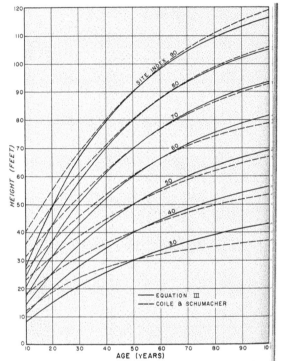

Figure 4.—*Equation III curves compared to site index curves of Coile and Schumacher.*

for poor sites, but the Miscellaneous Publication 50 curves tend to produce higher estimates of site index for sites above index 50 than do the equation III curves (fig. 5). At age 20 the difference in site index between the two sets of curves would be approximately 3 feet at site index 60, 5 feet at site index 70, and 8 feet at site index

8

nber

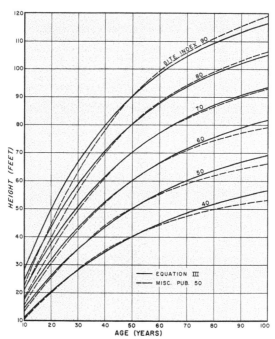

Figure 5.—*Equation III curves compared to Miscellaneous Publication 50 curves.*

e index curves of

on 50 curves tend
above index 50 f
the difference to
be approximatel
d 8 feet at site i

80. The difference between the two sets of curves becomes less with increasing age up to age 50. Due to the close agreement observed between the Miscellanous Publication 50 and equation III curves, predictive ability of each set was evaluated. Height-age data for testing the site curves could be obtained from two sources: an additional set of data from new plots or a random sample of the available plots. In the latter case, a new set of curves must be derived from the remaining data, then predictive ability of these

curves can be evaluated on the withdrawn data. Since it was not feasible to take new plots, 25 percent of the plots were withdrawn for the test. The new set of polymorphic curves based on 75 percent of the data were found to be nearly identical with the equation III curves presented in figure 1. Tree heights at ages 10, 20, and 50 (site index) were read directly from the height-age curves of each test plot. Site index values for each plot were predicted from heights at ages 10 and 20 years. In figure 6, the deviations of estimated site index from observed site index on each plot are compared for each set of curves. The deviations in 6 A and B are for estimates based on height at age 10, while 6 C and D represent the deviations of estimates based on height at age 20. Figure 6 A and C indicate that the errors of estimates for the Miscellaneous Publication 50 curves are strongly correlated with observed site index, with the magnitude of the errors increasing with increasing site index. A similar bias exists at age 30 but the magnitude of the errors is lower. The

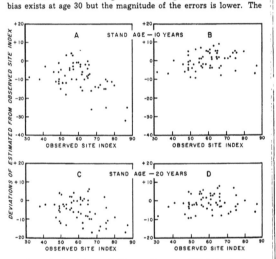

Figure 6.—*Miscellaneous Publication 50 curves yielded biased estimates of site index in stands at ages 10 (A) and 20 (C). Polymorphic curves produced unbiased estimates at these stand ages (B and D).*

Long-Term Contracts

For

nber

polymorphic curves allow for a change in growth pattern with the level of site index, produce unbiased estimates of site index, and reduce the magnitude of errors (fig. 6 B and D).

The results of this evaluation indicate that all three sets of curves produce fairly reliable estimates of site index in even-aged stands older than 40 years. However, site index estimates are most often required for younger stands. The Coile and Schumacher curves appear unreliable for all young stands in the study area. The Miscellaneous Publication 50 curves appear to be fairly reliable for young stands on poor sites, but probably produce biased site index estimates for very young stands on good sites. Since polymorphic curves derived from equation III produced unbiased and reliable estimates for all sites and ages, they are recommended for site index measurements in the Ouachita Mountain Province.

LITERATURE CITED

1. Beck, D. E.
   1971. Height-growth patterns and site index of white pine in the southern Appalachians. For. Sci. 17: 252-260.

2. Brickell, J. E.
   1968. A method of constructing site index curves from measurements of tree age and height—its application to inland Douglas-fir. USDA For. Serv. Res. Pap. INT-47, 23 p. Intermt. For. & Range Exp. Stn., Ogden, Utah.

3. Bruce, D.
   1926. A method of preparing timber-yield tables. J. Agric. Res. 32: 543-557.

4. Bull, H.
   1931. The use of polymorphic curves in determining site quality in young red pine plantations. J. Agric. Res. 43: 128.

5. Carmean, W. H.
   1972. Site index curves for upland oaks in the Central States. For. Sci. 18: 109-120.

6. Coile, T. S., and Schumacher, F. X.
   1953. Site index of young stands of loblolly and shortleaf pines in the Piedmont Plateau Region. J. For. 51: 432-435.

7. Cooper, C. F.
   1961. Equations for the description of past growth in even-aged stands of ponderosa pine. For. Sci. 7: 72-80.

8. Curtis, R. O.
    1964. A stem-analysis approach to site-index curves. For. Sci. 10: 241-256.

9. King, J. E.
    1966. Site index curves for Douglas-fir in the Pacific Northwest. Weyerhaeuser For. Pap. 8, 49 p. Centralia, Wash.

10. Richards, F. J.
    1959. A flexible growth function for empirical use. J. Exp. Bot. 10: 290-300.

11. Stage, A. R.
    1963. A mathematical approach to polymorphic site index curves for grand fir. For. Sci. 9: 167-180.

12. USDA Forest Service.
    1929. Volume, yield, and stand tables for second-growth southern pines. USDA Misc. Pub. 50, 202 p.

Polym
derimer
beyond

dex curves. For.

the Pacific North-
). Centralia Wash

trical use J Exp

oorphic site index
67-180.

for second-growth
0 202 p

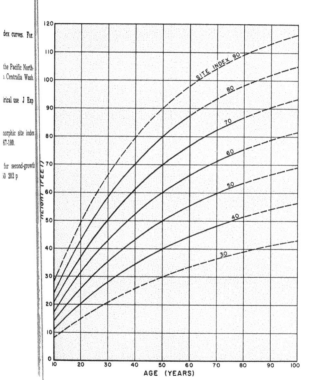

Polymorphic site index curves (base age 50) for shortleaf pine
derived from stem analysis data. Dashing indicates extrapolations
beyond the range of original data.

Long-Term Contracts

For

nber

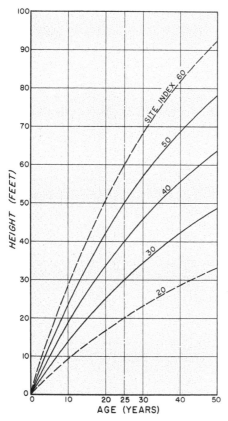

Polymorphic site index curves (base age 25) for shortleaf pine. Dashing indicates extrapolations beyond the original data ranges.

Long-Term Contracts

For

nber

Graney, David L., and Burkhart, Harold E.

1973. Polymorphic site index curves for shortleaf
pine in the Ouachita Mountains. South. For.
Exp. Stn., New Orleans, La. 12 p. (USDA
For. Serv. Res. Pap. SO-85)

Polymorphic site index curves for shortleaf pine at ages 25
and 50 were derived from stem analysis data collected in
the Ouachita Mountain Province. They produce unbiased
estimates for all ages and site qualities and reduce estimation errors.

Long-Term Contracts For

nber

U.S. Department of Agriculture
Forest Service Research Paper SO-86

18: 50-86

Field Evaluation
of
Two-Stage 3P Sampling

Dwane D. Van Hooser

Southern Forest Experiment Station
Forest Service
U.S. Department of Agriculture

1973

Long-Term Contracts

Field Evaluation of Two-Stage 3P Sampling

Dwane D. Van Hooser

Two-stage, point-3P sampling provided accurate and statistically sound midcycle estimates for 14 counties in central Mississippi. The 1967 Forest Survey was the base or first-stage sample. For a 1973 update, a 3P second-stage subsample was drawn, and predictions from it were compared to the base data. Since the volume estimates matched within 0.06 percent, the low-intensity subsample was deemed valid. Subsample trees were then remeasured in about 10 percent of the time required for the original inventory.

Additional keywords: Timber volume, timber growth, timber cut, Mississippi, forest survey.

The Forest Survey must provide estimates of timber volume in Southern States at approximately 10-year intervals, but demands to shorten the cycle have been growing. Multistage 3P samples may be ideal for midcycle volume estimates. In a computer simulation in southwestern Alabama, point-3P sampling predicted virtually the same volume as the standard cruise with a basal area factor of 37.5 (Van Hooser, 1972). What is more, 3P sample designs, which can be superimposed on existing inventory samples, promise tremendous savings in time and labor for remeasurement surveys. This paper describes a field evaluation of such a design in 14 counties of central Mississippi.

Since sampling with probability proportional to prediction (3P) was introduced by L. R. Grosenbaugh (1964), many studies have shown that 3P can be applied to various field situations. The sampling technique was designed to improve accuracy and efficiency in appraisal of standing timber for lump-sum sales; consequently, it was first applied primarily to timber sales in the West and South. Hartman (1967) found 3P to be accurate and efficient and recommended expansion of the 3P sales program to the BLM in the West. Johnson *et al.* (1967) found that 35 3P-sampled and dendrometered trees produced far better estimates than a standard cruise of 88 sampled trees. It would have been necessary, moreover, to sample 336 trees by the usual method to achieve a standard error equal to that of the 3P cruise. Hazard and Berger (1972) found that volume table biases were commonly injected where sample trees were not dendrometered.

The author thanks L. R. Grosenbaugh for assistance in this study and for the constructive comments on the manuscript.

Sharpnack (1965) found 3P to be efficient, with estimates free from detectable bias and well within the designed sampling error. He employed Monte Carlo computer simulations that repeatedly drew a set of sample-trees from a known population, and then compared the various sampling statistics. In comparisons with fixed and variable-plot methods, Bonnor (1972) found 3P to be the most efficient sampling technique in terms of variance and sampling time.

In the study reported here, the object was to evaluate point-3P sampling in midcycle inventories. The area selected was the Central Unit of Mississippi (fig. 1). The standard survey had been made in 1967, and the year of reinventory was 1973.

The above studies were all concerned with pure 3P; i.e., all elements of the population were visited but only the 3P-selected trees were measured in detail. This procedure is appropriate if the cruise is for a timber sale or condemnation proceedings, but how could 3P be applied to large-scale inventories such as those of the nationwide Forest Survey?

FIGURE 1.—*The 14-county study area in central Mississippi.*

2

nber

The question was answered when Grosenbaugh (1971) incorporated 3P techniques into multistage sampling designs. One such design is point-3P, in which the first-stage sample is selected with an angle gage and the second-stage sample is selected by 3P from guessed or measured tree heights. Thus, the second-stage sample is selected with probability proportional to D^2H. The program STX was also updated to handle a broad spectrum of multistage sampling designs.

Steber and Space (1972) found that a first-stage point sample of 406 locations would be required to determine population characteristics on an area of about 1 million acres. Only 221 3P-selected trees on 93 points were required, however, to estimate volume with a combined standard error of ± 7.1 percent. Of this 7.1 percent, 6.9 percent was attributable to the first-stage point sample.

In the Forest Survey study in southwest Alabama (Van Hooser, 1972), the standard cruise or first-stage sample required some 6,500 trees on 456 plots, while the second-stage 3P sample required only 259 trees on 170 locations. The standard cruise had a sampling error of 4.3 percent, while the point-3P cruise had a combined error of 4.5 percent. No estimates of time differential could be obtained because the point-3P sample was drawn by computer simulation.

Procedure

The Central Unit includes 14 counties which contained nearly 4 million acres of commercial forest land in 1967 (Van Sickle and Van Hooser, 1969). This acreage supported 8.5 billion cubic feet of growing stock, three-fifths of which was pine. Pine growth rates are among the best in the State; most growth takes place on sawtimber-size trees (Van Hooser, 1970). Average volume per acre is high and stands are well stocked.

In 1967, 1,056 plots were visited, of which 719 were found to be forested. These contained 10,400 trees at least 5.0 inches in d.b.h.

The procedure to select the initial 3P sample was the same as that described in detail for the southwest Alabama study (Van Hooser, 1972).

After analysis of the coefficients of variation (CV) of volume/D^2H for trees in study area, the size of the sample for dendrometry was set at 350 trees or an average of 25 trees per county. Within counties, CV for volume/D^2H ranged from 15 to 23 percent. With such a CV, a sample of 25 trees would yield a standard error of about ± 5 percent at the county level. Since the dendrometry CV for the Unit was 19 percent, expected dendrometry error would be slightly greater than ± 1.0 percent with a sample of 350 trees. This amount of variation would not increase the combined error by more than 0.2 percentage point.

An appropriate set of random numbers was generated, and a final sample of 342 trees for the unit was selected through computer simula-

NOXUBEE

KEMPER

LAUDERDALE

CLARKE

Mississippi

tion. The number of trees selected per county ranged from 16 to 34. This variation was due primarily to the proportion of total KPI (height) associated with each county. In all, the sample was only 3.3 percent of the total number of trees tallied during the first-stage prism cruise. The total 3P sample included 258 of the 719 first-stage locations (fig. 2) or only 36 percent of the initial cruise. The CV for sums of D²H among first-stage samples was about 70 percent.

FIGURE 2.—*Distribution of inventory locations in first-stage sample (top) and in second-stage 3P samples (bottom).*

4

Once the 3P sample was selected and written on tape, it had to be reformatted for processing by STX (Grosenbaugh, 1971). This step was accomplished by the programs ORDSTX and STXORD. The former uses regression to predict stump and bark measurements necessary for STX volume calculations, while the latter reformats and assigns strata and stratum weights as required by STX.

After the initial inventory was processed and summarized by STX, the sample trees were preprinted onto field record forms and the ground phase was ready to commence.

To avoid confusion in the discussion that follows, the Mississippi Forest Survey of 1967 (Van Sickle and Van Hooser, 1969) will be referred to as the standard. To insure that the 3P sample was adequate, 1967 volumes in the study area were computed from 1967 data on the 3P sample trees. The results were compared with the standard.

Results

Initial Inventory Comparisons

The point-3P sample of 342 trees predicted total 1967 cubic volume to within 0.06 percent of the standard, which had been computed by regression on 10,440 trees selected by prism and then measured (table 1). The 3P sample underestimated softwood cubic volume by 2.1 percent and overestimated hardwood cubic volume by 2.6 percent (table). These errors are much less than sampling error (table 2). At the county level, the 3P estimates of total volume varied from +6.1 to −5.9 percent. On an absolute basis, this variation was from +19.4 million cubic feet to −18.6 million cubic feet, with the same relative differences occurring by species. Variations both within counties and within species are to be expected because the sample was selected at the unit level, where species volumes tend to be proportional to the amount of the total KPI each accounts for.

In board-foot volume, the 3P estimate was within +0.7 percent of the standard (table 3). The number of trees for this estimate is less than that for growing stock because softwoods must be 9.0 inches and hardwoods 11.0 inches to qualify as sawtimber. The 3P estimates varied from the standard by +0.3 percent for softwoods and +1.5 percent for hardwoods. Again these volumes are well within the standard error for sawtimber.

Table 2, which lists the standard error attributable to each stage, shows that the 3P second-stage error does not appreciably increase overall variability. The combined error for the Unit was ±2.8 percent, while the error associated with the first-stage estimate alone was ±2.6 percent. Thus, a 1.0 percent second-stage error only increased total variation by 0.2 percentage point. The multistage error was computed

5

Table 1. *Standard and 3P estimates of 1967 growing-stock volume on commercial forest land for the 14--county test area, by species group*

County	All species			Softwood			Hardwood		
	37.5	3P	Difference	37.5	3P	Difference	37.5	3P	Difference
	- Million cubic feet -								
Attala	213.7	202.6	-11.1	99.6	120.7	+21.1	114.1	81.9	-32.2
Clarke	315.6	300.0	-15.6	190.4	160.8	-29.6	125.2	139.2	+14.0
Jasper	288.3	304.4	+16.1	187.1	198.0	+10.9	101.2	106.4	+ 5.2
Kemper	296.3	309.7	+13.4	166.8	185.9	+19.1	129.5	123.8	- 5.7
Lauderdale	352.7	334.1	-18.6	214.2	218.1	+ 3.9	138.5	116.0	-22.5
Leake	231.5	221.2	-10.3	125.7	103.1	-22.6	105.8	118.1	+12.3
Neshoba	204.2	195.7	- 8.5	110.0	104.6	- 5.4	94.2	91.1	- 3.1
Newton	210.2	214.9	+ 4.7	105.0	107.4	+ 2.4	105.2	107.5	+ 2.3
Noxubee	194.4	204.2	+ 9.8	90.4	77.5	-12.9	104.0	126.7	+22.7
Rankin	326.1	333.3	+ 7.2	176.1	193.6	+17.5	150.0	139.7	-10.3
Scott	231.7	234.8	+ 3.1	146.5	124.7	-21.8	85.2	110.1	+24.9
Simpson	173.3	163.1	-10.2	95.8	103.7	+ 7.9	77.5	59.4	-18.1
Smith	317.4	336.8	+19.4	213.9	199.5	-14.4	103.5	137.3	+33.8

Table 2. *Sampling error for each sampling stage for total volume in 14-county test area, 1967*

County	First-stage 37.5 BAF error	Second-stage 3P error	Combined error
	- - - - - - Percent - - - - - -		
Attala	10	2.4	10.2
Clarke	8	3.1	8.6
Jasper	11	2.8	11.3
Kemper	7	4.1	8.1
Lauderdale	8	3.9	8.9
Leake	8	3.6	8.8
Neshoba	10	4.8	11.1
Newton	10	4.4	10.9
Noxubee	10	4.4	10.9
Rankin	9	3.2	9.6
Scott	9	4.1	9.9
Simpson	13	4.0	13.6
Smith	10	3.8	10.7
Winston	12	4.3	12.7
All counties	2.6	1.0	2.8

by taking the square root of the sum of the squared errors for each stage; the covariance term, which is usually negligible, was ignored. Table 2 also shows that at the county level the increase in error due to the second stage is minor when compared to the magnitude of the first-stage error. It is obvious, therefore, that to decrease the overall variability the first-stage sample would have to be intensified.

Remeasurement Inventory

Of the 342 3P sample trees in the 1967 inventory, 246 trees (72 percent) survived (table 4). Eighty-one had been cut, 9 were present but dead, and 6 could not be relocated.

Table 3. *Standard and 3P estimates of 1967 sawtimber volume on the 14-county test area, by species*

County	All species			Softwood			Hardwood		
	37.5	3P	Difference	37.5	3P	Difference	37.5	3P	Difference
					Million board feet				
Attala	531.8	652.4	+120.6	306.7	549.1	+242.4	225.1	103.3	-121.8
Clarke	980.2	1,067.8	+ 87.6	723.3	650.8	- 72.5	256.9	417.0	+160.1
Jasper	953.5	944.1	- 9.4	748.0	720.0	- 28.0	205.5	224.1	+ 18.6
Kemper	799.0	1,024.1	+225.1	567.5	906.9	+339.4	231.5	117.2	-114.3
Lauderdale	1,076.4	905.8	-170.6	794.1	652.8	-141.3	282.3	253.0	- 29.3
Leake	712.1	907.6	+195.5	421.6	489.9	+ 68.3	290.5	417.7	+127.2
Neshoba	680.6	717.7	+ 37.1	453.4	577.8	+124.4	227.5	139.9	- 87.6
Newton	651.2	396.0	-255.2	350.1	138.9	-211.2	301.1	257.1	- 44.0
Noxubee	628.3	639.2	+ 10.9	404.2	429.5	+ 25.3	224.1	209.7	- 14.4
Rankin	1,011.1	810.9	-200.2	623.4	497.0	-126.4	387.7	313.9	- 73.8
Scott	763.1	676.0	- 87.1	525.0	436.1	- 88.9	238.1	239.9	+ 1.8
Simpson	482.9	546.6	+ 63.7	314.4	479.0	+164.6	168.5	67.6	-100.9
Smith	1,130.6	1,214.8	+ 84.2	919.5	730.4	-189.1	211.1	484.4	+273.3

8

Table 4. *Number of 1967 3P trees tallied in 1973 in the 14-county test area, by type*

County	1967 3P inventory	1973 3P inventory				
		Survived	Cut	Dead	Not found	Total
		- - - - - - - -Number of trees - - - - - - - -				
Attala	24	20	4	0	0	24
Clarke	29	17	12	0	0	29
Jasper	34	19	14	1	0	34
Kemper	30	21	6	1	2	30
Lauderdale	34	26	6	2	0	34
Leake	29	21	7	1	0	29
Neshoba	24	17	5	0	2	24
Newton	19	12	5	2	0	19
Noxubee	16	10	5	0	1	16
Rankin	24	20	4	0	0	24
Scott	18	15	2	1	0	18
Simpson	24	15	9	0	0	24
Smith	21	18	2	0	1	21
Winston	16	15	0	1	0	16
All counties	342	246	81	9	6	342

Growth of the 246 survivors indicated an increase of 9.0 percent in growing stock volume to 3,842.7 million cubic feet (table 5). Softwood growing stock increased 10 percent to 2,187.5 million cubic feet, while the hardwood gained 6 percent to 1,655.2 million cubic feet. Gains in sawtimber volume amounted to 1.4 million board feet for softwood and 1.5 million board feet for hardwood (table 6).

Changes in county inventory varied from —105.3 million cubic feet to +107.3 million. Accuracy of estimate within a county is affected not only by the intensity of the sample but also by the variation in number of sample trees per plot (clustering). When timber is removed from a plot—whether by cutting or beetle attack or some other cause—all sample trees on that plot are usually affected. For example, Jasper County was represented by 22 plots containing a total of 34 sample trees. Thirteen plots had one sample tree each, seven had two, one had three, and one had four. Cutting had occurred on eight plots, and all samples

9

Table 5. *Growing-stock volume on commercial forest land in the 14-county test area, 1973*

County	All species	Softwood	Hardwood
- - - *Million cubic feet* - - -			
Attala	217.8	126.1	91.7
Clarke	254.0	91.7	162.3
Jasper	199.1	125.2	73.9
Kemper	325.6	170.5	155.1
Lauderdale	402.6	276.4	126.2
Leake	242.9	92.9	150.0
Neshoba	227.5	107.3	120.2
Newton	257.6	168.9	88.7
Noxubee	162.7	121.3	41.4
Rankin	414.6	244.0	170.6
Scott	288.4	144.3	144.1
Simpson	144.0	81.6	62.4
Smith	444.1	281.1	163.0
Winston	261.8	156.2	105.6
All counties	3,842.7	2,187.5	1,655.2

had been removed from the plot containing four as well as from three plots with two samples. It is well known that the variability of an esti' mate based on cluster sampling is usually greater than that of ar estimate based on single-tree samples. To the extent that it is practical therefore, clustering should be minimized when designing low-intensity samples such as this.

The total 3P estimate of cut in the Unit is very close to the estimate of timber removals for 1966 (Hedlund and Earles, 1969) times 6, the number of years since the estimate. The early estimate times 6 is 801.'

Table 6. *Sawtimber volume on commercial forest land in the 14-county test area, 1973*

County	All species	Softwood	Hardwood
	- - - *Million board feet* - - -		
Attala	948.0	689.1	258.9
Clarke	718.2	325.1	393.1
Jasper	878.4	747.3	131.1
Kemper	1,113.4	955.7	157.7
Lauderdale	1,201.1	650.6	550.5
Leake	1,028.9	437.9	591.0
Neshoba	950.5	661.7	288.8
Newton	661.9	459.3	202.6
Noxubee	783.3	620.0	163.3
Rankin	1,122.8	653.3	469.5
Scott	1,169.3	586.4	582.9
Simpson	525.5	395.7	129.8
Smith	1,669.3	1,056.6	612.7
Winston	1,196.3	693.3	503.0
All counties	13,966.9	8,932.0	5,034.9

million cubic feet, and the 3P estimate of change is 763.8 million cubic feet. The difference is only 4.7 percent.

The final component of change—mortality—represents 101.8 million cubic feet or 3 percent of the 1967 inventory. To get average annual mortality this figure, which represents periodic mortality, must be divided by the measurement period. The result of this division is 16.9 million cubic feet. The average annual mortality for the 1967 standard survey was 17 million cubic feet. Thus, the 3P remeasurement inventory estimated mortality to within 0.1 percent of that determined by the full-scale cruise of 1967.

It should be pointed out here that the sample does not take into
account ingrowth at the time of remeasurement. During the first-stage
inventory by Forest Survey standards, only trees 5.0+ inches were
tallied. The number of stems growing into this class may be significant.
Therefore, on subsequent inventories a subsample should be taken on
every 3P location to tally only those trees which were below 5.0 inches
initially but have since equalled or exceeded this diameter.

Land Use Survey

Since the 3P sample was not designed to determine land use, an
additional survey was made to reveal shifts in land base since 1967. All
1,056 plots were field-checked to ascertain the current use. If a plot had
gone from forest to nonforest, the area associated with it in the 1967
survey was subtracted from the 1967 estimate of total forest land. If,
on the other hand, a nonforest plot had reverted to forest, its area-
equivalent was added to the 1967 forest area estimate.

The results indicate that forest area in the Unit increased 2 percent
or 96.8 thousand acres (table 7). Forest area in this Unit also increased
between the 1957 and 1967 surveys. An analysis of the change shows
that 19 plots classified as forest during the 1967 survey are currently
being used for nonforest purposes. These uses include pasture, cultivation, interstate highways, and urban development. In all, some 104,000
previously forested acres are currently classified as nonforest.

This shift from forest to nonforest is more than offset by the rever-
sion of 32 plots to forest. These plots represent more than 200,000 acres
that were classified as nonforest during the prior survey.

Eight counties had net gains in forest area; in four counties the
increases exceeded 20,000 acres. Of the remaining six counties all but
one lost forest area.

A comparison of the past and present area and volume estimates
reveals that stand density, in terms of volume, has increased from an
average of 896 cubic feet per acre (Van Sickle and Van Hooser, 1969)
to 953 cubic feet per acre, a gain of 6.4 percent.

Time Study

Field work for the entire survey was completed in 12 working days,
including a 1-day training session. To aid in assessing costs, manpower
requirements were recorded.

The field crew consisted of six two-man teams the first week and
five two-man teams thereafter. The 3P inventory took slightly more
than 12 team weeks to complete, while the 1967 standard inventory
required 106. No data are available to determine whether the 1967

12

Table 7. *Commercial forest area, 1967 and 1973 and change since 1967*

County	Commercial forest		Net change	Area shifts from:	
	1972	1967		Nonforest to forest	Forest to nonforest
			- - - - - - - - Thousand acres - - - - - - - - -		
Attala	309.3	313.2	- 3.9	12.4	16.3
Clarke	370.7	347.2	+23.5	29.1	5.6
Jasper	316.0	308.0	+ 8.0	13.6	5.6
Kemper	342.0	347.7	- 5.7	...	5.7
Lauderdale	340.2	345.6	- 5.4	...	5.4
Leake	259.1	230.0	+29.1	34.2	5.1
Neshoba	205.7	202.4	+ 3.3	7.7	4.4
Newton	268.2	245.0	+23.2	33.2	10.0
Noxubee	209.1	225.5	-16.4	...	16.4
Rankin	367.8	353.8	+14.0	25.5	11.5
Scott	265.7	261.0	+ 4.7	4.7	...
Simpson	248.0	248.0	...	...	...
Smith	276.9	278.4	- 1.5	10.2	11.7
Winston	277.6	253.7	+23.9	29.8	5.9
All counties	4,056.3	3,959.5	+96.8	200.4	103.6

survey would have required more or less time if it had been established as a two-stage point-3P sample.

The time study was in two parts. In the first, on one-half of the Unit area, the 3P portion was completed before the land-use survey was made, and times for each were maintained separately. Twenty-six team-days were required to locate and tally the sample trees on 138 plots; average daily production was 5.3 plots/team-day. An additional 9 team-days were required for the land-use survey. Total time thus was 35 team-days.

In the remaining seven counties, crews were instructed to conduct both the 3P and land-use surveys simultaneously. Here, 120 plots were cruised and 397 ground checks were made in 29 team-days. The com-bined sample thus was considerably more efficient, primarily because it avoided retravel.

As was anticipated, the majority of time was for travel. Nearly 80 percent of the total was for getting to and from 3P locations. Searching for plot center took 5.8 percent of the total time, while the actual measurement required 7.8 percent. Miscellaneous activities accounted for the remainder. It should be pointed out that on short-term projects such as this, efficiency of field crews tends to increase, especially when times are being recorded.

Conclusions

Two-stage point-3P sampling provided accurate and statistically sound estimates of volume for the Central Unit of Mississippi. The results of the full-scale 1967 survey and the point-3P predictions were virtually the same.

Remeasurement of the 3P sample, which took slightly more than 1/10 the time required to establish the first-stage point sample, revealed that the volume of growing stock had increased 9 percent. Twenty-one percent of the original stand had been cut and 3 percent died. Estimated cut, on an average annual basis, was quite close to that obtained by the 1967 survey.

The second-stage 3P sample did not appreciably affect overall variability; the increase in total error was 0.2 percentage point.

Two-stage point-3P sampling is one of the most feasible alternatives for obtaining quick and accurate estimates of midcycle volumes and components of change for Forest Survey. A comprehensive first-stage inventory with adequate samples of all the required strata must be made to provide detailed resource statistics. In the South, these surveys are conducted at approximately 10-year intervals. When a second-stage or a midcycle survey is desired, a 3P subsample can be selected at very low cost. By remeasuring this subsample and expanding it on the basis of the initial measurement frequencies, accurate and statistically sound estimates of updated volume can be obtained. This volume can serve not only as an absolute estimate at midcycle but as a benchmark for an additional updating.

The same basic technique can be repeated periodically until the original inventory base becomes outmoded. The life of the base will vary from situation to situation, but 10- and 20-year cycles with interim updating seem reasonable in the South. Of course, a drastic change in the base, as the result of some catastrophic event, would probably necessitate reestablishment in this as in any other sampling technique.

Literature Cited

Bonnor, G. M.
        1972. A test of 3-P sampling in forest inventories. For. Sci. 18 198-202.

Grosenbaugh, L. R.
   1964. Some suggestions for better sample-tree-measurement. Soc. Am. For. Proc. 1963: 36-42.

Grosenbaugh, L. R.
   1971. STX 1-11-71 for dendrometry of multistage 3P samples. USDA For. Serv. Publ. FS-277, 63 p.

Hartman, G. B.
   1967. Some practical experience with 3-P sampling and the Barr and Stroud dendrometer in timber sales. Soc. Am. For. Proc. 1966: 126-130.

Hazard, J. W., and Berger, J. M.
   1972. Volume tables vs. dendrometers for forest surveys. J. For. 70: 216-219.

Hedlund, A., and Earles, J. M.
   1969. Forest statistics for Mississippi counties. USDA For. Serv. Resour. Bull. SO-15, 24 p. South. For. Exp. Stn., New Orleans, La.

Johnson, F. A., Dahms, W. G., and Hightree, P. E.
   1967. A field test of 3P cruising. J. For. 65: 722, 724-726.

Sharpnack, D. A.
   1965. A computer trial of 3-P sampling. Soc. Am. For. Proc. 1964: 225-226.

Steber, G. D., and Space, J. C.
   1972. New inventory system sweeping the South. J. For. 70: 76-79.

Van Hooser, D. D.
   1970. Loblolly pine growth on forest survey plots in Mississippi. For. Sci. 16: 342-346.

Van Hooser, D. D.
   1972. Evaluation of two-stage 3P sampling for forest surveys. USDA For. Serv. Res. Pap. SO-77, 9 p. South. For. Exp. Stn., New Orleans, La.

Van Sickle, C. C., and Van Hooser, D. D.
   1969. Forest resources of Mississippi. USDA For. Serv. Resour. Bull. SO-17, 34 p. South. For. Exp. Stn., New Orleans, La.

15

Van Hooser, Dwane D.

 1973. Field evaluation of two-stage 3P sampling.
  South. For. Exp. Stn., New Orleans, La. 15 p.
  (USDA For. Serv. Res. Pap. SO-86)

Two-stage point-3P sampling provided accurate and
statistically sound volume estimates for 14 counties
in central Mississippi. The results of the 1967 full-scale
inventory and the point-3P predictions were virtually
the same. The 3P sample was remeasured in about 1/10
the time required for the original Forest Survey inventory. A 9-percent gain in growing-stock volume was
estimated for the study area.

Additional keywords: Timber volume, timber growth,
timber cut, Mississippi, forest survey.

Long-Term Contracts
For
mber

Long-Term Contracts
For
Forest Land and Timber
in the South

William C. Siegel

U. S. Department of Agriculture
Forest Service Research Paper SO-87

1973

Long-Term Contracts For orest Land and Timber in the South

William C. Siegel

U. S. Department of Agriculture
Forest Service Research Paper SO-87

1973

Southern Forest Experiment Station
Forest Service, U. S. Department of Agriculture

Contents

LONG-TERM CONTRACTS FOR FOREST LAND
AND TIMBER IN THE SOUTH

William C. Siegel

To help ensure a supply of raw material, the South's wood-using industries have negotiated long-term contracts with non-industrial private woodland owners. A study of the agreements of 54 firms revealed that, between 1967 and 1970, long-term contract acreage in the South increased from 6.0 to 6.7 million acres. More than half of the 54 companies have at least 50,000 acres under contracts of various types. Most of the agreements have been written in the last 20 years.

Southern pine is the most prevalent timber type on contract acreage, but there are also substantial volumes of hardwoods. Management ranges from very intensive to merely custodial. Two-thirds of the firms assume all management costs on contract woodlands.

Most agreements that provide for rental and stumpage payments contain adjustment indexes. A variety of arrangements are used to pay ad valorem taxes; to provide for losses by disaster, trespass, and theft; and to provide for condemnations and expropriations.

Additional keywords: Land leases, timber leases, long term cutting contracts.

Forest land values in the South have risen rapidly during the last two decades. In many areas external factors have supplanted timber production as the dominant consideration in pricing woodland. As a result, forest product firms striving to expand their holdings have often found prices to be beyond the payout capabilities of forestry. And large forest tracts are not often put on the open market.

Some landowners are becoming increasingly reluctant to sell because of existing or potential benefits from oil, gas, and mineral rights. Others tend to hold out for extraordinary gains in the indefinite future. Still others have no definite purpose but just do not want to give up their forest ownership. In our affluent society many landowners are free of pressing financial needs and, with out-of-pocket costs of proprietorship low, have no motive for selling.

For these reasons and others, few of the South's wood-using industries have acquired enough timberland in fee-simple ownership to supply their present and projected requirements for raw material. The problem has been partially solved by long-term contracts providing various degrees of resource control without change of ownership. These agreements, often a matter of necessity rather than choice, exist in infinite variety. They range from simple contracts between vendor and vendee to complicated arrangements that may involve lease of land; lease of timber; purchase of timber or cutting rights or both, and with or without retained economic interest; management services; and options to purchase land in fee simple. The agreements may extend to all timber on the land as of the date of contract, to trees of certain sizes at the present time or to achieve those sizes in future years, and possibly to timber yet to be grown. In recent years there has been considerable innovation in formulating long-term contracts. The changes have been motivated by a number of forestry, accounting, tax, legal, and economic considerations on the part of both timberland owners and users.

Long-term agreements hold promise for improving large segments of the southern forest resource. In the 12 Southern States 147 million acres of commercial forest (73 percent of the total) are held by 1.6 million private non-industrial owners. The vast majority of these owners under-manage their stands. Most of them have little or no technical forestry interest, knowledge, or skills. And some who have the ability are unable or unwilling to spend the time and effort required. For these persons, long-term contracts with industry will usually insure a periodic income with minimum effort and also provide good management for their holdings.

Industrial leasing of timberland in the South was initiated in the early 1930's by the Union Bag-Camp Paper Corporation (now Union-Camp Corporation) and the St. Regis Paper Company (Segur, 1960). In 1947 St. Regis began to supplement its leasing program with numerous long-term cutting contracts (Segur, 1960). Hundreds of thousands of southern acres are presently under contract to these two companies. Since the end of the Second World War, many other forest-product firms, large and small, have obtained leases and cutting agreements.

Today, close to 7 million acres of privately owned nonindustrial woodland are under long-term agreements in the South. This acreage is more than one-fifth of that owned by firms using these agreements. Thus long-term contracts have come to represent an important raw-material source, and supplement to land ownership, for many timber products companies.

In 1967 the Southern Forest Experiment Station sent questionnaires to primary wood-using firms drawing timber from the Southern States. Included were all multiproduct companies, all pulp and paper mills, all veneer and plywood manufacturers, sawmills with an annual output of at least 5 million board feet of lumber, and some miscellaneous-product firms. More than three-fourths of the companies responded—including virtually all of the large ones. A total of 92 respondents reported having long-term leases or cutting contracts of some type. Long-term was defined as 10 years or more. The basic statistics were published in 1968 (Siegel and Guttenberg, 1968).

THE STUDY

Of the 92 firms identified in 1967, a total of 64 reported 10,000 or more acres under con-

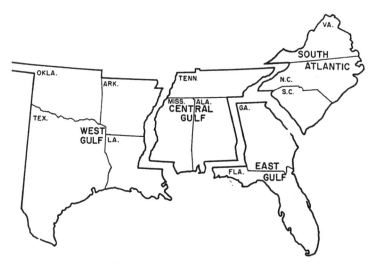

Figure 1.—*Forest regions of the South.*

Most existing contracts are recent. More an half of the 54 firms have made such arngements only since 1955. Just 10 entered o their first agreements before 1946. The cent increases appear related not only to the ircity of suitable land for purchase, but also the large expansion in wood-processing faities during the last decade. This period :nessed many mergers and purchases of exing firms.

The 6.7 million acres reported in 1970 are itrolled by 2,191 contracts for an average 2,725 acres each. Although the number of eements per firm averages 41, the range is m one to 304. As shown in table 1, more n half the companies had 10 or less. Only had more than 100.

tract Category

irtually all long-term agreements are writon an individual basis; that is, the particucircumstances of each situation are incorited into the document. However, most of

Table 1.—*Range in number of long-term agreements per firm*

Agreements (number)	Firms	
	Number	Percent
1-5	23	43
6-10	7	13
11-20	5	9
21-50	8	15
51-75	4	7
76-100	2	4
101-200	3	5
Over 200	2	4
Total	54	100

the agreements do fit into one of the broad categories shown in table 2.

Lump-sum agreements.—Slightly more than a third of the contracts stipulate a lump-sum payment that covers both land rental and timber purchase. Sometimes the entire payment is made at the beginning of the agreement, and sometimes it is made in two to five equal annual installments during the first few

3

Type of agreement	Number	Percent of total	Acres	Percent of total	Average contract acreage
One lump-sum payment that covers both land rental and timber purchase for term of contract	737	33.6	137,033	2.3	186
Specified timber cutting rights with payment on a volume basis as cut. No lease of land or timber. Lessee has management responsibilities	526	24.0	1,281,495	21.5	2,436
Same as above but with no management responsibilities	31	1.5	1,120,126	18.8	36,133
Lease of land plus lump-sum purchase of timber	230	10.5	861,823	14.4	3,747
Lease of bare land	71	3.2	190,487	3.2	2,683
Lease of land plus timber cutting rights with timber payments on a volume basis as cut	71	3.2	1,007,376	16.9	14,188
Lease of both land and timber with no other payment when timber is cut and with stipulation as to standing timber volume to be returned at contractor's end	274	12.6	520,058	8.7	1,825
Same as above but with no stipulation as to volume to be returned	12	.5	19,025	.3	1,585
Share-crop contract—company manages the tract, sells the timber to itself or on open market, and shares proceeds as agreed with landowner	8	.4	21,433	.3	2,679
Combinations or variations of preceding	231	10.5	812,506	13.6	3,517
Total	2,191	100.0	5,971,362	100.0	2,725

years. The timber to be returned at the end of the contract ranges from none to an amount equal to the original volume. Sometimes the provision for return of timber is expressed as a certain number of stems per acre. Sometimes, also, the lessee is required to prepare and plant the tract to trees before returning it. Many of the older contracts are of the lump-sum type. As landowners have become more sophisticated, however, and as land and timber values have risen, contracts more closely tied to changing economic conditions have been insisted upon. Although lump-sum agreements are still the most numerous, they represent only 2.3 percent of the total contract acreage and the average tract size is only 186 acres.

Cutting contracts.—Nearly 2½ million acres are controlled by long-term cutting contracts, making this type of agreement the most important in terms of acreage. Contracts with the lessee having management responsibilities account for almost one-fourth of the total agreements and slightly more than one-fifth of the total acreage. Similar arrangements, but with no management functions on the lessee's part, numbered only 31 but accounted for 19 percent of the total acreage. Thus the average contract in this subcategory is tremendously large—36,133 acres or 15 times the size of the average holding under cutting contracts with management responsibilities. More than 70 percent of the woodland under cutting contract is in the Central and West Gulf States. Only 300,000 acres are controlled by this means in Virginia and the Carolinas.

The landowners generally sell the timber growth at an agreed-upon unit price as it is removed. Alternatively, they may sell the right to remove a given quantity of timber with payment also made on a volume basis as cut. There is no lease of either the land or the timber. Many cutting agreements assure an annual minimum payment which may be designated as an advance or "cord credit" against future cuts. Thus the landowner is assured some annual income. In many instances, the initial volume of timber, unless independently purchased by specific lump-sum principal payment, must be returned with the land when the contract expires.

Land leases with timber purchases.—Three general types of agreement involve lease or rental of the land, coupled with various stipulations as to how payment for the timber is to be made. Under each type the lease payments are usually made annually, but some are made periodically at lesser intervals.

The most common contract in this category is lease of the land coupled with an initial lump-sum purchase of the timber. About 10 percent of the agreements—covering 14 percent of the acreage—are of this type. Such leases usually are not as strict as others in regard to the conditions of stocking when the land is returned to the lessor. In some instances the tracts can be clearcut just prior to lease termination. Since the socking is initially purchased outright rather than rented, the periodic payments are normally smaller than if both assets were rented. The basis for adjustment usually is the prorated principal and the interest credit accruing on the value of the purchased timber.

The other two general types of land-lease contracts each comprise 3 percent of the South's long-term agreements. These are: lease of bare land for timber-growing, and land leases coupled with timber payments to be made on a volume basis as cut. The former cover less than 200,000 acres. The latter, however, control more than a million acres, with the average contract acreage exceeding 14,000. Again, provisions regarding cutting rights and returnable timber differ widely. The bare-land leases, of course, involve initial stocking by the lessee.

Lease of both land and timber.—Contracts that provide for lease of both the land and the timber, with no other payments made when timber is cut, number 286—about 13 percent of the South's total. However, only slightly more than a half million acres, or 9 percent of the total, are controlled by this means. The average contract is for less than 2,000 acres. The most prevalent arrangement is for the lessee to have cutting rights to the growth and to return the original volume to the lessor at the close of the contract. Thus the timber land as a whole is actually rented, and in effect treated as a factory, with only the production being utilized. These contracts, which are perhaps the most sophisticated in terms of provisions and conditions, are used only by a small number of the largest firms—principally in the Southeast. All but 12 of the 286 reported have some stipulation as to the volume

of timber to be returned when the agreement ends.

Nearly half of the acreage under the five types of leases—as opposed to the other types of long-term contracts—is in Florida and Georgia. Another one-fourth is in Alabama, Mississippi, and Tennessee.

Share-crop contracts.—Only eight share-crop contracts were reported, all written by three companies in Louisiana and Tennessee. Under this type of agreement, the company manages the tract and has the option of making timber sales either to itself or on the open market at prevailing stumpage rates. The proceeds are then shared as agreed upon with the landowner. The first contract of this type was written in 1953.

Other contracts.—About 10 percent of the South's long-term agreements, covering more than 800,000 acres, do not fit into any of the categories that have been described. They are principally combinations of one or more types, but some are extreme variations of a particular category. The situation was typified by one woodland manager who said, "Our contracts are highly variable, and it would be next to impossible to separate them into meaningful categories."

Contract Acreage

What minimum acreage are the companies willing to put under contract? Almost half—25 firms—reported that they set no limit. They look at each tract as a separate entity and determine whether it can fit into their plans. Some of these companies have agreements for tracts of less than 200 acres.

The other 29 firms have set a minimum tract size below which no contract can be written regardless of how favorable other factors are. The range for these minimums is shown in table 3. Under optimum conditions, all but three of the 54 companies will consider holdings of 1,000 acres or less. And 38 of these take tracts smaller than 500 acres. Usually a woodland of less than 500 acres is placed under contract only if its productive potential is very high and it is near one of the firm's mills or else contiguous to other lands owned or controlled by the company.

2.3

21.5   2,436

18.8   38,132

14.4   3,747

2.2   2,683

16.9   14,108

8.7   1,826

3   1,585

3   2,878

13.6   3,517

2   100.0   2,725

subcategory is tr
acres or 15 times th
g under cutting con.
esponsibilities. Man
odland under cuttin
ral and West Gul
es are controlled b
the Carolinas.

ally sell the timb
n unit price as it i
they may sell ti
antity of timber wit
volume basis as co
r the land or the ti
ments assure an a
which may be desi
"cord credit" agai
andowner is assu
many instances, t
unless independen
p-sum principal p
with the land wi

er purchases.—Th
ient involve lease
ed with various st
t for the timber b
ported

5

Table 3.—*Minimum tract size required for long-term contracts*

Acres (number)	Firms	
	Number	*Percent*
No minimum	25	46
Less than 100	2	4
100-250	5	9
251-500	6	11
501-1,000	13	24
1,001-2,000	1	2
2,001-3,000	1	2
3,001-5,000	1	2
Total	54	100

The range in total acreage for the 54 firms is shown in table 4. Only nine companies reported less than 10,000 acres under contract. Some of these had reported more than 10,000 in 1967 and thus were included in the current study. Others of the nine had just begun utilizing long-term agreements and were not in the 1967 group. More than half the companies have 50,000 or more acres under contract. Generally, those with the most acreage are pulp and paper manufacturers in the Southeast.

Table 4.—*Range in total acreage under contract per firm*

Acres (number)	Firms	
	Number	*Percent*
Less than 10,000	9	16
10,000-24,999	5	9
25,000-49,999	10	19
50,000-99,999	10	19
100,000-199,999	13	24
200,000-499,999	6	11
Over 500,000	1	2
Total	54	100

Contract Length

About two-thirds of the companies have set a minimum length for their agreements, but only two out of five have set a maximum (table 5). A 10-year minimum, specified by one-fourth of the firms, is the most prevalent. The maximum used most often is 99 years—specified by nearly one firm in five. Some companies will waive their minimum if the terms include a purchase option.

Acreage	Contract term, in years										Totals	
	10-20		21-40		41-60		61-80		81-100			
	No.	Percent	No.	Percent	No.	Percent	No.	Percent	No.	Percent	No.	Percent
Less than 500	383	31.8	50	4.1	63	5.2	24	2.0	686	56.9	1,206	100.0
500-999	56	16.2	42	12.1	185	53.5	35	10.1	28	8.1	346	100.0
1,000-1,999	64	29.9	37	17.3	79	36.9	26	12.1	8	3.8	214	100.0
2,000-4,999	25	14.7	43	25.3	72	42.4	27	15.9	3	1.7	170	100.0
5,000-9,999	14	16.9	15	18.1	32	38.6	19	22.9	3	3.5	83	100.0
10,000 +	29	26.6	9	8.3	31	28.4	28	25.7	12	11.0	109	100.0
Total	571	26.8	196	9.2	462	21.7	159	7.5	740	34.8	2,128	100.0

stocking, and soon reach economic maturity. They can then be clearcut and the land returned to the owner. Shorter terms are possible for large natural stands, particularly those of 2,000 acres or more, because they tend to be self-sufficient units.

THE TIMBER AND ITS MANAGEMENT

Timber Types

Nearly two-thirds of the southern acreage under long-term contract can be classified as pine type (table 7). Upland hardwood is next most prevalent, and is followed in order by mixed pine-hardwood, bottom-land hardwood, and open land.

All but eight firms have some pine land under agreement, and nearly 60 percent some bottom-land hardwood. Less than half the companies, however, reported mixed pine-hardwood or upland hardwood. And only one-fourth have open land under contract. Six firms reported all five forest types, nine reported four, 10 specified three, 14 have two, and 15 have only one type.

Hardwood Contracts

A total of 156 hardwood contracts—7 percent of the total, and including both upland and bottom-land—were reported. Rather surprisingly, 21 companies—nearly 40 percent—have agreements for some tracts that are exclusively hardwood. Eight of these firms, all hardwood lumber producers, limit their agreements entirely to hardwood holdings.

The 33 firms with no hardwood contracts were asked if they would consider them if the opportunity arose. Twelve replied affirmatively. Most of the 21 who said no are either pine lumber producers or paper companies who require only a small percentage of hardwood pulp. The latter reported that sufficient hardwoods are obtainable from their own lands or on the open market.

The 12 firms who said yes either operate hardwood sawmills or are paper producers who use 20 percent or more hardwood pulp. Several are in both categories. Certain conditions, however, were specified as prerequisites. Most commonly mentioned were long terms,

Table 7.—Long-term contract acreage by general forest type

General forest type	Acres		Firms		Average acreage per firm
	Number	Proportion	Number	Proportion	
		Percent		Percent	
Pine (more than 50 percent pine)	3,826,726	64	46	85	83,168
Mixed pine-hardwood (50-75 percent hardwood and 25-50 percent pine)	594,559	10	24	44	24,773
Upland hardwood (more than 75 percent hardwoods)	899,846	15	¹ 24	44	37,494
Bottom-land hardwood	554,038	9	¹ 32	59	17,314
Open or nonstocked land	122,993	2	13	24	9,461
Total	5,998,162	100			

Eight firms lease entirely hardwood.

7

good sites, and extensive acreages. Four firms said, though, that average hardwood tracts would be acceptable. One specified a minimum 7-percent return.

Nonmerchantable Tracts

Fifteen companies will not contract for lands having only premerchantable lumber. These firms require either sawtimber or pulpwood-size stands at the beginning of the agreement. The other 39 will consider nonmerchantable tracts if certain standards are met. These firms generally evaluate site, stocking, location, and growth potential. One company, however, considers site only. Two firms would prefer to contract exclusively for precommercial stands.

Eleven of the 39 who will consider nonmerchantable stands limit such agreements to plantations. And one—a hardwood firm—will consider only cottonwood plantations. Another will contract only for pine plantations at least 10 years old. Most of the others will enter into agreements for plantings as young as 1 year. Three reported that they had tried but failed to negotiate plantation contracts.

The acreage of Soil Bank plantations in the South has been estimated at 1.9 million in commercial species (Kalmar, 1967). Many companies would welcome formal agreements for such plantations but few have been able to accumulate many. The most successful attempts have been in Georgia and Florida, where three pulp and paper companies reported a total of 275 Soil Bank leases. Firms in the Midsouth say opportunities are scarce. Several mentioned that many Soil Bank plantations in Alabama have been cut, and the land put into crops.

In general, leases would seem to be an attractive arrangement for plantation owners, particularly those with stands 10 years or more in age. Companies can pay well for such stands, because the period of high risk is past and the time to merchantability is short. Apparently, however, industry has been able to contract with relatively few plantation owners.

Management of Contract Lands

When the 54 firms were asked to compare management on their fee lands with manage-

Table 8.—*Regeneration treatments on contract acreage from which the commercial timber had been harvested*

Percent of acreage	Planned natural regeneration		Planted		Direct-seeded		Drained or bedded	
	No. firms	Percent	No. firms	Percent	No. firms	Percent	No. firms	Percent
0	17	32	14	26	41	76	39	72
1-25	15	28	17	32	10	20	5	9
26-50	6	11	6	11	2	3	4	8
51-75	1	1	1	1	1	1	0	0
76-100	15	28	16	30	0	0	6	11
Total	54	100	54	100	54	100	54	100

FINANCIAL CONSIDERATIONS

A major concern in writing long-term forestry contracts is that of financial stipulations. How should timber—both merchantable and nonmerchantable—standing at the beginning of the agreement be priced and how should growth be paid for? What indexes are the most suitable for adjusting payments? What types of periodic financial reviews are most appropriate to protect both the company's interest and that of the landowner? What provisions will be made for payment or assumption of management costs and ad valorem taxes? What kinds of revenue sharing and reimbursement should there be if contract land is condemned or appropriated? How should losses from natural disasters, fire, or theft be shared? Equitable answers to these questions are not easy to determine.

Landowners should recognize that companies contracting for their land and timber will usually have to make sizable investments to get the stands into full production—particularly during the critical first 10 years. This is especially true for long-term management leases as opposed to nonmanagement cutting contracts. At least 35 years are required to justify a program of intensive management. The financial risks are largely one-sided, since many years usually pass before a substantial payoff can be realized.

Rental Payments for Land and Timber

The 36 companies with agreements that embody periodic rental payments for land or timber, or both, were asked to discuss the bases for adjustments as economic conditions change. The existing rental contracts of 13 of these firms have no provisions for adjustment; the payment negotiated at the outset prevails throughout the term of the agreement, which sometimes is for 99 years. With a few exceptions, these 13 companies have little acreage under contract and most of their agreements are rather old. Representatives of several firms commented that any rental contracts negotiated today would have to contain an escalation clause, since landowners have been acquainted with the effects of inflation.

The 23 producers who do have contracts with rental escalation clauses utilize various indexes. Nineteen of them, however, base their adjustments for both land and timber rentals on the All-Commodities Wholesale Price Index issued by the Bureau of Labor Statistics of the U. S. Department of Labor.

Only two firms utilize BLS wholesale price indexes for specific products. One utilizes the index for "Paper except Newsprint, Paperboard, and Building Paper and Board," while the other relies on a consolidation of all the forest product indexes. The former is used for both land and timber rentals whereas the latter is applied only to land.

The U. S. Department of Labor (1967) considers the All-Commodities Index to have more statistical accuracy than the component group indexes. Perhaps this is the reason for its widespread usage in rental escalation clauses. It measures general price changes for more than 2,000 types of goods sold in U. S. primary markets. The intention is to measure pure price changes—that is, those not influenced by such factors as changes in quality, quantity, shipping terms, and product mix. The prices, insofar as possible, are FOB production point and refer to sales for immediate delivery. Whole-

9

sale price indexes are presently computed on the base that 1957-1959 equals 100. Wholesale is used to mean sales in large quantities and not necessarily prices received by wholesalers. Most of the quotations built into the index are the selling prices of representative manufacturers or producers.

Review intervals and provisions for price change vary among firms. Eight companies review the index yearly and seven of these adjust their payments proportionately for every 5-percent change. The eighth adjusts proportionately as the index changes, with no minimum time period or percentage interval required. The remaining 11 review at intervals of 3, 4, or 5 years; some adjust only in 5-percent increments, while others adjust with any amount of change.

Only one company reported that the All-Commodities Index has not proved satisfactory for rental payments. This firm's experience was that it did not properly reflect regional differences. No one index, of course, can be totally satisfactory. For one thing, stumpage values—which are closely reflected in timber rentals—are not likely to follow the index as closely as do changes in land values. However, contracts involving both types of rentals usually are written in terms of one consolidated payment for both items, and it is thus easier to use a single index for determining changes.

Seven other escalation guides are used for rental clauses, each by a different company. One is based on 10-cent increments in the average annual delivered price of pulpwood at the firm's mill; another is the BLS Cost of Living Index; another utilizes changes in the price of pulpwood per cord FOB railcar; the fourth is based on changes in the wholesale price of lumber sold by the company; the fifth involves a monthly review of the FOB truckwood price at the company's mill; the sixth entails an annual review of producers' prices of pulpwood delivered to the company's yard; and the seventh is an arbitrary increase per specified period of time.

Once a firm has determined a basis for rental escalations, it seldom changes. Only two companies were utilizing more than one index for agreements current at the time of the study. One was employing four guides and the other

all lumber producers—also rely on lumber prices, but of a more local nature. Three change stumpage rates in proportion to the retail prices of lumber that they sell. The fourth uses local lumber price quotations, which it reviews quarterly.

Stumpage prices are utilized directly by eight firms. Two adjust according to the "Louisiana Timber Products Quarterly Market Report." This report, issued by the State Department of Agriculture, covers a great variety of species and products. While the data are not necessarily typical for the South as a whole, they are nevertheless serviceable since they move up and down as economic conditions change. Four of these companies use local stumpage price quotations; another adjusts quarterly according to price variations in local USDA Forest Service sales; the eighth has an index based on the average open-market stumpage prices that it pays, with a specified minimum.

Two firms have developed an index based on percentage changes in the FOB price of pulpwood on railcars. Two others annually review FOB prices of timber delivered to their mills and adjust stumpage values proportionately. One producer uses the BLS Cost of Living Index and another has contracts with set increases for specified period of time.

Only four companies presently use more than one index for adjusting stumpage payments. Three have two indexes and the other four. Seven firms have adopted the same index for changes in both stumpage prices and rentals.

Management Costs

Two-thirds of the companies assume all the costs of management on their contract acreage. Payments to landowners are, of necessity, adjusted to reflect the extent to which management expenses are incurred by the company. In most instances the cost of management to the landowner will ultimately be less when the entire expense is borne directly by the contracting firm. The company can usually do the work more cheaply, since it can consolidate operations on both fee and contract acreage and can schedule work at cost-feasible times. Overhead can be kept constant, and men and machinery can be utilized efficiently.

Nevertheless, all the contracts of six companies require the landowner to assume the entire management cost. And four firms have some agreements with this stipulation.

The contracts of eight others all provide for cost-sharing. One firm pays every management cost but that of boundary surveys. Another pays all except road construction and maintenance. A third requires the landowner to pay site-preparation and road costs. Still another assumes only supervisory wages, and another only cruising and marking expenses. A sixth pays all costs except those of labor. The remaining two pay for all management as stipulated in the agreement, and the owner assumes the cost of any additional work that he undertakes.

Only one of the companies with contracts calling for owner assumption of management costs reported having a basis for adjustment as economic conditions change. This firm deducts an annual per-acre management fee from the payments it makes to lessors. The fee is adjusted annually in accordance with the average per-acre costs of management on the company's fee lands.

Ad Valorem Taxes

As with management expenses, there is little cost-sharing of ad valorem taxes. The contracts of 21 firms stipulate that the company will pay all such taxes; those of 15 others require the landowner to pay. Both types are being used by 12 firms. Contractual payments to landowners will, of course, usually reflect the allocation of taxes among the parties.

The other six manufacturers share land taxes with the owners. Some companies pay all ad valorem taxes after the first year. Others pay either the first 50 cents or $1 per acre, with the excess either assumed entirely by the owner or shared equally. In still others, the firm continues to pay the rate in force at the beginning of the contract and the owner absorbs either half or all of any subsequent increases. In some instances, taxes are shared equally. In a few contracts the firm assumes 95 percent and the owner 5 percent. Several stipulate that the company will pay the portion that exceeds a certain percentage of the annual land rental. In other agreements, the company will

11

pay the land taxes if it is given the hunting rights to the acreage.

A number of officials spoke of rapidly rising ad valorem taxes on contract acreage. They charged that some assessors deliberately raised assessments when it became known that land was under long-term agreement with industry and that the taxes were to be assumed by the company. The situation was termed most critical in Georgia, Texas, and Florida.

Natural Disasters, Trespass, and Theft

All but two companies stipulate who is to absorb timber damage and losses resulting from natural disasters, trespass, or theft. The agreements of 22 firms specify that the company will, in effect, bear all such losses—it is responsible to the landowner for all payments that he would normally receive if the loss had not occurred. These payments, of course, are adjusted for the landowner's recovery of any insurance proceeds or other compensation.

The contracts of nine companies require the landowner to bear all losses. Eight other firms have some of each type of agreement—that is, the entire loss is absorbed by the landowner in some contracts and by the company in others.

The remaining 13 producers utilize a variety of loss-sharing provisions. Some stipulate that they will bear all losses, but that if a disaster occurs during the first 10 years they are relieved from requirements relating to the timber volume to be left at the end of the agreement. Others also provide for company assumption of damage and losses, but state that after natural disasters the firm has the option of lowering the minimum cutting diameter by 2 inches and the stumpage price by $2 per thousand board feet. Two companies absorb only losses that are associated with timber after it has been cut. One firm assumes only fire losses and damage. Another company will bear the loss if less than 30 percent of the area is affected—otherwise the landowner assumes it all. And still another firm has the option of deducting specified proportions of losses from timber rent.

Condemnations and Expropriations

The contracts of 12 companies make no provision in the event of condemnation or expro-

agreements except cutting contracts. And still
another offers maximum payments only if it
receives a purchase option. A number of company spokesmen commented that many well-
informed landowners do not want purchase options in their agreements and, even if there
is one, do not want to sell. Of the 21 companies
who have no purchase options, all but two
would include such a stipulation if they could.

At the time of the study one firm had exercised all of its options, and eight had made
some purchases during the previous 10 years.
Acreage thus acquired ranged from 40 to 9,400
per firm but totaled only 23,000.

Adverse State Laws

Only two State laws were reported by company spokesmen to be deterimental to the formation of long-term forestry contracts. One
was Alabama's privilege tax, which is similar
to a license to do business and increases with
the value of the agreement. The representative of one firm said that plans for a large lease
had been cancelled because of the amount of
this tax. The other instance concerned the
Georgia State income tax laws and regulations,
which generally hold that income to lessors
under long-term leases is ordinary income even
though it may be a capital gain for Federal income tax purposes (see Williams v. Superior
Pine Products Co., Georgia Court of Appeals,
March 10, 1958).

INFORMAL AGREEMENTS

Numerous firms in the South work with
owners of small woodlands under informal,
verbal agreements—sometimes referred to as
tree farm family agreements. These usually
provide the company with first refusals of the
stumpage at prevailing rates. In a 1961 study
(Whaley and Guttenberg, 1962), at least 14
Midsouth firms were found to have such agreements. It was concluded that the arrangements
generally worked well and that the woodlands
were being managed substantially better than
"nonclient" lands.

The 54 companies with formal long-term contracts were asked if they also utilize informal
tree farm family agreements. Twenty-four said
yes, and one other planned to start this program soon. Another wanted to begin, but had
not received authorization from its home office.

Of the 30 firms who avoid these agreements,
however, most give free management advice
to landowners.

Twenty-three of the 24 users reported a
total of 994 informal contracts at the time of
the study, for an average of 43 each. Acreage
was reported by only 11 companies. The total
was 357,880 acres for an average of 32,535 per
firm and 848 acres per tract.

Only 13 of the 24 companies with informal
agreements are seeking more. The 11 who are
dissatisfied prefer formal contracts and feel
that the others have been too time-consuming
and troublesome for the returns. Several reported that many owners have backed out and
thus created supply problems for the firm.
The firms who would like additional informal
arrangements, however, reported generally
good success with them.

PLANS FOR MEETING TIMBER
REQUIREMENTS

Resource requirements of the 54 companies
are only partially met from long-term leases
and cutting contracts. In 1967, two-thirds of
the leasing firms drew timber from their contract woodlands and for most of these firms this
supply amounted to less than 20 percent of
total wood requirements. And only four of
them cut more than 60 percent of their wood
from leased acreage. Much of the woodland
is being developed for future supplies.

Lands under long-term cutting contracts
were drawn on more heavily than those under
lease. In 1967, more than nine out of 10 firms
with such contracts utilized them for part of
their timber supply, and more than two-fifths
received upwards of 40 percent of their raw
material from this source.

Forest Purchase Plans

Of the 54 companies, 35 stated that they were
actively seeking more purchases of timberland.
Most of these would eventually like enough
fee acreage to meet at least 50 percent of their
resource requirements, and a few are aiming
as high as 75 percent. One Georgia firm that is
seeking no new purchases reported that it plans
to remove all the timber from its fee lands
and put the acreage into agricultural use for
a greater return. This firm believes that its

timber requirements can be satisfied from the open market and long-term contracts.

Several manufacturers stipulated that they are only seeking purchases within specified distances of their mills. These distances range from 50 to 125 miles. Others wish to buy land only near new mill sites or to replace acreage lost through condemnation or to block out present holdings. A number of companies plan to buy all the woodlands they can, merely as an investment, and not necessarily to supply their own mills. Those who are not actively seeking more purchases generally cite high prices as the reason and thus plan to utilize long-term contracts to an even greater extent than at present.

Plans for Long-Term Contracts

When the companies were asked if they are actively seeking more long-term contracts, 28 of the 54 said yes. Many of those who answered in the negative cited legal problems with estates, elderly people, divided ownerships, and second-generation owners. Others mentioned income-tax problems associated with long-term agreements. Still others were discouraged by the reluctance of lessors to invest even minimally in their land.

The 28 seeking more contracts were asked the type they would prefer. Specified most often, and definitely preferred by a majority, were leases of both land and timber, and long-term cutting contracts with management responsibilities. Also mentioned were land leases coupled with lump-sum timber purchases, and land leases coupled with timber payments on an as-cut volume basis. One pulp and paper firm in the Southeast reported that it had turned down many good leases because management thought the money could be used more profitably elsewhere.

The 28 companies were also asked the proportion of their timber requirements they would eventually like to obtain from long-term contracts. The answers ranged from 100 percent (five firms) to 20 percent (two firms). The proportion cited most often was 50 percent; more than two-thirds of the companies are aiming for 50 percent or higher. One is seeking more long-term leases even though it presently has a sufficient land base, and foresees no problems in purchasing more if needed.

Siegel, William C.

    1973. Long-term contracts for forest land and timber in the South. South. For. Exp. Stn., New Orleans, La. 14 p. (USDA For. Serv. Res. Pap. SO-87)

By means of long-term contracts, southern wood-using industries control almost 7 million acres of privately owned, nonindustrial forest. While pine is the prevalent timber type under contract, there are also substantial acreages of hardwood. Nearly all agreements are written on an individual basis, but most fit into one of several broad categories.

Additional keywords: Land leases, timber leases, long term cutting contracts.

U. S. Department of Agriculture
Forest Service Research Paper SO-90

SO - 88

U. S. Department of Agriculture
Forest Service Research Paper SO-88

ng

Sawmill Manager Adapts To Change
With Linear Programming

George F. Dutrow

and

James E. Granskog

FEB

Southern Forest Experiment Station
Forest Service
U. S. Department of Agriculture

1973

A SAWMILL MANAGER ADAPTS TO CHANGE WITH LINEAR PROGRAMMING

George F. Dutrow and James E. Granskog

ng

Linear programming provides guidelines for increasing saw-mill capacity and flexibility and for determining stumpage-purchasing strategy. The operator of a medium-sized sawmill implemented improvements suggested by linear programming analysis; results indicate a 45 percent increase in revenue and a 36 percent hike in volume processed.

Additional keywords: Simulation, stumpage strategy, sawmill revenue, sawmill costs.

Independent sawmill owners have had to increase efficiency to survive. Now competition from new plywood plants and emerging processing complexes, rising labor costs, and growing environmental concerns demand continued improvements. Linear programming is an analytical technique helpful in meeting this challenge.

This paper shows how one independent operator prepared for the future. He travelled extensively to observe conditions and practices at mills that had installed improved equipment. Although he could eliminate some alternatives as inappropriate, he felt the need for formal analysis of those that appeared promising.

He was familiar with earlier research showing the value of linear programming for large sawmills (Row et al., 1965; McKillop and Hoyer-Nielsen, 1968; Sampson, 1972). Linear programming helped this independent operator expand his mill, evaluate potential revenues, and formulate stumpage strategies.

Operational Framework

Management must clearly define the constraints imposed by the mill and its operational environment. In this case, the sawmill is of medium size and processes mainly loblolly pine. Logs are supplied by contractors harvesting stumpage purchased on the open market by the company. Principal equipment includes a circular headsaw, a sash gangsaw, and a vertical resaw. The firm also has dry-kiln and finishing operations. Roughly 90 percent of the lumber output is 2-inch dimension. Residual material is chipped and sold to pulpmills.

The prevailing market, the geographical location of the mill, and the nature of the timber species limited production methods and possibilities for change. Market conditions favor maximum output of dimension lumber, and in this portion of the loblolly pine belt logs are sweepy. The canter-chipper, a profitable alternative in other southern locations, cannot make the most lumber from such timber.

Other limiting factors are the amount and quality of stumpage that can be purchased annually and the amount of capital that management is willing to risk in modifying a plant without a land base. Management must also evaluate additional requirements for working capital in the form of logs and lumber since these requirements can exceed the total value of the plant.

In light of these objectives and constraints, several stages of research were conducted. First, the mill was divided into major processing units, and productivity rates for each unit determined. Second, individual logs were measured before and after processing to correlate log size and values of final products. Third, a linear program model of the mill was constructed to provide current revenues and pinpoint machine limitations prohibiting increased log flow. Fourth, recommended changes were simulated under varying log input distributions to evaluate their effects. Finally, strategies for stumpage purchases were derived.

Data Collection

To gather information for the linear program, primary sawing operations were observed and time and yield studies conducted.

Mill layout is illustrated in figure 1. Tree-length timber is delivered to the log deck by forklifts. A conveyor feeds the raw material to a circular cutoff saw and then to a cambium-shear debarker. Logs are processed on two sides by the headrig chipper and circular saw; the cants are then conveyed to the sashgang, which is set to cut 2-inch dimension exclusively. When the gangsaw is down or its capacity exceeded, the headrig saws boards and timbers. Slabs from the gang are sent through a vertical resaw, and the resulting lumber plus other waney material from the gang or carriage is passed through the edger. Residual materials are chipped.

2

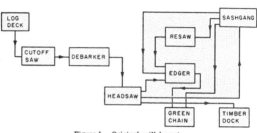

Figure 1.—*Original mill layout.*

Time Study

Machine rates were recorded to obtain equations for estimating processing requirements. Only the debarker, headrig, and gangsaw were timed since the resaw and edger units were obviously used far below their capacities.

Debarker.—The debarker and cutoff saw were treated as one unit. A batch of logs was clocked from the time the lead end of the first log reached a selected point until the log following the batch reached the same point. Batches consisted of one to three logs: diameter and length of each log were measured. The equation for estimating the time requirements of the debarker-cutoff unit is

$$Y = 7.71337 + 0.40132(L) + 11.09202(N)$$
$$R^2 = .57 \quad SE = 6.759$$

where

Y = processing time per batch (in seconds)
L = sum of length of logs (in feet)
N = number of logs in batch.

Headrig.—Processing on the headrig was measured from the time the first log of a batch hit the carriage until the first log following the batch touched the carriage. Batches ranged from two to five logs; diameter, length, and the number of cuts per log were recorded. The equation for estimating the time requirements of the headrig is

$$Y = -35.94915 + 1.10274 \ (L)$$
$$+ 5.28405(P) + 0.15390(D^2/N)$$
$$R^2 = .78 \quad SE = 18.038$$

3

where
- Y = processing time per batch (in seconds)
- L = sum of length of logs (in feet)
- P = total number of cuts on the carriage
- D = sum of the diameters (in inches)
- N = number of logs in batch.

The processing time per log can be obtained by dividing the estimated time per batch by the number of logs in the batch.

Sashgang.—Cants were timed through the gangsaw, beginning as the selected cant touched the infeed roller and ending when the following cant reached the same point. Cant thickness, width, and length were measured. The equation for estimating time requirements of the gangsaw is

$$Y = -10.31693 + 3.02561(L) + 1.83179(CT)$$
$$-0.25699(LD) + 0.01171(D^2L)$$
$$R^2 = .83 \qquad SE = 3.846$$

where
- Y = processing time per cant (in seconds)
- L = length of cant (in feet)
- CT = cant thickness (in inches)
- D = sum of diameters (in inches).

Since timing excluded stoppages, the equations represent the capacity of the respective machines. These equations were used to estimate processing times per thousand board feet of log input by diameter class and length for the log distributions used in the linear programming analysis.

Yield Study

Yields were determined for 81 logs ranging from 6 to 18 inches diameter inside bark and 12 to 20 feet long. Logs were sawn individually; volume and weight yields were recorded for each log. Boards and dimensions were tallied by width, thickness, length, and grade by a representative of the Southern Pine Inspection Bureau. Logs were weighed with and without bark, and all lumber and chippable material were weighed. Bark and sawdust weights were obtained by subtraction.

Logs were sawn to obtain the greatest possible volume of dimension lumber as this sawing pattern provides maximum output for the mills selected market.

Analysis of the yield data provided finalized regression equations for estimating weights, lumber volume, and total dollar value obtainable from a log (table 1). The products from each log were multiplied

4

by their current market prices, and the summed values supplied data for the dollar value equation.

Table 1.—*The yields*

	Equations	R^2	SE
Log weight (total lbs.)	$= 278.90935 - 39.64487(D)$ $+ 0.49885(D^2L)$ [1]	.97	96.581
Bark weight (lbs.)	$= -31.49903 + 0.52717(DL)$	.71	18.976
Sawdust weight (lbs.)	$= -65.83602 + 1.20237(DL)$	.76	38.214
Chip weight (lbs.)	$= -20.39227 + 1.01738(DL)$	.72	35.334
Lumber volume (M b.f.)	$= -0.01002 + 0.00332(L)$ $-0.00062(DL) + 0.00008(D^2L)$	.97	0.0122
Dollar value	$= 11.56702 - 1.40689(D)$	.97	2.232

[1] D = diameter in inches
L = length in feet

The Model

Linear programming is a mathematical technique for selecting the optimum level of production. The program consists of a mathematical expression of goals (objective function) and of limitations (constraints). The goal here was to maximize revenue received from sale of products. Machine rates and input volumes were the major constraints.

The linear program was formulated for a competitive firm since the mill is unable to control prices of inputs bought or products sold. Thus, with a goal of higher revenue, the only variables of choice were rates of consumption and production. Specifically, these variables included timber purchases, yields from different sawing patterns, and time availability on mill equipment.

Observations of log input, machine time, and yields were combined with product prices to formulate input-output relationships. The linear program was designed to maximize these revenue relationships in the objective function.

Table 2 presents the value coefficients for each log class when sawn or when sold instead as pulpwood or veneer bolts than processed. These values were derived by converting lumber-scale output to Doyle-rule log input. Each revenue is net of $100/M b.f. for the delivered timber.

The amount of time required to process a thousand board feet through the debarker, headrig, and gangsaw is given in table 3.

5

Table 2.—*Revenue values* [1] *for log classes—per M b.f. Doyle*

Diameter and length (inches and feet)	Revenue from dimension, boards, and chips	Revenue from pulpwood or veneer bolt sales
6 x 12	$256.67	—$30 (pulpwood)
8 x 12	315.83	— 10 (pulpwood)
10 x 12	168.89	20 (veneer)
12 x 12	131.46	20 (veneer)
14 x 12	120.27	30 (veneer)
6 x 14	220.00	— 30 (pulpwood)
8 x 14	312.14	— 10 (pulpwood)
10 x 14	177.81	20 (veneer)
12 x 14	147.14	20 (veneer)
14 x 14	134.43	30 (veneer)
6 x 16	191.88	— 30 (pulpwood)
8 x 16	309.38	— 10 (pulpwood)
10 x 16	192.22	20 (veneer)
12 x 16	159.06	20 (veneer)
14 x 16	147.30	30 (veneer)
16 x 16	142.92	30 (veneer)
10 x 18	196.34	20 (veneer)
12 x 18	168.33	20 (veneer)
14 x 18	157.50	30 (veneer)
16 x 18	151.36	30 (veneer)
10 x 20	206.22	20 (veneer)
12 x 20	175.75	20 (veneer)
14 x 20	163.60	30 (veneer)
16 x 20	158.11	30 (veneer)

[1] Net of delivered timber: $100/M b.f. Doyle.

Table 3.—*Time needed to process 1 M b.f. Doyle, by log class*

Diameter and length (inches and feet)	By debarker	By headrig	By gangsaw
	- - - - - - Minutes - - - - - -		
6 x 12	27.5	17.2	27.6
8 x 12	27.5	17.9	24.4
10 x 12	12.2	8.3	12.5
12 x 12	6.9	5.0	8.3
14 x 12	4.4	3.4	6.4
6 x 14	24.5	17.4	28.2
8 x 14	24.5	18.0	25.1
10 x 14	10.7	8.2	12.2
12 x 14	6.1	4.9	8.1
14 x 14	3.9	3.3	6.1
6 x 16	22.3	17.5	28.6
8 x 16	22.3	18.0	25.5
10 x 16	9.9	8.3	12.4
12 x 16	5.6	4.9	7.9
14 x 16	3.6	3.3	5.9
16 x 16	2.5	2.4	4.9
10 x 18	9.0	8.2	12.2
12 x 18	5.1	4.8	7.8
14 x 18	3.3	3.2	5.8
16 x 18	2.3	2.3	4.7
10 x 20	8.5	8.3	12.3
12 x 20	4.8	4.8	7.7
14 x 20	3.1	3.2	5.7
16 x 20	2.1	2.3	4.6

Maintenance-free operation time was calculated to be 55 minutes per hour for the headrig and gangsaw, and 57 minutes for the debarker. These processing and maintenance times served as constraints.

The final requirement was an estimate of the volume of each log class processed in an operating hour. These volumes were based on samples of incoming logs. The normal input distribution represents an hourly total of 6.023 M b.f. (table 4).

Tables 2, 3, and 4 provided the necessary data for optimizing hourly revenue with maintenance time, processing time, and volume constraints. Input volume was selected to just exceed the mill capacity, thus forcing the solutions to indicate which log classes should be sold rather than sawn and also which machines prohibited greater output.

Table 4.—

Normal log input pattern, by size class

Diameter and length (inches and feet)	M b.f. per hour
6 x 12	0.120
8 x 12	.120
10 x 12	.135
12 x 12	.125
14 x 12	.128
6 x 14	.210
8 x 14	.224
10 x 14	.224
12 x 14	.240
14 x 14	.194
6 x 16	.336
8 x 16	.264
10 x 16	.324
12 x 16	.320
14 x 16	.400
16 x 16	.288
10 x 18	.164
12 x 18	.432
14 x 18	.224
16 x 18	.486
10 x 20	.180
12 x 20	.400
14 x 20	.125
16 x 20	.360

Simulations

Mill Modifications

Initial linear programming solutions indicated an hourly revenue of $972.18 and marked the gangsaw as the obstacle to greater returns. When gangsaw capacity was exhausted, almost 20 minutes of unused capacity remained on the headrig and some 14 minutes on the debarker.

Simulated production with an additional gangsaw pointed to higher revenues and expanded output. Hourly revenues rose by about 31 percent to $1,279.05, and volume increased almost 13 percent to 7.4 M b.f. Unused capacity remained on the gangsaws and headrigs, but now the debarker became the limiting machine. Two alternatives were considered. The mill could add another cutoff saw and debarker complex or move the cutoff saw off line. The first alternative was rejected since it would require costly mill modifications. Reevaluation of the debarker complex showed that moving the cutoff saw off line would increase debarker capacity by about 10 percent.

7

The adjusted debarker arrangement was analyzed in a third simulation. Revenue increased to $1,414.90 per operational hour while processed volume rose to 8.229 M b.f. Relative increases were 11 percent for revenue and slightly less for log volume.

Of course, these gains in revenue cannot be viewed as increments to net profit. The cost of the second gangsaw, including additional labor to operate it, has to be deducted from revenue increases. Capacity and production boosts require substantial working capital for increased log supplies at the millyard and for additional stumpage purchases. Management weighed revenue gains against the added costs and made the suggested changes.

Certain log classes were sold by the model as pulpwood. An operational rule of thumb emerged: sell the long, small diameter logs rather than process them if mill capacity can be met by other log classes. Most of the 6-inch diameter logs were rejected by the linear-program solution as being unprofitable, even though the Doyle log rule permits a large overrun in small log classes.

Sawing Pattern Option

Although market considerations and managerial choice precluded detailed analyses of alternative sawing patterns, the option of cutting only 2 × 4's from butt logs was simulated. This option decreased revenues and absorbed more machine time; therefore, it was not a viable alternative.

Stumpage Guides

Stumpage-purchasing strategies can be derived from linear programs which calculate ranges of acceptability for the value coefficients in the objective function. Of primary interest is the low side of these ranges, i.e., the revenue value below which a given log class will be excluded from the optimal solution. Accordingly, those log classes most sensitive to falling prices are pinpointed.

Table 5 indicates the percentage decrease in prices which would make selling a log class as pulpwood or as veneer bolts more profitable than sawing it. For example, a 2 percent decrease in lumber prices would cause 6″ × 12′ logs to be sold as pulpwood; hence, the stumpage purchaser faces a very narrow price margin within which he can profitably buy small timber. On the other hand, a stand composed of larger trees provides substantially greater insulation from profit-eroding prices declines. Logs in the 12″ × 16′ class can absorb a 48 percent decline in market prices before sales to veneer plants become more profitable. Obviously, stumpage buyers can bid higher for stands of larger trees.

analyzed in a thi
er operational ho
lative increases we
volume.

viewed as incremer
including addition
revenue increas
tial working capi
additional stumpa
s against the adc

pulpwood. An ope
small diameter
be met by other
were rejected by
even though the De
g classes.

gerial choice precis
s, the option of cut
This option decreas
therefore, it was

rived from linear
y for the value con
interest is the low
which a given log
Accordingly, those
npointed.

in prices which
neer bolts more p
nt decrease in lo
s pulpwood; henc
e margin within
e other hand, a
ily greater insid
he 12" x 16' cla
before sales to vi
impage buyers ca

Table 5.—

*Product value declines per M
b.f. that delete a log class from
being sawn in the optimal solution*

Diameter and length (inches and feet)	Allowable value declines
	Percent
6 x 12	2
8 x 12	14
10 x 12	26
12 x 12	27
14 x 12	42
6 x 14	3
8 x 14	23
10 x 14	39
12 x 14	41
14 x 14	52
8 x 16	30
10 x 16	46
12 x 16	48
14 x 16	58
16 x 16	65
10 x 18	53
12 x 18	54
14 x 18	63
16 x 18	68
10 x 20	58
12 x 20	58
14 x 20	65
16 x 20	71

The question of how great a premium can be paid for bigger trees was answered by simulating different log input patterns. The comparison assumed that the purchaser can choose between tracts of relatively small-sized timber and stands of larger trees. His choice necessarily would be conditioned by comparative profitability and stumpage price. Our task was to determine how much revenue difference would occur and how much of a stumpage price differential could be justified.

The hypothetical log distributions in table 6 provided significantly different revenue values. A small-log volume of 4.957 M b.f. yielded an hourly revenue of $956.33; with larger logs, 11.902 M b.f. furnished $1,886.62. Comparing these volumes and values with existing ones at the mill indicates how much more or less could be paid for stumpage. At present, the normal log input for the modified mill is about 8.2 M b.f. per hour and yields $1,415 in revenue. Since operating costs are virtually the same for all three log distributions, direct comparisons are possible. If only small logs are purchased, a cost reduction of $90/M b.f., or a net savings of approximately $450 per operating hour, would be required for small-log revenue to equal existing revenues. The current delivered-stumpage price of $100/M b.f. would have to fall to $10 before the mill could profitably purchase these small-sized stands. Such a price is, of course, extremely unlikely. On the other hand, a premium of $40/M b.f. can be paid for the larger log input pattern before net revenue falls below current levels.

These log input patterns are only two of infinite possibilities. Also, the prices and the price blend are relevant to but one point in time. These considerations limit the general applicability of the results. The technique, however, enables the mill manager to

Table 6.—*Alternative log class input patterns*

Diameter and length (inches and feet)	Small logs	Large logs
	– – – – M b.f. – – – –	
6 x 12	0.134	0.036
8 x 12	.240	.048
10 x 12	.335	.270
12 x 12	.178	.528
14 x 12	.038	.300
6 x 14	.332	.042
8 x 14	.330	.056
10 x 14	.397	.384
12 x 14	.246	.840
14 x 14	.062	.528
6 x 16	.440	.048
8 x 16	.445	.080
10 x 16	.565	**.648**
12 x 16	.237	1.216
14 x 16	.090	.900
16 x 16	.000	.720
10 x 18	.287	.410
12 x 18	.108	.792
14 x 18	.056	.784
16 x 18	.000	.972
10 x 20	.292	.360
12 x 20	.120	.720
14 x 20	.025	.500
16 x 20	.000	.720
	$\Sigma = 4.957$ M b.f.	$\Sigma = 11.902$

analyze timber cruise data and determine how high he can bid f[
the tract.

Conclusions

Linear programming provided practical solutions to the problem
of a medium-sized sawmill. Guided by the analysis, manageme[
installed a second gangsaw, realigned the cutoff saw, and sold sma[
logs as pulpwood. Results indicated a 45 percent revenue increa[
and a 36 percent hike in volume processed. Log classes high[
sensitive to price changes were specified, and a stumpage-purchasi[
strategy was formulated.

Further benefits are derived from added machine capacity a[
the stumpage strategy. Increased production is attained witho[
overtime costs. Previously, gangsaw breakdown caused the mill [
shut down; now one gangsaw can continue operation while the oth[

s being repaired. Management can reevaluate purchasing strategies as the stumpage market changes. In fact, the current practice of oversupplying the mill and selling marginal logs provides for nearly automatic response to change.

Although this analysis applies only to one sawmill, linear programming does have general applicability. Other independent lumbermen can similarly review their circumstances and evaluate their opportunities. With linear programming they can also plan inventory levels and determine optimum sawing patterns. In any case, the technique is an important addition to the manager's survival kit.

Literature Cited

McKillop, W., and Hoyer-Nielsen, S.
    1968. Planning sawmill production and inventories using linear programming. For. Prod. J. 18(5): 83-88.

Row, C., Fasick, C., and Guttenberg, S.
    1965. Improving sawmill profits through operations research. USDA For. Serv. Res. Pap. SO-20, 26 p. South. For. Exp. Stn., New Orleans, La.

Sampson, G. R.
    1972. Determining maximum stumpage values for wood products mills with alternative processes. For. Prod. J. 22(8): 26-30.

Dutrow, George F., and Granskog, James E.

1973. A sawmill manager adapts to change with linear programming. South. For. Exp. Stn., New Orleans, La. 11 p. (USDA For. Serv. Res. Pap. SO-88)

Linear programming provides guidelines for increasing sawmill capacity and flexibility and for determining stumpage-purchasing strategy. The operator of a medium-sized sawmill implemented improvements suggested by linear programming analysis; results indicate a 45 percent increase in revenue and a 36 percent hike in volume processed.

Additional keywords: Simulation, stumpage strategy, sawmill revenue, sawmill costs.

U. S. Department of Agriculture
Forest Service Research Paper SO-90

ng

FEB 25

U. S. Department of Agriculture
Forest Service Research Paper SO-90

U. S. Department of Agriculture
Forest Service Research Paper SO-89

ng

South Carolina Best of 36 Loblolly Pine Seed Sources For Southern Arkansas

Hoy C. Grigsby

FEB 2

Southern Forest Experiment Station
Forest Service
U. S. Department of Agriculture

1973

U. S. Department of Agriculture
Forest Service Research Paper SO-90

SOUTH CAROLINA BEST OF
36 LOBLOLLY PINE SEED SOURCES
FOR SOUTHERN ARKANSAS

Hoy C. Grigsby [1]

ng

After 10 years, trees from 34 of 36 Pinus taeda L. *seed sources selected from throughout the range of the species grew faster than the local trees in southern Arkansas. Trees from South Carolina coastal sources grew fastest, but survival was best among trees from drier sites—especially western sources. Volume growth was correlated with summer rainfall at the seed source and to a lesser extent with growing-season rainfall. Fusiform rust occurred least on trees from western sources and most on sources from northern Florida, Georgia, and eastern Alabama. Rust decreased with an increase in latitude of seed source, particularly up the east coast from Georgia.*

Additional keywords: *Cronartium fusiforme*, growing season rainfall, summer rainfall, volume growth, *Pinus taeda*.

Judicious selection of seed source appears to be an important method for increasing loblolly pine *(Pinus taeda* L.) yield. Ten-year results of the Southwide Pine Seed Source Study (Wells and Wakeley, 1966), for example, show that in many locations certain provenances consistently grow fastest. Other provenances have superior survival and resistance to fusiform rust caused by *Cronartium fusiforme* Hedgc. and Hunt ex Cumm. That such differences persist was demonstrated in another study (Wakeley and Bercaw, 1965) in which performance of trees was reported through age 35.

The study described here was designed to discover which sources are best for planting in southern Arkansas, a major pine-producing region. Survival, growth, and rust resistance of trees from 36 geographic sources were compared in plantings in two counties. In the Southwide study, only 8 to 15 sources are being tested at each location.

The author is Associate Plant Geneticist, Institute of Forest Genetics, Gulfport, Mississippi, stationed at Crossett, Arkansas.

Methods

The complete range of loblolly pine from Delaware to Florida and west to Texas was sampled (fig. 1). Sources sampled that were not included in the Southwide study were: Delaware, Virginia, the Piedmont in North Carolina, the Coastal Plain in South Carolina, Florida, southern Alabama, southeastern Oklahoma, and the "Lost Pines" area of Texas.[3]

From each location a 4-ounce lot of routinely collected nonselect seed was obtained from a State, Federal, or private agency. Planting stock was grown in an Arkansas Forestry Commission nursery and outplanted in January 1957 in Hempstead and Cleveland Counties, Arkansas.

All plantings were arranged in randomized blocks. Plots, which contained trees from only one source, were 1/10-acre squares. Spacing was 6 by 6 feet, 121 trees per plot. To minimize border effects, only the 40 trees forming a square in the center of each plot were measured.

From three to eight replications of 1-0 seedlings from the 36 sources were planted in Hempstead County in eight separate locations.

Trees from 34 of the 36 sources were planted at one location in Cleveland County; 32 were replicated four times; one, three times; and one, twice.

In all, there were 356 plots. Thirty-six provenances were replicated a minimum of three times; 34, six times; 33, seven times; 32, eight times; 29, nine times; 23, 10 times; 19, 11 times; and 10, 12 times.

The Celeveland County plantation was sprayed with a herbicide at age 3 to control excessive competition from weeds and brush.

Field Data

Tree heights were measured immediately after planting and at the end of the first, second, fifth, and tenth growing seasons. Ten-year heights of every other tree were taken to the nearest ½ foot with a telescoping pole. Survival and incidence of fusiform rust were tallied at those times. Diameter breast height was taken on all trees to the nearest 1/10 inch with a diameter tape at age 10.

Analyses

Survival, diameter, height, volume, and rust infection were examined in analyses of variance which were confined to the 32 seed sources having at least eight complete blocks. Volumes were com-

[3] Islands of pines west of the continuous zone of loblolly pine, which ends in eastern Texas.

2

Delaware to Florid
es sampled that were
laware, Virginia, the
 in South Carolina,
oma, and the "Lo

ly collected nonselec
ate agency. Planting
mission nursery and
 Cleveland Counties.

 blocks. Plots, which
)-acre squares. Spac
 imize border effects
 er of each plot were

 edlings from the 3
 eight separate loca

 ed at one location
 es; one, three time

 venances were rep
 seven times; t
 nes; and 10, 12 time
 yed with a herbici
 reeds and brush.

 ifter planting and
 owing seasons. T
 the nearest ½ in
 ce of fusiform ru
 height was taken
 er tape at age 10.

 rust infection we
 nfined to the 3 s
 Volumes were co
 ne, which ends in eac

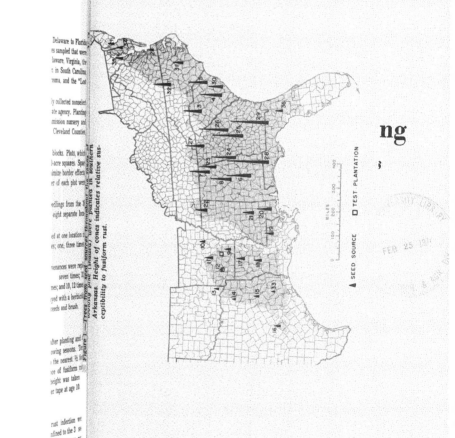

Figure 1.—Traces from seed sources where planted in southern Arkansas. Height of cones indicates relative susceptibility to fusiform rust.

MILES

0   100   200   300   400

▲ SEED SOURCE   ☐ TEST PLANTATION

ng

3

puted with Schmitt and Bower's (1970) formula for young loblolly
pines in plantations. Survival and rust data were transformed to
arc sin $\sqrt{\text{percentage}}$. Differences due to source were tested for
statistical significance at the 0.05 level with Duncan's multiple range
test.

Linear regressions were computed to show the relationships
between volume growth and summer rainfall, average January temperature, and distance from coast at seed source. Regressions were
tested for significance at the 0.05 level.

Results and Discussion

Survival

Although survival ranged from 93.4 to 64.4 percent (table 1),
only the four provenances with highest survivals differed significantly from the one with lowest survival after 10 growing seasons. In
general, trees from low rainfall areas survived best. Those from
high rainfall areas and mild climates, particularly with coastal influences, had the lowest survival. In previous studies in the Tennessee
Valley (Zarger, 1961), in Georgia (Kraus, 1967), and Southwide (Collins, 1964; Wells and Wakeley, 1966), inland or western sources have
survived better than coastal or more eastern sources.

Fusiform Rust

Incidence of rust varied from 2.4 percent for local trees to 34.3
percent for those from northwest Florida (table 1). Rust was greatest
on northwestern Florida, eastern Alabama, and Georgia trees and
decreased up the Atlantic Coast to about 5 percent on the trees from
Delaware, Maryland, and Virginia (fig. 1). The South Carolina
coastal provenances averaged 12 percentage points more infectious
than local trees, but they produced 27.3 percent more wood over
10-year period.

Diameter and Height

An analysis of variance and Duncan's test showed significant
differences among the provenances with largest and smallest diameters. Four of the five provenances with largest diameters were from
South Carolina. Trees with smallest diameters were from the western
edge of the loblolly pine range (table 1). Southeastern provenances,
particularly those nearest the coast, also grew fastest in diameter
in two Georgia studies (Collins, 1964; Kraus, 1967).

Height also increased as the origin of trees moved toward the
east and south. Greatest heights were attained by Coastal Plain
sources in North Carolina, South Carolina, Louisiana, and Virginia
(table 1). In most previous studies trees from Coastal Plain seed

4

Table 1.—*Volume, survival, height, and diameter by seed source 10 years after outplanting in Arkansas. Sources ranked by volume*

Origin of seed		Average volume per tree	Survival	Average height [1]	Average d.b.h.	Rust
		Cu. ft.[2]	Percent	Feet	Inches	Percent
S. C.	Lower Coastal Plain	1.98	77.7	30.2	5.52	11.4
S. C.	Lower Coastal Plain	1.87	77.3	28.7	5.51	17.2
S. C.	Upper Coastal Plain [3]	1.82	82.3	30.5	5.52	15.7
Fla.	Northwest	1.80	80.1	28.9	5.42	34.3
Miss.	Northwest	1.80	80.8	29.1	5.45	11.6
Fla.	Northeast	1.79	70.6	28.8	5.43	30.4
S. C.	Piedmont	1.79	75.5	27.6	5.46	15.0
La.	Southeast	1.77	74.2	29.7	5.33	7.0
Ala.	East Central	1.74	81.0	29.1	5.49	19.9
Va.	Southeast	1.73	80.6	29.3	5.31	5.3
Ga.	Upper Coastal Plain	1.73	70.7	27.6	5.41	24.4
N. C.	Lower Coastal Plain [3]	1.66	80.9	31.3	5.29	8.4
Ala.	Piedmont	1.66	81.8	27.3	5.35	17.0
N. C.	Piedmont	1.65	87.1	28.6	5.31	14.1
Miss.	Coastal Plain	1.62	88.5	28.5	5.29	9.0
Ala.	Coastal Plain	1.58	87.3	28.6	5.23	19.2
Ala.	Mountain	1.57	88.2	28.3	5.22	11.8
Tex.	Coastal Plain	1.56	84.2	27.5	5.32	4.2
Tex.	Lost Pines	1.54	86.2	27.3	5.13	4.9
Del.	Coastal Plain	1.53	93.4	27.9	5.76	5.0
Ga.	North	1.53	85.0	27.7	5.24	15.2
Fla.	Coastal Plain	1.52	64.4	25.7	5.20	6.1
La.	Coastal Plain	1.51	85.9	27.6	5.09	6.0
La.	Coastal Plain	1.49	83.5	26.2	5.21	6.9
Md.	Coastal Plain	1.49	82.3	27.1	5.16	5.5
Ga.	Central [3]	1.48	76.2	27.8	5.06	23.2
Ala.	Mountain	1.43	87.3	26.7	5.06	16.1
Tex.	Coastal Plain [3]	1.43	85.4	25.8	4.99	2.7
Ark.	Coastal Plain	1.42	89.8	26.7	5.14	6.6
Miss.	Coastal Plain	1.41	77.0	25.9	4.86	9.3
Ark.	Delta	1.40	84.2	26.0	5.04	4.8
Ark.	Coastal Plain	1.39	89.1	25.8	4.97	5.1
Okla.	Coastal Plain	1.39	88.6	25.4	5.10	3.8
Md.	Coastal Plain	1.39	93.4	28.2	5.12	4.1
Ark.	Coastal Plain	1.37	84.5	25.7	5.00	2.4
Tex.	Coastal Plain	1.21	80.1	26.4	4.72	2.5

Based on measurement of every other tree.

Volume means opposite the same line do not differ significantly at the 0.05 level of probability by Duncan's test.

Not included in analyses.

5

sources have grown fastest in height (Collins, 1964; Crow, 1964; Goggans and others, 1972; Wells and Wakeley, 1966). In a test of seed from six sources in the Cumberland Mountains of Tennessee, a North Carolina Piedmont provenance performed best (Thor, 1967), but this site was outside the natural range of loblolly pine and was too severe for trees of Coastal Plain origin.

Height growth relationships among provenances remained surprisingly constant as the plantings grew. Many fast and slow growers could be spotted as early as age 2.

Volume

The 36 provenance averages for volume varied from 1.98 to 1.21 cubic feet per tree (table 1). The three provenances with the greatest volume were all from South Carolina. The poorest performers were a Texas source, the local Arkansas trees, and trees from a Maryland source. The tallest trees, those from the North Carolina Coastal Plain, ranked 12th in volume. The three provenances with the greatest volume had the largest diameters. Thirty-four of the provenances had average volumes greater than the local trees, and 28 had volumes superior to all four Arkansas provenances.

Simple linear regressions revealed significant associations between tree volume and summer rainfall, average January temperature, and distance from seed source to a coast.

Average summer rainfall varied from 7.23 inches for a Texas provenance to 21.07 inches for one in Florida (U. S. Department of Agriculture, 1941), and accounted for 38 percent of the variation in volume (fig. 2). Average growing season (spring and summer) rainfall at seed source ranged from 18.96 to 35.00 inches for the same two provenances and in another regression was responsible for 2 percent of the variation. Summer rainfall normally has greater effect on loblolly pine growth than spring rainfall because winter rains leave the soil at field maximum. It is only after water deficits develop that trees respond to additional rainfall. Trees from seed originating in areas with heavy summer precipitation apparently have an inherent ability to make maximum use of available summer moisture. With few exceptions, trees producing the greatest volume were from areas where rainfall is at least 25 inches during the growing season, with 15 inches or more falling in the summer months.

There was a variation of 22.5°F. in average January temperatures at seed sources. The extremes were 35.1° at West Shore, Maryland, and 57.6° at Gainesville, Florida. The regression of volume on temperature was significant, and 10 percent of the variation in volume was associated with temperature at seed source (fig. 3). The

6

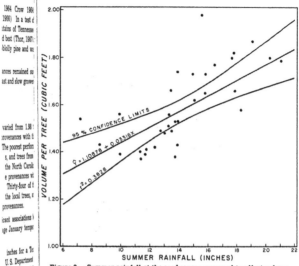

Figure 2.—*Summer rainfall at the seed source appeared to affect volume growth per tree in Arkansas.*

peak in volume occurred where the average January temperature at the seed source was about 50°.

Distance of seed source from coast was significantly correlated with volume. Sources closest to the Atlantic and Gulf Coasts produced the greatest volume. Growth varied inversely with distance from the coast. Extreme northern coastal provenances were not superior in growth, but trees of coastal origins from Virginia to Louisiana were within the top 12 of the 36 sources in volume production. Proximity of seed source to coast accounted for 29 percent of the variation in volume (fig. 4).

There was no significant interaction between planting site and seed source. This result is understandable because the sites are not widely separated. In the Southwide study, lack of interaction was also the general rule as trees from North Carolina and Louisiana coastal origins grew best in all but the two coldest locations.

7

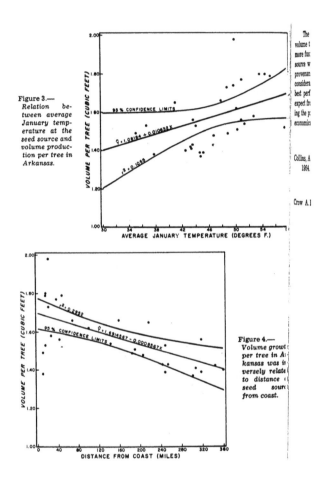

Figure 3.— *Relation between average January temperature at the seed source and volume production per tree in Arkansas.*

95% CONFIDENCE LIMITS

$\hat{Y} = 1.08184 + 0.010433X$

$r^2 = 0.1059$

VOLUME PER TREE (CUBIC FEET)

AVERAGE JANUARY TEMPERATURE (DEGREES F.)

VOLUME PER TREE (CUBIC FEET)

$r^2 = 0.2852$

$\hat{Y} = 1.6914267 - 0.0008567X$

95% CONFIDENCE LIMITS

DISTANCE FROM COAST (MILES)

Figure 4.— *Volume growth per tree in Arkansas was inversely related to distance of seed source from coast.*

The
volume t
more fus:
source w
provenan
considera
best perf
expect fr
ing the p:
economics

Collins, A
1964.

Crow A. l

The fastest growing provenance produced 30 percent more volume than the local trees in the test and had 9 percentage points more fusiform rust (11.4 percent). Survival for this South Carolina source was 77.7 percent, 6.8 percentage points less than the local provenance but still acceptable. Thus, based on 10-year performance, considerable improvement can be obtained by selecting one of the best performing provenances—more than tree breeders normally expect from F_1 progeny of the best plus trees in local stands. Selecting the proper seed source promises to be one of the simplest, most economical, and most reliable ways of improving tree growth.

Literature Cited

Collins, A. B., III.
1964. Tenth-year results of loblolly pine seed source planting in Georgia. USDA For. Serv. Res. Note SE-20, 4 p. Southeast. For. Exp. Stn., Asheville, N. C.

Crow, A. B.
1964. Ten-year results from a local geographic seed source study of loblolly pine in southeastern Louisiana. La. State Univ. Sch. For. and Wildl. Manage., LSU For. Notes 57, 4 p.

Goggans, J. F., Lynch, K. D., and Garin, G. I.
1972. Early results of a loblolly pine seed source study in Alabama. Auburn Univ. Agric. Exp. Stn. Circ. 194, 19 p.

Kraus, J. F.
1967. A study of racial variation in loblolly pine in Georgia—tenth-year results. Ninth South. For. Tree Improv. Conf. Proc., p. 78-85.

Schmitt, D., and Bower, D.
1970. Volume tables for young loblolly, slash, and longleaf pines in plantations in south Mississippi. USDA For. Serv. Res. Note SO-102, 6 p. South. For. Exp. Stn., New Orleans, La.

Thor, E.
1967. A ten-year-old loblolly pine seed source test in Tennessee. J. For. 65: 168-174.

U. S. Department of Agriculture.
1941. Climate and man. Yearb. of Agric. 1248 p.

Wakeley, P. C., and Bercaw, T. E.
1965. Loblolly pine provenance test at age 35. J. For. 63: 168-174.

9

Wells, O. O., and Wakeley, P. C.
1966. Geographic variation in survival, growth, and fu
rust infection of planted loblolly pine. For. Sci. N
11, 40 p.

Zarger, T. G.
1961. Ten year results on a cooperative loblolly pine seed
test. Sixth South. For. Tree Improv. Conf. Proc., p

Grigsby, Hoy C.

1973. South Carolina best of 36 loblolly pine seed
sources for southern Arkansas. South. For.
Exp. Stn., New Orleans, La. 10 p. (USDA
For. Serv. Res. Pap. SO-89)

After 10 years, trees from 34 of 36 *Pinus taeda* L. seed
sources selected from throughout the range of the species
grew faster than the local trees in southern Arkansas. Trees
from South Carolina coastal sources grew fastest, but survival was best among trees from drier sites—especially
western sources.

Additional keywords: *Cronartium fusiforme*, growing season rainfall, summer rainfall, volume growth, *Pinus taeda*.

Flowering and Fruiting

Of Southern Browse Species

L.K. Halls

FEB

Southern Forest Experiment Station
Forest Service
U. S. Department of Agriculture

1973

Flowering and Fruiting of Southern Browse Species

L. K. Halls [1]

Flowering and fruiting dates are reported for 14 browse species growing in the open and beneath trees in an east Texas pine-hardwood forest. Dates for individual species generally were not influenced by tree cover. Usually flowers bloomed earliest when March temperatures were highest. In the open, plants generally produced fruit more consistently and abundantly and at an earlier age than beneath the trees. American beautyberry and yaupon yielded the most fruit. Honeysuckle and yaupon fruits persisted longest through the winter.

Additional keywords: Phenology, fruit yields, *Pinus elliottii, P. taeda.*

Because plant fruits are a vital food source for wild birds and mammals, wildlife managers need to know when various plants bear fruits and how much they yield. This paper describes the flowering and fruiting characteristics of 14 common species of woody vines, shrubs, and small trees in southern forests. The information is based on observations made at the Stephen F. Austin Experimental Forest near Nacgdoches, Texas, from 1963 through 1972. Certain phases of the study were reported previously (Halls and Alcaniz, 1968 and 1972).

SITE TREATMENTS

A sawtimber-size stand of shortleaf *(Pinus chinata* Mill.) and loblolly *(P. taeda* L.) pines was thinned to a tree basal area of 70 sq. ft. per acre. Understory vegetation was cut or killed with chemicals before planting of 1-year-

[1] On the staff of the Wildlife Habitat and Silviculture Laboratory, which is maintained at Nacogdoches, Texas, by the Southern Forest Experiment Station in cooperation with the School of Forestry, Stephen F. Austin State University.

old seedlings. Nine plants of each of the 14 species were spaced equally within each of four contiguous ¼-acre blocks. Shrubs and vines were planted 5 feet apart; small trees, 10 feet apart.

The same planting schedule and arrangement were implemented in a nearby abandoned field. Here the land was disked before planting, and weeds were controlled by disking until the study plants were definitely established and thereafter by mowing.

Twenty-four plants of each species (three from each block in the two planting sites) were selected for flowering and fruiting observations. A few of these plants died during the study. The age when plants first bore mature fruits was noted. Thereafter, the seasonal development of flowers and fruits was recorded from beginning date of growth until final fruit fall the following winter. Samples of mature fruits were collected, dried at 70°C., and weighed to compute an average ovendry weight per fruit. This figure was multiplied by number of fruits per plant to obtain the total annual yield. Number of mature fruits per gram, moisture content, and ratio of pulp to seed were calculated when a sufficient sample was available from 1967 through 1972.

Temperatures were continuously recorded by hygrothermographs at both study locations, and daily rainfall was measured by a standard rain gage nearby.

WEATHER RECORDS

Rainfall.—Annual rainfall ranged from a low of 32 inches in 1967 to a high of 66 inches in 1968 (table 1). For 7 years out of 9 it was below the longtime average of 46 inches. Droughts of several weeks duration occurred in the summers of 1964 and 1965. Total rainfall in 1966 was very close to the yearly average. The 1967 rainfall was 14 inches below average, and soils were especially dry from June through November. In contrast, rainfall was much above average in 1968 and through May 1969. This wet period was in turn followed by a prolonged dry period that extended through June 1972. Heavy rains occurred in July, and rainfall was above average for the remainder of the year.

Table 1.—*Seasonal rainfall (inches) at Stephen F. Austin Experimental Forest*

Year	Winter	Spring	Summer	Fall	Total
1964	9.4	13.2	7.2	8.9	38.7
1965	10.9	11.9	7.2	11.7	41.7
1966	11.7	15.1	12.6	6.8	46.2
1967	6.6	11.2	6.3	8.0	32.1
1968	13.0	26.7	14.0	12.4	66.1
1969	15.1	13.6	5.2	8.8	42.7
1970	9.6	8.4	6.9	9.0	33.9
1971	4.8	9.1	7.0	16.0	36.9
1972	8.0	8.5	12.9	15.5	44.9

Temperature.—From 1964 through 1972 the average maximum monthly temperature was 78.7°F. in the open and 76.9°F. beneath the trees. Occasionally, however, the maximum temperature was higher under the trees. The widest differences occurred in the falls of 1966 and 1968 and in the summer and fall of 1969, when temperatures in the open were 4° to 8°F. above those in the shade. The highest maximum temperature, 95.0°F., was reported in August.

The average daily minimum temperature was 52.8°F. in the open and 53.4°F. beneath trees. The lowest mean minimum temperature, 35.6°F., occurred in February.

In comparison to longtime Weather Bureau records at nearby Nacogdoches, the following months were several degrees colder than normal: January in 1964, 1966, 1970; February from 1964 through 1968; March in 1965 and 1969; and July and August in 1968. Temperatures were considerably warmer than average in January and February of 1969, 1971, and 1972; March 1967 and 1972; April 1965 and

freeze varied from
in 1971, and the
'om October 19 in
72. The length of
m 205 days in 1971

IT OF FLOWERS

t species were not
tree cover. How
ry flowered a few
and flowering dog
on greenbrier flow
neath trees (fig. 1)

species was earlier
March temperature
lowers bloomed late
this relationship wa
for species blooming
and February temp
fect on flower initia
ecipitation. In a way
d flower early are
responses to the sam
ve been noted else

st flower emerges
rtest for rusty blac
and saw greenbri
t for flowering do
4 to 16 days). Oth
als of 6 to 10 da

: bloomed in the b
ering dogwood a
haw, red mulber
(3) yaupon and co
reenbrier, dwarf li
2, and Alabama su
nolia and muscad
eautyberry.

d sassafras form
eached their peak
ly April. Since fr
wers, only the b
ure fruits. Althou
ten discolored wl
ely damaged wl
rew to maturity

For rusty blackhaw, red mulberry, and smallflower pawpaw,[1] blooming usually peaked during the first or second week in April. Red mulberry blossoms were occasionally nipped by late freezes, and only a small proportion of all three species' flowers developed into mature fruits.

Yaupon and common greenbrier flowered late enough (mid-April) so that they were never damaged by a freeze. Yaupon produced abundant flowers; but common greenbrier, only a few.

Flowering peaked in late April for saw greenbrier, dwarf live oak, Japanese honeysuckle, and Alabama supplejack. Greenbrier flowers were sparse. Live oak continued to form new flowers several weeks after the peak of bloom; many of the flowers aborted and were replaced by others. Honeysuckle also formed a few flowers throughout the summer and fall if rainfall was above average; these late blooms seldom developed into mature fruits, however. Supplejack flowers were present only briefly in the spring, and most of them developed into mature fruit.

Sweetbay magnolia and muscadine grape flowered in mid-May. The magnolia's sparse flowers developed into one or two full-size cones and a few immature fruits. Grape flowers were also sparse; and many of the developing fruits aborted, especially in hot, dry weather.

Flowering of American beautyberry did not peak until June. Although flowers continued to form through summer and autumn when moisture was plentiful, late blooms seldom developed into mature fruits.

FRUITING

Age at first fruiting.—Whether planted in the open or under trees, American beautyberry first produced fruit at age 2 years. All other species except red mulberry produced mature fruit at an earlier age in open than beneath trees (table 2). In the open, sweetbay magnolia and common greenbrier bore fruit at age 6; and most other species, at 3 or 4 years. Beneath trees, the beginning fruit-bearing age ranged

[1] In this study the flowering and fruiting dates were earlier for smallflower pawpaw and red mulberry, and the fruiting dates were later for Alabama supplejack and rusty blackhaw than the dates presented in taxonomic texts (Vines, 1960; Correll and Johnston, 1970).

from 5 to 9 years. Sweetbay magnolia, dwarf live oak, saw greenbrier, rusty blackhaw, and muscadine grape plants beneath trees did not bear fruit during the study.

Table 2.—*Age at which species first bore mature fruit*

Species	Open	Beneath trees
		– – Years – –
Smallflower pawpaw	4	7
Alabama supplejack	3	8
American beautyberry	2	2
Flowering dogwood	4	9
Yaupon	4	7
Japanese honeysuckle	3	5
Sweetbay magnolia	6	(¹)
Red mulberry	4	2
Dwarf live oak	4	(¹)
Sassafras	4	8
Saw greenbrier	4	(¹)
Common greenbrier	6	9
Rusty blackhaw	5	(¹)
Muscadine grape	4	(¹)

[1] Plants beneath trees did not bear fruit by 1972.

The proportion of fruit-bearers was much higher for plants growing in the open than for those beneath trees (table 3). Usually this proportion increased as the plants grew older; however, this trend was not consistent for several species. On the average, American beautyberry and Japanese honeysuckle had the highest proportion of plants bearing fruit.

Season of fruiting.—Usually when a plant bloomed early, the fruit matured early; but for several species there was no relationship between dates when flowers appeared and dates when fruit matured. For example, in any given year a species might flower earlier than average but bear mature fruit later than usual.

Generally, the fruits of Alabama supplejack, American beautyberry, and red mulberry matured at least 10 days later on plants beneath trees than on those in the open. No consistent difference between locations was observed for other species.

Red mulberry was the only species that bore mature fruit in the spring (May), and the interval from first appearance to end of drop was only about 6 weeks. Smallflower pawpaw, sweetbay magnolia, sassafras, and muscadine grape produced mature fruit in summer or

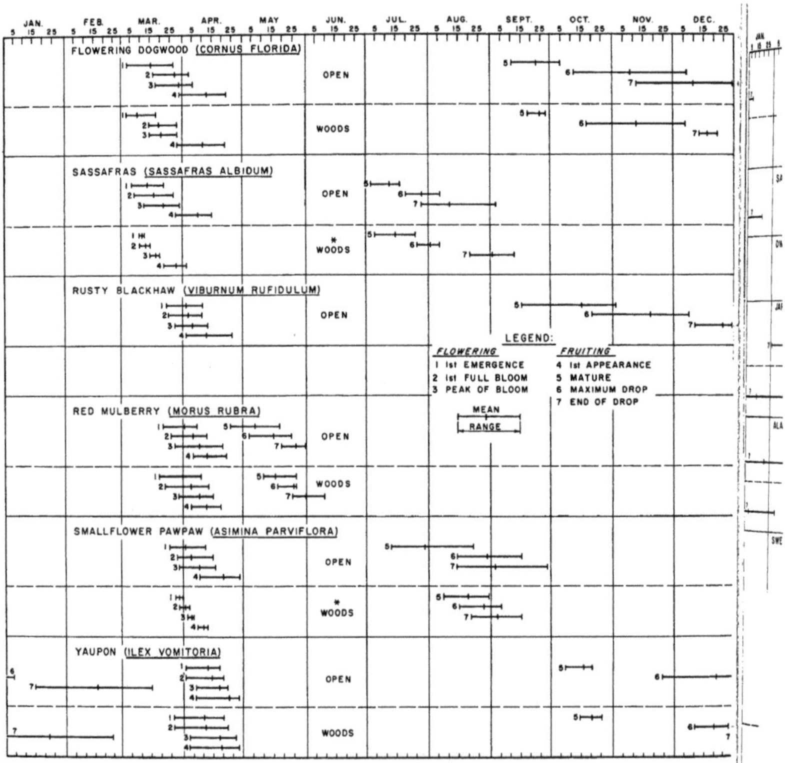

Figure 1.—*Flowering and fruiting chronology of plants growing in the open and beneath trees.*
(* indicates that data are available only for 2 years.)

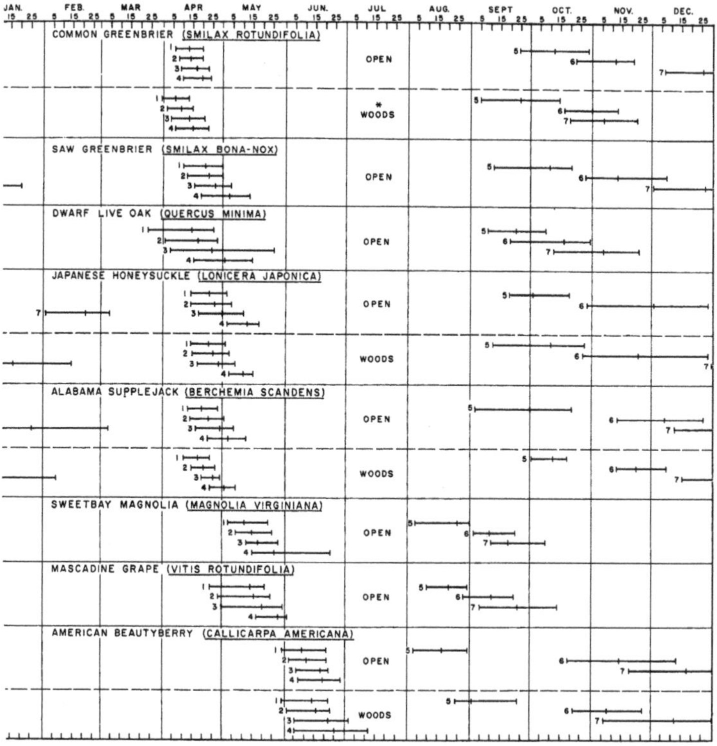

JAN.   FEB.   MAR   APR   MAY   JUN.   JUL   AUG.   SEPT   OCT.   NOV.   DEC.
15 25   5 15 25   5 15 25   5 15 25   5 15 25   5 15 25   5 15 25   5 15 25   5 15 25   5 15 25   5 15 25   5 15 25

COMMON GREENBRIER (SMILAX ROTUNDIFOLIA)

OPEN

WOODS

SAW GREENBRIER (SMILAX BONA-NOX)

OPEN

DWARF LIVE OAK (QUERCUS MINIMA)

OPEN

JAPANESE HONEYSUCKLE (LONICERA JAPONICA)

OPEN

WOODS

ALABAMA SUPPLEJACK (BERCHEMIA SCANDENS)

OPEN

WOODS

SWEETBAY MAGNOLIA (MAGNOLIA VIRGINIANA)

OPEN

MASCADINE GRAPE (VITIS ROTUNDIFOLIA)

OPEN

AMERICAN BEAUTYBERRY (CALLICARPA AMERICANA)

OPEN

WOODS

5

early fall; but almost all their fruit dropped by October. Fruit persisted through November or December on all other species and through most of the winter on yaupon and honeysuckle. In these two species, the end of drop was later for plants in the open, probably because they produced much more fruit than those in the woods.

Fruit yields.—In any year the yields were apt to be high for one species and low for another. The period of high production also varied considerably between species. Wright (1953) previously noted that there was little correlation among woody species in the occurrence of good seed years.

American beautyberry yielded more fruit than any other species during this study (table 3). In the open, plants grew rapidly the first few years and produced abundant fruit. Yields peaked at 1,722 grams per plant in 1966, when plants were 5 years old. The following year, however, the plants deteriorated; and many stems died back. This general decadence continued through 1969, and fruit yields declined sharply to 107 grams. In 1970 many of the old stems sprouted at the base, and the plants again produced a substantial fruit crop. During the next 2 years, severe dieback of stems diminished fruit yields. In the woods, the young plants developed more slowly; and they produced less fruit. Beginning in 1965 the yields tended to be relatively high and low on alternate years.

Yaupon had the second highest yield per fruiting plant; but because this species is dioecious—i.e., has separate staminate (male) and pistillate (female) plants—the proportion of total plants bearing fruit was relatively small. The ratio of fruit yields between open- and woods-grown plants was 19:1. After plants in the open began to bear fruit (age 4), their yields tended to be high or low on alternate years. In 1972 the yield per fruiting plant was the highest recorded for any species during the study.

The alternate high and low yields of American beautyberry growing beneath trees and of yaupon in the open were apparently unrelated to weather conditions. Rather, as noted in other studies (Kozlowski, 1971), the variance seemed to be a physiological characteristic of individual plants.

e 3.—*Fruit yields of browse plants, growing in the open and beneath pine trees*

pecies	Year	Open		Beneath trees	
		Yield per fruiting plant	Plants bearing fruit	Yield per fruiting plant	Plants bearing fruit
		Grams	*Percent*	*Grams*	*Percent*
llflower	1967	7	11	0	0
vpaw	1968	3	22	0	0
	1969	0	0	0	0
	1970	0	0	0	0
	1971	77	17	33	10
	1972	135	67	22	20
ama	1965	2	25	0	0
plejack	1966	129	33	0	0
	1967	141	33	0	0
	1968	196	33	0	0
	1969	173	33	0	0
	1970	165	33	1	10
	1971	874	42	6	10
	1972	258	42	0	0
rican	1963	49	91	1	26
utyberry	1964	578	91	2	75
	1965	1,554	100	17	25
	1966	1,772	100	65	91
	1967	585	100	29	58
	1968	231	100	133	100
	1969	107	100	51	58
	1970	834	100	240	58
	1971	342	100	18	58
	1972	183	83	26	83
vering	1965	1	25	0	0
wood	1966	13	58	0	0
	1967	5	58	0	0
	1968	4	17	0	0
	1969	150	58	0	0
	1970	9	50	9	42
	1971	269	100	3	25
	1972	788	100	38	83
pon	1965	1	17	0	0
	1966	93	25	0	0
	1967	36	25	0	0
	1968	339	42	19	8
	1969	150	42	28	8
	1970	336	42	5	42
	1971	270	42	39	42
	1972	2,212	42	88	50
nese	1965	4	100	0	0
eysuckle	1966	138	100	0	0
	1967	127	100	1	17
	1968	222	100	3	42
	1969	43	100	5	92
	1970	68	100	11	92
	1971	23	100	6	92
	1972	26	100	2	92
tbay	1968	20	42	0	0
nolia	1969	3	42	0	0
	1970	5	27	0	0
	1971	16	64	0	0
	1972	6	33	0	0

Table 3.—*Fruit yields of browse plants growing in the open and beneath pine trees* (Continued)

Species	Year	Open		Beneath trees	
		Yield per fruiting plant	Plants bearing fruit	Yield per fruiting plant	Plants bearing fruit
		Grams	*Percent*	*Grams*	*Percent*
Red mulberry	1965	0	0	2	8
	1966	0	0	2	8
	1967	7	50	0	0
	1968	22	58	3	17
	1969	17	58	1	8
	1970	10	50	3	25
	1971	6	42	0	¹0
	1972	10	50	0	0
Dwarf live oak	1967	5	75	0	0
	1968	50	75	0	0
	1969	67	88	0	0
	1970	60	100	0	0
	1971	201	100	0	0
	1972	45	100	0	0
Sassafras	1967	5	8	0	0
	1968	19	17	0	0
	1969	154	33	0	0
	1970	3	25	0	0
	1971	48	25	91	10
	1972	129	33	129	10
Saw greenbrier	1965	1	8	0	0
	1966	11	25	0	0
	1967	11	17	0	0
	1968	13	25	0	0
	1969	4	25	0	0
	1970	2	50	0	0
	1971	8	42	0	0
	1972	2	17	0	0
Common greenbrier	1967	1	33	0	0
	1968	2	67	0	0
	1969	7	100	0	0
	1970	10	100	1	8
	1971	1	67	0	0
	1972	6	58	2	17
Rusty blackhaw	1967	2	8	0	0
	1968	0	0	0	0
	1969	6	8	0	0
	1970	2	8	0	0
	1971	4	8	0	0
	1972	84	58	0	0
Muscadine grape	1967	56	17	0	0
	1968	107	18	0	0
	1969	388	18	0	0
	1970	239	27	0	0
	1971	300	36	0	0
	1972	149	36	0	0

¹All plants beneath trees had died by 1971.

Species	Fruit type	Color	Dimensions				Seed		
American beautyberry (Callicarpa americana)	Berry-like drupe	Reddish-purple	3-6 mm. long wide	71.6	64.9-79.7	80	75-84 4, about 1.5 mm. long	1:0.68	1:0.38-0.98
Flowering dogwood (Cornus florida)	Drupe	Bright red to yellow	1-1.5 cm. long	7.0	5.2-7.8	53	52-56 1-2, about 9 mm. long pale brown	1:0.96	1:0.85-1.05
Yaupon (Ilex vomitoria)	Drupe	Shiny red	About 6 mm. long	13.7	11.7-15.5	51	46-60 Usually 4 1-seeded stones, up to 4 mm. long	1:0.69	1:0.65-0.76
Japanese honeysuckle (Lonicera japonica)	Berry	Black	About 7 mm. in diameter	17.8	12.7-23.0	68	61-72 Several, irregularly ridged	1:0.35	1:0.27-0.45
Sweetbay magnolia (Magnolia virginiana)	A cone with drupe-like seeds	Red	4-5 cm. long	12.1	²9.3-15.0	41	²37-45 Each follicle with 1-2 seeds, about 7 mm. long, red	...	...
Red mulberry (Morus rubra)	An aggregate of small drupes	Dark purple	2-3 cm. long	8.7	6.2-11.5	78	72-81 1-2 mm. long	1:0.31	¹1:0.26-0.35
Dwarf live oak (Quercus minima)	Acorn	Glossy brown	Cup 10-15 mm. broad, 8-16 mm. high	0.8	²0.6-0.9	45	²37-54 15-20 mm. long, 8-12 mm. thick, ¾ to ⅞ included in cup	...	...
Sassafras (Sassafras albidum)	Drupe, borne on red pedicel	Lustrous blue	About 1 cm. long	6.2	6.0-6.9	47	43-50 Solitary stone, light brown, about 6 mm. long	1:0.73	1:0.60-0.81
Saw greenbrier (Smilax bona-nox)	Berry	Glaucous, black	About 6 mm. thick	12.7	8.4-16.5	64	59-69 Usually solitary, 4-5 mm. long, reddish brown with black basal disk	1:1.59	1:1.00-1.91
Common greenbrier (Smilax rotundifolia)	Berry	Blue-black	6-8 mm. thick	13.0	10.0-16.1	64	58-69 Usually 2-3, about 4-6 mm. thick	1:1.23	1:0.93-1.76
Rusty blackhaw (Viburnum rufidulum)	Drupe	Blue-black	8-15 mm. long	5.2	4.0-7.0	52	44-58 Solitary, flattened	1:0.50	1:0.41-0.56
Muscadine grape (Vitis rotundifolia)	Berry	Purple-black	12-25 mm. in diameter	1.4	1.2-1.6	81	78-84 2-3, 7-8 mm. long, 4-5 mm. thick	1:0.28	1:0.18-0.36

¹ Samples taken only 2 years.
² Includes only the seed portion of fruit.

9

The proportion of pulp to seed was highest for muscadine grape, red mulberry, and Japanese honeysuckle, and lowest for smallflower pawpaw, Alabama supplejack, and greenbriers. These ratios are significant in wildlife management since the nutrient content of seeds differs considerably from that of the pulpy and fleshy portions of fruit. Usually the seeds contain more crude protein than the pulp. The value of a fruit with a large proportion of seeds depends mainly on how well the animal digests the seeds (Wainio and Forbes, 1941).

LITERATURE CITED

Blaisdell, J. P.
    1958. Seasonal development and yield of native plants on the upper Snake River plains and their relation to certain climatic factors. USDA Tech. Bull. 1190, 68 p.

Correll, D. S., and Johnston, M. C.
    1970. Manual of the vascular plants of Texas. 1881 p. Renner, Tex.: Tex. Res. Found.

Halls, L. K., and Alcaniz, R.
    1968. Browse plants yield best in forest

Ha

Kc

Vii

Wa

Wr

Halls, L. K.

  1973. Flowering and fruiting of southern browse
    species. South. For. Exp. Stn., New Orleans,
    La. 10 p. (USDA For. Serv. Res. Pap. SO-90)

Flowering and fruiting dates are reported for 14 browse
species growing in the open and beneath trees in an east
Texas pine-hardwood forest. Dates for individual species
generally were not influenced by tree cover. In the open,
plants generally produced fruit more consistently and a-
bundantly and at an earlier age than beneath the trees.

U. S. Department of Agriculture

Forest Service Research Paper SO-91

HEDGING SOUTHERN PINE
THROUGH FUTURES TRADING

Lloyd C. Irland   and   James P. Olmedo, Jr.

SOUTHERN FOREST EXPERIMENT STATION
FOREST SERVICE
U. S. DEPARTMENT OF AGRICULTURE

HEDGING SOUTHERN PINE THROUGH FUTURES TRADING

Lloyd C. Irland [1]
and
James P. Olmedo, Jr. [2]

Futures trading in lumber and plywood offers an attractive method for southern pine manufacturers and wholesalers to protect themselves against price fluctuations and gain added marketing flexibility. Existing markets have sufficient liquidity to permit expanded trading. Price correlations with futures are close enough for effective hedging. Basis relations over time show discernible and predictable patterns. Hedging practices used by manufacturers and wholesalers are illustrated with examples.

Additional keywords: Prices, marketing, risk.

Fluctuating prices threaten profits and complicate the marketing of lumber and plywood (fig. 1). As a result, plywood and lumber have lost markets to products with more stable prices. For example, makers of aluminum studs reportedly stand ready to guarantee firm prices one year in advance. Hedging in futures markets offers a new technique for meeting these challenges.

This paper describes the technique for southern pine lumber and plywood manufacturers, wholesalers, and users. Following a general description of futures markets in forest products, the need for hedging and its benefits are discussed. Next, the feasibility of hedging southern pine products is demonstrated. Finally, specific hedging techniques are described.

The prices and charts shown here are general guides only, based on market average prices. They cannot be assumed to apply to all grades of southern products or to all areas of the South.

When the paper was written, Irland was Associate Economist with the Southern Forest Experiment Station, USDA Forest Service. He is now Assistant Professor of Forest Economics, Yale School of Forestry and Environmental Studies.
Research Director, Brascan International, New York, N. Y.

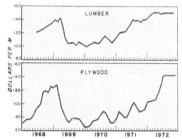

Figure 1.—*Recent trends in pine lumber and plywood prices. Source: Random Lengths Yearbook, 1972. Lumber: KD, 2x4, #2 12 ft. Plywood: ½ in. CDX, southern.*

Every firm differs in its marketing policies and exposure to risk. In addition, each firm must base hedging decisions on the prices it receives for its own production.

FUTURES MARKETS

History

Futures markets have long been used in the grain and produce industries to enable businessmen to manage the impact of price fluctuations on their firms. Until recently lumbermen had no such opportunity.

In the 1930's, a lumber futures market was proposed, with southern pine as the deliverable grade (Kulp, 1931). A group of western lumbermen actually started a short-lived National Lumber Exchange in the 1950's, but it quickly expired due to poor planning and the hostility of wholesalers. [3]

3 Schmelzlee, R. National lumber exchange in Portland. Unpub. MS thesis, Oregon State College. 1960.

In the late 1960's, officials of three major exchanges—the New York Mercantile Exchange, the Chicago Mercantile Exchange, and the Chicago Board of Trade—began studying opportunities for trading new commodities. Two products considered were plywood and lumber. The New York Mercantile Exchange opened plywood trading in September 1969, but that market has done little trading recently and is ignored in this paper. The Chicago Mercantile Exchange initiated lumber trading in October 1969. The Chicago Board of Trade plywood market opened in December 1969 (Sandor, 1972). In early winter of 1972, the Chicago Board of Trade began trading a lumber contract written for studs. Trading activity on these markets has grown rapidly (table 1).

Exchanges—Central Market Places

A commodity exchange is an association of firms and individuals which operates a central market for futures contracts and often also for cash commodities. The exchanges provide trading floors and price reporting and clearing services. The members trade on the exchange floor for their own accounts and offer brokerage services to the general public. Exchanges audit the accounts of members and enforce contract provisions and ethical standards of business practice. The exchanges, in turn, are supervised by the USDA's Commodity Exchange Authority. In addition, the Authority regulates trading in specified commodities, but plywood and lumber are not regulated commodities.

On the floor, commodities are traded in different pits. At each pit a price reporter records every transaction. All trades must be made by open outcry during prescribed hours. Price changes and market data are posted above the trading floor and electronically conveyed to outside wire services and brokerage houses. In addition, the floor itself receives a constant stream of market news and economic data from wire services, the CEA, and the exchanges' own statistical departments.

In its capacity as a clearinghouse, an exchange aids liquidity by clearing trades in much the same way the Federal Reserve System clears checks. The clearinghouse totals the transactions of all traders daily at the close of trading, sending each a statement of his net position. In this way, contracts lose their identification with

Table 1.—*The volume of trading and open interest, Chicago lumber and plywood futures markets, October 1969 to June 1973*

Month and year	Chicago Mercantile Exchange— lumber		Chicago Board of Trade— plywood	
	Volume (monthly total)	Open interest (last business day)	Volume (monthly total)	Open interest (last business day)
	— — — — — *Number of contracts* — — — — —			
1969				
October	416	277	. . .	. . .
November	218	369	. . .	. . .
December	110	404	393	208
Total	744		393	
1970				
January	367	526	593	406
February	283	536	1,434	797
March	817	436	2,132	1,059
April	528	404	3,628	1,652
May	519	313	3,232	1,555
June	403	336	3,613	1,784
July	411	305	4,344	1,895
August	18,240	3,142	3,454	2,309
September	21,613	4,325	5,949	2,931
October	13,300	3,490	7,261	2,612
November	12,187	4,770	5,050	2,847
December	16,845	6,748	6,539	3,625
Total	85,513		47,229	
1971				
January	13,653	1,890	14,362	4,610
February	9,563	2,248	25,008	5,384
March	7,633	2,220	35,692	5,540
April	4,628	2,523	33,271	5,586
May	7,421	2,573	20,260	4,909
June	7,358	2,581	18,203	4,381
July	6,926	3,145	18,058	3,760
August	10,202	2,637	17,669	3,351
September	9,293	2,243	6,185	2,163
October	6,472	2,396	9,185	3,805
November	7,456	2,823	9,623	2,791
December	9,544	2,861	11,340	3,139
Total	100,149		218,856	
1972				
January	3,815	1,844	12,678	3,496
February	3,082	1,936	13,232	3,124
March	3,498	2,234	8,288	3,348
April	2,629	1,892	9,866	3,053
May	1,722	1,479	7,670	3,292
June	4,714	2,048	19,463	4,009
July	4,963	1,854	22,250	3,724
August	5,780	1,576	28,613	3,916
September	5,964	1,971	31,122	3,030
October	8,601	2,422	26,860	4,204
November	11,092	2,525	20,007	4,150
December	9,624	2,922	16,828	4,494
Total	65,484		218,877	
1973				
January	10,380	1,523	21,462	4,069
February	12,185	2,076	20,904	3,954
March	10,657	1,988	21,630	3,450
April	9,881	2,170	14,757	3,431
May	17,071	2,304	26,495	4,341
June	12,173	2,417	20,557	4,384

Sources: Chicago Board of Trade; Chicago Mercantile Exchange; Merrill Lynch, Pierce, Fenner, and Smith, Inc.

3

Futures contracts are traded for specified delivery months in the future. In all three markets, contracts are traded for delivery in January, March, May, July, September, and November.

In August, the September contract would be referred to as the "near" or "nearby" future, and the others, as "far" or "distant" months.

Provisions of the plywood, lumber, and stud contracts are summarized in table 2. Details and further information are available from the exchanges and from commodity brokerage firms.

HEDGING

What Is Hedging?

Hedging is a marketing tool for increasing manufacturing or wholesaling profits. Hedging involves *using the futures market to offset positions taken in the cash market.*

A commitment to sell or deliver a car of lumber is called a *short commitment,* and the seller is said to be short one car of wood. The buyer of this car is holding a *long commitment.* So long as

Table 2.—*Provisions of forest products futures contracts*

Item	Plywood	Lumber	Studs
Exchange	Chicago Board of Trade	Chicago Mercantile Exchange	Chicago Board of Trade
Delivery grade	½ in. CDX western 4 or 5 ply	KD hem-fir R. L. 2 × 4, 8-20 ft. std. & better	2 × 4 × 8 ft. (92-5/8") stud grade lodgepole pine and other specified species
Delivery unit	36 banded units, 69,120 sq. ft. per car	100,000 bd. ft. in banded units	100,000 bd. ft. in banded units
Delivery method	Shipping certificate from regular mill or warehouse	On track	Shipping certificate
Delivery region	Oregon Washington Idaho Montana Northern California	California Idaho Montana Nevada Oregon Washington	Oregon Washington Idaho Montana Northern California British Columbia Alberta

he holds a binding obligation to take delivery, he is *long* in lumber whether actual delivery has occurred or not. Of course, most businessmen will be holding both long and short commitments at any one moment. They measure their exposure to price risk by their *net position.* The net position is the surplus of long commitments over short commitments or vice versa. Thus, a wholesaler owning 50 cars of wood and holding firm orders for 100 cars would be net short 50 cars.

In the same way, traders who have sold futures contracts are said to be short. Traders who have bought contracts are long. Traders may be

holding both long and short positions in the same commodity in different contract months, for example. Their net position is then the difference between their long and short commitments.

By taking an opposite position in the futures market, a firm can hedge its net position in the cash market. Depending on market conditions the position in futures may not be *equal* to the cash position. That is, the firm's overall net position—cash and futures combined—may not be zero. For a variety of reasons, a firm may choose to hold a cash commitment "open"—unhedged—or may actually hold futures contract as a temporary substitute for cash commitments

4

Potential Hedgers

As noted above, the firm's cash net position consists of *all of its current commitments*. These include:

Ownership of raw material inventory
Ownership of goods in process
Finished goods in storage or in transit
Raw materials contracted for at firm prices
Orders accepted for product at firm prices.

Producers, wholesalers, and wood users have different economic functions, different net positions in the cash market, and different hedging plans (Arthur, 1970).

Producers.—Producers typically hold inventories of logs at mills and have timber contracted for current logging. Log and timber prices are highly correlated with product prices. When product prices tumble, as they did in early 1969, millmen find themselves cutting high-priced logs into low-priced products. In addition, mills hold stocks of goods in process. These stocks are also subject to price risks. Offsetting these long commitments, however, is an order file covering several weeks' production. The order file represents the firm's short position. Typically, the quantity of timber on the log deck and under contract, and the goods in process compared to the file of unfilled orders, will leave the mill with a net long position in cash. This position can be covered by an appropriate short position in futures.

Wholesalers.—The wholesaler does not carry stocks of raw material or goods in process, but is constantly searching for low-cost purchases and high-price markets. He owns cars of wood on order or in transit and attempts to sell those cars at the most favorable prices. He may in addition hold warehouse stock for cash-and-carry service to local accounts. His file of firm orders and completed sales may or may not cover his current long position in cash. Wholesalers have often been substantial speculators in the cash markets. In other industries, it is the wholesalers who operate highly sophisticated hedging programs. They hedge to protect cash commitments, to enhance marketing flexibility, to earn carrying charges on inventory, and to aid in quoting firm orders farther out into the future.

Wood users.—Wood users tend to be net short in the cash market. A builder, for instance, may plan and price homes or apartment units in the winter, then order building materials for summer delivery. He wants material at his planned costs. Theoretically, the builder would be a long hedger in futures—he would buy futures to cover his short commitment in cash. In the flour, corn products, and soybean industries, futures trading has become an integral part of normal procurement practice. Millers and crushers use futures to reduce their net cost of raw material through hedging. In these industries the raw material is one item—grain—and accounts for a substantial part of total cost. In building, however, wood products are but one of many raw materials and account for a small portion of total cost. For these reasons, wood users have only recently become active participants in futures markets.

Gains from Hedging

Potential hedgers must evaluate the gains and costs which their firms may realize through futures trading.

The potential gains from hedging southern pine are:

1. Firms can cover their net positions in the cash market through hedging and thus reduce the impact of unfavorable price change on profits.

2. Firms can gain additional marketing flexibility by using futures as a temporary substitute for merchandising commitments. If a wholesaler views current prices as low, he can, in effect, inventory plywood by simply buying futures. A mill expecting prices to fall can, in effect, sell its production at current prices by selling futures. The low financial outlay required conserves working capital.

3. Wholesalers can use futures to protect inventory stocks, to reduce their net cost of wood, and to quote firm prices to users farther into the future than present practices permit.

4. Firms can obtain more favorable financing for hedged inventories than for unhedged stocks. By holding and turning over a higher volume of goods, firms can increase their profits.

5. Futures markets provide the entire industry with informed estimates of price trends. They also supply the first instantly available quotations of daily lumber and plywood prices.

6. The futures market can be used as an alternate source of supply in tight markets, by simply buying futures and taking delivery.

5

Costs of Hedging

In considering hedging programs, the potential benefits must be weighed against the following costs:

1. *Execution costs.*—These costs represent price changes unfavorable to the hedger which result from illiquid markets. The Chicago markets by now have sufficient liquidity that this cost is low. Additional market participation will raise volume, attract more informed pit trader and public speculation, and reduce execution costs still further.

2. *Interest on margin deposits.*—The buyer or seller of a futures contract must deposit a margin with the broker. The brokers, in turn, keep their margins up to date with the clearing-house on a daily basis. Margins normally amount to 10-15 percent of total contract value. Interest on this investment cost is part of the cost of hedging.

3. *Commissions.*—Commissions on futures trades are quite small in relation to the total contract value. They are more substantial, however, in relation to the value of margin deposited or to the net gains and losses on futures.

4. *Costs of management and information.*—As this paper points out, hedging programs require constant updating of data on prices, on the firm's cash net position, and on the results of current hedging activity. In addition, management time must be occupied in developing and implementing hedging policy. In many agribusiness organizations, these decisions are made at high management levels.

FEASIBILITY OF HEDGING SOUTHERN PINE

Southern pine can be profitably hedged only if the futures markets are liquid, if the peculiarities of the southern market are taken into account, if futures prices are correlated closely with cash prices, and if basis patterns can be defined.

Liquidity

Hedgers need to be certain that they can place and lift hedges as planned and at currently quoted market prices. Southern wholesalers and manufacturers, then. must ask themselves if the existing futures markets possess enough liquidity to offer a workable hedging medium.

market'
ling volu
ice relationship

l dimensions. Th

nding contracts.
rtant. Other, le

arized by the tra
of contracts trad
y, month, or ye
es that any giv
speculators are b
irket, buy or sell

n five different o

n each month so t
not or nearby are

er and plywood m

e controls. Howe
rated the capacity

contracts in Sept
arket topped 90

(figs. 2 and 3).

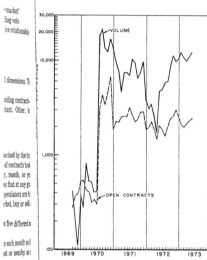

Figure 2.—*Trading activity in lumber futures, Chicago Mercantile Exchange. Trading volume, monthly totals; open interest as of last business day. Source: Table 1.*

 800 by mid-July. Thereafter, the open interest as slowly liquidated. By the close of trading, it reached 100 contracts, which represented the number of deliveries made.

In lumber, open interest jumped to more than 7,000 contracts in the trading flurry of late 1970 but since then has remained between 2,000 and 4,000. Plywood open interest reached its high of about 5,600 contracts early in 1971 and ended 1972 just above 4,000. The rising market in the second half of 1972 clearly benefitted open interest in both markets, though not as markedly as it aided trading volume.

Depth.—The depth of a market is determined by the extent to which pit traders are holding stop orders—orders to be executed at prices very near the ruling price at any moment. When a burst of buying or selling volume occurs, these stop orders are triggered; they aid in filling the orders with little disturbance of the price. The lumber and plywood markets have now developed substantial depth. At any given moment,

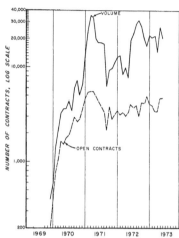

Figure 3.—*Trading activity in plywood, Chicago Board of Trade, December 1969 to June 1973. Trading volume, monthly totals; open interest as of last business day. Source: Table 1.*

therefore, these markets are more liquid than the trading volumes alone indicate.

Breadth.—Breadth refers to the diversity of the groups participating in trading. A market shared only by office wholesalers and floor traders lacks breadth. A broad market includes trading activity by manufacturers, office wholesalers, warehouse wholesalers, wood users, floor traders, and speculators from the general public. Today the markets are between these two extremes. Widespread trade participation developed in plywood quite rapidly, and both markets have continuously gained breadth.

Trading volume, open interest, depth, and breadth are such that lumber and plywood futures markets can offer southern pine hedgers workable hedging services. At times, however, trading in distant months may be undesirably light. Liquidity feeds itself. High volume and open interest tend to enhance the depth and breadth by assuring the most effective hedging opportunities and by attracting speculation and hedging from new groups. Increased participation by southern pine hedgers will make the futures markets better hedging media for everyone.

Differences Between Southern and Western Markets

Hedging needs and opportunities are affected by differences between southern and western markets.

Broadly speaking, southern pine is a regional species, while western items are the basis of the national market. The South has a large and growing regional market, and it is now roughly self-sufficient in softwood lumber and plywood. Thus, regional supply or demand disruptions such as bad weather or strikes in the South could leave production in other regions unaffected. National prices, however, would respond to disruption in the South.

Since southern pine is not shipped as far as the western species, southern pine wholesalers need not hold cars of wood as long as dealers of western species. In addition, the average hauling distance affects the modes of transportation used. Rail shipments account for only about 30 percent of southern pine production but for roughly 60 percent of western output. Since the delivery unit for futures contracts is a rail carload, truck shipments do not match contract specifications.

Seasonal patterns of production and demand differ in the South and in the West. The South's poorest logging conditions occur in the early spring; the Northwest has poorest conditions in the winter. Thus, the timing of potential supply interruptions and inventory needs is quite different in the two regions.

In addition, seasonal demand patterns in the South are distinctive. Due to its mild climate, the region has a longer building season and less fluctuation from quarter to quarter in building activity. In the South, construction is slack during the fourth quarter; elsewhere the slowest quarter is the first. The margin by which the high quarter exceeds the low quarter is also lowest in the South—26 percent as compared to 75 percent in the Midwest (table 3). These factors affect the seasonal patterns of orders and inventory holding (Row, 1961).

In the past, southern pine lumber prices were more stable than western prices (Robinson,

8

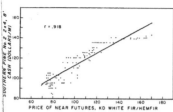

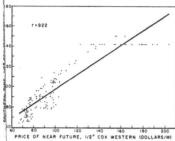

Figure 4.—*Scatter diagram of weekly cash prices versus futures prices for southern pine lumber, #2 2x4x8 ft., October 1969 to February 1973.*

Figure 5.—*Scatter diagram of weekly cash prices versus futures prices for southern pine ½ in. CDX plywood, December 1969 to February 1973.*

tatistics data for the period 1959-1968. He ᵇrrelated a series of species and grades with ᵗhite fir lumber, which was then the contract ᵣade. He found that Douglas-fir green 2×4's ᵇrrelated closely but that southern items corre-ᵗted less well (table 4). Intensive studies ᵛvering 1962-1968 with monthly averages from ᴸandom Lengths were carried out by the Harris ᵀrust Bank for the Chicago Mercantile Ex-ᵠange and summarized by Coy (1969) and ᴾrost (1970). These studies necessarily relied ᵘpon cash prices only.

An additional study was performed for this ᵖaper. The data period begins with the opening ᵈ trading in October 1969 for lumber and in ᴵecember 1969 for plywood, and ends with the ᶠrst week of February 1973. Weekly cash prices ᵃe compared with Friday closing prices of the ᵣar future. Comparison with the correlation of

Table 4.—*Correlation between futures and cash prices, selected lumber and plywood items*

Item	Correlation coefficient	
	This study	Robinson
Lumber		
White fir — hem-fir cash	0.975	1.00
Douglas fir std. & better		
R. L. green 2×4	.941	.974
Ponderosa pine #3COM 1×6	.926	.924
Southern pine #2, KD 2×4×8 ft.	.918	.939
Southern pine #2 & better		
boards, 1×6	.932	.921
Plywood		
½ in. CDX western — cash	.941	...
¼ in. AD interior western	.808	...
5/16 in. CDX southern	.877	...
½ in. CDX southern	.922	...
¾ in. CDX southern	.927	...

cash and futures prices *for the contract grades,* is the appropriate standard for deciding whether noncontract grades can be hedged.

For lumber, hem-fir cash prices correlate well with futures, with a coefficient of 0.975 (table 4). Douglas-fir comes relatively close to this level, while southern pine # 2 dimension is barely above 0.91.

Western plywood is currently being hedged with a correlation of only 0.941. Southern pine ½ in. CDX comes remarkably close to the western grades at 0.922. Due to the differences between sheathing and sanded markets, ¼ in. AD interior correlates only weakly with plywood futures.

Hedging prospects in the stud market must be determined using cash prices since the contract commenced trading only recently. Statistical studies suggest a high degree of correlation between lodgepole pine prices (the contract grade) and prices of other major stud species (table 5).

Table 5.—*Coefficient of determination (R²) in regression of stud species prices on lodgepole pine studs, cash prices. All grades are 8' 2×4 PET, std. & better*

Species	R²
Douglas-fir, green	0.97
Douglas-fir, KD	.97
Hem-fir, coastal KD	.97
Southern pine, KD	.80

Source: Merrill Lynch, Pierce, Fenner & Smith, Inc.

9

For southern pine, which contributes only about 9 percent of stud production, the correlation is only about 0.80. Southern stud producers may not be able to hedge in this market.

Basis Relationships

The price correlation only indirectly indicates hedging opportunity. A direct indicator is the pattern of basis relationships.

Basis is the difference between cash and futures prices at a given moment. It is commonly computed as follows:

$$\begin{bmatrix} \text{Price of near future} \\ \text{for contract grade} \end{bmatrix} - \begin{bmatrix} \text{Cash price for item} \\ \text{to be hedged} \end{bmatrix}$$
$$= \text{BASIS}$$

Hedgers must study basis patterns carefully because they powerfully affect the profitability of hedging.

Standard examples of hedging assume that the basis remains constant over the life of the contract. Consider the following short hedge:

December 1972		*May 1973*		*RESULTS*
CASH				
Buy 70 MSF[4] ¾ in. @$200 CDX		Sell plywood	@$180	−20
Futures				
Sell one contract May	@ 210	Cover hedge	@ 190	+20
BASIS	10		10	0

On this hedge, the wholesaler placed a car of plywood in storage at $200 and lost $20 per MSF or about $1,400 on selling in May. His futures transaction returned $1,400 (ignoring commissions and interest on margin) so that he was fully protected and broke even. Note that the basis—the price difference between cash and futures—remained the same. What would have happened had it changed? Suppose that futures had fallen faster than cash over the life of this hedge. The hedger would then gain more on his short sale than he lost on the cash plywood, and the basis would have narrowed. If futures had fallen to $180 by May, the gain on futures would have been $30; and the hedge would have turned a profit of $10 per MSF or $700. This profit is measured directly by the change in the basis from $10 to 0. By the same token, if futures fell

4 Thousand square feet

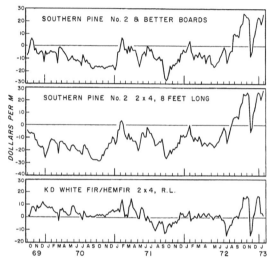

gure 6.—*Weekly lumber basis patterns, October 1969 to February 1973. Cash prices versus near future.*

These charts show that southern pine items ɔ possess basis patterns which are similar to ιose of the western contract grades. The pat-ɪrns are somewhat more marked and more able for plywood than for lumber.

mitations

Evidence assembled in this paper shows that ιuthern pine manufacturers, wholesalers, and ιers can profitably hedge in western contracts ɪen though they cannot deliver the cash com-odity. This general result, however, may not ιply to all firms because it is based on reported ɪerage prices for the items studied.

Price patterns, correlation coefficients, and sis patterns can vary between regions, be-ɪeen different grades and species, and over ɪe. Therefore, prospective hedgers will have study correlations and basis patterns using ɜ net realizations that apply to the products ɜy deal in and the markets they serve. Once ɜse studies are made, the information needed ˙ normal hedging can be utilized to update re-ts and compare with actual hedging outcomes.

Even after a firm has completely studied price ˙relation and basis patterns, obstacles remain.

These are due to the unrepresentative nature of price movements in the recent past. Since 1968 the demand for lumber and plywood has been highly unstable. Correlation and basis studies for this period may not be reliable guides to fu-ture market behavior. It is impossible to predict whether wide fluctuations will persist or whether markets will settle into the more placid pattern of the early 1960's. For this reason con-tinuous updating of correlation and basis studies and monitoring of hedging performance will be necessary.

Finally, price controls have distorted normal price relationships. These distortions appear clearly in the basis charts and impart an un-known bias to these results. Price history and therefore basis relationships are somewhat biased in that the market has been generally rising since the inception of trading.

Despite these difficulties, however, profitable hedging opportunities do exist for southern pine, as the following examples illustrate.

EXAMPLES OF HEDGING PRACTICE

The mechanics of hedging are best illustrated through examples (see also Gray, 1972; Radoll,

11

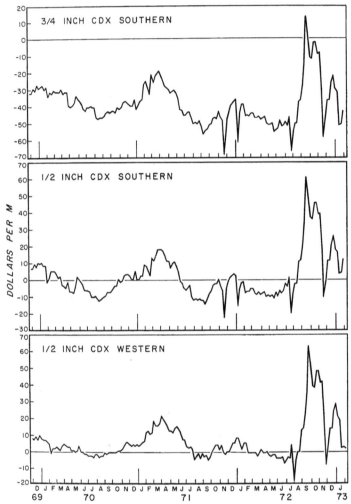

Figure 7.—*Weekly plywood basis patterns, December 1969 to February 1973. Cash prices versus near future.*

)72). In all of the examples given here, the im-
ıct of price fluctuations on reported profits de-
ʼnds upon practices used in accounting for raw
aterial costs and for changes in inventory
ılues.

Basis and correlation analyses provide the
·oad base from which to evaluate historical
:ice fluctuations and price ranges. The most
gnificant factor to evaluate, especially in non-
ʼntract grades, is expected price direction. It
price direction plus the basis and regression
ıtterns that guides hedging strategy. Our ex-
ʼnples illustrate the use of these concepts in
ʼdging practice.

For simplicity, all examples use southern ½
. CDX and southern pine # 2, 2×4×8 ft.
rices are taken from *Random Lengths* and
otted weekly in figures 8-11 to determine the
ısis. These prices are not discounted and could
ffer from actual net realizations.

ıy Hedge

In the second half of 1972, an Atlanta whole-
.ler plans his inventory buildup for the 1973
ıilding season. He plans to hedge in order to
ʼnefit from price protection and to hold inven-
ry without a large cash outlay by relying on a
ng position in futures. He considers the fol-

lowing factors:

Price direction —He expects prices to *rise*
next year due to increased construction activity.

Which contract —He needs the material before
month to use the end of March for spring
and summer building; so he
selects the March contracts.
The lumber contract expires
March 15 and the plywood on
March 23.

Option: a moving —Alternatively, he may elect
hedge to place a series of hedges in
intervening months through
the fall and winter, lifting
each one in turn and replacing
it with a more distant month—
"rolling over" his hedge. For
example, he might buy the
November, then sell at a favorable basis, and buy the March
contract to regain his hedged
position.

When to place —The wholesaler desires to
the hedge buy futures in expectation of a
favorable basis movement over
the life of the hedge. He makes
a trading profit if the basis

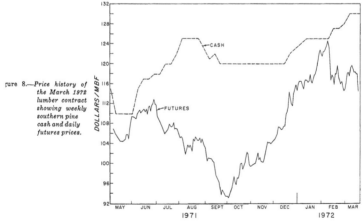

;ure 8.—*Price history of
the March 1972
lumber contract
showing weekly
southern pine
cash and daily
futures prices.*

13

Figure 9.—*Price history of the March 1972 plywood contract showing weekly southern CDX cash and daily futures prices.*

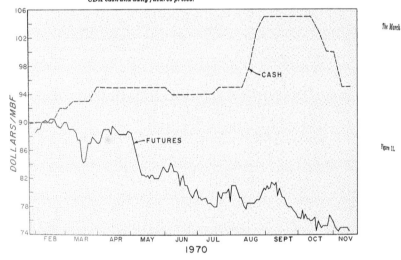

Figure 10.—*Price history of the November 1970 lumber contract showing weekly southern pine cash and daily futures prices.*

14

rises. He therefore wishes to place his hedge at a seasonal low in the basis and lift it (sell futures) when the basis is high.

mber:

The wholesaler would consider specific data r each item to be hedged. For lumber, he would ıd the following:

:mber price 1sonal	—Recent market fluctuations make a clear seasonal pattern difficult to identify, but there is a tendency for lows at year-end and for highs in August and September (fig. 1).
:ar term basis ttern	—The basis charts show futures generally below cash from late March to early fall and often reaching a premium in early winter (fig. 6). This pattern has prevailed since inception of trading, but it is biased by the generally rising market since that time.
ıe March basis r Southern pine	—Figure 8 for southern pine lumber shows cash and futures for March 1972 delivery.

March futures generally remain at a discount to cash from the beginning of trading. The discount increases in the late fall and early winter and narrows at yearend. Futures reach a premium over cash in the first months of the new year.

Plywood:

Plywood price seasonal	—Plywood prices tend to peak in midsummer, with lows during the winter (fig. 1).
Plywood near term basis	—In normal years futures move from a premium in winter or early spring, to discount by summer, and back to a premium in the fall (fig. 7).
The March 1972 plywood basis	—Futures remained at a premium to cash from inception of trading until July. Thereafter, futures were below cash until winter (fig. 9).

The same hedge analysis helps wood users, especially builders, protect themselves against price advances. They can thus fix material costs

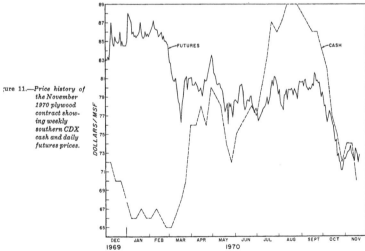

;ure 11.—*Price history of the November 1970 plywood contract showing weekly southern CDX cash and daily futures prices.*

15

and price homes more securely and competitively.

Example 1: Applying this strategy in the fall of 1972 for 1973 needs, the wholesaler made a profitable transaction.

Date	Cash transaction	Futures transaction
10/18/72	On October 18 arrange through normal supply channels for purchase of 1 million board feet of southern pine 2×4's for January delivery. The price will be determined by the prevailing market at time of delivery. Cash lumber is now $120.	Buy 10 March lumber futures at $98 per M b. f. through a commodity brokerage house.
1/24/73	Receive lumber at $125 per M b. f. or a cost of $125,000. He has paid $5,000 more for his lumber than he would have paid in early October.	Sell 10 lumber contracts at $120 per M b. f. He has a gross gain of $22,000 for his lumber.

Trading profits give him an average cost of $103 for his lumber, minus minimal commissions and interest on the margin deposited.

Example 2: The same wholesaler quotes a customer a price for plywood several months in advance using the hedge analysis described above.

Date	Cash transaction	Futures transaction
10/18/72	Current cash value for plywood is $88. Customer wants January delivery. Wholesaler quotes for future delivery of 700,000 sq. ft. of pine plywood at current prices. Customer accepts and wholesaler makes order through regular market channels for January delivery at the then prevailing price.	Wholesaler buys 10 March plywood futures contracts at $87 per MSF.

16

The millman effectively prices his plywood at $110 per MSF less small commission and interest costs.

REFERENCES

Arthur, H. B.
  1970. Commodity futures as a business management tool. 392 p. Boston: Harvard Bus. Sch.

Chicago Board of Trade.
  1972. Statistical annual 1971. 299 p. Chic. Board Trade.

Chicago Board of Trade.
  [n.d.] Hedging highlights. 18 p. Chic. Board Trade.

Chicago Mercantile Exchange.
  1972. Fundamental factors affecting lumber. 32 p. Chic. Merc. Exch.

Chicago Mercantile Exchange.
  1972. Yearbook, 1971-72. 363 p. Chic. Merc. Exch.

Coy, S.
  1969. Pricing, volume, and weighted averages. 14 p. Chicago: Harris Trust and Savings Bank.

Frost, R.
  1970. Lumber futures: first year review. In Futures trading in plywood and lumber—a profit margin insurance, p. 36-48. B.H. Box (ed.). LSU Coop. Ext. Serv. Pub. 1630.

Gray, R. W.
  1972. Trading in plywood futures. 26 p. Chic. Board Trade.

Hieronymus, T. A.
  1971. Economics of futures trading for commercial and personal profit. 338 p. New York: Commod. Res. Bur., Inc.

Irland, L. C.
  1973. Basis charts for southern lumber and plywood hedgers. USDA For. Serv. Res. Note SO-154, 5 p. South. For. Exp. Stn., New Orleans, La.

Irland, L. C.
  1973. Futures trading for southern pine manufacturers and wholesalers. South. Lumberman 227 (2816): 12, 14.

Kulp, C. A.
  1931. Possibilities of organized markets in various commodities. Ann. Am. Acad. Polit. and Soc. Sci. 155(1): 176-195.

Labys, W. C., and Granger, C. W. J.
  1970. Speculation, hedging, and commodity price forecasts. 320 p. Lexington, Mass.: D. C. Heath and Co.

Olmedo, J. P., Jr.
  1973. Understanding the lumber and plywood futures markets. In Commodity yearbook, 1973, p. 6-15. New York: Commod. Res. Bur., Inc.

Radoll, R. W.
  1972. Hedging: plywood futures. 39 p. Chic. Board Trade.

Random Lengths Publ. Co.
  1972. Random Lengths yearbook. 147 p. Eugene, Oreg.: Random Lengths Publ. Co.

Robinson, V. L.
  1970. Risk and the lumber futures market. For. Prod. J. 20(12): 11-16.

Row, C.
  1961. Seasons set pace for activity in the lumber business. Lumberman 88(1): 26-28.

17

Irland, Lloyd C., and Olmedo, James P., Jr.

    1973.   Hedging southern pine through futures
       trading. South. For. Exp. Stn., New
       Orleans, La. 17 p. (USDA For. Serv.
       Res. Pap. SO-91)

Futures trading in lumber and plywood enables
southern pine manufacturers and wholesalers to
protect themselves against price fluctuations and
gain added marketing flexibility. Existing
markets are sufficiently liquid, price correlations with futures are close enough, and basis
relations are predictible enough to allow effective hedging. Hedging practices are illustrated
with examples.

Additional keywords: Prices, marketing, risk.

woody plants valuable to wildlife in southern forests

U.S Department of Agriculture
Forest Service Research Paper SO-92

Simeon W. Oefinger, Jr and
Lowell K. Halls

Southern Forest Experiment Station
Forest Service
U S Department of Agriculture

USDA Forest Service
Research Paper SO-92

IDENTIFYING WOODY PLANTS
VALUABLE TO WILDLIFE IN SOUTHERN FORESTS

Simeon W. Oefinger, Jr.,
and
Lowell K. Halls

Southern Forest Experiment Station
Forest Service
U. S. Department of Agriculture

1974

Oefinger, Simeon W., Jr., and Halls, Lowell K.

1974.   Identifying woody plants valuable to wildlife
in southern forests. South. For. Exp. Stn.,
New Orleans, La. 76 p. (USDA For. Serv.
Res. Pap. SO-92)

Twigs, buds, and other key identification features are
illustrated in color for 70 browse species common to pine-
hardwood forests of the South.

FOREWORD

Trees, shrubs, and woody vines constitute the major source of food and cover for wildlife in southern forests. One must be able to identify these important plants in order to appraise the habitat. The task is difficult, especially during the winter when the more recognizable plant features such as leaves, flowers, and fruits are often absent.

This publication describes and illustrates some key identification features of 70 species. It stresses the characteristics of stems and buds in winter, since a description of these parts is rarely available in taxonomic references. As an aid in identification the species are grouped according to physical characters such as growth form (vines or shrubs), leaf, stem, and bud arrangement (alternate or opposite), and leaf persistence (deciduous or persistent).

The list consists mainly of woody understory species that are eaten by wildlife, with special emphasis on those preferred by white-tailed deer *(Odocoileus virginianus)*. The intention was to include sufficient species for a practical habitat appraisal almost anywhere in the loblolly-shortleaf pine-hardwood forest type. It is recognized, however, that the list is not all-inclusive and that other species may be important locally.

The dominant overstory trees such as oaks and hickories are extremely useful to wildlife. For the most part, however, they are omitted here because excellent guides to their identification are readily available elsewhere. Too, limitations of space have made it necessary to represent the large genera of *Rubus* and *Crataegus* by one species each.

References for Latin and common names, listed in order of preference, were:

Little, E. L., Jr.
    1953. Check list of native and naturalized trees of the United States (including Alaska). USDA Agric. Handb. 41, 472 p.
Fornald, M. L.
    1950. Gray's manual of botany. Ed. 8, 1,632 p. New York: Am. Book Co.
Correll, D. S., and Johnston, M. C.
    1970. Manual of the vascular plants of Texas. 1,881 p. Tex. Res. Foundation, Renner.
Kelsey, H. P., and Dayton, W. A.
    1942. Standardized plant names. Ed. 2, 675 p. Harrisburg, Pa.: J. Horace McFarland Co.

dlife
Stn.,
Serv.

are
pine-

ACKNOWLEDGMENTS

Daniel W. Lay, Texas Parks and Wildlife Department, Phil D. Goodrum, Bureau of Sport Fisheries and Wildlife, U. S. Department of the Interior, and Carroll J. Perkins, International Paper Company, helped in the selection of species to be illustrated.

Suggestions on content and organization were given by Stephen F. Austin State University faculty members Elray S. Nixon, Russell C. Faulkner, Jr., Creighton Delaney, and Reesman S. Kennedy.

Several biologists in private industry and government agencies supplied plant specimens for photographing.

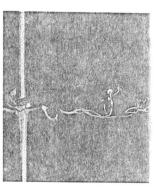

Bignonia capreolata L.—CROSSVINE

Tardily deciduous to evergreen vine with opposite bud
and leaf arrangement; climbs by tendrils. Leaves or scars
in two opposite pairs. Twigs glabrous, greenish to gray
with age; smooth when young. Old stems with rough, curled,
papery bark. Leaf-supporting stalks terminated by two
leaves or leaf scars and one branching tendril. Tendrils
have disk-like tips and are persistent on old growth. Lateral
buds small, glabrous to slightly puberulent; wider than tall;
with two scales. Leaf scars terete. One bundle scar; curved
line or circle. Cross section of larger stems with Maltese
cross pattern. Pith later absent.

Leaves two-foliate, tendril bearing.

Flowers red-orange externally, yellow internally, axillary
in 2 to 5 flowered cluster, April to June.

Fruit a many-seeded capsule.

3

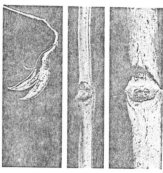

Campsis radicans (L.) Seem.
COMMON TRUMPETCREEPER

Deciduous climbing or creeping vine with opposite leaf scars. Aerial rootlets. Stems glabrous, green to reddish-tan, dying back in winter with no terminal bud. Leaf scars large, half round with continuous stipule scars. Bundle scars U-shaped to circular.

Leaves odd-pinnate, olive green, lustrous and glabrous above.

Flowers reddish-brown in terminal cluster, June to September.

Fruit a dry, brownish, two-celled capsule, with numerous winged seeds, pods long persistent.

4

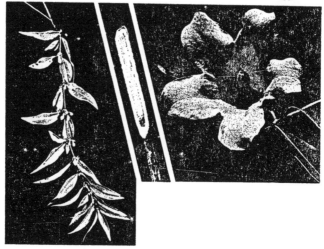

Gelsemium sempervirens (L.) Ait.—CAROLINA JESSAMINE

Evergreen vine with opposite leaves. Creeping to climbing. No tendrils or aerial rootlets. Stems hollow, glabrous, reddish-brown with occasional glaucous bloom. When stems are broken silvery hairlike strands can be seen at the edge of the break. Previous season's four-lobed fruit pods commonly persist through winter. Leaf scars half round to circular. One bundle scar.

Leaves slender, glabrous, entire, mostly three or more times longer than wide.

Flowers trumpet-shaped, deep yellow, fragrant and fragile, February to April.

Fruit a capsule with numerous winged seeds.

Lonicera japonica Thunb.—JAPANESE HONEYSUCKLE

Evergreen vine with opposite leaves. Climbing or creeping without tendrils or aerial rootlets. Stems pubescent, less so with age; becoming hollow. Leaf scars narrow, flat, and covered by remnants of leaf petioles. Bundle scars three with petiole remnant removed.

Leaves pubescent above and beneath, margins commonly entire but occasionally toothed to lobed.

Flowers fragrant, in pairs from leafy bracts, white or pink, later yellow, June to August.

Fruit a black pulpy berry with several seeds, September to March.

Rubus trivialis Michx.—SOUTHERN DEWBERRY

Deciduous trailing or creeping vine. New stems or primo-
canes with curved or hooked prickles and numerous reddish
glandular hairs. Stems becoming stiff and glabrous with
age. Prickles persistent. Stems commonly root at tips
and nodes.

Dead leaves occasionally persist through winter. Flori-
cane leaves usually 3-foliate and primocane leaves 5-foliate;
margins coarsely serrate to dentate, petioles with glandular
hairs.

Flowers white to pinkish on solitary, erect, armed pedicels,
March and April.

Fruit a large-seeded berry, black at maturity, sweet, juicy
and attractive, June and July.

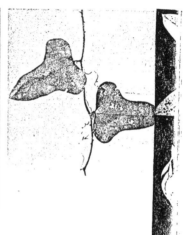

Smilax bona-nox L.—SAW GREENBRIER

Deciduous, spiny, climbing vine with tendrils. Alternate leaf scars. Young stems somewhat square, glabrous, green, and angled at joints. Buds and leaf scars covered by remnants of leaf bases. Pith absent, stems porous in cross section. Spines paired at nodes.

Leaves simple, usually thick and often blotched with white. Margins and main veins usually prickly.

Flowers borne in staminate and pistillate axillary umbels, April to June.

Fruit a shining or dull black berry, usually one-seeded, September to November.

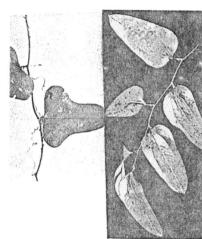
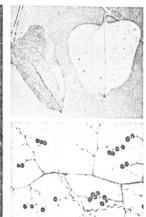

RIER

ils. Alternate
abrous, green,
vered by rem-
rous in cross

blotched with
ly.
cillary umbels,

y one-seeded,

Smilax glauca Walt.—CAT GREENBRIER

Tardily deciduous, spiny, climbing vine with or without tendrils. Alternate leaf scars. Young stems glabrous, green, and angled at joints. Leaf scars and buds covered with remnants of leaf bases. Nodes with or without spines. Pith absent, stems porous in cross section. Leaves simple, green to black-green above, glaucous white below.

Flowers small, greenish, borne in separate staminate and pistillate umbels, May and June.

Fruit a black berry with a bloom, ripens in first year, September and October.

9

Smilax laurifolia L.—LAUREL GREENBRIER

Evergreen climbing vine with tendrils and alternate leaves. Spines generally sparse, absent from nodes. Stems glabrous, waxy green. Pith absent; stems porous in cross section.

Leaves simple, shiny green, and glabrous, somewhat paler below; thick and leathery with three prominent veins.

Flowers greenish-white, borne in separate staminate and pistillate umbels, July and August.

Fruit a shiny black berry in clusters of 5 to 25, persistent, matures in October of second season after flowering.

RIER

ernate leaves.
. Stems gla-
rous in cross

Js, somewhat
ominent veins.
staminate and

25, persistent,
flowering.

Smilax rotundifolia L.—COMMON GREENBRIER

Deciduous, climbing, spiny vine with alternate leaf scars. Leaf scars and buds covered by remnants of leaf bases. Stems glabrous, green to yellowish-green. Pith absent, stems porous in cross section. Spines stout and scattered, absent at nodes.

Leaves simple, mostly round and smooth, tips abruptly pointed, green both sides, shiny beneath.

Flowers small greenish-yellow, borne in separate staminate and pistillate umbels, March to June.

Fruit a black berry, persistent, borne in rounded umbellate clusters, matures in 1 year.

11

Ampelopsis arborea (L.) Koehne—PEPPERVINE

Trailing or climbing vine with alternate buds and leaf scars. Tendrils branching and occasionally falling, with remaining scars appearing as opposite leaf scars. Twigs glabrous; grayish-tan to light brown in color. Lenticels numerous, especially so toward twig tips, light gray or tan in color. Pith soft, white, homogeneous but interrupted at nodes. Twigs swollen at nodes. Terminal buds absent. Lateral buds small, glabrous, pyramidal in shape. Leaf scars circular. Bundle scars several, arranged in a circle.

Leaves deciduous, bipinnate, coarsely toothed, incised or lobed.

Flowers greenish-white, small, inconspicuous, in cymes, June and July.

Berries shiny black at maturity, 1 to 4 seeds.

VINE

Berchemia scandens (Hill) K. Koch
ALABAMA SUPPLEJACK

s and leaf
lling, with
ars. Twigs
Lenticels
gray or tan
errupted at
ds absent.
Leaf scars
circle.
ed, incised

, in cymes,

Stout, deciduous vine with alternate leaf scars and buds. Climbing without aerial rootlets or tendrils. Very tight and smooth bark. Young stems glabrous and shiny green to red. Leaf scars elliptic, somewhat raised above the stem. One bundle scar, occasionally indistinctly broken into two or three parts. Buds with two scales, glabrous.

Leaves simple, with parallel veins.

Flowers small greenish-white in small panicles, March to June. Fruit a bluish-black drupe containing a two-celled stone, July to October.

13

  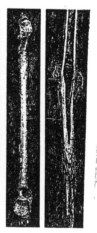

Parthenocissus quinquefolia (L.) Planch.
VIRGINIA CREEPER

Deciduous vine climbing by tendrils. Alternate buds and leaf scars. Twigs light purple-gray to tan-gray and puberulent; glabrous with age. Lenticels tan, large, and appearing split longitudinally with ridges remaining at sides. Leaf scars terete. Stipule scars present. Bundle scars four to six dots arranged in a circle. Terminal bud absent. Lateral buds in side-by-side pairs or occasionally absent; glabrous scales. Tendrils branched with terminal disk-like appendages. Tendrils opposite leaf scars. Pith variously chambered to hollow; not interrupted at nodes.

Leaves palmately compound with five leaflets, margins coarsely serrate-dentate except near base and apex.

Flowers in panicles of compound cymes, greenish, May to August.

Fruit a berry in spreading erect cymes, greenish-blue, seeds 2 to 4, ripen in fall.

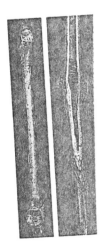

anch.

rnate buds and
ay and puberu-
, and appearing
at sides. Leaf
e scars four to
absent. Lateral
bsent; glabrous
ısk-like append-
ariously chamb-

leaflets, margins
and apex.
, greenish, May

s, greenish-blue,

Rhus radicans L.—COMMON POISON-IVY

Deciduous vine with alternate leaves. Climbing with aerial rootlets or creeping. Rootlets with disk-like appendages. Tendrils and thorns absent. Twigs dark reddish-brown to grayish-tan or brown. New growth pubescent, becoming glabrous. Leaf scars large, U-shaped. Bundle scars five. Buds naked.

Leaves trifoliate, margin entire, dentate or lobed.

Flowers small, greenish-white, fragrant in axillary panicles, May to July.

Fruit a persistent drupe, clustered, dull white, waxy, glabrous or sparsely pubescent, August to November.

15

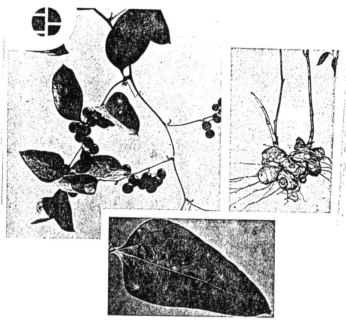

Smilax smallii Morong—LANCELEAF GREENBRIER

Evergreen vine with alternate leaves and branches. Tendrils present. Commonly climbs high in trees. Thorns uncommon, practically nonexistent on new growth. Stem porous in cross section.

Leaves leathery in texture, ovate to elongate ovate with acute tips, shiny green above and below, margins smooth, five veins, rarely seven.

Flowers yellowish-green, fragrant, staminate and pistillate borne separately in axillary umbellate clusters, April to July.

Fruit a maroon to blackish-red berry, in clusters, matures second year in June.

Underground stems firm, tuberous, up to 2 feet in length.

16

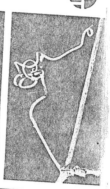

Vitis aestivalis Michx.—SUMMER GRAPE

Deciduous, thornless vine, climbing by forked tendrils. Leaf scars alternate. Stems reddish-brown, pubescent on new growth, becoming glabrous with bluish-gray waxy bloom on old growth. Buds rusty-brown pubescent underneath scales. Pith soft, brown, interrupted at nodes. Lenticels uncommon. Leaf scars triangular to half round. Bundle scars indistinct.

Leaves simple, margin of some irregularly toothed and unlobed, others shallowly to deeply 3 to 5 lobed.

Flowers small, clustered in a slender, loose panicle, May to July.

Fruit a globose, dark blue to black, persistent berry, September and October.

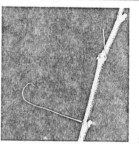

Vitis rotundifolia Michx.—MUSCADINE GRAPE

Deciduous, thornless vine climbing by tendrils. Tendrils not forked. Leaf scars alternate. Stems light rusty-tan to purplish-brown, glabrous. Buds rusty-brown underneath scales. Pith firm, light tan to greenish-tan, not interrupted at nodes. Lenticels profuse and lighter than twigs. Leaf scars half round to circular. Bundle scars indistinct.

Leaves simple, margin coarsely angular-dentate, seldom lobed.

Flowers small in dense short-branched panicles, May and June.

Fruit a purplish-black berry borne in short, loose, globular clusters, dropping singly as soon as ripe, July to September.

18

SHRUBS OR TREES WITH
STEMS, BUDS, AND LEAF
SCARS OPPOSITE,
SUB-OPPOSITE, OR
WHORLED; LEAVES
DECIDUOUS

Aesculus pavia L.—RED BUCKEYE

Deciduous shrub with opposite buds and leaf scars. Twigs greenish-gray to tan, puberulent, becoming glabrous. Buds large, reddish-gray. Terminal buds usually absent. Leaf scars large, deeply to shallowly U-shaped. Bundle scars three or roughly in three groups. Pith heterogeneous with irregular longitudinal chambers. Lenticels warty and numerous.

Leaves palmately compound.

Flowers red, in panicles, March to May.

Fruit a light brown capsule.

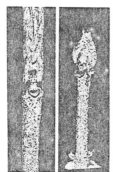

Callicarpa americana L.—AMERICAN BEAUTYBERRY

Deciduous ascending shrub with opposite leaf scars and buds. Generally with several stems arising from one rootstock. Remnants of fruit stalks at tips of stems remain through winter. Leaf scars circular with one U-shaped bundle scar. Current growth puberulent and gray to dark grayish-brown.

Leaves simple, white pubescent beneath, soft-textured.

Flowers small, bluish, clustered in leaf axils, June to November.

Fruit a purplish drupe with four seeds, August to November, persists to January.

Chionanthus virginicus L.—FRINGETREE

Deciduous shrub to small tree with opposite leaf ṣcars and buds. Twigs light gray to tan, puberulent and stout. Terminal and lateral buds many-scaled. Lenticels conspicuously warty. Pith soft, continuous, and white. Leaf scars half-round and somewhat protruding with one circular to horseshoe-shaped bundle scar.

Leaves simple, dark green and glabrous above, paler below with hairs on

Cephalanthus occidentalis L.—COMMON BUTTONBUSH

EREE

afcars and buds. Twigs
nd lateral buds many-
nnuous, and white. Leaf
ecircular to horseshoe-

paer below with hairs on

arh to June.
git to October.

Deciduous shrub typically branching near the ground. Buds and leaf scars opposite or whorled in threes. Stems upright at first, later bending. Glabrous, light tan to grayish twigs with opposite or whorled leaf scars. Commonly three circular leaf scars per node. Bundle scars circular to U-shaped. Stipules connecting to form circle around stem. Pith more than one-third of total stem diameter, soft and light tan in color. Lenticels large and warty. Commonly grows at the edge of or in shallow, slow-moving or still water.

Leaves simple, opposite or in whorls, glabrous above, sometimes pubescent on midrib.

Flowers numerous, white, clustered in spherical heads, June to September. Fruit a round cluster of reddish-brown nutlets, September and October.

23

Cornus florida L.—FLOWERING DOGWOOD

Deciduous shrub to small tree. Leaf scars and twigs with opposite arrangement. Twigs green to reddish with telescoping appearance at nodes. Stems with scattered appressed pubescence becoming nearly glabrous. Leaf scars from near linear to U-shaped to half round. Bundle scars three. Flower buds distinct from leaf buds, terminal and with long-stalked globose tip. Pith longitudinally parted with hard resinous glomerules frequently interspersed throughout.

Leaves simple, dark green above, light green and glaucous beneath, often clustered near end of branches.

Flower bracts snowy white or rarely pink, March to June.

Fruit a bright red, clustered drupe. September and October.

Euonymus americanus L.—BROOK EUONYMUS

Deciduous shrub with opposite branching. Leaf scars on current growth occasionally subopposite to alternate, slit-like to half round. One bundle scar. Stems remain green through several years, occasionally have reddish tinge. Fine longitudinal ridges connecting edges of succeeding pairs of leaf scars. Pith spongy. Terminal and lateral buds reddish.

Leaves simple and sessile.

Flowers in cymes, petals greenish to greenish-purple or reddish, May and June.

Fruit a warty capsule that turns purplish-red and splits at maturity to expose several reddish-orange, pulpy, seed-bearing appendages, September and October.

)OGWOOD

wigs with opposite arrange-
pearance at nodes. Stems
early glabrous. Leaf scars
scars three. Flower buds
alked globose tip. Pith lon-
les frequently interspersed

nd glaucous beneath, often

h to June.
and October.

25

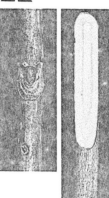

Fraxinus americana L.—WHITE ASH

Deciduous tree with opposite buds and leaf scars. Twigs greenish-gray to purple-gray, glabrous or with few scattered minute hairs towards tips. Leaf scars U-shaped, indented on upper edge. Bundle scar a curved line. Terminal buds punctate, two-scaled, and triangular in outline. Lateral buds smaller, terete, and punctate. Pith white, homogeneous, and interrupted at nodes.

Leaves odd-pinnately compound, usually seven leaflets.

Flowers with or before the leaves. Staminate clusters short and dense, pistillate clusters about 2 inches long and slender.

Fruit samaras in dense clusters, ripening in August and September.

Hydrangea arborescens L.—SMOOTH HYDRANGEA

Deciduous shrub with opposite leaf scars. Grows primarily in clumps. Twigs light tan, pubescent, becoming glabrous. Older stems with very thin, loose, shredding bark. Pith soft, continuous, white, occupying more than one-half of total stem diameter. Leaf scars triangular with tips connecting around stems. Bundle scars three. Flower head remnants commonly persist through winter.

Leaves dark green above, paler beneath.

Flowers borne in creamy white cyme which persists through winter, June and July.

Fruit a many-seeded capsule, October to December.

Rhododendron serrulatum (Small) Millais
HAMMOCKSWEET AZALEA

Deciduous shrub up to 20 feet; alternate leaf scars.
Twigs reddish-tan to grayish-brown, pubescent becoming
glabrous. Buds commonly clustered towards twig tips, giving rise to an almost whorled twig arrangement on some
stems. Scales of terminal buds with dark mucronate tips.
Leaf scars shield-shaped to triangular, with one bundle
scar.

Leaves simple, clustered at end of branches, lustrous
green on both sides, glabrous except for occasional scattered hairs on midrib beneath.

Flowers borne in few or many terminal clusters, corolla
white, sweet-scented, tube slender, cylindric, very viscid
glandular hairs, June to August.

Fruit a five-valved capsule.

28

Sambucus canadensis L.—AMERICAN ELDER

Deciduous shrub with many basal stems, rarely a small tree to 30 feet. Leaf scars opposite. Terminal buds absent. Twigs glabrous, gray to brownish-gray with small ridges towards tips. Pith soft, white and predominant, occupying over three-quarters of twig diameter. Lenticels light brown and numerous. Leaf scars large V-shaped or U-shaped with five or more dot-like bundle scars.

Leaves odd-pinnately compound with 5 to 7 leaflets, margin sharply serrate, upper surface lustrous bright green, lower surface paler.

Flowers white in conspicuous large terminal cyme, May to July.

Fruit a berry-like drupe in clusters, deep purple or black, four-celled, August and September.

Viburnum acerifolium L.—MAPLELEAF VIBURNUM

Deciduous shrub with opposite leaf scars and buds. Twigs finely puberulent, becoming glabrous in 1 to 2 years. Twigs gray to brownish-gray, becoming mottled gray. Leaf scars narrow, U-shaped to crescent shaped, connecting around twig. Bundle scars three. Buds yellowish to rusty, lightly pubescent along midrib and toward tips of scales only.

Leaves simple, margin coarsely dentate and often with three lobes, venation palmately three-ribbed.

Flowers creamy white in flattened or convex cymes, May to August.

Fruit a persistent, reddish- to purplish-black flattened drupe with shallow grooves on each side, July to October.

Viburnum dentatum L.—ARROWWOOD VIBURNUM

Deciduous shrub with opposite buds and leaf scars. Twigs reddish-tan to reddish-brown; gray on older growth; stellate pubescent at first, becoming glabrous in 1 to 2 years. Leaf scars narrow and V-shaped, continuous around twigs. Bundle scars three. Buds small, reddish-brown and glabrous except for fringe on edges of scale tips.

Leaves simple, margin serrate to dentate, the teeth rather triangular, many single or tufted hairs at fork of branched veins on underside.

Flowers white in cymes, June to August.

Fruit a bluish-black drupe with shallow grooves on one side, August to November.

OD VIBURNUM

nd leaf scars. Twigs
lder growth; stellate-
s in 1 to 2 years.
nuous around twigs.
ish-brown and gla-
ale tips.
late, the teeth rather
at fork of branched

ust.
low grooves on one

Viburnum nudum L.—POSSUMHAW VIBURNUM

Deciduous shrub with opposite buds and leaf scars. Twigs reddish-brown to purple-brown; glabrous or with sparse minute hairs. Older twigs gray and glabrous. Leaf scars U- to V-shaped and narrow, encircling young stems but later separated or only inconspicuously connecting. Bundle scars three. Terminal buds slender, elongate to one half-inch or more, with scales. Sides of buds nearly parallel. Roughness due to abundant rusty-brown scales and glands. Pubescence only along margins of scale or absent. Lateral buds similar in appearance but usually much smaller and somewhat more flattened. Pith white and soft.

Leaves simple, leathery, margin entire or obscurely serrulate, olive green to dark green and lustrous above, lower surface paler and with rusty-brown scales.

Flowers white in flat or round-topped cymes, April to June.

Fruit a drupe, pink at first but glaucous blue later, maturing in autumn.

33

Viburnum rufidulum Raf.—RUSTY BLACKHAW

Deciduous shrub with opposite leaf scars. Twigs glabrous, purple-brown with bluish-gray waxy bloom. Buds naked and covered with short rust-colored pubescence. Leaf scars U-shaped to narrowly V-shaped with three dot-like bundle scars. Pith white, soft, and one-third or less of total twig diameter.

Leaves simple, dark green, leathery, shiny above, paler below; red hairs on veins, margin finely serrate.

Flowers white in flat cymes, April and May.

Fruit a bluish-black glaucous drupe in drooping clusters, July to October.

Cliftonia monophylla (Lam.) Britt.—BUCKWHEAT-TREE

Evergreen shrub with alternate leaves. Twigs reddish-brown to gray. Pith at least one half the diameter of stem. Leaf scars shield-shaped. Bundle scars slit-like to U-shaped. Tip of largest scale on terminal bud overtops other scales with long curved tip.

Leaves entire, tapering acutely toward base, firm, green and shiny above, paler beneath.

Flowers fragrant, small, white or pinkish in slender spikes, erect when blooming, late winter or early spring.

Fruit a reddish-brown nut-like drupe with 2 or 3 light brown seeds, August and September.

Cyrilla racemiflora L.—SWAMP CYRILLA

Evergreen shrub to small tree with alternate leaf scars. Leaf scars half round to triangular with one circular to U-shaped bundle scar. Twigs glabrous, gray to brown with pith one-half or more of total stem diameter. Typically found in wet or swampy sites.

Leaves entire, glabrous, acutely tapered at bases and often clustered toward twig tips.

Flowers white in slender racemes opening in spring.

Fruit a dry, light yellowish-brown capsule, seeds minute, maturing in late summer.

36

37

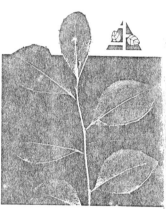

Ilex coriacea (Pursh) Chapm.—LARGE GALLBERRY

Evergreen shrub to small tree with alternate leaves. Pith one-third or less of total twig diameter. Stems puberulent, becoming glabrous, gray to greenish in color. Leaf scars half round to triangular, with one bundle scar.

Leaves glabrous above, puberulent to somewhat glaucous beneath. Tiny sharp spines sparingly borne on leaf margins from about midpoint to tips.

Flowers small and white, April and May.

Fruit a shiny black drupe, smooth, soft and pulpy, ripening in fall.

LA

ars. Leaf scars half
bundle scar. Twigs
total stem diameter.

and often clustered

ng.
minute, maturing in

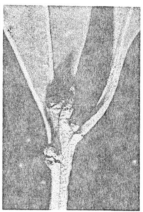

Ilex glabra (L.) Gray—INKBERRY

Evergreen shrub with alternate leaves. Twigs mottled green and light brown, very finely puberulent becoming glabrous. Pith one-third or less of total stem diameter. Leaf scars half round to flat oval and protruding. One bundle scar.

Leaves glabrous, green on top and bottom, somewhat lighter underneath. Margins entire or more commonly with few, small, blunt low teeth towards tips.

Pistillate flowers mostly solitary, staminate flowers in small pediceled clusters, petals white, February to July.

Fruit a black globose drupe, persistent, solitary or 2 to 3 together, matures in late autumn.

Twigs mottled
·ulent becoming
n diameter. Leaf
ng. One bundle

·ttom, somewhat
ɔ commonly with

inate flowers in
ɔruary to July.
solitary or 2 to 3

Ilex opaca Ait.—AMERICAN HOLLY

Evergreen tree. Stems stout, green to light brown or gray, glabrous or densely puberulent. Buds nearly globose, single bundle scar.

Leaves alternate, dark green above, paler beneath, ovate to oblong, stiff and coriaceous; margins wavy with sharp, stiff spines, sometimes spineless.

Staminate flowers small and white, 3 to 10 in cymes; pistillate flowers solitary or 2 to 3 together, April to June.

Fruit a globose or ellipsoid drupe, red, rarely yellow or orange; nutlets prominently ribbed, November and December.

39

Ilex vomitoria Ait.—YAUPON

Evergreen shrub with alternate leaves. Twigs stout, reddish-gray to gray, lightly puberulent becoming glabrous. Plants form dense shrubs under excessive utilization. Pith one-third or less of total twig diameter.

Leaves glabrous, shiny dark green above, light green below. Leaf bases rounded to a reddish petiole. Leaf scar small, half round to circular with one bundle scar.

Flowers small and white in warty sessile clusters on branches of previous year, April and May.

Fruit a bright red drupe maturing in October and persisting into winter.

eddish-gray to gray,
dense shrubs under
vig diameter.

n below. Leaf bases
und to circular with

branches of previous

ersisting into winter.

Juniperus virginiana L.—EASTERN REDCEDAR

Evergreen tree with alternate branching. Young stems covered with scales,. older stems with flaky and shreddy fibrous bark. Sap clear but sticky. Heartwood of larger stems purple. Characteristic cedar odor to stems when broken.

Leaves tiny, scale-like, overlapping, flat appressed and sharp-pointed.

Flowers small and cone-like, at end of short twigs.

Fruit a fleshy, sweet, pale blue berry-like cone with 1 or 2 seeds, September to December.

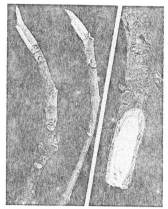

Magnolia virginiana L.—SWEETBAY

Tardily deciduous tree with alternate leaf scars and buds. Twigs mottled light and dark green to tan, becoming tan with age; softly pubescent on young stems. Terminal buds large, naked, light green with soft silvery pubescence. Leaf scars half round. Bundle scars 10 or more; arranged laterally on the face of the scar as a line or open ellipse paralleling the edges of the leaf scar. Lenticel scars completely encircling twigs. Pith homogeneous but diaphragmed.

Leaves simple, leathery, pungently aromatic; upper surface dark green, lower chalky white and often silky.

Flowers white and fragrant, May to July.

Fruits a cone-like aggregate of small follicles, yellow or reddish when ripe.

ars and buds.
becoming tan
erminal buds
scence. Leaf
ranged later-
ellipse paral-
rs completely
hragmed.
c; upper sur-
en silky.

es, yellow or

Persea borbonia (L.) Spreng.—REDBAY

Evergreen tree with alternate leaves and buds. Twigs dark green with gray woolly pubescence. Older twigs lighter brown, becoming glabrous. Buds small, generally globose with two outer scales gray woolly. Leaf scars terete. One bundle scar, dot-like. Pith near one-half of total twig diameter, firm but softer than surrounding wood, angular in cross section.

Leaves simple, bright green and elliptic to lanceolate, tapering to points at both ends, margins entire.

Flowers small, pale yellow, borne in axillary panicles, May to July.

Fruit an egg-shaped drupe, dark blue or deep purple, seed solitary.

Quercus minima (Sarg.) Small—DWARF LIVE OAK

Evergreen shrub with alternate leaves. Twigs reddish to grayish-brown. Pith star-shaped in cross section, terminal buds obtuse or rounded, clustered. Leaf scars triangular to circular with three bundle scars. No subtending bracts encircling buds.

Leaves simple, rounded or short-pointed at apex, often toothed, glabrous above, puberulent beneath.

Flowers staminate and pistillate, borne in separate catkins on same tree.

Acorns often long-stalked, solitary or several on a peduncle; cup hemispheric, nut avoid or ellipsoid, inner surface of shells hairless, July to November.

.-
n

ә,

y

әd

Quercus pumila Walt.—RUNNING OAK

Evergreen shrub with alternate leaves. Twigs reddish-brown to brown with star-shaped pith. Leaf scars triangular to half round with three bundle scars. Buds pointed, with long subtending bracts, especially towards twig tips.

Leaves simple, ending in a bristle-tip, usually entire and not toothed, glabrous above and beneath.

Acorns sessile or nearly so, cup saucer-shaped, inner surface of the shells densely hairy, July to November.

Vaccinium myrsinites Lam.—GROUND BLUEBERRY

OAK

eddish-brown to brown
round with three bundle
cts, especially towards

entire and not toothed,

d, inner surface of the

Evergreen shrub with alternate leaves. Stems green even on 2- and 3-year-old portions. Leaf scars small, half round to circular. One bundle scar, inconspicuous.

Leaves simple, tapering toward base, less than 1 inch long with serrulate margins, glabrous, green above, paler below with scattered, brown, gland-tipped, short hairs.

Flowers white to deep pink in umbel-like clusters.

Fruit a many-seeded black berry, May.

SHRUBS OR TREES WITH ALTERNATE STEMS, BUDS, AND LEAF SCARS; LEAVES DECIDUOUS

Asimina triloba (L.) Dunal—PAWPAW

Deciduous shrub to small tree, with alternate leaf scars and buds. Twigs dark gray to olive brown; pubescent at first, becoming glabrous with age. Leaf scars slightly raised on lower lip; U-shaped to V-shaped. Bundle scars five and dot-like. Terminal buds elongate and densely covered with dark-brown hairs. Lateral buds globose, covered with dark-brown hairs. Pith continuous but diaphragmed.

Leaves simple, rusty-pubescent when young, glabrous later.

Flowers purplish-green, solitary on rusty-hairy stems, appearing with or before the leaves.

Fruit banana-like with several seeds, maturing in autumn.

48

Carpinus caroliniana Walt.—AMERICAN HORNBEAM

Small deciduous tree. Glabrous reddish-brown twigs with alternate buds and leaf scars. True terminal buds absent. Bundle scars three. Trunk of larger specimens fluted. Bark smooth, gray, and tight.

Leaves simple, margin sharply double-serrate.

Staminate flowers green, borne in linear-cylindric catkins; pistillate catkins with arrowhead-like bracts that develop into a three-lobed involucre, April to June.

Fruit a nutlet, August to October.

49

Castanea pumila Mill.—ALLEGHENY CHINKAPIN

Deciduous thicket-forming shrub to tree with alternate leaf scars. Bundle scars several. Twigs pubescent at first, later glabrous. Pith angled or star-shaped in cross section. True terminal bud absent.

Leaves simple, margins coarsely serrate with pointed teeth.

Flowers appearing after the leaves, staminate catkins cylindric, slender, tomentose; pistillate flowers in threes or scattered toward base of catkins; involucre prickly.

Fruit a one-seeded nut enclosed in burs that form spike-like clusters.

Clethra alnifolia L.—SUMMERSWEET CLETHRA

Deciduous shrub with alternate buds and leaf scars. Stems ascending. Twigs pubescent, becoming glabrous. Gray to reddish-gray. Single bundle scar. Light tan to silvery gray terminal buds with villous valvate scales Remnants of previous season's fruit stalks persist until spring. Pith soft, pale green, and continuous.

Leaves simple, margins conspicuously serrate.

Flowers white, fragrant, borne in erect racemes, June to September.

Fruit a round three-sectioned capsule, erect, persistent.

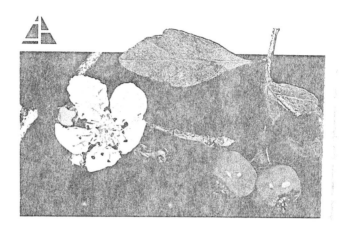

Crataegus opaca Hook. & Arn.—RIVERFLAT HAWTHORN

Deciduous shrub or small tree, with alternate leaf scars and buds. Twigs brown to reddish-brown, becoming gray with some mottling. Scattered white pubescence on young twigs, becoming glabrous. Spines up to 1.5 inches long on most twigs; dark-brown to reddish-brown, becoming brittle and mottled gray with age. On older growth the flower buds commonly occur on short, stout spurs. Buds small, round and glabrous, reddish. Leaf scars slender and flat to slightly U-shaped. Bundle scars three. Pith hard.

Leaves simple, dark green and usually glabrous above, densely rusty-brown pubescent below.

Flowers white, borne before leaves in February or March. Fruit a pome with 3 to 5 nutlets, April and May.

Fagus grandifolia Ehrh.—AMERICAN BEECH

Deciduous tree with alternate leaf scars and buds. Leaf scars small, flat oval to half round with one dot-like bundle scar. Buds elongate up to 1 inch; at least six times longer than wide. Bud scales reddish-tan with light gray to tan tips. Terminal buds present. Twigs reddish-brown on new growth, becoming light-gray mottled with darker gray. Light tan lenticels frequent on new growth. Pith hard. Stem cross section with radial rays.

Leaves simple, coarsely serrate on margin; when mature glabrous and dark green above, paler and pubescent beneath.

Flowers form after leaves unfold in April and May, staminate in globose heads pendant on hairy peduncles, pistillate in clusters of 2 to 4 borne on short hairy peduncles.

Fruit a pair of small brown three-angled sweet nuts in a bur-like involucre, full grown in midsummer but becoming brown and persistent on branches. September to November.

Gleditsia triacanthos L.—HONEYLOCUST

Deciduous tree with alternate buds and leaf scars. Twigs glabrous, grayish-brown to reddish-brown. Older twigs greenish-gray to dark gray with silver-gray longitudinal striations. Stout, sharp, thorns present at most nodes and positioned above leaf scars and buds. Thorns straight or branched, purple-red to purple-black. Terminal buds absent. Lateral buds scaly, small, and oval. U-shaped leaf scars appearing three-lobed with one bundle scar in each lobe. Pith homogeneous, light-colored, and softer than surrounding wood.

Leaves once- or twice-pinnate.

Flowers borne in axillary, dense, green racemes. Staminate flowers often clustered. Pistillate flowers few, usually solitary, May and June.

Fruit a legume borne on short peduncles, usually in twos or threes.

54

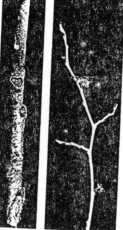

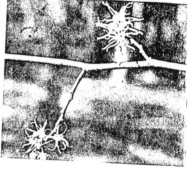

Hamamelis virginiana L.—WITCH-HAZEL

Deciduous shrub to small tree with alternate leaf scars. Twigs pubescent, becoming glabrous. Buds without scales. Lateral buds superposed. Terminal bud on short stalk and asymmetric. Leaf scars half round to triangular with three bundle scars. Pith firm, small, and darker than surrounding wood.

Leaves simple, wavy-toothed, usually glabrous above, somewhat pubescent beneath.

Flowers bright yellow in axillary or terminal clusters, usually surrounded by a scale-like three-part involucre,. September to November.

Fruit a woody capsule with two bony, shiny, black seeds; matures a year after flowering.

UST

Twigs glabrous, grayish-to dark gray with silver-ent at most nodes and it or branched, purple-s scaly, small, and oval. e bundle scar in each han surrounding wood.

Staminate flowers often and June.
in twos or threes.

55

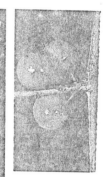

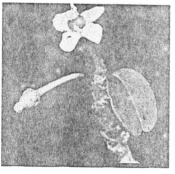

Ilex decidua Walt.—POSSUMHAW

Deciduous shrub to small tree with alternate buds and leaf scars. Twigs greenish-gray and puberulent, becoming mottled gray and glabrous with age, often with spur-like lateral twigs. Buds small with two scales. Leaf scars small, half-circular or occasionally narrow slits. One bundle scar. Pith homogeneous, white, and much softer than surrounding wood. Shiny red to red-orange spherical fruit 0.25 inch diameter commonly persistent long after leaves have fallen.

Leaves in crowded groups at end of short branchlets, thick and firm. Flowers small, whitish, March to May.

Fruit a globose drupe, orange to scarlet, ripens in fall.

gl
Le
ra
be
to

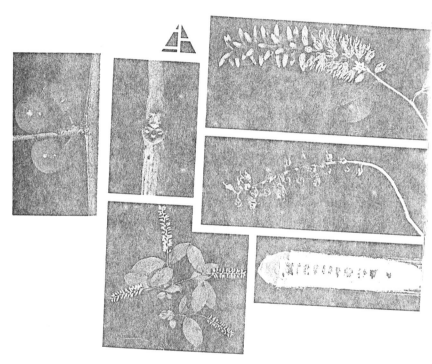

Itea virginica L.—VIRGINIA SWEETSPIRE

d leaf scars. Twigs
and glabrous with
h two scales. Leaf
One bundle scar.
inding wood. Shiny
ommonly persistent

ets, thick and firm.

ıll.

Deciduous shrub with alternate leaf scars. Twigs pubescent, becoming glabrous, reddish-brown. Pith interrupted at regular intervals by air spaces. Leaf scars triangular to crescent shaped, with three bundle scars. Terminal racemes persist through winter.

Leaves with finely serrated margin except near base, turn reddish-brown before falling.

Flowers white in conspicuous racemes that persist through winter, April to June.

Fruit a two-grooved capsule, July to September.

Liquidambar styraciflua L.—SWEETGUM

Deciduous tree with alternate buds and leaf scars. Twigs gray to tan, glabrous; occasionally with corky wings. Twigs and buds strongly aromatic if crushed when fresh. Buds shiny and glabrous except for short marginal pubescence on scales. Bud scales greenish-yellow with purple-brown on keel and margins. Leaf scars half circular. Bundle scars three. Pith homogeneous, white, and much softer than surrounding wood.

Leaves simple with 3 to 7 acuminate lobes.

Flowers very small, greenish; staminate flowers in terminal, erect, tomentose racemes; pistillate flowers in axillary, globose, long-peduncled, drooping heads, March to May.

Fruit an aggregate of many two-celled ovaries tipped by two-beaked or horn-like styles, September to November.

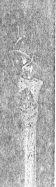

Liquidambar styraciflua L.—SWEETGUM

Deciduous tree with alternate buds and leaf scars. Twigs gray to tan, glabrous; occasionally with corky wings. Twigs and buds strongly aromatic if crushed when fresh. Buds shiny and glabrous except for short marginal pubescence on scales. Bud scales greenish-yellow with purple-brown on keel and margins. Leaf scars half circular. Bundle scars three. Pith homogeneous, white, and much softer than surrounding wood.

Leaves simple with 3 to 7 acuminate lobes.

Flowers very small, greenish; staminate flowers in terminal, erect, tomentose racemes; pistillate flowers in axillary, globose, long-peduncled, drooping heads, March to May.

Fruit an aggregate of many two-celled ovaries tipped by two-beaked or horn-like styles, September to November.

58

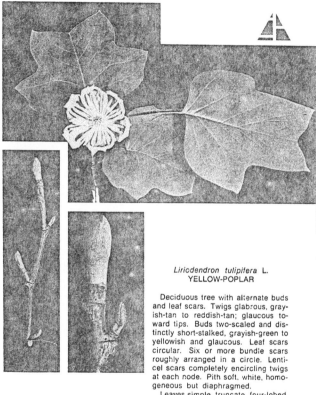

Liriodendron tulipifera L.
YELLOW-POPLAR

Deciduous tree with alternate buds and leaf scars. Twigs glabrous, grayish-tan to reddish-tan; glaucous toward tips. Buds two-scaled and distinctly short-stalked, grayish-green to yellowish and glaucous. Leaf scars circular. Six or more bundle scars roughly arranged in a circle. Lenticel scars completely encircling twigs at each node. Pith soft, white, homogeneous but diaphragmed.

Leaves simple, truncate, four-lobed. Flowers conspicuous, greenish-yellow, orange-banded at base, tulip-like April and May.

Fruit a samara-bearing cone, September to November.

59

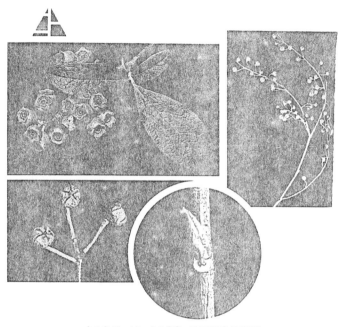

Lyonia ligustrina (L.) D C.—HE-HUCKLEBERRY

Deciduous shrub with alternate leaf scars and buds. Commonly 6 to 12 feet high. Leaf scars somewhat raised; half circular with one dot-like bundle scar. Buds light to dark red; acute tip. True terminal buds absent. New growth puberulent, later becoming glabrous. Stems light yellow-green to dark brown or gray; commonly with small black spots.

Leaves simple, margin entire or obscurely serrulate.

Flowers white in elongate panicles with umbel-like clusters of 2 to 6 flowers, May to July.

Fruit a dry capsule. Panicles remain on plant through winter.

Morus rubra L.—RED MULBERRY

nmonly 6 to 12
th one dot-like
al buds absent.
s light yellow-
k spots.

usters of 2 to 6

winter.

Deciduous tree with alternate leaf scars. Twigs glabrous, yellowish-tan to bluish-gray. Pith soft, continuous, and white. Inner bark of fresh twigs turns orange when outer bark is scraped away. Sap milky and sticky. Bud scales orange-tan with dark edges. Leaf scars half round to flat oval or circular. Bundle scars many in one central cluster.

Leaves simple, doubly serrate, some with numerous lobes, rough and glabrous above, soft pubescent beneath.

Flowers appear with leaves in spring, staminate spikes cylindric, pistillate spikes cylindric and sessile.

Fruit resembles a blackberry, red at first and then purplish-black, juicy; May to August.

61

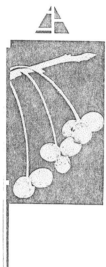
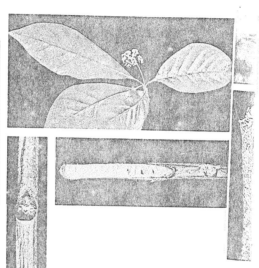

Nyssa sylvatica Marsh.—BLACKGUM

Deciduous tree with alternate branching and leaf scar arrangement. Twigs slightly appressed-pubescent, becoming glabrous, light tannish-gray. Pith continuous but with diaphragms. Leaf scars half round, somewhat protruding particularly towards twig tips. Bundle scars three. Visible terminal bud scales three, two large and one small.

Leaves simple, lustrous green above, paler and hairy below.

Flowers axillary; staminate flowers in long-peduncled capitate clusters, pistillate flowers in slender-peduncled clusters of two or more, April to June.

Fruit an egg-shaped drupe, acid, dark-blue to black, September and October.

—BLACKGUM

and leaf scar
scent, becom-
uous but with
hat protruding
three. Visible
 small.
ler and hairy

eduncled cap-
uncled clusters

 to black, Sep-

Ostrya virginiana (Mill.) K. Koch
EASTERN HOPHORNBEAM

Deciduous tree with alternate leaf scars. Twigs reddish to brown with white lenticels. Bud scales with fine longitudinal striations, yellow-green with purple-brown edges. Leaf scars half round to crescent shaped, with three bundle scars. Trunk bark on larger specimens broken into small elongate brownish-gray scales.

Leaves simple, margin sharply and doubly serrate, glabrous and yellowish-green above, hairy and paler below, turning yellow in autumn.

Staminate catkins 1 to 3 at end of branches, pistillate catkins small, usually solitary, slender; catkins open with leaves in spring.

Fruit a flattened-ovoid nutlet in cone-like imbricate clusters, nuts enclosed in the sac, September and October.

63

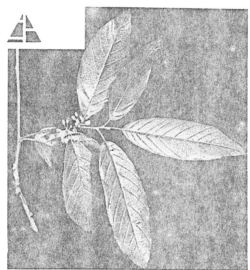

Rhamnus caroliniana Walt.—CAROLINA BUCKTHORN

Deciduous shrub with alternate buds and leaf scars. Twigs purple-gray to purple-brown; fine woolly pubescence near tips, becoming glabrous. Leaf scars half circular to oval with no lobes. Bundle scars three. Lateral and terminal buds woolly, tan, without scales. Terminal buds larger and sessile to twig tips. Stipules persistent, especially toward twig tips. Pith white, homogeneous, and soft. Inner bark of stems greenish-yellow.

Leaves simple, with prominent parallel veins

Flowers small greenish-yellow, solitary or 2 to 10 in peduncled umbels, May and June.

Fruit a drupe with 2 to 4 (usually 3) seeds, persistent, red at first but turning black and lustrous at maturity, August to October.

64

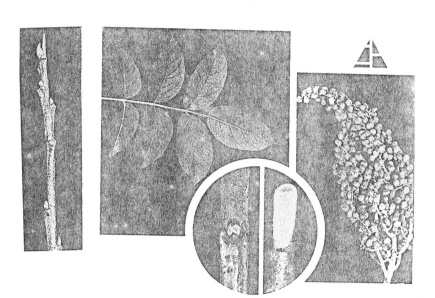

Rhus copallina L.—SHINING SUMAC

Deciduous shrub with one primary stem, occasionally reaching heights of 30 feet. Twigs velvety pubescent becoming glabrous, reddish-brown to gray. Buds silvery-tan, pubescent, and without scales. True terminal buds absent. Leaf scars horseshoe-shaped, encircling buds halfway. Bundle scars many. Pith dark tan, occupying well over one-half stem diameter. Sap milky and sticky. Lenticels dark rusty-brown and numerous.

Leaves alternate, pinnate with 7 to 17 leaflets, entire or with a few teeth, lustrous; rachis pubescent and broadly winged.

Flowers borne in a densely pubescent, compact terminal panicle, petals greenish-white, July to September.

Fruit a red, glandular-hairy drupe in compact panicles, erect or drooping; seed solitary, smooth, ripens in fall and stalks commonly persist through winter.

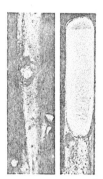

Rhus glabra L.—SMOOTH SUMAC

Deciduous shrub with one main stem up to 20 feet. Twigs glabrous, sparingly puberulent when young, tan to blotchy gray. Buds with tan to silvery dense pubescence. True terminal buds absent. Leaf scars encircle buds. Bundle scars many. Pith orange, occupying over one-half twig diameter. Sap milky and sticky. Lenticels light brown and scattered.

Leaves alternate, pinnate with 11-31 leaflets, sharply serrate, usually dark green above, lighter to white beneath.

Flowers in terminal compact panicles, petals white, June to August.

Fruit a drupe with short red-velvety hairs, one-seeded, smooth; ripens September and October.

66

H SUMAC

et. Twigs
to blotchy
nce. True
s. Bundle
-half twig
rown and

arply ser-
eneath.
hite, June

e-seeded,

Rhus vernix L.—POISON SUMAC

Deciduous shrub to small tree with one primary stem, 12 to 25 feet high. Terminal buds present. Leaf scars heart-shaped. Bundle scars several, arranged around edges of leaf scars. Twigs light orange-gray to dark gray, with dark gray to black mottling; glabrous but with numerous lenticels. Pith soft and homogeneous. Stem cross section revealing dark strands in outer edges of white pith.

Leaves alternate, odd-pinnate with 7 to 13 leaflets, entire, smooth, shining above and more or less pubescent beneath.

Flowers green in axillary panicles, April to July.

Fruit a greenish-white or gray drupe, August to November.

67

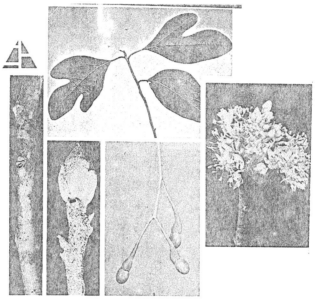

Sassafras albidum (Nutt.) Nees—SASSAFRAS

Deciduous tree with alternate leaf scars. Twigs light green, mottled with dark green and grayish-brown, pubescent, becoming glabrous. Stems and roots with strong spicy odor. Pith soft and white. Leaf scars half round to crescent-shaped. One bundle scar, straight to crescent-shaped.

Leaves simple, thin, aromatic, entire on the margin or divided into 2 to 3 mitten-shaped lobes, bright green above, glabrous and glaucous beneath.

Flowers greenish-yellow; loose open clusters appear before leaves unfold.

Fruit a blue spicy drupe borne erect on bright red club-shaped stalk, August to October.

68

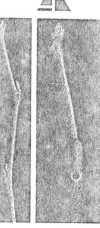

FRAS

light
ubes-
trong
nd to
scent-

ɡin or
above,

ar be-

club-

Styrax americana Lam.—AMERICAN SNOWBELL

Deciduous shrub to small tree with alternate buds and leaf scars. Twigs greenish, becoming purple-gray; stellate pubescence at first, glabrous on older stems. Leaf scars U-shaped with V-notch on upper margin. Scar lobes pointed. One bundle scar, curved and raised with short points on both sides. Buds two per joint and superposed. Old fruiting stems commonly persist through winter. Pith homogeneous, white, with a darker center.

Leaves simple, margin varying from entire to serrate or remotely toothed, upper surface dark green, lower surface paler.

Flower fragrant, on short lateral leafy branches; racemes axillary, subtended by small leafy bracts; corolla white; anthers bright yellow; May and June.

Fruit a persistent dry drupe, finely tomentose, seed usually solitary, September and October.

Symplocos tinctoria (L.) L'Her.—COMMON SWEETLEAF

Deciduous shrub arising from one main stem. Alternate leaf scars. Twigs sparingly pubescent, becoming glabrous, gray to reddish-brown. Terminal buds two-scaled, pubescence appressed. Leaf scars half round with one round to crescent-shaped bundle scar. Pith interrupted by air spaces.

Leaves simple, drooping on upcurved twigs, persistent in the South. Margin obscurely and remotely serrate or almost entire, thick and leathery, upper surface dark green, glabrous and lustrous, lower surface paler and pubescent.

Flowers fragrant, small yellowish-white in axillary clusters on branches of previous year, March to May.

Fruit orange-brown drupe, seed solitary, early autumn.

Ulmus alata Michx.—WINGED ELM

Deciduous shrub or small tree with alternate leaf scars. Twigs reddish-brown to gray-brown, pubescent to puberulent, becoming glabrous. Corky ridges on older growth. Buds fairly slender, acutely pointed, reddish-brown, and puberulent. Leaf scars half round to flat oval, with three dot-like bundle scars. Pith small, hard, and white.

Leaves simple, coarsely and doubly serrate.

Flowers borne in few-flowered drooping fascicles that appear before leaves in the spring, red to yellow.

Fruit a reddish or greenish samara, seed and wing hairy especially on margin, March to June.

Ulmus americana L.—AMERICAN ELM

Deciduous tree with alternate leaves. Twigs reddish- to grayish-brown, pubescent to puberulent, becoming glabrous. Buds acutely pointed, reddish to brown, and puberulent. Leaf scars half round to flat oval with three dot-like bundle scars. Pith small, white, and hard. Twigs without corky ridges.

Leaves simple, somewhat cordate on one side at base and rounder or cuneate on other side, margin coarsely and doubly serrate.

Flowers borne before leaves in spring, axillary in 3- to 4-flowered fascicles, calyx red to green.

Fruit a red to green samara consisting of a central flattened seed surrounded by a membranous wing, ciliate on the margin, March to June.

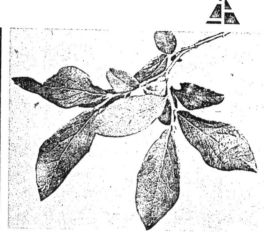

ddish- to
labrous.
perulent.
e bundle
ut corky

ase and
sely and

3- to 4-

ntral flat-
iliate on

Vaccinium vacillans Torr.—LOW BLUEBERRY

Deciduous shrub with alternate leaf scars. Twigs green
on one side and reddish on other, even when 2 to 3 years
old. Leaf scars small, narrow, crescent-shaped with one
bundle scar. Branchlets erratically bent or zigzag at nodes.
Pith hard, exceeding one-third total twig diameter.
Leaves simple, U-shaped at base, leathery when mature.
Flowers greenish to purplish on ends of branchlets or
from old axils, appear before leaves are fully expanded.
Fruit a blue to black globose berry, June to September.

Vaccinium stamineum L.—COMMON DEERBERRY

Deciduous shrub with alternate leaf scars. Twigs—even when 2 to 3 years old—commonly green on one side, red on the other. Twigs slender, branchlets straight or gently curved. Leaf scars small, half round to circular with one dot-like, inconspicuous bundle scar.

Leaves simple, margin entire and ciliate, rounded at base.

Flowers greenish-white to purple on specialized flowering branches subtended by leafy bracts, April to June.

Fruit a green or yellowish globose berry, July to September.

INDEX OF PLANT NAMES

Structural Pest Control
Regulations

Richard V. Smythe
and
Lonnie H. Williams

Southern Forest Experiment Station
Forest Service
U.S. Department of Agriculture

1974

ACKNOWLEDGMENTS

We thank the following people for providing information on their State's laws and regulations and for reviewing a draft of our manuscript: W. A. Ruffin, Alabama; J. N. Roney, Arizona; E. C. Sizemore and M. Bonner, California; D. M. Burchett, Colorado; F. R. Du Chanois, Florida; C. C. Yent, Hawaii; H. D. Garwood, Kansas; C. H. Gayle, Kentucky; R. Carlton, Louisiana; M. K. Fresvik, Minnesota; N. S. Weber, New Mexico; R. E. Howell, North Carolina; R. M. Rogers and H. H. Latham, Oklahoma; and C. Chapman and V. E. Walter, Texas.

We also thank G. King, R. W. Anderson, and M. Tucker, Arkansas; C. M. Scott, Jr., Georgia; R. McCarty, Mississippi; and H. L. Bruer, Tennessee; who not only reviewed our manuscript but gave unstintingly of their time on many occasions during the past 3 years and who made their offices and records available 24 hours a day, 7 days a week, to facilitate our data collection.

We acknowledge valuable review comments by T. A. Amburgey, Southern Forest Experiment Station; E. R. Baker, Environmental Protection Agency; J. R. Cook, Cook's Pest Control, Inc.; W. Ebeling, UCLA; C. J. Hromada, Bruce-Terminix Company; M. P. Levi, North Carolina State University; C. D. Mampe, National Pest Control Association; H. B. Moore, North Carolina State University; R. J. Pence, UCLA; J. C. Redd, Redd Pest Control; P. J. Spear, National Pest Control Association; and J. W. Steckel, Torco Termite and Pest Control Company.

Although we were helped by many people, the validity of the data and the expressed opinions are our responsibility and should not be ascribed to any reviewers unless expressly referenced in the text.

WARNING!

Pesticide recommendations mentioned in the appendices were in effect before the Federal Environmental Pesticide Act of 1972. They varied widely from State to State, and some had never been approved by USDA. Listing is for information only. Current status of the pesticides must be checked with responsible State or Federal authorities.

CAUTION: Pesticides can be injurious to humans, domestic animals, desirable plants, and fish or other wildlife—if they are not handled or applied properly. Use all pesticides selectively and carefully. Follow recommended practices for the disposal of surplus pesticides and pesticide containers.

Use Pesticides Safely
FOLLOW THE LABEL
U S DEPARTMENT OF AGRICULTURE

ii

Contents

iii

Appendices

Structural Pest Control Regulations

Richard V. Smythe and Lonnie H. Williams[1]

INTRODUCTION

Why would two research entomologists leave their studies of insects long enough to review State laws governing structural pest control (SPC)? As research scientists at the only laboratory with a research program devoted solely to wood products insects, our nationwide responsibility is the safe, effective, and economical control of insects that damage wood in use. Subterranean termite (ST) insecticidal controls in field tests conducted by this laboratory have remained 100 percent effective up to 24 years. The pesticides' environmental hazard is minimized and their effectiveness is maximized when they are properly applied. Because these insecticides are used worldwide for termite control and because suitable chemical substitutes are lacking, we have examined how effectively SPC laws implement current ST control recommendations. We have also noted practical problems in SPC and in law enforcement which require further research.

Many a homeowner does not know that his house contains construction faults conducive to attack by wood-destroying organisms (WDO). Nor does he know where to look for and how to recognize early signs of attack or how to treat his house if the attack has already occurred. Because WDO control is complex, it is largely the responsibility of commercial pest control operators (PCOs). The wood products protection services provided by the pest control (PC) industry influence the lives of millions of Americans and represent a business grossing several hundred million dollars yearly. For years the PC industry has been the major user of WDO research. The industry's importance will increase in the near future as fed-

eral law will soon require that insecticides for WDO control be sold only to certified applicators. Thus, homeowners will have to depend increasingly on the services of a PCO.

Eighteen States[2] have laws and regulations governing PCOs controlling structural pests (SP), which we define as wood-destroying arthropods and fungi. We are not considering laws governing either the licensing of other kinds of pesticide applicators (e.g., agricultural pest control applicators) or the sale, application, and disposal of pesticides, even though the latter laws in many States also regulate SPC applicators.[3] Most of these laws have been reviewed elsewhere (Environmental Protection Agency (EPA) 1971). We do not review legislation concerning fumigation, whether these laws are part of SPC or other PC, because many States lacking SPC laws have fumigation regulations. Further, fumigation is often used for pests other than SP.

The intent of this paper is to examine the laws and regulations now governing SPC. These laws and regulations vary greatly in their coverage and effectiveness and have never been brought together and analyzed in one source. We believe this compilation should be helpful both to the PC industry and to the States, whether or not they currently have SPC legislation. It should certainly benefit those States who soon will be developing

[1] The authors are Principal Entomologist and Entomologist, respectively, at the Wood Products Insect Laboratory, Southern Forest Experiment Station, USDA, Gulfport, Mississippi.

[2] Hawaii and Texas have not fully implemented their SPC legislation. Tennessee passed a new SPC law in 1972 which has already been amended several times and, at this writing, may be amended again. Therefore, we review and discuss Tennessee's former law which has existed since 1955.

[3] In addition to the 18 SPC laws reviewed in this paper, 18 other States license SPC applicators in their Pesticide Applicator's Laws. Seven other States have pending pesticide applicator's legislation which includes SPC operators. Maine and North Dakota have pesticide legislation which specifically excludes SPC operators. Five States have no pesticide legislation or SPC laws.

similar legislation or who will be revising existing legislation. The information may also be useful to the EPA in its efforts to implement the Federal Environmental Pesticide Control Act of 1972 (Public Law 92-516, approved October 21, 1972).

The recent passage of the Federal Environmental Pesticide Control Act has resulted in a greatly increased public concern for, and awareness of, environmental problems; has heightened SPC legislative activity in various States; and has increased efforts on the part of the National Pest Control Association (NPCA) to promote and insure better training and increased public confidence in its members. With continued increasing costs of purchasing and maintaining adequate

housing, we believe it is timely for a detailed at existing SPC legislation. We will evaluate effectiveness of this legislation for consumer s ice and protection on the basis of the avail data, which is admittedly limited, and we will cuss some of the future alternatives for obtai the maximum in quality service from the com cial control of SP—a service of vital import to the American homeowner and to the Amer economy.

Information for this paper was gathered ov period of 3 years. We contacted regulatory offi from the District of Columbia and every Stat cept Alaska. We consulted with PCOs, the NP and numerous homeowners.

WHY STRUCTURAL PEST CONTROL REGULATIONS HAVE BEEN DEVELOPED

Before we discuss the existing State laws we believe it would be useful to provide some background material. The damage and control of SP differ in several ways from those of other pests. Homeowners are seldom familiar with the appearance or habits of SP. Inadequate service is rarely evident to the homeowner and may not result in obvious damage for many years. Once damage is discovered, repairs may be costly. Because one of every five Americans moves each year (Toffler, 1970), a new owner may unknowingly acquire a damaged house that will soon require repairs. Usually the commodity protected is a house, a major lifetime investment. Most termite control measures require well-trained personnel for proper application, but the control is effective for many years (Johnston et al., 1972).

Conversely, most agricultural pests and household pests such as cockroaches and ants are easily detected. Likewise, even an untrained observer can detect the damage caused by such pests and he can observe the failure or success of most control measures which normally are easily applied and give only short-term protection to a commodity often of less value than a house.

Because the quality of SP control services v widely, and because poor work can result in c damage, the question of State or Federal rej tion of the PC industry often arises. And wher it does, it arouses strong feelings. There is middle ground; you're either for regulatio against it (Anon., 1956).

Many interested parties and observers w agree that some minimum treatment standar procedures are desirable to insure quality se and maximum consumer protection. The disa ments arise in how to achieve these goals. solutions are typically proposed, self-po through industry-imposed standards or re tion by State or Federal authorities. Propone both views have argued pro and con for : years. It is not our intention to advocate (position. We will present the available inf(tion, discuss it, and let the reader draw hi: conclusions.

For over 20 years there has been sent within the industry (probably a minority in favor of SPC legislation. At the 1951 Cali Pest Control Operators convention, most gates interviewed believed some kind of bi

2

vania PC Association Fall Conference (Anon.,
1960a) "... After consultation with Dr. Ralph E.
Heal of the NPCA and representatives from associations in neighboring Pennsylvania and New
Jersey, we soon realized legislation was not the
answer to questionable, fly-by-night, itinerant
pseudo PCOs. ..."

Legislation has been proposed and forestalled
in various States over the last 25 years. State regulation in Texas was proposed in the 1940s, but a
law was not passed until 1971. In a letter to the
editor, W. D. Bedingfield (1966), then a 20-year
member of the NPCA and a charter member of the
Texas PC Association which was organized to
fight the legislation, wrote,

"... thank goodness, there have been enough
level headed people around to put up such a
fight that state pest control regulatory laws
have never been passed in Texas. ... One will
succeed or fail on his own initiative whether
he be large or small. He can do all of this
without regulatory measures by a state board
which has no idea of the problems PCOs face
in their particular area."

A somewhat analogous situation occurred in Delaware (Anon., 1960a). The Delaware PC Association was developed in response to a public demand
for a law to regulate allegedly unscrupulous operators in the State. Less than 2 years after the association was formed, public opinion improved to
the extent that the demand for industry regulation
was withdrawn, and Delaware currently has no
SPC law.

It is not clear in the above instances whether the
lack of legislation was due more to lobbying and
advertising success or to the industry upgrading
its own operations. In either case, there is strong
sentiment within the industry to be self-policing
and to be allowed to upgrade the level of competence. Many individuals and associations have indicated the necessity of keeping the industry's
house in order to avert legislation (Bruer, 1960;
Lance, 1969).

In addition, there is a strong feeling within the
industry that the best and most desirable method
of consumer protection is the contract provided
the homeowners by the PC companies or the
NPCA Insured Termite Warranty Program.
Without question, these programs can provide a
valuable public service and we will discuss them
in more detail later in the paper.

EXISTING SPC LEGISLATION

Since SPC regulations vary widely in scope and depth, we established 17 criteria for their evaluation. Results of the evaluation are summarized in table 1. The criteria and the reasons for their selection are listed below:

#1 Dates of law passage and amendments.— The age of the law may influence the experience and expertise of the State Regulatory Agency. The number and dates of amendments may suggest the activity and improvement in regulatory work, PC activities, or both.

Louisiana and California passed the first two laws in 1930 and 1935, respectively. Mississippi and Arkansas followed in the same decade, and Alabama and Florida passed laws in the 1940s. Six States adopted legislation in the 1950s, three in 1955. New Mexico and Arizona both passed legislation in 1965; Minnesota and Colorado followed in 1970 and 1972, respectively. Texas and Hawaii passed laws in 1971 and 1972, but neither State has fully implemented this legislation.

#2 Financing of regulatory work. — The method by which a regulatory agency is financed can influence its enforcement authority and operational independence. Five States — Arizona, Arkansas, California, Louisiana, and Texas—support their SPC enforcement activities entirely by collected fees. Although Alabama, Georgia, Hawaii, Kentucky, Minnesota, Mississippi, New Mexico, and Tennessee collect some fees, the monies are deposited in the State general fund, and these States support their activities by legislative appropriation. The five remaining States use a combination of collected fees and legislative appropriation.

At the end of each fiscal year in Louisiana, all surplus funds in excess of $100,000 are given to Louisiana State University (LSU) for research on control of structural and household pests. Last year, LSU received $6,841 from the Louisiana SPC Commission (Anon., 1972a). Mississippi charges a 5 percent sales tax on all PC work. This tax, which totaled $296,806 in fiscal year 1971, generated more revenue from PC activities than was obtained in any of the 18 States, except California.

#3 Regulatory board.—The composition of the SPC board or its counterpart may indicate how well various segments of the public are represented. The Board's placement in the hierarchy of State government may indicate its authority or effectiveness. F. R. Du Chanois commented on Florida's Board (Anon., 1960b):

"... One feature of the law which has always disturbed me and many others, both in and out of the industry, is the provision for an all-industry regulatory body ... it is my personal conviction that the original intent and objectives of having a law would be best maintained by a body of men unshackled and unfettered with the natural bias associated with belonging to the industry. It is very understandable that no industry man alone wants to judge his fellow competitor."

Complete representation would include the PC industry, the State Regulatory Office, State university or extension entomology, and public health. Boards in many States have an entomologist either from the State Department of Agriculture or from the State university; Arizona, Colorado, Georgia, Hawaii, Minnesota, and Texas have one or more representatives from the Department of Public Health. PC industry representatives make up a majority on the Arizona, California, Georgia, Hawaii, and Texas Boards. Florida and Oklahoma have no well-defined Board with licensing and regulatory power, but the law is administered by a department of the State assisted by a technical advisory council. Minnesota's Board has only advisory powers and lacks regulatory authority.

#4 License requirements. — The education, training, experience, and financial responsibility of SPC licensees obviously influences the quality of work performed. Most regulations require licensees to have some education or experience and some liability insurance or bonding. Florida is the only State requiring each new licensee to have a high school diploma or its equivalent. California, Hawaii, Kansas, Minnesota, and Texas have no minimum educational requirements, and the latter four States also lack minimum experience requirements. Most other States demand a minimum of 2 years of college or 1 to 4 years of experience.

All States except Florida, Georgia, North Carolina, Oklahoma, and Texas require surety bonds or insurance as guarantees of financial liability. If a Georgia licensee advertises himself as bonded he must be able to (a) issue a separate bond for

4

Phoenix, Tucson, and Yuma. Mississippi has a 20-week course that meets 1 night per week. It, too, is given in different cities throughout the State. Of the 18 States with SPC laws, only California, Florida, and North Carolina have specific curricula in pest control (Ebeling, 1973). North Carolina has a 2-year curriculum leading to an associate degree with a major in urban and industrial pest control.

#7 Offices and licensed supervision. — Small companies usually have only one place of business. Larger companies operate with a main office and one or more sub, branch, sales, or call offices. Supervision of activities at these offices, which could influence the quality of work performed, varies by State.

Florida, Hawaii, Kentucky, Louisiana, Minnesota, Mississippi, North Carolina, Tennessee. and Texas require a licensed operator residing in the immediate (75 miles) vicinity to be in charge of every place of business. Colorado and Kansas do not require that branch office supervisors be licensed although they must pass an examination. The other nine States permit some offices to be supervised by licensed operators not residing nearby.

Some States require that all work from any office be supervised by personnel proven qualified by examination. Small, one-office companies have their treating crews under the supervision of a licensee, probably the owner. Larger companies with more than one branch office frequently have treating crews with no direct licensed supervision, even though they may be nominally trained and supervised by a licensee from the main office. When regulations require branch office supervision by licensees, companies with branch locations sometimes lack a licensee at every branch office because of personnel changes. So at times, the PC company is in violation of the State law.[1] While this does not necessarily mean that the branch office should be closed, regulatory officials from several States have commented that the lack of licensed supervisors at the branch offices of some of the less responsible companies has created problems in regulatory enforcement.

#8 Regulatory fee structure.—Some States finance part or all of their regulatory activities by assessing fees for job reporting, company registration, examinations, licenses, reinspection, and identification cards. Arkansas, California, Louisi-

---

[1] Paragraph based largely on review comments by John B. Cook, President, Cook's Pest Control, Inc., Decatur, Alabama

Table 1a.— Characteristics of structural pest control laws, Alabama through Kansas.

ITEM	ALABAMA	ARIZONA	ARKANSAS	CALIFORNIA
1. Date(s) of law and amendments	Law – 1940 Amended – 1943, 1951, 1953, 1965, 1969	Law – 1965 Amended – 1971	Law – 1939 Amended – 1951, 1953, 1965	Law – 1935 Amended – Many times
2 Financing of regulatory work	Legislative appropriation	Fees	Fees	Fees
3 Regulatory board	Composition 4 members Comm. Agric. and industries Entomologist–Agric. Exp. Stn. Horticulturist–Agric. Exp. Str Plant pathologist–Agric. Exp Stn. Authoritative agency Dep. Agric.–Licensee eXamining and regulatory powers Tenure Unlimited	Composition 5 members 2 members representing public interest from field of public health, pesticides, entomology, or SPC 3 members active in PC industry in past 5 years Authoritative agency. Dep. Agric–Licensee eXamining and regulatory powers Tenure. 5-year terms, limit 2 consecutive	Composition: 13 members Chairman, Dep. Entomol., Univ. Ark. Chairman, Dep. Plant Pathology, Univ. Ark. Representatives of: Feed Manuf. Assoc. Pesticides Assoc. Pest Control Assoc. Horticulturists Soc. Nurserymen's Assoc. Agric. Aviation Assoc. Rice Growers Assoc. Fertilizer Manuf. Assoc. Cotton Growers Assoc. Seed Growers Assoc. Seed Dealers Assoc. Authoritative agency: Dep Commerce–Licensee examining and regulatory powers Tenure: Unlimited	Composition: 6 members 4 members licensed for SPC 2 public members Authoritative agency: Dep. Consumer Affairs–Licensee examining and regulatory powers Tenure: 4-year terms, limit 3 consecutive
4 Licensee requirements	Education College degree in appropriate fields, or EXperience 1 year in appropriate field Financial: Bond–$25,000 public liability for fire and eXplosion during subterranean termite (ST) control Insurance– $5,000 property damage only required for ST control NOT for dry-wood termites or wood destroying beetles	Education: College degree with more than 10 hours in entomology, or EXperience. 2 years in past 10 for any field eXcept fumigation and weed control • Financial Deposit, insurance or bond Public liability–$25,000 Property damage–$25,000	Education. 2 years college with 1 course in entomology, or Experience 1 year in appropriate field Financial: Bond–$1,500 and Insurance–public liability $25,000/person $50,000/accident $5,000/property damage	Education: None Experience: Termite – 4 years Financing, Insurance, deposit, or surety bond Public liability–$25,000 Property damage–$25,000 Indemnity bond–$2,000 Restoration bond–$1,000–$8,000 if guilty of violation
5. Licensee eXaminations	Type: Written Classes: 1 Termites and powder-pest beetles (PPB) 2. General household pests 3. Fumigation 4. Wood decay	Type: Written, oral, or both Classes: 1. Includes all classes below 2. General pests 3. Wood-destroying organisms (WDO) 4. Fumigation 5. Weed control	Type: Written Classes: 1. Tree surgery 2. Spraying fruit trees 3. Spraying shade trees and shrubs 4. Termites and other structural pests 5. Household pests and rodents 6. Fumigation 7. Pecan pests 8. Fabric pests–rugs and carpeting 9. Upholstery materials	Type: Written Classes: 1. Fumigation 2. General pests 3. Termites and any combination of the above except 1 only
6. Training eXceeding 1 week	None	10–12 week servicemen's course, twice weekly, rotated between Phoenix, Tucson, and Yuma.	None	None, but does have university sponsored training
7. Offices and licensed supervision	Main office–certified operator in residence Branch office–3 or more service technicians, supervisor must be certified operator in residence or branch supervisor who cannot supervise main office	Each office providing treatments must be supervised by licensee	Main office–licensed operator in residence Branch office–licensed operator in residence	Principal office–supervised by licensee Branch office–any place of business other than princip office, usually supervised licensee
8. Fees	Reporting–none Company registration–none Examination–$10 License: Main–$50 Branch–$25 Sub–$10 Reinspection–none Identification–none	Reporting–none Company registration–$100 Examination–$50 License–$50 Reinspection–none Identification–none	Reporting–$2/job Company registration–none Examination–$30, re-exam $20 License–$10/operator, $5/solicitor, $5/agent New license $15/job for first 5 jobs which are state inspected Reinspection–$15 each Identification–none	Reporting: Inspection report filing–$. Certified copy of inspectic report–$2 Notice of work completed– Company registration: Principal office–$40 Branch office–$20 Examination: Operator–$25 Field representative–$10 License: Operator–$50 Field representative–$10 Duplicate–$2 Reinspection–none Identification–none

6

COLORADO	FLORIDA	GEORGIA	HAWAII	KANSAS
Law – 1972 Amended – No Fees plus legislative appropriation Composition: 6 members Member Dep. Entomol., Colo. State Univ. Chief, Div. Plant Ind., Colo. Dep. Agric. Member Dep. Health 3 members licensed for SPC Authoritative agency: Dep. Agric.–Licensee examining and regulatory powers Tenure: 3-year terms, unlimited	Law – 1947 Amended – 1950, 1953, 1955, 1959, 1963, 1965, 1967, 1969 Fees plus legislative appropriation Composition 5 members 3 members licensed for SPC representing, PC Assoc. Fla. Horticultural Spraymen's Assoc. PC Ind. Counc. Member Dep. Entomol., Univ. Fla. Member chemical manufacturing industry Authoritative agency. Dep. Health & Rehabilitation Services—Advisory only Tenure. 3-year terms, unlimited	Law – 1955 Amended–1957, 1960, 1961, 1966, 1969, 1971 Legislative appropriation Composition 6 members 3 members licensed for SPC Member Dep. Entomol., Univ. Ga. Director, Div. Entomol., Ga. Dep. Agric. Director, Dep. Public Health Authoritative agency. Dep. Agric.–Licensee examining and regulatory powers Tenure. 3-year terms, unlimited	Law – 1972 Amended – No Legislative appropriation Composition 7 members 4 PCOs, appointed by Governor Member Dep. Agric. Member Dep. Health Chairman, Dep. Entomol., Univ. Hawaii Authoritative agency Dep. Regulatory Agencies– Licensee examining and regulatory powers Tenure 4-year terms, unlimited	Law – 1953 Amended – 1969 Fees plus legislative appropriation Composition: 12 members Elected at annual convention affiliation unspecified Authoritative agency: Dep. Agric.–Licensee examining and regulatory powers Tenure: Staggered 3-year terms, unlimited
Education: College degree in entomology, sanitary or public health engineering, or related subjects plus sufficient practical experience, or Experience: 2 years in SPC, including 1 in last 5 years Financial: Bond—none required Insurance— $50,000/person $100,000/accident for bodily injury $50,000 property damage accident or other approved evidence of financial responsibility Type: Written Classes: 1. Household pests 2. Fumigation 3. WDO 4. Rodent	Education: All certified operators must be high school graduates or the equivalent or have college degree with advanced training or major in entomology, or Experience: 3 years in appropriate field, 1 year of which must have been in Fla. immediately preceding application Financial: No bond or insurance required Type: Written, may include practical demonstration Classes: 1. Fumigation 2. General household 3. Lawn and ornamental 4. Rodent 5. Termites and other WDO	Education: College degree in entomology, sanitary health or related subjects plus 1 year experience, or Experience: 2 years under licensee, 1 of which must be in last 5 years Financial: No performance bond required but licensee shall be able to: (A) issue separate bond for each job or (B) provide bond for minimum of 5% of previous year's gross sales or liability, whichever is higher, not to exceed $100,000 Type: Written, oral, or both Classes: 1. Fumigation 2. Household pests 3. WDO	Education: None Experience: None Financial: Bond—none required Insurance— $20,000/claim $50,000 for all claims/year Type: Written Classes: 1. Fumigation 2. General pests 3. Termites	Education: None Experience: None Financial: Bond—$2,000 for each place of business, or Insurance—$5,000/person/accident, $25,000/2 or more people/accident, $5,000 property damage/accident Type: Written Classes: 1. Fumigation 2. Rodent 3. Structural insects 4. Termites 5. Trees
None	None	None	None	None
Each office providing treatments must be supervised by licensee residing within 100 miles	Each office providing treatments must be supervised by licensed certified operator	Main or branch office—two or more employees, supervised by certified operator	Each office providing treatments must be supervised by licensee	Each office providing treatments must have a licensed technical representative
Reporting—none Company registration—none Examination—$25 License—$50 Reinspection—none Identification—none	Reporting—none Company registration—$25/ business location Examination—$25/application for each category License—$25/certified operator Reinspection—none Identification—$2/card	Reporting—none Company registration—$25 (called licensee fee) Examination—$10/category License—$25 for certification of operators Reinspection—none Identification—$2/card	Reporting—none Company registration—none Examination—none License—$25 Reinspection—none Identification—none	Reporting—none Company registration—$50 for each place of business Examination—none License—none Reinspection—none Identification—none

ITEM	ALABAMA	ARIZONA	ARKANSAS	CALIFORNIA
9. Identification	Personal—none Vehicle—none	Personal—I.D. cards for licensee only Vehicle—license number, company name, address	Personal—none Vehicle—name of operator or company, vehicles of solicitors and agents exempted	Personal—I.D. cards for licensee only Vehicle—license number, company name, address
10. Standard contracts or other forms related to treatment or inspection	Standard contract—no Other forms—monthly work activity reports	Standard contracts—no WDO damage must be described in writing Other forms—none	Standard contract—no, but contracts must be state approved Other forms—Appendices III-VI	Standard contracts—no, but contracts must be state approved Other forms—Appendices VII, VIII
11. Definition of structural pests restricted to WDO?	No	License class 3 is	Yes	No
12. Information reported to regulatory agency by each company	Monthly reports—list of jobs with dates and addresses of treated properties Copies of contracts filed—no	Monthly reports—none Copies of contracts filed—no	Monthly reports—list of jobs with name of property owner and address, date treated, years guaranteed Copies of contracts filed—yes, with scale drawing of property treated	Monthly reports—inspection reports (Appendix VII) filed within 5 days after inspection is made. Notices of work completed (Appendix VIII) on a contract made within 5 days of work completion Copies of contracts filed—no
13. State population and volume of work	Population: 1960—3,266,740 1970—3,444,165 Volume of work: 1972—43,015 Published licensee list—yes Work performance published—no	Population: 1960—1,302,161 1970—1,772,482 Volume of work: Unknown Published licensee list—yes Work performance published—no	Population: 1960—1,786,272 1970—1,923,295 Volume of work: 21-year mean—10,234 1971-72—26,458 Published licensee list—yes Work performance published—no	Population: 1960—15,712,204 1970—19,953,134 Volume of work: 1971-72 completion notices 219,431 Published licensee list—yes Work performance published—no
14. Minimum job requirements excluding those for fumigation	None	None	Appendix XVII	Appendix XIX
15. State inspections	Inspection authority—yes No. of inspectors: 1 Percent jobs inspected: Unknown	Inspection authority—no No. of inspectors: None Percent jobs inspected: Not applicable	Inspection authority—yes No. of inspectors: 5 Percent jobs inspected: 21-year mean—27.5 1971-72—20.6 Miscellaneous: First 5 jobs of each new licensee are inspected. If 15% or more jobs of any licensee are found substandard, regulatory office may give 60 days notice then inspect all jobs reported during previous year	Inspection authority—yes No. of inspectors: 7 plus trainees or aides within Division of Investigation Percent jobs inspected: Unknown
16. Penalties	Substandard work: No fines Delinquent license renewal: 10% of original fee Misdemeanor: Violating provisions of law	Substandard work: No fines Delinquent license renewal: 100% of original fee Misdemeanor: Violating provisions of law	Substandard work: $15 inspection fee for each substandard job; request for substandard treatment waiver will cost $30 if impractical situation does not exist Delinquent reporting fees: 100% of original fee Misdemeanor: Violating provisions of law	Substandard work: No fines Delinquent license renewal: 100% of original fee, over 3 months delinquent, processed as new applicant Misdemeanor: Violating provisions of law
17. Policy for FHA and VA financed houses	Letters of clearance: Not reported State inspected: Yes, only on complaint	Letters of clearance: Copies for FHA financed houses are received by state State inspected: No	Letters of clearance: Copies for FHA and VA financed houses are received by state State inspected: Yes, at random	Letters of clearance: Copies for FHA and VA financed houses are receiv by state State inspected: No

COLORADO	FLORIDA	GEORGIA	HAWAII	KANSAS
...onal—none ...cle—license number, ...pany name, address	Personal—I.D. cards for licensees, solicitors, servicemen Vehicle—company name, address	Personal—I.D. cards for licensee, servicemen and solicitors Vehicle—license number, company name, address	Personal—none Vehicle—none	Personal—none Vehicle—none
...dard contracts—no ...r forms—quarterly work ...vity reports	Standard contracts—no, but contracts must be state approved Other forms—Appendix IX	Standard contracts—no, but contracts must be state approved Other forms—Appendices X-XII	Standard contracts—no, but contracts must include whether guarantee offered, its terms and duration, WDOs treated and portions of structure covered Other forms—none	Standard contracts—no, but contracts must be state approved Other forms—none
...) category is	License class 5 is	License class 3 is	License class 3 is	License class 4 is
...rterly reports—common ...I technical names of ...micals and amounts used ...ies of contracts filed—no	Monthly reports—none Copies of contracts filed—no	Monthly reports—Appendix X Copies of contracts filed—no	Monthly reports—none Copies of contracts filed—no	Monthly reports—none Copies of contracts filed—no
...ulation: ...0—1,753,947 ...0—2,207,259 ...ume of work: ...nown ...lished licensee list—yes ... performance published—	Population: 1960—4,951,560 1970—6,789,443 Volume of work: Unknown Published licensee list—yes Work performance published—no Appendix XX	Population: 1960—3,946,116 1970—4,589,575 Volume of work: 16-year mean—34,062 1970—71—67,958 Published licensee list—yes Work performance published—no Appendix XXI	Population: 1960—632,772 1970—769,913 Volume of work: Unknown Published licensee list—yes Work performance published—no None	Population: 1960—2,178,611 1970—2,249,071 Volume of work: Unknown Published licensee list—yes Work performance published—no Appendix XXII
...ction authority—yes ...of inspectors: ...tomologists whose res- ...ibilities are not confined ...PC work ...ent jobs inspected: ...own	Inspection authority—yes No. of inspectors: 5 regional entomologists whose responsibilities are not confined to SPC work Percent jobs inspected: Unknown	Inspection authority—yes No. of inspectors: 2 full time, 2 part time Percent jobs inspected: 16-year mean—6.7 1970—71—5.7 Miscellaneous: Soil samples taken on a random selection of jobs after treatment	Inspection authority—yes No. of inspectors: 1 part time Percent jobs inspected: Unknown	Inspection authority—yes No. of inspectors: 7 regional entomologists whose responsibilities are not confined to SPC work Percent jobs inspected: Unknown Miscellaneous: Soil samples may be taken
...andard work: ...ines ...quent license renewal: ... of original fee ...meanor: ...ating provisions of law	Substandard work: No fines Delinquent license renewal: License cancelled if delinquent after grace period stated to company in writing Misdemeanor: Violating either provisions of the law or rules of the Division of Health	Substandard work: No fines Delinquent license renewal: 100% of original fee Misdemeanor: Violating provisions of law	Substandard work: No fines Delinquent license renewal: No fines Misdemeanor: Violating provisions of law	Substandard work: No fines Delinquent license renewal: Delinquent renewals processed as new applicants Misdemeanor: Violating provisions of law
...rs of clearance: ...eatment not required for ...and VA financed houses ... inspected:	Letters of clearance: Not reported State inspected: No	Letters of clearance: Copies for FHA and VA financed houses are received by state State inspected: Yes, only on complaint	Letters of clearance: Not reported State inspected: Yes, only on complaint	Letters of clearance: Not reported State inspected: No

Table 1b.—Characteristics of structural pest control laws, Kentucky through Texas.

ITEM	KENTUCKY	LOUISIANA	MINNESOTA	MISSISSIPPI
1 Date(s) of law and amendments	Law – 1954 Amended – 1960	Law – 1930 Amended – 1952	Law – 1970 Amended – No	Law – 1938 Amended – 1942, 1969, 1971
2. Financing of regulatory work	Fees plus legislative appropriation	Fees	Legislative appropriation	Legislative appropriation 5% sales tax on all work, revenue deposited in State General Fund
3. Regulatory board	Composition: 5 members 2 members licensed for SPC 2 members Dep. Entomol., Univ. Ky. Member Dep. Agric. Authoritative agency: Dep. Agric.–Regulatory powers Examining Bd.–Licensee examining Tenure: Staggered 1–and 2– year terms, unlimited	Composition: 5 members Comm. Agric. and Immigration State entomologist Member, Dep. Entomol., LSU 2 members licensed for SPC Authoritative Agency: Dep. Agric.– Licensing and regulatory powers Tenure: Unlimited number 2-year terms for latter 3, tenure unlimited for others	Composition: 13 members 3 members licensed for SPC 1 of which is designated by State Assoc. 3 representatives of State or Municipal Health Deps. Member food industry Member State Dep. Agric. Others not specified Authoritative Agency: Dep. Agric.–Advisory only Tenure: Terms not specified	Composition: 3 members Head, Dep. Entomol Miss. State Univ. Head, Dep. Plant Path. & Weed Science, Miss. State Univ. State Chemist Authoritative Agency: Dep. Agric. & Commerce– Licensing and regulatory powers Tenure: Unlimited
4. Licensee requirements	Education: 2 years college in appropriate field, or Experience: 2 years in appropriate field Financial: Bond–$50,000 or Insurance– $25,000 property damage $50,000 bodily injury/ person – $100,000 bodily injury/accident plus Workman's Compensation	Education: College graduate with major in entomology, or Experience: 4 years in appropriate field Financial: Bond–$2,000 Insurance–$25,000 injury/ person $50,000/accident for injuries $10,000 property damage	Education: None Experience: None Financial: Bond or insurance $100,000 bodily injury/ person $200,000 bodily injury/ 2 or more persons $10,000 property damage	Education: College graduate with courses in appropriate field, or 2 years college with special training in appropriat field, or Experience: High school graduate plus 1 year experience, or 2 years under licensee Financial: Bond–$1,000 or $2,500 for any one guaranteeing work for more than a year or who has service contracts of more than a year
5. Licensee examinations	Type: Written, oral, or both Classes: 1. Termites and other WDO 2. General pests and rodents 3. Fumigation	Type: Written or oral and may have pest identification quiz also Classes: 1. Entomological work a. Fumigation b. Household pests c. Termites and other wood destroying insects 2. Rodent work	Type: Written or oral and possibly also a practical demonstration Classes: 1. Licensed by state 2. Licensed by Minneapolis or St. Paul	Type: Written and oral Classes: 1. Domestic animals 2. Home pests 3. Horticultural and floricultural pests 4. Orchard pests 5. Ornamental plants, shade trees, and lawns 6. Termites and other structural pests 7. Tree surgery
6. Training exceeding 1 week	None	None	None	20 week course, once weekly for servicemen and managers held in various locations of the state
7. Offices and licensed supervision	Each office providing treatments must be supervised by licensee	Each office providing treatments must be supervised by licensee	Each office providing treatments must be supervised by licensee	Each office providing treatments must have 1 or more licensed operators
8. Fees	Reporting–none Company registration–$100/ place of business Examination. $15 for first exam in any or all fields $15 for subsequent exam in each additional field Re-examination–$5 License–none Reinspection–none Identification–none	Reporting–$3/contract Company registration–none Examination–none License–none Reinspection–none Identification–none	Reporting–none Company registration–$50 but $10 where county or municipality also licenses Examination–none License–$5 Reinspection–none Identification–none	Reporting–none Company registration–none Examination–none License–none Reinspection–none Identification–$.50/card

10

	NEW MEXICO	NORTH CAROLINA	OKLAHOMA	TENNESSEE	TEXAS
	Law – 1965 Amended – No Legislative appropriation	Law – 1955 Amended – 1957, 1967, 1971 Fees plus legislative appropriation	Law – 1953 Amended – 1955, 1957, 1967 Fees plus legislative appropriation	Law – 1955 Amended – 1961 Legislative appropriation	Law – 1971 Amended – No Fees
	Composition: 5 members Appointed by Governor, affiliation unspecified Authoritative agency: Environmental Improvement Agency—Licensing and and regulatory powers Tenure: Staggered 5-year terms, unlimited	Composition: 5 members Member State Board Agric. Member State Dep. Agric. Member Dep. Entomol., N.C. State Univ. 2 members licensed for SPC Authoritative agency: Dep. Agric.—Licensee examin- ing and regulatory powers Tenure: 4-year terms, PCO members cannot suceed themselves; unlimited number of terms for others	Composition: 4 members Head, Dep. Entomol., Okla. State Univ. Head, State Health Dep. 2 members State PC Assoc. Authoritative agency: Dep. Agric.—Advisory only Tenure: Unlimited	Composition: 5 members Comm. Agric. Director Entomol. & Plant Path. Entomologist of Exp. Stn. or Extension Staff, Univ. Tenn. 2 members licensed for SPC Authoritative agency: Dep. Agric.—Licensee examining and regulatory powers Tenure: 3-year terns, un- limited	Composition: 7 members 4 members licensed for SPC for past 5 years, appointed by Governor Member Dep. Agric. Member Dep. Entomol., Texas A&M Authoritative agency: Dep. Agric.—Licensee examining and regulatory powers Tenure: 2-year terms, un- limited
	Education: College degree in biological sciences, or Experience: 4 years in ap- propriate field Financial: Surety Bond– $5,000	Education: College degree in pertinent field plus some unspecified practical experience or, Experience: 2 years in ap- propriate field Financial: No bond or in- surance required	Education: 2 years college with major courses in en- tomalogy, or 1 year in SPC with 1 year experience, or Experience: 2 years in ap- propriate field Financial: No bond or in- surance required	Education: 2 years college in appropriate field Experience: 2 years in ap- propriate field Financial: Each office of licensee must have surety bond of $1,000	Education: None Experience: None Financial: No bond or in- surance required
	Type: Written Classes: None – 1 exam covers all classes	Type: Written, oral, or both Classes: 1. Fumigation 2. Household pests 3. WDO 4. Any combination of the above	Type: Written and practical Classes: 1. Termites 2. General pests 3. Fumigation 4. Any combination of the above	Type: Preliminary written exam Those passing take written and oral exam before ex- amining board. 100% score– 1st class license (recommend control measures, train new employees), other passing scores–2nd class license (supervise work crews) Classes: 1. Agric. pest and disease control (P&D)–aircraft applied 2. Agric. P&D control– general 3. Consulting entomol.– general 4. Consulting–Plant Path.– general 5. Fumigation 6. Horticultural spraying 7. Insects affecting man and animals 8. Rodents 9. Storage, warehouse and transportation pests 10. Structural and household pests	Type: Written Classes: 1. Termites and other WDO 2. Pests in homes, busi- nesses, and industries 3. Pests of ornamental plants, shade trees, and lawns 4. Fumigation 5. Weed control
	None	None, but N.C. State Univ. has a 2 year course in PC	None	None	None
	Each office providing treat- ments must be supervised by licensee	Each office providing treat- ments must be supervised by licensee residing within 75 miles	None specified	Home office–supervised by resident 1st class licensee Branch office–supervised by resident 1st or 2nd class licensee	Each office providing treat- ments must be supervised by licensee
	Reporting–none Company registration–none Examination–none License–$10 Reinspection–none Identification–none	Reporting–none Company registration–none Examination–$25/class License. $100 for 1 class $50 for each additional class Reinspection–$10 Identification–$20/card/ solicitor, estimator, salesman, or serviceman Miscellaneous: $5 duplicate license $10 delinquent license plus normal fee	Reporting–none Company registration–none Examination–$25/class License–$25/class Reinspection–none Identification–none	Reporting–$3/contract Company registration–none Examination–none for preliminary, $10 each exam after first License–$10/class regard- less of PC fields Reinspection–none Identification–none	Reporting–none Company registration–$50/ business location Examination–none License–none Reinspection–none Identification–$5/card

Table 1b—Characteristics of structural pest control laws, Kentucky through Texas. (Cont'd.)

ITEM	KENTUCKY	LOUISIANA	MINNESOTA	MISSISSIPPI
9. Identification	Personal—I.D. cards for all servicemen and solicitors Vehicle—license number, company name address	Personal—I.D. cards for licensee, servicemen, solicitors Vehicle—none	Personal—None Vehicle—none	Personal—I.D. cards for all licensees, servicemen, solicitors Vehicle—marked for easy identification
10. Standard contracts or other forms related to treatment or inspection	Standard contracts—no, but contracts must be state approved, must have treatment date, length of contract period, WDOs treated, total price, reinspection and retreatment conditions, and spot treatments shown on scale drawing	Standard contracts—yes, 2 kinds, 1 conventional and 1 for slab construction— Appendices I, II Other forms—Appendices XIII—XV	Standard contracts—no, but contracts must include name of person requesting service, property address, date, chemicals and control measures used Other forms—none	Standard contracts—no, but termite contracts cover beetles also, unless specifically eXcluded. Has to meet minimum requirements, guaranteed for 1 year, a complete diagram showing infestation and damaged areas, and out line of work performed Other forms—Appendix XVI
11. Definition of structural pests restricted to WDO?	License class 1 is	License class 1c 1s	No	License class 6 is
12. Information reported to regulatory agency by each company	Monthly reports—list of jobs with contract number, name and address and owner and property treated, and number of houses inspected under renewal contracts Copies of contracts filed—no	Monthly reports—list of jobs with dates, addresses of treated properties, and name and address of property owner Copies of contracts filed—no	Monthly reports—none Copies of contracts filed—no	Monthly reports—list of jobs with name and address of customer and kind of service rendered, i.e., ST or PPB Copies of contracts filed—no
13. State population and volume of work	Population : 1960—3,038,156 1970—3,219,311 Volume of work 1972—20,223 Published licensee list—yes Work performance published—no	Population: 1960—3,257,022 1970—3,643,180 Volume of work: 1971.72—29,409 Published licensee list—yes Work performance published—no, but semi-annual newsletters are circulated to PCOs listing, among other things, total number of jobs inspected and the number found substandard	Population: 1960—3,413,864 1970—3,804,971 Volume of work: Unknown Published licensee list—no Work performance published—no	Population: 1960—2,178,144 1970—2,216,912 Volume of work: 1972—31,301 Published licensee list—yes Work performance published—no
14. Minimum job requirements excluding those for fumigation	None When a building is found to be unsatisfactorily treated, 2 or 3 local companies are asked to check the building. If they agree that it is unsatisfactory, then the treatment has to be corrected	Appendix XXIII	None	Appendix XXIV
15. State inspections	Inspection authority—yes No. of inspectors: 4 regional entomologists whose responsibilities are not confined to SPC work Percent jobs inspected: Unknown, estimated 480—500 annually Miscellaneous. Whenever 30% of treatments of a licensee are found unsatisfactory, the licensee is either suspended or he must pay $50 per day until work is satisfactory	Inspection authority—yes No. of inspectors: 6 Percent jobs inspected: 1971—15-20 (estimated)	Inspection authority—no No. of inspectors: none Percent jobs inspected: not applicable	Inspection authority—yes No. of inspectors: 3 Percent jobs inspected: Unknown Miscellaneous: Whenever 25% of a licensee's work is found substandard, license is given 30 days to show cause why he should not be suspended
16. Penalties	Substandard work No fines Delinquent license renewal: Delinquent renewals processed as new applicants Misdemeanor: Violating provisions of law. Fines deposited with county where infraction occurred	Substandard work· No fines Delinquent license renewal: No license renewal necessary unless operator has been inactive for 4 years	Substandard work: No fines Delinquent license renewal: 50% of original fee; over 3 months delinquent, processed as new applicants	Substandard work: No fines Delinquent license renewal 50% of original fee Misdemeanor: Violating provisions of law
17. Policy for FHA and VA financed houses	Letters of clearance: Copies for FHA and VA financed houses are received by state State inspected: No	Letters of clearance: Not reported State inspected: Yes, if property is covered by standard contract	Letters of clearance: Not reported State inspected· No	Letters of clearance: Not reported State inspected No

NEW MEXICO	NORTH CAROLINA	OKLAHOMA	TENNESSEE	TEXAS
'ersonal—I.D. cards for licensee, servicemen, solicitors 'ehicle—license number, company name, address	Personal—I.D. cards for servicemen and solicitors, certification card for each licensee Vehicle—license number, company name, address	Personal—none Vehicle—license number, company name, address	Personal—I.D. cards for licensee, servicemen, solicitors Vehicle—none	Personal—I.D. cards for licensees and servicemen Vehicle—license number and license category
standard contracts—no, but contracts must include date of inspection, address of property, type of construction, sketch of infested areas, and chemicals used)ther forms—none	Standard contracts—no, but contracts must be state approved Other forms—Appendix XVII	Standard contracts—no Other forms—monthly work activity reports	Standard contract—no, but contracts must be state approved	Standard contracts—no Other forms—none
'o	License class 3 is	No	No	License class 1 is
'onthly reports—none :opies of contracts filed—no	Monthly reports—none Copies of contracts filed—no	Monthly reports—list of jobs with name of property owner, property address, kind of structure, and date work completed Copies of contracts filed—no	Monthly reports—list of jobs with name and address of treated property, date work completed, kind of termite treatment, kind of structure, and amount of contract Copies of contracts filed— law states they are to be submitted but only work orders are submitted	Monthly reports—none Copies of contracts filed—no
'opulation: 1960—951,023 1970—1,016,000 'olume of work: Jnknown 'ublished licensee list—yes 'ork performance published— 'o	Population: 1960—4,556,155 1970—5,082,059 Volume of work: Unknown Published licensee list—yes Work performance published— no, but it is reported to the appropriate local Better Business Bureaus, Chambers of Commerce, property control officers, building contractors, VA, and FHA	Population: 1960—2,328,284 1970—2,559,253 Volume of work: 1971.72—26,500 Published licensee list—yes Work performance published— no	Population: 1960—3,567,089 1970—3,924,164 Volume of work: 16-year mean—25,777 1970-71—41,624 Published licensee list—yes Work performance published— no, but statewide totals are published for number treated, number state inspected, numbers of suspensions and fines, and number of soil samples taken. No data listed by company	Population: 1960—9,579,677 1970—11,196,730 Volume of work: Unknown Published licensee list—yes Work performance published— no
one	Appendix XXV	Appendix XXVI	Under revision	None
spection authority—yes o. of inspectors 1 part time ercent jobs inspected: Jnknown	Inspection authority—yes No. of inspectors. 7 Percent jobs inspected: Unknown Miscellaneous. First 6 jobs of each new licensee are inspected. Soil samples taken on a random selection of jobs after treatment	Inspection authority—yes No. of inspectors. 1 plus 8 regional entomologists whose responsibilities are not confined to SPC work Percent jobs inspected: Less than 1	Inspection authority—yes No. of inspectors. 5 Percent jobs inspected: 2 8	Inspection authority—yes No. of inspectors. 5 Percent jobs inspected: Unknown Miscellaneous: Soil samples taken at inspector's discretion
bstandard work: No fines elinquent license renewal: Ielinquent renewals rocessed as new applicants isdemeanor: iolating provisions of law	Substandard work $10 for each necessary reinspection resulting from major discrepancies (Appendix XXVII), if a major discrepancy is found, PCO has 30 days to correct, then another state inspection, etc. until corrected Delinquent license renewal: $10 fine. license phase, over 1 year delinquent, processed as new applicant Misdemeanor: Violating provisions of law	Substandard work No fines Delinquent license renewal: Delinquent renewals processed as new applicants Misdemeanor: Violating provisions of law	Substandard work: No fines Delinquent reporting fees: $5 plus 10% of fees due Misdemeanor: Violating provisions of law	Substandard work: No fines Delinquent license renewal: Delinquent renewals processed as new applicants Misdemeanor: Violating provisions of law
tters of clearance: ot reported ite inspected: No	Letters of clearance: Copies for FHA and VA financed houses are received by state on state-supplied forms—Appendix XVII State inspected. Yes, at random	Letters of clearance: Not reported State inspected Yes, only on complaint	Letters of clearance: Copies for FHA and VA financed houses are received by state State inspected: Yes, at random	Letters of clearance: Not reported State inspected: No

ana, and Tennessee require each job to be reported to the State and charge a fee for each job reported. All of these States except Tennessee support their SPC regulatory work entirely by fees. Eight States with SPC legislation have a company registration fee, which varies from $20 for branch offices in California to $100 for companies in Arizona and Kentucky. Ten States charge examination fees ranging from $10 to $50. License fees vary from $5 in Minnesota to $100 in North Carolina. Kansas, Kentucky, Louisiana, Mississippi, and Texas have no license fees.

Arkansas and North Carolina charge for reinspecting a property reported as substandard by a State inspector. This fee enables Arkansas and North Carolina to immediately penalize PCOs for faulty treatments without revoking or suspending licenses.

Florida, Georgia, Mississippi, North Carolina, and Texas charge for issuing personal identification cards.

#9 Identification.—Identification of license holders and their service vehicles offers three advantages: (1) assures the public that work is done by licensed firms; (2) improves the image of the PC industry and of individual companies; and (3) gives the State Regulatory Office some publicity.

Only Hawaii, Kansas, and Minnesota require neither personal nor vehicle identification. At this writing, the Kansas legislature is considering a bill to require vehicle identification. Of the other States, Alabama, Arkansas, Colorado, and Oklahoma require no personal identification and Louisiana and Tennessee require no vehicle identification. All the other States require both personal and vehicle identification in some form. For example, Arizona and California require only the licensee, not servicemen or solicitors, to carry personal identification cards. Upon initiation of the new California State Association Certification Program, all certified servicemen in California will be identified by a shoulder patch. The Association is advertising this program to encourage the public to admit only certified servicemen into homes. Alabama, Arkansas, Florida, and Mississippi do not require vehicles to be marked with the company's license number.

#10 Standard contracts or other forms.—Many of the possible benefits of SPC legislation require that regulatory offices receive copies of contracts or forms containing accurately reported information. Also, regulatory enforcement is obviously

tics of treated properties, pest incidence, treatment types, etc. California gathers some similar information in its "Standard Structural Pest Control Inspection Reports." Tennessee's law requires the submission of contracts, but the State usually receives only work orders and monthly work reports.

#13 *State population and volume of work.*— These data indicate how much of a problem SP are in each State and, hence, the extent of legislation necessary to adequately regulate PC activities. Further, a knowledge of the work volume is essential for maximum enforcement efficiency.

Maintaining adequate records is expensive and most States do not know the dollar cost of SPC damage. For example, California stopped tabulating the dollar cost of SPC work in 1965 when 331,256 inspection reports and 130,407 completion notices were filed.

A list of licensees is available from all States except Minnesota. Until 1972, only Arkansas published company performance data in its Plant Board Annual Report. This report provided a list, by company, of the number of houses treated, the number of houses that were State-inspected, and the number of inspected houses that were found substandard. However, only about 200 copies of this report were distributed. Although the report is public information, it obviously was not widely disseminated. If it had been made more available, it could have been a valuable public service and could have offered effective advertising for the better companies. North Carolina compiles similar information and reports it to Better Business Bureaus, Chambers of Commerce, the North Carolina Property Control Officer, building contractors, FHA, and VA. Louisiana sends all PC companies a summary of statewide totals for treatments, inspections, and license suspensions. Tennessee lists statewide totals (not by company) for number of buildings treated, number of inspections, number of suspensions and fines, and number of soil samples taken, but not number of substandard treatments.

#14 *Minimum job requirements.*—Effective regulation requires definition of satisfactory performance. Ideally, some attention in definitions should be given to long-term protection rather than just to killing the pests now present. Frequently the design of a structure or its maintenance must be altered to prevent termite attack. For example, wooden debris under a house, soil fills outside the house above the sill line, any

ana, and Tennessee require each job to be reported to the State and charge a fee for each job reported. All of these States except Tennessee support their SPC regulatory work entirely by fees. Eight States with SPC legislation have a company registration fee, which varies from $20 for branch offices in California to $100 for companies in Arizona and Kentucky. Ten States charge examination fees ranging from $10 to $50. License fees vary from $5 in Minnesota to $100 in North Carolina. Kansas, Kentucky, Louisiana, Mississippi, and Texas have no license fees.

Arkansas and North Carolina charge for reinspecting a property reported as substandard by a State inspector. This fee enables Arkansas and North Carolina to immediately penalize PCOs for faulty treatments without revoking or suspending licenses.

Florida, Georgia, Mississippi, North Carolina, and Texas charge for issuing personal identification cards.

#9 *Identification.*—Identification of license holders and their service vehicles offers three advantages: (1) assures the public that work is done by licensed firms; (2) improves the image of the PC industry and of individual companies; and (3) gives the State Regulatory Office some publicity.

Only Hawaii, Kansas, and Minnesota require neither personal nor vehicle identification. At this writing, the Kansas legislature is considering a bill to require vehicle identification. Of the other States, Alabama, Arkansas, Colorado, and Oklahoma require no personal identification and Louisiana and Tennessee require no vehicle identification. All the other States require both personal and vehicle identification in some form. For example, Arizona and California require only the licensee, not servicemen or solicitors, to carry personal identification cards. Upon initiation of the new California State Association Certification Program, all certified servicemen in California will be identified by a shoulder patch. The Association is advertising this program to encourage the public to admit only certified servicemen into homes. Alabama, Arkansas, Florida, and Mississippi do not require vehicles to be marked with the company's license number.

#10 *Standard contracts or other forms.*—Many of the possible benefits of SPC legislation require that regulatory offices receive copies of contracts or forms containing accurately reported information. Also, regulatory enforcement is obviously aided by concise, clearly worded working forms.

Louisiana is the only State that issues standard contracts that must be used by all PCOs (Appendices I, II). Louisiana issues two kinds of contracts, one for slab and one for other types of construction. Both contracts are good for 2 years with guaranteed annual inspections. Any rider to the contract must be approved by the SPC Commission.

Arizona, Colorado, Oklahoma, and Texas neither have standard contracts nor require State approval of contracts. The other States with SPC laws either have minimum requirements that must be met by each contract or approve the contracts used by the individual companies.

Many other forms are used by States in administering their SPC laws. The most important forms not discussed under # 12 are mentioned here and are reproduced (partially modified) in the Appendices. If several States have similar forms, only one is discussed and illustrated. In Arkansas, each operator must fill out and sign a "Minimum Requirements Checklist" (Appendix III) for each job, attach it to the copy of the contract, and mail it to the State Regulatory Office. This office inspects 25 to 33 percent of all jobs performed in the State (see #15) and files a "Termite Inspection Report" (Appendix IV) for each job inspected. If the work is unsatisfactory, a "Report of Substandard Termite Treatment" form is filed (Appendix V). A "Request for Approval of Substandard Contract for Existing Structures" (Appendix VI) (see #14) is used also.

In California, when a PCO inspects a house to bid on a job, he files a "Standard Structural Pest Control Inspection Report" (Appendix VII) with the regulatory office. Within 5 days after completion of work under a contract a "Standard Notice of Work Completed and Not Completed" (Appendix VIII) is filed.

If a complaint is made in writing to the Florida State Board of Health, the State inspects the property and files a "Standard Structural Pest Control Investigation Report" (Appendix IX).

Georgia supplies monthly activity forms on which PCOs must report work performed (Appendix X). Georgia also has forms for State inspectors and for requesting State approval of substandard treatments (Appendices XI and XII).

Louisiana has routine inspection reports (Appendix XIII), reports of termite eradication (monthly activity reports) (Appendix XIV), and

14

tics of treated properties, pest incidence, treatment types, etc. California gathers some similar information in its "Standard Structural Pest Control Inspection Reports." Tennessee's law requires the submission of contracts, but the State usually receives only work orders and monthly work reports.

#13 State population and volume of work.— These data indicate how much of a problem SP are in each State and, hence, the extent of legislation necessary to adequately regulate PC activities. Further, a knowledge of the work volume is essential for maximum enforcement efficiency.

Maintaining adequate records is expensive and most States do not know the dollar cost of SPC damage. For example, California stopped tabulating the dollar cost of SPC work in 1965 when 331,256 inspection reports and 130,407 completion notices were filed.

A list of licensees is available from all States except Minnesota. Until 1972, only Arkansas published company performance data in its Plant Board Annual Report. This report provided a list, by company, of the number of houses treated, the number of houses that were State-inspected, and the number of inspected houses that were found substandard. However, only about 200 copies of this report were distributed. Although the report is public information, it obviously was not widely disseminated. If it had been made more available, it could have been a valuable public service and could have offered effective advertising for the better companies. North Carolina compiles similar information and reports it to Better Business Bureaus, Chambers of Commerce, the North Carolina Property Control Officer, building contractors, FHA, and VA. Louisiana sends all PC companies a summary of statewide totals for treatments, inspections, and license suspensions. Tennessee lists statewide totals (not by company) for number of buildings treated, number of inspections, number of suspensions and fines, and number of soil samples taken, but not number of substandard treatments.

#14 Minimum job requirements.—Effective regulation requires definition of satisfactory performance. Ideally, some attention in definitions should be given to long-term protection rather than just to killing the pests now present. Frequently the design of a structure or its maintenance must be altered to prevent termite attack. For example, wooden debris under a house, soil fills outside the house above the sill line, any

earth-wood contact, earth-filled porches not chemically treated, and many other correctable conditions make a house susceptible to termite attack.

Eight States have no specific minimum job requirements—Alabama, Arizona, Colorado, Hawaii, Kentucky, Minnesota, New Mexico, and Texas. The other States have specific minimum requirements that we have listed in Appendices XVIII to XXVI.

#15 State inspections.—No other feature of SPC legislation can do more to protect the public from unscrupulous practices. Experience of States that inspect show that although many companies in the PC industry render excellent professional service, some do not. The undesirable companies are eventually controlled by the normal flow of private enterprise, but the high-value, long term investments being protected and the hazardous (if misused) means of protection mitigate against such slow-acting controls. Inspections also identify the competent operator, thus benefiting both him and his industry.

Only Arizona and Minnesota have no inspection authority. Most of the remaining States have one or more full-time inspectors whose sole job is to check SPC work. Colorado, Florida, Kansas, and Kentucky have regular inspectors who check SPC work only as one of many duties. In these States the number of houses checked is small, and inspections usually are made only to follow up a complaint. Alabama, Hawaii, and New Mexico have only one SPC inspector, and he inspects only on complaint.

Arkansas regularly inspects a much higher percentage of jobs than any other State—24.3 percent of the work performed during fiscal year 1971. Arkansas' 21-year mean percentage of jobs inspected is 27.5 percent. The first five jobs of each new licensee are inspected to determine whether the operator has the capacity to perform satisfactorily. Whenever 15 percent or more of the jobs of any operator are found to be substandard, the regulatory office may give the operator 60 days' notice and then inspect all jobs reported during the previous year.

Georgia has inspected an average of 6.7 percent of all jobs over the past 6 years. In fiscal year 1970 the figure was 5.7 percent. Soil samples are taken at randomly selected properties. One-hundred ppm of insecticide is the minimum acceptable residue for termite control.

Kentucky checks about 500 properties annually. Whenever a building is found to be unsatisfactorily treated, representatives of two or three local companies are asked to check. If they agree that the treatment is unsatisfactory, it has to be corrected. Whenever 30 percent of the treatments of a licensee are found unsatisfactory, he is either fined or suspended until the work is corrected.

Prior to 1971, Mississippi had only regional entomologists to check treated properties on complaint. Since then, three full-time SPC inspectors have been hired, and Mississippi is now checking a much greater number of reported jobs. Whenever 25 percent of a licensee's work is found substandard, the licensee is given 30 days to show cause why his license should not be suspended.

In 1971, Louisiana inspected an estimated 15 to 20 percent of all reported jobs. From 1954 to 1970, Tennessee inspected an average of 9.5 percent of all reported jobs.

North Carolina inspects the first six jobs of each new licensee to determine whether the operator has the capacity to perform satisfactorily. Soil samples are taken routinely after termite treatment to test for the adequacy of insecticidal residues. These samples are categorized by location for both pre- and post-treatment jobs—e.g., inside and outside foundation walls, around pillars, under slabs, around pipes and conduits, and inside earth-filled porches.

Soil samples may also be taken at complaint sites in Kansas; 50 ppm of insecticide residue is considered adequate treatment.

#16 Penalties.—SPC regulations cannot be enforced without some penalties for offenders. Alabama, Arizona, Arkansas, Colorado, Georgia, North Carolina, and Tennessee impose fines for delinquent license renewals or for delinquent monthly reports.

Only Arkansas and North Carolina impose fines for substandard work. Arkansas charges $15 to inspect each substandard job. Further, any request for a substandard treatment waiver costs $30 if, in the opinion of the State, a standard treatment could have been performed. North Carolina imposes a $10 fee to inspect each substandard job resulting from a "major discrepancy" (Appendix XXVII).

In all States, a violation of the law is a misdemeanor. Thus, PCOs must be licensed and must conform to the rules and regulations authorized by the law or be penalized.

16

200 properties an-
found to be unsatis-
ves of two or three
heck. If they agree
actory, it has to be
t of the treatment
tactory, he is either
ork is corrected.

had only regional
l properties on com-
ime SPC inspectors
ppi is now checking
eted jobs. Whenever
t is found substand-
days to show cause
suspended.

ed an estimated 15
jobs. From 1954 to
i average of 9.5 per-

the first six jobs of
ne whether the oper-
rform satisfactorily.
tinely after termite
quacy of insecticidal
categorized by loca-
reatment jobs—e.g.
on walls, around pil-
es and conduits, and

taken at complaint
insecticide residue is
it.

lations cannot be en-
s for offenders. Also
Colorado, Georgia
see impose fines for
or for delinquent

Carolina impose fines
ate charges $15 to the
Further, any request
waiver costs $30 if, it
standard treatment
North Carolina in-
ith substandard job
repancy" (Appendix

f the law is a misde-
e licensed and must
gulations authorize

#17 *Policy for FHA and VA financed homes.*—
The NPCA (1973) reports that "In some areas of
the country where termite infestations are heavy,
the FHA requires certification of all subject properties. In other areas, a certification is required
only if the FHA inspector suspects a termite condition. The decision for determining whether or
not *all* properties will be inspected is made by the
regional FHA insuring office." For example,
neither existing nor new homes are required to be
examined or treated in Colorado. Obviously the
public has to fend for itself unless adequate standards are set and someone other than the buyer is
made liable for failure to detect problems.

Only in California, Arkansas, Kentucky, and
North Carolina must copies of letters of clearance
be sent to the State. Georgia requires that each
letter of clearance be placed on the monthly job
report. Of these five States, only Arkansas, Georgia, and North Carolina randomly inspect these
properties. North Carolina provides standard
forms for FHA and VA letters of clearance and has
long recognized the desirability (Smith, 1960) of
a standard form acceptable to all lending agencies.
If the properties are covered by a valid contract,
Louisiana randomly inspects them.

Recently a form (fig. 1, see instructions in Appendix XXVIII) has been prepared through the
cooperation of FHA and NPCA which has also
been approved by the VA. The intent of this form
is to provide a standardized inspection procedure
and report for SP in buildings involved in real estate transactions. States may either accept this
form or modify their own form to concur with it.
Otherwise, the States must submit both the FHA
form and their own form.

17

Figure 1.—*Wood infestation report (FHA Form No. 2053).*

FHA FORM NO 2053

U S DEPARTMENT OF HOUSING AND URBAN DEVELOPMENT
FEDERAL HOUSING ADMINISTRATION

Form Approved
OMB NO 63-R1395

WOOD INFESTATION REPORT

Company Name _____ Date _____

Address _____ Property Address _____

Phone Number _____ _____

Lic No (if any) _____ FHA Case Number _____

This is to certify that a qualified inspector employed by this Company has carefully inspected and sounded all accessible areas of the property located at the above address for termites or other wood destroying insects with the following findings

	Check One
	YES NO

1 There is active infestation of. (A) Termites .... ☐ ☐
  (B) Other wood destroying insects .... ☐ ☐

2. There is evidence of a previous infestation of: (A) Termites .... ☐ ☐
  (B) Other wood destroying insects .... ☐ ☐

3 There is evidence of conditions conducive to infestation (earth-wood contact. faulty grades, insufficient ventilation. etc.) If yes. describe on reverse side of form .... ☐ ☐

4 There is evidence of damage to structural items (columns, girders. sills, joists. plates. headers, stairs, porch supports. rafters. etc.) If yes. describe on reverse side of form .... ☐ ☐

5. There is evidence of damage to other construction (exterior porch floors and steps. door and window sills. Jambs. siding. subflooring. etc.) If yes. describe on reverse side of form .... ☐ ☐

6 The premises have been treated by the undersigned by the application of _____ on _____ (Date) and a one (1) year guarantee issued against reinfestation by subterranean termites This guarantee is transferable to any subsequent owner of this property during the life of this guarantee .... ☐ ☐

7 This Company has made inspections of these premises previously .... ☐ ☐

8 This Company has treated these premises previously .... ☐ ☐

I hereby certify that neither I nor the company for whom I am acting have had. presently have or contemplate having any interest in the property involved I do further certify that neither I nor the company for whom I am acting is associated in any way with any party to this transaction.

Inspector _____ Date _____

Authorized Agent _____ Date _____

Original to HUD/FHA

Copy To: Buyer
  Seller
  Mortgagee

18

REMARKS

USE THIS SPACE TO AMPLIFY STATEMENTS NUMBER 3, 4 AND 5. THIS SPACE CAN BE USED TO
CLARIFY ANY STATEMENT MADE. INCLUDE ITEM NUMBER WITH EACH EXPLANATION

cted and wounded all
rying insects with the

Check One

YES

HAS SPC LEGISLATION WORKED?

Insufficient data are available for accurately assessing how work quality is influenced by SPC laws. We have no independently gathered data from States lacking SPC laws to compare with that from States having SPC laws. Of the latter, only Arkansas, Georgia, North Carolina, and Tennessee had adequate data and there was considerable variation among these States in volume of work, minimum job requirements, percentage of work inspected, thoroughness of inspection, and the like. A job that is satisfactory in one of these States may not be in another. In the future, some indication of work activity and pest incidence will be available from States lacking SPC regulations because the original copy of the FHA Wood Infestation Report (fig. 1) will be retained by FHA.

The available data (figs. 2 to 4, tables 2, 3) indicate that State regulation may improve performance. Arkansas, which has strict standards of performance, reports that less than 10 percent of inspected jobs have been substandard over the last decade (fig. 2), despite an increase in mimimum acceptable standards during the early 1960s. At that time, the definition of substandard work, which only included treated homes with live termite infestations, was expanded to include insufficient preventive measures such as failure to remove wood debris from around foundations, failure to tunnel porches, and failure to correct

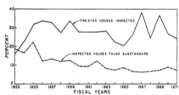

Figure 2.—Annual State inspection activity and work
quality in Arkansas.

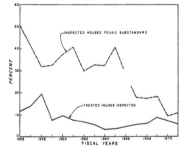

Figure 3.—Annual State inspection activity and work
quality in Georgia.

aders,

aders,

r and

rate-
ig the

have
l not

19

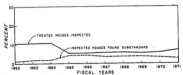

Figure 1.—Annual State inspection activity and work quality in Tennessee.

Table 3.—Adequacy of termite soil treatments based annual State post-treatment inspections North Carolina

Fiscal year	No. soil samples tested	Percen substand: sample
1963-64	58	100
1964-65	67	58
1965-66	127	42
1966-67	85	28
1967-68	497	14
1968-69	1,071	12
1969-70	800	12
1970-71	1,189	7

faulty grades. The majority of substandard work in recent years has been in these categories.

The Georgia data (fig. 3) suggests several interesting features. In 1956, one-half the State-inspected jobs were substandard. During the next 2 years the State increased the inspection rate from 11.8 to 19.3 percent of the treated houses. During this time the percentage of State-inspected substandard jobs decreased from 50.4 to 31.8 percent. For the next 3 years the State's inspection rate dropped and the incidence of substandard treatments increased. This cycle was essentially repeated between 1962-1965. Since 1965, the State-inspected percentage of substandard jobs has decreased from 40.6 to 10.9 percent.

We do not suggest that the improvements in

work quality in Georgia or any other State shou be attributed solely to the frequency of St inspections. The PC industry has taken ma steps on its own to improve performance. I example, early in 1967 the Georgia Pest Cont Association held its first serviceman's semin other examples of industry self-improvement v be discussed later. The most dramatic increase performance (fig. 3), however, occurred soon af Georgia's regulatory office began taking s samples and testing them for insecticide resid in 1964. From 1964 to 1967, the percentage of s

Table 2.—A summary of SPC work volume, work quality, and regulatory enforcement in Arkansas, Georgia, No Carolina, and Tennessee

State	Years of records	Number of houses treated	Number of treated houses inspected by State	Percent of treated houses inspected by State	Number of State inspected houses with substandard treatment	Percent of State inspect houses with substandar treatment
Arkansas	[1]21					
Mean		10,234	[3]2,816	27.5	[4]262	[4]9.3
1970-71		21,500	[3]5,217	24.3	375	7.2
Georgia	16					
Mean		34,062	2,268	6.7	596	26.3
1970-71		55,291	3,124	5.7	340	10.9
North Carolina	16					
Mean			[3]808	. . .[2]	175	21.7
1970-71			[3]1,811	. . .[2]	346	19.1
Tennessee	10					
Mean		34,037	2,398	7.0	67	2.8
1970-71		41,624	2,963	7.1	82	2.8

[1] Records for fiscal year 1952 are missing.
[2] North Carolina Pest Control Regulatory Office has no record of total volume of work within the State.
Does not include reinspection of substandard treatments.
[3] Substandard work during the 1950s included only treated houses with live infestations. Beginning in the e 1950s and to the present, substandard treatments include many other deficiencies (e.g., debris, porches not tunn etc.). Therefore, the mean numbers and percentages of substandard treatments would be much higher if they had based on the current definition of substandard work.

his agency's experience,

> "We've spent a lot of time recently just watching various firms perform pretreats, and the picture on pretreats over the entire State has changed substantially over the past year. We used to have some operators pretreating entire houses of 1500 to 2000 ft^2 with a 50-gallon drum of insecticide. Now they're doing the job right."

McCarty estimated that 50 percent of the work prior to 1971 was substandard; from more than 70 houses we checked in that period, his estimate appears accurate.

An interesting contrast to this estimate is provided by the NPCA, which reports[5] that its members have an average of 1.5 percent callbacks on termite jobs. This figure does not necessarily reflect a 98.5 percent satisfactory treatment rate, for the public has no way to judge the adequacy of many termite jobs. What the NPCA figure does indicate is that 1.5 percent of its members' termite work is, rightly or wrongly, considered inadequate by customers.

Adequate State inspection is essential for the maximum benefit of any SPC legislation. A law without enforcement is like an empty gun—a threat soon discovered to be meaningless. State inspections should not result from specific customer complaints; as Bruer (1960) observed, ". . . more satisfactory results will be obtained if routine inspections are made of jobs which did not draw complaints. This is the only way to secure an overall picture of the work being performed." According to Bruer, the better operators in Tennessee welcome State inspection because they feel it affords a valuable check on the work of their servicemen.

Several PCOs have told us that work quality goes out the window when a competitive bid is required. Reliable companies are undersold by competitors who bid prices below the cost of materials and labor. For this reason, some PCOs would like to see more State inspections made.

Many problems with SPC laws and with State inspections result from inadequate financing. Bruer (1960) commented,

> "A law under-financed cannot be efficient and is never satisfactory to anyone. TOs in states having termite control laws are just as interested in having adequate appropriations

[5] Personal Communication Telephone conversation between C D Mampe, Director of Technical Service, National Pest Control Association and R V Smythe, July 1972

for the operation and enforcement of those laws as is the regulatory agency charged with such responsibility."

Some support for Bruer's statement is suggested by the fact that many States have sufficient funds only for inspecting after complaints.

In July 1972, we asked a small independent operator in southwestern Mississippi what he would do if the Mississippi SPC law was repealed.

"I'd probably sell my business," he replied.

"Why?" we asked.

"Well, our law protects the honest operators and makes it harder on the dishonest ones. Some people say the dishonest operators will put themselves out of business, anyway. Some will, but it takes too long and they do too many lousy jobs while they're in business.

I've got a good reputation here in _____. It's taken me 15 years of hard work to build it and with no law it would be just that much harder for me to make a living—I'm all for it."

The above quote is typical of many operators with whom we've talked. We've also talked with many PCOs who are opposed either to their State's legislation if their State has a SPC law or to S legislation in general. For example, in June 19 one PCO from southern California, the owner a relatively large business, commented as follc when we asked his opinion on SPC legislation:

"Who needs it! We can all do without it Let's face it, although its theoretically pos sible to write a good SPC law, none has evei been written. All legislation does is put re strictions on the good, honest operators Legislation doesn't run the crooks out There's not a State with legislation that doesn't have its share of crooks."

One reason for this divergence of opinion is marked lack of uniformity among SPC laws b in licensee requirements and in enforceme Without adequate training, examination, and forcement, SPC regulations can have the effect licensing unscrupulous PCOs to commit fra' Present PC advertising which stresses that a fi is licensed and bonded gives the public no inforn tion on professional expertise because all firms most regulated States are licensed and bonded law. If PC advertising accentuated emplo' training and experience, the public could bet determine that all licensed firms were not equa professional.

SPC LAWS AND REGULATIONS SHOULD BE UNIFORM

Although SPC regulations vary greatly from State to State, attempts have been made to achieve uniformity and reciprocity. In 1961, regulatory officials met for this purpose.[6] The conference led to an organization of administrative officials—National Association of Pest Control Regulatory Officials—which meets annually. Although sentiment for reciprocity was favorable even at their first meeting, the wide variation from State to State in license examinations and minimum standards for termite control prevented any substantial progress toward reciprocal agreements. For example, the percentage of PCO applicants passing license examinations ranged from 15 percent in one State to 90 percent in another and

averaged less than 50 percent. No reciproc agreements have yet been made.

The lack of reciprocity and uniformity in l and regulations is a serious problem. More : more companies are operating in more than State and have to contend with a variety of lic ing, reporting, supervising, and treating requ ments, different fee schedules, and other incor tencies. This unnecessary and undesirable bur has resulted from a maze of existing laws regulations which have been developed over past 42 years in response to a variety of press and in an attempt to meet the needs of homec ers in 18 different States ranging from Florid Minnesota to California.

Largely as a result of these difficulties, var people have proposed the formulation of a m law. Although there has been no nationwide c

[6] Unpublished history of National Association of State Pest Control Regulatory Officials prepared by F. R. Du Chanois, presented at the 8th annual organizational meeting held at Hot Springs, Arkansas, in 1967

weed out the unscrupulous operators. We have noted sharp disagreements on this view.

A conscientious PCO representing a large firm pointed out to us that certain companies are placed in an unfavorable competitive position in some States because company policy dictates more rigid treatment standards than does the State. A Florida State regulatory official commented (Du Chanois, 1961), ". . . The only startling thing to me is the complacency of the industry in Florida in tolerating . . . some of the shoddy, substandard so-called termite control work foisted on the public from time to time." Du Chanois pointed out that even though the performance standards of an individual PCO may equal or exceed those imposed by regulation, the failure on the part of competitors to meet standards ". . . reflects directly or indirectly, sooner or later, on you and the industry as a whole. Over the long pull, you suffer from lack of the general public acceptance, confidence and trust. Whether such performance standards should be self-imposed industry-wide in some way, or state-imposed, is not for me to say."

Eighteen of 50 States now have SPC laws and similar legislation is pending in four other States. The fact that these laws exist indicates that some people have felt the need for legislative protection to assure a minimum level of acceptable service from the PC industry. The general movement toward increased Federal protection of consumers leads us to believe that some form of general legislation, in addition to the Federal Environmental Control Act of 1972, governing SPC may soon be enacted. Whether the regulations or the enforcement necessary to implement the legislation will also be promulgated is, we feel, less certain. We believe that the concept of a model law should be carefully considered. Aside from the obvious advantage of creating uniformity, such a law would have an added attraction for those industry members favoring self-regulation. The PC industry could help formulate the legislation and the legislation could provide for substantial industry self-regulation. It is to be hoped that such legislation would not only avoid ill-conceived or unnecessary requirements but would also increase the difficulty for special interest groups or unknowledgeable persons to amend the legislation when such amendments lack industry or regulatory office approval.

RELATED BENEFITS FROM
ADEQUATE SPC LEGISLATION

The main purpose for regulatory legislation is consumer protection. This objective has been consistently discussed or implied throughout the paper. There are some additional benefits that need mentioning. In the discussion of these benefits, numbers in parentheses refer to the numbered criteria under which State laws are summarized in table 1.

To many people associated with the PC industry, the benefits that could result from proper SPC legislation will seem unrealistic. Hope for adequate industry self-regulation, however, may seem equally unrealistic. By either approach there will always be some fraudulent SPC work. It should be recognized that most of the benefits of strong legislation are being provided now by all ethical, well-trained PCOs. But the unknowledgeable consumer may find it difficult to identify ethical operators. Furthermore, some States would need only a few changes in their SPC laws and little, if any, additional financing to achieve most of the discussed benefits.

Regulatory Personnel Intermediate Between The PC Industry and Customers

Recourse for improper PC service before resorting to court action is provided by State inspections (15), minimum work standards (14), and penalties for substandard work (16). Even in States without all these provisions, court expenses and delays are often avoided, benefiting both the customer and the PC firm. Often by correcting substandard work, the PC firm avoids risk of greater expense in a damage suit.

Regulatory personnel can also protect PC firms against unjustified complaints. Homeowners frequently either do not understand their SPC contracts or have structurally modified their homes since receiving treatment, thus nullifying contract guarantees. The homeowner who does not keep his SPC contract in force loses his guaranteed protection.

Although pretreatments and some other contracts offer 1- to 5-year damage guarantees, SP damage may develop very slowly or escape detection for more than 5 years. The home may be sold before damage appears. A lawsuit is often the only recourse for a homeowner who has received poor service, but legal action is frequently impractical. Often the time and means of termite entry or beetle attack are difficult to determine. This difficulty, frequent transfers of PC office managers, and legal expenses relative to most damage costs usually discourage homeowner court action. However, such damage seldom occurs with proper initial treatment, and responsible PC firms are anxious to correct errors they may make in treatments.

Regulatory Agencies Can Inform the Public About PCO Capabilities

Regulatory personnel can publish certain information on all SPC companies licensed in the State. For example, companies may be listed by size (13) (the number of jobs performed), by work quality (12, 14, 15, 16), by service capacity (5) (license classifications, i.e., household pests, wood-destroying organisms, fumigation, etc.), and by availability (office addresses). Although only North Carolina currently makes this information reasonably available to the public, this practice would benefit both consumers and reliable PC firms.

For example, company size and services vary greatly. Frequently, very conscientious work is performed by smalltown, small-volume operators. These individuals depend on their reputation with neighbors and cannot afford the advertising and sales campaign of larger companies. Small companies that are members of the NPCA's Warranty Program can offer the same damage insurance as large firms. Thus, a customer need not sacrifice quality by getting small, routine services from a small firm if there is a reliable indicator of performance. For people with large or unusual jobs, information could be provided about high technical competence or sophisticated equipment possessed by large companies.

Providing such information could benefit the industry as a whole, even though the idea would not be popular with many companies. By this practice, the work of superior companies could be recognized, individual companies could compare their service with that of their competitors, companies giving consistently poor service would be

tion of existing regulations or passage of new regulations (12, 13, 14, 15, 16) to insure that the necessary data were obtained in usable form. If a model law were drafted, it could easily incorporate the necessary requirements. It is not reasonable or necessary for every State to have equally detailed reporting regulations, but 10 to 15 States with considerable SP damage and adequate geographic representation could adopt similar enough regulations so that data verified by competent inspectors could be obtained.

With the cooperation of the NPCA and various PC companies in selected States, much of the same information could be obtained. It is doubtful, however, whether identification, treating, and reporting of SP would be sufficiently uniform under industry reporting so that comparable data could be accumulated from many companies in various States. Collection through regulatory activities would insure that firms of all sizes and professional capabilities supplied comparable data. Data collection, whether by State regulatory personnel or the PC industry, is expensive. For example, a representative of one of the country's largest pest control firms recently told us that he was interested in the percentage of wood-destroying beetle jobs his company had performed over the past 5 years. Although the company's work records are computerized, the cost of obtaining such information has been prohibitive.

People responsible for SPC and the public they serve need such data. Effective controls are available for many SP, but we do not know whether they are being applied. The PC industry and researchers like ourselves need to know how many homes are properly protected and which pest(s) are of greatest concern. A case in point is "powder-post beetles." During World War II and the decade thereafter, beetles of the family Lyctidae (true powder-post beetles) were the most important of the five families called powder-post beetles. This statement is no longer true for the Southern and Southeastern United States, yet most PCOs, 17 years later, still are not aware that beetles of the family Anobiidae are the most prevalent. The problem of beetle families is not academic; beetles of the different families vary widely in their damage capabilities and frequently require different methods of control.

25

INDUSTRY SELF-REGULATION

Perhaps the most important deterrent to SPC legislative regulation has been long-standing industry opposition. Many PCOs believe that laws and regulations improve neither the lot of the PCO nor the quality of his work. Instead of legislative regulation, they favor self-regulation. What are the possibilities for effective self-regulation? A look at the industry indicates that strong self-regulation would be hard to achieve.

The industry is composed of more than 6,000 firms (2,600 of them offering SPC services) employing approximately 26,000 servicemen.[5] Some service personnel are highly trained, others have only an elementary knowledge of pest insects. Company size varies from one man to nationwide corporations; the range of services also varies. The industry is relatively young and is striving to improve its image (Spitz, 1971). As it expands and diversifies, justification for training and service standards increases.

Training.—Osmun (1959) stated:

"Few businesses have strived harder for higher educational standards than the pest control industry in the last 25 years. Sincere intellectual ambition has been the theme. The period has been characterized by the birth of conferences, short courses, college courses, excellent association-developed technical literature, and company training programs. This is vast—but it is only the beginning."

The impressive efforts made have not always been equal to the public needs. Bruer (1960) noted that:

"Violations of termite law stem from three causes: ignorance, arrogance, and fraudulent intent. It is surprising the number of operators who learned a method of termite control 20 years ago and who continue to follow the same methods today, ignorant of new chemicals and techniques. The problem of insufficiently trained operators has plagued this industry down through the years. . . . I think the most promising approach is through PCO conferences. It is regrettable more operators

do not avail themselves of these opportunities. I feel the termite control industry can serve itself by promoting attendance at such conferences."

Legislation could assist PCO training efforts by requiring attendance. Minnesota is the only State presently requiring that licensees attend at least one short course annually for license renewal.

Although many supervisors and managers are licensed and highly trained, many, if not most, homes are inspected and treated for SP by servicemen. Most servicemen are never examined for technical competence, a major failing of present SPC legislation. Many are self-taught or have received only cursory on-the-job training. While the value of experience cannot be discounted, new chemicals, new control techniques, and new information on pest habits all emphasize the importance of continuous training.

Has training changed in the past decade? Are adequate training programs available? Pi Chi Omega, a professional fraternity of the PC industry, in cooperation with Purdue University, sponsored a 2-day workshop in July 1969 to answer this question (Killough, 1969). Dr. Robert Snetsinger and David E. Schneider of Pennsylvania State University documented many types of training opportunities for PCOs. Much of the more formalized training is not being used, and Killough emphasized that certain PC courses at some universities might be canceled for lack of students. Daily workloads, the lack of company training programs for most PCOs, and the few specialized training courses sponsored by universities or regulatory offices (table 1) prevent formalized training for most servicemen.

The industry has recognized the need for certification and training for years, but financing and priorities of the NPCA did not permit action.[9] In 1969, the dues structure of the NPCA was changed and the Board of Directors authorized hiring of an Education and Training Director and developing a Professional Pest Control Operators

⁹ Personal communication. Telephone conversation between C. D Mampe, Director of Technical Service, National Pest Control Association and T. H. Williams, September 1972.

SPC industry and has asked EPA for permission to prepare its own standards for certification. If EPA accepts this proposal, NPCA will develop a certification institute which will promulgate certification standards, prepare training programs, and administer the certification program. The program may be operated through existing educational bodies such as extension services, junior colleges, and high school adult education courses. Such a program would be open to anyone wishing to be certified in SPC.

The serviceman's training program that the NPCA is currently developing is not expected to meet the needs for EPA certification. NPCA plans to continue this program anyway, probably making it part of another type of industry certification program. This certification may be analagous to that for public accountants. Various types and levels of certification may be included. The entire program might be coordinated with EPA-approved certification programs in the future.[10]

Minimum work performance standards.—No minimum standards for SPC are planned under industry self-regulation, but additional Approved Reference Procedures are anticipated. We have illustrated the minimum standards prescribed by States with SPC laws (Appendices XVIII-XXVI). We have mentioned, also, the need for certain standards while discussing the uniformity of regulation.

Can the PC industry itself be responsible for developing and maintaining uniform standards? There is ample evidence that individual companies can. An excellent example is Rentokil, the largest PC company outside the United States. Rentokil operated in 37 countries and had gross sales of $49 million in 1971.[11] Ted Buchan, co-director, commented (Anon., 1965b), "Whether it's a matter of morality, or just sound business, and there isn't much difference between them, . . . we have to do the very best possible job for our customers and stand behind what we do. It may cost a bit more in the beginning, but it pays off in the long haul." As a result of this policy, Rentokil has reduced from 5 to 0 the percentage of gross sales held in reserve for contingency claims.

In several States lacking SPC regulation, PC associations have adopted minimum standards (Anon., 1961; Anon., 1964). Dempsey Sapp, Flor-

10 Personal communication Letter from C D Mampe, Director of Technical Service, National Pest Control Association, to R V Smythe, March 1973

11 Personal communication Letter from J W Rodgers, Director, Rentokil Laboratories, Ltd to R V Smythe, August 1972

ida Pest Control and Chemical, felt that minimum standards were the direct responsibility of the PC organizations and said (Anon., 1963), "If you don't set up governing rules for your own industry, someone else will . . . do it for you.. . ." Sapp also said minimum standards were by definition "rules established by authority using standards set by research and experience. . . . Minimum standards . . . would set down in writing those rules for performance of termite control. . . . At present the public is ignorant of the difference between an $80 job and a $300 job which claim to do the same thing."

Frequently the objection to standards is that they stifle the initiative and independence of action of ethical operators (Du Chanois,1961). Unrealistic or unnecessarily intricate standards can, but Du Chanois continued,

". . . on the basis of our present knowledge, there are certain procedures which are undeniably and incontestably essential to every subterranean termite prevention or control job that was ever done or ever will be done . . . to stop the termites successfully."

The Approved Reference Procedures for Subterranean Termite Control of the NPCA are partial testimony to that statement. Although Dr. Ralph Heal. former Executive Secretary, NPCA, was asked at the 1963 meeting of the National Association of Pest Control Regulatory Officials to formulate model minimum standards, he declined to do so at their 1964 meeting. He decided against writing national minimum standards because of variation among States and because he felt that most States have built-in devices to get around minimum standards. The Approved Reference Procedures were suggested as a substitute, not as minimum standards but as good practice manual. These reference procedures are excellent and should be required reading for all PCOs doing SPC work. It is necessary to realize, however, that there is no penalty for doing less than what is described.

The contracts provided to homeowners by responsible PC firms and the NPCA Insured Termite Warranty Program are forms of self-imposed performance standards. A typical contract by a responsible company provides for any necessary retreatment or repair, provided the company has the opportunity to reinspect the property at least once annually. Two plans of the NPCA's Warranty Program are available to participating firms. A 1-year "corrective" warranty may be issued for

work on existing buildings. It provides for any necessary retreatment and may include the repair of damage. The warranty may be extended for 4 more years upon annual inspection and customer payment of annual fees. "Pretreat" warranties for termite prevention in new construction cover retreatments only for 5 years without annual inspection or fee. Damage repairs may be added if the owner pays for annual inspections. After 5 years a new warranty may be issued for an additional 5 years, without retreatment, if the PC firm considers the property a sound risk.

Valuable as these contracts are, certain problems persist. A contract is valid only if the homeowner keeps it in force every year. All too frequently the homeowner pays the annual reinspection fee for 2 to 3 years, then, thinking that continued payment is a poor investment, drops the contract, thereby relieving the company of any further obligations and canceling the protection he formerly had. Since one of five Americans move every year, long term protection seems unnecessary to many homeowners. Furthermore, it is often difficult (or impossible) to maintain a termite contract on a house that is sold. The NPCA Warranty Program has similar drawbacks, and only about 50,000 NPCA warranties are now in effect.[12] This is valuable protection for these 50,000 properties but the United States has approximately 68.7 million housing units (U.S. Dep. Commer., 1971).

Advertising.—An uninformed public is always a threat to a reputable business. Recognizing this, the NPCA, State PC associations, and individual firms have recently increased their advertising efforts to educate the public about WDOs and to help the public find responsible SPC service. Two examples of such efforts are the NPCA's brochure, "Termites and All About Them" and their information sheet, "Suggestions and Termite Control Services."

In 1962, entomologists at the University of Delaware and the Delaware PC Association staged a complete termite control operation on a large fraternity house on the University of Delaware campus (Webb, 1962). The public and news media were invited. The demonstration was staged to educate the public about termites and professional termite control methods, to overcome the lack of communication between researchers, PCOs, and homeowners, and to thwart the bad

---

12 Personal communication. Letter from P. J. Spear, Senior Director, Research, National Pest Control Association, to T. A. Harrington, Assistant Director, Southern Forest Experiment Station. March 1973.

publicity caused by the occasional fly-by-night operator. Newspaper, radio, and television gave the demonstration wide publicity. Dr. Dale F. Bray, University of Delaware, considered the demonstration an unqualified success and commented, "The demonstration has done more than anything else we have tried to drive out the fly-by-nighters and encourage the use of association members. This has helped maintain high standards in the profession and insure homeowners the most termite control per dollar invested." This excellent method of constructive advertising, although well received in Delaware, apparently has not been widely copied.

Many State PCO associations have become increasingly aware of the need for greater professional expertise among their members and of the need to inform the public about this expertise. Probably the most extensive association effort is that of the PCOs of California, Inc. (Anon., 1972b). The California association has: (1) employed a State Educational Chairman; (2) developed a five-course training program (Principles of Construction and Termite Repair, General Pest Control, Wood-Destroying Organisms, Fumigation, and Safe Chemical Handling) for owners and servicemen; (3) obtained the counsel of a commercial public relations firm; and (4) promoted an aggressive education program through various group meetings such as PTAs, service clubs, and chambers of commerce.

Evaluation.—How will performance be determined with self-regulation? An official of the NPCA has commented that while policing or inspection of work and penalties for violations have not been planned as part of compliance with the Federal Environmental Pesticide Control Act of 1972, whatever steps become necessary for regulation will be taken.[9] How effective (i.e., random, thorough, and objective) any required in-house evaluations will be is, obviously, unknown.

Financing.—Training, certification or licensing, and evaluation of work cost money. We have shown how SPC laws are financed. Presumably, training and certification under the auspices of the NPCA would be through dues and fees. In either case, financing must be adequate, and presumably PCOs would pass their additional costs on to their customers. How these costs would compare with the costs of legislative regulation is unknown.

Ethics.—Self-regulation of the PC industry will be assured if the industry's membership truly subscribes to the ethics of a profession. Maintaining high ethical standards is often the soundest business practice. For example, Rentokil Ltd. of England has made a policy of letting its services sell themselves and spends very little on advertising (Anon., 1965b). The PC industry has taken many steps to upgrade its image, particularly with regard to ethics. Wayland A.Tonning (1963), Professor of Marketing, Memphis State University, stated:

"... To be sure, the industry subscribes to the harsh requirements of a profession, and many firms qualify on many counts. Professionalism is a product of ethics, energy, and education! Ethics is a combination of moral principles, quality of work performed, and a science of ideal human character."

The NPCA and most State associations have adopted a code of ethics which clearly sets forth the behavior expected of the members.

29

CONCLUSION

How effective industry self-regulation can be in improving SPC work is an open question. The next few years will probably be the last in which the industry has opportunities to regulate itself. Success of industry self-regulation will hinge on the standards specified for NPCA certification, on the effectiveness of the NPCA's serviceman's training program, and on other training programs aimed at improving the professionalism of the serviceman. We believe the leadership of the NPCA will be a crucial factor in the PC industry's ability to develop the expertise and public esteem necessary for adequate performance. Because only 20 percent of the PC firms currently are NPCA members, membership obviously will have to be made more attractive. The threat of further SPC legislation in addition to the EPA's 1976 deadline for certification of pesticide applicators may make NPCA membership, certification, and training extremely desirable to many PCOs. The rapidly increasing number of new NPCA membership applications in the current year indicates this effect. Another important factor is the NPCA's vigorous campaign for new members.

Seven years ago an editorial in Pest Control (Anon., 1966) said: ". . . The industry could defend itself against unjust, restrictive legislation; could have a nationwide public relations and sales promotion campaign; . . . could have certification and standards to be sure everyone practicing the art is a fully trained craftsman; . . . could have better intraindustry cooperation. . . .

"The list of 'coulds' goes on and on.

". . . The plain truth is, PCOs are investing all too little in the welfare of their industry's future. There are too many associations, and too few who support any of them. . . . If, instead, . . . full support were given to the National, most of the 'ills' could be corrected. . . . For an industry with a sales volume of nearly $500 million, the shameful truth is that less than 1/2 of 1/10th of 1% is invested in a nationally directed program."

One of the major difficulties in judging the effectiveness of SPC legislation is the lack of comparable information about SPC work in States without legislation. The limited data available suggest that some laws have helped to improve SPC service. A common argument is that the cost and burden of implementing such legislation is too high. We leave that judgment to the reader.

We stress the importance of considering the desirability of a model law that is developed through the cooperation of the NPCA, PC industry, State regulatory personnel, EPA, FHA, VA, and other interested groups. This law could help upgrade PC services to a more professional level and, perhaps of equal or greater importance, could provide various mechanisms for educating the consumers of PC services.

Only recently has it been widely recognized that our Nation's resources are limited. Improved SPC would extend the service life of wood and reduce an important drain on natural and human resources. It would also promote user preference for wood, a renewable resource, over nonrenewable substitutes. The environment would be protected by eliminating retreatments caused by faulty work and by minimizing misuse of pesticides by poorly trained servicemen. Financial institutions (FHA, VA, commercial banks, and mortgage institutions) would benefit by obtaining better protection of their investments. There is considerable evidence to indicate that such goals could be reached through strong legislation. A strong nationwide certification or training program might achieve the same results.

Roughly half of our Nation's houses are older than 30 years, a tribute to the "over-engineering" of early builders and the durability of early building materials (U.S. Department of Commerce, 1972). With widespread use of insect- and decay-prone sapwood and changes in construction practices since World War II, will today's new homes last 30 to 50 years? Or will houses be like automobiles—used, abused, and discarded. The value of houses and the housing needs of the Nation are too high for such practices. SPC is an important part of building maintenance, and service expertise must be assured either through legislation or industry self-regulation.

Anonymous, 1964.

Anonymous, 1964.

Anonymous, 1965.

Anonymous, 1966a.

Anonymous, 1966b.

Anonymous, 1961.

Anonymous, 1962.

Anonymous, 1964.

Anonymous, 1966a.

Anonymous, 1966b.

Anonymous, 1966.

Anonymous.

REFERENCES CITED

Anonymous.
1954. "Public Hearing" provision essential in legislation Oser tells Ohio PCO meeting. Pest Control 22(9) : 45-46.

Anonymous.
1956. Legislation controversy. Pest Control 24(3) : 5.

Anonymous.
1959. Texas drafts PCO legislation to keep fly-by-nighters out. Pest Control 27(5) : 64.

Anonymous.
1960a. Dr. Bray tells Pa. PCA fall conference how association action forestalled legislation. Pest Control 28(10) : 78, 80, 82.

Anonymous.
1960b. Non-PCO structural pest control commissions are more effective, according to Florida inspector. Pest Control 28(10) : 84, 86.

Anonymous.
1961. Virginia state PCA fashions minimum standard list for termite control. Pest Control 29(5) : 66, 67.

Anonymous.
1963. Termite inspection charges, reports debated at Purdue PCO conference. Pest Control 31(3) : 72, 74, 76, 78.

Anonymous.
1964. Ohio PCA approves minimum termite standards during '64 summer meeting. Pest Control 32(9) : 56, 58.

Anonymous.
1965a. What's ahead in termite control? Pest Control 33(2) : 18, 19, 22, 25, 50.

Anonymous.
1965b. Rentokil. Portrait of a company on the go. Pest Control 33(9) : 12-16, 18, 26, 28, 30, 32, 36-37.

Anonymous.
1966. Too many and too few. Pest Control 34(10) : 5-6.

Anonymous.
1971. Industry responds to proposed legislation. Pest Control 39(5) : 13-14, 16.

Anonymous.
1972a. What's news in the industry (picture caption). Pest Control 40(5) : 34.

Anonymous.
1972b. Pest Control Operators of California launches unprecedented training program. Nat. Pest Control Oper. News 32(10) : 23-27.

Bedingfield, W. D.
1966. The legislative holocaust (letter to the editor). Pest Control 34(10) : 6.

Bennett, G. W., and Walker, L. L.
1971. Where do we go from here? Indiana PCOs look at legislation and training. Pest Control 39(11) : 9, 10, 13, 14, 16.

Bruer, H. L.
1960. Finances, personnel, operations stifle enforcement of termite laws. Pest Control 28(10) : 66, 68, 70, 72.

Chamberlin, W. J.
1960. Insects affecting forest products and other materials. 159 p. O. S. C. Coop. Assoc., Corvallis, Oregon.

Davis, J. J.
1956. Legislation pro and con (letter to the editor). Pest Control 24(5) : 6.

Du Chanois, F. R.
1961. Shortcomings in termite control work in Florida. Pest Control 29(11) : 42, 44, 46, 55.

Ebeling, W.
1973. Urban entomology. Univ. Calif. Off. of Agric. Publ., Berkeley, Calif. (in press)

Environmental Protection Agency.
1971. Digest of State pesticide use and application laws. 52 p.

Johnston, H. R., Smith, V. K., and Beal, R. H.
1972. Subterranean termites, their prevention and control in buildings. USDA Home Gard. Bull. 64 (rev.). 30 p.

Joseph, J.
1951. Legislation chief topic at California PCO convention. Pest Control 19(4) : 26-28, 42

Killough, R. A.
1969. Crisis in pest control education. Pest Control 37(12) : 18, 20, 22, 24.

Lance, W.
1969. The answer is training. Pest Control 37(10) : 22, 24, 75.

31

National Pest Control Association.
1973. FHA termite certifications for existing construction. Serv. Lett. 1267, 4 p.

Osmun. J. V.
1959. What's ahead in the next 25 years for the pest control industry. Pest Control 27(1) : 14, 16, 17.

Rodgers, J. W.
1972. Cloth cap to white coat. Speculation on the future of pest control. Pest Control 40(1) : 15, 16, 40, 41, 42, 43.

Showalter, R. L.
1956. About legislation (letter to the editor). Pest Control 24(4) : 6.

Smith, C. F.
1960. North Carolina finds 3-year old inspection form still acceptable. Pest Control 28(9) : 40-41.

Spitz, W. J.
1971. "Do we join consumerism?" Pest Control 39(12) : 36-37.

Toffler, A.
1970. Future shock. 561 p. Bantam Books, New York.

Tonning, W. A.
1963. Is pest control a profession . . . or a trade? Pest Control 31(5) : 14, 16, 18, 20.

U. S. Department of Commerce.
1971. General housing characteristics. U. S Summary. HC(1)-A1.

U. S. Department of Commerce.
1972. Detailed housing characteristics. U. S Summary. HC(1)-B1.

Webb, J.
1962. TOs, researchers demonstrate termite control to public in Delaware project. Pest Control 30(10) : 84, 86, 88.

Louisiana Standard Contract for Control
of Subterranean Termites in Conventional
Type Construction with Crawl Space

This contract made this_____day of_____, 19_____, at_____, Louisiana, by and between

_____, Licensed under Act 488 of 1962 herein represented by _____

hereunto duly authorized, hereinafter called the Termite Contractor, and _____
                                                                              Name of Building Owner

of age and resident of the city of_____, Louisiana, who declares himself to
be the owner of the property named, and hereinafter called the OWNER.
WITNESSETH:

For and in consideration of the price and sum of_____dollars

($), to be paid by the OWNER as follows:_____
The Termite Contractor agrees to furnish subterranean termite control service to the property located at

for a period of 2 years from this date under the following terms and conditions:

1. Should any reinfestation occur in any portion of the building covered by the contract during its duration,
the Termite Contractor agrees to treat such infested portions within 30 days of discovery of such reinfestation without additional charge to the Owner.

2. The Termite Contractor agrees to exercise care in applying the treatment in order to avoid damage to
shrubs or vegetation but under no circumstances or conditions shall the Termite Contractor be responsible
for damage to shrubs or vegetation involved in this job, or stains or discolorations to any part of the structure except those resulting from gross negligence on the part of the Termite Contractor. The Termite Contractor shall be responsible only for new damage to visible portions of the foundations, sills, joists, plates and
sub floor where an active termite infestation is discovered 6 months or more after completion of initial treatment

3. The Termite Contractor agrees to use first class materials and workmanship, and agrees to inspect the
premises annually without charge during the next two years.

4. The Owner warrants full co-operation with the Termite Contractor during the life of this contract, and
agrees to maintain the area treated free from any factor contributing to infestation such as wood, trash, lumber, or direct wood contact in the area treated, and agrees not to make any alterations or additions to the
structure without notifying the Termite Contractor; and agrees to immediately eliminate faulty plumbing,
leaks and dampness from drains, condensation or leaks from the roof or otherwise into, onto, or under said
area treated At no time will damage caused to any portion of the structure, even by active termite infestation, be the responsibility of the Termite Contractor in areas where any of the conditions in this paragraph
exist.

The Termite Contractor will commence work_____and prosecute same in
thorough and workmanlike manner under proper and adequate supervision, and complete the work without
delay. The Termite Contractor will report this termite job to the Louisiana Structural Pest Control Commission
s having been treated.

No fences, garage, or other outbuildings at above address are included in this agreement unless specified in
writing in this contract.

It is further agreed and understood that in the event of default by the Owner in the payment of the contract
price, or any part thereof under this contract, the Termite Contractor shall be released from further inspections
r reservicing, as herein provided. Such failure to make payments nullifies all guarantee under this contract.

No replacements, alterations or repairs, other than those specified herein shall be made by the Termite

ontractor. The Termite Contractor agrees to_____

_____

_____

This contract may continue in force on a service basis after the expiration date, at the option of both parties.
ccepted:_____

I 2439 (R 11/70)   Owner   By:_____   _____

APPENDIX II

Terms of Louisiana Standard Contract
for Control of Subterranean Termites in
Concrete Slab Foundation Type Construction

This contract made this_____day of_____, 19____, at_____, Louisiana, by and betwe

_____, Licensed under Act 488 of 1962 herein represented by _____

hereunto duly authorized, hereinafter called the Termite Contractor, and _____
Name of Building Owner

of age and resident of the city of_____, Louisiana, who declares himself
be the owner of the property named, and hereinafter called the OWNER.
WITNESSETH:
　For and in consideration of the price and sum of_____dolla

($　　　), to be paid by the OWNER as follows:_____
　The Termite Contractor agrees to furnish subterranean termite control service to the property located

for a period of 2 years from this date under the following terms and conditions:

1. Should any reinfestation occur in any portion of the building covered by the contract during its duration, t Termite Contractor agrees to treat such infested portions within 30 days of discovery of such reinfestati without additional charge to the Owner.

2. The Termite Contractor agrees to exercise care in applying the treatment in order to avoid damage shrubs or vegetation but under no circumstances or conditions shall the Termite Contractor be responsil for damage to shrubs or vegetation involved in this job, or stains or discolorations to any part of the stru ture except those resulting from gross negligence on the part of the Termite Contractor. Due to extrer hazards encountered in slab type construction and the possibilities of hidden infestations, the Termite Co tractor cannot be held responsible for any damage to the structure or its contents caused by termite a tivity.

3. The Termite Contractor agrees to use first class materials and workmanship, and agrees to inspect t premises annually without charge during the next two years.

4. The Owner warrants full co-operation with the Termite Contractor during the life of this contract, a agrees to maintain the area treated free from any factor contributing to infestation such as wood, tra lumber, or direct wood contact in the area treated; and agrees not to make any alterations or additions the structure without notifying the Termite Contractor; and agrees to immediately eliminate faulty plu ing, leaks and dampness from drains, condensation or leaks from the roof or otherwise into, onto, or un said area treated.

　The Termite Contractor will commence work_____and prosecute same a thorough and workmanlike manner under proper and adequate supervision, and complete the work with delay. The Termite Contractor will report this termite job to the Louisiana Structural Pest Control Commiss as having been treated.

　No fences, garage, or other outbuildings at above address are included in this agreement unless specified writing in this contract.

　It is further agreed and understood that in the event of default by the Owner in the payment of the contr price, or any part thereof under this contract, the Termite Contractor shall be released from further inspecti or reservicing, as herein provided. Such failure to make payments nullifies all guarantee under this contract.

　No replacements, alterations or repairs, other than those specified herein shall be made by the Tern

Contractor. The Termite Contractor agrees to

_____

　This contract may continue in force on a service basis after the expiration date, at the option of both par
Accepted:_____

AI 2439 (R 11/70)　　　Owner　　　　　　By:_____　　_____

34

FORM 905

Arkansas Minimum Requirements Checklist

TO ALL TERMITE OPERATORS

20M—6-71—63886—PP&SCo.

This form is to be signed by the operator in charge and attached to the copy of contract on each termite job, and mailed to the STATE PLANT BOARD.

		Yes	No
1.	Does the house have access openings to admit a large man?		
2.	Have you used sufficient amount of approved chemicals?		
3.	Have you removed all debris and form boards, including stumps?		
4.	Have you made proper clearance under all parts of building as required?		
5.	Have you brushed down all shelter tubes?		
6.	Have you serviced all piers and stiff legs properly?		
7.	Have you serviced all wood on concrete floors properly?		
8.	Have you serviced all wood steps properly?		
9.	Have you serviced all pipes properly?		
10.	Have you made all practical corrections to prevent dampness or if impractical to correct have you stated this in your contract?		
11.	Have you provided all needed ventilation?		
12.	Have you made all necessary replacements or shown on your diagram and contract the location and description of existing damage that you are not to replace?		
13.	Have you made proper clearance for outside grade?		
14.	Have you serviced all skirting and lattice work properly, including fences?		
15.	Have you serviced all steps, porches, ground slabs and similar structures as required?		
16.	Have you treated all stucco as required?		
17.	Have you treated ground area inside as necessary?		
18.	Have you treated ground area outside as necesary?		
19.	Have you drilled and treated all masonry as required?		
20.	Have you treated for powderpost beetles or notified the owner in writing that they were not to be treated if present?		
21.	Have you drawn a diagram indicating the location of infestation or shown on the diagram that the treatment was preventive if no infestation was found?		

For effective termite control, all of the above 21 minimum requirements should be met. If this cannot be done, prior approval for substandard work must be obtained from the Plant Board before any

work is done. State here your reason for substandard work, if any:_____

_____

_____

_____

_____

To the best of my knowledge, I have complied with the MINIMUM REQUIRE-MENTS for termite work, as stated in the revised Circular 6, SECTION III, A, B & C, issued by the STATE PLANT BOARD.

Amount Charged $_____

_____at_____
(Owner's Signature)   (Address)

TREATED BY_____ Signed_____
(Name of Company)   (Signature of Operator in Charge)

Date_____

35

APPENDIX IV

10M—3-71—82374—I

Arkansas Termite Inspection Report

Address of Property	Inspector	Date Inspected

Owner_____Address_____

TREATED BY_____**DATE**_____

Constr. of house_____Constr. of foundation_____

Was infestation as reported by operator?_____

How close (originally) was outside grade to siding or top of foundation at closest point, and had there been infestati

this point? _____

Termites present _____ 9. Pipes_____ 16. Stucco _____

1. Access opening _ ...._____ 10. Dampness_____   Wood supports _____
2. Method used _____ 11. Ventilation_____ ... 17. Gnd. treatment (inside) _____
3. Debris_____ 12. Replacements_____ _____   Meth. 1 (complete) _____
4. C'earance .. _____ 13. Outside grade_____   Meth. 2 (partial) _____
5. Shelter tubes _____   Siding_____ 18. Gnd. treatment (outside) _____
6. Stiff legs _____ 14. Skirting_____ 19. Masonry _____
    Piers (under 8") _____ 15. Steps_____   Drill hole depth _____
7. Wood on concrete floors_____   Porches_____ 20. Powderpost_____
    _____   Dirt fills_____   _____
8. Wood steps_____   Ground slabs _____

How does this job fail to meet minimum standards? Were these exceptions stated and agreed to in operator's contract, and does property owner understand them?

To what condition was original damage due? _____

FORM 909
2M—9—71—65532—PP&SCo.

APPENDIX V

Arkansas Report of Substandard Termite Treatment

First Inspection_____

Second Inspection_____

Third Inspection_____

_____
Date Treated

REPORT OF SUB-STANDARD TERMITE TREATMENT

To_____Date_____
(Operator Who Treated Property)

The property belonging to_____

located at_____

was inspected by an inspector of the Plant Board on_____

and the treatment was found to be sub-standard in the following respects:

COPY OF THIS SUB-STANDARD REPORT SHOWING
CORRECTIONS MADE, TOGETHER WITH $15
REINSPECTION FEE, SHOULD BE RETURNED.

BY_____ _____
Chief Inspector

TO THE OPERATOR: Please bring the treatment up to standard with respect to the items noted above. In addition, please make a thorough examination and should you find that the inspector has overlooked any other defects in the work, please correct these, also. Then, return this sheet to the board, after having noted in the space below, in detail, just what you have done, not later than 15 days after the date of this notice, keeping duplicate for your files. The Board will then reinspect as soon as possible.

Date_____   Signed_____ _____
(Operator)

37

Form 920
5M—3-65—58496—Ark.P.&L Co

APPENDIX VI

Arkansas Request for Approval of Substandard
Contract for Existing Structures

(Submit in Duplicate)

Note of Explanation to Property Owner or Prospective Buyer:

The State Plant Board has established a set of minimum standards which a licensed operator must meet in treating a property for termites and other structural pests if it is to be accepted by the Plant Board as "Standard Job." Anything less than these minimum standards in the opinion of the Plant Board does not give the homeowner the maximum protection. There are a few instances when because of physical obstructions, or due to the way the property is constructed, etc., it would not be feasible to comply with all of the minimum standards. In such cases this form can be used to obtain Prior Approval from the Plant Board for a "Sub Standard Contract, which is required by the Plant Board of the licensed operator. The Plant Board requires that the licensed operator give the customer a copy of this agreement.

This contract does not relieve the operator of any part of the Arkansas Pest Control Law as is set forth in Circular No. 6, "The Law and Regulations on Licensing Pest Control Operators."

Owner or Prospective Buyer's Name_____

Complete Mailing Address _____

Location of Property _____

List the Exception points by which this property does not meet the Plant Board's Minimum Standards, and for which Prior Approval is requested:

_____

_____

_____

_____

I hereby certify that I have read and understand the information on this request, and that I am in agreement as to the points listed above, and hereby request Prior Approval for a "Sub-Standard" job on this property.

| Date | Signed (Owner or Prospective Buyer) | Date |

For Office Use only

$30.00 Received _____ Date _____   Name of Company

Inspection Made _____ Date _____

State Inspector _____   Address

Approved_____
Date

Head, Pest Control Division

38

California Standard Structural
Pest Control Inspection Report

This is an inspection report only--not a Notice of Completion

ADDRESS OF PROPERTY INSPECTED	BLDG. NO.	STREET	CITY		DATE OF INSPECTION
			CO. CODE		

FIRM NAME AND ADDRESS

Affix stamp here on Board copy only
A LICENSED PEST CONTROL OPER-
ATOR IS AN EXPERT IN HIS FIELD.
ANY QUESTIONS RELATIVE TO THIS
REPORT SHOULD BE REFERRED TO
HIM.

FIRM LICENSE NUMBER	CO. REPORT NO. (if any)	STAMP NUMBER

Inspection Ordered by (Name and Address)
Report Sent to (Name and Address)
Owner's Name and Address
Name and Address of a Party in Interest
INSPECTED BY: LICENSE NO.   Original Report   Supplemental Report   Number of Pages

YES CODE SEE DIAGRAM BELOW   YES CODE SEE DIAGRAM BELOW   YES CODE SEE DIAGRAM BELOW   YES CODE SEE DIAGRAM BELOW
S--Subterranean Termites   B--Beetles--Other Wood Pests   Z--Dampwood Termites   EM--Excessive Moisture Condition
K--Dry-Wood Termites   FG--Faulty Grade Levels   SL--Shower Leaks   IA--Inaccessible Areas
F--Fungus or Dry Rot   EC--Earth-wood Contacts   CD--Cellulose Debris   FI--Further Inspection Recom.

1. SUBSTRUCTURE AREA (soil conditions, accessibility, etc.)
2. Was Stall Shower water tested?   Did floor coverings indicate leaks?
3. FOUNDATIONS (Type, Relation to Grade, etc.)
4. PORCHES . . . STEPS . . . PATIOS
5. VENTILATION (Amount, Relation to Grade, etc.)
6. ABUTMENTS . . . Stucco walls, cloumns, arches, etc.
7. ATTIC SPACES (accessibility, insulation, etc.)
8. GARAGES (Type, accessibility, etc.)
9. OTHER

DIAGRAM AND EXPLANATION OF FINDINGS (This report is limited to structure or structures shown on diagram.)

General Description

Signature

APPENDIX VIII

California Standard Notice of
Work Completed and Not Completed

NOTICE - All recommendations may not have been completed - See Below - Recommendations not complet

This form is prescribed by the Structural Pest Control Board, with whom a copy must be filed by licensee within 5 days after completion of work under a contract.

THIS IS A NOTICE OF COMPLETION ONLY, NOT AN INSPECTION REPORT.

ADDRESS OF PROPERTY INSPECTED	BLDG. NO.	STREET	CITY CO. CODE	DATE OF COMPLETION:

LICENSEE FIRM NAME AND ADDRESS:

AFFIX STAMP TO
BOARD COPY ONLY

FIRM LICENSE NO.	COMPLETION STAMP NO.

Notice of Completion
        Sent To and Date: _____

Owner's Name and Address: _____
Buyer's Name and Address: _____

        This is to certify that the recommendations on the above designated property, as outlined in STANDARD INSPECTION REPORT NO. _____, dated _____ _____, REGISTRATION STAMP NO._____, have been completed with exceptions as noted below:

        The following recommendations were not completed by this Company:

_____
_____
_____

ESTIMATED COST OF WORK RECOMMENDED
BUT NOT COMPLETED BY THIS COMPANY - $

        The following recommendations have been completed and are considered secondary me under Section 1992 of the Structural Pest Control Board's Rules and Regulations:

_____
_____

REMARK

_____ ___
_____

Signature

40

APPENDIX IX

Florida Standard Structural Pest
Control Investigation Report

:ral Information:
perty owner's (agent's) name:_____

              (last)   (first)   (middle initial)

perty address:_____

        (street)   (city)   (county)   (phone No.)

)lainee (if other than property owner/agent):_____

ress(home and mailing):_____ ,vs.

nsee or non-licensed person involved:_____

ress:_____Cert. Oper. In Charge:_____

No.:_____Identification Card No.:_____Other:_____

itification Card holders involved (if known):

:!_____

    last   first   initial   address   Iden. Card No. date iss. date ter.

plaint Information: written_____verbal_____date submitted_____

:ific complaint(s) (describe briefly):_____

tract or Agreement Data: Contract issued: yes_____no_____Contract No._____

< Order issued: yes_____no_____Work Order No._____date of Work Order_____

:ctive date of contract:_____Date of treatment:_____

ictural pest(s) covered:_____

a of contract: _____years, renewable for _____years, Frequency of reinspections:_____

iitions under which retreatments to be made:_____

itment cost: $_____Repairs cost: $_____Contract in force: yes_____no_____

ial renewal fee: $_____Contract signed by_____

iir or replacement of damage covered by contract: yes_____no_____

:ract furnished prior to performance of work: yes____no____Contract in order: yes____no____

ients:_____

iection Data: Inspection made: yes_____no_____date_____

icting official:_____Title:_____

istation status: active_____inactive (old)_____none visible_____other_____

i(s) living organism found:_____

ive: severe_____moderate_____slight_____none visible_____

tion (specify):_____

r data or comments:_____

41

Treatment Data: Adequate (std. all respects)_____inadequate (substd. certain respects)____
Exact discrepancies noted and requiring attention: (reinspect, retreat, adjustment, structural,
1._____
2._____
3._____
4._____
5._____
Enforcement Data: Actual or apparent violations involved, if any (Exact Section of Law or Rule,
known, or describe):_____
_____
_____
_____

Action taken by investigating official and (or) Bureau of Entomology:
1. Verbal instructions given to (name(s):_____
    date:_____
2. Written instructions given to (name(s):_____
    date:_____
3. Registered letter to certified operator (Rule 305.0): yes_____no_____date_____
    reply date_____date previous 30-day notice (if any)_____
4. Report submitted to Bureau of Entomology for: further action_____information only_____
    other_____date_____
5. Violations, if any, reported to State Health Officer and Florida Structural Pest Control Co
    sion: yes_____no_____date of report
6. Instructions given to certified operator,
    etc:

Additional Information, Comments or Recommendations:

_____   _____   _____
Date of report   Investigating official's signature   Ageno

42

APPENDIX X

Georgia Monthly Job Report
on Wood-Destroying Organisms

e of Company_____L.P.C.O. No._____

ess_____

ature of Designated Certified Operator_____

Company Have An Approved Bond? Yes_____ No_____

Number Jobs Completed (Including Form 2) _____.

The following work was completed during the month of_____in the field of Wood Des-
ng Organisms. A report must be made on each job for which a fee is charged and a service rendered. Such reports
be mailed monthly, not later than the 10th of the month following the completion of the job. If no reportable
are done in any month, you must so notify the Department of Agriculture. Any jobs listed on this form that do not
ly in all respects to the required State Minimum Standards and/or bond requirements must also be on a Form II.

NAME AND COMPLETE ADDRESS OF PROPERTY TREATED	DATE JOB COMPLETED	Clearance Letter	WOOD DESTROYING ORGANISMS				STRUCTURE			BOND TO BE ISSUED ***	EXCEP- TIONS— Form II Attached
			Sub. Termites	Powder Post Beetles	Fungus	Other*	Crawl Space	Slab Con- struction **	New Con- struction		

me of wood destroying organisms treated for
es whether treated by rodding (R) or drilling (D)
f company does not have an approved bond.

(OVER)

APPENDIX XI

Georgia Termite Inspection Report

_____19___

COMPANY_____

NAME OF OCCUPANT OR OWNER_____

A. THE SECTIONS CHECKED BELOW REPRESENT APPARENT VIOLATIONS OF THE STRUCTURAL PEST CONTR
REGULATIONS.

1 · ☐ ALL DEBRIS NOT REMOVED. (620-7-.03A)

2 · ☐ WOODEN CONTACTS NOT REMOVED OR INSULATED. (620-7-.03B)

3 · ☐ TERMITE TUNNELS NOT REMOVED. (620-7-03C)

4 · ☐ TRENCHES NOT ADEQUATELY TREATED (620-7-03D)

        ☐ INSIDE   ☐ OUTSIDE   ☐ PIERS

5 · ☐ VOIDS NOT ADEQUATELY TREATED. 620-7-03E)

        ☐ FOUNDATION   ☐ PIERS   ☐ CHIMNEY   ☐ OTHER

6 · ☐ EARTH FILLED PORCHES NOT ADEQUATELY TREATED (620-7-.03F)

7 · ☐ CONTINUOUS POURED SLAB (620-7-.03G)

8 · ☐ VENTILATION NOT ADEQUATE (620-7-.03H)

9 · ☐ CRAWL SPACE NOT ADEQUATE (620-7-03I)

10 · ☐

     ☐

THE MINIMUM STANDARDS HAVE APPARENTLY BEEN MET IN ALL SECTIONS NOT CHECKED.

B — IN SPACE BELOW INDICATE IF ALL VIOLATIONS REPORTED ABOVE HAVE BEEN CORRECTED TO M
MINIMUM STANDARDS.

        ☐ YES   ·   ☐ NO

IF ABOVE CHECKED <u>NO</u>, LIST BY SECTION NUMBER REASONS FOR NOT DOING THIS WORK. U
REVERSE SIDE IF NECESSARY.

MAIL TO: (<u>Within 30 Days</u>)

Georgia Department of Agriculture
Entomology Division
Capitol Square
Atlanta, Georgia · 30334

Date Returned:

_____

SIGNATURE OF LICENSEE
_____

COMPANY
_____

ADDRESS
_____

AG 15-027-093
REV. 11/21/72

44

APPENDIX XII

Georgia Termite Control Work Reported
on Monthly Job Report in Which Exceptions
are Made to Minimum Standards

ame of Company _____   Location of Job _____

ddress _____   Address _____

gnature of Designated
ertified Operator _____   L.P.C.O. No. _____

wner _____   Occupant _____

ddress _____   Date Job Completed _____

• •

COMPLETED FORM TO: GEORGIA DEPARTMENT OF AGRICULTURE
DIVISION OF ENTOMOLOGY
CAPITOL SQUARE - ATLANTA , GA. 30334

A CHECK MARK IN THE "YES" COLUMN INDATES WORK DONE TO BRING JOB UP
TO MINIMUM STANDARDS FOR REASONABLY EFFECTIVE TERMITE CONTROL. IF
A CHECK MARK APPEARS IN "NO" COLUMN THIS ITEM MUST BE EXPLAINED UNDER
REMARKS AND THIS REPORT SIGNED BY THE CUSTOMER.

	YES	NO
1 - CONTRACT ISSUED (620-7.02)	☐	☐
2 - ALL DEBRIS REMOVED (620-7-.03a)	☐	☐
3 - ALL WOOD TO EARTH CONTACTS REMOVED (620-7-.03b)	☐	☐
4 - ALL VISIBLE TERMITE TUNNELS REMOVED (620-7-.03c)	☐	☐
5 - TRENCHED INSIDE FOUNDATIONS AND AROUND PIERS, CHIMNEY (620-7-.03d)	☐	☐
6 - TRENCHED OR RODDED OUTSIDE FOUNDATION	☐	☐
7 - HOLLOW WALLS, PIERS, CHIMNEY (S) DRILLED & FLOODED (620.7-.03e)	☐	☐
8 - EARTH FILLED PORCHES VOIDED; DRILLED OR RODDED (620-7-.03f)	☐	☐
9 - ON-GRADE SLAB(S) DRILLED OR RODDED (620-7-.03g)	☐	☐
10-VENTILATION ADEQUATE OR ADDED (620-7-.03h)	☐	☐
11-CRAWL SPACE ADEQUATE OR DUG (620.7-.03i)	☐	☐

BY   SIGNING BELOW THE CUSTOMER SIGNIFIES THAT HE UNDERSTANDS THAT HE
WILL RECEIVE LESS THAN THE MINIMUM TERMITE CONTROL JOB AS SPECIFIED BY
THE STATE OF GEORGIA.

CUSTOMER _____
Signature

Address _____

56-085
/72

45

APPENDIX XIII

Louisiana Termite Inspection Report

NAME OF OWNER

ADDRESS OF OWNER

ADDRESS OF PROPERTY INSPECTED

SOURCE OF REQUEST OR INFORMATION FOR MAKING THIS INSPECTION

NAME OF COMPANY ISSUED CONTRACT

ADDRESS OF COMPANY

TYPE OF STRUCTURE: ☐ Slab ☐ Pier ☐ Other

ACTIVE INFESTATION: ☐ Is Present ☐ Is Not Present

EXPLAIN:

IF PRESENT, LOCATION IS:

CORRECTIONS TO BE MADE:

DATE — INSPECTOR

AI 2438

Louisiana Report of Termite Eradication

Please type or print

Name of Firm _____ Report for the Month of _____ Page _____

Address _____ Licensee _____

NO.	NAME AND ADDRESS OF PROPERTY OWNER	ADDRESS OF PROPERTY TREATED	CITY PROPERTY LOCATED	DATE WORK BEGAN	DATE CONTRACT ISSUED

I hereby certify that the above listed properties represent all properties on which work has been started and/or completed for the eradication of termites by the above named operator during the above named month.

Attached hereto is check in the amount shown below for fees due on the number of contracts* shown below.

*Each contract as herein set out shall cover only one unit or one individual property treated for eradication or prevention of termites, provided that a garage appurtenant to one unit or one individual property may be included in such contract. Failure to make this payment subjects the licensee to revocation of his license.

Amount of Check	No. of Contracts	Signature of "LICENSEE"	Date	
$				☐ Check ack ☐ Trans. ☐ Cert. (Do Not Write in this Space)

47

APPENDIX XV

Louisiana State Inspection Complaint Report

COMPLAINT FILED BY _____ Letter

Address _____ Telepho

_____ Other

DATE OF COMPLAINT _____

NATURE OF COMPLAINT

RESULT O

ACTION T

REMARKS

INVESTIGATED BY

DATE OF REPORT

P L E A S E   T Y P E   O R   P R I N T

Send Original to Office   Employee Retains C

APPENDIX XVI

Mississippi Termite and Other
Structural Pest Inspection Report

_____Date Inspected_____

ss_____Location if Rental_____

ne____ Requested____ Complaint____ Termite____ Beetle____ Other_____

act Date_____Treated for: Termites____ Beetles____ Other_____

received copy of: Contract____ Specifications____ Plat____ Other____ Not Determined____

ny_____Address_____

REPORT ON FINDINGS BY (x) IN PROPER SPACE

Type of Construction: Full Basement____ Crawl Space____ Slab____ Combination_____

SUBTERRANEAN TERMITES: Active____ No visible activity____.

POWDER POST BEETLES: Active____ No visible activity____ Not applicable____.

OTHER:_____ Active____ Inactive____ Not applicable____.

Old Termite infestation: Tubes____ Damaged wood____ No evidence found____.

Cellulóse debris: None____ Small amount____ Large amount____.

Clearance between soil and wood: Less than 6"____ More than 6"____ Contacts soil____.

Foundation walls: (A) Double or triple masonry walls. Present____ Not Present____;

Drilled____ Not Drilled____ Partly Drilled____. (B) Hollow concrete block walls.

Present____ Not Present____; Drilled____ Not Drilled____ Partly Drilled____.

Pillars: (A) Double brick or larger. Present____ Not Present____; Drilled____

Not Drilled____ Partly Drilled____. (B) Hollow concrete blocks. Present____

Not Present____; Drilled____ Not Drilled____ Partly Drilled____.

Chimney (s): Present____ Not Present____; Drilled____ Not Drilled____ Partly Drilled____;

Excavated____ Not Excavated____.

Dirt-filled concrete porch (es): Present____ Not Present____; (A) Tunneled____

Not Tunneled____; Satisfactory____ Not Satisfactory____; (B) Other Treatment____

Satisfactory____ Not Satisfactory____.

Trenching of soil: (A) Inside Foundation Walls - Yes____ No____ Partly____; (B)

Pillars - Yes____ No____ Partly____; (C) Chimney (s) - Yes____ No____ Partly____.

Remarks:

Company representative present. Yes____ No____. _____

                                                        Inspector

APPENDIX XVII

North Carolina Statement Concerning the Absence of Wood-Destroying Insects from an FHA or VA Financed Building or Structure

Two copies to be submitted directly to:

Date: _____

File No. _____

THERE IS NOW NO VISIBLE TERMITE OR OTHER WOOD–DESTROYING INSECT INFESTATION IN THE BELOW–DESCRI... PROPERTY; AND, IF SUCH INFESTATION PREVIOUSLY EXISTED, IT IS NOT NOW ACTIVE AND ANY DAMAGE DUE TO SUCH INI... TATION HAS BEEN CORRECTED OR, ALTERNATIVELY, IS FULLY DISCLOSED HEREIN.

Location of Property _____

Buyer _____

Lender _____

Seller _____

TYPE OF SALE
FHA _____ VA _____ Conventional Loan _____ Loan Assumption _____ Cash Sale _____

THE INFORMATION CONTAINED IN THIS STATEMENT IS BASED ON CAREFUL VISUAL INSPECTION OF ACCESSIBLE AREAS AND ON SOUNDING OF ACCESSIBLE STRUCTURAL MEMBERS.

INFESTATION
Was there evidence of infestation of wood–destroying insects in this property? Yes _____ No _____ . If answer is "Yes"

a) Specify: Termites _____; Powder–Post Beetles _____; Others (specify) _____

b) Did evidence include live insects? Yes _____ No _____ . If "No", specify the evidence found: _____

DAMAGE
Were there any insect–damaged timbers in this property? Yes _____ No _____ . If answer is "Yes",

a) Specify: Termites _____; Powder–Post Beetles _____; Others (specify) _____

b) Were they damaged sufficiently to cause structural weakness? Yes _____ No _____ If answer is "Yes", were damaged timbe...

areas repaired to correct structural weakness? Yes _____ No _____ . If "No" explain _____

If timb...

Explain _____

TREATMENT
Was this property treated for wood–destroying insects by your company? Yes...
(1) If answer is "Yes"

a) Specify Termites _____, Powder–Post Beetles _____; Others (specify) _____

b) Specify: Date of Treatment _____ .

c) Specify: Name of Pesticide used: _____ .

(2) If answer is "No", explain
If no treatment of the property was found necessary, the undersigned agrees to treat, free of charge, for any wood–destroying i...
which are discovered within 90 days of date of this statement.
If treated, this property is covered by a written guarantee against insects for which treatment was made. The guarantee extends a mini...
of (1) one year from date of this statement and is renewable annually for a minimum of (4) four years for $_____ ann...

CHARGES
A fee of $_____ was charged for this statement. A fee of $_____ was charged for this statement, treat...
and repairs (if any) (It has been estimated that a minimum charge of $15.00 for buildings 1,000 square feet and under and $20.00 for buil...
over 1,000 square feet is a reasonable fee for a statement when no treatment is involved. Such fees should be borne by the seller.)

By _____

Firm _____

Address _____

| NORTH CAROLINA STRUCTURAL PEST |
| CONTROL LICENSE NO. _____ |

(City)   (State)

Form No. 1, Rev. (December 11, 1970)

50

APPENDIX XVIII

Arkansas Minimum Requirements for Structural Pest Control

SECTION IIIA. MINIMUM REQUIREMENTS FOR STRUCTURAL PEST WORK ON EXISTING STRUCTURES

1. Access Openings at least 14 in. high and 16 in. wide should be provided. to permit inspection under all parts of building

2. Chemicals. The term chemical when used hereafter shall refer to any chemical listed for use under termite work or powder-st beetles in Section IIIC—Materials.

3. Debris and Form Boards. (a) Remove all wood (including stumps and dead roots) and other debris from under the build-;. Large stumps, if their removal is impractical, may be trenched and treated with chemical, provided they are not in contact with within 5 inches of foundation timbers. (b) Remove all unnecessary form boards

Comment: Remove all pieces of wood which can be caught by a rake, both on and near surface of ground

4. Clearance Under Buildings. Remove all soil which is within 12 inches of bottom edge of floor joists or within 5 inches of sills.

Note: Adequate clearance must be provided for passage of a large man to make complete inspection under all parts of building

5. Shelter Tubes. Brush all termite shelter tubes from piers, walls, sills, joists, sub-floors, pipes, and other parts of the under-ucture.

6. Piers and Stiff-legs. (a) Stiff-legs or other wood supports must have concrete or metal-capped bases extending at least 4 in ove the ground. (b) Piers under sills or subsills, if less than 8 in high, must be concrete or metal-capped

7. Wood on Concrete Floors. Where wood parts such as posts, doorframes, partitions, or stair-carriages (a) have been attacked termites working up thru concrete, or (b) are set down into concrete, said wood parts must be cut off and set on metal or con-te bases raised at least 1 inch above the floor level.

8. Wood Steps. Place all wood steps on concrete bases, which should extend several inches above ground level, and prefer-ly several inches beyond the steps in all directions. See Fig 5.

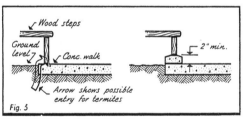

Fig. 5

FIGURE 5 EXPLAINS SECTION 8. In the left-hand drawing, the steps are set on a concrete sidewalk which is level with the ground. With this construction, dirt may wash against steps and provide favorable conditions for termites. To correct this, steps should be placed on a concrete base (with sloping shoulders for drainage), to raise the wood at least 2 inches above ground level, as shown in right-hand drawing. Trench and treat heavily around the concrete base.

9. Pipes. Packing around pipes, if not removed, should be saturated with chemical after breaking contact with ground. Pipes uld also be trenched and treated.

10. Dampness. Dampness favors the development of termites and wood rots. If water can run under the building through ac- s opening, ventilators, or other openings in or under side walls or skirting, this condition must be remedied. Seepage through or ler walls should be prevented. If it is impracticable to prevent seepage, owner must be notified in the contract

If condensation of soil moisture upon wood of the substructure is a problem, it can be prevented by constructing a vapor bar- ·. An adequate barrier can be made by covering the soil under the building completely with roofing paper or polyethylene sheet, owing an overlap of 4 to 6 inches between pieces.

11. Ventilation. Inadequate ventilation also favors the development of termites and wood rots. Provide ventilation at the rate it least one 8 x 16 inch opening (or equivalent) for each 25 linear feet of foundation wall. Provide ventilation for all dead air kets.

12 Replacements Wood which has been damaged or substantially weakened by termites, powder-post beetles or decay fungi (rot) should be replaced Replacements may be made by the operator or the property owner, or both. Replacements for which the operator is not to be responsible must be specifically excluded in writing by, (1) Describing the excluded replacements in the contract and stating therein that they are not to be made by the operator, (2) Showing their location on the diagram of the structure, and (3) Noting these replacements as an exception to Item 12 on Form 905.

13 Outside Grade. See Fig. 1

A Wood Construction.

Top of foundation wall and all exterior wood such as siding and framing must be at least 3 inches above outside grade. To lower grade, soil next to wall should be removed to necessary depth and a retaining wall built, or a concrete gutter installed after heavy application of chemical is made.

B Masonry Veneer Construction

In lieu of a retaining wall or concrete gutter as specified in A above, soil against masonry veneer walls may be treated with a heavy application of chemical.

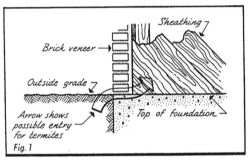

Sheathing

Brick veneer →

Outside grade

Arrow shows possible entry for termites

Top of foundation

Fig. 1

FIGURE 1 EXPLAINS SECTION 13. In this drawing, the top of the foundation is level with the outside grade. This is a hazardous condition, for 2 reasons: (1) wood at or near ground level is often damp, a condition favorable to termites, (2) termites can easily enter from the outside, unseen, by going thru cracks in the first mortar joint and thence into the sheathing. Unless the outside grade is at least 3 inches below the wood parts of the house (including the siding in a frame house), it must be treated as described in Sec. 13.

14. Skirting and Lattice Work. These should rest on solid concrete or cemented brick extending at least 3 inches above outside grade, unless suspended, in which case there should be at least 3 in. clearance above grade. Or, contact must be broken between the building and any lattice which may rest on or in the soil.

15. Steps, Porches, Ground Slabs* and Similar Structures.

I. Dirt Filled or Ground Slab

A When such structures are built against masonry veneer walls they must be either:

1. Drilled on not more than 24 inch centers, or rodded, and the soil thoroughly flooded with chemical at Juncture of structure and wall, (If long-rodded, route must be shown on diagram of building), or

2. Tunneled and treated as described in Section III below.

Note 1: Neither 1 nor 2 is required if the top of the structure is below the bottom of the sill and no infestation is associated with the fill or slab.

Note 2: When masonry veneer rests on a slab whose top is even with or above the bottom of the sill, method 1 or 2 must be used This also applies if top of slab is below the sill line but an infestation is associated with the fill.

B. When such structures are built against the foundation, the top of the foundation wall must be above the top surface of the structure to provide exposure of termites working to wood.

When an infestation is associated with a structure built in this manner, the structure must be treated by method 1 or method under IA above.

If the top of the structure is even with or above the bottom of the sill, providing hidden access to wood, the structure must be tunneled and treated as described in Item III below. See Fig. 2.

*"Ground Slab" as used in this item refers to such structures as concrete or masonry walks, gutters, patios, etc., laid directly upon the ground rather than upon a dirt fill.

C. Siding or other exterior wood which touches the top of a dirt filled or ground slab structure must be either.

    1. Drilled through slab or foundation on not more than 24 inch centers and the soil flooded with chemical, or

    2. Tunneled and treated as described in Section III below, or

    3. Remedied by cutting off a strip from the bottom of the siding to create space enough between siding and slab that termites must expose themselves getting to wood

II. Solid (Not Filled or Ground Slab)

A. Solid structures built against the foundation wall must be made accessible for thorough treating or inspection by either:

    1. Drilling through foundation on 9 inch centers and flooding the void between the foundation and the back of the structure with chemical, or

    2. Opening the foundation sufficient for visual examination all the way across the adjacent structure.

    Note: When possible to do so, removal of sill or plate to expose back of adjacent structure is permissible.

B. Solid structures built against masonry veneer walls must be drilled and flooded or opened for inspection as in A-1 and 2, Item II above, unless there is adequate space between the top of the structure and any wood of the building, such as the threshold, to cause termites to expose themselves getting to the wood.

Exception to I and II. Appurtenant structures, dirt-filled, ground slab or solid, need not be drilled and flooded, opened, or tunneled and treated if both of the following apply:

    1. Construction is such that termites working to wood will be forced to expose themselves in the open, and

    2. No infestation, present or previous, is associated with the structure.

III Tunneling and Treating. See Figures 2 and 4.

When required or elected in I-A, B or C above, a horizontal tunnel may be created by excavating the soil at the junction of slab and wall. The tunnel must extend the length of the fill and be at least 12 inches deep (or down to grade) and 12 inches wide. Dirt of the tunnel should be saturated with chemical at all points of contact with wall and slab. Supports for the slab should be erected in the tunnel if needed. Tunnel must be well ventilated, preferably at the ends.

Exception Size of tunnel under ground slabs is not specified

Note 1. Where the sill or other wood extends below the bottom of the concrete slab, it will be necessary to make the tunnel more than 12 inches deep next to the foundation wall, in order to provide at least 12 inches clearance from the bottom of the wood to the bottom of the tunnel.

Note 2. Where the bottom of the tunnel is above ground level under the house, the trench for applying the chemical in the tunnel should, if practicable, be dug down to the level of the trench on the opposite side of the wall, unless the wall is of solid, uncracked concrete. If this is impracticable, drilling and injection of the wall should be done at the level of the bottom of the tunnel, rather than at the ground level.

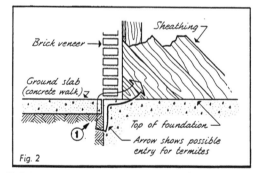

FIGURE 2 EXPLAINS SECTION 15-IB AND 15-III (IN PART). In this drawing, the top of the concrete walk (ground slab) is level with top of the foundation wall. Termites can come up unseen thru the joint between foundation and slab, then thru cracks in the first mortar joint, and thence into the sheathing. To correct this condition, a tunnel must be made at 1 and chemical applied, as described in Section 15-III.

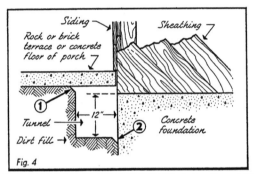

Fig. 4

FIGURE 4 EXPLAINS SECTION 15-III (IN PART). In tunneling a dirt-filled porch (or other dirt-filled structure), the bottom of the tunnel must be at least 12 inches below the sill or other wood parts of the building. Apply chemical heavily at 1 and 2.

Note 3. If it appears that fireplace hearth would fall in if tunneled. chimney base may be drilled and treated as stated in Par. 19.

16) Stucco (a) Wood supports for stucco must be at least 6 in. above outside ground level. or 4 in. above level of ground under an adjacent slab. (b) Where stucco extends to or below grade. chemical should be applied heavily in trenches dug below and under the edge of the stucco. so as to assure saturation of ground beneath. This is in addition to ground treatment under building (17). (c) Where ground slabs prevent the trenching required under (b) the ground may be saturated by flooding thru the void between stucco and inner walls.

17. Ground Treatment Under the Building. Chemical should be applied in narrow trenches 3 to 6 in. deep (but not lower than top of footing), dug in contact with and around foundation walls. pipes. chimneys and piers. Either of the following methods may be used: Method 1 Apply 1 to 2 gal. of chemicals per 10 lineal ft. entirely around inside of foundation wall. and around all pipes. chimneys. and piers, regardless of location of infestation. Method 2. Treatment may be limited to infested areas (except for stucco. Par. 16). as follows Near points at which tubes or damage are found. apply 2 to 4 gal. per 10 lineal ft. of trench. The application should extend along the walls for at least 6 feet in each direction past farthest extension of tubes or damage on walls. sills. or floor. Similar applications should be made around piers. chimneys. and pipes which are within 6 ft of farthest extension of tubes or damage

18. Ground Treatment Outside the Building. Heavy applications of chemical should be made in trenches outside the foundation wall (a) opposite points of termite damage. and (b) where there are tubes on the outside of the foundation wall. The trenches should extend 6 ft. or more past tubes or points of damage along wall in either direction and should be dug on the level with the inside trench (Fig. 3). where possible (unless the wall is solid. uncracked concrete).

Example. If the outer ground level is one foot above the inner ground level. the outside trench should be dug one foot deep at this point. to put it on a level with the inside trench. The dirt used to refill this deep outside trench should be saturated with chemical as it is put back in the trench. thus providing a barrier which should prevent termites from entering from the outside thru the mortar joints The chemical in the inside trench prevents termites from coming up the inside face of the wall. The only remaining way by which termites could come up the wall would be by coming from below thru crevices in a brick or stone footing. and the drilling and injection of the first mortar joint above the trench (Sec. 19 (b)) should prevent this.

19. Chemical Treatment of Masonry. Chemicals should be applied as follows to foundation walls. piers. and chimneys which are within 6 ft. of termite damage or tubes. (a) Flood all cracks in concrete (b) Drill mortar joints on 9-inch centers and flood all cracks and voids. in the horizontal layer of masonry which is adjacent to the treated ground at the bottom of the trench (Fig. 3) (c) Flood voids between walls. as in brick-veneer and stucco construction. (d) Flood between top of masonry and the sills or other timbers resting thereon. When the foregoing appears insufficient. the top of the wall or piers should be capped with concrete or metal

Note· When foundation wall is of hollow masonry blocks. and these blocks are not capped with metal or solid concrete. drilling should be done at each void. and heavy application of chemical applied. This should be done even though there is no visible sign of infestation

Note When footing is of solid concrete. with no cracks. and trenches on both sides of wall extend down to it. drilling and injection of mortar joints in wall is ordinarily not necessary

54

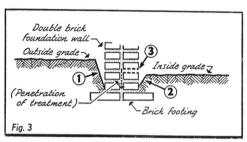

Fig. 3

FIGURE 3 EXPLAINS SECTION 18 (IN PART), AND SECTION 19 (b). To apply chemical around foundation walls, outside and inside trenches 1 and 2 should be dug to the same level, and the first horizontal mortar joint above the trench at 3 should be drilled and treated at 9-inch intervals. Drill the full width of the brick, so chemical will reach and penetrate down into the vertical mortar joint, as shown.

20. Powder-post Beetles If powder-post beetles are present in the understructure and are not to be treated, the owner must be so informed in writing before the contract is signed. Treatment, if given, should be applied to all wood showing signs of infestation. This can be done by spraying, brushing or mopping heavily onto wood surfaces one of the following: 5% pentachlorophenol in oil, pentachlorophenol emulsion paste, or oil or water mixtures of any of the materials listed for termite work in Section IIIC, Item A. Two or more applications must usually be made at intervals to achieve control. Special formulations are required for treating flooring and furniture to avoid damage to the finish

21. Slab Construction. This type of construction prevents adequate inspection Treatments are made when swarms occur or damage is found, or as a preventative measure. Treatment is accomplished by drilling holes in the slab or foundation through which sufficient chemical can be pumped with a power sprayer to thoroughly flood the soil. Long-rodding can sometimes be accomplished without drilling.

When the grade permits and drilling is necessary, holes should be drilled through the outside foundation at the horizontal level of the bottom edge of the slab. This will avoid damage to floors and floor coverings which may occur if drilling is done inside

When the slab is at or very near grade level the only route beneath it in many cases is by drilling through from inside This is also true when treating junctures of the slab with partitioning foundations. Care must be exercised to avoid damage to utility pipes. wiring or heating systems embedded in the concrete Building plans should be consulted before drilling, if available.

SECTION IIIB. MINIMUM REQUIREMENTS FOR STRUCTURAL PEST WORK ON NEW CONSTRUCTION (PRE-TREAT)

Conventional Construction (Crawl space)

New structures which are treated for termite control before or during construction must meet all of the minimum requirements given for existing structures in Section IIIA. except 13B and 16 through 21. 15 is also suspended except Part IB See Figure 2. Treatments should be made as directed in the "Chemical Treatment" section below.

Slab Construction.

Termite prevention in slab-construction buildings can be accomplished if: (a) All debris is removed, including grade stakes. before pouring the slab, and (b) The soil in the area where the slab is to be poured is pre-treated heavily with chemical. Treatments should be made as directed in the "Chemical Treatment" section below.

Chemical Treatment.

1. Apply 2-4 gallons of chemical to each 10 linear feet of trench around the inside and outside of foundations. pipes. ductwork. piers, etc., after soil has been leveled.

2. Treat all soil surface to be covered by structure and adjacent to it with 1 gallon of chemical to every 10 square feet.

3 Apply same treatment as in 2 above to soil under and adjacent to steps, porches, garage floors, carport slabs. or any other structure adjoining the building.

55

SECTION IIIC. MATERIALS

Chemical and Strength Commonly Used For.	Amount To Use Per 100 Gallons of Water or Diesel Fuel
A Termite Work	
Aldrin @ 0 5%	2 gallons of 2 lb. Aldrin e.c. or
or	1 gallon of 4 lb. Aldrin e.c.
Chlordane @ 1.0%	2 gallons of 4 lb. Chlordane e.c. or
or	1 gallon of 8 lb. Chlordane e.c.
Dieldrin @ 0.5%	2-2/3 gallons of 1.5 lb. Dieldrin e.c.
or	
Heptachlor @ 0.5%	2 gallons of 2 lb. Heptachlor e.c.
B. Powder-post Beetles	
Pentachlorophenol @ 5.0%	10 gallons of 40% material
or	
Pentachlorophenol Emulsion Paste	Not Applicable
or	
Any material Listed in A above	Same as A Above

APPENDIX XIX

California Minimum Requirements
for Structural Pest Control

Article 6. Wood Destroying Organisms

1990 Report Requirements Under Section 8516(b) 1-9 Inclusive.

All reports shall be completed as prescribed by the board. Copies filed with the board shall be clear and legible. All reports must supply information and indicate conditions existing as required by paragraphs 1-9, inclusive, of subdivision (b) of Section 8516 of the code and contain or describe the following

(a) Date of the inspection.

(b) Name and Structural Pest Control License number of person making the inspection.

(c) Name of person or firm ordering the report.

    (1) Name and address of any person whom the operator may designate as a party in interest pursuant to Section 8516 (b) 3 of the code.

(d) Address of location of the property.

(e) General description of building and premises inspected.

(f) A foundation diagram or sketch of the structure or structures or portions of such structure or structures inspected. The diagram shall indicate thereon the approximate location of any infested or infected areas evident and the parts of the structure where conditions exist which would ordinarily subject such parts to attack by wood destroying pests or organisms

(g) Infestations, infections or evidence thereof.

(h) Conditions usually deemed likely to lead to infestation or infection including, but not limited to·

    (1) Faulty Grade Level. A faulty grade level exists when the top of the foundation is less than 2 inches above the adjacent earth, or when the drainage is such that an adverse moisture condition is evident on the substructure timbers. The existing earth level shall be considered grade.

    (2) Inaccessible subareas or portions thereof and areas where there is less than 12 inches clear space between the bottom of the floor joists and the unimproved ground area.

    (3) Excessive cellulose debris. (This is defined as any cellulose debris of a size that can be raked or larger Stumps and wood imbedded in footings shall be reported.)

    (4) Information regarding the substructure, foundation walls and footings, porches, patios and steps, stairways, air vents, abutements, stucco walls, columns, or other parts of a structure normally subject to attack by wood-destroying pests or organisms.

    (5) Structural members found to be structurally weakened by wood destroying pests or organisms.

    (6) Earth-wood contacts.

    (7) Commonly controllable moisture conditions which would foster the growth of a fungus infection materially damaging to woodwork, including conditions resulting from stucco leaks, from first floor stall showers when tested in the manner required by Section 1991 (a) (13) of these regulations, and from shower baths

    (8) Woodwork which is adjacent to or in contact with a concrete slab floor.

(i) Areas normally subject to damage which are inaccessible or for other reasons not inspected. The Board recognizes the following areas as inaccessible for the purposes of this subdivision: inaccessible attics or portions thereof; the interior of hollow walls; space between a floor or porch deck and the ceiling or soffit below, such structural segments as porte cocheres, enclosed bay windows, buttresses, and similar areas to which there is no access without defacing or tearing out lumber, masonry, or finished work; areas behind stoves, refrigerators, and built-in cabinet work, floors beneath coverings, areas where storage conditions or locks make inspection impracticable.

(j) The report shall indicate or describe any area which is not inspected because it is inaccessible, or which is not inspected for any other reason, and shall either recommend inspection of such areas if inspection is practicable or state the reason inspection is not practicable.

Note: Authority cited for Sections 1990 through 1994: Sections 8516 (b) and 8525, Business and Professions Code.

History: 1. Amendment filed 11-18-57, effective thirtieth day thereafter (Register 57, No. 20).

    2. Amendment filed 9-16-65; effective thirtieth day thereafter (Register 59, No. 16).

    3. Amendment filed 5-19-64; designated effective 9-1-64 (Register 64, No. 11).

1991. Report Requirements Under Section 8516(b)9. (a) Recommendations for corrective measures for the conditions found shall be made as required under paragraph 9 of subdivision (b) of Section 8516 of the code to accomplish the following·

    (1) Break all contacts between wood members and the soil.

    (2) Clean the subarea of excessive cellulose debris. This excludes shavings or other cellulose too small to be raked. Stumps and wood imbedded in footings shall be treated if removal is impractical.

(3) Correct foundations or piers when faulty grade levels exist. Piers with 4 inches or more of exposed concrete are acceptable. In lieu of raising the foundation, flash walls of a minimum of 3 inch thickness may be installed provided they are properly bonded to the foundation and insure proper drainage.

Exterior grade may be lowered if natural exterior drainage is maintained away from structure.

When evidence of moisture, infestation or infection exists as a result of loose stucco, a recommendation shall be made to correct the condition.

(4) Separate earth in planters from all woodwork by a minimum of 3 inches of concrete or masonry or by a waterproof container.

(5) Separate earth fills such as under porches or other paving from all woodwork by concrete, masonry, or good quality cement plaster, so there is no earth-wood contact. Chemical treatment of infestations arising from earth fills is considered adequate if fill is separated from woodwork by concrete, masonry or good quality cement plaster.

(6) Replace or reinforce structural members visibly structurally weakened by wood-destroying pests or organisms. Wood structurally weakened by fungus shall be removed. Wood which is surface-infected by fungus may be treated by a wood preservative if no structural damage is present. Correct the adverse moisture condition.

(7) Provide adequate ventilation. It shall be considered inadequate when lack of ventilation has contributed to the growth of wood-destroying pests or organisms Vents shall be above grade or protected by a vent well which will not allow roof drainage or other water to enter the subarea.

(8) Correct conditions in frame and stucco walls and similar construction so that the wood framing is separated from the main structure by a complete concrete or masonry plug with no voids that will allow infestations or infections to enter the structure from the wall. If there is no plug, the foundation shall be 2 inches or more above the grade levels and at least as high as the adjoining slabs or a 4 inch concrete barrier seal off installed.

(9) Exterminate all reported infestations of drywood termites (kalotermes), subterranean termites (reticulitermes), dampwood termites (zootermopsis), or other wood-destroying pests. If evidence indicates that an infestation of drywood termites (kalotermes), dampwood termites (zootermopsis) or wood boring beetles extends into an inaccessible wall or area, recommendation shall be made to fumigate or to expose the infestation for local chemical treatment. Where accessible, termite pellets shall be removed, covered, or masked over after treatment. Subterranean termite tubes shall be removed where accessible.

(10) Treat an infested area under the structure when subterranean termite tubes are found connected to the ground or when active infestations are found in the ground.

(11) Make all unimproved areas of the subarea accessible for inspection. Twelve inches or more of clearance under floor joists shall be considered adequate except that such clearance shall not be necessary when ratproofing or other paving is present, or where the subarea soil is of such a nature as to prevent excavation or where excavation would create a hazard from shifting soil or other causes These conditions must be fully set forth in the report. Recommendations shall be made to provide crawl spaces so that any part is sufficiently visible to enable the inspector to comply with Section 8516 (b), Items 6, 7, 8, and 9.

(12) Correct any excessive moisture condition that is commonly controllable. When there is reasonable evidence to believe a fungus infection exists in a concealed wall or area, recommendations shall be made to open the wall or area.

(13) Repair a stall shower if it is found to leak when water tested for a minimum of 15 minutes after the shower drain has been plugged and the base filled to a minimum of 2 inches. Showers over finished ceilings are excluded unless water stains are evident on the ceiling, in which case recommendations shall be made for further testing

(b) When recommendations are made to correct conditions set forth as required by paragraphs 6 to 8 inclusive of subdivision (b) of Section 8516 of the Code, such recommendations shall not be labeled other than corrective Under no circumstances shall the same be labeled as preventive.

History 1. Amendment filed 9-16-59: effective thirtieth day thereafter (Register 59, No. 16)

2. Amendment filed 5-19-64: designated effective 9-1-64 (Register 64, No 11).

3 Amendment to subsections (a) (3), (9) and new (b) filed 1-2-68; effective thirtieth day thereafter (Register 68, No. 1).

1992. Secondary Recommendations. In addition to the recommendations required by Section 1991, the report may suggest secondary recommendations with a full explanation of why they are made and that they are secondary measures. If secondary recommendations are performed as allowed under Section 1992, any letter of completion, billing or other document referring to the work completed, must state specifically which recommendations were secondary

History. 1. New section filed 9-16-59: effective thirtieth day thereafter (Register 59, No. 16).

2 Amendment filed 5-19-64: designated effective 9-1-64 (Register 64, No. 11).

1993. Supplementary Reports. All supplementary reports, as required under Section 8516 (b) 8 of the Code shall indicate the absence or presence of wood-destroying pests or organisms, or conditions conducive thereto and shall be on the form prescribed by and filed with the Board with stamps affixed.

History 1 New section filed 9-16-59: effective thirtieth day thereafter (Register 59, No. 16).

2 Amendment filed 5-19-64: designated effective 9-1-64 (Register 64, No. 11).

1994. Limited Inspection Reports. If a report is made on part of a building then it must be designated as a limited report. The report shall have a diagram of the area inspected and shall specifically indicate which portions of the building were inspected and the name of the person or agency requesting a limited report.

History 1. New section filed 9-16-59; effective thirtieth day thereafter (Register 59, No. 16).

2. Amendment filed 5-19-64; designated effective 9-1-4 (Register 64, No. 11).

1995. Limitation of Report. The licensee shall limit the inspection report to buildings shown on the diagram only and identified the report.

History. 1. New section filed 5-19-64; designated effective 9-1-64 (Register 64, No. 11)

1996. Requirements for Reporting All Inspections Under Section 8516(b). (a) A written inspection report conforming to Section 16 (b) of the code shall be prepared and delivered to the person requesting the inspection or his designated agent whether or not : licensee has offered to perform the inspection without charge. The granting of permission to make an inspection shall be deem- a request to make an inspection.

(b) A copy of each such report shall be filed with the Board at or within the time specified by Section 8516 (b) of the code

(c) After commencing any inspection the failure or refusal of the person for whom the inspection was undertaken or his des- lated agent to pay for the inspection or a written inspection report conforming to Section 8516 (b), whether or not any payment is agreed on, shall not excuse the licensee from preparing and delivering a report and filing a copy thereof, as required by this :tion.

History. 1. New section filed 5-19-64; designated effective 9-1-64 (Register 64, No. 11).

1997. Filing Fees—Inspection Reports and Completion Notices Pursuant to the provisions of Section 8674 of the Business and ofessions Code, the following filing fee is determined, set and established (1) Inspection report filing—$0.50 (2) Notice of work mpleted filing—$2.00.

History. 1. New section filed 8-26-65; designated effective 10-1-65 (Register 65, No. 15).

2. Amendment filed 1-19-67; designated effective 3-1-67 (Register 67, No. 3).

3. Amendment filed 2-7-68; designated effective 3-15-68 (Register 68, No. 6).

APPENDIX XX

Florida Minimum Requirements
for Structural Pest Control

We would like to remind all licensees and certified operators that the Pest Control Act, Chap. 482.051 FS, authorizes this agency to promulgate rules, requiring that pesticides, fumigants, and rodenticides be used only in accordance with label requirements or as otherwise *accepted* by federal or state pesticide regulating agencies. *Acceptance* is only by label registration for purposes of this Act.

This has been implemented in the rules of the State Board of Health, Pest Control Regulations, Chap. 1701-2.04(1) and (2); 2.10(1), (2), (6), (7), (9), and (10), 2 14(2)(a), (4)(a), and (5)(a); and 2.15(2)(c) FAC. In effect the label becomes the regulation and has the force of law. Label directions and precautions become "Minimum Standards", so to speak, encompassing active ingredients, claims for pests to be controlled, selection, mixing, storage, use directions and safety precautions for all pesticides.

The purpose of this memorandum is to notify licensees and certified operators of "minimum treatment standards" for subterranean termite control in the light of registered labels, and that the law and rules will be enforced accordingly. Accepted chemicals, together with use concentration, most widely recommended and commonly used for subterranean termite control are as follows

Aldrin		0.5%
BHC[1]		0.8% Gamma Isomer (Lindane)
Chlordane		1.0%
Dieldrin		0.5%
Heptachlor		0 5%

The registered commercial labels for these compounds are very uniform in their directions for use in subterranean termite control These labels are the "Minimum Treatment Standards for Subterranean Termite Prevention and Control" under present Florida law and regulations.

The directions for treating various types of construction in accordance with MINIMUM LABEL TREATMENT STANDARDS are as follows

To control termites dilute the required amount of the registered product being used with sufficient water to provide a mixture containing the required finished concentration of the registered product being used and apply at the following rates:

(A) BUILDINGS WITH CRAWL SPACES

(1) Apply 2 gallons per 5 linear feet to critical area only under the house, such as along the inside of foundation walls, around piers, sewer pipes and conduits, etc Trench according to directions in part (2) below.

(2) Along the outside of foundation walls, dig a narrow trench, such trench to be dug no deeper than the top of the footings. If the trench is less than 15 inches in depth to the top of the footings, apply 1 gallon per 5 linear feet. Replace the soil and apply another 1 gallon per 5 linear feet to the back fill. Cover the back fill with a thin layer of soil. If the trench is more than 15 inches in depth to the top of the footings, apply 2 gallons per 5 linear feet. Replace the soil and apply another 2 gallons per 5 linear feet to the back fill. Cover the back fill with a thin layer of soil. A trench 30 inches deep is a maximum depth required alongside those foundations where the top of the footings is greater than 30 inches deep. In lieu of trenching to a 30" depth, make the trench 12" to 15" deep and rod to footing, spacing the holes about 1 foot apart

(3) Apply 1 gallon per 10 square feet of soil surface under attached porches, entrance platforms, utility entrances, and similar situations where slab or fill is at the grade level Where crawl spaces exist, treat as described in part (1) preceding.

(4) Treat all voids in hollow masonry units of the foundations at the rate of at least 1 gallon per 5 linear feet of wall. It is best to apply the chemical near the footing.

(B) BUILDINGS WITH CONCRETE SLAB ON THE GROUND

(1) Apply 1 gallon per 10 square feet as an over-all treatment under the slab as well as under attached porches, entrance platforms, utility entrances and similar situations where slab or fill is at grade level. In case the fill is washed gravel, cinders, or similar coarse material, increase the dosage by at least one-half

(2) Apply 2 gallons per 5 linear feet to critical areas only under the slab, such as along the inside of foundation walls, around sewer pipes, conduits, etc Trench as in part (3) below.

(3) Along the outside of foundation walls, dig a narrow trench no deeper than the top of the footings, but no deeper than 15 inches, unless the footing is much deeper in some places on account of the slope of the land, then treat as under (A) (2) preceding

(4) Treat all voids in hollow masonry units of the foundations at the rate of at least 1 gallon per 5 linear feet of wall. It is best to apply the chemical near the footings.

· [1] This compound is a mixture of several isomers, and the label must be carefully read and followed in order to obtain the correct amount of the Gamma Isomer for termite control

C) BUILDINGS WITH BASEMENTS

(1) Apply 1 gallon per 10 square feet as an over-all treatment under the basement flooring, as well as under attached porches, entrance platforms, utility entrances, and similar situations where slab fill is at the grade level. In case fill is washed gravel, cinders, or similar coarse material, increase the dosage by at least one-half. Where crawl spaces exist, treat as described in part (2) below

(2) Apply 2 gallons per 5 linear feet to critical areas only under the basement floorings, as well as porches and entrances having crawl spaces, such as along the inside of foundation walls, around sewer pipes, conduits, piers, etc. Trench according to directions in part (3) below.

(3) Along the outside of foundation walls, dig a narrow trench, such trench to be dug no deeper than the top of the footings. If the trench is less than 15 inches in depth to the top of the footings, apply 1 gallon per 5 linear feet. Replace the soil and apply another 1 gallon per 5 linear feet to the back fill. Cover the back fill with a thin layer of soil. If the trench is more than 15 inches in depth to the top of the footings, apply 2 gallons per 5 linear feet. Replace the soil and apply another 2 gallons per 5 linear feet to the back fill. Cover the back fill with a thin layer of soil. A trench 30 inches deep is a maximum depth required alongside foundations where the top of the footings is greater than 30 inches deep. In lieu of trenching to a 30" depth, make the trench 12" to 15" deep and rod to footing, spacing the holes about 1 foot apart

(4) Treat all voids in hollow masonry units of the foundation at the rate of 1 gallon per 5 linear feet of wall. It is best to apply the chemical near the footing.

****

These are the minimum treatment standards for subterranean termite prevention and control in accordance with the recommendations of all registered labels so far as pesticide usage is concerned. It is the responsibility of licensees and certified operators llow the label directions in their work

61

APPENDIX XXI

Georgia Minimum Requirements
for Structural Pest Control

620-7-.03 CONTROL MEASURES

(1) Control measures used shall be appropriate to the type of termites and termite damage, or the type of other Wood Destroying Organisms present as previously determined by inspection, and in accordance with the written agreement or contract

(2) The chemicals permitted in the control of termites and/or other Wood Destroying Organisms shall be one or more (either alone or in combination) of those that are registered with the Georgia Economic Poisons Law. These chemicals shall be used in the proper proportions as specified by the formulator and in the quantities and in the manner directed herein

(3) The chemicals recommended in the control of subterranean termites are those chemicals and concentrations listed below (arranged alphabetically)

Aldrin, 0.5% in water emulsion or oil solution

Benzene hexachloride, 0.8% gamma isomer in water emulsion or oil solution

Chlordane, 1.0% in water emulsion or oil solution

Dieldrin, 0.5% in water emulsion or oil solution

Heptachlor, 0.5% in water emulsion or oil solution.

NOTE: Oil no heavier than No. 2 fuel should be used. The above percentages should be used at specified rates (620-7-.03(4)/ or if any deviation from these percentages is made then at least the equivalent amount of toxic materials should be applied.

(4) On each Minimum Adequate Subterranean Termite Control Treatment the Designated Certified Operator shall give the following minimum service.

(a) Remove from underneath the building or immediately adjacent thereto all cellulosic debris (wood, paper, etc.) and other debris which would interfere with effective treatment and inspections. Remove form boards which are less than six (6) inches above grade

(b) Remove all wooden contacts between buildings and soil, both outside and inside. Wooden supports under buildings other than those resting on pillars, concrete slabs or walls shall rest on a masonry footing. The top of the footing shall be not less than six (6) inches above the ground, then capped with concrete or metal shield if openings in, or between parts of the footing allow for termite access

(c) Scrape off all visible and accessible termite tunnels

(d) Cut trenches in exposed soil on inside of foundation, at least six (6) inches in depth, but not lower than the top of the footing in soil in contact with masonry around all foundation walls, pillars and supports. Apply four (4) gallons of permitted insecticid or solid equivalent of recommended insecticide per ten (10) linear feet to trenches and backfill per foot of depth to top of footing with heavier applications where termite tunnels enter the soil Trenching and backfilling or rodding is required along the outsid foundation walls using an equivalent amount of insecticide The soil immediately around the point where pipes enter the soil sha be saturated with an effective chemical. When pipes are covered with insulating material, care shall be taken that the soil treatment above mentioned is sufficient to insure penetration of the soil below the depth to which such covering extends.

(e) Drill and flood, at not more than sixteen (16) inch intervals, the cavities in hollow pillars, tile, brick, concrete block or oth cavity type walls, chimneys or any other structure likely to be penetrated by termites by injecting at least one (1) gallon of permitt chemicals per five (5) linear feet. If walls are uncapped, flooding from the top is acceptable.

(f) Treat earth fills beneath concrete slabs in contact with building:

(1) Wherever practicable by cutting access openings and removing soil adjacent to the foundation the length of the fill at lea six (6) inches deep below the bottom of the sill or other wood framing, and six (6) inches wide poisoning as in (d) above.

(2) Or by drilling and poisoning as in (d) above from the top of the slab at not more than sixteen (16) inch intervals parall to and not more than twelve (12) inches away from the foundation wall.

(3) Or by rodding and applying the permitted chemical beneath the slab in a continuous barrier not more than six (6) inch from foundation walls

(g) Treat grade level slabs in contact with structure, but where there is no wood to earth contact, by drilling slab from abo at not more than sixteen (16) inch intervals parallel to and not more than twelve (12) inches from the structure of by rodding und slab and flooding with permitted chemical.

(h) Provide adequate ventilation (under circumstances approximately two (2) square feet of opening for each twenty-five (linear feet of wall), making certain that no "dead ends" or corners are left unventilated.

(i) Provide, by excavation, sufficient space for the application of proper control measures between wood super structure a the soil. In any case, minimum clearance between wood and soil shall be eight (8) inches.

(j) In treating structures on concrete slab on ground, soil beneath all points of potential termite entry shall be saturated w termite chemical by treating from above or by rodding beneath the slab.

62

(k) In new construction treatment, the permitted chemical may be applied to cavities in pillars, tile, brick or concrete block ills, chimneys, or other cavities likely to be penetrated by Wood Destroying Organisms by flooding the voids before they are covered, if practical. Soil surfaces to be covered by slabs may be treated before slab is poured, otherwise, drilling will be necessary; he overall treatment shall be not less than the Minimum Standards as herein before set forth.

0-7-04 INSPECTIONS

The Enforcement Agency, may in its descretion, inspect properties treated for termites and/or other Wood Destroying Organisms for the purpose of determining the efficiency of the remedial measures applied and to determine if treatment (s) comply th contract and Minimum Standards. When requested by the Enforcement Agency, the Designated Certified Operator or Licensee all furnish for analysis, sufficient samples of pesticides used in treatment.

ithority Ga. L. 1955, pp. 564, 568; Ga. L. 1957, pp. 299, 301; Ga. L. 1960, pp. 813, 817

(1) All Designated Certified Operators shall file reports with the Enforcement Agency on forms to be provided by the Enforcent Agency, 304 State Agriculture Building, Atlanta, Georgia, 30334, by the tenth (10) of each month, for each Wood Destroying rganisms job on which treatment was completed during the previous month or where an initial inspection was made for a fee. licating the contract number or the date the work was completed, the name and address of the property so treated, and the type work performed. Any omissions or deviations from the Minimum Standards and/or bond requirements shall be indicated and plained on the provided report form and shall be signed by the property owner or his agent. The form on which omissions and viations from the Minimum Standards and/or bond requirements shall be submitted to the Enforcing Agency with the Monthly port. Submission after inspection by the Enforcing Agency is not acceptable.

(2) If deviations from the Minimum Standards or from the contract agreement are found by the inspector, the Enforcement gency shall notify the Designated Certified Operator of such deficiencies on an Inspection Report. The correction of any defiencies shall be indicated on the Inspection Report and the Inspection Report returned to the Enforcement Agency within thirty 0) days.

ithority Ga. L. 1955, pp. 564, 572; Ga. L. 1960, pp. 813, 823.

APPENDIX XXII

Kansas Minimum Requirements
for Structural Pest Control

4-12-5. SUBTERRANEAN TERMITE CONTROL RECOMMENDATIONS: Each pest control operator shall use (a) insecticide, (b) the rates of application, and (c) the applications procedures, which have been approved and required by Kansas State Board of Agriculture regulation, and such shall be used on each treatment which he performs for the control of subterranean termites: Provided, That when the use of an insecticide, rate of application, or application procedure, different from those approved by such regulation is requested or agreed to by the customer, and when in the opinion of the pest control operator such insecticide, rate of application, or application procedure, would be effective, such different insecticide, rate of application, or application procedure, may be used.

When a different insecticide, a different rate of application, or a different application procedure, is used, the pest control operator shall furnish adequate control, and shall state on the required written statement the kind of insecticide requested and used, the rate of application used, and the application procedure used.

A. *Definition of Terms*

(a) "Treat Soil", "Soil Treat", and "Soil Treatment" means the application of a required insecticide to the soil in the area treated at the required rate of application.

(b) "Treat Voids", "Void Treat", and "Void Treatment" means the complete flushing of the void area surfaces treated with a required insecticide at the required rate of application.

(c) "Flood" and "Flooding" means the complete flushing of the entire surface of the area treated with a required insecticide at the required rate of application.

(d) "Rod" and "Rodding" means the practice of inserting a required insecticide at the required rate of application into an area treated by use of a hollow tube or pipe through which the chemical is pumped under pressure.

(e) "Wood Treat" and "Wood Treatment" means the complete impregnation of the wood in the area treated with a required insecticide.

B. *Required Insecticide Solutions:* One of the following insecticides shall be used for subterranean termite control and shall be applied in a solution strength not less than the following: .5% aldrin, .8% BHC, .8% lindane, 1% chlordane, .3% dieldrin, .5% heptachlor, 5% pentachlorophenol.

C. *Required Rates of Application.* Insecticide solutions shall be applied at the following rates of application when treating for the control of subterranean termites:

(a) Not less than 2 gallons of insecticide solution per 5 linear feet along the outside of foundation walls, including entrance platforms and porches, where the foundation is not more than 18 inches deep. If foundation is over 18 inches deep, apply not less than 4 gallons of insecticide solution per 5 linear feet of foundation wall.

(b) Not less than 2 gallon of insecticide solution per 5 linear feet of wall or other space to be soil treated along inside of foundation walls, on both sides of cross walls and around piers and other structural components under the building being treated.

(c) Not less than 1 gallon of insecticide solution per 5 linear feet of void space in foundations, piers, and other void containing areas.

(d) Not less than 1 gallon of insecticide solution per 10 linear square feet of area for soil treated under concrete slabs and other areas under which it is impossible to crawl.

D. *Required Application Procedures.* The following application procedures shall be used in treatment for the control of subterranean termites:

(a) All wood scrap, paper scrap, cardboard scrap and other scrap and debris containing cellulose shall be removed from the spaces under the building being treated.

(b) All points of contact where the wood of the building being treated rests on or in the ground shall be removed or treated. Wooden steps, support piers, window frames, trellises, lattice work and other such wooden parts of the building shall be set on a concrete or other base which is impervious to termites, or shall be otherwise altered so that they are not in direct contact with the ground, unless, because of financial or other considerations, the customer does not consent to the removal of such point of contact, in which event both wood treatment and soil treatment at the point of contact, shall be used.

(c) If the building being treated has no basement, (1) soil treatment shall be applied against both the inside and outside of the foundation walls, including cross walls, but excluding the outside of poured concrete outside foundation walls which are free of cracks, and which are completely exposed; and soil treatment shall be applied completely around all porch piers, support piers, pipes, and other such points of contact with the ground under and on the outside of the building. Where the soil against a foundation wall is covered by a concrete slab, it shall be reached for soil treatment and shall be soil treated by drilling holes through the slab from 18 to 24 inches apart; it shall be reached and shall be treated by making a tunnel under the slab adjacent to the foundation wall and by soil treating thereunder; it shall be reached and shall be treated by drilling holes through the head board from 12 to 24 inches apart; or it shall be reached and shall be treated by rodding adjacent to the foundation wall. (2) Voids and cracks shall be void treated in all outside foundation walls, cross walls, piers, chimneys, fireplace bases, and all other

64

such void containing structural components which might be used by termites to enter the building, but excluding voids in rock or stone foundation walls when the top of the foundation, under the sill plate, is flooded with an approved insecticide, and excluding voids in rock, stone, or brick veneer foundation walls or in rock or stone foundation walls when the veneer or entire foundation wall rests on solid concrete which extends at least 8 inches above the level of the soil against the foundation wall.

(d) If the building being treated has a basement, treatment shall be the same as a building with no basement, except as follows: soil treatment need not be applied to the soil in the basement area unless furred walls, dry walls or wooden partitions are present; if a furred wall, dry wall, or wooden partition is present in a basement, wood treatment shall be applied to the base of such wall or partition, or the basement floors next to such wall or partition shall be drilled every 18 to 24 inches and soil treatment shall be used next to the wall, partition or foundation.

(e) Buildings or parts of buildings with inadequate crawl space shall be soil treated by application of a required insecticide at the required rate to the entire area under the building, by rodding approximately every 24 inches apart, or by drilling through the floor approximately every 12 inches square.

(f) Buildings or parts of buildings built on concrete slabs, shall be treated by the application of a required insecticide at the required rate of application to the areas under the slab which provide potential termite entry, by rodding beneath the slab, or by inserting the insecticide through drilled holes, expansion joints, utility openings or cracks in the slab. (Authorized by KSA 2-2403; Compiled Jan. 1, 1966.)

-12-6. POWDER-POST BEETLE CONTROL RECOMMENDATIONS: Each pest control operator shall treat infestations of power post beetles by wood treating, or by spraying or painting, the infested area with a kerosene solution containing not less than % DDT, 2% chlordane, .5% lindane, .5% dieldrin or 5% pentachlorophenol: Provided, That when the use of an insecticide other than those listed, or a solvent other than kerosene, is requested or agreed to by the customer, and when in the opinion of the pest control operator such insecticide or solvent would be effective, and when the use of such insecticide or solvent is stated on the required written statement, such requested or agreed to insecticide or solvent may be used. (Authorized by KSA 2-2403; Compiled, Jan. 1, 1966.)

APPENDIX XXIII

Louisiana Minimum Requirements
for Structural Pest Control

MINIMUM REQUIREMENTS FOR TERMITE CONTROL WORK IN LOUISIANA
Adopted by Louisiana Pest Control Operators' Association
and approved by Structural Pest Control Commission
Effective February 1. 1958
(Amended November 1. 1962)

ACCEPTABLE CHEMICALS

A The chemical to be used shall be one which is accepted by the Structural Pest Control Commission or the U. S. Department of Agriculture. Division of Insecticides and Fungicides as having prolonged effectiveness as a toxicant against subterranean termites. In no event shall the anticipated effective duration of the termite chemical be for less than two years. The chemical shall be applied at the dosage rate recommended by the manufacturer and acceptable to the Structural Pest Control Commission.

B. The following chemicals are approved by the Structural Pest Control Commission for termite control work.

1 Aldrin 0.5 per cent in water emulsion or oil solution.

2. Benzene Hexachloride (Technical) To be used at a concentration of 0.8 per cent gamma isomer. applied in oil solution or water emulsion.

3 Chlordane To be used at a concentration of 1.0 per cent. applied in oil solution or water emulsion.

4 Dieldrin To be used at a concentration of 0.5 per cent. applied in oil solution or water emulsion.

5 DDT To be used at a concentration of 8.0 per cent in oil solution only.

6 Heptachlor 0.5 per cent in water emulsion or oil solution

C There are certain Proprietary Materials which are being used as soil poisons for subterranean termite control. These are acceptable. provided they contain one or more of the above - named chemicals in the concentrations specified. and providing, further. that they possess acceptable compatibility. Materials that do not conform to these requirements will be evaluated on the basis of field performance according to standards approved by the Structural Pest Control Commission. Any other chemicals which in the future may be recommended by the Southern Forest Experiment Station's Forest Insect Laboratory at Gulfport, Mississippi. will be acceptable by the Structural Pest Control Commission.

II. TREATMENT

A Pier-Type Construction

1. Access Openings Provide suitable access openings to partially excavated areas and to any other areas requiring inspection or treatment for the presence of termites.

2. Sanitation Remove all cellulose-bearing debris such as scrapwood. wood chips. paper. stumps. dead roots. etc. from underneath building Large stumps or roots that are too sound to be removed may be trenched. rodded and treated provided they are 6 inches or more from foundation timbers

3. Eliminate all direct contact of wood with ground as follows:

a Piers and Still Legs Still legs or other wood supports shall have concrete or metal capped bases extending at least 4 in. above the ground Pressure treated piling foundations are excepted from this requirements.

b Wood on Concrete Floors Where wood parts such as posts. doorframes. or stair carriages have been attacked by termites working up through concrete. or are set down into concrete. said wood parts shall be cut off and set on metal or concrete bases raised at least 1 inch above the floor level.

c. Wood Steps Place all wood steps on concrete bases which shall extend several inches above ground level and preferably several inches beyond the steps in all directions.

4 All temporary form boards which may have been left in place shall be removed.

5 Pipes Packing around pipes. if not removed. shall be saturated with chemical after breaking contact with ground. Pipes shall also be trenched and treated.

6. Skirting and Lattice Work These shall rest on solid concrete or cemented brick extending at least 3 inches above outside grade. unless suspended. in which case there shall be at least 3 in. clearance above grade.

7 Stucco (a) Where stucco extends to or below grade. chemical shall be applied heavily in trenches dug below and under the edge of the stucco. so as to assure saturation of ground beneath. This is in addition to ground treatment under building. as set forth in item #9 (b) Where ground slabs prevent the trenching required under (a) the ground may be saturated by flooding through the void between stucco and inner walls.

8 Dirt Fills When a foundation wall is in contact with a dirt-filled structure, such as a concrete slab porch or steps, the dirt must be excavated so as to leave a horizontal tunnel at junction of slab and foundation wall. The tunnel shall extend the length of the fill and shall be saturated with chemical at all points of contact with wall and slab. Supports for the slab shall be erected in the tunnel if necessary. Tunnel should be well ventilated but care shall be taken to assure that water does not run into these tunnels

EXCEPTION. If the construction is such that termites in working from fill to woodwork of house will be forced into the open. and if in addition there has been no previous infestation from fill. then fill need not be tunneled. If foundation wall is in contact with a solid structure such as porches and steps. the void made in conjunction with such structure shall be made accessible for thorough treating and inspection at all points. EXCEPTION. If due to construction it is impractical to

66

break into and excavate dirt filled areas, a method of drilling, flooding, and rodding may be employed with the permission of the Secretary of the Louisiana Pest Control Commission. Other dirt filled structure such as chimneys, porch columns, etc. will also be excavated.

Note 1: Where the sill or other wood extends below the bottom of the concrete slab, it will be necessary to make the tunnel more than 12 inches deep next to the foundation wall in order to provide sufficient clearance from the bottom of the wood to the bottom of the tunnel.

Note 2: Where the bottom of the tunnel is above the ground level under the house, the trench for applying the chemical in the tunnel shall, if practicable, be dug down to the level of the trench on the opposite side of the wall, unless the wall is solid, uncracked concrete, if this is impracticable, drilling and injection of the wall should be done at the level of the bottom of the tunnel, rather than at the ground level.

9. Ground Treatment. This rate of application shall be in accordance with the recommendations of the manufacturer but in no case will this rate be less than 1 gal. per 5 lineal feet of trench.

10. Chemical Treatment of Masonry. Chemical shall be applied to porous areas, cracks and accessible voids in foundation walls, piers, chimneys, step buttresses, etc. as follows.

   a. Flood all cracks in concrete.

   b. Drill mortar joints on all 2 course brick formations such as piers, foundation walls, chimneys, step buttresses, etc., in a horizontal line at sufficient intervals to provide thorough saturation of wall voids but in no case shall the distance between holes exceed 24 inches. Holes shall be deep enough to reach the center mortar joint and shall be flooded under sufficient pressure to flood all cracks and voids therein. Drilling shall not be required when solid concrete footing extends above grade level or when wall is capped with solid concrete.

   c. Drill mortar joints on all brick formations with 3 or more courses of brick on each side of formation at the end of every other brick but with the locations of the holes on each side of the formation alternating as much as is practicable and flood under pressure all cracks and voids therein. Where the outside finish of a 3 course brick wall makes drilling from each side of wall impractical, this wall can be drilled from one side by extending holes two bricks deep.

   d. Drill into the center vertical core of each of a complete row of hollow concrete (or other light weight aggregate) blocks in construction using this type of building material and apply chemical into the openings at rate sufficient to flood the area at the bottom of each block. In hollow concrete block construction, drilling will not be required where accessibility to the opening is already available through construction.

Existing Slab-Type Construction

1. Trench and trest the entire perimeter of the slab foundation. Rate of application shall be 1 gallon per 5 linear feet

2. Treat all traps and other openings in the slab.

3. Treat all expansion joints, cracks and other voids in slab by rodding under or drilling through slab and thoroughly saturating the area beneath the slab where the above stated conditions exist. When the slab is drilled, the holes must not be more than 3 feet apart along the above stated areas.

4. Treat all other obvious points of entry according to the best of the individual operator's ability.

Pre-Treatment of Slabs

1. Chemical shall be applied to the area under the inside the foundation of the building after the final grade has been reached but either before or after the gravel fill has been spread at the following rates:

   a. Overall treatment of area under slab at the rate of 10 gallons per 100 sq. ft.

   b. Along inside of exterior foundation wall on monolithic slabs at rate of 1 gal. per 5 linear ft

   c. Along all expansion and/or construction joints at the rate of 2 gals. per 5 linear ft.

   d. Critical areas under slab such as around plumbing, electrical conduits, air conditioning vents, chimney bases, etc., at the rate of 1 gal. per 10 sq. ft.

   NOTE: Items b, c, and d are in addition to overall treatment outlined in item a.

2. Soil under open slabs such as porches, carports, walkways, etc., attached to building shall be pre-treated at the rate of 15 gals. of chemical per 100 sq. ft. along a strip extending at least 3 feet from wall of building.

3. Enclosed garages, breezeways, sun rooms, etc., shall be treated as outlined in item #1

4. After building has been completed and the yard filled and leveled so that the final grade has been reached along the outside of the foundation wall, this area shall be trenched and treated at the rate of 1 gal. per 5 linear ft. of wall.

"Spot" Treatment

1. "Spot" treatment shall not be done on pier-type construction except with permission of the Secretary of the Pest Control Commission.

2. "Spot" treatment of existing slab-type construction is permissible when it is impractical to treat the entire slab and where the property owner requests this type treatment.

Responsibility of Operator to Property Owner and Commission

1. Any wood which is not in sound structural condition shall be promptly brought to the attention of the property owner, so that replacement, reconditioning, or repairing can be provided for. This, of course, pertains only to those visible portions of the building that it is possible to inspect and locate unsound wood.

2. In special cases, where it is apparent that these specifications are either insufficient or more than sufficient to insure adequate protection, the operator shall consult the Structural Pest Control Commission for advice.

3. The operator shall provide for air space on the water hose used in supplying the water to the chemical tank

67

APPENDIX XXIV

Mississippi Minimum Requirements
for Structural Pest Control

SECTION 11. *APPROVED CHEMICALS — MINIMUM REQUIREMENTS*

1. Acceptable chemicals—
   (a) Persons licensed in accordance with these Regulations shall use all pesticides in a safe manner and in accordance with procedures acceptable to the Division.
   (b) The following chemicals are approved by the Division for termite control work:
   Aldrin 0.5 percent concentration in water emulsion or oil solution
   Chlordane 1.0 percent concentration in water emulsion or oil solution
   Dieldrin 0.5 percent concentration in water emulsion or oil solution
   Heptachlor 0.5 percent concentration in water emulsion or oil solution
   (c) Chemicals approved for powder-post beetles to be used as liquids are those given in paragraph (b) of this section and Pentachlorophenol 5.0 percent concentration. "Poison Gas" fumigants may also be used for powder-post beetles.
   (d) There are certain Proprietary Materials which are being used as soil poisons for subterranean termite control. These are acceptable provided they contain *one* or *more* of the chemicals given in paragraph (b) of this section in the concentrations specified, and providing, further, that they possess acceptable compatibility. Any other chemicals which in the future may be recommended by the Southern Forest Experiment Station's Forest Insect Laboratory at Gulfport, Mississippi, will be acceptable by the Division.

2. *TREATMENT REQUIREMENTS — Pier-Type Construction*
   (a) Remove all cellulose-bearing debris such as scrapwood, wood chips, paper, stumps, dead roots, etc. from underneath building. Large stumps or roots that are too sound to be removed may be trenched, drilled or rodded and treated provided they are 6 inches or more from foundation timbers.
   (b) Remove all wooden contacts between building and soil, both inside and outside. Wooden supports under building must rest on concrete footing or on brick capped with concrete, the top of the brick or footing not less than 6 inches above the ground. This includes but is not limited to wood steps, skirting and lattice work, form boards, piers and stiff legs. (Pressure treated piling foundations are excepted from this requirement.)
   (c) Termite tunnels—Scrape off all termite tunnels from foundation walls and pillars.
   (d) Trenches—Cut trenches a minimum of 4 inches wide and deep, but not below top of footing, in contact with masonry around all inside foundation walls and pillars. A minimum of 4 gallons of approved repellent liquid shall be applied per 10 linear feet of trench. When it is necessary to trench more than 4 inches deep the rate of approved repellent liquid shall not be less than 4 gallons per 10 linear feet for each foot of depth. If trenches are back filled, the back fill shall be saturated with approved chemical.
   (e) Pipes—Remove all non-metal packing from pipes underneath the house which extends from the ground into the building or saturate packing with approved chemical after breaking direct contact with soil. Pipes shall also be trenched and treated.
   (f) Treatment of Masonry and Voids—Approved chemical shall be applied to porous areas, cracks and voids in foundation walls, piers, chimneys, step buttresses and other structures likely to be penetrated by termites. (1) Flood all cracks in concrete, (2) Drill mortar joints on all 2 course brick formations such as piers, foundation walls, chimneys, step buttresses, etc., in a horizontal line at sufficient intervals to provide thorough saturation of wall voids but in no case shall the distance between holes exceed 24 inches. Holes shall be deep enough to reach the center mortar joint and shall be flooded under sufficient pressure to flood all cracks and voids therein. Drilling shall not be required when solid concrete footing extends above grade level or when wall is capped with solid concrete. (3) Drill mortar joints on all brick formations with 3 or more courses of brick on each side of formation at the end of every other brick but with the locations of the holes on each side of the formation alternating as much as is practicable and flood under pressure all cracks and voids therein. Where the outside finish of a 3 course brick wall makes drilling from each side of wall impractical, this wall can be drilled from one side by extending holes two bricks deep. (4) Drill into the center vertical core of each of a complete row of hollow concrete (or other light weight aggregate) blocks in construction using this type of building material and apply chemical into the openings at rate sufficient to flood the area at the bottom of each block. In hollow concrete block construction, drilling will not be required where accessibility to the opening is already available through construction.
   (g) Dirt Fills—All dirt filled structures such as concrete slab porches, steps, chimneys, porch columns, etc. shall be treated by excavating, trenching and applying chemicals in same manner as around pillars and foundation. EXCEPTION. If due to construction it is impractical to break into and excavate dirt filled areas, a method acceptable to the Division of drilling, flooding, and rodding may be employed.
   (h) Beetles—Approved controls must be applied for beetles in timbers, walls and flooring, if beetles are present, unless contract states that protection against beetle injury is not included.

3. *EXISTING SLAB-TYPE CONSTRUCTION*
   (a) Trench and treat the entire perimeter of the slab foundation. Rate of application shall be 1 gallon per 5 linear feet.
   (b) Treat all traps and other openings in the slab.
   (c) Treat all expansion joints, cracks and other voids in slab by rodding under or drilling through slab and thoroughly saturating the area beneath the slab where the above stated conditions exist. When the slab is drilled, the holes must not be more than 3 feet apart along the above stated areas.

68

(d) Treat all other obvious points of entry according to the best of the individual operator's ability.

4. *PRETREATMENT OF SLABS*

In pretreatment of soil prior to the pouring of a concrete slab for "Slab-on-the-ground" construction, an over-all treatment of the soil beneath the proposed slab must be made by using chemicals and application rates which will meet the minimum requirements set forth by the FHA in "Minimum Property Standards For One and Two Living Units."

5. *SPOT TREATMENT*

(a) "Spot" treatment shall not be done on pier-type construction except with permission of the Division.

(b) "Spot" treatment of existing slab-type construction is permissible when it is impractical to treat the entire slab and where the property owner requests this type of treatment. The contract shall specify "Spot" treatment and clearly define area treated.

6. *SPECIAL CASES*

In special cases. where it is apparent that these specifications are either insufficient or more than sufficient to insure adequate protection. the operator shall consult the Division for advice before treatment is started.

SECTION 12. *INSPECTION BY DIVISION*

At such times as it may deem desirable the Division shall examine properties treated for termites for the purpose of determining the efficiency of the treatment given. On or about January 1, and July 1, of each year the Division shall determine the number of properties it has examined during the previous six (6) months for each operator. listing those receiving satisfactory treatments, and separately those receiving unsatisfactory treatments. Whenever the number of properties receiving unsatisfactory treatments (as evidenced by the presence of active termites in the foundation or super structure. or failure to carry out effectively the requirements of Section 11) shall exceed 25% of the total examined for any operator during a period, he shall be notified in writing by the Entomologist of the Division that effective 30 days from the date of notice, his license will be suspended unless he shall in the meantime show to the satisfaction of the Division why such action should not be taken. While his license is suspended the operator shall have the privilege of examining and retreating all properties on which he has current contracts but shall not solicit any new business on pain of immediate cancellation of his license. He shall notify the Division of the dates of all re-examinations and retreatments. When all properties previously reported in an unsatisfactory condition have been re-examined and retreated. the operator shall request the Division to reinspect the properties. The Division shall then make the reinspections at its earliest convenience but any property where termites were actively infesting the building at the last previous inspection by the Division shall not be reinspected earlier than 3 months after the retreatment. If the Division on reinspection shall find all the properties in satisfactory condition the suspension shall be removed. otherwise the license of the operator shall be cancelled permanently, PROVIDED, that the license of an operator may be cancelled immediately for gross neglect of contracts and general failure to give satisfactory service.

APPENDIX XXV

North Carolina Minimum Requirements
for Structural Pest Control

ARTICLE 5. CHEMICALS AND REQUIREMENTS FOR
POWDER-POST BEETLE AND SUBTERRANEAN TERMITE
TREATMENTS; REPORTING WOOD-DESTROYING ORGANISMS
DAMAGE AND INFESTATION.

§20-29. One or more of the following chemicals and at the concentrations specified shall be used for powder-post beetle control — (a) Chlordane — to be used on unfinished wood surfaces at a concentration of 2 0 percent applied in water emulsion.

(b) Dieldrin — to be used on unfinished wood surfaces at a concentration of 0.5 percent applied in water emulsion.

(c) Dieldrin — to be used on finished wood surfaces at a concentration of 0.5 percent applied in oil solution.

§20-30. One or more of the following chemicals and at the concentrations specified shall be used for subterranean termite control — (a) Chemicals and concentrations

(1) Aldrin — to be used at a concentration of 0.5 percent or if used at a lower concentration, a sufficient volume of the lower concentration shall be used so that the actual aldrin toxicant applied will be equal to that used when a 0.5 percent concentration is used at the specified rate of application, applied in a water emulsion or oil solution.

(2) Chlordane — to be used at a concentration of 1.0 percent or if used at a lower concentration, a sufficient volume of the lower concentration shall be used so that the actual chlordane toxicant applied will be equal to that used when a 1.0 percent concentration is used at the specified rate of application; applied in a water emulsion or oil solution.

(3) Dieldrin — to be used at a concentration of 0.5 percent, or if used at a lower concentration, a sufficient volume of the lower concentration shall be used so that the actual dieldrin toxicant applied will be equal to that used when a 0.5 percent concentration is used at the specified rate of application: applied in a water emulsion or oil solution.

(4) Heptachlor — to be used at a concentration of 0.5 percent, or if used at a lower concentration, a sufficient volume of the lower concentration shall be used so that the actual heptachlor toxicant applied will be equal to that used when a 0.5 percent concentration is used at the specified rate of application; applied in a water emulsion or oil solution.

(b) Proprietary materials — there are certain proprietary materials which are being used as soil poisons. Such materials may be approved if:

(1) They meet a 5-year field test conducted by an agency approved by the North Carolina Structural Pest Control Committee;

(2) They contain one or more of the above-named chemicals in the concentrations recommended; and

(3) Proof is provided that no toxic effects to humans or to beneficial plant and animal life will result from their use.

§20-31. Minimum requirements for the chemical control of subterranean termites in buildings after they are constructed. — (a) Basement or crawl-space construction:

(1) Access openings shall be provided to permit inspection of all basement and crawl-space areas of a building.

(2) Clean up and remove all wood debris and cellulose material, such as wood, paper, cloth, etc., contacting soil in all crawl-space areas. This excludes shavings or other cellulose material too small to be raked with the tines of an ordinary garden rake. Remove all visible stumps from all crawl-space areas. Should the property owner waiver their removal, the soil around stumps not removed shall be trenched or rodded and treated with toxic chemical at the rate of one gallon per 2-1/2 lineal feet per foot of depth. Remove all visible form boards in contact with soil.

(3) Thoroughly saturate with toxic chemical areas under the structure where termite tubes or tunnels are present or where active termite infestations are found in or on the soil.

(4) Remove all earth which is within 12 inches of the bottom edges of floor joists or within 8 inches of the bottom edges of subsills or supporting girders, but not below footings of foundation walls If foundation footings are less than 12 inches below the bottom edges of joists or subsills or supporting girders, a bank of soil 12 inches to 18 inches wide shall be left adjacent to footing for the purpose of support. Clearance shall be adequate to provide passage of a man to all crawl-space areas of a building.

(5) All visible termite tubes or tunnels on pillars, pilasters, foundation walls, chimneys, step buttresses, sill, pipe and other structures below the sill line shall be removed.

(6) Eliminate all wooden parts between the building and soil, both outside and inside.

a. No wood of any access opening shall be in contact with the soil.

b. Where wood parts such as door frames, partition walls, posts, stair carriages or others can be reasonably ascertained to be making direct soil contact through concrete or where there is evidence of termite activity or damage they shall be cut off above the ground or floor level the wood removed from the concrete; the hole filled with concrete or covered with a metal plate, after the point of contact has been flooded with toxic chemical at the rate of one quart per square foot of soil.

c Where wood parts such as vertical wood supports or other wood parts under a building or steps outside a building are not resting on solid masonry or concrete bases extending at least 2 inches above the soil surface or are in direct soil contact and such supports or steps are not removed, the supports and steps shall be cut off and set on a solid masonry or concrete footing extending at least 2 inches above the ground after the point of contact has been flooded with toxic chemical at the rate of one quart per square foot of soil

d. When wood skirting and lattice work are suspended, there shall be at least a 2-inch clearance between the top of the soil and the bottom edges of the wood skirting or lattice work. If the 2-inch clearance is not acceptable to the property owner, it may be closed with solid masonry or concrete but a minimum clearance of 1/4 inch shall be provided between the masonry and wood.

70

(7) Drill above highest soil line on either side of wall and treat all voids in multiple masonry foundation and bearing walls.

a. Drill walls every 16 lineal inches and treat with toxic chemical at the rate of one gallon per 5 lineal feet, or until there is an overlap of chemical.

b. If wall is hollow concrete blocks drill blocks, except those which are 4 inches thick, at each void in one row above highest soil line on either side of wall unless blocks are accessible from top or unless blocks have solid masonry cap consisting of solid poured concrete. Treat voids with toxic chemical at the rate of one gallon per 5 lineal feet or until there is an overlap of chemical.

c. Toxic chemical shall be applied to all voids in walls under sufficient pressure to flood all cracks and voids therein below the level of application.

(8) Drill voids in all multiple masonry pillars, pilasters, chimneys and step buttresses, and any void created by their placement, except hollow concrete block pillars and pilasters which have solid masonry caps consisting of concrete blocks with the wide faces turned down or solid poured concrete, and treat all voids with toxic chemical at the rate of one gallon per 5 lineal feet or until there is an overlap of chemical.

a. The distance between drill holes shall not exceed 16 inches and holes shall be deep enough to reach the center mortar joint

b. Drill holes shall be treated with toxic chemical under sufficient pressure to cover all cracks and voids therein below the level of application.

c. Drilling shall not be required if solid concrete masonry footings of pillars, or pilasters, or chimneys, or step buttresses extend 4 inches or more above top of soil surface.

(9) If foundation walls of dirt-filled concrete porches, or concrete slabs over dirt-filled areas, have voids, drill and treat all voids, herein as specified for foundation walls, for a distance of 3 feet from the main foundation wall. Drill holes shall be treated with toxic chemical under sufficient pressure to cover all cracks and voids therein below the level of application

(10) Where concrete slabs over dirt-filled areas are at the level of, above the level of, or in contact with, wood foundation members rod treat dirt-filled areas.

a. Drill vertically 1/2 inch or larger holes in the slab, no more than 8 inches from the house foundation, at 15 inch intervals and rod treat soil below slab with toxic chemical at the rate of one gallon per lineal foot

b. Alternatively drill horizontally 1/2 inch or larger holes in the foundation wall of the concrete slab no more than 8 inches from the house foundation and immediately below the bottom of the slab and rod treat soil adjacent to house foundation with toxic chemical at the rate of one gallon per lineal foot.

(11) Trench and/or rod soil adjacent to, but not more than 8 inches from, all pillars, pilasters, chimneys, step buttresses inside of foundation walls, outside of foundation walls if sill line is less than 12 inches above surface and the outside of foundation walls of concrete slabs over dirt-filled areas for a distance of three feet from the main foundation wall, and treat from top of soil surface to top of footing with toxic chemical at the rate of one gallon per 2-1/2 lineal feet per foot of depth to footing. Depth of trench and distance between rod holes will depend upon type of soil and depth of footing. Where concrete slabs adjacent to the foundation prevent trenching of soil, drill 1/2 inch or larger holes not more than 18 inches apart and within a foot of the foundation wall, through slabs or through adjoining foundation wall, and treat soil below slabs with toxic chemical at the rate given above. The soil immediately around pipes and other utility conduits making contact with soil and wood of structure, shall be trenched and/or rodded and treated with toxic chemical at the rate of one gallon per lineal foot of soil.

(12) Packing around pipes, if not removed, shall be saturated with toxic chemical after breaking contact between the packing and the ground.

(13) Drill at 12-inch intervals and treat with toxic chemical, under pressure, all known termite-infested timbers below the sub-floor.

(14) Where stucco or wood or similar type materials extend to or below grade, trench soil to a depth below and under the edge of the stucco or similar type materials and treat soil with toxic chemical at the rate of one gallon per 2-1/2 lineal feet per foot of depth from top of soil surface to top of foundation footing. After the soil has been treated a masonry barrier wall may be erected to hold back the soil from making direct contact with the stucco or similar type materials. Where slabs on grade adjacent to foundation prevent trenching of soil, drill 1/2 inch or larger holes not more than 18 inches apart and within a foot of the foundation wall through slabs or through adjoining foundation wall, and treat soil below slabs with toxic chemical, at the rate given above

(b) Slab-on-ground construction:

(1) Trench and/or rod soil adjacent to, but not more than 8 inches from, the wall around the entire perimeter of the slab foundation wall, including foundation walls of concrete slabs over dirt-filled areas for a distance of three feet from the main foundation wall, and treat from top of soil surface to top of footing with toxic chemical at the rate of one gallon per 2-1/2 lineal feet per foot of depth to footing. Depth of trench and distance between rod holes will depend upon type of soil and depth of footing. Where concrete slabs adjacent to the foundation prevent trenching of soil, drill 1/2 inch or larger holes not more than 18 inches apart and within a foot of the foundation wall, through slabs or through adjoining foundation wall, and treat soil below slabs with toxic chemical at the rate given above.

(2) Treat soil with toxic chemical in, under, and around, all traps, openings and utility conduits in the slab at the rate of one gallon per 2-1/2 lineal feet of soil.

(3) Drill at all visible or known expansion and construction joints, cracks, and other crevices in slab at 16 inch intervals and treat soil below slab with toxic chemical at the rate of one gallon per 2-1/2 lineal feet of soil. Where wooden structural members are in contact with concrete or masonry floors which have joints or cracks beneath the wooden structural members, the concrete or masonry shall be drilled and treated to flood the soil beneath them

(4) Eliminate all wooden parts between the building and soil, both outside and inside

a. Where wood parts, such as door frames, partition walls, posts, stair carriages, or others can be reasonably ascertained to be making direct soil contact through slab or where there is evidence of termite activity or damage, they shall be cut off above the ground or floor level; the wood removed from the slab, the hole filled with concrete or covered with a metal plate, after the point of contact has been flooded with toxic chemical at the rate of one quart per square foot of soil.

71

b. Where wood steps or other wood parts contacting building and soil are not resting on solid masonry or concrete bases at least 2 inches above the soil surface or are in direct contact with soil and such step or other wood parts are not removed, the steps or other wood parts shall be cut off and set on a solid masonry or concrete footing extending at least 2 inches above the ground after the point of contact has been flooded with toxic chemical at the rate of one quart per square foot of soil

c When wood skirting and lattice work are suspended, there shall be at least a 2-inch clearance between the top of the soil and the bottom edges of the wood skirting or lattice work If the 2-inch clearance is not acceptable to the property owner, it may be closed with solid masonry or concrete but a minimum clearance of 1/4 inch shall be provided between the masonry and wood.

(5) Drill above highest soil line on either side of wall and treat all voids in multiple masonry foundation and bearing walls.

a Drill every 16 lineal inches, and treat with toxic chemical at the rate of one gallon per 5 lineal feet or until there is an overlap of chemical, all multiple masonry foundation and partition walls which are at the level of, in contact with, or extending below the soil

b. If wall is hollow concrete blocks drill blocks, except those which are 4 inches thick, at each void in one row above highest soil line on either side of wall unless blocks are accessible from top or unless blocks have solid masonry cap consisting of solid poured concrete. Treat void with toxic chemical at the rate of one gallon per 5 lineal feet or until there is an overlap of chemical.

c. Toxic chemical shall be applied to all voids in walls under sufficient pressure to flood all cracks and voids therein below the level of application

(6) Drill, at or near the footing, voids of all multiple masonry chimneys, pillars, pilasters, and step buttresses adjacent to foundation walls, or which penetrate concrete slabs, and treat voids with toxic chemical at the rate of one gallon per 5 lineal feet.

a. The distance between drill holes shall not exceed 16 inches and holes shall be deep enough to reach the center mortar joint.

b Drill holes shall be treated with toxic chemical under sufficient pressure to cover all cracks and voids therein below the level of application

c Drilling shall not be required if solid concrete masonry footing of structures extends to or above top of slab.

(7) If foundation walls of dirt-filled concrete porches, or concrete slabs over dirt-filled areas, have voids, drill and treat all voids, therein as specified for partition and foundation walls, for a distance of 3 feet from the main foundation wall. Drill holes shall be treated with toxic chemical under sufficient pressure to cover all cracks and voids therein below the level of application.

(8) Where concrete slabs over dirt-filled areas are at the level of, above the level of, or in contact with, wood foundation members rod treat dirt-filled areas.

a. Drill vertically 1/2 inch or larger holes in the slab, no more than 8 inches from the house foundation, at 15 inch intervals and rod treat soil below slab with toxic chemical at the rate of one gallon per lineal foot.

b. Alternatively drill horizontally 1/2 inch or larger holes in the foundation wall of the concrete slab no more than 8 inches from the house foundation and immediately below the bottom of the slab and rod treat soil adjacent to house foundation with toxic chemical at the rate of one gallon per lineal foot

(9) Where stucco on wood or similar type materials extend to or below grade, trench soil to a depth below and under the edge of the stucco or similar type materials and treat soil with toxic chemical at the rate of one gallon per 2-1/2 lineal feet per foot of depth from top of soil surface to top of foundation footing. After the soil has been treated a masonry barrier wall may be erected to hold back the soil from making direct contact with the stucco or similar type materials. Where concrete slabs adjacent to the foundation prevent trenching of soil, drill 1/2 inch or larger holes not more than 18 inches apart and within a foot of the foundation wall, through slabs or through the adjoining foundation wall, and treat soil below slabs with toxic chemical at the rate given above.

(10) Section 20-31, (a) (1), (2), (3), (4), (5), (6) a, e, (11), (12) and (13), of these rules and regulations shall be followed, if any or all of the above, apply to existing slab-on-ground construction.

§20-32. Reporting wood-destroying organisms damage and infestation or areas inaccessible to inspection.—(a) All wood members of a structure which can be ascertained, by visual inspection, to be structurally weakened or damaged by wood-destroying organisms shall be promptly brought to the attention of the property owner or his authorized agent and shall also be indicated in writing, in the contract or agreement, by the licensee or his authorized agent.

(b) The licensee or his authorized agent, shall indicate in writing, in the contract or agreement, whether or not he is responsible for the replacement, repair or re-inforcement of any or all of the wood members which were ascertained by visual inspection to be structurally weakened or damaged by wood-destroying organisms.

(c) Any evidence of infestation of wood-destroying organisms in, on, under, or in contact with, a structure shall be promptly brought to the attention of the property owner or his authorized agent and shall be specified, in writing, in the contract or agreement, by the licensee or his authorized agent.

(d) In buildings with basements or crawl-spaces, the licensee or his authorized agent shall indicate and describe, in writing, on the contract or agreement, any area of the basement or crawl-space of the building which has not been inspected and give the reasons for not making such inspection.

§20-33. Minimum requirements for the chemical control or prevention of subterranean termites for buildings under construction.—(a) Basement or crawl-space construction:

(1) Apply toxic chemical at the rate of one gallon per 2-1/2 lineal feet per foot of depth, from top of soil surface to top of footing along, and not more than 6 inches away from' the inside of the main foundation wall; the entire perimeter of all multiple masonry chimney bases, pillars, pilasters, and piers, and both sides of partition or inner walls

(2) After a building or structure has been completed and the excavation filled and leveled, so that the final grade has been reached along the outside of the main foundation wall, trench and/or rod soil adjacent to, along, and not more than 6 inches from, the outside of the main foundation wall and treat with toxic chemical at the rate of one gallon per 2-1/2 lineal feet per foot of depth, from top of final grade to top of footing, or to top of soil previously treated.

(3) Apply toxic chemical to all voids of unit masonry foundation walls, piers, pillars, pilasters, chimneys, and other supporting

72

or attached unit masonry structures, at the rate of one gallon per 5 lineal feet or until there is an overlap of chemical, except where voids are to be capped with a solid masonry cap consisting of solid concrete or masonry. Toxic chemical shall be applied to voids, under sufficient pressure, to flood all cracks and voids therein below the level of application.

(4) Apply toxic chemical, at the rate of one gallon per 10 square feet of surface area, to soil within 3 feet of the main foundation, under slabs, such as patios, walkways, driveways, terraces, gutters, etc., attached to the building. Toxic chemical shall be applied before slab is poured, but after fill material has been spread. If fill under slab is gravel or other course absorbent material, toxic chemical shall be applied at the rate of one gallon per 7 square feet of surface area.

(5) Apply an over-all treatment of toxic chemical under the entire surface of floor slabs, such as basements, porches, entrance platforms, garages, carports, breezeways, sun room, etc. Apply toxic chemical under floor slabs at the rate of one gallon per 10 square feet of surface area, except that if fill under such slabs is gravel or other course absorbent material, apply toxic chemical at the rate of one gallon per 7 square feet of surface area. Toxic chemical shall be applied before slab is poured but after fill material has been spread.

(6) In addition to treatment outlined above, apply toxic chemical, at the rate of one quart per square foot of surface area, to and around all critical areas, such as expansion and construction joints, electrical conduits, air conditioning vents, heating and plumbing outlets, pipes, utility lines, etc. at their point of penetration of the slab or floor.

(b) Slab-on-ground construction.— All parts of Section 20-33 (a) of these rules and regulations shall be followed, if applicable, in treating slab-on-ground construction.

(c) Treatment shall not be made when the soil or fill is excessively wet or immediately after heavy rains, to avoid surface flow of toxicant from the application side. Unless the treated areas are to be immediately covered, precautions shall be taken to prevent disturbance of the treatment by human or animal contact with the treated surface.

(Adopted: August 6, 1970; Amended February 24, 1971.)

(Authority: G. S. 106-65.29.)

APPENDIX XXVI

Oklahoma Minimum Requirements
for Structural Pest Control

3-412 MINIMUM REQUIREMENTS FOR TERMITE WORK

    A The "minimum requirements" are the irreducible minimum measures deemed necessary to bring worthwhile termite control service to the public.

    1. ACCESS

        All new access openings shall be a minimum of 14" x 16" to permit entrance under all crawl space

        No wood of any access opening shall be in contact with the soil.

    2. DEBRIS AND FORM BOARDS

        Remove all wood or other cellulose material on the ground in unexcavated areas that can be caught with a rake. Remove visible form boards. If form boards are buried in concrete, break out as much as practical. Visible stumps must be trenched and treated if they cannot be removed.

    3. CLEARANCE UNDER BUILDINGS

        Remove all earth which is within 12" of the bottom edge of floor joist or within 5" of stringers, but not below footing of foundation walls. If foundation footings are less than 12" below bottoms of joist, a bank of soil 12" to 18" wide should be left adjacent to the footing for the purpose of support.

    4. SHELTER TUBES

        Brush all visible tubes from piers, walls, sills, joist, subfloor, pipes and all other parts of the understructure.

    5 PIERS AND STIFF-LEGS

        Stiff-legs or other wood supports must have concrete or other termite impervious bases extending at least 4 inches above the ground, or resting on unbroken concrete bases which must extend 4 inches beyond the support.

    6. OUTSIDE WOOD STEPS

        Place all outside wood steps on concrete bases which extend above ground level and preferably beyond the steps in all directions.

    7. DAMPNESS

        If ventilators or access opening in foundation or veneer walls are so constructed in relation to outside grade, or within 2 inches of exterior grade so as to allow normal rainfall to run under the house, this condition shall be corrected by the construction of a permanent masonry or metal water barrier, or wall so constructed as to stop the water from running under the structure through the opening.

    8. REPLACEMENTS

        If visible termite damaged material is not to be replaced or repaired at the time of the original contract, it must be noted in the contract

    9. SKIRTING AND LATTICE WORK

        These should rest on concrete or other continuous, termite impervious material extending several inches above the outside grade. If they are suspended, there must be at least 1 inch clearance above grade.

    10. FRAME SIDING

        On homes where the frame siding runs below grade, the soil must be treated the same as for stucco, or preferably a barrier wall erected to hold back the soil from contact with structural timbers.

    11 STUCCO OR SIMILAR TYPE MATERIALS

        Where stucco extends to or below grade, chemical must be applied heavily in trenches dug below and under the edge of stucco, so as to assure saturation of ground underneath. This is in addition to ground treating under the building Where ground slabs prevent trenching, the ground must be saturated by flooding through holes drilled in the slab.

        The erection of a barrier wall to hold back the soil from making contact with the bottom of the stucco is preferred.

    12. DIRT FILLS

        Dirt filled areas must be treated in one of three ways, drilled, rodded, or tunneled. If tunneled, the tunnel must be a minimum of 12" x 12" and thoroughly treated with chemical and ventilators placed at both ends, if practicable. If drilled, holes shall not be in excess of 24" apart.

    13. GROUND TREATMENT UNDER BUILDING

        In the unexcavated areas of structures (without basements but with under floor area) chemical shall be applied in narrow trenches 3" to 6" deep, dug in contact with and around walls and footings, pipes, chimneys, and piers except where trenches cannot be dug because of rock formation. Back filling of the trench is permitted.

    14 VOID TREATMENT

        The ground or hollow area in cracks and voids in foundation walls, piers, between foundation walls, and masonry veneer, and in chimneys shall be drilled or otherwise entered and treated. Drill holes shall not be farther apart than 36 inches.

        CONCRETE BLOCKS

            1. Voids in a given row of concrete block wall must be drilled or otherwise entered and treated Drilling can follow the grade contour.

2. Where concrete block walls with termite shield are capped by solid concrete or solid concrete blocks, they need not be drilled.

3. Where outside concrete block walls are behind brick or other veneer, the treatment of the block voids may be effected by trenching the inside grade to the footing and flooding with chemicals. This will also apply to brick stem walls.

15. GROUND TREATMENT OUTSIDE OF BUILDING

The exterior open ground along the foundation or veneer wall shall be trenched and treated to the footing, or 12" deep where the footing is deeper, except solid concrete where foundation is exposed more than 12" above the outside grade. Treat the back fill.

Where trenching is not possible due to presence of masonry walls, gutters, terraces, etc., holes must be drilled at not more than 24" intervals and the ground beneath thoroughly treated adjacent to the structure, or treating may be done by rodding or tunneling.

When concrete slab garage floors show a termite infestation, the cracks or crevices shall be drilled or otherwise treated.

16. WOOD PARTS-CONCRETE SLABS

Where wooden parts are resting on cracked concrete, holes must be drilled diagonally on both sides of the wooden part and the soil below thoroughly treated. Where wood parts extend into or through concrete slab or adjacent to an expansion joint, drill the slab adjacent to the wood part at 24" intervals and pressure treat. If wood parts rest on footing, drill both sides. If practical, wood parts may be cut off and placed on termite impervious material. Wood parts resting on top of unbroken concrete are excluded from this provision, provided there is no visible evidence of termites.

17. CONCRETE SLABS

Pretreating- Apply under the floor slab, including porch floors, entrance platforms and garage floors, at the rate of one gallon per ten square feet of dirt filled. If fill is gravel or other coarse absorbent material, apply at the rate of one gallon per seven square feet.

1. Apply to areas along outside perimeter of the structure, 1 gallon per 2-1/2 linear feet. Should footing exceed 18 inches in depth, apply more material proportionally.

2. To voids below grade on unit masonry foundation wall, apply at rate of one gallon per five linear feet.

3. Treatment should not be made when soil or fill is excessively wet or immediately after heavy rains. Unless the treated area is to be immediately covered, precaution should be taken to prevent disturbance of the treated area by human or animal contact.

Existing slabs—In so far as possible approved chemicals at the recommended percentages given in these regulations and in the gallonage amounts listed under 17 (a), shall be introduced to the area under the slab, as well as to the outside perimeter and to voids. A drawing or statement must be on the contract showing how and where treatment was applied.

13 APPROVED CHEMICALS FOR TERMITE CONTROL

A. The following chemicals are approved for termite control work within the State of Oklahoma: Sodium arsenite, pentachlorophenol, orthodichlorobenzene, trichlorobenzene, nitrochlortoluene, creosote, tetrachlorophenol, betanaphthol, tar acid oil, monochlornaphthalene, sodium fluoride, dinitrophenol, sodium arsenate, dichloro-diphenyl-trichloroethane (DDT), ethylene dibromide, copper naphthenate, chlordane, dieldrin, aldrin, heptachlor, BHC and lindane.

B. Upon proper request the Board will grant to any licensed operator an "experimental permit" for the testing and use of any new or unlisted chemical or method that offers promise in the field of termite control. The request for approval and permit shall be treated as restricted information, upon request of the applicant, so as to allow him ample time to collect data and perfect a patent, if applicable.

C. The treatment of soil for slab construction shall be with the following materials at the dilutions given, or the equivalent amount of technical chemical; aldrin-0.5% in oil or water, BHC-0.8% in oil or water, chlordane-1.0% in oil or water, dieldrin-0.5% in oil or water, DDT-8.0% in oil only, lindane-0.8% in oil or water, heptachlor-1.0% in oil or water, sodium arsenite-10.0% in water, trichlorobenzene-1 part to 3 parts in oil, orthodichlorobenzene 1 part to 10 parts in oil.

APPENDIX XXVII

North Carolina Major Discrepancies
in Structural Pest Control

1. State structural pest control license number not displayed on service vehicle after 30 days from the date licensee acquired vehicle. Article 3, Section 20-20.
2. Property owner or authorized agent not furnished a written agreement signed by licensee or his authorized agent before treatment is started and licensee does not have copy of said agreement. Article 6, Sections 20-34 and 20-38 (a)
3. Active subterranean termites present in any part of the building(s) covered by the agreement or on anything in contact with the building(s) after 30 days from the original treatment date or anytime thereafter as long as agreement remains effective. Article 1, Section 20-1 (b), (c)
4. Active powder-post beetles present in any part of the building(s) covered by the agreement after 10 months from the original treatment date or anytime thereafter as long as agreement remains effective. Article 1, Section 20-1 (b), (c)
5. Access opening not provided to permit inspection of basement or crawlspace area and property owner not furnished waiver on same. Article 5, Sections 20-31 (a) and 20-31 (b) (10) and Article 6, Section 20-36 (a)
6. Less than 12-inch clearance between soil and bottom edge of floor joist and property owner not furnished waiver on same. Article 5, Sections 20-31 (a) (4) and 20-31 (b) (10) and Article 6, Section 20-36 (a)
7. Less than 8-inch clearance between soil and bottom edge of subsill or supporting girder and property owner not furnished waiver on same. Article 5, Sections 20-31 (a) (4) and 20-31 (b)(10) and Article 6, Section 20-36 (a)
8. Wood part making direct soil contact and property owner not furnished waiver on same. Article 5, Sections 20-31 (a) (6) and 20-31 (b)(4) and Article 6, Section 20-36 (a)
9. Multiple masonry foundation or bearing wall not drilled and property owner not furnished waiver on same. Article 5, Sections 20-31 (a)(7) and 20-31 (b) (5) and Article 6, Section 20-36 (a)
10. Chimney not drilled and property owner not furnished waiver on same. Article 5, Sections 20-31 (a) (8) and 20-31 (b) (6) and Article 6, Section 20-36 (a)
11. Concrete slab over dirt-filled area not drilled (through slab or through foundation wall of slab) and property owner not furnished waiver on same. Article 5, Sections 20-31 (a) (10) and 20-31 (b) (8) and Article 6, Section 20-36 (a)
12. Soil not trenched and/or rodded inside of foundation wall and property owner not furnished waiver on same. Article 5, Sections 20-31 (a) (11) and 20-31 (b) (10) and Article 6, Section 20-36 (a)
13. Soil not trenched and/or rodded outside of foundation wall and property owner not furnished waiver on same. Article 5, Sections 20-31 (a) (11) and 20-31 (b) (1) and Article 6, Section 20-36 (a)
14. Soil deficient in toxic chemical. Article 5, Sections 20-31 (a), 20-31 (b) and 20-33
15. Vehicle compartment containing pesticidal concentrate or acutely toxic poison bait not locked. Article 7, Section 20-39 (b)
16. Pesticidal container not labeled as specified. Article 4, Section 20-28 (a), (b) and Article 7, Section 20-39 (c)
17. Covered bait station used for acutely toxic rodenticidal bait not locked and marked as specified. Article 7, Section 20-39 (d)

*Effective January 1, 1972, any minor discrepancy which is not corrected within 30 days after the date of the initial inspection notice shall become a major discrepancy and be handled in accordance with Article 11, Section 20-62 (d) of the Committee rules and regulations.

APPENDIX XXVIII

U.S. Department of Housing
and Urban Development Instructions
for Use of HUD Wood Infestation Report

1. This Notice Transmits:
   FHA Form 2053, Wood Infestation Report.

2. Explanation of Material Transmitted:
   The Wood Infestation Report provides a standard form for reporting
   inspections for infestation in existing houses.

3. Requirement: Form FHA 2800, "Application for Property Appraisal and
   Commitment", requires under Specific Commitment Conditions, Item 2,
   2800-5, that when checked, a certificate shall be furnished on an
   existing house by a recognized termite control operator. The Wood
   Infestation Report is the Certificate and the house, when accepted
   for mortgage insurance, must show no evidence of active termite
   infestation.

4. Instructions: When Item 2 of Form 2800-5 has been checked, the
   termite control operator will, when requested to do so, complete the
   Wood Infestation Report as follows:

   a. He completes the heading and includes his license number.

   b. He certifies that he has carefully inspected and sounded all
      accessible areas for termites and other wood destroying insects.

   c. He completes Items 1 through 8 and checks boxes either "yes" or
      "no" as applicable.

   d. Item 3. If there is evidence of conditions conducive to infestation
      he describes these conditions on the reverse side of the form.

   e. Item 4. If there is structural damage he describes this on the
      reverse side of the form.

   f. Item 5. If there is damage to construction (non-structural mem-
      bers) he describes the damage and the location on the reverse side
      of the form.

g. Item 6. If the premises have been treated by the company making the inspection, the chemical used and the date of treatment are entered in the spaces provided. Additional information, such as the rate of application, is entered on the reverse side of the form.

h. The inspector and the company he represents certify that they have no interest, present or prospective, in the property or parties involved in the transact on. The inspector and an agent authorized to sign for the compai y sign and date the form.

i. The original of the form is sent to the appropriate HUD/FHA field office and a copy is sent to the buyer, seller and mortgagee.

If the Wood-Infestation Report shows evidence of active termite infestation, FHA will refuse to issue mortgage insurance until a subsequent report is issued showing no active infestation.

If the property is located in a state having an approved wood-infestation form, the HUD Wood-Infestation Report must be completed for HUD purposes in addition to the state approved form unless it is acceptable to the state to modify their form to conform.

Attachment

Smythe, Richard V., and Williams, Lonnie H.

    1973.   Structural pest control regulations. South. For.
         Exp. Stn., New Orleans, La. 78 p. (USDA For.
         Serv. Res. Pap. SO-93)

A compilation of existing structural pest control legislation evaluated by 17 criteria and an examination of possible alternative—industry self-regulation.

Additional keywords: Structural pest control, structural pest control legislation, industry self-regulation, wood-destroying organisms.

U.S. Department of Agriculture
Forest Service Research Paper SO-95

U.S. Department of Agriculture
Forest Service Research Paper SO-94

Growth of Pine Plantations In North Mississippi

ons

Hamlin L. Williston
and
T.R. Dell

Southern Forest Experiment Station
Forest Service
U.S. Department of Agriculture

1974

U.S. Department of Agriculture
Forest Service Research Paper SO.95

Growth Of Pine Plantations
In North Mississippi

Hamlin L. Williston and T. R. Dell[1]

Performance of loblolly and shortleaf pines planted by the Civilian Conservation Corps in northern Mississippi provides guides for managing flood control plantings made in the same area since World War II. Thinning may cause losses to Fomes annosus *on certain soils and is likely to reduce total pulpwood yields. Thinning is advisable where sawtimber is the desired product.*

Additional keywords: *Pinus taeda, P. echinata,* plantation management, plantation growth, plantation yields, *Fomes annosus.*

More than ½ million acres of pine plantations have been established n north Mississippi by the Yazoo-Little Tallahatchie Flood Prevention Project (Y-LT) since 1947. These stands were established to control erosion, but they are reaching ages where decisions must be made about commercial thinning, product objectives, and management accommodations to the threat of disease. Data on which to base management prescriptions are badly needed, and plantings made in the same area between 1934 and 1942 by the Civilian Conservation Corps (CCC) are probably the best source of such information.

This paper reports the 10-year results of a study installed in 1959 n the CCC plantings. Records available in 1959 indicated that at least 22,575 acres had been planted with pine by the CCC in north Mississippi to control erosion. Some of these plantations were on the Holly Springs National Forest; some were on private land. The original study was designed to determine how many acres of CCC plantations were still in existence and to obtain information on the early development of erosion control plantations.

METHODS

Y-LT foresters supervised by Southern Forest Experiment Station scientists established 199 permanent sampling points during 1959 in plantations 16 to 26 years old (Grosenbaugh, 1958). Ninety-nine of the points were in loblolly (*Pinus taeda* L.), 88 were in shortleaf (*P. echinata* Mill.), and 12 were in slash pine (*P. elliottii* Engelm.). Since the plots were selected by superimposing a grid on aerial photographs, numbers of plots are representative of the total area covered by each species. There were too few slash pine plots for any valid data analysis.

[1] Respectively, Softwood Management Specialist, Forest Resource Management Group, Southeastern Area State and Private Forestry, Alexandria, Louisiana, and Mathematical Statistican, Southern Forest Experiment Station, USDA Forest Service, New Orleans, Louisiana.

Site indices (base age 50) according to Coile and Schumacher's curves (1953) ranged from 30 to 105 for loblolly and 20 to 97 for short leaf. Basal areas were from 10 to 220 square feet per acre.

At each sample point trees for measurement were selected with a prism having a basal area factor of 10. D.b.h., form class ($\frac{\text{d.i.b. 17 feet}}{\text{d.o.b. at d.b.h.}}$), merchantable height (3-inch top d.i.b. for pulpwood and 6-inch top d.i.b. to nearest half log for sawtimber), and total height were recorded for each selected tree. The plots—minus some that were destroyed in the interim—were remeasured in 1964 and 1969 by the same procedures.

Soils were identified and the physical characteristics of the soil described for each plot by a soil scientist from the Soil Conservation Service. The soils within the study area are extremely varied. Of the original loblolly plots, 70 are on loessial soils and 20 are on Coastal Plain soils. Of the original shortleaf plots, 59 are on loessial soils and 28 on Coastal Plain soils. Nineteen plots are on mixed aluvium. The soil series most frequently encountered on the plots were Loring, 26 plots; Providence, 34; Grenada, 21; and Ruston, 19.

Volumes in rough cords per acre (in trees 4.6 inches and larger at d.b.h.) were determined from Minor's merchantable length table (1950). Point sample volumes were expanded to estimates per acre. The heights of the five trees of largest d.b.h. on each plot were used in determining site index when there were five or more. If there were fewer than five, all trees were used. Sawtimber volumes were determined for trees 9.6 inches and larger with tables (International ¼ inch rule) and procedures given by Mesavage and Girard (1946).

Growth analysis.—Since the major objectives were to survey rather than study growth, management practices were not restricted on the plots between measurements. The data, therefore, were not ideal for analysis of growth as a function of site, age, and residual density. However, it was reasoned that a subset of the data might be used in such an analysis of loblolly and shortleaf growth.

Data for a given period on a plot were included in these analyses if no significant cutting or catastrophic mortality occurred between the measurements defining the period. Serious mortality was usually associated with outbreaks of *Fomes annosus* (Fr.) Cke., and it seemed desirable to study the growth relations with this disrupting influence eliminated. To further restrict the disruption of trends by extraneous influences, only data from surviving trees were utilized. Values for volume growth, therefore, do not express a comprehensive penalty for mortality and represent results when significant mortality is avoided. Thus, the growth values are higher than net growth, but the analyses allow some evaluation of general trends otherwise clouded by mortality.

2

Periodic annual increment in cords for each of the two growth periods was related to site index and to functions of age and residual basal area at the beginning of the 5-year period. Data for a plot in a given period were included if there was no large-scale cutting or mortality between measurements. Only trees surviving the 5-year period were used in establishing the basal area and growth. Trees less than 4.6 inches d.b.h. were treated as having no merchantable volume.

There were 96 observations included for loblolly and 98 for shortleaf. The ranges in values of age and basal area of all trees at the beginning of the period along with site index were:

	LOBLOLLY		SHORTLEAF	
	Minimum	*Maximum*	*Minimum*	*Maximum*
Age (A)	18	31	17	29
Site index (S)	33	103	31	82
Basal area (D)	10	150	10	170

Analysis was done separately for each species. Regression equations involving all possible combinations of 11 independent variables (S, A, D, D², SA, SD, SD², AD, AD², SAD, SAD²) were screened (Grosenbaugh, 1967). The equation of best fit for loblolly included only three independent variables and explained 63 percent of the variation. Graphs of this equation indicated the trends were reasonable. For shortleaf pulpwood growth the equation of best fit included seven of the independent variables, explained 65 percent of the variation, but expressed illogical trends for low basal areas on poor sites. It was rejected. The equation form, with three variables, adopted for loblolly when fitted to the shortleaf data explained nearly as much of the variation (56 percent) as did the seven variable equation, and it was selected for presentation in this report.

Only the data of the second 5-year period of observation were considered in analyzing sawtimber growth because the stands were not of sufficient size in the first period. Again, only survivors were considered and plots were discarded for reasons previously mentioned. In addition, shortleaf plots on the poorest sites were ignored because trees had not reached sawtimber size (9.6 inches).

The resulting data sets included 46 observations for loblolly and 29 for shortleaf with the following extremes in site, age, and basal area:

	LOBLOLLY		SHORTLEAF	
	Minimum	*Maximum*	*Minimum*	*Maximum*
Age (A), years	23	30	23	29
Site index (S), feet	33	103	54	82
Basal area (D), square feet	20	150	20	160

3

The basal area is that of the trees which survived and was measured at the beginning of the 5-year period on all trees regardless of siz

Analyses for each species followed the same general pattern as tha for pulpwood growth. For sawtimber growth the equation of best f via the combinatorial screening process was accepted for each of th two species. The loblolly equation explained 56 percent of the variation and the shortleaf equation 68 percent.

SURVEY RESULTS

Survival.—In 1959 only 8,005 acres of the 22,575 acres of CCC pir plantations were still in existence. Possible explanations for the lo include wildfires, overgrazing, and clearing for pasture or row crop Eighty-eight percent of the loblolly plots and 92 percent of the shortlea had been planted at a spacing of 6 by 6 feet. Survival in 1959 for a spacings in uncut stands, 16 to 26 years old, averaged 38 percent fo loblolly and 48 percent for shortleaf. In 1959 in the plantations estab lished at 6- by 6-foot spacing, the average number of trees before thi ning was 494 for loblolly and 633 for shortleaf.

Ninety of the plots were on private land; of this number, 15 plo were destroyed during each of the two 5-year growth periods—a 10-ye loss of 33 percent. Most of the 30 plots that were clearcut were convert to pasture.

For the plots of this study it was impossible to detect a relationsh between site index and depth of the A horizon, depth to the least perm able layer, or texture of the least permeable layer. The most obvio explanation is that varying degrees of erosion confused the pictur Uniformity of soil, topography, and stand were not requirements of th sampling design. It is also possible that the site index curves used a not entirely suitable for young stands.

Stand condition.—Most of the study plantations were establishe in eroding fields on which only a few hardwood sprouts were growin As the years passed, hardwoods invaded the understory in many plot particularly those in the minor bottoms and on lower slopes. On son plots a luxuriant growth of kudzu, muscadine, or cow-itch vines deve oped. In 1959, 19 percent of the plots needed release from hardwo competition, and in 1964, 16 percent of the plots needed release.

In 1959, 42 percent of the plots had basal areas of 110 square fe per acre or more. Pulpwood markets improved and in 1964 only 31 pe cent had basal areas of 110 or more and in 1969 but 25 percent. Throug out the study, 22 to 26 percent of the plots have had basal areas too lo to adequately protect or fully utilize the site. Many of the understock plantations were on severely eroded sites where initial survival h been poor.

4

Twenty-nine percent of the study plots had been thinned prior to 1959; the work was often haphazard. Loblolly stands were thinned 4 or 5 years sooner than shortleaf. Thinning on some plots was from above and on others from below. Between 1959 and 1964 the average thinning removed 38 square feet of basal area or 6.2 cords per acre. During the next 5 years, 32 percent of the stands were thinned, with the average cut removing 39 square feet of basal area or 8.1 cords per acre. Some plots have yet to be thinned.

GROWTH AND DEVELOPMENT

Individual Tree Development

Stand prescriptions are frequently based upon assumptions about growth responses of the individual trees. Ten-year growth records are available on 444 loblolly and 541 shortleaf pines growing in the plantations. Some of these trees were affected by thinning. It should be understood that the species growth differences are confounded by earlier and more extensive thinning in the loblolly than in the shortleaf pine plantations. Table 1 shows average performance by diameter class and species.

Table 1.—*Ten-year growth of loblolly and shortleaf pines in CCC plantings in northern Mississippi*

Species and 1959 d.b.h. (inches)	D.b.h. growth	Volume increase
	Inches	*Percent*
Loblolly pine		
4	0.9	142
5	1.0	164
6	1.7	254
7	1.9	294
8	2.0	217
9	2.2	210
10	2.6	240
11	2.6	260
12	2.6	222
Shortleaf pine		
4	0.8	325
5	1.0	196
6	1.2	199
7	1.5	196
8	1.8	247
9	1.6	235
10	1.9	189
11	2.0	190

5

The 4- and 5-inch loblolly and shortleaf pines were largely in t̶ suppressed and intermediate crown classes and increased in diamet̶ very slowly. Even the larger trees, which were in the dominant crov class, did not increase in diameter at a rapid rate. Volumes of trees all sizes increased markedly during the 10-year period.

Table 2 shows the improvement in form class, merchantable heig̶ and total height during the 10-year period.

In 1959 the average loblolly was 6.6 inches in d.b.h.; the avera̶ shortleaf, 5.3 inches. In 10 years these trees were 8.5 and 6.3 inches d.b.h., respectively. Form class of the average loblolly increased 7 pe̶ centage points, merchantable height 13 feet, and total height 16 fe̶ Corresponding increases for the average shortleaf were 13 percenta̶ points in form class, 9 feet in merchantable height, and 15 feet in tot̶ height. Each increase of 1 percentage point in form class results in ̶ increase of approximately 2 percent in volume.

Due to thinning and mortality as well as growth, the average me̶ chantable loblolly in 1969 was 8.8 inches in d.b.h. with a form class of 7̶ merchantable height of 47 feet, and total height of 65 feet. Similarly, t̶ average merchantable shortleaf in 1969 was 7.2 inches in d.b.h. with

Table 2.—*Ten-year improvement in form class and height*

D.b.h.	Form class		Merchantable height		Total height	
1959 (inches)	1959	1969	1959	1969	1959	196̶
	— — Percent — —		— — Feet — —		— — Feet — —	
LOBLOLLY PINE						
4	53	59	13	18	28	38
5	58	65	19	24	32	44
6	64	72	25	35	38	53
7	69	76	29	42	43	59
8	70	77	32	45	46	62
9	71	77	34	47	48	65
10	74	79	36	51	50	68
11	73	80	38	51	51	69
12	76	80	38	52	54	72
13	71	77	39	51	53	69
14-16	70	78	39	53	53	70
SHORTLEAF PINE						
4	49	64	8	16	29	40
5	54	67	15	24	31	46
6	61	72	21	32	35	51
7	64	74	25	36	38	54
8	67	76	28	39	41	56
9	69	77	30	40	42	56
10	72	79	32	41	44	59
11	67	78	32	41	44	58

6

form class of 74, merchantable height of 36 feet, and total height of 54 feet. On quite similar sites, the average loblolly was 11 feet taller at age 31 than the average shortleaf.

Growth Analysis

Pulpwood growth.—The two equations selected for periodic annual increment (PAI) of pulpwood were:

Loblolly

$$\text{PAI (cords/acre)} = 0.147 + 3.5653\,(SD)\,(10^{-4}) - 6.9474\,(D^2)\,(10^{-5}) - 1.8156\,(AD)\,(10^{-4})$$

Shortleaf

$$\text{PAI (cords/acre)} = 0.043 + 1.8827\,(SD)\,(10^{-4}) - 8.4918\,(D^2)\,(10^{-6}) + 1.3440\,(AD)\,(10^{-4})$$

Table 3 expresses these equations for several ages, stand densities, and site indices. Figures 1 and 2 depict the equations at initial age 20 for various sites and initial basal areas. Each of the lines is concave downward but only in the case of loblolly on the poorer sites does the growth rate relative to basal area culminate within the range of densities likely to be realized. Similar graphs across the range of ages included in the data show the same general pattern but a decline in growth rate with age for loblolly and an increase for shortleaf with increasing age. This difference may be caused by shortleaf expressing a surge of

Table 3.—*Periodic annual increment of pulpwood (cords/acre/year)*

Site index and age	Basal area (sq. ft.)											
	Loblolly pine						Shortleaf pine					
	40	60	80	100	120	140	40	60	80	100	120	140
20	0.461	0.535	0.553	0.515	0.422	0.274	0.438	0.626	0.806	0.980	1.147	1.307
25	.425	.480	.480	.424	.313	.146	.465	.666	.860	1.047	1.228	1.401
30	.388	.426	.408	.334	.204	.019	.492	.706	.914	1.114	1.308	1.495
20	.604	.749	.838	.872	.850	.773	.514	.738	.957	1.168	1.373	1.571
25	.567	.694	.765	.781	.741	.646	.540	.779	1.010	1.235	1.454	1.665
30	.531	.640	.693	.690	.632	.518	.567	.819	1.064	1.303	1.534	1.759
20	.746	.962	1.123	1.228	1.279	1.272	.589	.852	1.107	1.356	1.599	1.834
25	.710	.908	1.051	1.138	1.169	1.145	.616	.982	1.161	1.424	1.680	1.928
30	.674	.854	.978	1.047	1.060	1.018	.642	.932	1.215	1.491	1.760	2.022
20	.889	1.176	1.408	1.585	1.706	1.771	.664	.964	1.258	1.545	1.825	2.098
25	.853	1.122	1.336	1.494	1.597	1.644	.691	1.005	1.312	1.612	1.905	2.192
30	.816	1.068	1.263	1.403	1.488	1.517	.718	1.045	1.366	1.679	1.986	2.286
20	1.032	1.390	1.694	1.941	2.134	2.270	.739	1.077	1.409	1.733	2.051	2.362
25	.995	1.336	1.621	1.851	2.025	2.143	.766	1.118	1.462	1.800	2.131	2.456
30	.959	1.281	1.548	1.760	1.916	2.016	.793	1.158	1.516	1.867	2.212	2.550
20	1.174	1.604	1.979	2.298	2.561	2.769						
25	1.138	1.550	1.906	2.207	2.452	2.642						
30	1.102	1.495	1.834	2.116	2.344	2.515						
20	1.317	1.818	2.264	2.654	2.989	3.268						
25	1.280	1.764	2.192	2.564	2.880	3.141						
30	1.244	1.709	2.119	2.473	2.771	3.014						

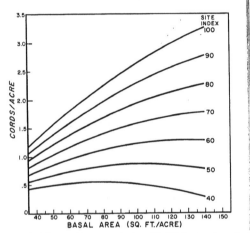

Figure 1.—*Periodic annual increment of pulpwood by loblolly pine at age 20 on various sites.*

ingrowth over the merchantability threshold during the measuremen interval. Loblolly reached this stage at an earlier age. Shortleaf appear to be a "late bloomer," able to tolerate higher densities than loblolly.

One is forced to conclude from these results that pulpwood growth between ages 20 and 35 years, was enhanced by retaining very high densities. Partial cutting may reduce growth as well as increase the risk of losses to *Fomes annosus*.

Sawtimber growth.—The two equations of best fit were:

Loblolly
$$PAI \text{ (board feet/acre)} = 163.23 + 1.3968\,(D^2) - 9.0365\,(SD)\,(10^{-1})$$
$$- 1.0659\,(SD^2)\,(10^{-2}) - 6.1462\,(AD^2)\,(10^{-2})$$
$$+ 3.9331\,(SAD)\,(10^{-2}) + 4.804\,(SAD^2)\,(10^{-4})\,)$$

Shortleaf
$$PAI \text{ (board feet/acre)} = -241.37 - 9.201\,(D) + 1.8106\,(SD^2)\,(10^{-3})$$
$$+ 1.0931\,(AD) - 9.5749\,(AD^2)\,(10^{-3})$$

These equations are shown in figures 3 and 4 for an initial age o 28 years. Graphs for other ages within the range of the data have the same general appearance. The equations are also expressed in table 4 for several ages, stand densities, and site indices.

8

ons

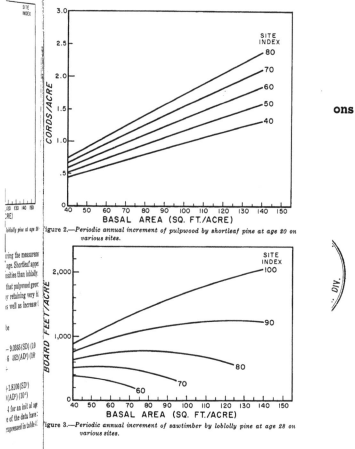

Figure 2.—*Periodic annual increment of pulpwood by shortleaf pine at age 20 on various sites.*

Figure 3.—*Periodic annual increment of sawtimber by loblolly pine at age 28 on various sites.*

9

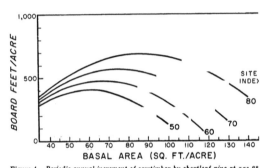

Figure 4.—*Periodic annual increment of sawtimber by shortleaf pine at age 28 various sites.*

Table 4.—*Periodic annual increment of sawtimber (board feet/acre/year)*

Site index and age		Basal area (sq. ft.)					
		40	60	80	100	120	
LOBLOLLY PINE							
40	25	154	...	...	...	...	
	30	130	...	...	...	...	
50	25	207	142	...	...	...	
	30	301	58	...	...	...	
60	25	260	239	170	...	...	
	30	471	359	69	...	...	
70	25	314	335	320	269	...	
	30	642	660	531	254	...	
80	25	367	431	470	484	472	
	30	812	961	992	906	702	
90	25	421	528	620	698	762	
	30	983	1,262	1,453	1,557	1,574	1
100	25	474	624	771	713	1,052	1
	30	1,154	1,563	1,914	2,209	2,446	2
SHORTLEAF PINE							
50	25	246	310	256	83	...	
	30	389	466	387	151	...	
60	25	275	376	372	264	51	
	30	417	531	503	332	18	
70	25	303	441	488	445	312	
	30	446	596	619	513	278	
80	25	332	506	604	626	573	
	30	474	662	734	694	539	

10

PAI for sawtimber relative to basal area culminates in almost all situations for both species, because of the more stringent merchantability limits for sawtimber than for pulpwood. Failure to thin dense stands results in sawtimber growth rates considerably below the maximums in the age span considered.

The culmination pattern is more dramatic for shortleaf. For both species culmination occurs at a higher density as site index increases.

When a large number of trees is established to stabilize the soil and sawtimber production at an early age is demanded, there obviously must be some thinning. This cutting will concentrate growth on fewer trees. Higher residual densities should be retained on the better sites.

The age variable was included in these sawtimber equations. However, the plots were nearly all in the 25 to 30 year age range at the beginning of the 5-year growth period. Extrapolation much beyond this range is not recommended. Within the age span of interest, the growth rate associated with the optimum density for a given site tends to be greater as age increases. Also, for a given site there is a trend of realizing the maximum growth at a lower density as age progresses. This property will have to be investigated with a more comprehensive data set. With the exception of the age variable, the general pattern of the growth-site-density relationship for both species is identical to that depicted by Nelson et al. (1963) for loblolly. The equation given here for loblolly predicts growth values higher than Nelson et al.'s. This result is not surprising in that those authors were considering net growth and their data set emphasized older age classes and naturally established stands. Their formulation specified a decline in growth with age except in young stands with many trees just crossing the merchantability limit. The trend toward increasing growth with age in our results probably expresses such a surge of trees crossing the merchantability threshold.

DISCUSSION

Thinning is of doubtful economic benefit when pulpwood is the only foreseeable product, but if the owner insists upon an early return from his pine plantation a single thinning is certainly preferable to liquidation of the stand. A commercial thinning becomes feasible at age 15-20 for loblolly and 20-25 for shortleaf pine in the area. Thinning should be primarily from below, and about one-third of the basal area and up to half the stems will probably have to be removed to make the return attractive. It should be understood that such a thinning may reduce pulpwood growth by as much as ½ cord per acre per year, but it will result in larger, more cheaply logged trees for the final harvest. If a market for small saw logs develops before the final harvest, the presence of large, well-formed trees would considerably improve the stumpage price received.

Board-foot volume growth culminates within the range of basal areas sampled except for loblolly pine on site 100. If the primary objective is to grow sawtimber, therefore, basal areas will have to be limited through thinning. In the process, some pulpwood yields will be sacrificed.

Where sawtimber is the goal of management, it appears advisable to thin loblolly plantations to 120 square feet of basal area per acre on sites of 95 or better. On sites 80 and 90 the basal area after thinning should approximately equal the site index, and on sites 60 and 70 basal area equal to the site index minus 10 should be left. On shortleaf pine sites of 60 to 90, cutting back to a basal area equal to the site index is recommended at age 25 to 30. These prescriptions will tend to minimize investment in growing stock, maximize board-foot volume growth, protect the site, and maintain an acceptable rate of diameter growth.

For small landowners, such as those on the Y-LT area, once pine plantations are established it is highly practical to prolong the rotation as long as a good rate of growth is maintained—500 board feet or 1.5 cords per acre per year.

The fact that incidence of *Fomes annosus* in the study plots was high enough to limit the plots available for growth analysis is evidence that this disease must be considered a serious threat to pine plantations in north Mississippi. During the 1964 measurements, 41 percent of the thinned plots had tree mortality attributable to *Fomes* root rot. The picture in 1969 was clouded by the fact that several of the plots with previous infections had been clearcut or lost. Virtually all of the infected plots were on deep sands or deep sandy loams. Those soils with clay near the surface had little *Fomes* infection.

We suggest the development of a classification system similar to that employed by the Virginia Division of Forestry (Morris and Frazier, 1966); root rot potential of the sites in north Mississippi could be at least roughly defined. High risk areas would be assigned to the production of pulpwood, low risk areas to the production of sawtimber and the associated interim cuttings. Prediction of root rot potential would also allow more selective prescription of preventive treatment.

Borax sprinkled on the fresh stump surface immediately after a tree is cut has given the best and most consistent control of *Fomes annosus* infection on artificially and naturally inoculated stumps of slash, loblolly, and white pine (Phelps et al., 1971). It is difficult to get cutters to use borax, but a premium could be paid for treating stumps.

Thinning in the summer has also reduced infection by *Fomes* because the spore production approaches 0 when the maximum temperature approaches 90°F and the mean minimum-maximum temperature approaches 70°F (Ross, 1973). These findings suggest that on *Fomes* prone sites, such as sandy loams and loamy sands with A horizons 8 t

12

n the range of basal
If the primary obje
ill have to be limite
yields will be sacri

, it appears advisabl
asal area per acre e

l area after thinning
sites 60 and 70 has
ft. On shortleaf pir
al to the site index
will have to minimiz
: volume growth, ye
iameter growth.

Y-LT area, once pr
) prolong the rotatir
500 board feet or l

the study plots w
h analysis is eviden
at to pine plantatio
nts, 41 percent of tl
Fomes root rot. Th
eral of the plots wt
ially all of the infec
se soils with clay st

ion system similar
try (Morris and Fr
h Mississippi could
e assigned to the p
tion of sawtimber a
ot rot potential wer
ive treatment.

: immediately after
ent control of Fm
inoculated stumps
). It is difficult tol
for treating stump

ifection by Fomes
ie maximum temper
iximum temperat
iggest that on Fm
s with A horizons f

2 inches deep, thinning be done in summer.

In northern Mississippi, loblolly pine is now regarded as more desirable than shortleaf pine for planting. Loblolly grows well on a wider range of sites, casts more needle litter to hold soil in place, and grows faster when young. This does not mean that owners of shortleaf pine plantations should consider liquidation and conversion to loblolly pine.

It is true that at a given location the site index of loblolly pine is generally higher than that of shortleaf. The difference may range from 3 to 20 feet at age 50, depending upon whose formula is accepted (Bassett and Beene, 1967; Coile, 1948; Turner, 1938; Zahner, 1958). Curves in USDA Miscellaneous Publication 50 (U. S. Department of Agriculture, 1929) show, however, that at the same site index shortleaf pine produces more cubic volume growth than loblolly pine after age 20. If the loblolly site index is 10 points higher than that of shortleaf, the mean and periodic annual cubic-foot increments of shortleaf will be higher than those of loblolly from age 25 on (U. S. Department of Agriculture, 1929). The superior cubic volume growth of shortleaf is due largely to its propensity to carry more stems and higher basal areas per acre. Our data confirm this observation. Shortleaf pine grows so well after age 20 that a landowner cannot afford to clearcut it and replace it with loblolly pine prior to age 30-35.

The implications in the loss of one-third of the privately owned CCC plantations on erosion-prone soils in 10 years are serious. All advisory agencies should stress the need for sound land-use planning by owners. Row crops should be confined to flat lands and pastures to the gentle slopes. Steep slopes should be kept in trees. The majority of sites planted to pine with Federal Government assistance on the Yazoo-Little Tallahatchie watersheds are unsuitable for pasture because of poor soil or steep slope.

The hydrologic consequences of removing pine plantations from severely eroded lands after one rotation are largely unknown. Research has demonstrated that the protective forest floor dissipates rapidly after the crown cover is removed. Until the combination of this effect and that of soil disturbance are quantified in terms of water quality, water quantity, and subsequent erosion, recommended practices established elsewhere should be applied. Logging during summer and fall when soils are normally driest will reduce formation of ruts and minimize surface disturbances that could develop into gullies. Truck roads should be located on ridges, skidding should be on contour or uphill, and the length of skid trails should be minimized. Planting trees as soon as possible after final harvest will minimize the time soils are left exposed and will reduce competition from hardwoods and minor vegetation that may seed on the newly logged area. The erosion hazard can be further reduced by constructing waterbars on roads and skid trails and by

13

fertilizing and seeding disturbed areas with a temporary cover crop. Healed gullies and drainage channels should not be disturbed.

LITERATURE CITED

Bassett, J. R., and Beene, J. T.
  1967. Loblolly pine site quality on four loess soils. USDA For. Serv. Res. Note SO-68, 6 p. South. For. Exp. Stn., New Orleans, La.

Coile, T. S.
  1948. Relation of soil characteristics to site index of loblolly and shortleaf pines in the lower Piedmont Region of North Carolina. Duke Univ. Sch. For. Bull. 13, 78 p.

Coile, T. S., and Schumacher, F. X.
  1953. Site index of young stands of loblolly and shortleaf pines in the Piedmont Plateau Region. J. For. 51: 432-435.

Grosenbaugh, L. R.
  1958. Point-sampling and line-sampling: probability theory, geometric implications, synthesis. USDA For. Serv. South. For. Exp. Stn. Occas. Pap. 160, 34 p.

Grosenbaugh, L. R.
  1967. REX—Fortran-4 system for combinatorial screening or conventional analysis of multivariate regressions. USDA For. Serv. Res. Pap. PSW-44, 47 p. Pac. Southwest For. and Range Exp. Stn., Berkeley, Calif.

Mesavage, C., and Girard, J. W.
  1946. Tables for estimating board-foot volume of timber. USDA For. Serv., 94 p.

Minor, C. O.
  1950. Form class volume tables for use in southern pine pulpwood timber estimating. La. Agric. Exp. Stn. Bull. 445, 39 p.

Morris, C. L., and Frazier, D. H.
  1966. Development of a hazard rating for *Fomes annosus* in Virginia. Plant Dis. Rep. 50: 510-511.

Nelson, T. C., Trousdell, K. B., Brender, E. V., and Lotti, T.
  1963. Board-foot growth of loblolly pine as related to age, site, and stand density. J. For. 61: 120-123.

Phelps, W. R., Wolfe, R. D., and Laird, P. P.
  1971. Evaluation of stump treatments for control of *Fomes annosus*. USDA For. Serv., Southeast. Area State and Priv. For., 13 p.

14

U.S. Department of Agriculture
Forest Service Research Paper SO-95

W.
Fomes annosus in the Southeastern United States: relation of
environmental and biotic factors to stump colonization and
losses in the residual stand. USDA Tech. Bull. 1459, 26 p.

L. M.
Some profile characteristics of the pine-growing soils of the
Coastal-Plain region of Arkansas. Ark. Agric. Exp. Stn. Bull.
361, 52 p.

epartment of Agriculture
Volume, yield, and stand tables for second-growth southern
pines. USDA Misc. Publ. 50, 202 p.

ons

R.
Site-quality relationships of pine forests in southern Arkansas
and northern Louisiana. For. Sci. 4: 162-176.

Williston, Hamlin L., and Dell, T. R.

    1974. Growth of pine plantations in north Missis-
        sippi. South. For. Exp. Stn., New Orleans, La.
        15 p. (USDA For. Serv. Res. Pap. SO-94)

Performance of loblolly and shortleaf pines planted by the Civilian
Conservation Corps in northern Mississippi provides guides for
managing flood control plantings made in the same area since
World War II.

Additional keywords: *Pinus taeda, P. echinata*, plantation management, plantation growth, plantation yields, *Fomes annosus.*

U.S. Department of Agriculture
Forest Service Research Paper SO-95

.ons

Recurring Southern Pine Beetle Infestations
Near Oakdale, Louisiana

Peter L. Lorio, Jr., and William H. Bennett

CLEMSON UNIVERSITY LIBRARY
OCT 20 1975
S^l., TECH. & AGR. DIV.

Southern Forest Experiment Station
Forest Service
U.S. Department of Agriculture
1974

Acknowledgment

The authors thank the Southwestern Improvement Company for its cooperation and Mr. Alvin Laird for his assistance.

Recurring Southern Pine Beetle Infestations Near Oakdale, Louisiana

Peter L. Lorio, Jr., and William H. Bennett[1]

Recurring southern pine beetle (Den-roctonus frontalis *Zimm.) infestations were associated with high stand density, a large proportion of pine sawtimber, nearness to an earlier infestation, and lightning strikes. Intensive chemical control applied on 10,095 trees in 2,959 infestations apparently maintained an endemic situation over a 5-year period. Following abandonment of control, epidemic status developed within 2 years.*

Additional keywords: Dendroctonus fron-talis, *D.* terebrans, *stand density, stand composition, soil drainage.*

Attempts to control the southern pine beetle (SPB) (*Dendroctonus frontalis* Zimm.) have generally been sporadic and ineffectual; some researchers even suggest that chemical control disrupts SPB populations and encourages epidemics (Williamson, 1971; Williamson and Vité, 1971). Control would be facilitated if beetle attacks can be linked to specific forest conditions and environmental factors. Several potential factors were identified during the present study in which direct chemical controls were conscientiously applied for 5 years.

Relative isolation from other known SPB populations, interest of the major landowner in attempts to control infestations, ready access to control records, and availability of a recently completed forest type map of the study area facilitated this study.

Methods

Some 31,500 acres belonging to the Southwest-Improvement Company were selected for the study. The land is near Oakdale, Louisiana, and includes portions of the Calcasieu River bottom,

Lorio is Principal Soil Scientist, and Bennett, now retired, was Principal Entomologist at the Southern Forest Experiment Station, Pineville, Louisiana 71360. Robert C. Thatcher, Principal Entomologist at the Southern Forest Experiment Station, helped to plan the study.

East Bay, and West Bay. Both East and West Bays are flat and poorly drained; widespread inundation is common after heavy rains. Both areas are low, fluviatile terraces outcropping on the coastwise Montgomery terrace (Holland et al., 1952). The predominant soils are Caddo and Guyton silt loams, Typic Glossaqualfs, on flat intermound areas, and Messer silt loam, a Typic Glossudalf, on low circular mounds. In the Calcasieu River bottom, the Bibb series, a Typic Fluvaquent, predominates. Further information on the soils and water regime is provided by Lorio and Hodges (1971). Lorio et al. (1972) described the characteristics of pine tree root systems on these sites.

Early in the century the study area supported mostly hardwoods. In 1927-28 the merchantable trees were clearcut by railway logging (personal communication from Alvin Laird, formerly with Southwestern Improvement Co., now retired). Few hardwood trees of any size remained standing, and natural seeding from pines converted large portions of this forest to pure pine and pine-hardwood mixtures. However, as late as 1940 pines were hardly prominent (Holland et al., 1952). By the fall of 1963, about 49 percent of the area was in pine sawtimber; but only 16 percent had a density $\geq$ 5 M b.f. per acre (designated as Type P11 in the forest type map) (tables 1 and 2).

The area has a history of black turpentine beetle (*D. terebrans* (Oliv.)) problems, beginning in the late 1950's. The first SPB infestation was discovered in the West Bay Game Management Area in October 1963; at that time the nearest known SPB activity was about 75 miles to the west in southeast Texas.

An intensive control program began in April 1964 and continued through March 1969. Every green infested tree was cut, and the entire bole was sprayed with a 0.5 percent solution of the gamma isomer of benzenehexachloride (BHC)

Table 1.—*Forest types represented in area in 1963*[1]

Size and type	Percent of total area	
Sawtimber	92	
Pine		49
Pine-hardwood		36
Hardwood		7
Large pulpwood	4	
Pine		3
Pine-hardwood		<1
Hardwood		<1
Other	4	
Total	100	

[1] Based on type map compiled by Pomeroy and McGowin, Monticello, Arkansas.

Table 2.—*Distribution of sawtimber by type and density classes in 1963*[1]

Type	Density class (M b.f.)				
	<1.5	1.5-3	3-5	≥5	Total
	— — — *Percent of total acreage* — — —				
Pine	5	14	14	16	49
Pine-hardwood	10	25	1	0	36
Hardwood	4	3	0	0	7
Total	19	42	15	16	92

[1] Based on type map compiled by Pomeroy and McGowin, Monticello, Arkansas.

in diesel oil; this was the procedure then recommended by the Division of Forest Pest Control, Southeastern Area, State and Private Forestry Branch.

Infestations were located in biweekly flights by Louisiana Forestry Commission personnel from April until fall coloration of broadleaved species made aerial detection impractical. Field checks and ground observations were made throughout the year.

The primary units for locating infestations were 40-acre blocks. With a dot grid and a forest type map,[2] the number of acres in each forest type-size density class was obtained. The following information was recorded for each of 785 blocks:

1. Identification and location.
2. Total dot count for block (1.1 acres per dot).

[2] A forest type map of Southwestern Improvement Company lands was prepared in 1963 by Pomeroy and McGowin of Monticello, Arkansas.

3. Dot count for each forest type-size-density class in block.
4. Distance, in 20-chain units, from the original infestation.
5. Monthly totals of controlled infestations and of infested trees from April 1964 through March 1969.
6. Number of controlled infestations associated with lightning (lightning-strike spots) in each month from April 1965 through March 1969.

Results

A total of 10,095 trees in 2,959 infestations was treated during the 5-year period. Of the total, 2,811 infestations occurred in West Bay, 98 in East Bay, and 50 in the Calcasieu River bottom. The greatest number of infestations were treated in July, August, June, and October, in that order (fig. 1). The total number of trees treated peaked in June and July.

About 75 percent of all infestations occurred in 238 of the blocks; 219 blocks had none at all (fig. 2). The highest frequency of infestations per block occurred in Section 7, T3S, R4W in West Bay. Here two blocks with 77 percent of their areas in Type P11 had 27 infestations each —the maximum recorded.

Lightning was associated with 29 percent of the infestations treated from April 1965 through March 1969. Lightning-strike spots were most frequent in August and nearly absent in the winter and early spring (fig. 3); their distribution was similar to that of all infestations.

Correlation analyses yielded only one apparent relationship between number of infestations and forest type-size-density classes: percent of block area in Type P11 was significantly correlated with number of infestations per acre (r= 0.43). The coefficient was not improved by deleting infestations associated with lightning strikes.

Several regressions were developed with number of infestations per acre (excluding lightning-strike spots) as the dependent variable (y) and percent of block area in pine or pine-hardwood type and distance from original infestation as the independent variables. The regression yielding the largest coefficient of determination (R^2=0.33) and a standard error of estimate of 0.069, (P=0.01), was

2

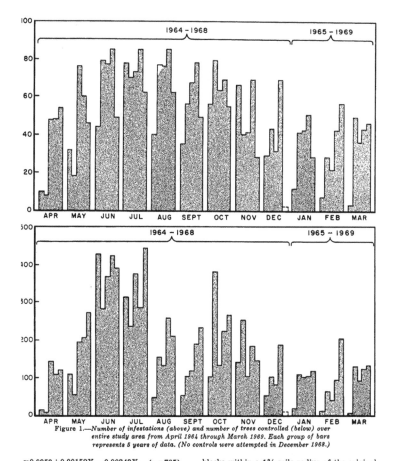

Figure 1.—*Number of infestations (above) and number of trees controlled (below) over entire study area from April 1964 through March 1969. Each group of bars represents 5 years of data. (No controls were attempted in December 1968.)*

=0.0953+0.00152X₁−0.00343X₂ (n=785), ere X₁=percent of block area in Type P11 '₂=distance from original infestation. 'o further explore the relationship between icks and stand characteristics, the average aber of infestations per acre was plotted inst percent of block area in Type P11 for a year (fig. 4). Data were limited to 161

blocks within a 1¾-mile radius of the original infestation. A linear regression of infestations per acre on percent of block area in Type P11 for the 5 years of data was not significant; but when the regression was computed only for those blocks with 41 percent or more of their areas in Type P11, it was significant (P=.05, n=34). The first 2 years of data seemed to contribute

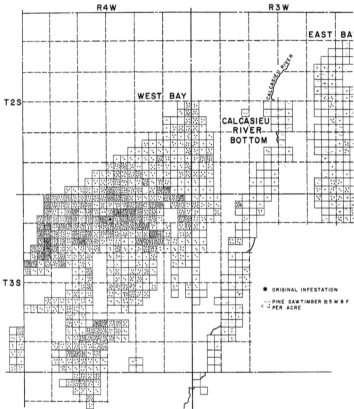

Figure 2.—*Distribution of total infestations from April 1964 through March 1969 and location of stands of pure pine sawtimber >5 M b.f. per acre. Each dot represents one infestation.*

most to this relationship (fig. 4).

For the same 161 blocks, average number of infestations per acre was plotted against distance from original infestation (fig. 5). In the first 2 years, infestations tended to recur near the origin, but by the third year this tendency was beginning to disappear.

Discussion

Stand composition (pine, pine-hard hardwood), size, and density appeared to portant factors in recurring SPB infesta Usually, blocks with a high frequency of in tions had large percentages of their areas i

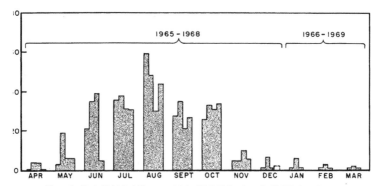

Figure 3.—*Controlled infestations associated with lightning from April 1965 through March 1969. Each group of bars represents 4 years of data. (No controls were attempted in December 1968.)*

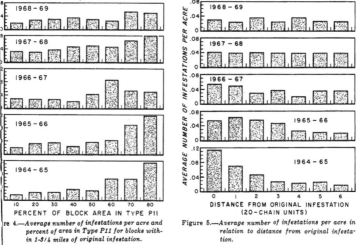

re 4.—*Average number of infestations per acre and percent of area in Type P11 for blocks within 1-3/4 miles of original infestation.*

Figure 5.—*Average number of infestations per acre in relation to distance from original infestation.*

sawtimber ≥ 5 M b.f. per acre.
odges and Pickard (1971) studied trees in same area and concluded that trees struck lightning provide a favorable attack and d environment for southern pine and *Ips*

beetles. In this study struck trees were associated with 29 percent of the infestations over a 4-year period, and in August of 1965 a high of 77 percent of the infestations treated included struck trees.

The very low average number of trees per infestation (3.4) and the maintenance of endemic conditions during the study period apparently resulted from the intensity of the control effort. Chemical control was abandoned in 1969; by November 1971 surveys indicated at least 49 infestations in West Bay with an average of 34 trees each (personal communication from B. F. Griffin, District Forester, La. For. Comm., Oberlin, La.). The two largest infestations contained 600 and 300 trees. This activity marked the beginning of an epidemic that developed through the mild 1971-72 winter.

These findings differ markedly from those of Williamson (1971) and Williamson and Vité (1971). They concluded that chemical control actually dispersed SPB populations and encouraged epidemics in east Texas in the 1960's. However, the study area contained millions of acres owned by many different people; perhaps the recommended controls were not being consistently applied by all landowners. In addition, the Toledo Bend and Sam Rayburn Reservoirs were developed during this time, and the drastic change in water regime contributed significantly to the outbreak of beetle attacks. In contrast, our study concerned lands owned by only one company, which had committed itself to a program of intensive control. Full records of the control operations were available, and no environmental disruptions interfered with our findings.

Literature Cited

Hodges, J. D., and Pickard, L. S.
  1971. Lightning in the ecology of the southern pine beetle *Dendroctonus frontalis* (Coleoptera: Scolytidae). Can. Entomol. 103: 44-51.

Holland, W. C., Hough, L. W., and Murray, G. E.
  1952. Geology of Beauregard and Allen Parishes. La. Dep. Conserv., Geol. Surv. Bull. 27, 224 p.

Lorio, P. L., Jr., and Hodges, J. D.
  1971. Microrelief, soil water regime, and loblolly pine growth on a wet, mounded site. Soil Sci. Soc. Am. Proc. 35: 795-800.

Lorio, P. L., Jr., Howe, V. K., and Martin, C. N.
  1972. Loblolly pine rooting varies with microrelief on wet sites. Ecology 53: 1134-1140.

Williamson, D. L.
  1971. Management to reduce pine beetle infestations. For. Farmer 30 (4): 6-7, 18.

Williamson, D. L., and Vité, J. P.
  1971. Impact of insecticidal control on the southern pine beetle population in east Texas. J. Econ. Entomol. 64: 1440-1444.

Lorio, Peter L., Jr., and Bennett, William H.

    1974. Recurring southern pine beetle infestations near
        Oakdale, Louisiana. South. For. Exp. Stn., New
        Orleans, La. 6 p. (USDA For. Serv. Res. Pap.
        SO-95)

Recurring southern pine beetle (*Dendroctonus frontalis*
Zimm.) infestations were associated with high stand density, a large proportion of pine sawtimber, nearness to
an earlier infestation, and lightning strikes. Intensive
chemical control applied on 10.095 trees in 2.959 infestations apparently maintained an endemic situation over a
5-year period. Following abandonment of control, epidemic
status developed within 2 years.

Yield Tables and Stand Structure For LOBLOLLY PINE Plantations In Tennessee, Alabama, and Georgia Highlands

GLENDON W. SMALLEY
AND
ROBERT L. BAILEY

Southern Forest Experiment Station
Forest Service
U. S. Department of Agriculture

1974

Yield Tables and Stand Structure For Loblolly Pine Plantations In Tennessee, Alabama, and Georgia Highlands

Glendon W. Smalley[1]
and
Robert L. Bailey[2]

᾽ than 300,000 acres of abandoned fields
᾽en planted with loblolly pine (*Pinus taeda*)
.he four highland regions (fig. 1) of Ten-
Alabama, and Georgia. To assist forest-
l landowners in making decisions about
.o harvest these plantations we present
;howing expected cubic-foot yields, basal
verage height, and number of stems per
᾽ 1-inch diameter classes for all combina-
᾽:

anting density—500 to 2,500 trees per
acre in increments of
250

te index (base age 25)—40 to 70 feet in
steps of 10

ᵧe from seed—10 to 40 years, by intervals
of 5.

ᵢods of collecting the data and construct-
tables are discussed in the sections imme-
following. The tables were derived
h a system of equations to predict number
ᵢiving stems per acre, diameter distribu-
ite index (Smalley and Bower 1971), tree
, and volume per tree (Smalley and
1968).

ᵢers interested only in the tables and a
ᵢion of growth trends should turn directly
section on Stand and Stock Tables, p. 7.
for combinations of variables not de-
᾽—or for combinations of surviving trees
᾽e, site, and age—can be obtained from the
ᵣn Forest Experiment Station, Biometrics
ᵢ, T-10210 Federal Building, 701 Loyola
᾽, New Orleans, Louisiana 70113. Users
mputer facilities may wish to acquire the
ms, which are written in FORTRAN.

ntist at the Silviculture Laboratory, maintained at Sewanee,
᾽e, by the Southern Forest Experiment Station in cooperation
University of the South.
.tical Statistician, Biometrics Branch, Southern Forest Ex-
. Station, New Orleans, La.

A similar publication is available for shortleaf
pine (Smalley and Bailey 1974).

PLANTATION MEASUREMENTS

Diameter, height, and age data were recorded
in 302 loblolly plantations distributed almost
evenly in the four physiographic regions.

All plantations sampled were at least 10 years
old from seed and had not been thinned, burned,
or pruned. Survival and tree distribution were
judged reasonably good, and no reinforcement
planting had been done. The plantations bore no
evidence of severe damage by insects or disease,
and ice or snow had not deformed more than 25
percent of the trees. Most of the stands contained
some competing vegetation, but quantity of competition was not a criterion except as it influenced pine survival and distribution.

One plot, usually 0.05 acre in size, was established in each plantation. The diameter at breast
height (D_i) of each tree on the plot was measured to the nearest 0.1 inch. The total height of
every fourth tree in each 1-inch diameter class
was measured to the nearest foot. Average total
height of dominants and codominants (H) was
calculated for each plot. Age from seed (A) was
determined by counting annual rings at ground
level on a felled tree. Mortality was estimated
from the presence of standing dead trees, rotten
stumps, fallen trees, and obvious gaps in the
planting rows. This estimate was added to the
count of live trees to obtain the number of planted trees.

Data from 267 of the 302 plots were used to
screen and fit mathematical models to estimate
survival, heights of individual trees (H_i), and
relative diameter distributions as functions of
age, number of trees planted per acre (T_p), number of trees surviving per acre (T_s), diameter
class midpoints, the midpoint of the largest

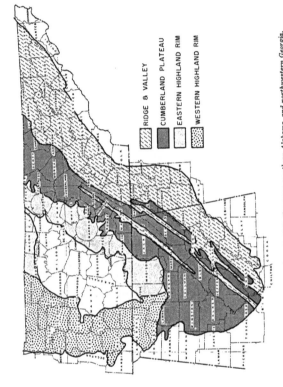

Figure 1.—*Highland regions of middle Tennessee, northern Alabama, and northwestern Georgia.*

RIDGE & VALLEY

CUMBERLAND PLATEAU

EASTERN HIGHLAND RIM

WESTERN HIGHLAND RIM

2

eter class with nonzero predicted frequency
cre (DMAX), and site index at base age 25
s (S). No attempt was made to develop sepa-
predictors for each physiographic region,
rior experience (Smalley and Bower 1968,
) had indicated that there was little chance
gression coefficients being significantly dif-
it.

or model validation, a randomly selected sub-
f 32 plots was withheld at the rate of 10 per-
per physiographic region. Data for the
lining three plots had unresolvable errors
were discarded.

The distribution of sample plots by site index,
age, and planting density is shown in table 1.
Ages ranged from 10 to 31 years, average heights
of dominant and codominants from 22 to 72 feet,
site indexes from 31 to 89, number of trees planted from 305 to 3,460 per acre, and number of surviving trees from 202 to 2,240. A chi-square test
showed no relationship between age and site index (all statistical tests were at the 0.05 level).

An initial screening was made to determine
which of six height-on-diameter models to use.
Since two of the models involved a logarithmic
transformation of the dependent variable, the in-

1.—*Distribution of loblolly plantations by age, site index, and planting density*

Planting density (trees per acre)	Age from seed					Total
	10-14	15-19	20-24	25-29	30-34	
	— — — — — — — — — — *Number of plots* — — — — — — — — — —					
<500	1		2		1	4
501-1000	31	5	6	11	2	55
1001-1500	72	46	24	30	3	175
1501-2000	10	4	2	2		18
2001-2500	4	4	1			9
2501-3000		3	2			5
>3000	1					1
Total	119	62	37	43	6	267

n rs)	Site index													Total
	30	35	40	45	50	55	60	65	70	75	80	85	90	
	— — — — — — — — — — *Number of plots* — — — — — — — — — —													
10-14		1	4	8	19	42	25	15	3	1			˙	119
15-19			1	6	14	18	12	7	4					62
20-24				2	11	11	8	3	1					37
25-29					13	19	7	4						43
30-34					1	2	1	2						6
otal	1	1	5	16	57	91	54	30	10	1			1	267

	Site index												Total
	35	40	45	50	55	60	65	70	75	80	85	90	
	— — — — — — — — — — *Number of plots* — — — — — — — — — —												
<500				1		2	1						4
01-1001	1		1	5	24	13	9	2					55
01-1500		4	11	39	59	39	17	5					175
01-2000			2	4	6	2		2	1				18
01-2500		1	2	3	1		2						9
01-3000				4	1								5
>3000				1									1
otal	1	1	5	16	57	91	54	30	10	1		1	267

dex to fit suggested by Furnival (1961) was taken as the basis of comparison. A test for homogeneity of variance and a one-way analysis of variance was performed on the "I" values (untransformed standard errors). The variance was homogeneous and the models were not significantly different. Principally on the recommendations of Curtis (1967), therefore, we chose log H, on D_i^{-1}. Seventy percent of the plots and 88 percent of the randomly selected subset of 32 plots had $r^2 \geq .70$. These regressions and the tree-volume regressions of Smalley and Bower (1968) gave equations for tree volumes as functions of diameter only. Plot volumes were computed with these equations.

MODEL DEVELOPMENT AND EVALUATION

Survival

The basic structure chosen for a survival model was:

$$\log (T_p/T_s) = Af(A, H, T_p) .$$

Merits of this particular form are that when $A = 0$, $\log (T_p/T_s) = 0$. Thus $T_p = T_s$ and the equation predicts 100-percent survival. For any value of the right-hand side a nonnegative (T_p/T_s) ratio results.

The function, $f(A,H,T_p)$, was defined by screening linear, nonlinear, and interaction transformations of A, H, T_p with REX (Grosenbaugh 1967). Graphical evaluation (for example, see fig. 2) was the final step in screening. The model chosen was:

$$\log (T_p/T_s) = A\{0.0130 \log (T_p) + 0.0009 H - 0.0109 \sqrt{H}\} \quad (1)$$
$$R^2 = 0.84; \ S^2_{y \cdot x} = 0.0042; \ \overline{Y} = 0.1329; \ S^2_y = 0.0081$$

There were no discernible trends in residuals.

Number of surviving trees was calculated for each plot of the randomly selected subset by solving for T_s. Sixty-six percent of the estimates were within ±10 percent of the observed values, and the algebraic average error of +0.02 percent corresponded to a difference of less than one tree per acre for the stand of mean stocking.

Diameter Distributions

The three-parameter Weibull function (Bailey 1972, Bailey and Dell 1973) was used to model diameter distributions. Maximum-likelihood estimation gave excellent results. By the Kolmo-

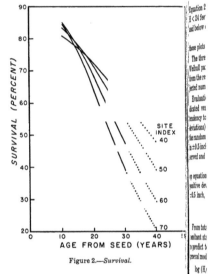

Figure 2.—*Survival.*

gorov-Smirnov test (Massey 1951), none of th? fitted distributions differed significantly and th? maximum difference between fitted and ol? served quadratic mean diameter was —0.07 inc? The average of all differences was 0.004 inch.

Regressions were fitted to predict the max?? mum-likelihood estimates from stand variabl?? A, H, and T_s :

Location parameter
$$a = -1.5254 + 0.0639 H , \quad (?)$$
$$R^2 = 0.28, \ S^2_{y \cdot x} = 1.3762, \ \overline{Y} = 1.188?$$
$$S^2_y = 1.9146;$$

Scale parameter
$$a + b = -6.6951 - 0.0008 T_p + 7.51?$$
$$\log H + 1{,}036.0133/T_s , \quad (?)$$
$$R^2 = 0.92 , \ S^2_{y \cdot x} = 0.1599, \ \overline{Y} - 5.9?$$
$$S^2_y - 1.985 ;$$

Shape parameter
$$c = 3.3542 + 0.0002A + 0.3416 \log T_s ,$$
$$R^2 - 0.09 , \ S^2_{y \cdot x} = 1.602, \ \overline{Y} - 4.06 ,$$
$$S^2_y - 1.756.$$

4

ation 2 will predict negative values for a if 24 feet (age 14 and below on site 40, age 10 below on site 45). This situation only occur-twice in our data. We specified $a = 0$ for e plots and $a \geq 0$ in all predictions.

ie three equations were used for predicting oull parameters. Stand tables were obtained 1 the resulting relative distributions and pro-:d numbers of surviving stems per acre.

valuations against observed distributions in-ted very good fits. Although there was a ency to underestimate (71 percent negative ations), the mean diameter of 81 percent of ·andomly selected subset was predicted with-0.5 inch. For 78 percent of these plots the ob-ed and predicted distributions were not sig-:antly different. With the data used to devel-quations there were 143 negative and 124 tive deviations; of these, 84 percent within 5 inch, and the average was —0.01 inch.

hts

rom total height and diameter pairs and con-itant stand data, an equation was developed redict total height of individual trees. The :ral model was:

$$\log (H/H_1) = b_0 + (1/D_1 - 1/DMAX) \\ g (A,H,T_s).$$

characteristics are discussed by Lenhart {8).

he form of the function $g(A,H,T_s)$ and para-:r estimates were obtained through the ening process described previously.

$$\log (H/H_1) = — 0.0187 \\ + (1/D_1 - 1/DMAX) \\ \{2.3357 + 0.7849 \, A \, T_s (10^{-5}) \\ + 0.0029 \, (T_s/A) \\ — 1.1990 \log (T_s/A) \\ + 0.3919 \log (H/A) \} \quad (5)$$

$R^2 = 0.72, \, S^2_{y \cdot x} = 0.0016, \, \overline{Y} = 0.0358,$
$S^2_y = 0.0056$

Predicted Yields

Equations 2 through 5 and those for tree volume (Smalley and Bower 1968) made possible a yield prediction system in which age, average dominant height, and number of surviving trees were the variables.

Number of trees and average heights by diameter classes were used to calculate volumes by classes, which were then summed for expected yields per acre. When this approach was evaluated against observed yields on the 267 plots, R^2's for all volume categories were in excess of 0.90. Precision was best for 4-inch top volumes ($R^2 = 0.91$). Average differences between predicted and observed yields per acre ranged between —48 cubic feet for volume to a 4-inch top (i.b.) and 26 cubic feet for total volume (o.b.); differences were within ±10 percent of observed 36 percent of the time.

When evaluated with the randomly selected subset, this system explained 87 percent or more of the variation about mean volume for all categories (table 2). Approximately one-third of the predicted values were within ±10 percent of the

Table 2.—*Evaluation of predicted per-acre yields of all trees with d.b.h.* > 4.5 inches

Volume category	R^2	Predicted values in relation to observed			Predicted minus observed		T-value for test of zero percent difference
		±10 percent	±25 percent	±500 ft³	Negative	Positive	
		— — Number — —					
Total							
Outside bark	0.87	25	68	72	14	18	1.82*
Inside bark	.87	31	68	78	14	18	1.63
To 2-inch top							
Outside bark	.87	25	69	72	14	18	1.78*
Inside bark	.87	34	69	78	14	18	1.59
To 3-inch top							
Outside bark	.87	31	69	75	14	18	1.63
Inside bark	.88	31	66	78	15	17	1.40
To 4-inch top							
Outside bark	.88	31	66	75	16	16	1.14
Inside bark	.89	34	60	81	19	13	.84

*Significant at 0.05 level.

5

observed and-two-thirds were within ±25 percent. Prediction seemed to be better for inside-bark volumes. Precision improved as top diameter increased. Average differences between predicted and observed yields ranged from —0.25 to 47 cubic feet and were almost equally divided between positive and negative. A two-tailed t-test for average difference (tabular value 1.70) was significant for only two categories.

Simple correlation coefficients between residuals and observations were not significant for any volume category (table 3). Correlations with age averaged 0.01 and were lower for merchantable than for total volumes. The reverse trend occurred for site index, where the correlation between residuals and surviving stems averaged 0.30 and was lowest (0.27) for volume inside bark to a 4-inch top.

Table 3.—*Correlation coefficients between residuals (predicted minus observed per-acre yield) and stand characteristics. Data are for all trees with d.b.h. > 4.5 inches*

Category	Correlation[1] with—			
	Age	Site index	Surviving stems per acre	Average dominant height
Total				
Outside bark	0.02	−0.11	0.32	−0.02
Inside bark	.01	− .11	.31	− .03
To 2-inch top				
Outside bark	.01	− .11	.31	− .02
Inside bark	.01	− .12	.30	− .03
To 3-inch top				
Outside bark	.01	− .11	.31	− .03
Inside bark	.01	− .12	.30	− .04
To 4-inch top				
Outside bark	.01	− .13	.28	− .04
Inside bark	.00	− .13	.27	− .05

[1] Critical value at 0.05 level with 30 degrees of freedom is 0.35.

Comparisons

The algebraic average error of +4 percent over all categories in this study compares favorably with tests of diameter-distribution yield prediction systems based on the beta function. For old-field loblolly plantations in the Georgia Piedmont predicted yields exceeded observed yields by 6 percent (Lenhart and Clutter 1971), and in the interior West Gulf Coastal Plain by 5 percent (Lenhart 1972). Burkhart (1971) reported an average under-prediction of 1 percent

for slash pine plantations in south Georgia and north Florida. For natural stands of yellow-poplar (*Liriodendron tulipifera* L.) in the southern Appalachians, the average over-prediction was 0.5 percent (Beck and Della-Bianca 1970).

LITERATURE CITED

Bailey, R. L.
1972. Development of unthinned stand of *Pinus radiata* in New Zealand. Ph.D. thesis. Univ. Ga., Athens. 73 p. [Diss. Abstr. Int. 33(9): 4061-B, 1973.]

Bailey, R. L., and Dell, T. R.
1973. Quantifying diameter distributions with the Weibull function. For. Sci. 19: 97-104.

Beck, D. E., and Della-Bianca, L.
1970. Yield of unthinned yellow-poplar. USDA For. Serv. Res. Pap. SE-58, 20 p. Southeast. For. Exp. Stn., Asheville, N. C.

Burkhart, H. E.
1971. Slash pine plantation yield estimates based on diameter distribution: an evaluation. For. Sci. 17: 452-453.

Curtis, R. O.
1967. Height-diameter and height-diameter-age equations for second-growth Douglas-fir. For. Sci. 13: 365-375.

Furnival, G. M.
1961. An index for comparing equations used in constructing volume tables. For. Sci. 7: 337-341.

Grosenbaugh, L. R.
1967. REX—FORTRAN-4 system for combinatorial screening or conventional analysis of multivariate regressions USDA For. Serv. Res. Pap. PSW-44 47 p. Pac. Southwest For. and Range Exp. Stn., Berkeley, Calif.

Lenhart, J. D.
1968. Yield of old-field loblolly pine plantations in the Georgia Piedmont. Ph.D. thesis. Univ. Ga., Athens. 10 p. [Diss. Abstr. Int. 30(8): 3450-E 1970.]

Lenhart, J. D.
1972. Cubic-foot yields for unthinned old-field loblolly pine plantations in the interior West Gulf Coastal Plain. Tex. For. Pap. 14, 46 p.

Lenhart 1971.

Massey 1961.

Smalley 1974.

Smalley, 1968.

Smalley, 1971

Lenhart, J. D., and Clutter, J. L.
  1971. Cubic-foot yield tables for old-field loblolly pine plantations in the Georgia Piedmont. Ga. For. Res. Counc. Rep. 22, Ser. 3, 13 p.

Massey, F. J., Jr.
  1951. The Kolmogorov-Smirnov test for goodness of fit. J. Am. Stat. Assoc. 46: 68-78.

Smalley, G. W., and Bailey, R. L.
  1974. Yield tables and stand structure for shortleaf pine plantations in Tennessee, Alabama, and Georgia highlands. USDA For. Serv. Res. Pap. SO-97, 57 p. South. For. Exp. Stn., New Orleans, La.

Smalley, G. W., and Bower, D. R.
  1968. Volume tables and point-sampling factors for loblolly pines in plantations on abandoned fields in Tennessee, Alabama, and Georgia highlands. USDA For. Serv. Res. Pap. SO-32, 13 p. South. For. Exp. Stn., New Orleans, La.

Smalley, G. W., and Bower, D. R.
  1971. Site index curves for loblolly and shortleaf pine plantations on abandoned fields in Tennessee, Alabama, and Georgia highlands. USDA For. Serv. Res. Note SO-126, 6 p. South. For. Exp. Stn., New Orleans, La.

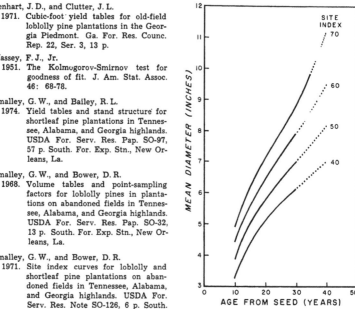

Figure 3.—*Quadratic mean diameter.*

STAND AND STOCK TABLES

The tables that follow are believed to be representative of the natural development of undisturbed loblolly pine plantations on abandoned fields throughout the Interior Highlands of Tennessee, Alabama, and Georgia. To enhance their utility, the tables have been extended 9 years beyond the oldest sampled stands. We recognize that extrapolation is risky, but all relationships appear biologically valid. Some trends are worthy of comment. For conciseness, graphical examples in figures 2-8 are limited to a planting density of 1,250 trees per acre, or about a 6- by 6-foot spacing. Where discussions relate to changes in planting density, trends can be verified from the tables.

Survival.—On all sites survival percentage decreased as planting density and age increased.

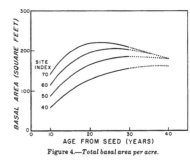

Figure 4.—*Total basal area per acre.*

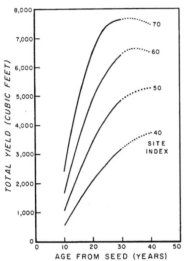

Figure 5.—*Total yield per acre of entire stem (outside bark) of all trees.*

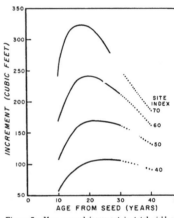

Figure 7.—*Mean annual increment in total yield per acre.*

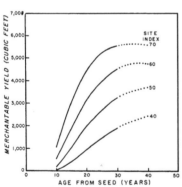

Figure 6.—*Merchantable yield (inside bark) per acre to a 4-inch top (outside bark) in stems with d.b.h. > 4.5 inches.*

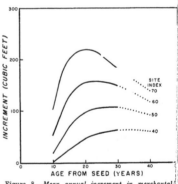

Figure 8.—*Mean annual increment in merchantable yield per acre.*

With an increase in site, however, survival w slightly better at early ages and worse at old ages—a consequence of intensified competiti on better sites as age increased.

Quadratic mean diameter.—As planting de sities increased, mean diameter declined for . ages and sites, but improvement in site alwa

ansed dia sities, dia past age 20

Basal ar sizes) for a 40 for all pl tion was at acre.

occur at abc

Yie yields incre but the effe Yield increa sities on site 0,000 to 2,50 yield culmin

8

aused diameter increases. For all sites and densities, diameter growth was essentially linear past age 20.

Basal area.—Total basal area (trees of all sizes) for sites 60 and 70 culminated before age 0 for all planting densities. On site 50, culmination was at densities greater than 1,000 trees per cre. On poor sites culmination will probably ccur at about age 50.

Yields.—Total and merchantable cubic-foot yields increased with site and planting density, but the effect of density was small on poor sites. Yield increased with age for all planting densities on sites 40 and 50. For planting densities of ,000 to 2,500 trees per acre on sites 60 and 70, yield culminated at 30 to 35 years. In effect, the

loss of volume from mortality began to exceed both total and merchantable growth on the remaining trees.

Mean annual increment.—For total volume, mean annual increment culminated for all sites and planting densities. Age at culmination decreased from 25 years on poor sites to about 19 years on the best sites, regardless of planting density. Increment at culmination increased with planting density up to 2,000 trees per acre on all sites.

Merchantable-volume increment culminated on all sites at all planting densities. However, the culmination occurred at older ages than for total-volume, e.g., age 35 to 40 for site 40 and age 20 to 23 for site 70.

9

SI 40

500 stems per acre

Age	Av. dom. ht.	D.b.h.	Stems per acre	Basal area	Av. ht.	All trees, total stem		5-inch class and greater to o.b. tops of—					
								2 inches		3 inches		4 inches	
						O. b.	I. b.	O. b.	I. b.	O. b.	I. b.	O. b.	I. b.
Yrs	Ft	In	No.	Ft 2	Ft	– – – – – – – – – – – – Ft 3 – – – – – – – – – – – –							
10	21	1	2	.0	5	0	0						
		2	19	.4	12	3	2						
		3	62	3.0	15	24	17						
		4	120	10.5	17	92	67						
		5	139	19.0	19	185	134	185	120	159	98	92	47
		6	88	17.3	20	176	128	175	121	157	107	116	75
		7	25	6.7	21	71	52	70	50	65	47	53	38
		8	3	1.0	21	11	8	11	8	10	8	9	7
			458	57.9		562	408	441	299	391	260	270	167
		MEAN DIA. 4.8 INCHES, WEIBULL PARAMETERS A= .0, B= 5.10, C= 4.27											
15	29	1	1	.0	5	0	0						
		2	6	.1	14	1	1						
		3	27	1.3	19	13	10						
		4	66	5.8	23	68	49						
		5	111	15.1	25	193	140	193	130	171	113	118	73
		6	125	24.5	27	337	244	330	237	306	218	248	174
		7	88	23.5	28	334	242	324	239	308	227	268	196
		8	34	11.9	29	174	127	168	127	162	122	148	110
		9	6	2.7	30	40	29	39	29	38	29	35	27
			464	84.9		1160	842	1054	762	985	709	817	580
		MEAN DIA. 5.8 INCHES, WEIBULL PARAMETERS A= .3, B= 5.83, C= 4.27											
20	35	2	2	.0	15	0	0						
		3	14	.7	22	8	6						
		4	41	3.6	26	48	35						
		5	82	11.2	29	165	120	164	113	148	101	109	71
		6	117	23.0	31	361	262	353	256	331	239	277	198
		7	112	29.9	33	500	363	485	361	464	346	415	308
		8	67	23.4	34	402	292	388	292	376	285	348	263
		9	22	9.7	35	172	125	165	125	162	124	153	117
		10	4	2.2	36	40	29	38	29	37	29	36	28
			461	103.7		1696	1232	1593	1176	1518	1124	1338	985
		MEAN DIA. 6.4 INCHES, WEIBULL PARAMETERS A= .7, B= 6.09, C= 4.27											
25	40	2	1	.0	15	0	0						
		3	8	.4	23	5	3						
		4	27	2.4	28	34	24						
		5	60	8.2	32	133	97	132	92	120	83	92	62
		6	99	19.4	34	335	243	327	239	308	224	263	190
		7	113	30.2	36	550	400	532	399	512	383	463	345
		8	89	31.1	38	597	434	575	434	559	426	523	397
		9	42	18.6	39	365	266	351	266	344	264	328	252
		10	11	6.0	40	121	88	116	88	114	88	111	85
		11	1	.7	41	14	10	13	10	13	10	13	10
			451	116.9		2154	1565	2046	1528	1970	1478	1793	1341
		MEAN DIA. 6.9 INCHES, WEIBULL PARAMETERS A= 1.0, B= 6.26, C= 4.27											

Age	Av. dom. ht.	D.b.h.	Stems per acre	Basal area	Av. ht.	All trees, total stem		5-inch class and greater to o.b. tops of—					
								2 inches		3 inches		4 inches	
						O. b.	I. b.	O. b.	I. b.	O. b.	I. b.	O. b.	I. b.
Yrs	Ft	In	No.	Ft²	Ft	— — — — — — — — — — — — Ft³ — — — — — — — — — — — —							
30	44	3	4	.2	24	2	2						
		4	17	1.5	30	23	17						
		5	44	6.0	34	104	75	102	72	94	66	73	50
		6	80	15.7	37	295	214	287	211	272	199	235	172
		7	108	28.9	40	584	424	564	424	545	410	498	375
		8	98	34.2	42	726	528	698	528	682	521	642	490
		9	59	26.1	43	566	411	543	411	534	411	512	394
		10	21	11.5	44	254	185	244	185	241	185	233	181
		11	4	2.6	46	61	45	59	45	58	45	57	44
			435	126.6		2615	1901	2497	1876	2426	1837	2250	1706
			MEAN DIA. 7.3 INCHES, WEIBULL PARAMETERS A= 1.3, B= 6.42, C= 4.26										
35	48	3	2	.1	24	1	1						
		4	11	1.0	30	15	11						
		5	32	4.4	35	78	56	76	54	70	49	55	38
		6	63	12.4	39	244	178	238	176	226	166	197	145
		7	92	24.6	41	510	371	492	371	476	359	437	329
		8	99	34.6	44	768	558	738	558	722	553	683	522
		9	72	31.8	45	722	525	693	525	682	525	655	505
		10	33	18.0	47	427	310	409	310	404	310	393	305
		11	9	5.9	48	144	105	137	105	136	105	134	104
		12	1	.8	49	19	14	19	14	18	14	18	14
			414	133.5		2928	2129	2802	2113	2734	2081	2572	1962
			MEAN DIA. 7.7 INCHES, WEIBULL PARAMETERS A= 1.5, B= 6.58, C= 4.25										
40	51	3	1	.0	24	1	0						
		4	8	.7	31	11	8						
		5	23	3.1	36	57	42	56	40	52	37	41	28
		6	48	9.4	40	191	139	185	137	177	130	155	114
		7	77	20.6	44	458	333	442	333	428	324	396	299
		8	92	32.1	46	746	542	717	542	702	538	666	510
		9	78	34.5	48	835	607	800	607	789	607	761	587
		10	44	24.0	50	605	440	579	440	573	440	559	434
		11	15	9.9	51	255	185	243	185	242	185	237	185
		12	3	2.4	52	62	45	59	45	59	45	58	45
			389	136.7		3221	2341	3081	2329	3022	2306	2873	2202
			MEAN DIA. 8.0 INCHES, WEIBULL PARAMETERS A= 1.7, B= 6.76, C= 4.25										

SI 40

750 stems per acre

Age	Av. dom. ht.	D.b.h.	Stems per acre	Basal area	Av. ht.	All trees, total stem		5-inch class and greater to o.b. tops of—					
								2 inches		3 inches		4 inches	
						O. b.	I. b.	O. b.	I. b.	O. b.	I. b.	O. b.	I. b.
Yrs	Ft	In	No.	Ft 2	Ft					Ft 3			
10	21	1	7	.0	7	0	0						
		2	54	1.2	14	10	7						
		3	162	8.0	17	72	51						
		4	239	20.9	18	195	140						
		5	155	21.1	20	216	157	216	142	187	118	113	61
		6	33	6.5	20	66	48	65	45	59	40	43	28
		7	2	.5	21	6	4	6	4	5	4	4	3
			652	58.2		565	407	287	191	251	162	160	92
			MEAN DIA. 4.0 INCHES, WEIBULL PARAMETERS A= .0, B= 4.28, C= 4.32										
15	29	1	1	.0	7	0	0						
		2	16	.3	16	3	2						
		3	64	3.1	21	35	25						
		4	147	12.8	24	158	115						
		5	201	27.4	26	364	264	362	246	322	216	226	143
		6	153	30.0	28	427	310	419	301	389	278	318	224
		7	54	14.4	29	212	154	206	152	196	145	172	126
		8	7	2.4	30	37	27	36	27	35	26	32	24
			643	90.6		1236	897	1023	726	942	665	748	517
			MEAN DIA. 5.1 INCHES, WEIBULL PARAMETERS A= .3, B= 5.07, C= 4.32										
20	35	2	5	.1	17	1	1						
		3	30	1.5	23	18	13						
		4	88	7.7	28	110	80						
		5	160	21.8	30	333	242	330	229	298	205	223	147
		6	183	35.9	32	583	424	570	414	535	387	451	323
		7	117	31.3	34	538	391	521	389	500	373	448	333
		8	36	12.6	35	222	162	214	162	208	158	193	146
		9	4	1.8	36	32	23	31	23	30	23	29	22
			623	112.6		1837	1336	1666	1217	1571	1146	1344	971
			MEAN DIA. 5.8 INCHES, WEIBULL PARAMETERS A= .7, B= 5.38, C= 4.31										
25	40	2	2	.0	17	0	0						
		3	15	.7	25	10	7						
		4	54	4.7	30	72	52						
		5	116	15.8	33	266	193	262	184	239	167	185	125
		6	165	32.4	36	591	429	576	423	544	399	470	342
		7	148	39.6	38	760	553	735	553	708	532	644	483
		8	74	25.8	39	509	370	490	370	477	364	447	340
		9	17	7.5	40	152	110	146	110	143	110	136	105
		10	2	1.1	41	23	16	22	16	21	16	21	16
			593	127.7		2363	1730	2231	1656	2132	1568	1903	1411
			MEAN DIA. 6.3 INCHES, WEIBULL PARAMETERS A= 1.0, B= 5.60, C= 4.31										

Age	Av. dom. ht.	D.b.h.	Stems per acre	Basal area	Av. ht.	All trees, total stem		5-inch class and greater to o.b. tops of—					
								2 inches		3 inches		4 inches	
						O. b.	I. b.	O. b.	I. b.	O. b.	I. b.	O. b.	I. b.
Yrs	Ft	In	No.	Ft 2	Ft	------------- Ft 3 -------------							
30	44	2	1	.0	17	0	0						
		3	8	.4	25	5	4						
		4	33	2.9	31	46	33						
		5	82	11.2	36	205	148	201	143	185	131	147	101
		6	136	26.7	39	528	383	513	379	487	359	426	313
		7	152	40.6	41	842	612	813	612	786	593	721	543
		8	102	35.6	43	773	562	744	562	727	556	686	524
		9	37	16.3	44	363	264	348	264	343	264	329	253
		10	6	3.3	46	76	55	73	55	72	55	70	54
			557	137.0		2838	2061	2692	2015	2600	1958	2379	1788
		MEAN DIA. 6.7 INCHES, WEIBULL PARAMETERS A= 1.3, B= 5.80, C= 4.30											
35	48	3	4	.2	25	3	2						
		4	21	1.8	32	30	22						
		5	56	7.6	37	144	104	141	101	130	92	104	72
		6	105	20.6	40	418	303	406	300	386	285	339	249
		7	136	36.3	43	790	574	762	574	739	558	681	514
		8	117	40.8	45	928	675	892	675	873	669	827	633
		9	59	26.1	47	618	450	593	450	584	450	563	434
		10	16	8.7	48	211	154	202	154	200	154	195	151
		11	2	1.3	49	33	24	31	24	31	24	30	24
			516	143.6		3175	2308	3027	2278	2943	2232	2739	2077
		MEAN DIA. 7.1 INCHES, WEIBULL PARAMETERS A= 1.5, B= 6.00, C= 4.29											
40	51	3	2	.1	26	1	1						
		4	13	1.1	33	19	14						
		5	38	5.2	38	100	73	98	70	91	64	73	51
		6	78	15.3	42	326	237	316	235	302	224	267	197
		7	114	30.5	45	693	504	668	504	648	491	601	454
		8	117	40.8	47	969	705	931	705	912	700	867	664
		9	76	33.6	49	830	604	796	604	784	604	758	585
		10	29	15.8	51	407	296	389	296	386	296	376	292
		11	6	4.0	52	104	76	99	76	99	76	97	76
			473	146.4		3449	2510	3297	2490	3222	2455	3039	2319
		MEAN DIA. 7.5 INCHES, WEIBULL PARAMETERS A= 1.7, B= 6.22, C= 4.28											

13

SI 40

1,000 stems per acre

Age	Av. dom. ht.	D.b.h.	Stems per acre	Basal area	Av. ht.	All trees, total stem		5-inch class and greater to o.b. tops of—					
								2 inches		3 inches		4 inches	
						O. b.	I. b.	O. b.	I. b.	O. b.	I. b.	O. b.	I. b.
			No.	Ft²	Ft	— — — — — — — — — — — Ft³ — — — — — — — — — — —							
10	21	1	15	.1	8	1	0						
		2	111	2.4	15	21	15						
		3	298	14.6	18	140	100						
		4	312	27.2	19	268	193						
		5	96	13.1	20	134	97	134	88	116	73	70	38
		6	5	1.0	21	10	8	10	7	9	6	7	5
			837	58.4		574	413	144	95	125	79	77	43
			MEAN DIA. 3.6 INCHES, WEIBULL PARAMETERS A= .0, B= 3.79, C= 4.35										
15	29	1	2	.0	8	0	0						
		2	29	.6	17	6	4						
		3	116	5.7	22	66	47						
		4	243	21.2	25	273	197						
		5	268	36.5	26	485	351	483	328	429	287	302	190
		6	130	25.5	28	363	263	356	256	331	236	270	190
		7	21	5.6	29	82	60	80	59	76	56	67	49
		8	1	.3	30	5	4	5	4	5	4	5	3
			810	95.6		1280	926	924	647	841	583	644	432
			MEAN DIA. 4.7 INCHES, WEIBULL PARAMETERS A= .3, B= 4.61, C= 4.35										
20	35	2	9	.2	18	2	1						
		3	52	2.6	24	32	23						
		4	146	12.7	28	183	132						
		5	240	32.7	31	516	375	511	356	463	320	350	234
		6	217	42.6	33	713	518	696	508	654	476	555	400
		7	92	24.6	34	423	307	410	306	393	293	353	262
		8	14	4.9	35	86	63	83	63	81	61	75	57
		9	1	.4	36	8	6	8	6	8	6	7	5
			771	120.7		1963	1425	1708	1239	1599	1156	1340	958
			MEAN DIA. 5.4 INCHES, WEIBULL PARAMETERS A= .7, B= 4.95, C= 4.34										
25	40	2	3	.1	18	1	0						
		3	25	1.2	26	17	12						
		4	87	7.6	31	121	87						
		5	178	24.3	34	420	305	414	292	379	265	295	201
		6	222	43.6	37	817	594	795	586	754	553	653	477
		7	152	40.6	39	801	582	774	582	747	562	682	512
		8	48	16.8	40	339	246	326	246	318	242	298	227
		9	6	2.7	41	55	40	53	40	52	40	49	38
			721	136.8		2571	1866	2362	1746	2250	1662	1977	1455
			MEAN DIA. 5.9 INCHES, WEIBULL PARAMETERS A= 1.0, B= 5.20, C= 4.34										

14

Age	Av. dom. ht.	D.b.h.	Stems per acre	Basal area	Av. ht.	All trees, total stem		5-inch class and greater to o.b. tops of—					
								2 inches		3 inches		4 inches	
						O. b.	I. b.	O. b.	I. b.	O. b.	I. b.	O. b.	I. b.
Yrs	Ft	In	No.	Ft 2	Ft	─ ─ ─ ─ ─ ─ ─ ─ ─ ─ ─ Ft 3 ─ ─ ─ ─ ─ ─ ─ ─ ─ ─ ─ ─							
30	44	2	1	.0	18	0	0						
		3	12	.6	26	8	6						
		4	52	4.5	32	74	54						
		5	123	16.8	36	307	223	302	214	278	196	220	152
		6	188	36.9	39	729	530	709	524	674	497	589	432
		7	177	47.3	41	981	713	947	713	916	691	840	632
		8	89	31.1	43	675	491	649	491	634	485	599	457
		9	20	8.8	44	196	143	188	143	185	143	178	137
		10	2	1.1	46	25	.18	24	18	24	18	23	18
			664	147.1		2995	2178	2819	2103	2711	2030	2449	1828
			MEAN DIA. 6.4 INCHES, WEIBULL PARAMETERS A= 1.3, B= 5.42, C= 4.32										
35	48	3	6	.3	27	4	3						
		4	31	2.7	33	46	33						
		5	83	11.3	38	218	159	215	153	198	141	160	111
		6	147	28.9	41	599	435	582	432	555	410	489	360
		7	169	45.2	44	1005	730	969	730	939	711	868	656
		8	118	41.2	46	957	696	920	696	900	690	854	654
		9	43	19.0	48	460	335	441	335	435	335	419	324
		10	7	3.8	49	94	69	90	69	89	69	87	68
			604	152.4		3383	2460	3217	2415	3116	2356	2877	2173
			MEAN DIA. 6.8 INCHES, WEIBULL PARAMETERS A= 1.5, B= 5.64, C= 4.31										
40	51	3	3	.1	26	2	1						
		4	18	1.6	33	27	19						
		5	54	7.4	38	142	103	140	100	129	92	104	72
		6	107	21.0	42	447	325	434	322	414	307	366	270
		7	146	39.0	45	888	645	856	645	830	629	769	582
		8	129	45.0	48	1091	794	1048	794	1028	789	978	750
		9	67	29.6	50	747	543	716	543	706	543	682	527
		10	18	9.8	51	253	184	242	184	239	184	234	181
		11	2	1.3	52	35	25	33	25	33	25	32	25
			544	154.9		3632	2639	3469	2613	3379	2569	3165	2407
			MEAN DIA. 7.2 INCHES, WEIBULL PARAMETERS A= 1.7, B= 5.89, C= 4.30										

15

SI 40

1,250 stems per acre

Age	Av. dom. ht.	D.b.h.	Stems per acre	Basal area	Av. ht.	All trees, total stem		5-inch class and greater to o.b. tops of—					
								2 inches		3 inches		4 inches	
						O. b.	I. b.	O. b.	I. b.	O. b.	I. b.	O. b.	I. b.
Yrs	Ft	In	No.	Ft²	Ft	— — — — — — — — — — — — — Ft³ — — — — — — — — — — — — —							
10	21	1	27	.1	9	1	1						
		2	197	4.3	16	39	28						
		3	449	22.0	18	210	151						
		4	305	26.6	20	275	199						
		5	38	5.2	21	56	40	56	37	48	31	30	17
			1016	58.3		581	419	56	37	48	31	30	17
			MEAN DIA. 3.2 INCHES, WEIBULL PARAMETERS A= .0, B= 3.44, C= 4.38										
15	29	1	4	.0	9	0	0						
		2	47	1.0	18	10	7						
		3	183	9.0	23	108	78						
		4	345	30.1	26	402	291						
		5	296	40.4	27	556	403	552	378	494	333	353	226
		6	89	17.5	29	257	167	252	182	235	168	194	137
		7	6	1.6	30	24	18	24	18	23	17	20	15
			970	99.6		1357	984	828	578	752	518	567	378
			MEAN DIA. 4.3 INCHES, WEIBULL PARAMETERS A= .3, B= 4.28, C= 4.38										
20	35	2	14	.3	19	3	2						
		3	79	3.9	25	51	36						
		4	215	18.8	29	279	202						
		5	313	42.7	32	695	504	686	481	625	433	478	321
		6	222	43.6	34	751	546	733	536	691	503	589	426
		7	61	16.3	35	289	210	280	209	268	201	242	180
		8	5	1.7	36	32	23	31	23	30	23	28	21
			909	127.3		2100	1523	1730	1249	1614	1160	1337	948
			MEAN DIA. 5.1 INCHES, WEIBULL PARAMETERS A= .7, B= 4.64, C= 4.37										
25	40	2	4	.1	19	1	1						
		3	37	1.8	27	26	18						
		4	125	10.9	31	173	125						
		5	242	33.0	35	587	426	578	409	531	373	418	287
		6	262	51.4	37	965	701	939	691	889	653	771	562
		7	138	36.9	39	727	529	703	529	678	510	619	464
		8	28	9.8	40	198	144	190	144	185	141	174	133
		9	2	.9	41	18	13	18	13	17	13	16	13
			838	144.8		2695	1957	2428	1786	2300	1690	1998	1459
			MEAN DIA. 5.6 INCHES, WEIBULL PARAMETERS A= 1.0, B= 4.91, C= 4.36										

16

Age	Av. dom. ht.	D.b.h.	Stems per acre	Basal area	Av. ht.	All trees, total stem		5-inch class and greater to o.b. tops of—					
								2 inches		3 inches		4 inches	
						O. b.	I. b.	O. b.	I. b.	O. b.	I. b.	O. b.	I. b.
Yrs	Ft	In	No.	Ft 2	Ft	– – – – – – – – – – – Ft 3 – – – – – – – – – – –							
30	44	2	1	.0	19	0	0						
		3	18	.9	28	13	9						
		4	72	6.3	33	106	77						
		5	168	22.9	37	431	313	423	302	391	276	312	217
		6	236	46.3	40	939	682	912	675	868	641	762	560
		7	186	49.7	42	1056	767	1019	767	986	745	907	684
		8	70	24.4	44	543	395	522	395	510	391	483	369
		9	10	4.4	46	103	75	98	75	97	75	93	72
			761	155.0		3191	2318	2974	2214	2852	2128	2557	1902

MEAN DIA. 6.1 INCHES, WEIBULL PARAMETERS A= 1.3, B= 5.15, C= 4.34

Age	Av. dom. ht.	D.b.h.	Stems per acre	Basal area	Av. ht.	O. b.	I. b.	O. b.	I. b.	O. b.	I. b.	O. b.	I. b.
35	48	3	8	.4	27	6	4						
		4	42	3.7	34	64	46						
		5	111	15.1	38	292	212	287	205	265	188	214	149
		6	186	36.5	42	777	564	754	560	719	533	637	470
		7	194	51.8	44	1153	838	1112	838	1078	816	997	753
		8	110	38.4	46	892	649	857	649	839	643	796	610
		9	29	12.8	48	310	226	297	226	293	226	283	218
		10	3	1.6	49	40	29	39	29	38	29	37	29
			683	160.4		3534	2568	3346	2507	3232	2435	2964	2229

MEAN DIA. 6.6 INCHES, WEIBULL PARAMETERS A= 1.5, B= 5.39, C= 4.33

Age	Av. dom. ht.	D.b.h.	Stems per acre	Basal area	Av. ht.	O. b.	I. b.	O. b.	I. b.	O. b.	I. b.	O. b.	I. b.
40	51	3	4	.2	27	3	2						
		4	24	2.1	34	36	26						
		5	71	9.7	39	192	139	188	135	174	124	141	99
		6	136	26.7	42	568	413	551	409	526	390	465	343
		7	171	45.7	45	1039	756	1002	756	973	737	901	681
		8	133	46.4	48	1125	818	1081	818	1059	813	1008	773
		9	55	24.3	50	613	446	587	446	579	446	560	433
		10	10	5.5	51	140	102	134	102	133	102	130	101
		11	1	.7	52	17	13	17	13	16	13	16	13
			605	161.2		3733	2715	3560	2679	3460 .	2625	3221	2443

MEAN DIA. 7.0 INCHES, WEIBULL PARAMETERS A= 1.7, B= 5.65, C= 4.31

SI 40

1,500 stems per acre

Age	Av. dom. ht.	D.b.h.	Stems per acre	Basal area	Av. ht.	All trees, total stem		5-inch class and greater to o.b. tops of—					
								2 inches		3 inches		4 inches	
						O.b.	I.b.	O.b.	I.b.	O.b.	I.b.	O.b.	I.b.
Yrs	Ft	In	No.	Ft 2	Ft	------------Ft 3------------							
10	21	1	44	.2	10	2	1						
		2	316	6.9	16	63	44						
		3	587	28.8	19	290	208						
		4	234	20.4	20	211	152						
		5	10	1.4	21	15	11	15	10	13	8	8	4
			1191	57.7		581	416	15	10	13	8	8	4
						MEAN DIA. 3.0 INCHES, WEIBULL PARAMETERS A= .0, B= 3.15, C= 4.41							
15	29	1	5	.0	10	0	0						
		2	70	1.5	19	16	11						
		3	266	13.1	23	157	113						
		4	445	38.8	26	519	375						
		5	286	39.0	27	537	389	534	365	477	322	341	218
		6	50	9.8	29	145	105	142	102	132	95	109	77
		7	1	.3	30	4	3	4	3	4	3	3	2
			1123	102.5		1378	996	680	470	613	420	453	297
						MEAN DIA. 4.1 INCHES, WEIBULL PARAMETERS A= .3, B= 4.02, C= 4.40							
20	35	1	1	.0	9	0	0						
		2	19	.4	20	5	3						
		3	112	5.5	26	75	54						
		4	291	25.4	29	378	273						
		5	377	51.4	32	837	607	827	579	753	522	575	387
		6	205	40.3	34	694	504	677	495	638	465	544	393
		7	35	9.4	35	166	120	160	120	154	115	139	103
		8	1	.3	36	6	5	6	5	6	5	6	4
			1041	132.7		2161	1566	1670	1199	1551	1107	1264	887
						MEAN DIA. 4.8 INCHES, WEIBULL PARAMETERS A= .7, B= 4.39, C= 4.39							
25	40	2	6	.1	20	1	1						
		3	50	2.5	28	36	26						
		4	168	14.7	32	240	174						
		5	307	41.9	36	766	556	753	535	694	489	550	380
		6	287	56.4	38	1085	788	1055	778	1001	737	872	638
		7	115	30.7	40	622	452	601	452	580	437	531	399
		8	15	5.2	41	108	79	104	79	102	78	96	73
			948	151.4		2858	2076	2513	1844	2377	1741	2049	1490
						MEAN DIA. 5.4 INCHES, WEIBULL PARAMETERS A= 1.0, B= 4.68, C= 4.38							

Age	Av. dom. ht.	D.b.h.	Stems per acre	Basal area	Av. ht.	All trees, total stem		5-inch class and greater to o.b. tops of—					
								2 inches		3 inches		4 inches	
						O. b.	I. b.	O. b.	I. b.	O. b.	I. b.	O. b.	I. b.
Yrs	Ft	In	No.	Ft 2	Ft	------------Ft 3-------------							
30	44	2	2	.0	19	0	0						
		3	23	1.1	28	16	12						
		4	95	8.3	34	144	104						
		5	214	29.2	37	549	398	539	384	498	352	398	276
		6	275	54.0	40	1094	795	1063	787	1011	747	888	653
		7	185	49.4	43	1075	781	1037	781	1005	760	926	699
		8	52	18.2	44	403	293	388	293	379	290	359	274
		9	5	2.2	46	51	37	49	37	48	37	47	36
			851	162.4		3332	2420	3076	2282	2941	2186	2618	1938
		MEAN DIA. 5.9 INCHES, WEIBULL PARAMETERS A= 1.3, B= 4.93, C= 4.36											
35	48	2	1	.0	18	0	0						
		3	11	.5	28	8	6						
		4	53	4.6	34	80	58						
		5	139	19.0	38	366	266	359	257	332	236	268	187
		6	224	44.0	42	935	680	908	674	866	642	767	566
		7	208	55.6	44	1236	899	1193	899	1156	875	1069	807
		8	98	34.2	46	795	578	764	578	748	573	709	543
		9	19	8.4	48	203	148	195	148	192	148	185	143
		10	1	.5	49	13	10	13	10	13	10	12	10
			754	166.9		3636	2645	3432	2566	3307	2484	3010	2256
		MEAN DIA. 6.4 INCHES, WEIBULL PARAMETERS A= 1.5, B= 5.19, C= 4.34											
40	51	3	5	.2	28	4	3						
		4	30	2.6	35	47	34						
		5	87	11.9	40	241	175	236	170	219	157	179	126
		6	163	32.0	44	713	518	691	515	661	492	589	436
		7	193	51.6	47	1225	891	1181	891	1147	871	1067	809
		8	132	46.1	49	1140	829	1095	829	1074	825	1023	785
		9	45	19.9	51	512	372	490	372	484	372	468	362
		10	6	3.3	52	86	62	82	62	81	62	79	62
			661	167.5		3968	2884	3775	2839	3666	2779	3405	2580
		MEAN DIA. 6.8 INCHES, WEIBULL PARAMETERS A= 1.7, B= 5.46, C= 4.33											

SI 40

1,750 stems per acre

Age	Av. dom. ht.	D.b.h.	Stems per acre	Basal area	Av. ht.	All trees, total stem		5-inch class and greater to o.b. tops of—					
								2 inches		3 inches		4 inches	
						O.b.	I.b.	O.b.	I.b.	O.b.	I.b.	O.b.	I.b.
Yrs	Ft	In	No.	Ft²	Ft	— — — — — — — — — — — — Ft³ — — — — — — — — — — — —							
10	21	1	70	.4	10	4	2						
		2	472	10.3	16	94	66						
		3	377	33.2	19	334	239						
		4	142	12.4	20	128	93						
		5	1	.1	21	1	1	1	1	1	1	1	0
			1362	56.4		561	401	1	1	1	1	1	0
			MEAN DIA. 2.8 INCHES, WEIBULL PARAMETERS A= .0, B= 2.91, C= 4.43										
15	29	1	7	.0	10	0	0						
		2	99	2.2	19	23	16						
		3	364	17.9	24	225	161						
		4	530	46.3	27	641	464						
		5	248	33.8	28	483	350	479	330	430	292	312	202
		6	24	4.7	30	72	52	70	51	66	47	55	39
			1272	104.8		1444	1043	549	381	496	339	367	241
			MEAN DIA. 3.9 INCHES, WEIBULL PARAMETERS A= .3, B= 3.80, C= 4.42										
20	35	1	1	.0	9	0	0						
		2	26	.6	21	7	5						
		3	152	7.5	27	105	76						
		4	374	32.6	30	502	363						
		5	422	57.5	33	966	701	953	670	870	606	672	455
		6	174	34.2	35	606	440	591	433	558	408	479	347
		7	18	4.8	36	88	64	85	64	82	61	74	55
			1167	137.2		2274	1649	1629	1167	1510	1075	1225	857
			MEAN DIA. 4.6 INCHES, WEIBULL PARAMETERS A= .7, B= 4.19, C= 4.41										
25	40	2	8	.2	20	2	1						
		3	66	3.2	28	47	34						
		4	216	18.8	33	318	231						
		5	367	50.0	36	916	665	901	640	829	585	658	454
		6	298	58.5	38	1127	818	1096	808	1040	765	906	662
		7	90	24.1	40	487	354	470	354	454	342	415	312
		8	7	2.4	41	51	37	49	37	48	36	45	34
			1052	157.3		2948	2140	2516	1839	2371	1728	2024	1462
			MEAN DIA. 5.2 INCHES, WEIBULL PARAMETERS A= 1.0, B= 4.49, C= 4.39										

Age	Av. dom. ht.	D.b.h.	Stems per acre	Basal area	Av. ht.	All trees, total stem		5-inch class and greater to o.b. tops of—					
								2 inches		3 inches		4 inches	
						O. b.	I. b.	O. b.	I. b.	O. b.	I. b.	O. b.	I. b.
Yrs	Ft	In	No.	Ft²	Ft	— — — — — — — — — — — — — Ft³ — — — — — — — — — — — — —							
30	44	2	2	.0	20	0	0						
		3	29	1.4	28	21	15						
		4	120	10.5	34	182	132						
		5	260	35.5	38	684	497	672	480	622	441	501	349
		6	310	60.9	40	1233	896	1198	887	1140	842	1001	736
		7	175	46.8	43	1017	739	981	739	950	719	876	661
		8	37	12.9	44	287	209	276	209	270	207	255	195
		9	2	.9	46	21	15	20	15	19	15	19	14
			935	168.8		3445	2503	3147	2330	3001	2224	2652	1955
		MEAN DIA. 5.8 INCHES, WEIBULL PARAMETERS A= 1.3, B= 4.76, C= 4.37											
35	48	2	1	.0	19	0	0						
		3	13	.6	29	10	7						
		4	66	5.8	35	103	75						
		5	169	23.0	40	468	340	459	330	426	304	348	245
		6	258	50.7	43	1103	801	1070	796	1022	759	908	671
		7	217	58.0	46	1348	980	1300	980	1262	957	1172	887
		8	85	29.7	48	719	523	691	523	677	520	644	494
		9	12	5.3	49	131	95	126	95	124	95	120	92
			821	173.1		3882	2821	3646	2724	3511	2635	3192	2389
		MEAN DIA. 6.2 INCHES, WEIBULL PARAMETERS A= 1.5, B= 5.03, C= 4.36											
40	51	3	6	.3	28	4	3						
		4	36	3.1	35	56	41						
		5	104	14.2	40	288	209	282	203	262	187	214	151
		6	189	37.1	44	827	601	801	597	767	570	683	506
		7	209	55.9	47	1327	965	1279	965	1243	943	1156	876
		8	128	44.7	49	1105	804	1062	804	1041	800	992	761
		9	36	15.9	51	409	298	392	298	387	298	374	289
		10	4	2.2	52	57	42	55	42	54	42	53	41
			712	173.3		4073	2963	3871	2909	3754	2840	3472	2624
		MEAN DIA. 6.7 INCHES, WEIBULL PARAMETERS A= 1.7, B= 5.31, C= 4.34											

SI 40

2,000 stems per acre

Age	Av. dom. ht.	D.b.h.	Stems per acre	Basal area	Av. ht.	All trees, total stem		5-inch class and greater to o.b. tops of—					
								2 inches		3 inches		4 inches	
						O.b.	I.b.	O.b.	I.b.	O.b.	I.b.	O.b.	I.b.
Yrs	Ft	In	No.	Ft 2	Ft	– – – – – – – – – – – – – Ft 3 – – – – – – – – – – – – –							
10	21	1	108	.6	10	6	3						
		2	668	14.6	17	141	99						
		3	689	35.8	19	340	244						
		4	65	5.7	21	62	44						
			1530	54.7		549	390	0	0	0	0	0	0

MEAN DIA. 2.6 INCHES, WEIBULL PARAMETERS A= .0, B= 2.70, C= 4.44

Age	Av. dom. ht.	D.b.h.	Stems per acre	Basal area	Av. ht.	O.b.	I.b.	O.b.	I.b.	O.b.	I.b.	O.b.	I.b.
15	29	1	10	.1	10	1	0						
		2	136	3.0	20	33	23						
		3	477	23.4	24	294	212						
		4	591	51.6	27	715	517						
		5	193	26.3	28	376	272	373	257	335	227	243	157
		6	9	1.8	30	27	20	26	19	25	18	20	15
			1416	106.1		1446	1044	399	276	360	245	263	172

MEAN DIA. 3.7 INCHES, WEIBULL PARAMETERS A= .3, B= 3.60, C= 4.43

Age	Av. dom. ht.	D.b.h.	Stems per acre	Basal area	Av. ht.	O.b.	I.b.	O.b.	I.b.	O.b.	I.b.	O.b.	I.b.
20	35	1	1	.0	10	0	0						
		2	35	.8	21	9	6						
		3	198	9.7	27	137	99						
		4	461	40.2	31	639	463						
		5	448	61.1	33	1026	744	1012	711	924	644	713	483
		6	137	26.9	35	477	347	465	341	439	321	377	273
		7	8	2.1	36	39	28	38	28	36	27	33	24
			1288	140.8		2327	1687	1515	1080	1399	992	1123	780

MEAN DIA. 4.5 INCHES, WEIBULL PARAMETERS A= .7, B= 4.02, C= 4.42

Age	Av. dom. ht.	D.b.h.	Stems per acre	Basal area	Av. ht.	O.b.	I.b.	O.b.	I.b.	O.b.	I.b.	O.b.	I.b.
25	40	2	10	.2	21	3	2						
		3	83	4.1	28	59	43						
		4	268	23.4	33	395	286						
		5	426	58.1	36	1063	771	1045	743	962	679	763	527
		6	295	57.9	38	1115	810	1085	800	1029	757	896	656
		7	67	17.9	40	362	263	350	263	338	255	309	232
		8	3	1.0	41	22	16	21	16	20	16	19	15
			1152	162.6		3019	2191	2501	1822	2349	1707	1987	1430

MEAN DIA. 5.1 INCHES, WEIBULL PARAMETERS A= 1.0, B= 4.33, C= 4.40

Age	Av. dom. ht.	D.b.h.	Stems per acre	Basal area	Av. ht.	All trees, total stem		5-inch class and greater to o.b. tops of—					
								2 inches		3 inches		4 inches	
						O. b.	I. b.	O. b.	I. b.	O. b.	I. b.	O. b.	I. b.
Yrs	Ft	In	No.	Ft 2	Ft	– – – – – – – – – – – – – Ft 3 – – – – – – – – – – – –							
30	44	2	3	.1	20	1	1						
		3	36	1.8	28	26	19						
		4	147	12.8	34	223	162						
		5	307	41.9	38	808	587	793	567	734	521	591	412
		6	336	66.0	41	1370	995	1330	986	1268	938	1118	823
		7	160	42.8	43	930	676	897	676	869	657	801	605
		8	25	8.7	44	194	141	186	141	182	140	172	132
		9	1	.4	46	10	7	10	7	10	7	9	7
			1015	174.4		3562	2588	3216	2377	3063	2263	2691	1979

MEAN DIA. 5.6 INCHES, WEIBULL PARAMETERS A= 1.3, B= 4.61, C= 4.39

Age	Av. dom. ht.	D.b.h.	Stems per acre	Basal area	Av. ht.	O. b.	I. b.	O. b.	I. b.	O. b.	I. b.	O. b.	I. b.
35	48	2	1	.0	19	0	0						
		3	16	.8	29	12	9						
		4	79	6.9	35	123	89						
		5	199	27.1	40	551	400	540	388	502	358	410	288
		6	288	56.5	43	1231	895	1194	888	1141	847	1013	749
		7	220	58.8	46	1367	994	1318	994	1280	971	1188	899
		8	72	25.1	48	609	443	585	443	574	440	546	418
		9	8	3.5	49	87	64	84	64	83	64	80	62
			883	178.8		3980	2894	3721	2777	3580	2680	3237	2416

MEAN DIA. 6.1 INCHES, WEIBULL PARAMETERS A= 1.5, B= 4.89, C= 4.37

Age	Av. dom. ht.	D.b.h.	Stems per acre	Basal area	Av. ht.	O. b.	I. b.	O. b.	I. b.	O. b.	I. b.	O. b.	I. b.
40	51	3	7	.3	28	5	4						
		4	42	3.7	35	66	48						
		5	122	16.6	40	338	245	331	238	308	220	251	177
		6	214	42.0	44	936	680	907	676	868	646	773	573
		7	223	59.6	47	1416	1029	1364	1029	1326	1007	1233	935
		8	121	42.2	49	1045	760	1004	760	984	756	938	719
		9	28	12.4	51	318	232	305	232	301	232	291	225
		10	2	1.1	52	29	21	27	21	27	21	26	21
			759	178.0		4153	3019	3938	2956	3814	2882	3512	2650

MEAN DIA. 6.6 INCHES, WEIBULL PARAMETERS A= 1.7, B= 5.18, C= 4.35

SI 40

2,250 stems per acre

Age	Av. dom. ht.	D.b.h.	Stems per acre	Basal area	Av. ht.	All trees, total stem		5-inch class and greater to o.b. tops of—					
								2 inches		3 inches		4 inches	
						O.b.	I.b.	O.b.	I.b.	O.b.	I.b.	O.b.	I.b.
Yrs	Ft	In	No.	Ft 2	Ft	— — — — — — — — — — — — Ft 3 — — — — — — — — — — — —							
10	21	1	162	.9	10	8	5						
		2	897	19.6	16	179	125						
		3	615	30.2	19	303	217						
		4	21	1.8	21	20	14						
			1695	52.5		510	361	0	0	0	0	0	0
		MEAN DIA. 2.4 INCHES, WEIBULL PARAMETERS A= .0, B= 2.51, C= 4.46											
15	29	1	14	.1	11	1	0						
		2	183	4.0	20	45	31						
		3	604	29.6	24	373	268						
		4	619	54.0	27	749	542						
		5	134	18.3	28	261	189	259	178	232	158	169	109
		6	3	.6	30	9	7	9	6	8	6	7	5
			1557	106.6		1438	1037	268	184	240	164	176	114
		MEAN DIA. 3.5 INCHES, WEIBULL PARAMETERS A= .3, B= 3.43, C= 4.45											
20	35	1	1	.0	10	0	0						
		2	45	1.0	21	11	8						
		3	251	12.3	27	173	125						
		4	550	48.0	31	762	552						
		5	455	62.0	33	1042	756	1027	722	938	654	724	491
		6	100	19.6	35	348	253	339	249	321	234	275	199
		7	3	.8	36	15	11	14	11	14	10	12	9
			1405	143.8		2351	1705	1380	982	1273	898	1011	699
		MEAN DIA. 4.3 INCHES, WEIBULL PARAMETERS A= .7, B= 3.86, C= 4.43											
25	40	2	12	.3	21	3	2						
		3	103	5.1	28	74	53						
		4	325	28.4	33	479	347						
		5	478	65.2	36	1193	866	1173	833	1080	761	857	591
		6	281	55.2	38	1062	772	1033	762	980	721	854	624
		7	47	12.6	40	254	185	245	185	237	179	217	163
		8	1	.3	41	7	5	7	5	7	5	6	5
			1247	166.9		3072	2230	2458	1785	2304	1666	1934	1363
		MEAN DIA. 5.0 INCHES, WEIBULL PARAMETERS A= 1.0, B= 4.19, C= 4.42											

Age	Av. dom. ht.	D.b.h.	Stems per acre	Basal area	Av. ht.	All trees, total stem		5-inch class and greater to o.b. tops of—					
								2 inches		3 inches		4 inches	
						O. b.	I. b.	O. b.	I. b.	O. b.	I. b.	O. b.	I. b.
Yrs	Ft	In	No.	Ft 2	Ft				Ft 3				
50	44	2	3	.1	21	1	1						
		3	44	2.2	30	34	24						
		4	175	15.3	35	273	198						
		5	355	48.4	39	959	696	941	674	872	621	707	495
		6	354	69.5	42	1478	1074	1435	1066	1369	1015	1211	894
		7	142	37.9	44	844	614	814	614	789	598	730	551
		8	17	5.9	46	138	100	132	100	130	99	123	94
			1090	179.3		3727	2707	3322	2454	3160	2333	2771	2034

MEAN DIA. 5.5 INCHES, WEIBULL PARAMETERS A= 1.3, B= 4.48, C= 4.40

Age	Av. dom. ht.	D.b.h.	Stems per acre	Basal area	Av. ht.	O. b.	I. b.	O. b.	I. b.	O. b.	I. b.	O. b.	I. b.
35	48	2	1	.0	20	0	0						
		3	19	.9	29	14	10						
		4	93	8.1	36	149	108						
		5	229	31.2	40	634	460	622	447	578	412	471	331
		6	316	62.0	43	1351	982	1310	975	1252	930	1112	822
		7	218	58.3	46	1355	985	1306	985	1268	962	1177	891
		8	60	20.9	48	508	369	488	369	478	367	455	349
		9	5	2.2	49	55	40	52	40	52	40	50	38
			941	183.8		4066	2954	3778	2816	3628	2711	3265	2431

MEAN DIA. 6.0 INCHES, WEIBULL PARAMETERS A= 1.5, B= 4.77, C= 4.38

Age	Av. dom. ht.	D.b.h.	Stems per acre	Basal area	Av. ht.	O. b.	I. b.	O. b.	I. b.	O. b.	I. b.	O. b.	I. b.
40	51	3	8	.4	28	6	4						
		4	48	4.2	35	75	54						
		5	139	19.0	40	385	279	377	271	351	250	286	201
		6	238	46.7	44	1041	756	1009	752	965	718	860	637
		7	234	62.5	47	1485	1080	1432	1080	1391	1056	1294	981
		8	114	39.8	49	984	716	946	716	927	712	883	678
		9	22	9.7	51	250	182	240	182	236	182	229	177
		10	1	.5	52	14	10	14	10	14	10	13	10
			804	182.9		4240	3081	4016	3011	3884	2928	3565	2684

MEAN DIA. 6.5 INCHES, WEIBULL PARAMETERS A= 1.7, B= 5.07, C= 4.35

SI 40

2,500 stems per acre

Age	Av. dom. ht.	D.b.h.	Stems per acre	Basal area	Av. ht.	All trees, total stem		5-inch class and greater to o.b. tops of—					
								2 inches		3 inches		4 inches	
						O. b.	I. b.	O. b.	I. b.	O. b.	I. b.	O. b.	I. b.
Yrs	Ft	In	No.	Ft²	Ft	-------------- Ft³ --------------							
10	21	1	240	1.3	10	12	8						
		2	1142	24.9	16	228	160						
		3	472	23.2	19	233	167						
		4	4	.3	21	4	3						
			1858	49.7		477	338	0	0	0	0	0	0
15	29	1	18	.1	11	1	1						
		2	241	5.3	20	59	41						
		3	740	36.3	24	456	328						
		4	612	53.4	27	741	536						
		5	83	11.3	28	162	117	160	110	144	98	105	68
		6	1	.2	30	3	2	3	2	3	2	2	2
			1695	106.6		1422	1025	163	112	147	100	107	70
20	35	1	2	.0	10	0	0						
		2	56	1.2	21	14	10						
		3	312	15.3	27	216	155						
		4	640	55.9	31	887	642						
		5	441	60.1	33	1010	732	996	700	909	634	702	476
		6	68	13.4	35	237	172	231	169	218	159	187	136
		7	1	.3	36	5	4	5	4	5	3	4	3
			1520	146.1		2369	1715	1232	873	1132	796	893	615
25	40	2	15	.3	21	4	3						
		3	125	6.1	28	89	64						
		4	385	33.6	33	567	411						
		5	523	71.3	36	1305	947	1283	912	1182	833	937	647
		6	259	50.9	38	979	711	952	702	904	665	787	576
		7	31	8.3	40	168	122	162	122	156	118	143	108
		8	1	.3	41	7	5	7	5	7	5	6	5
			1339	170.9		3119	2263	2404	1741	2249	1621	1873	1336

MEAN DIA. 2.2 INCHES, WEIBULL PARAMETERS A= .0, B= 2.33, C= 4.47

MEAN DIA. 3.4 INCHES, WEIBULL PARAMETERS A= .3, B= 3.27, C= 4.46

MEAN DIA. 4.2 INCHES, WEIBULL PARAMETERS A= .7, B= 3.72, C= 4.44

MEAN DIA. 4.8 INCHES, WEIBULL PARAMETERS A= 1.0, B= 4.06, C= 4.43

Age	Av. dom. ht.	D.b.h.	Stems per acre	Basal area	Av. ht.	All trees, total stem		5-inch class and greater to o.b. tops of—					
								2 inches		3 inches		4 inches	
						O. b.	I. b.	O. b.	I. b.	O. b.	I. b.	O. b.	I. b.
Yrs	Ft	In	No.	Ft 2	Ft	Ft 3							
30	44	2	4	.1	21	1	1						
		3	52	2.6	30	40	29						
		4	206	18.0	35	322	233						
		5	401	54.7	39	1083	786	1063	761	985	701	799	559
		6	366	71.9	42	1528	1111	1483	1102	1415	1049	1253	924
		7	123	32.9	44	731	532	705	532	684	518	632	477
		8	11	3.8	46	89	65	86	65	84	64	80	61
			1163	183.9		3794	2757	3337	2460	3168	2332	2764	2021
				MEAN DIA.	5.4 INCHES, WEIBULL PARAMETERS A= 1.3, B= 4.37, C= 4.41								
35	48	2	1	.0	20	0	0						
		3	22	1.1	29	16	12						
		4	107	9.3	36	172	124						
		5	260	35.5	40	720	523	706	507	656	468	535	376
		6	342	67.2	43	1462	1062	1418	1055	1355	1006	1203	890
		7	213	56.9	46	1323	962	1276	962	1239	940	1150	871
		8	49	17.1	48	415	301	398	301	390	300	371	285
		9	3	1.3	49	33	24	31	24	31	24	30	23
			997	188.4		4141	3008	3829	2849	3671	2738	3289	2445
				MEAN DIA.	5.9 INCHES, WEIBULL PARAMETERS A= 1.5, B= 4.67, C= 4.39								
40	51	3	9	.4	29	7	5						
		4	55	4.8	36	88	64						
		5	156	21.3	40	432	314	424	304	393	281	321	226
		6	260	51.1	44	1137	826	1103	821	1055	784	940	696
		7	241	64.4	47	1530	1112	1475	1112	1433	1088	1333	1010
		8	106	37.0	49	915	666	879	666	862	662	821	630
		9	17	7.5	51	193	141	185	141	185	141	177	137
		10	1	.5	52	14	10	14	10	14	10	13	10
			845	187.0		4316	3138	4080	3054	3940	2966	3605	2709
				MEAN DIA.	6.4 INCHES, WEIBULL PARAMETERS A= 1.7, B= 4.98, C= 4.36								

SI 50

500 stems per acre

Age	Av. dom. ht.	D.b.h.	Stems per acre	Basal area	Av. ht.	All trees, total stem		5-inch class and greater to o.b. tops of—					
								2 inches		3 inches		4 inches	
						O. b.	I. b.	O. b.	I. b.	O. b.	I. b.	O. b.	I. b.
Yrs	Ft	In	No.	Ft 2	Ft	— — — — — — — — — — — — — — Ft 3 — — — — — — — — — — — — — —							
10	26	1	1	.0	6	0	0						
		2	10	.2	14	2	1						
		3	38	1.9	19	19	13						
		4	86	7.5	21	81	59						
		5	129	17.6	23	207	150	207	138	181	118	119	71
		6	121	23.8	24	290	210	286	202	262	184	205	140
		7	66	17.6	25	224	162	218	159	205	150	175	127
		8	18	6.3	26	83	60	80	60	77	57	69	51
		9	2	.9	27	12	9	12	9	11	9	10	8
			471	75.7		918	664	803	568	736	518	578	397

MEAN DIA. 5.4 INCHES, WEIBULL PARAMETERS A= .1, B= 5.63, C= 4.27

Age	Av. dom. ht.	D.b.h.	Stems per acre	Basal area	Av. ht.	O. b.	I. b.	O. b.	I. b.	O. b.	I. b.	O. b.	I. b.
15	36	2	2	.0	17	0	0						
		3	13	.6	23	8	6						
		4	40	3.5	28	50	36						
		5	82	11.2	30	171	124	169	117	153	105	114	76
		6	119	23.4	33	391	284	382	278	359	261	305	219
		7	116	31.0	34	533	388	517	386	496	370	445	331
		8	71	24.8	35	438	319	423	319	410	311	381	288
		9	24	10.6	36	193	140	185	140	181	139	172	132
		10	4	2.2	37	41	30	39	30	38	30	37	29
			471	107.3		1825	1327	1715	1270	1637	1216	1454	1075

MEAN DIA. 6.5 INCHES, WEIBULL PARAMETERS A= .8, B= 6.06, C= 4.27

Age	Av. dom. ht.	D.b.h.	Stems per acre	Basal area	Av. ht.	O. b.	I. b.	O. b.	I. b.	O. b.	I. b.	O. b.	I. b.
20	44	3	5	.2	26	3	2						
		4	20	1.7	31	28	20						
		5	51	7.0	35	124	90	122	86	112	79	88	60
		6	90	17.7	38	340	247	331	244	314	231	273	200
		7	116	31.0	40	627	456	606	456	585	441	535	402
		8	100	34.9	42	741	538	712	538	696	532	655	500
		9	55	24.3	43	527	384	506	384	497	383	477	367
		10	17	9.3	44	206	150	197	150	195	150	189	146
		11	2	1.3	45	30	22	29	22	28	22	28	22
			456	127.4		2626	1909	2503	1880	2427	1838	2245	1697

MEAN DIA. 7.2 INCHES, WEIBULL PARAMETERS A= 1.3, B= 6.28, C= 4.27

Age	Av. dom. ht.	D.b.h.	Stems per acre	Basal area	Av. ht.	O. b.	I. b.	O. b.	I. b.	O. b.	I. b.	O. b.	I. b.
25	50	3	2	.1	27	1	1						
		4	10	.9	34	15	11						
		5	31	4.2	38	82	59	80	57	74	53	60	42
		6	64	12.6	42	267	194	259	193	247	183	219	162
		7	96	25.7	44	571	415	551	415	534	404	493	373
		8	105	36.7	46	851	619	818	619	801	614	760	582
		9	77	34.0	48	824	599	790	599	778	599	751	580
		10	35	19.1	50	461	350	461	350	456	350	445	345
		11	9	5.9	51	153	111	146	111	145	111	142	111
		12	1	.8	52	21	15	20	15	20	15	19	15
			430	139.9		3266	2374	3125	2359	3055	2329	2889	2210

MEAN DIA. 7.7 INCHES, WEIBULL PARAMETERS A= 1.7, B= 6.48, C= 4.26

Age	Av. dom. ht.	D.b.h.	Stems per acre	Basal area	Av. ht.	All trees, total stem		5-inch class and greater to o.b. tops of—					
								2 inches		3 inches		4 inches	
						O.b.	I.b.	O.b.	I.b.	O.b.	I.b.	O.b.	I.b.
Yrs	Ft	In	No.	Ft²	Ft	— — — — — — — — — — — Ft³ — — — — — — — — — —							
30	55	3	1	.0	28	1	1						
		4	5	.4	35	8	6						
		5	16	2.5	41	51	37	50	36	47	33	38	27
		6	42	8.2	45	188	137	182	136	174	130	156	116
		7	72	19.2	48	467	339	450	339	437	332	408	309
		8	93	32.5	51	836	608	802	608	788	606	752	578
		9	86	38.0	53	1016	739	973	739	961	739	931	721
		10	54	29.5	54	802	584	767	584	760	584	744	579
		11	21	13.9	56	391	285	374	285	372	285	366	285
		12	4	3.1	57	90	66	86	66	86	66	85	66
			396	147.3		3850	2802	3684	2793	3625	2775	3480	2681

MEAN DIA. 8.3 INCHES, WEIBULL PARAMETERS A= 2.0, B= 6.70, C= 4.25

Age	Av. dom. ht.	D.b.h.	Stems per acre	Basal area	Av. ht.	O.b.	I.b.	O.b.	I.b.	O.b.	I.b.	O.b.	I.b.
35	60	4	3	.3	36	5	3						
		5	10	1.4	42	29	21	28	21	27	19	22	16
		6	27	5.3	47	126	92	122	91	117	88	105	78
		7	50	13.4	50	338	245	325	245	317	241	296	225
		8	73	25.5	53	682	496	654	496	643	495	615	473
		9	81	35.8	55	993	722	951	722	939	722	912	707
		10	64	34.9	57	1003	730	959	730	952	730	933	726
		11	35	23.1	59	687	500	656	500	653	500	644	500
		12	12	9.4	60	285	207	272	207	271	207	269	207
		13	2	1.8	62	58	42	55	42	55	42	55	42
			357	150.8		4206	3058	4022	3054	3974	3044	3851	2974

MEAN DIA. 8.8 INCHES, WEIBULL PARAMETERS A= 2.3, B= 6.98, C= 4.23

Age	Av. dom. ht.	D.b.h.	Stems per acre	Basal area	Av. ht.	O.b.	I.b.	O.b.	I.b.	O.b.	I.b.	O.b.	I.b.
40	63	4	1	.1	36	2	1						
		5	6	.8	42	17	13	17	12	16	11	13	9
		6	16	3.1	48	76	55	74	55	71	53	64	48
		7	33	8.8	52	232	168	223	168	217	166	204	155
		8	53	18.5	55	514	374	493	374	485	374	465	358
		9	67	29.6	57	851	619	815	619	806	619	783	607
		10	65	35.5	60	1073	780	1025	780	1018	780	1000	778
		11	46	30.4	61	933	679	892	679	887	679	877	679
		12	22	17.3	63	549	399	524	399	522	399	518	399
		13	7	6.5	64	208	151	198	151	198	151	197	151
		14	1	1.1	66	36	26	34	26	34	26	34	26
			317	151.6		4491	3265	4295	3263	4254	3258	4155	3210

MEAN DIA. 9.4 INCHES, WEIBULL PARAMETERS A= 2.5, B= 7.34, C= 4.22

SI 50

750 stems per acre

Age	Av. dom. ht.	D.b.h.	Stems per acre	Basal area	Av. ht.	All trees, total stem		5-inch class and greater to o.b. tops of—					
								2 inches		3 inches		4 inches	
						O.b.	I.b.	O.b.	I.b.	O.b.	I.b.	O.b.	I.b.
Yrs	Ft	In	No.	Ft²	Ft	— — — — — — — — — — — Ft³ — — — — — — — — — — —							
10	26	1	3	.0	8	0	0						
		2	28	.6	16	6	4						
		3	98	4.8	20	51	36						
		4	193	16.8	23	200	144						
		5	211	28.6	24	353	256	352	237	310	204	209	128
		6	113	22.2	25	282	205	277	197	255	180	202	140
		7	23	6.1	26	81	59	79	58	75	55	64	47
		8	1	.3	27	5	3	5	3	4	3	4	3
			670	79.7		978	707	713	495	644	442	475	318
			MEAN DIA. 4.7 INCHES, WEIBULL PARAMETERS A= .1, B= 4.83, C= 4.32										
15	36	2	5	.1	19	1	1						
		3	31	1.5	25	20	14						
		4	92	8.0	29	119	86						
		5	169	23.0	32	375	272	371	260	337	234	258	173
		6	194	38.1	34	657	477	640	468	603	440	515	372
		7	122	32.6	35	577	420	559	418	537	401	483	360
		8	36	12.6	36	229	166	220	166	214	163	199	151
		9	4	1.8	37	33	24	32	24	31	24	30	23
			653	117.7		2011	1460	1822	1336	1722	1262	1485	1079
			MEAN DIA. 5.7 INCHES, WEIBULL PARAMETERS A= .8, B= 5.31, C= 4.32										
20	44	2	1	.0	20	0	0						
		3	11	.5	28	8	6						
		4	44	3.8	33	65	47						
		5	106	14.5	37	272	197	267	190	246	174	197	137
		6	165	32.4	39	640	465	622	460	591	436	517	379
		7	166	44.4	41	920	669	888	669	859	648	788	593
		8	94	32.8	43	713	518	685	518	670	512	632	483
		9	26	11.5	44	255	186	245	186	241	186	231	178
		10	3	1.6	45	37	27	36	27	35	27	34	26
			616	141.6		2910	2115	2743	2050	2642	1983	2399	1796
			MEAN DIA. 6.5 INCHES, WEIBULL PARAMETERS A= 1.3, B= 5.57, C= 4.31										
25	50	3	4	.2	29	3	2						
		4	21	1.8	36	34	24						
		5	62	8.5	40	172	125	168	121	156	112	128	90
		6	120	23.6	43	513	373	498	370	475	353	422	312
		7	156	41.7	46	969	705	935	705	907	688	842	638
		8	128	44.7	48	1083	788	1040	788	1020	783	970	744
		9	60	26.5	49	655	477	628	477	619	477	598	462
		10	14	7.6	51	196	143	188	143	186	143	182	141
		11	1	.7	52	17	13	17	13	16	13	16	13
			566	155.2		3642	2650	3474	2617	3379	2569	3158	2400
			MEAN DIA. 7.1 INCHES, WEIBULL PARAMETERS A= 1.7, B= 5.80, C= 4.30										

Age	Av. dom. ht.	D.b.h.	Stems per acre	Basal area	Av. ht.	All trees, total stem		5-inch class and greater to o.b. tops of—					
								2 inches		3 inches		4 inches	
						O. b.	I. b.	O. b.	I. b.	O. b.	I. b.	O. b.	I. b.
Yrs	Ft	In	No.	Ft 2	Ft				Ft 3				
30	55	3	1	.0	30	1	1						
		4	10	.9	37	16	12						
		5	35	4.8	43	104	76	102	74	95	69	79	56
		6	78	15.3	47	364	265	353	264	339	253	304	227
		7	122	32.6	50	824	599	793	599	772	588	722	549
		8	132	46.1	52	1209	880	1161	880	1140	878	1090	838
		9	89	39.3	54	1071	779	1026	779	1013	779	983	761
		10	34	18.5	56	524	381	501	381	497	381	487	379
		11	6	4.0	57	114	83	109	83	108	83	107	83
			507	161.5		4227	3076	4045	3060	3964	3031	3772	2893
		MEAN DIA. 7.6 INCHES, WEIBULL PARAMETERS A= 2.0, B= 6.04, C= 4.26											
35	60	4	5	.4	38	8	6						
		5	19	2.6	44	58	42	57	41	53	38	44	31
		6	48	9.4	48	229	166	222	166	213	159	192	143
		7	86	23.0	52	604	439	581	439	567	432	532	405
		8	111	38.7	55	1076	782	1032	782	1015	782	974	750
		9	99	43.7	57	1258	915	1204	915	1190	915	1158	897
		10	57	31.1	59	925	673	884	673	878	673	861	671
		11	18	11.9	60	359	261	343	261	341	261	337	261
		12	3	2.4	62	74	54	70	54	70	54	70	54
			446	163.2		4591	3338	4393	3331	4327	3314	4168	3212
		MEAN DIA. 8.2 INCHES, WEIBULL PARAMETERS A= 2.3, B= 6.34, C= 4.27											
40	63	4	2	.2	38	3	2						
		5	10	1.4	44	30	22	30	22	28	20	23	17
		6	28	5.5	49	136	99	132	99	127	95	115	86
		7	55	14.7	53	394	286	379	286	369	282	347	265
		8	82	28.6	56	809	588	776	588	763	588	733	565
		9	90	39.8	59	1183	861	1133	861	1120	861	1091	846
		10	70	38.2	61	1174	854	1122	854	1114	854	1095	853
		11	36	23.8	63	754	549	720	549	717	549	709	549
		12	11	8.6	64	279	203	266	203	265	203	263	203
		13	2	1.8	66	61	45	58	45	58	45	58	45
			386	162.5		4823	3509	4616	3507	4561	3497	4434	3429
		MEAN DIA. 8.8 INCHES, WEIBULL PARAMETERS A= 2.5, B= 6.70, C= 4.25											

SI 50

1,000 stems per acre

Age	Av. dom. ht.	D.b.h.	Stems per acre	Basal area	Av. ht.	All trees, total stem		5-inch class and greater to o.b. tops of—					
						O.b.	I.b.	2 inches		3 inches		4 inches	
								O.b.	I.b.	O.b.	I.b.	O.b.	I.b.
Yrs	Ft	In	No.	Ft 2	Ft	― ― ― ― ― ― ― ― ― ― ― Ft 3 ― ― ― ― ― ― ― ― ― ― ―							
10	26	1	6	.0	10	0	0						
		2	54	1.2	17	11	8						
		3	184	9.0	21	100	72						
		4	311	27.1	23	322	232						
		5	238	32.5	25	414	300	413	279	366	243	252	157
		6	64	12.6	26	166	121	163	116	151	107	121	84
		7	4	1.1	27	15	11	14	10	13	10	12	9
			861	83.5		1028	744	590	405	530	360	385	250
			MEAN DIA. 4.2 INCHES, WEIBULL PARAMETERS A= .1, B= 4.34, C= 4.36										
15	36	2	9	.2	21	2	2						
		3	55	2.7	27	38	27						
		4	160	14.0	31	222	161						
		5	263	35.9	33	602	437	594	418	542	378	419	284
		6	232	45.6	35	808	587	787	577	744	543	638	463
		7	92	24.6	36	448	325	433	325	417	312	377	281
		8	12	4.2	37	78	57	75	57	73	56	68	52
			823	127.0		2198	1596	1889	1377	1776	1289	1502	1080
			MEAN DIA. 5.3 INCHES, WEIBULL PARAMETERS A= .8, B= 4.85, C= 4.35										
20	44	2	2	.0	22	1	0						
		3	19	.9	30	15	10						
		4	75	6.5	35	117	85						
		5	171	23.3	38	450	327	442	316	409	290	329	229
		6	237	46.5	41	966	702	938	696	894	662	788	581
		7	183	48.9	43	1063	773	1026	773	994	751	916	691
		8	66	23.0	44	512	372	492	372	481	368	455	348
		9	9	4.0	45	90	66	87	66	85	66	82	63
			762	153.3		3214	2335	2985	2223	2863	2137	2570	1912
			MEAN DIA. 6.1 INCHES, WEIBULL PARAMETERS A= 1.3, B= 5.14, C= 4.34										
25	50	3	6	.3	31	5	3						
		4	35	3.1	37	58	42						
		5	100	13.6	41	284	206	278	200	259	185	212	150
		6	179	35.1	45	801	582	776	579	743	553	664	493
		7	201	53.7	47	1276	928	1230	928	1195	907	1111	842
		8	125	43.6	49	1079	785	1037	785	1017	781	969	743
		9	37	16.3	50	412	300	395	300	390	300	377	291
		10	4	2.2	52	57	42	55	42	54	42	53	41
			687	168.0		3972	2888	3771	2834	3658	2768	3386	2560
			MEAN DIA. 6.7 INCHES, WEIBULL PARAMETERS A= 1.7, B= 5.38, C= 4.33										

Age	Av. dom. ht.	D.b.h.	Stems per acre	Basal area	Av. ht.	All trees, total stem		5-inch class and greater to o.b. tops of—					
								2 inches		3 inches		4 inches	
						O. b.	I. b.	O. b.	I. b.	O. b.	I. b.	O. b.	I. b.
Yrs	Ft	In	No.	Ft²	Ft	‑ ‑ ‑ ‑ ‑ ‑ ‑ ‑ ‑ ‑ ‑ ‑ Ft³ ‑ ‑ ‑ ‑ ‑ ‑ ‑ ‑ ‑ ‑ ‑ ‑							
50	44	2	3	.1	21	1	1						
		3	44	2.2	30	34	24						
		4	175	15.3	35	273	198						
		5	355	48.4	39	959	696	941	674	872	621	707	495
		6	354	69.5	42	1478	1074	1435	1066	1369	1015	1211	894
		7	142	37.9	44	844	614	814	614	789	598	730	551
		8	17	5.9	46	138	100	132	100	130	99	123	94
			1090	179.3		3727	2707	3322	2454	3160	2333	2771	2034
					MEAN DIA. 5.5 INCHES, WEIBULL PARAMETERS A= 1.3, B= 4.48, C= 4.40								
35	48	2	1	.0	20	0	0						
		3	19	.9	29	14	10						
		4	93	8.1	36	149	108						
		5	229	31.2	40	634	460	622	447	578	412	471	331
		6	316	62.0	43	1351	982	1310	975	1252	930	1112	822
		7	218	58.3	46	1355	985	1306	985	1268	962	1177	891
		8	60	20.9	48	508	369	488	369	478	367	455	349
		9	5	2.2	49	55	40	52	40	52	40	50	38
			941	183.8		4066	2954	3778	2816	3628	2711	3265	2431
					MEAN DIA. 6.0 INCHES, WEIBULL PARAMETERS A= 1.5, B= 4.77, C= 4.38								
40	51	3	8	.4	28	6	4						
		4	48	4.2	35	75	54						
		5	139	19.0	40	385	279	377	271	351	250	286	201
		6	238	46.7	44	1041	756	1009	752	965	718	860	637
		7	234	62.5	47	1485	1080	1432	1080	1391	1056	1294	981
		8	114	39.8	49	984	716	946	716	927	712	883	678
		9	22	9.7	51	250	182	240	182	236	182	229	177
		10	1	.5	52	14	10	14	10	14	10	13	10
			804	182.9		4240	3081	4018	3011	3884	2928	3565	2684
					MEAN DIA. 6.5 INCHES, WEIBULL PARAMETERS A= 1.7, B= 5.07, C= 4.35								

SI 40

2,500 stems per acre

Age	Av. dom. ht.	D.b.h.	Stems per acre	Basal area	Av. ht.	All trees, total stem		5-inch class and greater to o.b. tops of—					
								2 inches		3 inches		4 inches	
						O. b.	I. b.	O. b.	I. b.	O. b.	I. b.	O. b.	I. b.
Yrs	Ft	In	No.	Ft 2	Ft	– – – – – – – – – – – – Ft 3 – – – – – – – – – – – –							
10	21	1	240	1.3	10	12	8						
		2	1142	24.9	16	228	160						
		3	472	23.2	19	233	167						
		4	4	.3	21	4	3						
			1858	49.7		477	338	0	0	0	0	0	0
					MEAN DIA. 2.2 INCHES, WEIBULL PARAMETERS A= .0, B= 2.33, C= 4.47								
15	29	1	18	.1	11	1	1						
		2	241	5.3	20	59	41						
		3	740	36.3	24	456	328						
		4	612	53.4	27	741	536						
		5	83	11.3	28	162	117	160	110	144	98	105	68
		6	1	.2	30	3	2	3	2	3	2	2	2
			1695	106.6		1422	1025	163	112	147	100	107	70
					MEAN DIA. 3.4 INCHES, WEIBULL PARAMETERS A= .3, B= 3.27, C= 4.46								
20	35	1	2	.0	10	0	0						
		2	56	1.2	21	14	10						
		3	312	15.3	27	216	155						
		4	640	55.9	31	887	642						
		5	441	60.1	33	1010	732	996	700	909	634	702	476
		6	68	13.4	35	237	172	231	169	218	159	187	136
		7	1	.3	36	5	4	5	4	5	3	4	3
			1520	146.1		2369	1715	1232	873	1132	796	893	615
					MEAN DIA. 4.2 INCHES, WEIBULL PARAMETERS A= .7, B= 3.72, C= 4.44								
25	40	2	15	.3	21	4	3						
		3	125	6.1	28	89	64						
		4	385	33.6	33	567	411						
		5	523	71.3	36	1305	947	1283	912	1182	833	937	647
		6	259	50.9	38	979	711	952	702	904	665	787	576
		7	31	8.3	40	168	122	162	122	156	118	143	108
		8	1	.3	41	7	5	7	5	7	5	6	5
			1339	170.9		3119	2263	2404	1741	2249	1621	1873	1336
					MEAN DIA. 4.8 INCHES, WEIBULL PARAMETERS A= 1.0, B= 4.06, C= 4.43								

Age	Av. dom. ht.	D.b.h.	Stems per acre	Basal area	Av. ht.	All trees, total stem		5-inch class and greater to o.b. tops of—					
								2 inches		3 inches		4 inches	
						O. b.	I. b.	O. b.	I. b.	O. b.	I. b.	O. b.	I. b.
Yrs	Ft	In	No.	Ft 2	Ft	------------ Ft 3 ------------							
30	44	2	4	.1	21	1	1						
		3	52	2.6	30	40	29						
		4	206	18.0	35	322	233						
		5	401	54.7	39	1083	786	1063	761	985	701	799	559
		6	366	71.9	42	1528	1111	1483	1102	1415	1049	1253	924
		7	123	32.9	44	731	532	705	532	684	518	632	477
		8	11	3.8	46	89	65	86	65	84	64	80	61
			1163	183.9		3794	2757	3337	2460	3168	2332	2764	2021

MEAN DIA. 5.4 INCHES, WEIBULL PARAMETERS A= 1.3, B= 4.37, C= 4.41

35	48	2	1	.0	20	0	0						
		3	22	1.1	29	16	12						
		4	107	9.3	36	172	124						
		5	260	35.5	40	720	523	706	507	656	468	535	376
		6	342	67.2	43	1462	1062	1418	1055	1355	1006	1203	890
		7	213	56.9	46	1323	962	1276	962	1239	940	1150	871
		8	49	17.1	48	415	301	398	301	390	300	371	285
		9	3	1.3	49	33	24	31	24	31	24	30	23
			997	188.4		4141	3008	3829	2849	3671	2738	3289	2445

MEAN DIA. 5.9 INCHES, WEIBULL PARAMETERS A= 1.5, B= 4.67, C= 4.39

40	51	3	9	.4	29	7	5						
		4	55	4.8	36	88	64						
		5	156	21.3	40	432	314	424	304	393	281	321	226
		6	260	51.1	44	1137	826	1103	821	1055	784	940	696
		7	241	64.4	47	1530	1112	1475	1112	1433	1088	1333	1010
		8	106	37.0	49	915	666	879	666	862	662	821	630
		9	17	7.5	51	193	141	185	141	183	141	177	137
		10	1	.5	52	14	10	14	10	14	10	13	10
			845	187.0		4316	3138	4080	3054	3940	2966	3605	2709

MEAN DIA. 6.4 INCHES, WEIBULL PARAMETERS A= 1.7, B= 4.98, C= 4.36

SI 50

500 stems per acre

Age	Av. dom. ht.	D.b.h.	Stems per acre	Basal area	Av. ht.	All trees, total stem		5-inch class and greater to o.b. tops of—					
								2 inches		3 inches		4 inches	
						O. b.	I. b.	O. b.	I. b.	O. b.	I. b.	O. b.	I. b.
Yrs	Ft	In	No.	Ft 2	Ft	$\leftarrow$				Ft 3			$\rightarrow$
10	26	1	1	.0	6	0	0						
		2	10	.2	14	2	1						
		3	38	1.9	19	19	13						
		4	86	7.5	21	81	59						
		5	129	17.6	23	207	150	207	138	181	118	119	71
		6	121	23.8	24	290	210	286	202	262	184	205	140
		7	66	17.6	25	224	162	218	159	205	150	175	127
		8	18	6.3	26	83	60	80	60	77	57	69	51
		9	2	.9	27	12	9	12	9	11	9	10	8
			471	75.7		918	664	803	568	736	518	578	397
		MEAN DIA. 5.4 INCHES, WEIBULL PARAMETERS A= .1, B= 5.63, C= 4.27											
15	36	2	2	.0	17	0	0						
		3	13	.6	23	8	6						
		4	40	3.5	28	50	36						
		5	82	11.2	30	171	124	169	117	153	105	114	76
		6	119	23.4	33	391	284	382	278	359	261	305	219
		7	116	31.0	34	533	388	517	386	496	370	445	331
		8	71	24.8	35	438	319	423	319	410	311	381	288
		9	24	10.6	36	193	140	185	140	181	139	172	132
		10	4	2.2	37	41	30	39	30	38	30	37	29
			471	107.3		1825	1327	1715	1270	1637	1216	1454	1075
		MEAN DIA. 6.5 INCHES, WEIBULL PARAMETERS A= .8, B= 6.06, C= 4.27											
20	44	3	5	.2	26	3	2						
		4	20	1.7	31	28	20						
		5	51	7.0	35	124	90	122	86	112	79	88	60
		6	90	17.7	38	340	247	331	244	314	231	273	200
		7	116	31.0	40	627	456	606	456	585	441	535	402
		8	100	34.9	42	741	538	712	538	696	532	655	500
		9	55	24.3	43	527	384	506	384	497	383	477	367
		10	17	9.3	44	206	150	197	150	195	150	189	146
		11	2	1.3	45	30	22	29	22	28	22	28	22
			456	127.4		2626	1909	2503	1880	2427	1838	2245	1697
		MEAN DIA. 7.2 INCHES, WEIBULL PARAMETERS A= 1.3, B= 6.28, C= 4.27											
25	50	3	2	.1	27	1	1						
		4	10	.9	34	15	11						
		5	31	4.2	38	82	59	80	57	74	53	60	42
		6	64	12.6	42	267	194	259	193	247	183	219	162
		7	96	25.7	44	571	415	551	415	534	404	493	373
		8	105	36.7	46	851	619	818	619	801	614	760	582
		9	77	34.0	48	824	599	790	599	778	599	751	580
		10	35	19.1	50	481	350	461	350	456	350	445	345
		11	9	5.9	51	153	111	146	111	145	111	142	111
		12	1	.8	52	21	15	20	15	20	15	19	15
			430	139.9		3266	2374	3125	2359	3055	2329	2889	2210
		MEAN DIA. 7.7 INCHES, WEIBULL PARAMETERS A= 1.7, B= 6.48, C= 4.26											

28

Age	Av. dom. ht.	D.b.h.	Stems per acre	Basal area	Av. ht.	All trees, total stem		5-inch class and greater to o.b. tops of—					
								2 inches		3 inches		4 inches	
						O. b.	I. b.	O. b.	I. b.	O. b.	I. b.	O. b.	I. b.
Yrs	Ft	In	No.	Ft 2	Ft	─ ─ ─ ─ ─ ─ ─ ─ ─ ─ ─ Ft 3 ─ ─ ─ ─ ─ ─ ─ ─ ─ ─							
30	55	3	1	.0	28	1	1						
		4	5	.4	35	8	6						
		5	18	2.5	41	51	37	50	36	47	33	38	27
		6	42	8.2	45	188	137	182	136	174	130	156	116
		7	72	19.2	48	467	339	450	339	437	332	408	309
		8	93	32.5	51	836	608	802	608	788	606	752	578
		9	86	38.0	53	1016	739	973	739	961	739	931	721
		10	54	29.5	54	802	584	767	584	760	584	744	579
		11	21	13.9	56	391	285	374	285	372	285	366	285
		12	4	3.1	57	90	66	86	66	86	66	85	66
			396	147.3		3850	2802	3684	2793	3625	2775	3480	2681
					MEAN DIA. 8.3 INCHES, WEIBULL PARAMETERS A= 2.0, B= 6.70, C= 4.25								
35	60	4	3	.3	36	5	3						
		5	10	1.4	42	29	21	28	21	27	19	22	16
		6	27	5.3	47	126	92	122	91	117	88	105	78
		7	50	13.4	50	338	245	325	245	317	241	296	225
		8	73	25.5	53	682	496	654	496	643	495	615	473
		9	81	35.8	55	993	722	951	722	939	722	912	707
		10	64	34.9	57	1003	730	959	730	952	730	933	726
		11	35	23.1	59	687	500	656	500	653	500	644	500
		12	12	9.4	60	285	207	272	207	271	207	269	207
		13	2	1.8	62	58	42	55	42	55	42	55	42
			357	150.8		4206	3058	4022	3054	3974	3044	3851	2974
					MEAN DIA. 8.8 INCHES, WEIBULL PARAMETERS A= 2.3, B= 6.98, C= 4.23								
40	63	4	1	.1	36	2	1						
		5	6	.8	42	17	13	17	12	16	11	13	9
		6	16	3.1	48	76	55	74	55	71	53	64	48
		7	33	8.8	52	232	168	223	168	217	166	204	155
		8	53	18.5	55	514	374	493	374	485	374	465	358
		9	67	29.6	57	851	619	815	619	806	619	783	607
		10	65	35.5	60	1073	780	1025	780	1018	780	1000	778
		11	46	30.4	61	933	679	892	679	887	679	877	679
		12	22	17.3	63	549	399	524	399	522	399	518	399
		13	7	6.5	64	208	151	198	151	198	151	197	151
		14	1	1.1	66	36	26	34	26	34	26	34	26
			317	151.6		4491	3265	4295	3263	4254	3258	4155	3210
					MEAN DIA. 9.4 INCHES, WEIBULL PARAMETERS A= 2.5, B= 7.34, C= 4.22								

SI 50

750 stems per acre

Age	Av. dom. ht.	D.b.h.	Stems per acre	Basal area	Av. ht.	All trees, total stem		5-inch class and greater to o.b. tops of—					
								2 inches		3 inches		4 inches	
						O. b.	I. b.	O. b.	I. b.	O. b.	I. b.	O. b.	I. b.
Yrs	Ft	In	No.	Ft 2	Ft	— — — — — — — — — — — — — Ft 3 — — — — — — — — — — — — —							
10	26	1	3	.0	8	0	0						
		2	28	.6	16	6	4						
		3	98	4.8	20	51	36						
		4	193	16.8	23	200	144						
		5	211	28.8	24	353	256	352	237	310	204	209	128
		6	113	22.2	25	282	205	277	197	255	180	202	140
		7	23	6.1	26	81	59	79	58	75	55	64	47
		8	1	.3	27	5	3	5	3	4	3	4	3
			670	79.7		978	707	713	495	644	442	479	318
		MEAN DIA. 4.7 INCHES, WEIBULL PARAMETERS A= .1, B= 4.83, C= 4.32											
15	36	2	5	.1	19	1	1						
		3	31	1.5	25	20	14						
		4	92	8.0	29	119	86						
		5	169	23.0	32	375	272	371	260	337	234	258	173
		6	194	38.1	34	657	477	640	468	603	440	515	372
		7	122	32.6	35	577	420	559	418	537	401	483	360
		8	36	12.6	36	229	166	220	166	214	163	199	151
		9	4	1.8	37	33	24	32	24	31	24	30	23
			653	117.7		2011	1460	1822	1336	1722	1262	1485	1079
		MEAN DIA. 5.7 INCHES, WEIBULL PARAMETERS A= .8, B= 5.31, C= 4.32											
20	44	2	1	.0	20	0	0						
		3	11	.5	28	8	6						
		4	44	3.8	33	65	47						
		5	106	14.5	37	272	197	267	190	246	174	197	137
		6	165	32.4	39	640	465	622	460	591	436	517	379
		7	166	44.4	41	920	669	888	669	859	648	788	593
		8	94	32.8	43	713	518	685	518	670	512	632	483
		9	26	11.5	44	255	186	245	186	241	186	231	178
		10	3	1.6	45	37	27	36	27	35	27	34	26
			616	141.6		2910	2115	2743	2050	2642	1983	2399	1796
		MEAN DIA. 6.5 INCHES, WEIBULL PARAMETERS A= 1.3, B= 5.57, C= 4.31											
25	50	3	4	.2	29	3	2						
		4	21	1.8	36	34	24						
		5	62	8.5	40	172	125	168	121	156	112	128	90
		6	120	23.6	43	513	373	498	370	475	353	422	312
		7	156	41.7	46	969	705	935	705	907	688	842	638
		8	128	44.7	48	1083	788	1040	788	1020	783	970	744
		9	60	26.5	49	655	477	628	477	619	477	598	462
		10	14	7.6	51	196	143	188	143	186	143	182	141
		11	1	.7	52	17	13	17	13	16	13	16	13
			566	155.2		3642	2650	3474	2617	3379	2569	3158	2400
		MEAN DIA. 7.1 INCHES, WEIBULL PARAMETERS A= 1.7, B= 5.80, C= 4.30											

Age	Av. dom. ht.	D.b.h.	Stems per acre	Basal area	Av. ht.	All trees, total stem		5-inch class and greater to o.b. tops of—					
								2 inches		3 inches		4 inches	
						O. b.	I. b.	O. b.	I. b.	O. b.	I. b.	O. b.	I. b.
Yrs	Ft	In	No.	Ft²	Ft				Ft³				
30	55	3	1	.0	30	1	1						
		4	10	.9	37	16	12						
		5	35	4.8	43	104	76	102	74	95	69	79	56
		6	78	15.3	47	364	265	353	264	339	253	304	227
		7	122	32.6	50	824	599	793	599	772	588	722	549
		8	132	46.1	52	1209	880	1161	880	1140	878	1090	838
		9	89	39.3	54	1071	779	1026	779	1013	779	983	761
		10	34	18.5	56	524	381	501	381	497	381	487	379
		11	6	4.0	57	114	83	109	83	108	83	107	83
			507	161.5		4227	3076	4045	3060	3964	3031	3772	2893

MEAN DIA. 7.6 INCHES, WEIBULL PARAMETERS A= 2.0, B= 6.04, C= 4.28

35	60	4	5	.4	38	8	6						
		5	19	2.6	44	58	42	57	41	53	38	44	31
		6	48	9.4	48	229	166	222	166	213	159	192	143
		7	86	23.0	52	604	439	581	439	567	432	532	405
		8	111	38.7	55	1076	782	1032	782	1015	782	974	750
		9	99	43.7	57	1258	915	1204	915	1190	915	1158	897
		10	57	31.1	59	925	673	884	673	878	673	861	671
		11	18	11.9	60	359	261	343	261	341	261	337	261
		12	3	2.4	62	74	54	70	54	70	54	70	54
			446	163.2		4591	3338	4393	3331	4327	3314	4168	3212

MEAN DIA. 8.2 INCHES, WEIBULL PARAMETERS A= 2.3, B= 6.34, C= 4.27

40	63	4	2	.2	38	3	2						
		5	10	1.4	44	30	22	30	22	28	20	23	17
		6	28	5.5	49	136	99	132	99	127	95	115	86
		7	55	14.7	53	394	286	379	286	369	282	347	265
		8	82	28.6	56	809	588	776	588	763	588	733	565
		9	90	39.8	59	1183	861	1133	861	1120	861	1091	846
		10	70	38.2	61	1174	854	1122	854	1114	854	1095	853
		11	36	25.8	63	754	549	720	549	717	549	709	549
		12	11	8.6	64	279	203	266	203	265	203	263	203
		13	2	1.8	66	61	45	58	45	58	45	58	45
			386	162.5		4823	3509	4616	3507	4561	3497	4434	3429

MEAN DIA. 8.8 INCHES, WEIBULL PARAMETERS A= 2.5, B= 6.70, C= 4.25

SI 50

1,000 stems per acre

Age	Av. dom. ht.	D.b.h.	Stems per acre	Basal area	Av. ht.	All trees, total stem		5-inch class and greater to o.b. tops of—					
								2 inches		3 inches		4 inches	
						O. b.	I. b.	O. b.	I. b.	O. b.	I. b.	O. b.	I. b.
Yrs	Ft	In	No.	Ft 2	Ft	----------- Ft 3 -----------							
10	26	1	6	.0	10	0	0						
		2	54	1.2	17	11	8						
		3	184	9.0	21	100	72						
		4	311	27.1	23	322	232						
		5	238	32.5	25	414	300	413	279	366	243	252	157
		6	64	12.6	26	166	121	163	116	151	107	121	84
		7	4	1.1	27	15	11	14	10	13	10	12	9
			861	83.5		1028	744	590	405	530	360	385	250
						MEAN DIA. 4.2 INCHES, WEIBULL PARAMETERS A= .1, B= 4.34, C= 4.36							
15	36	2	9	.2	21	2	2						
		3	55	2.7	27	38	27						
		4	160	14.0	31	222	161						
		5	263	35.9	33	602	437	594	418	542	378	419	284
		6	232	45.6	35	808	587	787	577	744	543	636	463
		7	92	24.6	36	448	325	433	325	417	312	377	281
		8	12	4.2	37	78	57	75	57	73	56	68	52
			823	127.0		2198	1596	1889	1377	1776	1289	1502	1080
						MEAN DIA. 5.3 INCHES, WEIBULL PARAMETERS A= .6, B= 4.85, C= 4.35							
20	44	2	2	.0	22	1	0						
		3	19	.9	30	15	10						
		4	75	6.5	35	117	85						
		5	171	23.3	38	450	327	442	316	409	290	329	229
		6	237	46.5	41	966	702	938	696	894	662	788	581
		7	183	48.9	43	1063	773	1026	773	994	751	916	691
		8	66	23.0	44	512	372	492	372	481	368	455	348
		9	9	4.0	45	90	66	87	66	85	66	82	63
			762	153.3		3214	2335	2985	2223	2863	2137	2570	1912
						MEAN DIA. 6.1 INCHES, WEIBULL PARAMETERS A= 1.3, B= 5.14, C= 4.34							
25	50	3	6	.3	31	5	3						
		4	35	3.1	37	58	42						
		5	100	13.6	41	284	206	278	200	259	185	212	150
		6	179	35.1	45	801	582	776	579	743	553	664	493
		7	201	53.7	47	1276	928	1230	928	1195	907	1111	842
		8	125	43.6	49	1079	785	1037	785	1017	781	969	743
		9	37	16.3	50	412	300	395	300	390	300	377	291
		10	4	2.2	52	57	42	55	42	54	42	53	41
			687	168.0		3972	2888	3771	2834	3658	2768	3386	2560
						MEAN DIA. 6.7 INCHES, WEIBULL PARAMETERS A= 1.7, B= 5.38, C= 4.33							

Age	Av. dom. ht.	D.b.h.	Stems per acre	Basal area	Av. ht.	All trees, total stem O. b.	I. b.	2 inches O. b.	I. b.	3 inches O. b.	I. b.	4 inches O. b.	I. b.
Yrs	Ft	In	No.	Ft 2	Ft								
30	66	4	2	.2	42	4	3						
		5	11	1.5	48	37	27	36	26	34	24	29	21
		6	34	6.7	53	179	130	173	130	167	126	152	114
		7	69	18.4	57	531	386	510	386	499	382	472	360
		8	102	35.6	60	1078	784	1034	784	1018	784	982	758
		9	106	46.8	62	1464	1065	1401	1065	1387	1065	1354	1051
		10	74	40.4	64	1302	948	1245	948	1236	948	1217	948
		11	30	19.8	66	659	479	629	479	626	479	620	479
		12	7	5.5	67	186	135	177	135	177	135	176	135
		13	1	.9	69	32	23	31	23	31	23	30	23
			436	175.8		5472	3980	5236	3976	5175	3966	5032	3889

MEAN DIA. 8.6 INCHES, WEIBULL PARAMETERS A= 2.7, B= 6.32, C= 4.26

Age	Av. dom. ht.	D.b.h.	Stems per acre	Basal area	Av. ht.	All trees, total stem O. b.	I. b.	2 inches O. b.	I. b.	3 inches O. b.	I. b.	4 inches O. b.	I. b.
35	72	4	1	.1	42	2	1						
		5	4	.5	49	14	10	13	10	12	9	11	8
		6	15	2.9	55	82	60	79	60	76	58	70	53
		7	36	9.6	59	287	208	276	208	270	207	255	195
		8	62	21.6	63	688	500	659	500	650	500	626	486
		9	80	35.3	65	1159	843	1108	843	1098	843	1074	834
		10	79	43.1	68	1477	1075	1411	1075	1403	1075	1383	1075
		11	53	35.0	70	1234	898	1178	898	1174	898	1164	898
		12	23	18.1	71	646	470	617	470	615	470	612	470
		13	6	5.5	73	203	148	194	148	194	148	194	148
		14	1	1.1	74	40	29	38	29	38	29	38	29
			360	172.9		5832	4242	5573	4241	5530	4237	5429	4196

MEAN DIA. 9.4 INCHES, WEIBULL PARAMETERS A= 3.0, B= 6.79, C= 4.23

Age	Av. dom. ht.	D.b.h.	Stems per acre	Basal area	Av. ht.	All trees, total stem O. b.	I. b.	2 inches O. b.	I. b.	3 inches O. b.	I. b.	4 inches O. b.	I. b.
40	76	5	1	.1	49	3	2	3	2	3	2	3	2
		6	6	1.2	55	33	24	32	24	31	23	28	21
		7	16	4.3	60	130	94	125	94	122	93	116	89
		8	32	11.2	64	361	262	346	262	341	262	330	255
		9	49	21.6	67	731	532	700	532	693	532	679	528
		10	63	34.4	70	1213	882	1158	882	1152	882	1137	882
		11	57	37.6	72	1365	993	1303	993	1299	993	1288	993
		12	40	31.4	74	1171	853	1117	853	1116	853	1111	853
		13	19	17.5	76	671	488	639	488	639	488	639	488
		14	6	6.4	78	252	183	240	183	240	183	240	183
		15	1	1.2	79	49	36	47	36	47	36	47	36
			290	167.0		5979	4349	5710	4349	5683	4347	5618	4330

MEAN DIA. 10.3 INCHES, WEIBULL PARAMETERS A= 3.3, B= 7.45, C= 4.20

SI 60

1,000 stems per acre

Age	Av. dom. ht.	D.b.h.	Stems per acre	Basal area	Av. ht.	All trees, total stem		5-inch class and greater to o.b. tops of—					
								2 inches		3 inches		4 inches	
						O.b.	I.b.	O.b.	I.b.	O.b.	I.b.	O.b.	I.b.
Yrs	Ft	In	No.	Ft 2	Ft	—	—	—	—	Ft 3	—	—	—
10	31	1	1	.0	11	0	0						
		2	24	.5	20	6	4						
		3	110	5.4	25	71	51						
		4	250	21.8	27	302	219						
		5	300	40.9	29	604	438	599	414	540	369	398	260
		6	159	31.2	30	475	345	465	336	434	313	361	257
		7	29	7.8	31	122	88	118	88	113	84	100	74
		8	1	.3	32	6	4	5	4	5	4	5	4
			874	108.0		1586	1149	1187	842	1092	770	864	595

MEAN DIA. 4.8 INCHES, WEIBULL PARAMETERS A= .5, B= 4.57, C= 4.36

Age	Av. dom. ht.	D.b.h.	Stems per acre	Basal area	Av. ht.	O.b.	I.b.	O.b.	I.b.	O.b.	I.b.	O.b.	I.b.
15	43	2	2	.0	24	1	0						
		3	24	1.2	31	19	14						
		4	93	8.1	36	149	108						
		5	204	27.8	39	551	400	541	387	501	357	406	285
		6	262	51.4	41	1068	776	1037	769	988	732	872	642
		7	176	47.0	42	999	726	964	726	933	705	858	647
		8	51	17.8	44	396	288	380	288	372	285	352	269
		9	5	2.2	45	50	36	48	36	47	36	46	35
			817	155.6		3233	2348	2970	2206	2841	2115	2534	1878

MEAN DIA. 5.9 INCHES, WEIBULL PARAMETERS A= 1.2, B= 5.00, C= 4.35

Age	Av. dom. ht.	D.b.h.	Stems per acre	Basal area	Av. ht.	O.b.	I.b.	O.b.	I.b.	O.b.	I.b.	O.b.	I.b.
20	52	3	5	.2	34	4	3						
		4	33	2.9	40	59	43						
		5	101	13.8	45	314	228	307	223	288	208	242	173
		6	190	37.3	48	906	659	877	657	843	630	760	566
		7	218	58.3	50	1472	1070	1418	1070	1380	1051	1291	981
		8	138	48.2	52	1264	920	1214	920	1192	917	1140	876
		9	40	17.7	53	473	344	453	344	447	344	433	335
		10	4	2.2	54	59	43	57	43	56	43	55	43
			729	180.5		4551	3310	4326	3257	4206	3193	3921	2974

MEAN DIA. 6.7 INCHES, WEIBULL PARAMETERS A= 1.8, B= 5.26, C= 4.34

Age	Av. dom. ht.	D.b.h.	Stems per acre	Basal area	Av. ht.	O.b.	I.b.	O.b.	I.b.	O.b.	I.b.	O.b.	I.b.
25	60	3	1	.0	35	1	1						
		4	11	1.0	43	21	15						
		5	44	6.0	48	146	106	142	104	134	98	114	82
		6	108	21.2	52	558	405	539	405	520	391	473	355
		7	170	45.4	55	1262	918	1214	918	1186	906	1118	853
		8	169	59.0	57	1697	1234	1628	1234	1602	1234	1540	1187
		9	95	42.0	59	1249	909	1196	909	1183	909	1152	894
		10	25	13.6	61	419	305	401	305	398	305	391	305
		11	3	2.0	62	62	45	59	45	59	45	58	45
			626	190.2		5415	3938	5179	3920	5082	3888	4846	3721

MEAN DIA. 7.5 INCHES, WEIBULL PARAMETERS A= 2.3, B= 5.53, C= 4.31

Age	Av. dom. ht.	D.b.h.	Stems per acre	Basal area	Av. ht.	All trees, total stem O. b.	All trees, total stem I. b.	2 inches O. b.	2 inches I. b.	3 inches O. b.	3 inches I. b.	4 inches O. b.	4 inches I. b.
Yrs	Ft	In	No.	Ft^2	Ft	— — — — — — — — — — — — — Ft^3 — — — — — — — — — — — —							
30	66	4	3	.3	44	6	4						
		5	18	2.5	50	62	45	61	44	57	42	49	36
		6	53	10.4	55	289	210	280	210	270	204	247	186
		7	104	27.6	58	814	592	783	592	766	586	724	554
		8	140	48.9	61	1504	1094	1442	1094	1421	1094	1371	1059
		9	122	53.9	64	1740	1266	1664	1266	1649	1266	1611	1252
		10	62	33.8	66	1125	819	1075	819	1068	819	1052	819
		11	16	10.6	67	357	259	340	259	339	259	336	259
		12	2	1.6	69	55	40	52	40	52	40	52	40
			520	189.6		5952	4329	5697	4324	5622	4310	5442	4205

MEAN DIA. 8.2 INCHES, WEIBULL PARAMETERS A= 2.7, B= 5.87, C= 4.29

35	72	4	1	.1	44	2	1						
		5	7	1.0	51	25	18	24	18	23	17	20	14
		6	23	4.5	56	128	93	123	93	119	90	110	82
		7	54	14.4	61	445	323	427	323	418	321	397	304
		8	89	31.1	64	1003	730	961	730	948	730	917	709
		9	105	46.4	67	1567	1140	1499	1140	1486	1140	1454	1131
		10	85	46.4	69	1613	1173	1541	1173	1532	1173	1511	1173
		11	43	28.4	71	1015	739	969	739	966	739	958	739
		12	12	9.4	73	347	252	331	252	330	252	329	252
		13	2	1.8	74	69	50	66	50	65	50	65	50
			421	163.5		6214	4519	5941	4518	5887	4512	5761	4454

MEAN DIA. 8.9 INCHES, WEIBULL PARAMETERS A= 3.0, B= 6.33, C= 4.26

40	76	5	2	.3	51	7	5	7	5	6	5	6	4
		6	9	1.8	57	51	37	49	37	48	36	44	33
		7	24	6.4	62	201	146	193	146	189	145	180	138
		8	46	16.1	66	535	389	512	389	506	389	490	379
		9	68	30.0	69	1045	761	1000	761	991	761	971	756
		10	77	42.0	72	1524	1109	1456	1109	1448	1109	1430	1109
		11	61	40.3	74	1501	1093	1433	1093	1428	1093	1418	1093
		12	33	25.9	76	993	722	947	722	945	722	942	722
		13	11	10.1	77	393	286	375	286	375	286	375	286
		14	2	2.1	79	85	62	81	62	81	62	81	62
			333	175.0		6335	4610	6053	4610	6017	4608	5937	4582

MEAN DIA. 9.8 INCHES, WEIBULL PARAMETERS A= 3.3, B= 6.95, C= 4.22

SI 60

1,250 stems per acre

Age	Av. dom. ht.	D.b.h.	Stems per acre	Basal area	Av. ht.	All trees, total stem		5-inch class and greater to o.b. tops of—					
								2 inches		3 inches		4 inches	
						O. b.	I. b.	O. b.	I. b.	O. b.	I. b.	O. b.	I. b.
Yrs	Ft	In	No.	Ft 2	Ft	------------- Ft 3 -------------							
10	31	1	2	.0	12	0	0						
		2	41	.9	21	10	7						
		3	181	8.9	26	121	87						
		4	372	32.5	28	466	338						
		5	345	47.0	30	719	521	711	494	644	442	480	318
		6	112	22.0	31	346	251	338	245	317	229	265	189
		7	8	2.1	32	35	25	34	25	32	24	29	21
			1061	113.4		1697	1229	1083	764	993	695	774	528
			MEAN DIA. 4.4 INCHES, WEIBULL PARAMETERS A= .5, B= 4.22, C= 4.39										
15	43	2	3	.1	25	1	1						
		3	37	1.8	32	30	22						
		4	145	12.7	36	233	169						
		5	295	40.2	39	797	578	782	560	725	516	588	411
		6	319	62.6	41	1301	945	1263	936	1203	891	1061	781
		7	153	40.9	42	868	631	838	631	811	613	746	562
		8	25	8.7	44	194	141	186	141	182	140	172	132
		9	1	.4	45	10	7	10	7	9	7	9	7
			978	167.5		3434	2494	3079	2275	2930	2167	2576	1893
			MEAN DIA. 5.6 INCHES, WEIBULL PARAMETERS A= 1.2, B= 4.67, C= 4.38										
20	52	3	7	.3	35	6	4						
		4	49	4.3	41	89	65						
		5	149	20.3	45	464	337	453	329	425	307	356	255
		6	261	51.2	48	1245	905	1205	902	1157	866	1043	778
		7	255	68.1	50	1722	1252	1658	1252	1614	1229	1510	1148
		8	118	41.2	52	1081	786	1038	786	1019	784	975	749
		9	20	8.8	53	236	172	226	172	223	172	217	168
		10	1	.5	54	15	11	14	11	14	11	14	11
			860	194.9		4858	3532	4594	3452	4452	3369	4115	3109
			MEAN DIA. 6.4 INCHES, WEIBULL PARAMETERS A= 1.8, B= 4.95, C= 4.36										
25	60	3	1	.0	36	1	1						
		4	15	1.3	44	29	21						
		5	65	8.9	49	220	160	215	157	202	148	173	125
		6	152	29.8	52	785	571	759	571	732	550	666	499
		7	220	58.8	55	1633	1188	1571	1188	1534	1173	1446	1104
		8	185	64.6	58	1890	1375	1813	1375	1785	1375	1717	1324
		9	76	33.6	59	999	727	956	727	946	727	921	715
		10	13	7.1	61	218	159	208	159	207	159	203	158
		11	1	.7	62	21	15	20	15	20	15	19	15
			728	204.8		5796	4217	5542	4192	5426	4147	5145	3940
			MEAN DIA. 7.2 INCHES, WEIBULL PARAMETERS A= 2.3, B= 5.22, C= 4.34										

Age	Av. dom. ht.	D.b.h.	Stems per acre	Basal area	Av. ht.	All trees, total stem		5-inch class and greater to o.b. tops of—					
								2 inches		3 inches		4 inches	
						O.b.	I.b.	O.b.	I.b.	O.b.	I.b.	O.b.	I.b.
Yrs	Ft	In	No.	Ft 2	Ft	------------ Ft 3 -------------							
30	66	4	4	.3	45	8	6						
		5	25	3.4	52	90	65	87	64	83	61	71	52
		6	75	14.7	56	417	303	403	303	389	294	357	269
		7	140	37.4	60	1134	824	1089	824	1067	818	1012	775
		8	172	60.0	63	1909	1388	1829	1388	1804	1388	1743	1347
		9	125	55.2	65	1810	1317	1732	1317	1716	1317	1677	1304
		10	47	25.6	67	866	630	827	630	822	630	810	630
		11	8	5.3	69	184	134	175	134	175	134	173	134
			596	202.1		6418	4667	6142	4660	6056	4642	5843	4511

MEAN DIA. 7.9 INCHES, WEIBULL PARAMETERS A= 2.7, B= 5.56, C= 4.31

35	72	4	1	.1	45	2	1						
		5	9	1.2	53	33	24	32	24	30	22	26	19
		6	33	6.5	58	190	138	183	138	178	134	164	123
		7	73	19.5	62	611	444	587	444	575	441	547	419
		8	115	40.1	66	1337	972	1281	972	1264	972	1224	947
		9	123	54.3	68	1863	1356	1782	1356	1767	1356	1730	1346
		10	84	45.8	71	1640	1193	1566	1193	1558	1193	1538	1193
		11	32	21.1	73	777	565	742	565	739	565	734	565
		12	6	4.7	74	176	128	168	128	167	128	167	128
			476	193.4		6629	4821	6341	4820	6278	4811	6130	4740

MEAN DIA. 8.6 INCHES, WEIBULL PARAMETERS A= 3.0, B= 6.00, C= 4.28

40	76	5	3	.4	52	11	8	10	8	10	7	9	6
		6	13	2.6	58	75	54	72	54	70	53	64	49
		7	33	8.8	62	276	201	265	201	260	199	247	189
		8	61	21.3	66	709	516	679	516	670	516	649	502
		9	85	37.6	69	1307	951	1250	951	1239	951	1214	944
		10	85	46.4	72	1683	1224	1607	1224	1599	1224	1579	1224
		11	59	38.9	74	1452	1057	1386	1057	1382	1057	1372	1057
		12	25	19.6	76	752	547	717	547	716	547	714	547
		13	6	5.5	77	215	156	205	156	204	156	204	156
		14	1	1.1	79	43	31	41	31	41	31	41	31
			371	182.2		6523	4745	6232	4745	6191	4741	6093	4705

MEAN DIA. 9.5 INCHES, WEIBULL PARAMETERS A= 3.3, B= 6.61, C= 4.24

SI 60

1,500 stems per acre

Age	Av. dom. ht.	D.b.h.	Stems per acre	Basal area	Av. ht.	All trees, total stem		5-inch class and greater to o.b. tops of—					
								2 inches		3 inches		4 inches	
						O. b.	I. b.	O. b.	I. b.	O. b.	I. b.	O. b.	I. b.
Yrs	Ft	In	No.	Ft²	Ft	------------ Ft 3 ------------							
10	31	1	3	.0	12	0	0						
		2	63	1.4	21	16	11						
		3	273	13.4	26	182	131						
		4	499	43.5	28	626	453						
		5	342	46.6	30	712	517	705	490	638	438	476	315
		6	62	12.2	31	191	139	187	136	175	127	147	105
		7	2	.5	32	9	6	8	6	8	6	7	5
			1244	117.7		1736	1257	900	632	821	571	630	425
			MEAN DIA. 4.2 INCHES, WEIBULL PARAMETERS A= .5, B= 3.94, C= 4.41										
15	43	2	5	.1	26	2	1						
		3	55	2.7	33	46	33						
		4	208	18.2	37	343	249						
		5	390	53.2	40	1080	784	1059	761	984	702	803	564
		6	349	68.5	42	1457	1059	1414	1051	1350	1001	1194	881
		7	116	31.0	43	674	490	650	490	630	476	581	438
		8	10	3.5	45	79	58	76	58	75	57	71	54
			1133	177.2		3681	2674	3199	2360	3039	2236	2649	1937
			MEAN DIA. 5.4 INCHES, WEIBULL PARAMETERS A= 1.2, B= 4.41, C= 4.40										
20	52	3	10	.5	36	9	7						
		4	69	6.0	42	129	94						
		5	204	27.8	46	649	471	634	461	595	431	501	360
		6	331	65.0	49	1611	1171	1559	1169	1499	1123	1355	1012
		7	271	72.4	51	1866	1357	1797	1357	1751	1334	1640	1248
		8	91	31.8	53	850	618	816	618	801	617	767	590
		9	9	4.0	54	108	79	104	79	102	79	99	77
			985	207.5		5222	3797	4910	3684	4748	3584	4362	3287
			MEAN DIA. 6.2 INCHES, WEIBULL PARAMETERS A= 1.8, B= 4.70, C= 4.38										
25	60	3	2	.1	38	2	1						
		4	21	1.8	45	42	30						
		5	87	11.9	50	301	218	293	215	277	202	237	172
		6	199	39.1	54	1067	776	1031	776	995	750	910	684
		7	265	70.8	57	2039	1483	1961	1483	1917	1467	1811	1384
		8	186	64.9	59	1933	1406	1854	1406	1826	1406	1758	1357
		9	57	25.2	61	775	564	741	564	734	564	716	556
		10	6	3.3	62	102	74	98	74	97	74	95	74
			823	217.1		6261	4552	5978	4518	5846	4463	5527	4227
			MEAN DIA. 7.0 INCHES, WEIBULL PARAMETERS A= 2.3, B= 4.98, C= 4.36										

Age	Av. dom. ht.	D.b.h.	Stems per acre	Basal area	Av. ht.	All trees, total stem		5-inch class and greater to o.b. tops of—					
								2 inches		3 inches		4 inches	
						O. b.	I. b.	O. b.	I. b.	O. b.	I. b.	O. b.	I. b.
Yrs	Ft	In	No.	Ft²	Ft	$-------------Ft^3-------------$							
30	66	4	6	.5	46	12	9						
		5	34	4.6	52	122	89	119	87	113	82	97	71
		6	98	19.2	57	555	403	535	403	518	391	476	359
		7	176	47.0	60	1425	1036	1370	1036	1341	1028	1272	974
		8	195	68.1	63	2164	1574	2074	1574	2045	1574	1976	1527
		9	120	53.0	65	1738	1264	1663	1264	1647	1264	1610	1251
		10	34	18.5	67	626	456	598	456	595	456	586	456
		11	4	2.6	69	92	67	88	67	87	67	87	67
			667	213.7		6734	4898	6447	4887	6346	4862	6104	4705

MEAN DIA. 7.7 INCHES, WEIBULL PARAMETERS A= 2.7, B= 5.32, C= 4.32

Age	Av. dom. ht.	D.b.h.	Stems per acre	Basal area	Av. ht.	O. b.	I. b.	O. b.	I. b.	O. b.	I. b.	O. b.	I. b.
35	72	4	1	.1	46	2	1						
		5	12	1.6	53	44	32	43	31	41	30	35	26
		6	42	8.2	58	242	176	233	176	226	171	208	157
		7	93	24.9	63	791	575	759	575	744	572	708	543
		8	140	48.9	66	1627	1184	1559	1184	1539	1184	1491	1153
		9	135	59.6	68	2045	1488	1956	1488	1939	1488	1899	1477
		10	77	42.0	71	1503	1094	1436	1094	1428	1094	1410	1094
		11	23	15.2	73	558	406	533	406	531	406	527	406
		12	3	2.4	74	88	64	84	64	84	64	83	64
			526	202.9		6900	5020	6603	5018	6532	5009	6361	4920

MEAN DIA. 8.4 INCHES, WEIBULL PARAMETERS A= 3.0, B= 5.76, C= 4.29

Age	Av. dom. ht.	D.b.h.	Stems per acre	Basal area	Av. ht.	O. b.	I. b.	O. b.	I. b.	O. b.	I. b.	O. b.	I. b.
40	76	5	4	.5	53	15	11	14	10	14	10	12	9
		6	16	3.1	59	94	68	90	68	88	66	81	61
		7	42	11.2	64	363	264	348	264	342	262	325	250
		8	76	26.5	68	910	662	872	662	861	662	835	646
		9	100	44.2	71	1582	1151	1512	1151	1500	1151	1471	1145
		10	92	50.2	73	1846	1344	1764	1344	1755	1344	1733	1344
		11	54	35.6	75	1347	980	1285	980	1282	980	1273	980
		12	18	14.1	77	549	399	523	399	522	399	521	399
		13	3	2.8	79	110	80	105	80	105	80	105	80
			405	188.3		6816	4959	6513	4958	6469	4954	6356	4914

MEAN DIA. 9.2 INCHES, WEIBULL PARAMETERS A= 3.3, B= 6.35, C= 4.25

SI 60

1,750 stems per acre

Age	Av. dom. ht.	D.b.h.	Stems per acre	Basal area	Av. ht.	All trees, total stem		5-inch class and greater to o.b. tops of—					
								2 inches		3 inches		4 inches	
						O. b.	I. b.	O. b.	I. b.	O. b.	I. b.	O. b.	I. b.
Yrs	Ft	In	No.	Ft 2	Ft	- - - - - - - - - - - - Ft 3 - - - - - - - - - - - -							
10	31	1	5	.0	12	0	0						
		2	93	2.0	22	25	17						
		3	389	19.1	26	259	187						
		4	612	53.4	29	794	575						
		5	296	40.4	31	637	462	630	439	572	394	432	288
		6	27	5.3	32	86	62	84	61	79	57	67	48
			1422	120.2		1801	1303	714	500	651	451	499	336

MEAN DIA. 3.9 INCHES, WEIBULL PARAMETERS A= .5, B= 3.70, C= 4.43

Age	Av. dom. ht.	D.b.h.	Stems per acre	Basal area	Av. ht.	O. b.	I. b.	O. b.	I. b.	O. b.	I. b.	O. b.	I. b.
15	43	2	7	.2	26	2	2						
		3	76	3.7	33	64	46						
		4	283	24.7	37	467	338						
		5	486	66.3	40	1346	977	1320	948	1226	875	1000	703
		6	350	68.7	42	1462	1062	1418	1054	1353	1003	1198	884
		7	77	20.6	43	447	325	432	325	418	316	386	291
		8	3	1.0	45	24	17	23	17	22	17	21	16
			1282	185.2		3812	2767	3193	2344	3019	2211	2605	1894

MEAN DIA. 5.1 INCHES, WEIBULL PARAMETERS A= 1.2, B= 4.19, C= 4.42

Age	Av. dom. ht.	D.b.h.	Stems per acre	Basal area	Av. ht.	O. b.	I. b.	O. b.	I. b.	O. b.	I. b.	O. b.	I. b.
20	52	3	14	.7	37	13	9						
		4	92	8.0	43	176	128						
		5	265	36.1	47	861	626	841	613	790	574	669	481
		6	395	77.6	49	1923	1397	1861	1395	1789	1340	1617	1207
		7	269	71.9	51	1852	1347	1784	1347	1738	1324	1628	1238
		8	65	22.7	53	607	441	583	441	572	441	548	421
		9	4	1.8	54	48	35	46	35	46	35	44	34
			1104	218.8		5480	3983	5115	3831	4935	3714	4506	3381

MEAN DIA. 6.0 INCHES, WEIBULL PARAMETERS A= 1.8, B= 4.49, C= 4.40

Age	Av. dom. ht.	D.b.h.	Stems per acre	Basal area	Av. ht.	O. b.	I. b.	O. b.	I. b.	O. b.	I. b.	O. b.	I. b.
25	60	3	2	.1	38	2	1						
		4	27	2.4	45	54	39						
		5	113	15.4	50	391	284	381	279	359	262	308	223
		6	248	48.7	54	1330	967	1285	967	1240	934	1134	852
		7	304	81.2	57	2339	1701	2249	1701	2199	1682	2078	1588
		8	178	62.1	59	1850	1346	1774	1346	1747	1346	1683	1298
		9	40	17.7	61	544	396	520	396	515	396	502	390
		10	2	1.1	62	34	25	33	25	32	25	32	25
			914	228.7		6544	4759	6242	4714	6092	4645	5737	4376

MEAN DIA. 6.8 INCHES, WEIBULL PARAMETERS A= 2.3, B= 4.79, C= 4.37

Age	Av. dom. ht.	D.b.h.	Stems per acre	Basal area	Av. ht.	All trees, total stem		5-inch class and greater to o.b. tops of—					
								2 inches		3 inches		4 inches	
						O. b.	I. b.	O. b.	I. b.	O. b.	I. b.	O. b.	I. b.
Yrs	Ft	In	No.	Ft 2	Ft	------------ Ft 3 ------------							
30	66	4	2	.2	42	4	3						
		5	11	1.5	48	37	27	36	26	34	24	29	21
		6	34	6.7	53	179	130	173	130	167	126	152	114
		7	69	18.4	57	531	386	510	386	499	382	472	360
		8	102	35.6	60	1078	784	1034	784	1018	784	982	758
		9	106	46.8	62	1464	1065	1401	1065	1387	1065	1354	1051
		10	74	40.4	64	1302	948	1245	948	1236	948	1217	948
		11	30	19.8	66	659	479	629	479	626	479	620	479
		12	7	5.5	67	186	135	177	135	177	135	176	135
		13	1	.9	69	32	23	31	23	31	23	30	23
			436	175.8		5472	3980	5236	3976	5175	3966	5032	3889

MEAN DIA. 8.6 INCHES, WEIBULL PARAMETERS A= 2.7, B= 6.32, C= 4.26

Age	Av. dom. ht.	D.b.h.	Stems per acre	Basal area	Av. ht.	All trees, total stem		2 inches O.b.	2 inches I.b.	3 inches O.b.	3 inches I.b.	4 inches O.b.	4 inches I.b.
35	72	4	1	.1	42	2	1						
		5	4	.5	49	14	10	13	10	12	9	11	8
		6	15	2.9	55	82	60	79	60	76	58	70	53
		7	36	9.6	59	287	208	276	208	270	207	255	195
		8	62	21.6	63	688	500	659	500	650	500	628	486
		9	80	35.3	65	1159	843	1108	843	1098	843	1074	834
		10	79	43.1	68	1477	1075	1411	1075	1403	1075	1383	1075
		11	53	35.0	70	1234	898	1178	898	1174	898	1164	898
		12	23	18.1	71	646	470	617	470	615	470	612	470
		13	6	5.5	73	203	148	194	148	194	148	194	148
		14	1	1.1	74	40	29	38	29	38	29	38	29
			360	172.9		5832	4242	5573	4241	5530	4237	5429	4196

MEAN DIA. 9.4 INCHES, WEIBULL PARAMETERS A= 3.0, B= 6.79, C= 4.23

Age	Av. dom. ht.	D.b.h.	Stems per acre	Basal area	Av. ht.	All trees, total stem		2 inches O.b.	2 inches I.b.	3 inches O.b.	3 inches I.b.	4 inches O.b.	4 inches I.b.
40	76	5	1	.1	49	3	2	3	2	3	2	3	2
		6	6	1.2	55	33	24	32	24	31	23	28	21
		7	16	4.3	60	130	94	125	94	122	93	116	89
		8	32	11.2	64	361	262	346	262	341	262	330	255
		9	49	21.6	67	731	532	700	532	693	532	679	528
		10	63	34.4	70	1213	882	1158	882	1152	882	1137	882
		11	57	37.6	72	1365	993	1303	993	1299	993	1288	993
		12	40	31.4	74	1171	853	1117	853	1116	853	1111	853
		13	19	17.5	76	671	488	639	488	639	488	639	488
		14	6	6.4	78	252	183	240	183	240	183	240	183
		15	1	1.2	79	49	36	47	36	47	36	47	36
			290	167.0		5979	4349	5710	4349	5683	4347	5618	4330

MEAN DIA. 10.3 INCHES, WEIBULL PARAMETERS A= 3.3, B= 7.45, C= 4.20

SI 60

1,000 stems per acre

Age	Av. dom. ht.	D.b.h.	Stems per acre	Basal area	Av. ht.	All trees, total stem		5-inch class and greater to o.b. tops of—					
								2 inches		3 inches		4 inches	
						O. b.	I. b.	O. b.	I. b.	O. b.	I. b.	O. b.	I. b.
Yrs	Ft	In	No.	Ft²	Ft	– – – – – – – – – – Ft³ – – – – – – – – – –							
10	31	1	1	.0	11	0	0						
		2	24	.5	20	6	4						
		3	110	5.4	25	71	51						
		4	250	21.8	27	302	219						
		5	300	40.9	29	604	438	599	414	540	369	398	260
		6	159	31.2	30	475	345	465	336	434	313	361	257
		7	29	7.8	31	122	88	118	88	113	84	100	74
		8	1	.3	32	6	4	5	4	5	4	5	4
			874	108.0		1586	1149	1187	842	1092	770	864	595
		MEAN DIA. 4.8 INCHES, WEIBULL PARAMETERS A= .5, B= 4.57, C= 4.36											
15	43	2	2	.0	24	1	0						
		3	24	1.2	31	19	14						
		4	93	8.1	36	149	108						
		5	204	27.8	39	551	400	541	387	501	357	406	285
		6	262	51.4	41	1068	776	1037	769	988	732	872	642
		7	176	47.0	42	999	726	964	726	933	705	858	647
		8	51	17.8	44	396	288	380	288	372	285	352	269
		9	5	2.2	45	50	36	48	36	47	36	46	35
			817	155.6		3233	2348	2970	2206	2841	2115	2534	1878
		MEAN DIA. 5.9 INCHES, WEIBULL PARAMETERS A= 1.2, B= 5.00, C= 4.35											
20	52	3	5	.2	34	4	3						
		4	33	2.9	40	59	43						
		5	101	13.8	45	314	228	307	223	288	208	242	173
		6	190	37.3	48	906	659	877	657	843	630	760	566
		7	218	58.3	50	1472	1070	1418	1070	1380	1051	1291	981
		8	138	48.2	52	1264	920	1214	920	1192	917	1140	876
		9	40	17.7	53	473	344	453	344	447	344	433	335
		10	4	2.2	54	59	43	57	43	56	43	55	43
			729	180.5		4551	3310	4326	3257	4206	3193	3921	2974
		MEAN DIA. 6.7 INCHES, WEIBULL PARAMETERS A= 1.8, B= 5.26, C= 4.34											
25	60	3	1	.0	35	1	1						
		4	11	1.0	43	21	15						
		5	44	6.0	48	146	106	142	104	134	98	114	82
		6	108	21.2	52	558	405	539	405	520	391	473	355
		7	170	45.4	55	1262	918	1214	918	1186	906	1118	853
		8	169	59.0	57	1697	1234	1628	1234	1602	1234	1540	1187
		9	95	42.0	59	1249	909	1196	909	1183	909	1152	894
		10	25	13.6	61	419	305	401	305	398	305	391	305
		11	3	2.0	62	62	45	59	45	59	45	58	45
			626	190.2		5415	3938	5179	3920	5082	3888	4646	3721
		MEAN DIA. 7.5 INCHES, WEIBULL PARAMETERS A= 2.3, B= 5.53, C= 4.31											

Age	Av. dom. ht.	D.b.h.	Stems per acre	Basal area	Av. ht.	All trees, total stem		5-inch class and greater to o.b. tops of—					
								2 inches		3 inches		4 inches	
						O. b.	I. b.	O. b.	I. b.	O. b.	I. b.	O. b.	I. b.
Yrs	Ft	In	No.	Ft 2	Ft	– – – – – – – – – – – – – Ft 3 – – – – – – – – – – – – –							
30	66	4	3	.3	44	6	4						
		5	18	2.5	50	62	45	61	44	57	42	49	36
		6	53	10.4	55	289	210	280	210	270	204	247	186
		7	104	27.8	58	814	592	783	592	766	586	724	554
		8	140	48.9	61	1504	1094	1442	1094	1421	1094	1371	1059
		9	122	53.9	64	1740	1266	1664	1266	1649	1266	1611	1252
		10	62	33.8	66	1125	819	1075	819	1068	819	1052	819
		11	16	10.6	67	357	259	340	259	339	259	336	259
		12	2	1.6	69	55	40	52	40	52	40	52	40
			----	-----		----	----	----	----	----	----	----	----
			520	189.6		5952	4329	5697	4324	5622	4310	5442	4205

MEAN DIA. 8.2 INCHES, WEIBULL PARAMETERS A= 2.7, B= 5.87, C= 4.29

Age	Av. dom. ht.	D.b.h.	Stems per acre	Basal area	Av. ht.	All trees, total stem		2 inches O.b.	2 inches I.b.	3 inches O.b.	3 inches I.b.	4 inches O.b.	4 inches I.b.
35	72	4	1	.1	44	2	1						
		5	7	1.0	51	25	18	24	18	23	17	20	14
		6	23	4.5	56	128	93	123	93	119	90	110	82
		7	54	14.4	61	445	323	427	323	418	321	397	304
		8	89	31.1	64	1003	730	961	730	948	730	917	709
		9	105	46.4	67	1567	1140	1499	1140	1486	1140	1454	1131
		10	85	46.4	69	1613	1173	1541	1173	1532	1173	1511	1173
		11	43	28.4	71	1015	739	969	739	966	739	958	739
		12	12	9.4	73	347	252	331	252	330	252	329	252
		13	2	1.8	74	69	50	66	50	65	50	65	50
			----	-----		----	----	----	----	----	----	----	----
			421	183.5		6214	4519	5941	4518	5887	4512	5761	4454

MEAN DIA. 8.9 INCHES, WEIBULL PARAMETERS A= 3.0, B= 6.33, C= 4.26

Age	Av. dom. ht.	D.b.h.	Stems per acre	Basal area	Av. ht.	All trees, total stem		2 inches O.b.	2 inches I.b.	3 inches O.b.	3 inches I.b.	4 inches O.b.	4 inches I.b.
40	76	5	2	.3	51	7	5	7	5	6	5	6	4
		6	9	1.8	57	51	37	49	37	48	36	44	33
		7	24	6.4	62	201	146	193	146	189	145	180	138
		8	46	16.1	66	535	389	512	389	506	389	490	379
		9	68	30.0	69	1045	761	1000	761	991	761	971	756
		10	77	42.0	72	1524	1109	1456	1109	1448	1109	1430	1109
		11	61	40.3	74	1501	1093	1433	1093	1428	1093	1418	1093
		12	33	25.9	76	993	722	947	722	945	722	942	722
		13	11	10.1	77	393	286	375	286	375	286	375	286
		14	2	2.1	79	85	62	81	62	81	62	81	62
			----	-----		----	----	----	----	----	----	----	----
			333	175.0		6335	4610	6053	4610	6017	4608	5937	4582

MEAN DIA. 9.8 INCHES, WEIBULL PARAMETERS A= 3.3, B= 6.95, C= 4.22

SI 60

1,250 stems per acre

Age	Av. dom. ht.	D.b.h.	Stems per acre	Basal area	Av. ht.	All trees, total stem		5-inch class and greater to o.b. tops of—					
								2 inches		3 inches		4 inches	
						O. b.	I. b.	O. b.	I. b.	O. b.	I. b.	O. b.	I. b.
Yrs	Ft	In	No.	Ft 2	Ft	‑ ‑ ‑ ‑ ‑ ‑ ‑ ‑ ‑ ‑ ‑ ‑ Ft 3 ‑ ‑ ‑ ‑ ‑ ‑ ‑ ‑ ‑ ‑ ‑ ‑							
10	31	1	2	.0	12	0	0						
		2	41	.9	21	10	7						
		3	181	8.9	26	121	87						
		4	372	32.5	28	466	338						
		5	345	47.0	30	719	521	711	494	644	442	480	318
		6	112	22.0	31	346	251	338	245	317	229	265	189
		7	8	2.1	32	35	25	34	25	32	24	29	21
			1061	113.4		1697	1229	1083	764	993	695	774	528
			MEAN DIA. 4.4 INCHES, WEIBULL PARAMETERS A= .5, B= 4.22, C= 4.39										
15	43	2	3	.1	25	1	1						
		3	37	1.8	32	30	22						
		4	145	12.7	36	233	169						
		5	295	40.2	39	797	578	782	560	725	516	588	411
		6	319	62.6	41	1301	945	1263	936	1203	891	1061	781
		7	153	40.9	42	868	631	838	631	811	613	746	562
		8	25	8.7	44	194	141	186	141	182	140	172	132
		9	1	.4	45	10	7	10	7	9	7	9	7
			978	167.5		3434	2494	3079	2275	2930	2167	2576	1893
			MEAN DIA. 5.6 INCHES, WEIBULL PARAMETERS A= 1.2, B= 4.67, C= 4.38										
20	52	3	7	.3	35	6	4						
		4	49	4.3	41	89	65						
		5	149	20.3	45	464	337	453	329	425	307	356	255
		6	261	51.2	48	1245	905	1205	902	1157	866	1043	778
		7	255	68.1	50	1722	1252	1658	1252	1614	1229	1510	1148
		8	118	41.2	52	1081	786	1038	786	1019	784	975	749
		9	20	8.8	53	236	172	226	172	223	172	217	168
		10	1	.5	54	15	11	14	11	14	11	14	11
			860	194.9		4858	3532	4594	3452	4452	3369	4115	3109
			MEAN DIA. 6.4 INCHES, WEIBULL PARAMETERS A= 1.8, B= 4.95, C= 4.36										
25	60	3	1	.0	36	1	1						
		4	15	1.3	44	29	21						
		5	65	8.9	49	220	160	215	157	202	148	173	125
		6	152	29.8	52	785	571	759	571	732	550	666	499
		7	220	58.8	55	1633	1188	1571	1188	1534	1173	1446	1104
		8	185	64.6	58	1890	1375	1813	1375	1785	1375	1717	1324
		9	76	33.6	59	999	727	956	727	946	727	921	715
		10	13	7.1	61	218	159	208	159	207	159	203	158
		11	1	.7	62	21	15	20	15	20	15	19	15
			728	204.8		5796	4217	5542	4192	5426	4147	5145	3940
			MEAN DIA. 7.2 INCHES, WEIBULL PARAMETERS A= 2.3, B= 5.22, C= 4.34										

Age	Av. dom. ht.	D.b.h.	Stems per acre	Basal area	Av. ht.	All trees, total stem		5-inch class and greater to o.b. tops of—					
								2 inches		3 inches		4 inches	
						O. b.	I. b.	O. b.	I. b.	O. b.	I. b.	O. b.	I. b.
Yrs	Ft	In	No.	Ft 2	Ft	———————————— Ft 3 ————————————							
30	66	4	4	.3	45	8	6						
		5	25	3.4	52	90	65	87	64	83	61	71	52
		6	75	14.7	56	417	303	403	303	389	294	357	269
		7	140	37.4	60	1134	824	1089	824	1067	818	1012	775
		8	172	60.0	63	1909	1388	1829	1388	1804	1366	1743	1347
		9	125	55.2	65	1810	1317	1732	1317	1716	1317	1677	1304
		10	47	25.6	67	866	630	827	630	822	630	810	630
		11	8	5.3	69	184	134	175	134	175	134	173	134
			596	202.1		6418	4667	6142	4660	6056	4642	5843	4511
				MEAN DIA. 7.9 INCHES, WEIBULL PARAMETERS A= 2.7, B= 5.56, C= 4.31									
35	72	4	1	.1	45	2	1						
		5	9	1.2	53	33	24	32	24	30	22	26	19
		6	33	6.5	58	190	138	183	138	178	134	164	123
		7	73	19.5	62	611	444	587	444	575	441	547	419
		8	115	40.1	66	1337	972	1281	972	1264	972	1224	947
		9	123	54.3	68	1863	1356	1782	1356	1767	1356	1730	1346
		10	84	45.8	71	1640	1193	1566	1193	1558	1193	1538	1193
		11	32	21.1	73	777	565	742	565	739	565	734	565
		12	6	4.7	74	176	128	168	128	167	128	167	128
			476	193.4		6629	4821	6341	4820	6278	4811	6130	4740
				MEAN DIA. 8.6 INCHES, WEIBULL PARAMETERS A= 3.0, B= 6.00, C= 4.28									
40	76	5	3	.4	52	11	8	10	8	10	7	9	6
		6	13	2.6	58	75	54	72	54	70	53	64	49
		7	33	8.8	62	276	201	265	201	260	199	247	189
		8	61	21.3	66	709	516	679	516	670	516	649	502
		9	85	37.6	69	1307	951	1250	951	1239	951	1214	944
		10	85	46.4	72	1683	1224	1607	1224	1599	1224	1579	1224
		11	59	38.9	74	1452	1057	1386	1057	1382	1057	1372	1057
		12	25	19.6	76	752	547	717	547	716	547	714	547
		13	6	5.5	77	215	156	205	156	204	156	204	156
		14	1	1.1	79	43	31	41	31	41	31	41	31
			371	182.2		6523	4745	6232	4745	6191	4741	6093	4705
				MEAN DIA. 9.5 INCHES, WEIBULL PARAMETERS A= 3.3, B= 6.61, C= 4.24									

SI 60

1,500 stems per acre

Age	Av. dom. ht.	D.b.h.	Stems per acre	Basal area	Av. ht.	All trees, total stem		5-inch class and greater to o.b. tops of—					
								2 inches		3 inches		4 inches	
						O.b.	I.b.	O.b.	I.b.	O.b.	I.b.	O.b.	I.b.
Yrs	Ft	In	No.	Ft²	Ft	---------- Ft³ ----------							
10	31	1	3	.0	12	0	0						
		2	63	1.4	21	16	11						
		3	273	13.4	26	182	131						
		4	499	43.5	28	626	453						
		5	342	46.6	30	712	517	705	490	638	438	476	315
		6	62	12.2	31	191	139	187	136	175	127	147	105
		7	2	.5	32	9	6	8	6	8	6	7	5
			1244	117.7		1736	1257	900	632	821	571	630	425

MEAN DIA. 4.2 INCHES, WEIBULL PARAMETERS A= .5, B= 3.94, C= 4.41

Age	Av. dom. ht.	D.b.h.	Stems per acre	Basal area	Av. ht.	O.b.	I.b.	O.b.	I.b.	O.b.	I.b.	O.b.	I.b.
15	43	2	5	.1	26	2	1						
		3	55	2.7	33	46	33						
		4	208	18.2	37	343	249						
		5	390	53.2	40	1080	784	1059	761	984	702	803	564
		6	349	68.5	42	1457	1059	1414	1051	1350	1001	1194	881
		7	116	31.0	43	674	490	650	490	630	476	581	438
		8	10	3.5	45	79	58	76	58	75	57	71	54
			1133	177.2		3681	2674	3199	2360	3039	2236	2649	1937

MEAN DIA. 5.4 INCHES, WEIBULL PARAMETERS A= 1.2, B= 4.41, C= 4.40

Age	Av. dom. ht.	D.b.h.	Stems per acre	Basal area	Av. ht.	O.b.	I.b.	O.b.	I.b.	O.b.	I.b.	O.b.	I.b.
20	52	3	10	.5	36	9	7						
		4	69	6.0	42	129	94						
		5	204	27.8	46	649	471	634	461	595	431	501	360
		6	331	65.0	49	1611	1171	1559	1169	1499	1123	1355	1012
		7	271	72.4	51	1866	1357	1797	1357	1751	1334	1640	1248
		8	91	31.8	53	850	618	816	618	801	617	767	590
		9	9	4.0	54	108	79	104	79	102	79	99	77
			985	207.5		5222	3797	4910	3684	4748	3584	4362	3287

MEAN DIA. 6.2 INCHES, WEIBULL PARAMETERS A= 1.8, B= 4.70, C= 4.38

Age	Av. dom. ht.	D.b.h.	Stems per acre	Basal area	Av. ht.	O.b.	I.b.	O.b.	I.b.	O.b.	I.b.	O.b.	I.b.
25	60	3	2	.1	38	2	1						
		4	21	1.8	45	42	30						
		5	87	11.9	50	301	218	293	215	277	202	237	172
		6	199	39.1	54	1067	776	1031	776	995	750	910	684
		7	265	70.8	57	2039	1483	1961	1483	1917	1467	1811	1384
		8	186	64.9	59	1933	1406	1854	1406	1826	1406	1758	1357
		9	57	25.2	61	775	564	741	564	734	564	716	556
		10	6	3.3	62	102	74	98	74	97	74	95	74
			823	217.1		6261	4552	5978	4518	5846	4463	5527	4227

MEAN DIA. 7.0 INCHES, WEIBULL PARAMETERS A= 2.3, B= 4.98, C= 4.36

Age	Av. dom. ht.	D.b.h.	Stems per acre	Basal area	Av. ht.	All trees, total stem		5-inch class and greater to o.b. tops of—					
								2 inches		3 inches		4 inches	
						O.b.	I.b.	O.b.	I.b.	O.b.	I.b.	O.b.	I.b.
Yrs	Ft	In	No.	Ft²	Ft	— — — — — — — — — — — — Ft³ — — — — — — — — — — — —							
30	66	4	6	.5	46	12	9						
		5	34	4.6	52	122	89	119	87	113	82	97	71
		6	98	19.2	57	555	403	535	403	518	391	476	359
		7	176	47.0	60	1425	1036	1370	1036	1341	1028	1272	974
		8	195	68.1	63	2164	1574	2074	1574	2045	1574	1976	1527
		9	120	53.0	65	1738	1264	1663	1264	1647	1264	1610	1251
		10	34	18.5	67	626	456	598	456	595	456	586	456
		11	4	2.6	69	92	67	88	67	87	67	87	67
			667	213.7		6734	4898	6447	4887	6346	4862	6104	4705

MEAN DIA. 7.7 INCHES, WEIBULL PARAMETERS A= 2.7, B= 5.32, C= 4.32

Age	Av. dom. ht.	D.b.h.	Stems per acre	Basal area	Av. ht.	O.b.	I.b.	O.b.	I.b.	O.b.	I.b.	O.b.	I.b.
35	72	4	1	.1	46	2	1						
		5	12	1.6	53	44	32	43	31	41	30	35	26
		6	42	8.2	58	242	176	233	176	226	171	208	157
		7	93	24.9	63	791	575	759	575	744	572	708	543
		8	140	48.9	66	1627	1184	1559	1184	1539	1184	1491	1153
		9	135	59.6	68	2045	1488	1956	1488	1939	1488	1899	1477
		10	77	42.0	71	1503	1094	1436	1094	1428	1094	1410	1094
		11	23	15.2	73	558	406	533	406	531	406	527	406
		12	3	2.4	74	88	64	84	64	84	64	83	64
			526	202.9		6900	5020	6603	5018	6532	5009	6361	4920

MEAN DIA. 8.4 INCHES, WEIBULL PARAMETERS A= 3.0, B= 5.76, C= 4.29

Age	Av. dom. ht.	D.b.h.	Stems per acre	Basal area	Av. ht.	O.b.	I.b.	O.b.	I.b.	O.b.	I.b.	O.b.	I.b.
40	76	5	4	.5	53	15	11	14	10	14	10	12	9
		6	16	3.1	59	94	68	90	68	88	66	81	61
		7	42	11.2	64	363	264	348	264	342	262	325	250
		8	76	26.5	68	910	662	872	662	861	662	835	646
		9	100	44.2	71	1582	1151	1512	1151	1500	1151	1471	1145
		10	92	50.2	73	1846	1344	1764	1344	1755	1344	1733	1344
		11	54	35.6	75	1347	980	1285	980	1282	980	1273	980
		12	18	14.1	77	549	399	523	399	522	399	521	399
		13	3	2.8	79	110	80	105	80	105	80	105	80
			405	186.3		6816	4959	6513	4958	6469	4954	6356	4914

MEAN DIA. 9.2 INCHES, WEIBULL PARAMETERS A= 3.3, B= 6.35, C= 4.25

SI 60

1,750 stems per acre

Age	Av. dom. ht.	D.b.h.	Stems per acre	Basal area	Av. ht.	All trees, total stem		5-inch class and greater to o.b. tops of—					
								2 inches		3 inches		4 inches	
						O.b.	I.b.	O.b.	I.b.	O.b.	I.b.	O.b.	I.b.
Yrs	Ft	In	No.	Ft²	Ft	— — — — — — — — — — — — Ft³ — — — — — — — — — — — —							
10	31	1	5	.0	12	0	0						
		2	93	2.0	22	25	17						
		3	389	19.1	26	259	187						
		4	612	53.4	29	794	575						
		5	296	40.4	31	637	462	630	439	572	394	432	288
		6	27	5.3	32	86	62	84	61	79	57	67	48
			1422	120.2		1801	1303	714	500	651	451	499	336

MEAN DIA. 3.9 INCHES, WEIBULL PARAMETERS A= .5, B= 3.70, C= 4.43

15	43	2	7	.2	26	2	2						
		3	76	3.7	33	64	46						
		4	283	24.7	37	467	338						
		5	486	66.3	40	1346	977	1320	948	1226	875	1000	703
		6	350	68.7	42	1482	1062	1418	1054	1353	1003	1198	884
		7	77	20.6	43	447	325	432	325	418	316	386	291
		8	3	1.0	45	24	17	23	17	22	17	21	16
			1282	165.2		3812	2767	3193	2344	3019	2211	2605	1894

MEAN DIA. 5.1 INCHES, WEIBULL PARAMETERS A= 1.2, B= 4.19, C= 4.42

20	52	3	14	.7	37	13	9						
		4	92	8.0	43	176	128						
		5	265	36.1	47	861	626	841	613	790	574	669	481
		6	395	77.6	49	1923	1397	1861	1395	1789	1340	1617	1207
		7	269	71.9	51	1852	1347	1784	1347	1738	1324	1628	1238
		8	65	22.7	53	607	441	583	441	572	441	548	421
		9	4	1.8	54	48	35	46	35	46	35	44	34
			1104	218.8		5480	3983	5115	3831	4935	3714	4506	3381

MEAN DIA. 6.0 INCHES, WEIBULL PARAMETERS A= 1.8, B= 4.49, C= 4.40

25	60	3	2	.1	38	2	1						
		4	27	2.4	45	54	39						
		5	113	15.4	50	391	284	381	279	359	262	308	223
		6	248	48.7	54	1330	967	1285	967	1240	934	1134	852
		7	304	81.2	57	2339	1701	2249	1701	2199	1682	2078	1588
		8	178	62.1	59	1850	1346	1774	1346	1747	1346	1683	1298
		9	40	17.7	61	544	396	520	396	515	396	502	390
		10	2	1.1	62	34	25	33	25	32	25	32	25
			914	228.7		6544	4759	6242	4714	6092	4645	5737	4376

MEAN DIA. 6.8 INCHES, WEIBULL PARAMETERS A= 2.3, B= 4.79, C= 4.37

Age	Av. dom. ht.	D.b.h.	Stems per acre	Basal area	Av. ht.	All trees, total stem		5-inch class and greater to o.b. tops of—					
								2 inches		3 inches		4 inches	
						O. b.	I. b.	O. b.	I. b.	O. b.	I. b.	O. b.	I. b.
Yrs	Ft	In	No.	Ft²	Ft	Ft³							
30	66	4	7	.6	47	15	11						
		5	43	5.9	53	157	114	153	113	145	107	126	92
		6	123	24.2	57	696	506	672	506	650	491	598	451
		7	211	56.4	60	1709	1243	1642	1243	1608	1232	1525	1167
		8	214	74.7	63	2375	1727	2276	1727	2244	1727	2169	1676
		9	110	48.6	65	1593	1159	1524	1159	1510	1159	1476	1147
		10	23	12.5	67	424	308	405	308	402	308	397	308
		11	2	1.3	69	46	33	44	33	44	33	43	33
			733	224.2		7015	5101	6716	5089	6603	5057	6334	4874
				MEAN DIA. 7.5 INCHES, WEIBULL PARAMETERS A= 2.7, B= 5.13, C= 4.34									
35	72	4	2	.2	46	4	3						
		5	15	2.0	53	55	40	53	39	51	37	44	32
		6	52	10.2	59	305	221	294	221	285	215	263	198
		7	113	30.2	63	961	699	923	699	905	695	861	660
		8	162	56.5	66	1883	1370	1804	1370	1781	1370	1725	1334
		9	142	62.7	69	2183	1588	2088	1588	2070	1588	2028	1578
		10	69	37.6	71	1347	980	1287	980	1280	980	1263	980
		11	16	10.6	73	388	283	371	283	370	283	367	283
		12	1	.8	74	29	21	28	21	28	21	28	21
			572	210.9		7155	5205	6848	5201	6770	5189	6579	5086
				MEAN DIA. 8.2 INCHES, WEIBULL PARAMETERS A= 3.0, B= 5.57, C= 4.30									
40	76	5	5	.7	53	18	13	18	13	17	12	15	11
		6	20	3.9	59	117	85	113	85	109	83	101	76
		7	51	13.6	64	440	320	423	320	415	319	395	303
		8	90	31.4	68	1078	784	1032	784	1020	784	989	766
		9	112	49.5	71	1771	1289	1694	1289	1680	1289	1648	1283
		10	95	51.8	73	1907	1387	1821	1387	1812	1387	1790	1387
		11	48	31.7	75	1197	871	1143	871	1139	871	1131	871
		12	13	10.2	77	396	288	378	288	377	288	376	288
		13	2	1.8	79	73	53	70	53	70	53	70	53
			436	194.7		6997	5090	6692	5090	6639	5086	6515	5038
				MEAN DIA. 9.0 INCHES, WEIBULL PARAMETERS A= 3.3, B= 6.14, C= 4.26									

SI 60

2,000 stems per acre

Age	Av. dom. ht.	D.b.h.	Stems per acre	Basal area	Av. ht.	All trees, total stem		5-inch class and greater to o.b. tops of—					
								2 inches		3 inches		4 inches	
						O. b.	I. b.	O. b.	I. b.	O. b.	I. b.	O. b.	I. b.
Yrs	Ft	In	No.	Ft 2	Ft	— — — — — — — — — — — Ft 3 — — — — — — — — — —							
10	31	1	r	.0	12	0	0						
		2	134	2.9	22	36	25						
		3	530	26.0	26	353	254						
		4	694	60.6	29	901	652						
		5	223	30.4	31	480	348	474	331	431	297	325	217
		6	9	1.8	32	29	21	28	20	26	19	22	16
			1597	121.7		1799	1300	502	351	457	316	347	253

MEAN DIA. 3.7 INCHES, WEIBULL PARAMETERS A= .5, B= 3.49, C= 4.45

Age	Av. dom. ht.	D.b.h.	Stems per acre	Basal area	Av. ht.	All trees, total stem		2 inches		3 inches		4 inches	
15	43	2	9	.2	27	3	2						
		3	103	5.1	33	86	62						
		4	370	32.3	37	610	442						
		5	574	78.3	40	1590	1154	1559	1119	1448	1034	1182	831
		6	325	63.8	42	1357	986	1317	978	1257	932	1112	821
		7	46	12.3	43	267	194	258	194	250	189	230	174
		8	1	.3	45	8	6	8	6	7	6	7	5
			1428	192.3		3921	2846	3142	2297	2962	2161	2531	1831

MEAN DIA. 5.0 INCHES, WEIBULL PARAMETERS A= 1.2, B= 3.99, C= 4.43

Age	Av. dom. ht.	D.b.h.	Stems per acre	Basal area	Av. ht.	All trees, total stem		2 inches		3 inches		4 inches	
20	52	3	18	.9	37	17	12						
		4	120	10.5	43	230	166						
		5	333	45.4	47	1082	786	1056	770	993	721	841	605
		6	452	88.7	49	2200	1599	2129	1596	2047	1533	1850	1381
		7	252	67.3	51	1735	1262	1671	1262	1628	1240	1525	1160
		8	42	14.7	53	392	285	376	285	370	285	354	272
		9	1	.4	54	12	9	12	9	11	9	11	9
			1218	228.0		5668	4119	5244	3922	5049	3788	4581	3427

MEAN DIA. 5.9 INCHES, WEIBULL PARAMETERS A= 1.8, B= 4.32, C= 4.41

Age	Av. dom. ht.	D.b.h.	Stems per acre	Basal area	Av. ht.	All trees, total stem		2 inches		3 inches		4 inches	
25	60	3	3	.1	39	3	2						
		4	34	3.0	46	70	50						
		5	141	19.2	51	497	361	484	355	457	335	394	286
		6	299	58.7	54	1603	1165	1549	1165	1496	1126	1367	1027
		7	332	88.7	57	2554	1857	2456	1857	2401	1837	2269	1734
		8	163	56.9	59	1694	1232	1624	1232	1600	1232	1541	1189
		9	27	11.9	61	367	267	351	267	348	267	339	263
		10	1	.5	62	17	12	16	12	16	12	16	12
			1000	239.1		6805	4946	6480	4888	6318	4809	5926	4511

MEAN DIA. 6.6 INCHES, WEIBULL PARAMETERS A= 2.3, B= 4.63, C= 4.38

Age	Av. dom. ht.	D.b.h.	Stems per acre	Basal area	Av. ht.	All trees, total stem		5-inch class and greater to o.b. tops of—					
								2 inches		3 inches		4 inches	
						O. b.	I. b.	O. b.	I. b.	O. b.	I. b.	O. b.	I. b.
Yrs	Ft	In	No.	Ft 2	Ft	------------ Ft 3 ------------							
30	66	4	9	.8	47	19	14						
		5	53	7.2	53	194	141	189	139	179	131	155	113
		6	148	29.1	57	837	609	808	609	782	591	719	542
		7	245	65.5	61	2017	1467	1938	1467	1898	1456	1802	1381
		8	225	78.5	63	2497	1816	2393	1816	2360	1816	2280	1762
		9	98	43.3	65	1419	1033	1358	1033	1345	1033	1315	1022
		10	16	8.7	67	295	214	282	214	280	214	276	214
		11	1	.7	69	23	17	22	17	22	17	22	17
			795	233.8		7301	5311	6990	5295	6866	5258	6569	5051

MEAN DIA. 7.3 INCHES, WEIBULL PARAMETERS A= 2.7, B= 4.97, C= 4.35

Age	Av. dom. ht.	D.b.h.	Stems per acre	Basal area	Av. ht.	O. b.	I. b.	O. b.	I. b.	O. b.	I. b.	O. b.	I. b.
35	72	4	2	.2	47	4	3						
		5	18	2.5	54	67	49	65	48	62	46	54	39
		6	63	12.4	59	369	268	356	268	345	261	318	240
		7	134	35.8	63	1139	828	1094	828	1073	824	1021	783
		8	180	62.8	66	2092	1522	2004	1522	1979	1522	1916	1483
		9	146	64.5	69	2244	1633	2146	1633	2128	1633	2085	1622
		10	60	32.7	71	1171	852	1119	852	1113	852	1098	852
		11	11	7.3	73	267	194	255	194	254	194	252	194
		12	1	.8	74	29	21	28	21	28	21	28	21
			615	218.9		7382	5370	7067	5366	6982	5353	6772	5234

MEAN DIA. 8.1 INCHES, WEIBULL PARAMETERS A= 3.0, B= 5.40, C= 4.31

Age	Av. dom. ht.	D.b.h.	Stems per acre	Basal area	Av. ht.	O. b.	I. b.	O. b.	I. b.	O. b.	I. b.	O. b.	I. b.
40	76	5	6	.8	54	22	16	22	16	21	15	18	13
		6	24	4.7	60	143	104	138	104	134	101	124	93
		7	60	16.0	64	518	377	498	377	488	375	465	357
		8	104	36.3	68	1245	906	1193	906	1178	906	1143	885
		9	124	54.8	71	1961	1427	1875	1427	1860	1427	1825	1420
		10	95	51.8	73	1907	1387	1821	1387	1812	1387	1790	1387
		11	42	27.7	76	1062	773	1013	773	1010	773	1003	773
		12	9	7.1	77	274	200	262	200	261	200	260	200
		13	1	.9	79	37	27	35	27	35	27	35	27
			465	200.2		7169	5217	6857	5217	6799	5211	6663	5155

MEAN DIA. 8.9 INCHES, WEIBULL PARAMETERS A= 3.3, B= 5.97, C= 4.27

SI 60

2,250 stems per acre

Age	Av. dom. ht.	D.b.h.	Stems per acre	Basal area	Av. ht.	All trees, total stem		5-inch class and greater to o.b. tops of—					
								2 inches		3 inches		4 inches	
						O. b.	I. b.	O. b.	I. b.	O. b.	I. b.	O. b.	I. b.
Yrs	Ft	In	No.	Ft 2	Ft	$-------------Ft 3 --------------$							
10	31	1	10	.1	12	1	0						
		2	187	4.1	22	50	35						
		3	695	34.1	26	463	333						
		4	733	64.0	29	951	689						
		5	143	19.5	31	308	223	304	212	276	191	209	139
		6	2	.4	32	6	5	6	5	6	4	5	4
			1770	122.1		1779	1285	310	217	282	195	214	143
				MEAN DIA. 3.6 INCHES, WEIBULL PARAMETERS A= .5, B= 3.29, C= 4.47									
15	43	2	12	.3	27	4	3						
		3	136	6.7	34	117	85						
		4	471	41.1	38	798	578						
		5	647	88.2	41	1836	1333	1799	1295	1674	1199	1375	970
		6	280	55.0	43	1197	870	1161	864	1109	824	985	728
		7	24	6.4	45	146	106	141	106	137	103	126	96
			1570	197.7		4098	2975	3101	2265	2920	2126	2486	1794
				MEAN DIA. 4.8 INCHES, WEIBULL PARAMETERS A= 1.2, B= 3.82, C= 4.45									
20	52	3	23	1.1	38	22	16						
		4	151	13.2	44	295	214						
		5	406	55.4	48	1348	979	1314	960	1237	900	1052	759
		6	499	78.0	51	2528	1837	2444	1837	2354	1767	2138	1601
		7	225	60.1	53	1610	1171	1549	1171	1511	1153	1420	1082
		8	26	9.1	54	247	180	237	180	233	180	224	172
			1330	236.9		6050	4397	5544	4148	5335	4000	4834	3614
				MEAN DIA. 5.7 INCHES, WEIBULL PARAMETERS A= 1.8, B= 4.16, C= 4.43									
25	60	3	3	.1	40	3	2						
		4	42	3.7	47	88	64						
		5	172	23.5	52	618	449	602	442	569	417	491	358
		6	350	68.7	56	1946	1415	1879	1415	1817	1371	1668	1255
		7	355	94.9	58	2779	2021	2672	2021	2613	2001	2473	1891
		8	144	50.3	60	1522	1107	1459	1107	1438	1107	1386	1070
		9	17	7.5	62	235	171	225	171	222	171	217	169
			1083	248.6		7191	5229	6837	5156	6659	5067	6235	4743
				MEAN DIA. 6.5 INCHES, WEIBULL PARAMETERS A= 2.3, B= 4.48, C= 4.40									

Age	Av. dom. ht.	D.b.h.	Stems per acre	Basal area	Av. ht.	All trees, total stem		5-inch class and greater to o.b. tops of—					
								2 inches		3 inches		4 inches	
						O. b.	I. b.	O. b.	I. b.	O. b.	I. b.	O. b.	I. b.
Yrs	Ft	In	No.	Ft 2	Ft	------------Ft 3------------							
30	66	4	11	1.0	48	23	17						
		5	63	8.6	54	235	171	229	169	217	159	188	138
		6	176	34.6	59	1031	749	995	749	964	729	889	671
		7	277	74.0	62	2318	1685	2227	1685	2182	1674	2074	1590
		8	232	81.0	65	2656	1932	2545	1932	2511	1932	2430	1880
		9	85	37.6	67	1269	923	1214	923	1203	923	1177	915
		10	10	5.5	69	190	138	181	138	180	138	178	138
			854	242.1		7722	5615	7391	5596	7257	5555	6936	5332

MEAN DIA. 7.2 INCHES, WEIBULL PARAMETERS A= 2.7, B= 4.84, C= 4.36

Age	Av. dom. ht.	D.b.h.	Stems per acre	Basal area	Av. ht.	O. b.	I. b.	O. b.	I. b.	O. b.	I. b.	O. b.	I. b.
35	72	4	3	.3	48	6	5						
		5	21	2.9	55	80	58	78	57	74	54	64	47
		6	74	14.5	60	441	320	425	320	412	312	381	288
		7	154	41.2	64	1330	967	1277	967	1253	962	1193	916
		8	199	69.5	68	2383	1734	2283	1734	2254	1734	2187	1693
		9	146	64.5	70	2277	1656	2177	1656	2159	1656	2117	1647
		10	52	28.4	72	1029	749	983	749	978	749	966	749
		11	7	4.6	74	172	125	164	125	164	125	163	125
			656	225.8		7718	5614	7387	5608	7294	5592	7071	5465

MEAN DIA. 7.9 INCHES, WEIBULL PARAMETERS A= 3.0, B= 5.27, C= 4.32

Age	Av. dom. ht.	D.b.h.	Stems per acre	Basal area	Av. ht.	O. b.	I. b.	O. b.	I. b.	O. b.	I. b.	O. b.	I. b.
40	76	5	7	1.0	55	27	19	26	19	25	18	21	16
		6	28	5.5	61	170	123	163	123	159	120	147	111
		7	69	18.4	66	614	447	590	447	579	445	553	424
		8	117	40.8	69	1422	1034	1362	1034	1345	1034	1305	1011
		9	133	58.8	72	2133	1552	2040	1552	2024	1552	1986	1546
		10	94	51.3	75	1938	1410	1851	1410	1842	1410	1821	1410
		11	37	24.4	77	948	690	904	690	902	690	896	690
		12	7	5.5	79	219	159	209	159	208	159	208	159
			492	205.7		7471	5434	7145	5434	7084	5428	6937	5367

MEAN DIA. 8.8 INCHES, WEIBULL PARAMETERS A= 3.3, B= 5.83, C= 4.28

SI 60

2,500 stems per acre

Age	Av. dom. ht.	D.b.h.	Stems per acre	Basal area	Av. ht.	All trees, total stem O.b.	I.b.	5-in 2 inches O.b.	I.b.	3 inches O.b.	I.b.	4 inches O.b.	I.b.
Yrs	Ft	In	No.	Ft 2	Ft	Ft 3							
10	31	1	14	.1	12	1	1						
		2	258	5.6	22	68	48						
		3	882	43.3	27	609	439						
		4	709	61.9	30	951	689						
		5	77	10.5	32	171	124	169	118	154	107	117	79
			1940	121.4		1800	1301	169	118	154	107	117	79
			MEAN DIA. 3.4 INCHES, WEIBULL PARAMETERS A= .5, B= 3.11, C= 4.48										
15	43	2	15	.3	27	5	3						
		3	176	8.6	34	152	110						
		4	585	51.1	38	991	718						
		5	699	95.3	41	1984	1440	1944	1399	1809	1295	1485	1048
		6	223	43.8	43	953	693	925	688	883	656	785	580
		7	11	2.9	45	67	49	64	49	63	47	58	44
			1709	202.1		4152	3013	2933	2136	2755	1998	2328	1672
			MEAN DIA. 4.7 INCHES, WEIBULL PARAMETERS A= 1.2, B= 3.66, C= 4.46										
20	52	2	1	.0	29	0	0						
		3	28	1.4	38	27	19						
		4	187	16.3	44	366	265						
		5	484	66.0	48	1606	1167	1567	1144	1475	1073	1254	904
		6	531	104.3	51	2690	1955	2601	1954	2505	1881	2276	1703
		7	192	51.3	53	1374	999	1322	999	1290	984	1212	924
		8	14	4.9	54	133	97	128	97	126	97	120	93
			1437	244.2		6196	4502	5618	4194	5396	4035	4862	3624
			MEAN DIA. 5.6 INCHES, WEIBULL PARAMETERS A= 1.8, B= 4.02, C= 4.44										
25	60	3	4	.2	40	4	3						
		4	51	4.5	47	107	77						
		5	205	28.0	52	737	535	717	527	679	497	586	426
		6	403	79.1	56	2241	1629	2164	1629	2092	1578	1920	1446
		7	367	98.1	58	2873	2089	2762	2089	2702	2068	2556	1955
		8	123	42.9	60	1300	946	1246	946	1228	946	1184	914
		9	10	4.4	62	138	101	132	101	131	101	128	99
			1163	257.2		7400	5380	7021	5292	6832	5190	6374	4840
			MEAN DIA. 6.4 INCHES, WEIBULL PARAMETERS A= 2.3, B= 4.36, C= 4.41										

Age	Av. dom. ht.	D.b.h.	Stems per acre	Basal area	Av. ht.	All trees, total stem		5-inch class and greater to o.b. tops of—					
								2 inches		3 inches		4 inches	
						O. b.	I. b.	O. b.	I. b.	O. b.	I. b.	O. b.	I. b.
Yrs	Ft	In	No.	Ft 2	Ft	------------ Ft 3 ------------							
30	66	4	13	1.1	48	28	20						
		5	74	10.1	54	276	201	269	198	255	187	221	162
		6	204	40.1	59	1195	869	1153	869	1117	845	1031	778
		7	307	82.0	62	2569	1868	2468	1868	2418	1856	2299	1762
		8	234	81.7	65	2679	1949	2567	1949	2533	1949	2451	1896
		9	72	31.8	67	1075	782	1028	782	1019	782	997	775
		10	7	3.8	69	133	97	127	97	126	97	124	97
			911	250.6		7955	5786	7612	5763	7468	5716	7123	5470

MEAN DIA. 7.1 INCHES, WEIBULL PARAMETERS A= 2.7, B= 4.72, C= 4.37

Age	Av. dom. ht.	D.b.h.	Stems per acre	Basal area	Av. ht.	O. b.	I. b.	O. b.	I. b.	O. b.	I. b.	O. b.	I. b.
35	72	4	3	.3	48	6	5						
		5	25	3.4	55	95	69	92	68	88	65	76	56
		6	86	16.9	61	521	379	502	379	487	369	451	341
		7	174	46.5	65	1526	1110	1465	1110	1438	1105	1371	1052
		8	214	74.7	68	2563	1864	2455	1864	2424	1864	2351	1820
		9	144	63.6	70	2246	1634	2147	1634	2130	1634	2088	1625
		10	44	24.0	72	871	634	832	634	828	634	817	634
		11	5	3.3	74	123	90	117	90	117	90	116	90
			695	232.7		7951	5785	7610	5779	7512	5761	.7270	5618

MEAN DIA. 7.8 INCHES, WEIBULL PARAMETERS A= 3.0, B= 5.15, C= 4.33

Age	Av. dom. ht.	D.b.h.	Stems per acre	Basal area	Av. ht.	O. b.	I. b.	O. b.	I. b.	O. b.	I. b.	O. b.	I. b.
40	76	4	1	.1	47	2	2						
		5	7	1.0	55	27	19	26	19	25	18	21	16
		6	32	6.3	61	194	141	187	141	181	137	168	127
		7	79	21.1	66	703	512	675	512	663	510	633	486
		8	130	45.4	70	1602	1166	1535	1166	1516	1166	1473	1141
		9	141	62.3	73	2293	1668	2192	1668	2175	1668	2136	1663
		10	92	50.2	75	1697	1380	1812	1380	1803	1380	1782	1380
		11	31	20.5	77	794	578	758	578	756	578	751	578
		12	5	3.9	79	156	114	149	114	149	114	148	114
			518	210.7		7668	5580	7334	5578	7268	5571	7112	5505

MEAN DIA. 8.6 INCHES, WEIBULL PARAMETERS A= 3.3, B= 5.70, C= 4.29

SI 70

500 stems per acre

Age	Av. dom. ht.	D.b.h.	Stems per acre	Basal area	Av. ht.	All trees, total stem		2 inches		3 inches		4 inches	
						O. b.	I. b.	O. b.	I. b.	O. b.	I. b.	O. b.	I. b.
Yrs	Ft	In	No.	Ft 2	Ft	— — — — — — — — — — — — Ft 3 — — — — — — — — — — — —							
10	36	2	2	.0	19	0	0						
		3	13	.6	25	8	6						
		4	42	3.7	29	55	39						
		5	85	11.6	32	169	137	186	131	170	118	130	87
		6	122	24.0	33	401	291	391	285	368	267	312	225
		7	118	31.5	35	559	406	541	405	519	388	468	348
		8	71	24.8	36	451	328	435	328	422	321	393	298
		9	23	10.2	37	190	138	182	138	179	137	170	130
		10	4	2.2	37	41	30	39	30	38	30	37	29
			480	108.6		1894	1375	1774	1317	1696	1261	1510	1117

MEAN DIA. 6.4 INCHES, WEIBULL PARAMETERS A= .8, B= 6.02, C= 4.27

15	50	3	2	.1	31	2	1						
		4	12	1.0	37	20	14						
		5	36	4.9	41	102	74	100	72	93	67	76	54
		6	72	14.1	44	315	229	305	227	292	217	260	193
		7	107	28.6	46	665	483	641	483	622	472	578	437
		8	111	38.7	48	939	683	902	683	884	679	841	645
		9	77	34.0	49	841	612	806	612	795	612	767	593
		10	31	16.9	50	426	310	408	310	404	310	394	306
		11	7	4.6	51	119	86	114	86	113	86	111	86
		12	1	.8	52	21	15	20	15	20	15	19	15
			456	143.9		3450	2507	3296	2488	3223	2458	3046	2329

MEAN DIA. 7.6 INCHES, WEIBULL PARAMETERS A= 1.7, B= 6.32, C= 4.27

20	61	4	3	.3	41	5	4						
		5	13	1.8	47	42	31	41	30	39	28	33	24
		6	35	6.9	51	177	129	171	129	165	124	150	112
		7	66	17.6	54	481	350	463	350	452	345	425	324
		8	93	32.5	56	918	667	880	667	866	667	832	641
		9	96	42.4	58	1241	903	1188	903	1175	903	1143	887
		10	65	35.5	60	1073	780	1025	780	1018	780	1000	778
		11	28	18.5	61	568	413	543	413	540	413	534	413
		12	7	5.5	62	172	125	164	125	163	125	162	125
		13	1	.9	63	29	21	28	21	28	21	28	21
			407	161.8		4706	3423	4503	3418	4446	3406	4307	3325

MEAN DIA. 8.5 INCHES, WEIBULL PARAMETERS A= 2.4, B= 6.58, C= 4.27

25	70	4	1	.1	43	2	1						
		5	4	.5	50	14	10	13	10	13	9	11	8
		6	15	2.9	55	82	60	79	60	76	58	70	53
		7	33	8.8	59	263	191	253	191	247	189	234	179
		8	57	19.9	62	622	453	597	453	588	453	568	439
		9	76	33.6	65	1101	801	1053	801	1043	801	1020	793
		10	75	40.9	67	1382	1005	1320	1005	1312	1005	1293	1005
		11	52	34.3	69	1193	868	1139	868	1135	868	1125	868
		12	24	18.8	70	665	484	634	484	633	484	630	484
		13	7	6.5	71	231	168	220	168	220	168	220	168
		14	1	1.1	73	39	29	37	29	37	29	37	29
			345	167.5		5594	4070	5345	4069	5304	4064	5208	4026

MEAN DIA. 9.4 INCHES, WEIBULL PARAMETERS A= 2.9, B= 6.96, C= 4.23

Age	Av. dom. ht.	D.b.h.	Stems per acre	Basal area	Av. ht.	All trees, total stem		5-inch class and greater to o.b. tops of—					
								2 inches		3 inches		4 inches	
						O.b.	I.b.	O.b.	I.b.	O.b.	I.b.	O.b.	I.b.
Yrs	Ft	In	No.	Ft²	Ft	—	—	—	— Ft³ —	—	—	—	—
30	77	.5	1	.1	52	4	3	3	3	3	2	3	2
		6	5	1.0	58	29	21	28	21	27	20	25	19
		7	14	3.7	62	117	85	113	85	110	85	105	80
		8	28	9.8	66	325	237	312	237	308	237	298	231
		9	45	19.9	69	692	503	662	503	656	503	643	500
		10	58	31.6	72	1148	835	1097	835	1091	835	1077	835
		11	56	37.0	74	1378	1003	1315	1003	1311	1003	1302	1003
		12	42	33.0	76	1263	919	1205	919	1203	919	1199	919
		13	22	20.3	77	787	573	750	573	750	573	749	573
		14	8	8.6	79	340	248	324	248	324	248	324	248
		15	2	2.5	80	99	72	94	72	94	72	94	72
			281	167.4		6182	4499	5903	4499	5877	4497	5819	4482

MEAN DIA. 10.5 INCHES, WEIBULL PARAMETERS A= 3.4, B= 7.55, C= 4.20

Age	Av. dom. ht.	D.b.h.	Stems per acre	Basal area	Av. ht.	O.b.	I.b.	O.b.	I.b.	O.b.	I.b.	O.b.	I.b.
35	83	6	2	.4	58	12	8	11	8	11	8	10	7
		7	5	1.3	63	43	31	41	31	40	31	38	29
		8	11	3.8	68	132	96	126	96	125	96	121	94
		9	20	8.8	71	316	230	302	230	300	230	294	229
		10	31	16.9	74	631	459	602	459	599	459	592	459
		11	39	25.7	77	999	727	953	727	950	727	944	727
		12	41	32.2	79	1282	933	1222	933	1221	933	1218	933
		13	34	31.3	81	1279	931	1219	931	1219	931	1219	931
		14	22	23.5	83	983	716	937	716	937	716	937	716
		15	11	13.5	84	571	416	544	416	544	416	544	416
		16	4	5.6	85	239	174	228	174	228	174	228	174
		17	1	1.6	87	69	50	66	50	66	50	66	50
			221	164.8		6556	4771	6251	4771	6240	4771	6211	4765

MEAN DIA. 11.7 INCHES, WEIBULL PARAMETERS A= 3.8, B= 8.46, C= 4.16

Age	Av. dom. ht.	D.b.h.	Stems per acre	Basal area	Av. ht.	O.b.	I.b.	O.b.	I.b.	O.b.	I.b.	O.b.	I.b.
40	89	7	2	.5	63	17	12	16	12	16	12	15	12
		8	4	1.4	68	48	35	46	35	45	35	44	34
		9	8	3.5	72	128	93	123	93	122	93	119	93
		10	13	7.1	76	272	198	259	198	258	198	255	198
		11	18	11.9	79	473	344	451	344	450	344	447	344
		12	24	18.8	81	769	560	734	560	733	560	731	560
		13	27	24.9	83	1041	757	992	757	992	757	992	757
		14	26	27.8	85	1190	866	1134	866	1134	866	1134	866
		15	21	25.8	87	1129	822	1076	822	1076	822	1076	822
		16	14	19.5	88	866	631	825	631	825	631	825	631
		17	8	12.6	90	572	416	544	416	544	416	544	416
		18	3	5.3	91	243	177	231	177	231	177	231	177
		19	1	2.0	92	91	66	87	66	87	66	87	66
			169	161.2		6839	4977	6518	4977	6513	4977	6500	4976

MEAN DIA. 13.2 INCHES, WEIBULL PARAMETERS A= 4.1, B= 9.81, C= 4.12

SI 70

750 stems per acre

Age	Av. dom. ht.	D.b.h.	Stems per acre	Basal area	Av. ht.	All trees, total stem O.b.	I.b.	2 inches O.b.	I.b.	3 inches O.b.	I.b.	4 inches O.b.	I.b.
Yrs	Ft	In	No.	Ft 2	Ft	—	—	—	Ft 3	—	—	—	—
10	36	2	5	.1	21	1	1						
		3	34	1.7	27	23	17						
		4	101	8.8	31	140	101						
		5	184	25.1	33	421	306	415	292	379	264	293	198
		6	204	40.1	34	691	502	673	492	635	463	542	391
		7	121	32.3	36	589	428	570	427	548	410	495	370
		8	32	11.2	37	209	152	201	152	196	149	182	139
		9	3	1.3	37	25	18	24	18	23	18	22	17
			584	120.6		2099	1525	1883	1381	1781	1304	1534	1115

MEAN DIA. 5.7 INCHES, WEIBULL PARAMETERS A= .8, B= 5.22, C= 4.32

Age	Av. dom. ht.	D.b.h.	Stems per acre	Basal area	Av. ht.	O.b.	I.b.	O.b.	I.b.	O.b.	I.b.	O.b.	I.b.
15	50	3	5	.2	34	4	3						
		4	28	2.4	39	49	35						
		5	80	10.9	43	238	173	233	168	218	157	181	128
		6	150	29.5	46	686	498	664	496	637	475	571	424
		7	181	48.4	48	1173	853	1131	853	1099	835	1024	777
		8	131	45.7	50	1154	840	1108	840	1088	836	1037	797
		9	49	21.6	51	557	405	534	405	527	405	510	394
		10	8	4.4	52	114	83	109	83	108	83	106	82
			632	163.2		3975	2890	3779	2845	3677	2791	3429	2602

MEAN DIA. 6.9 INCHES, WEIBULL PARAMETERS A= 1.7, B= 5.55, C= 4.31

Age	Av. dom. ht.	D.b.h.	Stems per acre	Basal area	Av. ht.	O.b.	I.b.	O.b.	I.b.	O.b.	I.b.	O.b.	I.b.
20	61	4	7	.6	44	14	10						
		5	29	4.0	49	98	71	96	70	90	66	77	56
		6	75	14.7	53	395	287	382	287	368	277	336	252
		7	130	34.7	55	965	702	928	702	907	693	855	652
		8	149	52.0	58	1522	1107	1460	1107	1437	1107	1363	1067
		9	108	47.7	60	1444	1050	1382	1050	1368	1050	1333	1034
		10	42	22.9	61	705	513	673	513	669	513	657	512
		11	8	5.3	62	165	120	158	120	157	120	155	120
		12	1	.8	63	25	18	24	18	24	18	24	18
			549	182.7		5333	3878	5103	3867	5020	3844	4820	3711

MEAN DIA. 7.8 INCHES, WEIBULL PARAMETERS A= 2.4, B= 5.81, C= 4.29

Age	Av. dom. ht.	D.b.h.	Stems per acre	Basal area	Av. ht.	O.b.	I.b.	O.b.	I.b.	O.b.	I.b.	O.b.	I.b.
25	70	4	1	.1	46	2	1						
		5	9	1.2	52	32	23	31	23	30	22	26	19
		6	31	6.1	57	175	128	169	128	164	124	151	114
		7	68	18.2	61	560	407	538	407	527	404	500	383
		8	106	37.0	64	1195	869	1145	869	1129	869	1092	845
		9	116	51.2	66	1706	1241	1632	1241	1617	1241	1562	1230
		10	81	44.2	68	1514	1102	1447	1102	1438	1102	1418	1102
		11	34	22.4	70	792	576	756	576	753	576	747	576
		12	7	5.5	71	197	143	188	143	187	143	186	143
		13	1	.9	73	34	25	32	25	32	25	32	25
			454	186.9		6207	4515	5938	4514	5877	4506	5734	4437

MEAN DIA. 8.7 INCHES, WEIBULL PARAMETERS A= 2.9, B= 6.15, C= 4.27

Age	Av. dom. ht.	D.b.h.	Stems per acre	Basal area	Av. ht.	All trees, total stem		5-inch class and greater to o.b. tops of—					
								2 inches		3 inches		4 inches	
						O. b.	I. b.	O. b.	I. b.	O. b.	I. b.	O. b.	I. b.
Yrs	Ft	In	No.	Ft²	Ft	— — — — — — — — — — — Ft³ — — — — — — — — — — —							
30	77	5	2	.3	54	7	5	7	5	7	5	6	4
		6	11	2.2	60	66	48	53	48	61	46	57	43
		7	29	7.8	64	250	182	241	182	236	181	225	172
		8	55	19.2	68	659	479	631	479	523	479	604	468
		9	79	34.9	71	1250	909	1195	909	1185	909	1162	905
		10	85	46.4	73	1706	1241	1629	1241	1621	1241	1601	1241
		11	61	40.3	76	1542	1122	1471	1122	1467	1122	1457	1122
		12	29	22.8	77	884	643	843	643	842	643	839	643
		13	8	7.4	79	294	214	280	214	280	214	280	214
		14	1	1.1	80	43	31	41	31	41	31	41	31
			360	182.1		6701	4874	6401	4874	6363	4871	6272	4843

MEAN DIA. 9.6 INCHES, WEIBULL PARAMETERS A= 3.4, B= 6.68, C= 4.23

Age	Av. dom. ht.	D.b.h.	Stems per acre	Basal area	Av. ht.	All trees, total stem		2 in O.b.	2 in I.b.	3 in O.b.	3 in I.b.	4 in O.b.	4 in I.b.
35	83	5	1	.1	54	4	3	4	3	3	3	3	2
		6	3	.6	61	18	13	18	13	17	13	16	12
		7	10	2.7	66	89	65	85	65	84	65	80	62
		8	22	7.7	71	275	200	263	200	260	200	253	196
		9	39	17.2	74	643	468	615	468	610	468	599	467
		10	53	28.9	77	1122	816	1071	816	1067	818	1055	816
		11	58	38.3	80	1543	1123	1472	1123	1469	1123	1460	1123
		12	47	36.9	82	1525	1110	1454	1110	1453	1110	1450	1110
		13	28	25.8	84	1092	795	1041	795	1041	795	1041	795
		14	11	11.8	85	504	366	480	366	480	366	480	366
		15	3	3.7	87	161	117	154	117	154	117	154	117
			275	173.7		6976	5076	6657	5076	6638	5076	6591	5066

MEAN DIA. 10.8 INCHES, WEIBULL PARAMETERS A= 3.8, B= 7.48, C= 4.19

Age	Av. dom. ht.	D.b.h.	Stems per acre	Basal area	Av. ht.	All trees, total stem		2 in O.b.	2 in I.b.	3 in O.b.	3 in I.b.	4 in O.b.	4 in I.b.
40	89	6	1	.2	60	6	4	6	4	6	4	5	4
		7	3	.8	66	27	19	26	19	25	19	24	18
		8	7	2.4	71	88	64	84	64	83	64	80	62
		9	14	6.2	75	234	170	224	170	222	170	218	170
		10	23	12.5	78	493	359	471	359	469	359	464	359
		11	32	21.1	81	862	627	822	627	821	627	816	627
		12	36	29.8	84	1263	919	1204	919	1204	919	1201	919
		13	35	32.3	86	1398	1017	1332	1017	1332	1017	1332	1017
		14	27	28.9	88	1280	931	1219	931	1219	931	1219	931
		15	16	19.6	89	880	641	838	641	838	641	838	641
		16	7	9.8	91	448	326	427	326	427	326	427	326
		17	2	3.2	92	146	106	139	106	139	106	139	106
			205	166.6		7125	5183	6792	5183	6785	5183	6763	5180

MEAN DIA. 12.2 INCHES, WEIBULL PARAMETERS A= 4.1, B= 8.69, C= 4.15

SI 70

1,000 stems per acre

Age	Av. dom. ht.	D.b.h.	Stems per acre	Basal area	Av. ht.	All trees, total stem		5-inch class and greater to o.b. tops of—					
								2 inches		3 inches		4 inches	
						O. b.	I. b.	O. b.	I. b.	O. b.	I. b.	O. b.	I. b.
Yrs	Ft	In	No.	Ft²	Ft	------------ Ft³ ------------							
10	36	2	10	.2	23	3	2						
		3	63	3.1	29	47	34						
		4	182	15.9	32	260	188						
		5	293	40.0	34	691	501	681	480	623	436	486	332
		6	239	46.9	35	833	605	811	595	766	560	657	477
		7	82	21.9	36	399	290	386	290	371	278	336	251
		8	9	3.1	37	59	43	57	43	55	42	51	39
			878	131.1		2292	1663	1935	1408	1815	1316	1530	1099
				MEAN DIA. 5.2 INCHES, WEIBULL PARAMETERS A= .8, B= 4.74, C= 4.36									
15	50	3	9	.4	35	8	6						
		4	49	4.3	40	87	63						
		5	139	19.0	44	423	307	414	300	387	279	323	230
		6	235	46.1	47	1097	798	1063	795	1020	762	917	683
		7	230	61.5	48	1491	1084	1437	1084	1397	1062	1302	988
		8	112	39.1	50	987	718	948	718	930	715	887	681
		9	22	9.7	51	250	182	240	182	236	182	229	177
		10	1	.5	52	14	10	14	10	14	10	13	10
			797	180.6		4357	3168	4116	3089	3984	3010	3671	2769
				MEAN DIA. 6.4 INCHES, WEIBULL PARAMETERS A= 1.7, B= 5.09, C= 4.35									
20	61	3	1	.0	38	1	1						
		4	11	1.0	46	22	16						
		5	50	6.8	51	176	128	172	126	162	119	140	101
		6	124	24.3	54	665	483	642	483	620	467	567	426
		7	196	52.4	57	1508	1097	1450	1097	1418	1085	1340	1024
		8	184	64.2	59	1912	1391	1834	1391	1806	1391	1739	1342
		9	92	40.6	61	1251	910	1197	910	1185	910	1155	897
		10	20	10.9	62	341	248	326	248	324	248	318	248
		11	2	1.3	63	42	30	40	30	40	30	39	30
			680	201.7		5918	4304	5661	4285	5555	4250	5298	4068
				MEAN DIA. 7.4 INCHES, WEIBULL PARAMETERS A= 2.4, B= 5.35, C= 4.33									
25	70	4	2	.2	48	4	3						
		5	15	2.0	54	56	41	54	40	52	38	45	33
		6	51	10.0	59	299	217	288	217	279	211	258	195
		7	108	28.9	63	918	668	882	668	865	664	823	631
		8	153	53.4	65	1751	1274	1678	1274	1656	1274	1603	1240
		9	135	59.6	68	2045	1488	1956	1488	1939	1488	1899	1477
		10	68	37.1	70	1309	952	1250	952	1243	952	1227	952
		11	17	11.2	71	401	292	383	292	382	292	379	292
		12	2	1.6	73	58	42	55	42	55	42	55	42
			551	204.0		6841	4977	6546	4973	6471	4961	6289	4862
				MEAN DIA. 8.2 INCHES, WEIBULL PARAMETERS A= 2.9, B= 5.68, C= 4.30									

Age	Av. dom. ht.	D.b.h.	Stems per acre	Basal area	Av. ht.	All trees, total stem		5-inch class and greater to o.b. tops of—					
								2 inches		3 inches		4 inches	
						O.b.	I.b.	O.b.	I.b.	O.b.	I.b.	O.b.	I.b.
Yrs	Ft	In	No.	Ft 2	Ft	— — — — — — — — — — — — Ft 3 — — — — — — — — — — — —							
30	77	5	4	.5	56	15	11	15	11	14	11	12	9
		6	18	3.5	62	111	81	107	81	104	79	96	73
		7	46	12.3	66	410	298	393	298	386	297	368	283
		8	85	29.7	70	1048	762	1003	762	991	762	963	746
		9	111	49.0	73	1805	1313	1726	1313	1713	1313	1681	1309
		10	96	52.4	75	1980	1440	1890	1440	1881	1440	1860	1440
		11	52	34.3	77	1332	969	1271	969	1267	969	1259	969
		12	15	11.8	79	469	341	447	341	447	341	445	341
		13	2	1.8	80	74	54	71	54	71	54	71	54
			429	195.4		7244	5269	6923	5269	6874	5266	6755	5224

MEAN DIA. 9.1 INCHES, WEIBULL PARAMETERS A= 3.4, B= 6.16, C= 4.26

Age	Av. dom. ht.	D.b.h.	Stems per acre	Basal area	Av. ht.	O.b.	I.b.	O.b.	I.b.	O.b.	I.b.	O.b.	I.b.
35	83	5	1	.1	55	4	3	4	3	4	3	3	2
		6	5	1.0	62	31	22	30	22	29	22	27	20
		7	16	4.3	67	145	105	139	105	136	105	130	100
		8	35	12.2	71	438	318	419	318	414	318	402	312
		9	58	25.6	75	969	705	926	705	920	705	903	704
		10	74	40.4	77	1567	1140	1496	1140	1489	1140	1473	1140
		11	67	44.2	80	1783	1297	1701	1297	1697	1297	1687	1297
		12	43	33.8	82	1395	1015	1331	1015	1329	1015	1326	1015
		13	18	16.6	84	702	511	669	511	669	511	669	511
		14	4	4.3	85	183	133	174	133	174	133	174	133
		15	1	1.2	87	54	39	51	39	51	39	51	39
			322	183.7		7271	5288	6940	5288	6912	5288	6845	5273

MEAN DIA. 10.2 INCHES, WEIBULL PARAMETERS A= 3.8, B= 6.90, C= 4.22

Age	Av. dom. ht.	D.b.h.	Stems per acre	Basal area	Av. ht.	O.b.	I.b.	O.b.	I.b.	O.b.	I.b.	O.b.	I.b.
40	89	6	1	.2	62	6	4	6	4	6	4	5	4
		7	5	1.3	68	46	33	44	33	43	33	41	32
		8	12	4.2	73	154	112	148	112	146	112	142	110
		9	22	9.7	77	377	275	361	275	358	275	352	274
		10	35	19.1	80	770	560	735	560	732	560	725	560
		11	44	29.0	83	1215	884	1159	884	1156	884	1150	884
		12	46	36.1	85	1547	1126	1475	1126	1474	1126	1472	1126
		13	37	34.1	87	1495	1088	1425	1088	1425	1088	1425	1088
		14	22	23.5	89	1054	767	1005	767	1005	767	1005	767
		15	9	11.0	91	506	368	482	368	482	368	482	368
		16	3	4.2	92	194	141	185	141	185	141	185	141
			236	172.6		7364	5358	7025	5358	7012	5358	6984	5354

MEAN DIA. 11.6 INCHES, WEIBULL PARAMETERS A= 4.1, B= 8.01, C= 4.17

SI 70

1,250 stems per acre

Age	Av. dom. ht.	D.b.h	Stems per acre	Basal area	Av. ht.	All trees, total stem		5-inch class and greater to o.b. tops of—					
								2 inches		3 inches		4 inches	
						C. b.	I. b.	O. b.	I. b.	O. b.	I. b.	O. b.	I. b.
Yrs	Ft		No.	Ft²						Ft 3			
10	36	2	17	.4	24	5	3						
		3	105	5.2	29	70	56						
		4	285	24.9	32	407	295						
		5	389	53.0	34	917	666	904	638	828	579	645	440
		6	227	44.6	35	791	574	770	565	728	532	624	453
		7	42	11.2	37	210	153	203	153	196	147	177	133
		8	2	.7	37	13	9	13	9	12	9	11	9
			1067	139.9		2421	1756	1890	1365	1764	1267	1457	1035
						MEAN DIA. 4.9 INCHES, WEIBULL PARAMETERS A= .8, B= 4.38, C= 4.39							
15	50	3	13	.8	37	12	9						
		4	77	8.7	42	144	104						
		5	210	28.6	45	854	475	639	464	599	433	502	359
		6	320	62.8	48	1526	1109	1477	1106	1419	1061	1279	954
		7	247	86.0	50	1668	1213	1608	1213	1564	1191	1462	1112
		8	79	27.6	51	710	516	682	516	669	515	639	491
		9	8	3.5	52	93	67	89	67	88	67	85	66
			954	195.9		4807	3493	4493	3366	4339	3267	3967	2982
						MEAN DIA. 6.1 INCHES, WEIBULL PARAMETERS A= 1.7, B= 4.76, C= 4.37							
20	61	3	1	.0		2	1						
		4	17	1.5		38	26						
		5	76	10.4	52	273	198	266	195	252	184	217	158
		6	181	35.5	56	1008	732	972	732	940	709	862	649
		7	255	68.1	58	1996	1452	1919	1452	1877	1437	1776	1358
		8	198	68.4	60	2072	1507	1986	1507	1957	1507	1886	1456
		9	68	30.0	62	959	683	899	683	890	683	869	674
		10	6	4.4	63	159	101	132	101	132	101	129	101
			802	218.4		6462	4700	6174	4670	6048	4621	5739	4396
						MEAN DIA. 7.1 INCHES, WEIBULL PARAMETERS A= 2.4, B= 5.02, C= 4.35							
25	70	4	3	.3	50	7	5						
		5	23	3.1	56	89	65	86	64	82	61	72	53
		6	75	14.7	61	454	330	438	330	425	322	393	297
		7	151	40.4	64	1304	948	1252	948	1228	944	1170	898
		8	192	67.0	67	2265	1648	2170	1648	2143	1648	2077	1607
		9	140	61.9	69	2152	1566	2058	1566	2041	1566	2000	1556
		10	50	27.3	71	976	710	932	710	927	710	915	710
		11	7	4.5	73	170	124	162	124	162	124	160	124
			641	219.2		7417	5396	7098	5390	7008	5375	6787	5245
						MEAN DIA. 7.9 INCHES, WEIBULL PARAMETERS A= 2.9, B= 5.34, C= 4.32							

Age	Av. dom. ht.	D.b.h.	Stems per acre	Basal area	Av. ht.	All trees, total stem		5-inch class and greater to o.b. tops of—					
								2 inches		3 inches		4 inches	
						O. b.	I. b.	O. b.	I. b.	O. b.	I. b.	O. b.	I. b.
Yrs	Ft	In	No.	Ft 2	Ft	---	---	---	Ft 3	---	---	---	---
30	77	5	6	.8	57	24	17	23	17	22	16	19	14
		6	26	5.1	62	160	116	154	116	150	114	139	105
		7	66	17.6	67	597	434	573	434	562	433	537	413
		8	115	40.1	70	1418	1031	1357	1031	1341	1031	1303	1009
		9	134	59.2	73	2179	1585	2083	1585	2067	1585	2030	1580
		10	97	52.9	75	2000	1455	1910	1455	1901	1455	1879	1455
		11	39	25.7	77	999	727	953	727	950	727	944	727
		12	7	5.5	79	219	159	209	159	208	159	208	159
		13	1	.9	80	37	27	35	27	35	27	35	27
			491	208.0		7633	5551	7297	5551	7236	5547	7094	5489

MEAN DIA. 8.8 INCHES, WEIBULL PARAMETERS A= 3.4, B= 5.81, C= 4.28

Age	Av. dom. ht.	D.b.h.	Stems per acre	Basal area	Av. ht.	O. b.	I. b.	O. b.	I. b.	O. b.	I. b.	O. b.	I. b.
35	83	5	1	.1	57	4	3	4	3	4	3	3	2
		6	7	1.4	64	44	32	43	32	42	32	39	29
		7	23	6.1	69	214	156	205	156	202	155	193	149
		8	49	17.1	73	630	458	603	458	596	458	580	450
		9	77	34.0	76	1304	948	1246	948	1237	948	1216	948
		10	91	49.6	79	1976	1438	1887	1438	1879	1438	1860	1438
		11	70	46.2	81	1886	1372	1799	1372	1795	1372	1785	1372
		12	35	27.5	83	1150	837	1096	837	1095	837	1093	837
		13	10	9.2	85	395	287	376	287	376	287	376	287
		14	1	1.1	87	47	34	45	34	45	34	45	34
			364	192.4		7650	5565	7304	5565	7271	5564	7190	5546

MEAN DIA. 9.8 INCHES, WEIBULL PARAMETERS A= 3.8, B= 6.50, C= 4.24

Age	Av. dom. ht.	D.b.h.	Stems per acre	Basal area	Av. ht.	O. b.	I. b.	O. b.	I. b.	O. b.	I. b.	O. b.	I. b.
40	89	6	2	.4	63	13	9	12	9	12	9	11	8
		7	7	1.9	68	64	47	62	47	61	47	58	45
		8	16	5.6	73	206	150	197	150	195	150	189	147
		9	31	13.7	77	532	387	508	387	505	387	496	387
		10	46	25.1	80	1012	736	966	736	962	736	952	736
		11	54	35.6	83	1491	1085	1422	1085	1419	1085	1412	1085
		12	50	39.3	85	1682	1224	1604	1224	1603	1224	1600	1224
		13	34	31.3	87	1374	1000	1309	1000	1309	1000	1309	1000
		14	16	17.1	89	767	558	731	558	731	558	731	558
		15	5	6.1	91	281	205	268	205	268	205	268	205
		16	1	1.4	92	65	47	62	47	62	47	62	47
			262	177.5		7487	5448	7141	5448	7127	5448	7088	5442

MEAN DIA. 11.1 INCHES, WEIBULL PARAMETERS A= 4.1, B= 7.54, C= 4.19

SI 70

1,500 stems per acre

Age	Av. dom. ht.	D.b.h.	Stems per acre	Basal area	Av. ht.	All trees, total stem		5-inch class and greater to o.b. tops of—					
								2 inches		3 inches		4 inches	
						O. b.	I. b.	O. b.	I. b.	O. b.	I. b.	O. b.	I. b.
Yrs	Ft	In	No.	Ft 2	Ft	———————————— Ft 3 ————————————							
10	36	1	1	.0	13	0	0						
		2	25	.5	24	7	5						
		3	159	7.8	30	122	88						
		4	407	35.5	33	600	434						
		5	459	62.6	35	1114	808	1096	776	1007	707	792	543
		6	182	35.7	36	652	474	635	466	601	440	518	377
		7	17	4.5	37	85	62	82	62	79	59	72	54
			1250	146.7		2580	1871	1813	1304	1687	1206	1382	974
			MEAN DIA. 4.6 INCHES, WEIBULL PARAMETERS A= .8, B= 4.10, C= 4.41										
15	50	2	1	.0	29	0	0						
		3	19	.9	37	18	13						
		4	112	9.8	42	209	152						
		5	292	39.8	45	909	660	888	645	832	602	698	499
		6	394	77.4	48	1879	1366	1819	1362	1747	1307	1575	1174
		7	237	63.3	50	1600	1163	1541	1163	1501	1142	1403	1067
		8	48	16.8	51	431	314	414	314	407	313	388	298
		9	2	.9	52	23	17	22	17	22	17	21	16
			1105	208.9		5069	3685	4684	3501	4509	3381	4085	3054
			MEAN DIA. 5.9 INCHES, WEIBULL PARAMETERS A= 1.7, B= 4.49, C= 4.40										
20	61	3	2	.1	41	2	1						
		4	24	2.1	48	51	37						
		5	106	14.5	53	388	282	378	278	358	263	310	226
		6	242	47.5	56	1345	978	1299	978	1256	948	1153	868
		7	307	82.0	58	2403	1748	2311	1748	2260	1730	2138	1635
		8	189	66.0	61	2031	1477	1947	1477	1919	1477	1851	1429
		9	45	19.9	62	622	452	595	452	589	452	575	446
		10	3	1.6	63	52	38	50	38	49	38	49	38
			918	233.7		6894	5013	6580	4971	6431	4908	6076	4642
			MEAN DIA. 6.8 INCHES, WEIBULL PARAMETERS A= 2.4, B= 4.77, C= 4.37										
25	70	4	4	.3	50	9	6						
		5	31	4.2	57	122	89	119	88	113	83	99	73
		6	102	20.0	61	618	449	596	449	578	438	535	405
		7	196	52.4	64	1693	1231	1626	1231	1594	1225	1519	1165
		8	222	77.5	67	2619	1905	2509	1905	2477	1905	2401	1858
		9	133	58.8	69	2044	1487	1955	1487	1939	1487	1900	1478
		10	34	18.5	71	664	483	634	483	631	483	622	483
		11	3	2.0	73	73	53	70	53	69	53	39	53
			725	233.8		7842	5703	7509	5696	7401	5674	7145	5515
			MEAN DIA. 7.7 INCHES, WEIBULL PARAMETERS A= 2.9, B= 5.09, C= 4.34										

Age	Av. dom. ht.	D.b.h.	Stems per acre	Basal area	Av. ht.	All trees, total stem		5-inch class and greater to o.b. tops of—					
								2 inches		3 inches		4 inches	
						O. b.	I. b.	O. b.	I. b.	O. b.	I. b.	O. b.	I. b.
Yrs	Ft	In	No.	Ft 2	Ft	— — — — — — — — — — — Ft 3 — — — — — — — — — — —							
30	77	4	1	.1	51	2	2						
		5	8	1.1	58	32	23	31	23	30	22	26	19
		6	35	6.9	64	222	162	214	162	208	158	194	147
		7	88	23.5	68	807	587	775	587	761	586	728	560
		8	144	50.3	72	1826	1328	1748	1328	1728	1328	1680	1303
		9	151	66.7	74	2489	1811	2380	1811	2362	1811	2320	1807
		10	92	50.2	77	1948	1417	1860	1417	1851	1417	1831	1417
		11	27	17.8	79	709	516	677	516	675	516	671	516
		12	3	2.4	80	95	69	91	69	90	69	90	69
			549	218.9		8130	5915	7776	5913	7705	5907	7540	5838
			MEAN DIA. 8.6 INCHES, WEIBULL PARAMETERS A= 3.4, B= 5.54, C= 4.30										
35	83	5	2	.3	59	8	6	8	6	8	6	7	5
		6	10	2.0	65	64	47	62	47	60	46	56	43
		7	31	8.3	70	293	213	281	213	276	213	264	204
		8	64	22.3	74	834	607	798	607	790	607	769	596
		9	96	42.4	78	1668	1214	1594	1214	1583	1214	1558	1214
		10	99	54.0	81	2204	1604	2105	1604	2096	1604	2076	1604
		11	68	44.9	83	1877	1366	1791	1366	1787	1366	1778	1366
		12	27	21.2	85	908	661	866	661	865	661	864	661
		13	5	4.6	87	202	147	193	147	193	147	193	147
			402	200.0		8058	5865	7698	5865	7658	5864	7565	5840
			MEAN DIA. 9.5 INCHES, WEIBULL PARAMETERS A= 3.8, B= 6.20, C= 4.25										
40	89	6	2	.4	64	13	9	12	9	12	9	11	8
		7	9	2.4	70	85	62	82	62	80	62	77	59
		8	21	7.3	75	277	202	265	202	263	202	256	198
		9	39	17.2	79	686	499	656	499	652	499	641	499
		10	56	30.5	82	1262	919	1205	919	1200	919	1189	919
		11	64	42.2	84	1788	1301	1706	1301	1702	1301	1694	1301
		12	52	40.8	87	1790	1303	1707	1303	1706	1303	1704	1303
		13	30	27.7	89	1240	902	1182	902	1182	902	1182	902
		14	11	11.8	91	539	392	514	392	514	392	514	392
		15	2	2.5	92	114	83	108	83	108	83	108	83
			286	182.8		7794	5672	7437	5672	7419	5672	7376	5664
			MEAN DIA. 10.8 INCHES, WEIBULL PARAMETERS A= 4.1, B= 7.19, C= 4.20										

SI 70

1,750 stems per acre

Age	Av. dom. ht.	D.b.h.	Stems per acre	Basal area	Av. ht.	All trees, total stem		5-inch class and greater to o.b. tops of—					
								2 inches		3 inches		4 inches	
						O. b.	I. b.	O. b.	I. b.	O. b.	I. b.	O. b.	I. b.
Yrs	Ft	In	No.	Ft 2	Ft	– – – – – – – – – – – Ft 3 – – – – – – – – – – –							
10	36	1	1	.0	13	0	0						
		2	37	.8	24	11	8						
		3	230	11.3	30	176	127						
		4	543	47.4	33	800	580						
		5	491	66.9	35	1191	865	1173	830	1077	757	847	581
		6	123	24.2	36	441	320	429	315	406	297	350	255
		7	5	1.3	37	25	18	24	18	23	17	21	16
			1430	151.9		2644	1918	1626	1163	1506	1071	1218	852
			MEAN DIA. 4.4 INCHES, WEIBULL PARAMETERS A= .8, B= 3.86, C= 4.43										
15	50	2	1	.0	30	0	0						
		3	27	1.3	38	26	19						
		4	154	13.4	43	295	214						
		5	384	52.4	47	1248	906	1218	888	1145	832	969	697
		6	454	89.1	49	2210	1606	2139	1603	2056	1540	1858	1388
		7	206	55.1	51	1419	1031	1366	1031	1331	1014	1247	948
		8	25	8.7	52	229	167	220	167	216	166	206	159
			1251	220.1		5427	3943	4943	3689	4748	3552	4280	3192
			MEAN DIA. 5.7 INCHES, WEIBULL PARAMETERS A= 1.7, B= 4.27, C= 4.42										
20	61	3	2	.1	41	2	1						
		4	32	2.8	48	68	50						
		5	141	19.2	53	516	375	503	370	476	349	412	301
		6	308	60.5	56	1712	1245	1654	1245	1599	1206	1468	1105
		7	348	93.0	59	2771	2015	2664	2015	2606	1997	2469	1889
		8	170	59.3	61	1827	1329	1751	1329	1726	1329	1665	1286
		9	27	11.9	62	373	271	357	271	353	271	345	268
		10	1	.5	63	17	13	17	13	16	13	16	13
			1029	247.4		7286	5299	6946	5243	6776	5165	6375	4862
			MEAN DIA. 6.6 INCHES, WEIBULL PARAMETERS A= 2.4, B= 4.56, C= 4.39										
25	70	4	5	.4	51	11	8						
		5	41	5.6	57	161	117	157	116	149	110	131	96
		6	131	25.7	61	793	577	765	577	742	562	687	520
		7	240	64.1	65	2105	1531	2021	1531	1983	1524	1891	1452
		8	245	85.5	67	2891	2103	2769	2103	2734	2103	2650	2051
		9	119	52.6	69	1829	1331	1749	1331	1735	1331	1700	1322
		10	22	12.0	71	429	312	410	312	408	312	403	312
		11	1	.7	73	24	18	23	18	23	18	23	18
			804	246.6		8243	5997	7894	5988	7774	5960	7485	5771
			MEAN DIA. 7.5 INCHES, WEIBULL PARAMETERS A= 2.9, B= 4.89, C= 4.35										

Age	Av. dom. ht.	D.b.h.	Stems per acre	Basal area	Av. ht.	All trees, total stem		5-inch class and greater to o.b. tops of—					
								2 inches		3 inches		4 inches	
						O. b.	I. b.	O. b.	I. b.	O. b.	I. b.	O. b.	·I. b.
Yrs	Ft	In	No.	Ft 2	Ft	– – – – – – – – – – – Ft 3 – – – – – – – – – – – –							
30	77	4	1	.1	52	2	2						
		5	10	1.4	59	41	30	40	29	38	28	33	24
		6	44	8.6	64	279	203	269	203	262	199	243	185
		7	110	29.4	69	1024	745	983	745	965	743	924	711
		8	173	60.4	72	2193	1596	2100	1596	2076	1596	2019	1565
		9	164	72.5	75	2740	1993	2619	1993	2600	1993	2555	1990
		10	83	45.3	77	1757	1279	1678	1279	1670	1279	1652	1279
		11	18	11.9	79	473	344	451	344	450	344	447	344
		12	1	.8	80	32	23	30	23	30	23	30	23
			604	230.3		8541	6215	8170	6212	8091	6205	7903	6121
		MEAN DIA. 8.4 INCHES, WEIBULL PARAMETERS A= 3.4, B= 5.33, C= 4.31											
35	83	5	2	.3	59	8	6	8	6	8	6	7	5
		6	13	2.6	66	85	62	82	62	80	61	74	56
		7	39	10.4	71	374	272	358	272	352	272	338	260
		8	79	27.6	75	1043	759	999	759	988	759	962	747
		9	112	49.5	78	1946	1416	1860	1416	1847	1416	1817	1416
		10	107	58.4	81	2383	1734	2275	1734	2266	1734	2244	1734
		11	63	41.6	83	1739	1265	1659	1265	1656	1265	1647	1265
		12	20	15.7	85	673	490	641	490	641	490	640	490
		13	3	2.6	87	121	88	116	88	116	88	116	88
			438	208.7		8372	6092	7998	6092	7954	6091	7845	6061
		MEAN DIA. 9.3 INCHES, WEIBULL PARAMETERS A= 3.6, B= 5.96, C= 4.26											
40	89	6	3	.6	65	19	14	19	14	18	14	17	13
		7	11	2.9	70	104	76	100	76	98	75	94	72
		8	27	9.4	75	357	259	341	259	338	259	329	255
		9	48	21.2	79	845	615	807	615	802	615	789	615
		10	66	36.0	82	1488	1083	1420	1083	1415	1083	1402	1083
		11	69	45.5	85	1950	1419	1861	1419	1857	1419	1849	1419
		12	51	40.1	87	1756	1278	1674	1278	1673	1278	1671	1278
		13	25	23.0	89	1033	752	985	752	985	752	985	752
		14	7	7.5	91	343	250	327	250	327	250	327	250
		15	1	1.2	92	57	41	54	41	54	41	54	41
			308	187.5		7952	5787	7588	5787	7567	5786	7517	5778
		MEAN DIA. 10.6 INCHES, WEIBULL PARAMETERS A= 4.1, B= 6.92, C= 4.21											

SI 70

2,000 stems per acre

Age	Av. dom. ht.	D.b.h.	Stems per acre	Basal area	Av. ht.	All trees, total stem		5-inch class and greater to o.b. tops of—					
								2 inches		3 inches		4 inches	
						O.b.	I.b.	O.b.	I.b.	O.b.	I.b.	O.b.	I.b.
Yrs	Ft	In	No.	Ft 2	Ft	Ft 3							
10	36	1	1	.0	13	0	0						
		2	53	1.2	24	15	11						
		3	321	15.8	30	246	177						
		4	684	59.7	33	1008	730						
		5	476	64.9	35	1155	838	1137	805	1044	733	821	564
		6	70	13.7	36	251	182	244	179	231	169	199	145
		7	1	.3	37	5	4	5	4	5	3	4	3
			1606	155.5		2680	1942	1386	988	1280	905	1024	712
				MEAN DIA. 4.2 INCHES, WEIBULL PARAMETERS A= .8, B= 3.65, C= 4.45									
15	50	2	1	.0	30	0	0						
		3	36	1.8	39	36	26						
		4	205	17.9	43	392	284						
		5	484	66.0	47	1573	1143	1535	1119	1443	1048	1222	879
		6	491	96.4	49	2390	1737	2313	1734	2224	1665	2010	1501
		7	164	43.8	51	1129	821	1087	821	1059	807	992	755
		8	12	4.2	52	110	80	106	80	104	80	99	76
			1393	230.1		5630	4091	5041	3754	4830	3600	4323	3211
				MEAN DIA. 5.5 INCHES, WEIBULL PARAMETERS A= 1.7, B= 4.08, C= 4.43									
20	61	3	3	.1	43	3	2						
		4	42	3.7	49	91	66						
		5	180	24.5	54	672	488	653	461	619	455	538	393
		6	377	74.0	57	2133	1551	2059	1551	1993	1505	1832	1381
		7	375	100.2	60	3037	2208	2918	2208	2857	2190	2710	2075
		8	144	50.3	62	1573	1144	1507	1144	1486	1144	1435	1108
		9	15	6.6	63	211	153	201	153	200	153	195	151
			1136	259.5		7720	5612	7338	5537	7155	5447	6710	5108
				MEAN DIA. 6.5 INCHES, WEIBULL PARAMETERS A= 2.4, B= 4.39, C= 4.40									
25	70	4	7	.6	52	16	12						
		5	51	7.0	58	204	148	198	147	189	140	166	122
		6	163	32.0	63	1019	741	962	741	954	724	886	671
		7	284	75.9	66	2529	1839	2428	1839	2383	1832	2275	1747
		8	260	90.8	69	3159	2298	3026	2298	2969	2298	2901	2247
		9	102	45.1	71	1613	1174	1543	1174	1530	1174	1501	1168
		10	13	7.1	73	261	190	249	190	248	190	245	190
			880	258.4		8801	6402	8426	6389	8293	6358	7974	6145
				MEAN DIA. 7.3 INCHES, WEIBULL PARAMETERS A= 2.9, B= 4.72, C= 4.37									

Age	Av. dom. ht.	D.b.h.	Stems per acre	Basal area	Av. ht.	All trees, total stem		5-inch class and greater to o.b. tops of—					
								2 inches		3 inches		4 inches	
						O. b.	I. b.	O. b.	I. b.	O. b.	I. b.	O. b.	I. b.
Yrs	Ft	In	No.	Ft 2	Ft	— — — — — — — — — — — — — Ft 3 — — — — — — — — — — — — —							
30	77	4	1	.1	52	2	2						
		5	12	1.6	59	49	36	47	35	45	33	40	29
		6	55	10.8	65	355	258	342	258	332	253	310	235
		7	133	35.5	69	1238	900	1188	900	1167	899	1117	859
		8	199	69.5	72	2523	1835	2416	1835	2388	1835	2322	1800
		9	170	75.1	75	2840	2066	2715	2066	2695	2066	2648	2063
		10	72	39.3	77	1524	1109	1455	1109	1449	1109	1433	1109
		11	12	7.9	79	315	229	301	229	300	229	298	229
		12	1	.8	80	32	23	30	23	30	23	30	23
			655	240.6		8878	6458	8494	6455	8406	6447	8198	6347
			MEAN DIA. 8.2 INCHES, WEIBULL PARAMETERS A= 3.4, B= 5.16, C= 4.32										
35	83	5	2	.3	60	8	6	8	6	8	6	7	5
		6	15	2.9	66	98	71	95	71	92	70	86	65
		7	47	12.6	71	450	327	432	327	425	327	407	313
		8	95	33.2	75	1254	913	1201	913	1188	913	1157	898
		9	129	57.0	78	2241	1631	2142	1631	2126	1631	2093	1631
		10	112	61.1	81	2494	1815	2381	1815	2372	1815	2349	1815
		11	56	37.0	83	1546	1125	1475	1125	1472	1125	1464	1125
		12	14	11.0	85	471	343	449	343	449	343	448	343
		13	1	.9	87	40	29	39	29	39	29	39	29
			471	215.9		8602	6260	8222	6260	8173	6259	8050	6224
			MEAN DIA. 9.2 INCHES, WEIBULL PARAMETERS A= 3.8, B= 5.77, C= 4.27										
40	89	6	4	.8	65	26	19	25	19	24	18	23	17
		7	13	3.5	71	125	91	119	91	117	91	113	87
		8	32	11.2	75	423	307	404	307	400	307	390	302
		9	57	25.2	79	1003	730	959	730	952	730	937	730
		10	76	41.5	82	1713	1247	1636	1247	1629	1247	1614	1247
		11	73	48.2	85	2064	1502	1969	1502	1965	1502	1956	1502
		12	46	37.7	87	1653	1203	1576	1203	1575	1203	1573	1203
		13	20	18.4	89	827	602	788	602	788	602	788	602
		14	5	5.3	91	245	178	233	178	233	178	233	178
		15	1	1.2	92	57	41	54	41	54	41	54	41
			329	192.9		8136	5920	7763	5920	7737	5919	7681	5909
			MEAN DIA. 10.4 INCHES, WEIBULL PARAMETERS A= 4.1, B= 6.69, C= 4.22										

SI 70

2,250 stems per acre

Age	Av. dom. ht.	D.b.h.	Stems per acre	Basal area	Av. ht.	All trees, total stem		5-inch class and greater to o.b. tops of—					
								2 inches		3 inches		4 inches	
						O. b.	I. b.	O. b.	I. b.	O. b.	I. b.	O. b.	I. b.
Yrs	Ft	In	No.	Ft 2	Ft	– – – – – – – – – – – – – Ft 3 – – – – – – – – – – – – –							
10	36	1	2	.0	13	0	0						
		2	74	1.6	25	22	16						
		3	436	21.4	30	334	240						
		4	818	71.4	34	1241	899						
		5	416	56.7	36	1038	753	1021	725	940	663	745	514
		6	33	6.5	37	121	88	118	87	112	82	97	71
			1779	157.6		2756	1996	1139	812	1052	745	842	585
						MEAN DIA. 4.0 INCHES, WEIBULL PARAMETERS A= .8, B= 3.45, C= 4.47							
15	50	2	1	.0	30	0	0						
		3	48	2.4	39	47	34						
		4	266	23.2	44	521	377						
		5	590	80.4	47	1918	1393	1872	1364	1759	1278	1489	1071
		6	504	99.0	49	2453	1783	2374	1780	2283	1709	2063	1540
		7	119	31.8	51	819	596	789	596	769	586	720	548
		8	4	1.4	52	37	27	35	27	35	27	33	25
			1532	238.2		5795	4210	5070	3767	4846	3600	4305	3184
						MEAN DIA. 5.3 INCHES, WEIBULL PARAMETERS A= 1.7, B= 3.91, C= 4.45							
20	61	3	3	.1	43	3	2						
		4	53	4.6	50	118	85						
		5	225	30.7	54	839	610	817	602	774	569	673	491
		6	444	87.2	58	2556	1858	2467	1858	2389	1805	2201	1660
		7	390	104.2	60	3158	2297	3035	2297	2971	2278	2818	2158
		8	116	40.5	62	1267	921	1214	921	1197	921	1156	893
		9	8	3.5	63	112	82	107	82	106	82	104	81
			1239	270.9		8053	5855	7640	5760	7437	5655	6952	5283
						MEAN DIA. 6.3 INCHES, WEIBULL PARAMETERS A= 2.4, B= 4.23, C= 4.41							
25	70	4	8	.7	53	19	14						
		5	63	8.6	59	257	186	249	185	238	176	209	154
		6	197	38.7	63	1232	895	1187	895	1153	875	1071	811
		7	326	87.1	66	2903	2111	2787	2111	2735	2103	2611	2006
		8	266	92.9	69	3232	2351	3095	2351	3058	2351	2968	2299
		9	85	37.6	71	1344	978	1286	978	1275	978	1251	973
		10	8	4.4	73	161	117	153	117	153	117	151	117
			953	269.9		9148	6652	8757	6637	8612	6600	8261	6360
						MEAN DIA. 7.2 INCHES, WEIBULL PARAMETERS A= 2.9, B= 4.58, C= 4.38							

Age	Av. dom. ht.	D.b.h.	Stems per acre	Basal area	Av. ht.	All trees, total stem		5-inch class and greater to o.b. tops of—					
								2 inches		3 inches		4 inches	
						O. b.	I. b.	O. b.	I. b.	O. b.	I. b.	O. b.	I. b.
Yrs	Ft	In	No.	Ft 2	Ft	— — — — — — — — — — — — — Ft 3 — — — — — — — — — — — — —							
30	77	4	1	.1	54	2	2						
		5	15	2.0	61	63	46	61	46	59	43	52	38
		6	66	13.0	66	432	314	416	314	405	308	378	287
		7	157	42.0	70	1463	1078	1423	1078	1398	1077	1339	1031
		8	223	77.8	74	2906	2114	2781	2114	2751	2114	2678	2077
		9	173	76.4	76	2929	2131	2799	2131	2780	2131	2732	2129
		10	61	33.3	78	1308	952	1249	952	1244	952	1231	952
		11	8	5.3	80	213	155	203	155	203	155	201	155
			704	249.9		9336	6792	8932	6790	8840	6780	8611	6669
			MEAN DIA. 8.1 INCHES, WEIBULL PARAMETERS A= 3.4, B= 5.01, C= 4.33										
35	83	5	3	.4	60	12	9	12	9	12	9	10	8
		6	18	3.5	66	118	86	114	86	110	84	103	78
		7	56	15.0	71	536	390	515	390	506	390	485	373
		8	110	38.4	75	1453	1057	1390	1057	1376	1057	1340	1039
		9	142	62.7	78	2467	1795	2358	1795	2342	1795	2304	1795
		10	113	61.6	81	2516	1831	2402	1831	2393	1831	2370	1831
		11	49	32.3	83	1353	984	1290	984	1288	984	1281	984
		12	10	7.9	85	336	245	321	245	321	245	320	245
		13	1	.9	87	40	29	39	29	39	29	39	29
			502	222.8		8831	6426	8441	6426	8387	6424	8252	6382
			MEAN DIA. 9.0 INCHES, WEIBULL PARAMETERS A= 3.6, B= 5.61, C= 4.28										
40	89	6	4	.8	66	26	19	25	19	25	19	23	17
		7	16	4.3	72	155	113	149	113	147	113	141	108
		8	38	13.3	77	515	375	493	375	488	375	476	369
		9	66	29.2	80	1176	856	1124	856	1117	856	1099	856
		10	84	45.8	84	1940	1411	1852	1411	1845	1411	1829	1411
		11	76	50.2	86	2174	1582	2073	1582	2070	1582	2061	1582
		12	45	35.3	88	1567	1140	1494	1140	1493	1140	1492	1140
		13	16	14.7	90	669	487	637	487	637	487	637	487
		14	3	3.2	92	149	108	142	108	142	108	142	108
			348	196.8		8371	6091	7989	6091	7964	6091	7900	6078
			MEAN DIA. 10.2 INCHES, WEIBULL PARAMETERS A= 4.1, B= 6.50, C= 4.23										

SI 70

2,500 stems per acre

Age	Av. dom. ht.	D.b.h.	Stems per acre	Basal area	Av. ht.	All trees, total stem		5-inch class and greater to o.b. tops of—					
								2 inches		3 inches		4 inches	
						O. b.	I. b.	O. b.	I. b.	O. b.	I. b.	O. b.	I. b.
Yrs	Ft	In	No.	Ft 2	Ft	------------- Ft 3 -------------							
10	36	1	2	.0	13	0	0						
		2	102	2.2	25	30	22						
		3	577	26.3	30	441	318						
		4	931	81.2	34	1413	1023						
		5	326	44.5	36	813	590	800	568	737	519	584	403
		6	12	2.4	37	44	32	43	32	41	30	35	26
			1950	156.6		2741	1985	843	600	778	549	619	429
			MEAN DIA. 3.9 INCHES, WEIBULL PARAMETERS A= .8, B= 3.27, C= 4.48										
15	50	2	2	.0	30	1	1						
		3	62	3.0	39	61	44						
		4	339	29.6	44	663	481						
		5	695	94.8	47	2259	1641	2205	1607	2072	1505	1755	1262
		6	490	96.2	49	2385	1734	2308	1731	2219	1662	2006	1498
		7	79	21.1	51	544	396	524	396	510	389	478	364
		8	1	.3	52	9	7	9	7	9	7	8	6
			1668	245.1		5922	4304	5046	3741	4810	3563	4247	3130
			MEAN DIA. 5.2 INCHES, WEIBULL PARAMETERS A= 1.7, B= 3.75, C= 4.46										
20	61	3	4	.2	43	4	3						
		4	66	5.8	50	147	106						
		5	275	37.5	54	1026	745	998	736	946	696	822	601
		6	513	100.7	58	2954	2147	2851	2147	2760	2086	2543	1918
		7	390	104.2	60	3158	2297	3035	2297	2971	2278	2818	2158
		8	88	30.7	62	961	699	921	699	908	699	877	677
		9	4	1.8	63	56	41	54	41	53	41	52	40
			1340	280.9		8306	6038	7859	5920	7638	5800	7112	5394
			MEAN DIA. 6.2 INCHES, WEIBULL PARAMETERS A= 2.4, B= 4.09, C= 4.43										
25	70	4	10	.9	53	24	17						
		5	76	10.4	59	310	225	301	223	287	212	253	186
		6	233	45.7	63	1457	1059	1404	1059	1364	1035	1267	959
		7	367	98.1	66	3268	2377	3138	2377	3079	2368	2939	2258
		8	266	92.9	69	3232	2351	3095	2351	3058	2351	2968	2299
		9	68	30.0	71	1076	783	1028	783	1020	783	1001	779
		10	4	2.2	73	80	58	77	58	76	58	75	58
			1024	280.1		9447	6870	9043	6851	8884	6807	8503	6539
			MEAN DIA. 7.1 INCHES, WEIBULL PARAMETERS A= 2.9, B= 4.45, C= 4.39										

Yield Tables and Stand Structure
For SHORTLEAF PINE Plantations
In Tennessee, Alabama, and
Georgia Highlands

Glendon W. Smalley

and

Robert L. Bailey

Southern Forest Experiment Station
Forest Service
U. S. Department of Agriculture

1974

ACKNOWLEDGMENTS

For the basic data in this report the Southern Forest Experiment Station is indebted to the 31 private landowners and 16 industries, agencies, and other organizations who made their plantations available for study.

The authors thank Kenneth Pierce, Southern Forest Experiment Station, and John C. Allen, formerly with the Tennessee Valley Authority, now retired, for help in gathering data.

CONTENTS

Yield Tables and Stand Structure
For Shortleaf Pine Plantations
In Tennessee, Alabama, and Georgia Highlands

Glendon W. Smalley[1]
and
Robert L. Bailey[2]

ore than 60,000 acres of abandoned fields
been planted with shortleaf pine (*Pinus
ιata* Mill.) in the four highland regions (fig.
Tennessee, Alabama, and Georgia. To assist
ıters and landowners in making decisions
t when to harvest these plantations we pre-
tables showing expected cubic-foot yields,
l area, average height, and number of stems
ιcre by 1-inch diameter classes for all com-
tions of:

Planting density—750 to 2,000 trees per
acre in increments of
250

Site index (base age 25)—30 to 60 feet in
steps of 10

Age from seed—10 to 40 years, by intervals
of 5.

ethods of collecting the data and construct-
.he tables are discussed in the sections im-
ately following. The tables were derived
ıgh a system of equations to predict number
ırviving stems per acre, diameter distribu-
, site index (Smalley and Bower 1971), tree
ıts, and volume per tree (Smalley and
er 1968).

Readers interested only in the tables and a
.ssion of growth trends should turn directly
e section on Stand and Stock Tables, p. 7.
ıs for combinations of variables not de-
d—or for combinations of surviving trees
ιcre, site, and age—can be obtained from
iouthern Forest Experiment Station, Bio-
ics Branch, T-10210 Federal Building, 701

cientist at the Silviculture Laboratory, maintained at Sewanee,
ıssee, by the Southern Forest Experiment Station in cooperation
·he University of the South.

ınatical Statistician, Biometrics Branch, Southern Forest
·iment Station, New Orleans, La.

Loyola Avenue, New Orleans, Louisiana 70113.
Users with computer facilities may wish to
acquire the programs, which are written in
FORTRAN.

A similar publication is available for loblolly
pine (Smalley and Bailey 1974).

PLANTATION MEASUREMENTS

Diameter, height, and age data were recorded
in 104 shortleaf plantations, approximately one-
half of which occurred in the Ridge and Valley
region. The remainder were distributed almost
evenly in the other three regions.

All plantations sampled were at least 10 years
old from seed and had not been thinned, burned,
or pruned. Survival and tree distribution were
judged reasonably good, and no reinforcement
planting had been done. The plantations bore no
evidence of severe damage by insects or disease.
Most of the stands contained some competing
vegetation, but quantity of competition was not
a criterion except as it influenced pine survival
and distribution.

One plot, usually 0.05 acre in size, was established in each plantation. The diameter at breast
height (D,) of each tree on the plot was measured to the nearest 0.1 inch. The total height of
every fourth tree in each 1-inch diameter class
was measured to the nearest foot. Average total
height of dominants and codominants (H) was
calculated for each plot. Age from seed (A) was
determined by counting annual rings at ground
level on a felled tree. Mortality was estimated
from the presence of standing dead trees, rotten
stumps, fallen trees, and obvious gaps in the
planting rows. This estimate was added to the
count of live trees to obtain the number of
planted trees.

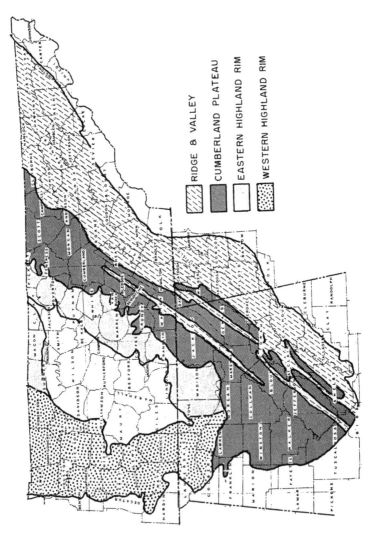

RIDGE & VALLEY

CUMBERLAND PLATEAU

EASTERN HIGHLAND RIM

WESTERN HIGHLAND RIM

2

Data from the 104 plots were used to screen, and evaluate mathematical models to esti-...e survival, heights of individual trees (H_1), relative diameter distributions as functions ...ge, number of trees planted per acre (T_p), ...ber of trees surviving per acre (T_s), diam-...: class midpoints, the midpoint of the largest ...neter class with nonzero predicted fre-...ncy per acre (DMAX), and site index at base age 25 years (S). No attempt was made to develop separate predictors for each physiographic region, as prior experience (Smalley and Bower 1968, 1971) had indicated that there was little chance of regression coefficients being significantly different.

The distribution of sample plots by site index, age, and planting density is shown in table 1. Ages ranged from 11 to 35 years, average heights

le 1.—*Distribution of shortleaf plantations by age, site index, and planting density*

nting sity es per re)	Age from seed						Total
	10-14	15-19	20-24	25-29	30-34	35-39	
				Number of plots			
<500					1		1
.-1000	8	1	2	15	1	1	28
1-1500	9	2	11	28	8	1	59
1-2000	2	2	1	3		1	9
.-2500	1						1
1-3000	1	1					2
>3000		4					4
otal	21	10	14	46	10	3	104

? a l rs)	Site index								Total
	25	30	35	40	45	50	55	60	
				Number of plots					
4				2	8	7	4		21
9				4	3	2	1		10
4			1	1	4	6	1		14
9		1	1	8	12	15	8	1	46
4					5	2	2	1	10
9				2		1			3
tal	1	1	2	17	32	33	16	2	104

nting sity es per re)	Site index								Total
	25	30	35	40	45	50	55	60	
				Number of plots					
<500							1		1
.-1000				6	4	10	7		28
.-1500			1	7	22	19	8	2	59
.-2000			1	1	2	4			9
.-2500					1				1
-3000				1	1				2
>3000				2	2				4
tal	1	1	2	17	32	33	16	2	104

3

of dominant and codominants from 20 to 66 feet, site indexes from 26 to 58, number of trees planted from 400 to 4,500 per acre, and number of surviving trees from 320 to 3,400. A chi-square test showed no relationship between age and site index (all statistical tests were at the 0.05 level).

Height-on-diameter regressions were fitted for each plot with a log H, on D,[1] relationship. On 65 percent of the plots $r^2 > 0.70$. These regressions and the tree-volume regressions of Smalley and Bower (1968) gave equations for tree volumes as functions of diameter only. Plot volumes were computed with these equations.

MODEL DEVELOPMENT AND EVALUATION

Survival

The basic structure chosen for a survival model was:

$$\log (T_p/T_s) = Af(A, H, T_p).$$

Merits of this particular form are that when $A = 0$, $\log (T_p/T_s) = 0$. Thus $T_p = T_s$ and the equation predicts 100-percent survival. For any value of the right-hand side a nonnegative (T_p/T_s) ratio results.

The function. $f(A,H,T_p)$, was defined by screening linear, nonlinear, and interaction transformations of A, H, T_p with REX (Grosenbaugh 1967). Graphical evaluation (for example, see fig. 2) was the final step in screening. The model chosen was:

$$\log (T_p/T_s) = A\{0.0100 \log (T_p) \\ + 0.0008\ H \\ - 0.0091\sqrt{H}\} \quad (1)$$

$$R^2 = 0.86;\ S^2_{y\cdot x} = 0.0040;\ \overline{Y} = 0.1424;$$
$$S^2_y = 0.0075$$

There were no discernible trends in residuals.

Number of surviving trees was calculated for each plot by solving for T_s. Fifty-nine percent of the estimates were within ±10 percent of the observed values, and the algebraic average error of —1.02 percent corresponded to a difference of 10 trees per acre for the stand of mean stocking.

Diameter Distributions

The three-parameter Weibull function (Bailey 1972, Bailey and Dell 1973) was used to model diameter distributions. Regressions were fitted to predict the maximum-likelihood estimates of

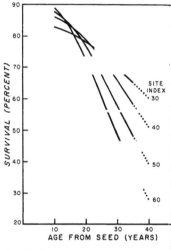

Figure 2.—*Survival.*

Weibull parameters from stand variables A, H and T_s:

Location parameter
$$a = -1.9492 + 0.0757\ H,$$
$$R^2 = 0.33,\ S^2_{y\cdot x} = 1.5425,\ \overline{Y} = 1.3010,$$
$$S^2_y = 2.2780;$$

Scale parameter
$$a + b = -5.2352 - 0.0003\ T_s + 6.2046\ \\ \log (H) + 1{,}195.4687/T_s,$$
$$R^2 = 0.88,\ S^2_{y\cdot x} = 0.2198,\ \overline{Y} = 5.9785,$$
$$S^2_y = 1.8343;$$

Shape parameter
$$c = 6.0560 - 0.0391\ H - 0.0006\ T_s,$$
$$R^2 = 0.11,\ S^2_{y\cdot x} = 1.3620,\ \overline{Y} = 3.7564,$$
$$S^2_y = 1.5089.$$

Equation 2 will predict negative values for $H < 26$ feet. When this situation occurs a sh be made zero.

These three equations were used for predic Weibull parameters. Stand tables were obta from the resulting relative distributions and jected numbers of surviving stems per acre.

4

Evaluations against observed distributions indicated very good fits. There were 56 negative and 48 positive differences between predicted and observed mean diameters. Of these, 80 percent were within ±0.5 inch, and the average difference was 0.001 inch. For 86 percent of the lots the observed and predicted distributions were not significantly different by the Kolmogorov-Smirnov test (Massey 1951).

ghts

From total height and diameter pairs and concomitant stand data, an equation was developed to predict total height of individual trees. The general model was:

$$\log(H/H_t) = b_o + (1/D_t - 1/DMAX) \cdot g(A,H,T_s).$$

characteristics are discussed by Lenhart (68).

The form of the function $g(A,H,T_s)$ and parameter estimates were obtained through the screening process described previously.

$$\log(H/H_t) = -0.0207 + (1/D_t - 1/DMAX)$$
$$\{1.9022 - 0.3861 \, A \, T_s \, (10^{-6})$$
$$+ 0.0016 \, (T_s/A)$$
$$- 0.8970 \log(T_s/A)$$
$$+ 0.9578 \log(H/A)\} \quad (5)$$
$$R^2 = 0.81, \; S^2_{y \cdot x} = 0.0012, \; \bar{Y} = 0.0423,$$
$$S^2_y = 0.0060$$

dicted Yields

Equations 2 through 5 and those for tree volume (Smalley and Bower 1968) made possible a

yield prediction system in which age, average dominant height, and number of surviving trees were the variables.

Number of trees and average heights by diameter classes were used to calculate volumes by classes, which were then summed for expected yields per acre. When evaluated against observed yields on the 104 plots, this system explained 85 percent of the variation about mean volume for all categories (table 2). Approximately one-fourth of the predicted values were within ±10 percent of the observed and two-thirds were within ±25 percent. Prediction seemed to be better for outside-bark volumes. Precision improved as top diameter decreased. Average differences between predicted and observed yields ranged from —60 to 7 cubic feet and were almost equally divided between positive and negative. The mean difference across all categories was —1 percent of the average observed. A two-tailed t-test for average difference (tabular value 1.98) was not significant for any category.

Simple correlation coefficients between residuals and observations were not significant for any volume category (table 3). Lowest correlations were with site index and averaged —0.03. Correlations were lower for merchantable than for total volumes and those for inside-bark were lower than for outside-bark categories

Comparisons

The average error of —1 percent in this study compares favorably with that in other tests of diameter-distribution systems for predicting yield. For old-field loblolly (*P. taeda* L.) planta-

le 2.—*Evaluation of predicted per-acre yields of all trees with d.b.h. > 4.5 inches*

Volume category	R²	Predicted values in relation to observed			Predicted minus observed		T-value for test of zero percent difference
		±10 percent	±25 percent	±500 ft³	Negative	Positive	
		— — — — — Percent — — — — —			— — —Number — — —		
Il							
utside bark	0.85	27	66	54	50	54	1.08
side bark	.85	27	64	67	51	53	1.12
-inch top							
utside bark	.85	27	66	55	50	54	1.08
side bark	.85	26	62	67	51	53	1.12
-inch top							
utside bark	.85	26	65	57	49	55	1.10
side bark	.85	24	61	67	52	52	1.14
-inch top							
utside bark	.85	25	62	59	52	52	1.12
side bark	.85	26	65	69	52	52	1.18

Table 3.—*Correlation coefficients between residuals (predicted minus observed per-acre yield) and stand characteristics. Data are for all trees with d.b.h. > 4.5 inches*

Volume category	Correlation[1] with			
	Age	Site index	Surviving stems per acre	Average dominant height
Total				
Outside bark	−0.20	−0.04	0.20	−0.20
Inside bark	− .19	− .03	.18	− .19
To 2-inch top				
Outside bark	− .20	− .04	.20	− .20
Inside bark	− .19	− .03	.18	− .19
To 3-inch top				
Outside bark	− .20	− .04	.19	− .19
Inside bark	− .18	− .03	.17	− .18
To 4-inch top				
Outside bark	− .18	− .03	.17	− .18
Inside bark	− .17	− .03	.15	− .16

[1] Critical value at 0.05 level with 100 degrees of freedom is 0.24.

tions, predicted yields exceeded observed yields by 4-6 percent in three studies (Smalley and Bailey 1974, Lenhart and Clutter 1971, Lenhart 1972). Burkhart (1971) reported an average under-prediction of 1 percent for slash pine (*P. elliottii* Engelm.) plantations in south Georgia and north Florida. For natural stands of yellow-poplar (*Liriodendron tulipifera* L.) in the southern Appalachians, the average over-prediction was 0.5 percent (Beck and Della-Bianca 1970).

LITERATURE CITED

Bailey, R. L.
    1972. Development of unthinned stand of *Pinus radiata* in New Zealand. Ph.D. thesis. Univ. Ga., Athens. 73 p. [Diss. Abstr. Int. 33(9): 4061-B, 1973.]

Bailey, R. L., and Dell, T. R.
    1973. Quantifying diameter distributions with the Weibull function. For. Sci. 19: 97-104.

Beck, D. E., and Della-Bianca, L.
    1970. Yield of unthinned yellow-poplar. USDA For. Serv. Res. Pap. SE-58, 20 p. Southeast. For. Exp. Stn., Asheville, N. C.

Burkhart, H. E.
    1971. Slash pine plantation yield estimates based on diameter distribution: an evaluation. For. Sci. 17: 452-453.

Grosenbaugh, L. R.
    1967. REX—FORTRAN-4 system for combinatorial screening or conventional analysis of multivariate regressions. USDA For. Serv. Res. Pap. PSW-44, 47 p. Pac. Southwest For. and Range Exp. Stn., Berkeley, Calif.

Lenhart, J. D.
    1968. Yield of old-field loblolly pine plantations in the Georgia Piedmont. Ph.D. thesis. Univ. Ga., Athens. 107 p. [Diss. Abstr. Int. 30(8): 3450-B, 1970.]

Lenhart, J. D.
    1972. Cubic-foot yields for unthinned old-field loblolly pine plantations in the interior West Gulf Coastal Plain. Tex. For. Pap. 14, 46 p.

Lenhart, J. D., and Clutter, J. L.
    1971. Cubic-foot yield tables for old-field loblolly pine plantations in the Georgia Piedmont. Ga. For. Res. Counc. Rep. 22, Ser. 3, 13 p.

Massey, F. J., Jr.
    1951. The Kolmogorov-Smirnov test for goodness of fit. J. Am. Stat. Assoc. 46: 68-78.

Smalley, G. W., and Bailey, R. L.
    1974. Yield tables and stand structure for loblolly pine plantations in Tennessee, Alabama, and Georgia highlands. USDA For. Serv. Res. Pap. SO-96, 81 p. South. For. Exp. Stn., New Orleans, La.

Smalley, G. W., and Bower, D. R.
    1968. Volume tables and point-sampling factors for shortleaf pines in plantations on abandoned fields in Tennessee, Alabama, and Georgia highlands. USDA For. Serv. Res. Pap. SO-31, 13 p. South. For. Exp. Stn., New Orleans, La.

Smalley, G. W., and Bower, D. R.
    1971. Site index curves for loblolly and shortleaf pine plantations on abandoned fields in Tennessee, Alabama, and Georgia highlands. USDA For. Serv. Res. Note SO-126, 6 p. South. For. Exp. Stn., New Orleans, La.

6

STAND AND STOCK TABLES

The tables that follow are believed to be representative of the natural development of undisturbed shortleaf pine plantations on abandoned fields throughout the Interior Highlands of Tennessee, Alabama, and Georgia. To enhance their utility, the tables have been extended 5 years beyond the oldest sampled stands. We recognize that extrapolation is risky, but all relationships appear biologically valid. Some trends are worthy of comment. For conciseness, graphical examples in figures 2-8 are limited to a planting density of 1,250 trees per acre, or about a 6- by 6-foot spacing. Where discussions relate to changes in planting density, trends can be verified from the tables.

-4 system for ea
ting or convention
ivariate regression
Res. Pap. PSW-4
west For. and Ran
ley, Calif.

d loblolly pine pla
Georgia Piedmon
iiv. Ga., Athens. I
Int. 30(8): 3458

s for unthinned e
ne plantations in t
Gulf Coastal Pla
14, 46 p.

er, J. L.
l tables for old-fie
antations in the Ge
Ga. For. Res. Cour
, 13 p.

ov-Smirnov test j
J. Am. Stat. Ass

ey, R. L.
d stand structure f
antations in Tennes
Georgia highlan
v. Res. Pap. SO-96
Exp. Stn. New C

ver, D. R.
and point-sampl
pines in plan
ned fields in Tenn
nd Georgia highlan
rv. Res. Pap. SO-
t. Exp. Stn., New

ver, D. R.
rves for lob
plantations on al
Tennessee, Alaba
ighlands. USDA F
e SO-126, 6 p. Sov
New Orleans, La

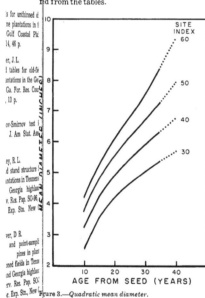

Figure 3.—Quadratic mean diameter.

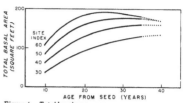

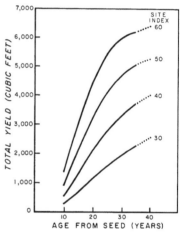

Figure 4.—Total basal area per acre.

Figure 5.—Total yield per acre of entire stem (outside bark) of all trees.

Survival.—On all sites survival percentage decreased as planting density and age increased. With an increase in site, however, survival was slightly better at early ages and worse at older ages—a consequence of intensified competition on better sites as age increased.

Quadratic mean diameter.—As planting densities increased, mean diameter declined for all ages and sites, but improvement in site always caused diameter increases. For all sites and densities, diameter growth was essentially linear past age 20.

Basal area.—Total basal area (trees of all sizes) for sites 50 and 60 culminated before age 40 for all planting densities. On site 40, culmination was at densities greater than 1,250 trees per acre. On poor sites culmination will probably occur at about age 50.

Yields.—Total and merchantable cubic-foot yields increased with site and planting density,

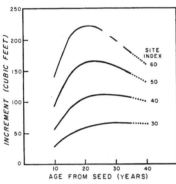

Figure 6.—*Mean annual increment in total yield per acre.*

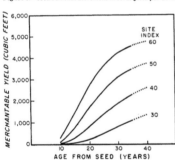

Figure 7.—*Merchantable yield (inside bark) per acre to a 4-inch top (outside bark) in stems with > 4.5 inches.*

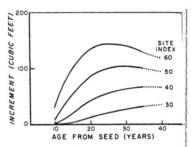

Figure 8.—*Mean annual increment in merchantable yield per acre.*

but the effect of density was small on poor sites. Yield increased with age for all planting densities on sites 30, 40, and 50. For a planting density of 2,000 trees per acre on site 60, yield culminated at about age 35. In effect, the loss of volume from mortality began to exceed both total and merchantable growth on the remaining trees.

Mean annual increment.—For total volume, mean annual increment culminated for all sites and planting densities. Age at culmination decreased from 30 years on poor sites to about 25 years on the best sites, regardless of planting density. Increment at culmination increased with planting density up to 2,000 trees per acre on all sites.

Merchantable-volume increment culminated on sites 40, 50, and 60 at all planting densities. However, the culmination occurred at older ages than for total-volume, e.g., age 35 to 40 for site 40 and age 25 for site 60.

STAND AND
STOCK TABLES

SI 30
750 stems per acre

Age	Av. dom. ht.	D.b.h.	Stems per acre	Basal area	Av. ht.	All trees, total stem		5-inch class and greater to o.b. tops of—					
								2 inches		3 inches		4 inches	
						O.b.	I.b.	O.b.	I.b.	O.b.	I.b.	O.b.	I.b.
Yrs	Ft	In	No.	Ft 2	Ft	----------- Ft 3 ------------							
10	14	1	9	.0	5	0	0						
		2	103	2.2	10	15	9						
		3	313	15.4	12	105	70						
		4	215	18.6	13	132	91						
		5	15	2.0	14	15	11	13	6	10	4	3	0
			655	38.5		267	181	13	6	10	4	3	0
		MEAN DIA. 3.3 INCHES, WEIBULL PARAMETERS A= .0, B= 3.47, C= 5.12											
15	20	1	3	.0	6	0	0						
		2	33	.7	12	8	3						
		3	129	6.3	16	56	38						
		4	245	21.4	18	205	143						
		5	197	26.9	19	267	189	238	135	197	105	108	39
		6	46	9.0	20	94	67	87	56	77	49	56	34
		7	2	.5	21	6	4	5	4	5	3	4	3
			655	64.9		634	444	330	195	279	157	168	76
		MEAN DIA. 4.3 INCHES, WEIBULL PARAMETERS A= .0, B= 4.51, C= 4.87											
20	26	1	2	.0	6	0	0						
		2	20	.4	14	4	2						
		3	76	3.7	19	39	27						
		4	169	14.7	21	164	115						
		5	213	29.0	23	348	247	316	193	272	160	176	89
		6	134	26.3	25	340	242	318	216	292	196	231	152
		7	32	8.6	26	114	82	109	78	103	73	88	62
		8	2	.7	26	9	7	9	7	9	6	8	6
			648	83.5		1018	722	752	494	676	435	503	309
		MEAN DIA. 4.9 INCHES, WEIBULL PARAMETERS A= .0, B= 5.16, C= 4.66											
25	30	1	1	.0	6	0	0						
		2	11	.2	16	2	2						
		3	49	2.4	21	27	19						
		4	123	10.7	24	136	95						
		5	191	26.0	26	352	250	323	204	284	175	197	111
		6	172	33.8	28	488	348	459	316	425	293	349	235
		7	74	19.8	29	295	211	281	203	267	193	235	168
		8	12	4.2	30	64	46	62	46	60	44	55	40
		9	1	.4	31	7	5	7	5	7	5	6	5
			634	97.6		1371	976	1132	776	1043	710	642	559
		MEAN DIA. 5.3 INCHES, WEIBULL PARAMETERS A= .3, B= 5.30, C= 4.50											

Age	Av. dom. ht.	D.b.h.	Stems per acre	Basal area	Av. ht.	All trees, total stem		5-inch class and greater to o.b. tops of—					
								2 inches		3 inches		4 inches	
						O. b.	I. b.	O. b.	I. b.	O. b.	I. b.	O. b.	I. b.
Yrs	Ft	In	No.	Ft 2	Ft	------------- Ft 3 -------------							
30	34	2	6	.1	17	1	1						
		3	33	1.6	23	20	14						
		4	92	6.0	27	114	80						
		5	162	22.1	29	332	236	307	199	274	175	201	121
		6	179	35.1	31	561	401	531	373	496	347	416	287
		7	108	28.9	33	489	350	469	343	449	328	401	292
		8	30	10.5	34	182	131	176	131	171	128	158	118
		9	3	1.3	35	24	17	23	17	23	17	21	16
			613	107.7		1723	1230	1506	1063	1413	995	1197	834
		MEAN DIA. 5.7 INCHES, WEIBULL PARAMETERS A= .6, B= 5.41, C= 4.37											
35	37	2	3	.1	17	1	0						
		3	22	1.1	24	14	10						
		4	69	6.0	28	88	62						
		5	133	18.1	31	291	207	270	178	243	158	183	114
		6	169	33.2	33	564	403	535	378	502	354	427	298
		7	129	34.5	35	619	443	594	437	571	419	515	377
		8	52	18.2	36	335	240	323	240	315	236	292	219
		9	9	4.0	37	75	54	73	54	72	54	68	52
		10	1	.5	38	11	8	10	8	10	8	10	8
			587	115.6		1998	1427	1805	1295	1713	1229	1495	1068
		MEAN DIA. 6.0 INCHES, WEIBULL PARAMETERS A= .8, B= 5.50, C= 4.26											
40	40	2	2	.0	18	0	0						
		3	15	.7	25	10	7						
		4	51	4.5	30	70	49						
		5	107	14.6	33	249	177	232	155	211	139	162	103
		6	152	29.8	36	552	395	526	376	497	355	430	304
		7	137	36.6	38	713	511	686	508	662	490	602	444
		8	72	25.1	39	502	360	486	360	474	357	443	334
		9	19	8.4	40	172	123	167	123	164	123	156	119
		10	2	1.1	41	23	16	22	16	22	16	21	16
			557	120.9		2291	1638	2119	1538	2030	1480	1814	1320
		MEAN DIA. 6.3 INCHES, WEIBULL PARAMETERS A= 1.1, B= 5.61, C= 4.17											

SI 30

1,000 stems per acre

Age	Av. dom. ht.	D.b.h.	Stems per acre	Basal area	Av. ht.	All trees, total stem		5-inch class and greater to o.b. tops of—					
								2 inches		3 inches		4 inches	
						O. b.	I. b.	O. b.	I. b.	O. b.	I. b.	O. b.	I. b.
Yrs	Ft	In	No.	Ft 2	Ft	— — — — — — — — — — — — — Ft 3 — — — — — — — — — — — — —							
10	14	1	26	.1	6	1	1						
		2	256	5.6	11	40	25						
		3	470	23.1	13	169	114						
		4	96	8.4	14	63	44						
			848	37.2		273	184	0	0	0	0	0	0
				MEAN DIA. 2.8 INCHES, WEIBULL PARAMETERS A= .0, B= 3.00, C= 5.01									
15	20	1	7	.0	7	0	0						
		2	72	1.6	14	14	9						
		3	246	12.1	17	113	77						
		4	347	30.3	19	305	214						
		5	152	20.7	20	216	153	194	113	163	89	94	38
		6	12	2.4	21	26	18	24	16	21	14	16	10
			836	67.0		674	471	218	129	184	103	110	48
				MEAN DIA. 3.8 INCHES, WEIBULL PARAMETERS A= .0, B= 4.06, C= 4.76									
20	26	1	4	.0	7	0	0						
		2	39	.9	16	8	5						
		3	140	6.9	20	75	52						
		4	267	23.3	23	282	199						
		5	256	34.9	24	436	309	397	245	345	207	229	121
		6	99	19.4	25	251	179	235	160	215	145	171	112
		7	11	2.9	26	39	28	37	27	35	25	30	21
			816	88.3		1091	772	669	432	595	377	430	254
				MEAN DIA. 4.5 INCHES, WEIBULL PARAMETERS A= .0, B= 4.73, C= 4.56									
25	30	1	1	.0	8	0	0						
		2	20	.4	17	4	3						
		3	87	4.3	22	51	35						
		4	200	17.5	25	229	162						
		5	262	35.7	27	500	356	461	294	407	255	289	167
		6	170	33.4	29	499	356	471	327	437	303	361	246
		7	44	11.8	30	181	130	173	126	165	119	146	105
		8	3	1.0	31	17	12	16	12	16	12	14	11
			787	104.1		1481	1054	1121	759	1025	689	810	529
				MEAN DIA. 4.9 INCHES, WEIBULL PARAMETERS A= .3, B= 4.89, C= 4.41									

Age	Av. dom. ht.	D.b.h.	Stems per acre	Basal area	Av. ht.	All trees, total stem		5-inch class and greater to o.b. tops of—					
								2 inches		3 inches		4 inches	
						O. b.	I. b.	O. b.	I. b.	O. b.	I. b.	O. b.	I. b.
Yrs	Ft	In	No.	Ft 2	Ft	– – – – – – – – – – – – Ft 3 – – – – – – – – – – – –							
30	34	2	11	.2	18	3	2						
		3	57	2.8	24	36	25						
		4	148	12.9	27	183	129						
		5	231	31.5	30	489	348	454	297	407	263	303	185
		6	202	39.7	32	653	467	619	437	580	408	490	340
		7	86	23.0	33	389	279	373	273	357	261	320	233
		8	14	4.9	34	85	61	82	61	80	60	74	55
		9	1	.4	35	8	6	8	6	8	6	7	5
			750	115.4		1846	1317	1536	1074	1432	998	1194	818

MEAN DIA. 5.3 INCHES, WEIBULL PARAMETERS A= .6, B= 5.01, C= 4.29

Age	Av. dom. ht.	D.b.h.	Stems per acre	Basal area	Av. ht.	All trees, total stem		2 in O.b.	2 in I.b.	3 in O.b.	3 in I.b.	4 in O.b.	4 in I.b.
35	37	2	6	.1	19	1	1						
		3	38	1.9	26	26	18						
		4	110	9.6	30	151	106						
		5	193	26.3	33	449	320	419	280	380	251	293	186
		6	206	40.4	35	728	520	692	494	653	464	562	396
		7	120	32.1	36	592	424	569	419	547	403	495	363
		8	32	11.2	37	212	152	205	152	199	150	186	139
		9	3	1.3	38	26	18	25	18	25	18	23	18
			708	122.9		2185	1559	1910	1363	1804	1286	1559	1102

MEAN DIA. 5.6 INCHES, WEIBULL PARAMETERS A= .8, B= 5.12, C= 4.19

Age	Av. dom. ht.	D.b.h.	Stems per acre	Basal area	Av. ht.	All trees, total stem		2 in O.b.	2 in I.b.	3 in O.b.	3 in I.b.	4 in O.b.	4 in I.b.
40	40	2	3	.1	19	1	0						
		3	25	1.2	26	17	12						
		4	80	7.0	31	113	80						
		5	156	21.3	34	374	266	349	235	318	211	248	159
		6	193	37.9	36	701	501	668	478	631	450	545	386
		7	141	37.7	38	734	526	706	523	681	504	620	457
		8	54	18.8	39	376	270	364	270	355	268	333	250
		9	9	4.0	40	81	58	79	58	78	58	74	56
		10	1	.5	41	11	8	11	8	11	8	11	8
			662	128.5		2408	1721	2177	1572	2074	1499	1831	1316

MEAN DIA. 6.0 INCHES, WEIBULL PARAMETERS A= 1.1, B= 5.24, C= 4.11

13

SI 30

1,250 stems per acre

Age	Av. dom. ht.	D.b.h.	Stems per acre	Basal area	Av. ht.	All trees, total stem		5-inch class and greater to o.b. tops of—					
								2 inches		3 inches		4 inches	
						O. b.	I. b.	O. b.	I. b.	O. b.	I. b.	O. b.	I. b.
Yrs	Ft	In	No.	Ft 2	Ft	– – – – – – – – – – – – – Ft 3 – – – – – – – – – – – – –							
10	14	1	59	.3	7	3	1						
		2	468	10.2	11	74	45						
		3	484	23.6	13	174	117						
		4	26	2.3	14	17	12						
			1037	36.6		268	175	0	0	0	0	0	0
					MEAN DIA. 2.5 INCHES, WEIBULL PARAMETERS A= .0, B= 2.68, C= 4.89								
15	20	1	14	.1	8	1	0						
		2	126	2.7	14	24	15						
		3	376	18.5	17	173	118						
		4	393	34.3	19	346	242						
		5	99	13.5	20	141	100	127	73	106	58	61	25
		6	3	.6	21	6	5	6	4	5	3	4	2
			1011	69.7		691	480	133	77	111	61	65	27
					MEAN DIA. 3.6 INCHES, WEIBULL PARAMETERS A= .0, B= 3.76, C= 4.65								
20	26	1	8	.0	8	0	0						
		2	65	1.4	16	14	9						
		3	213	10.5	21	119	82						
		4	353	30.8	23	373	263						
		5	265	36.1	24	451	320	411	254	357	214	237	125
		6	68	13.4	26	179	128	168	115	155	105	124	82
		7	4	1.1	26	14	10	14	10	13	9	11	8
			976	93.3		1150	812	593	379	525	328	372	215
					MEAN DIA. 4.2 INCHES, WEIBULL PARAMETERS A= .0, B= 4.44, C= 4.47								
25	30	1	3	.0	9	0	0						
		2	33	.7	18	8	5						
		3	133	6.5	23	81	56						
		4	277	24.2	26	330	233						
		5	306	41.7	28	606	431	559	360	497	314	358	212
		6	152	29.8	29	446	318	421	293	391	271	323	220
		7	26	6.9	30	107	77	102	74	97	71	86	62
		8	1	.3	31	6	4	5	4	5	4	5	4
			931	110.3		1584	1124	1087	731	990	660	772	498
					MEAN DIA. 4.7 INCHES, WEIBULL PARAMETERS A= .3, B= 4.61, C= 4.32								

Age	Av. dom. ht.	D.b.h.	Stems per acre	Basal area	Av. ht.	All trees, total stem		5-inch class and greater to o.b. tops of—					
								2 inches		3 inches		4 inches	
						O. b.	I. b.	O. b.	I. b.	O. b.	I. b.	O. b.	I. b.
Yrs	Ft	In	No.	Ft ^{2}	Ft	------------- Ft ^{3} -------------							
30	34	1	1	.0	9	0	0						
		2	18	.4	19	4	3						
		3	85	4.2	25	56	39						
		4	208	18.2	29	275	195						
		5	287	39.1	31	628	447	583	385	525	342	396	246
		6	207	40.6	33	690	493	655	464	615	434	523	365
		7	64	17.1	34	298	214	286	210	275	201	247	180
		8	7	2.4	35	44	31	42	31	41	31	38	29
			877	122.0		1995	1422	1566	1090	1456	1008	1204	820
						MEAN DIA. 5.1 INCHES, WEIBULL PARAMETERS A= .6, B= 4.74, C= 4.21							
35	37	2	9	.2	20	2	2						
		3	56	2.7	26	38	27						
		4	153	13.4	30	209	148						
		5	247	33.7	33	575	409	536	358	486	321	375	239
		6	228	44.8	35	806	576	766	546	723	514	622	438
		7	105	28.1	36	518	371	498	367	479	353	433	318
		8	20	7.0	37	132	95	128	95	125	94	116	87
		9	1	.4	38	9	6	8	6	8	6	8	6
			819	130.2		2289	1634	1936	1372	1821	1288	1554	1088
						MEAN DIA. 5.4 INCHES, WEIBULL PARAMETERS A= .8, B= 4.86, C= 4.12							
40	40	2	5	.1	20	1	1						
		3	36	1.8	28	26	18						
		4	111	9.7	32	162	115						
		5	202	27.5	35	498	355	466	315	425	285	335	217
		6	223	43.8	37	833	595	793	569	751	538	653	464
		7	136	36.3	39	727	520	700	519	675	501	617	456
		8	40	14.0	40	286	205	277	205	270	204	253	191
		9	4	1.8	41	37	27	36	27	35	27	34	26
			757	135.0		2570	1836	2272	1635	2156	1555	1892	1354
						MEAN DIA. 5.7 INCHES, WEIBULL PARAMETERS A= 1.1, B= 4.98, C= 4.05							

SI 30

1,500 stems per acre

Age	Av. dom. ht.	D.b.h.	Stems per acre	Basal area	Av. ht.	All trees, total stem		5-inch class and greater to o.b. tops of—					
								2 inches		3 inches		4 inches	
						O.b.	I.b.	O.b.	I.b.	O.b.	I.b.	O.b.	I.b.
Yrs	Ft	In	No.	Ft²	Ft	– – – – – – – – – – – – – Ft³ – – – – – – – – – – – – –							
10	14	1	111	.6	7	6	3						
		2	703	15.3	11	111	68						
		3	403	19.8	13	145	98						
		4	5	.4	14	3	2						
			1222	36.2		265	171	0	0	0	0	0	0
				MEAN DIA. 2.3 INCHES, WEIBULL PARAMETERS A= .0, B= 2.45, C= 4.78									
15	20	1	23	.1	9	1	1						
		2	195	4.3	15	39	25						
		3	504	24.7	17	232	159						
		4	398	34.7	19	350	245						
		5	60	8.2	20	85	60	77	44	64	35	37	15
		6	1	.2	21	2	2	2	1	2	1	1	1
			1181	72.2		709	492	79	45	66	36	38	16
				MEAN DIA. 3.3 INCHES, WEIBULL PARAMETERS A= .0, B= 3.54, C= 4.55									
20	26	1	12	.1	9	1	0						
		2	96	2.1	17	22	14						
		3	293	14.4	21	164	113						
		4	425	37.1	23	449	316						
		5	255	34.8	25	452	321	413	258	361	220	246	134
		6	46	9.0	26	121	86	114	78	105	71	84	56
		7	2	.5	26	7	5	7	5	6	5	6	4
			1129	98.0		1216	855	534	341	472	296	336	194
				MEAN DIA. 4.0 INCHES, WEIBULL PARAMETERS A= .0, B= 4.23, C= 4.38									
25	30	1	4	.0	10	0	0						
		2	48	1.0	19	12	8						
		3	183	9.0	24	116	81						
		4	351	30.6	27	433	306						
		5	334	45.5	29	684	487	633	411	565	361	414	249
		6	131	25.7	30	398	284	376	263	350	244	292	200
		7	16	4.3	31	68	49	65	47	62	45	55	40
			1067	116.2		1711	1215	1074	721	977	650	761	489
				MEAN DIA. 4.5 INCHES, WEIBULL PARAMETERS A= .3, B= 4.41, C= 4.24									

Age	Av. dom. ht.	D.b.h.	Stems per acre	Basal area	Av. ht.	All trees, total stem		5-inch class and greater to o.b. tops of—					
								2 inches		3 inches		4 inches	
						O. b.	I. b.	O. b.	I. b.	O. b.	I. b.	O. b.	I. b.
Yrs	Ft	In	No.	Ft 2	Ft	— — — — — — — — — — — — Ft 3 — — — — — — — — — — — —							
30	34	1	1	.0	9	0	0						
		2	25	.5	20	6	4						
		3	117	5.7	26	80	56						
		4	267	23.3	29	353	250						
		5	333	45.4	31	728	518	677	446	610	397	459	285
		6	202	39.7	33	674	481	639	452	600	423	510	356
		7	49	13.1	34	228	164	219	161	210	154	189	138
		8	3	1.0	35	19	13	18	13	18	13	16	12
			997	128.8		2088	1486	1553	1072	1438	987	1174	791

MEAN DIA. 4.9 INCHES, WEIBULL PARAMETERS A= .6, B= 4.54, C= 4.14

Age	Av. dom. ht.	D.b.h.	Stems per acre	Basal area	Av. ht.	O. b.	I. b.	O. b.	I. b.	O. b.	I. b.	O. b.	I. b.
35	37	2	13	.3	21	3	2						
		3	76	3.7	27	54	38						
		4	198	17.3	31	280	198						
		5	293	40.0	33	682	485	635	425	577	381	445	283
		6	238	46.7	35	841	601	800	570	754	536	649	457
		7	90	24.1	36	444	318	427	314	410	302	371	273
		8	13	4.5	37	86	62	83	62	81	61	75	57
		9	1	.4	38	9	6	8	6	8	6	8	6
			922	137.0		2399	1710	1953	1377	1830	1286	1548	1076

MEAN DIA. 5.2 INCHES, WEIBULL PARAMETERS A= .8, B= 4.66, C= 4.06

Age	Av. dom. ht.	D.b.h.	Stems per acre	Basal area	Av. ht.	O. b.	I. b.	O. b.	I. b.	O. b.	I. b.	O. b.	I. b.
40	40	2	7	.2	21	2	1						
		3	49	2.4	28	36	25						
		4	143	12.5	32	208	148						
		5	245	33.4	35	604	430	565	382	516	346	406	264
		6	243	47.7	37	907	649	865	620	818	586	711	505
		7	127	33.9	39	678	486	653	484	631	467	576	426
		8	29	10.1	40	207	149	201	149	196	148	184	139
		9	2	.9	41	19	13	18	13	18	13	17	13
			845	141.1		2661	1901	2302	1648	2179	1560	1894	1347

MEAN DIA. 5.5 INCHES, WEIBULL PARAMETERS A= 1.1, B= 4.79, C= 4.00

SI 30

1,750 stems per acre

Age	Av. dom. ht.	D.b.h.	Stems per acre	Basal area	Av. ht.	All trees, total stem		5-inch class and greater to o.b. tops of—					
								2 inches		3 inches		4 inches	
						O. b.	I. b.	O. b.	I. b.	O. b.	I. b.	O. b.	I. b.
Yrs	Ft	In	No.	Ft 2	Ft	— — — — — — — — — Ft 3 — — — — — — — — —							
10	14	1	188	1.0	8	11	5						
		2	923	20.1	11	145	89						
		3	292	14.3	13	105	71						
		4	1	.1	14	1	0						
			1404	35.6		262	165	0	0	0	0	0	0
		MEAN DIA. 2.2 INCHES, WEIBULL PARAMETERS A= .0, B= 2.27, C= 4.67											
15	20	1	36	.2	9	2	1						
		2	277	6.0	16	59	38						
		3	623	30.6	18	302	207						
		4	375	32.7	20	346	243						
		5	35	4.8	21	52	37	47	28	40	23	24	11
			1346	74.3		761	526	47	28	40	23	24	11
		MEAN DIA. 3.2 INCHES, WEIBULL PARAMETERS A= .0, B= 3.37, C= 4.45											
20	26	1	18	.1	10	1	1						
		2	133	2.9	18	31	20						
		3	375	18.4	21	210	145						
		4	483	42.1	23	511	359						
		5	237	32.3	25	420	298	384	240	336	204	228	125
		6	31	6.1	26	82	58	77	52	71	48	57	38
		7	1	.3	26	4	3	3	2	3	2	3	2
			1278	102.2		1259	864	464	294	410	254	288	165
		MEAN DIA. 3.8 INCHES, WEIBULL PARAMETERS A= .0, B= 4.06, C= 4.29											
25	30	1	6	.0	10	0	0						
		2	66	1.4	20	17	11						
		3	238	11.7	24	151	105						
		4	418	36.5	27	516	364						
		5	348	47.5	29	713	507	659	428	569	376	432	260
		6	112	22.0	30	340	243	321	224	299	208	249	171
		7	10	2.7	31	43	30	41	30	39	28	34	25
			1198	121.7		1780	1260	1021	682	927	612	715	456
		MEAN DIA. 4.3 INCHES, WEIBULL PARAMETERS A= .3, B= 4.25, C= 4.16											

18

Age	Av. dom. ht.	D.b.h.	Stems per acre	Basal area	Av. ht.	All trees, total stem		5-inch class and greater to o.b. tops of—					
								2 inches		3 inches		4 inches	
						O. b.	I. b.	O. b.	I. b.	O. b.	I. b.	O. b.	I. b.
Yrs	Ft	In	No.	Ft 2	Ft	– – – – – – – – – – – – Ft 3 – – – – – – – – – – – –							
30	34	1	2	.0	10	0	0						
		2	34	.7	21	9	6						
		3	152	7.5	26	104	72						
		4	325	28.4	29	430	304						
		5	367	50.0	31	803	571	746	492	672	437	506	314
		6	192	37.7	33	640	457	607	430	570	403	485	339
		7	37	9.9	34	172	123	165	121	159	116	143	104
		8	2	.7	35	13	9	12	9	12	9	11	8
			1111	134.9		2171	1542	1530	1052	1413	965	1145	765

MEAN DIA. 4.7 INCHES, WEIBULL PARAMETERS A= .6, B= 4.38, C= 4.07

Age	Av. dom. ht.	D.b.h.	Stems per acre	Basal area	Av. ht.	All trees, total stem		2 inches O.b.	I.b.	3 inches O.b.	I.b.	4 inches O.b.	I.b.
35	37	2	18	.4	22	5	3						
		3	97	4.8	28	71	50						
		4	242	21.1	32	353	250						
		5	335	45.7	34	803	572	750	504	683	454	532	342
		6	241	47.3	36	876	626	834	596	788	562	661	482
		7	77	20.6	37	390	280	375	277	362	267	328	242
		8	9	3.1	38	61	44	59	44	56	43	54	40
			1019	143.0		2559	1825	2018	1421	1891	1326	1595	1106

MEAN DIA. 5.1 INCHES, WEIBULL PARAMETERS A= .8, B= 4.51, C= 4.00

Age	Av. dom. ht.	D.b.h.	Stems per acre	Basal area	Av. ht.	All trees, total stem		2 inches O.b.	I.b.	3 inches O.b.	I.b.	4 inches O.b.	I.b.
40	40	2	9	.2	22	3	2						
		3	62	3.0	29	47	33						
		4	175	15.3	33	263	186						
		5	284	38.7	36	720	513	674	458	618	416	490	321
		6	257	50.5	38	985	705	940	676	891	640	778	555
		7	117	31.3	39	625	448	602	446	581	431	530	392
		8	22	7.7	40	157	113	152	113	149	112	139	105
		9	1	.4	41	9	7	9	7	9	7	8	6
			927	147.1		2809	2007	2377	1700	2248	1606	1945	1379

MEAN DIA. 5.4 INCHES, WEIBULL PARAMETERS A= 1.1, B= 4.64, C= 3.95

SI 30

2,000 stems per acre

Age	Av. dom. ht.	D.b.h.	Stems per acre	Basal area	Av. ht.	All trees, total stem		5-inch class and greater to o.b. tops of—					
								2 inches		3 inches		4 inches	
						O. b.	I. b.	O. b.	I. b.	O. b.	I. b.	O. b.	I. b.
Yrs	Ft	In	No.	Ft 2	Ft	— — — — — — — — — — — Ft 3 — — — — — — — — — — —							
10	14	1	295	1.6	8	17	7						
		2	1099	24.0	12	185	115						
		3	190	9.3	14	73	50						
			1584	34.9		275	172	0	0	0	0	0	0

MEAN DIA. 2.0 INCHES, WEIBULL PARAMETERS A= .0, B= 2.12, C= 4.57

15	20	1	53	.3	10	3	2						
		2	371	8.1	16	79	51						
		3	723	35.5	18	351	240						
		4	340	29.7	20	314	220						
		5	21	2.9	21	31	22	28	17	24	14	14	6
			1508	76.4		778	535	28	17	24	14	14	6

MEAN DIA. 3.0 INCHES, WEIBULL PARAMETERS A= .0, B= 3.22, C= 4.36

20	26	1	25	.1	11	2	1						
		2	175	3.8	19	43	28						
		3	459	22.5	22	269	186						
		4	525	45.8	24	578	407						
		5	216	29.5	25	383	272	350	219	306	186	208	114
		6	22	4.3	26	58	41	54	37	50	34	40	27
			1422	106.1		1333	935	404	256	356	220	248	141

MEAN DIA. 3.7 INCHES, WEIBULL PARAMETERS A= .0, B= 3.92, C= 4.20

25	30	1	8	.0	11	1	0						
		2	67	1.9	20	22	15						
		3	295	14.5	25	195	135						
		4	479	41.8	27	592	418						
		5	354	48.3	29	725	516	671	436	599	383	439	264
		6	95	18.7	30	288	206	272	190	254	177	212	145
		7	7	1.9	31	30	21	28	21	27	20	24	17
			1325	127.0		1853	1311	971	647	880	580	675	426

MEAN DIA. 4.2 INCHES, WEIBULL PARAMETERS A= .3, B= 4.11, C= 4.09

Age	Av. dom. ht.	D.b.h.	Stems per acre	Basal area	Av. ht.	All trees, total stem		5-inch class and greater to o.b. tops of—					
								2 inches		3 inches		4 inches	
						O. b.	I. b.	O. b.	I. b.	O. b.	I. b.	O. b.	I. b.
Yrs	Ft	In	No.	Ft²	Ft	------------- Ft³ -------------							
30	34	1	2	.0	11	0	0						
		2	45	1.0	21	12	8						
		3	189	9.3	27	134	93						
		4	361	33.2	30	521	369						
		5	393	53.6	32	887	632	825	548	746	490	569	358
		6	180	35.3	33	600	429	569	403	535	377	455	317
		7	29	7.8	34	135	97	130	95	124	91	112	82
		8	1	.3	35	6	4	6	4	6	4	5	4
			1220	140.5		2295	1632	1530	1050	1411	962	1141	761

MEAN DIA. 4.6 INCHES, WEIBULL PARAMETERS A= .6, B= 4.26, C= 4.01

Age	Av. dom. ht.	D.b.h.	Stems per acre	Basal area	Av. ht.	All trees, total stem		2 inches		3 inches		4 inches	
35	37	1	1	.0	11	0	-- 0						
		2	23	.5	22	6	4						
		3	120	5.9	28	88	61						
		4	286	25.0	32	417	295						
		5	370	50.5	34	866	632	828	556	754	502	587	378
		6	240	47.1	36	872	623	830	594	784	560	678	480
		7	66	17.6	37	335	240	322	238	310	229	281	207
		8	6	2.1	38	41	29	39	29	38	29	36	27
			1112	148.7		2645	1884	2019	1417	1886	1320	1582	1092

MEAN DIA. 5.0 INCHES, WEIBULL PARAMETERS A= .8, B= 4.38, C= 3.95

Age	Av. dom. ht.	D.b.h.	Stems per acre	Basal area	Av. ht.	All trees, total stem		2 inches		3 inches		4 inches	
40	40	2	12	.3	22	3	2						
		3	76	3.7	29	58	40						
		4	207	18.1	33	311	220						
		5	317	43.2	36	804	573	753	511	690	464	547	359
		6	266	52.2	38	1020	729	973	700	922	663	805	574
		7	108	28.9	39	577	413	556	412	536	398	490	362
		8	17	5.9	40	121	87	118	87	115	87	108	81
		9	1	.4	41	9	7	9	7	9	7	8	6
			1004	152.7		2903	2071	2409	1717	2272	1619	1958	1382

MEAN DIA. 5.3 INCHES, WEIBULL PARAMETERS A= 1.1, B= 4.52, C= 3.90

SI 40

750 stems per acre

Age	Av. dom. ht.	D.b.h.	Stems per acre	Basal area	Av. ht.	All trees, total stem		5-inch class and greater to o.b. tops of—					
								2 inches		3 inches		4 inches	
						O.b.	I.b.	O.b.	I.b.	O.b.	I.b.	O.b.	I.b.
Yrs	Ft	In	No.	Ft 2	Ft	— — — — — — — — — — — — Ft 3 — — — — — — — — — — — —							
10	18	1	4	.0	5	0	0						
		2	48	1.0	11	8	5						
		3	160	8.8	15	74	50						
		4	286	25.0	17	226	158						
		5	146	19.9	18	188	133	167	92	136	70	70	21
		6	14	2.7	19	27	19	25	16	22	14	16	9
			678	57.5		523	365	192	108	158	84	86	30
				MEAN DIA. 3.9 INCHES, WEIBULL PARAMETERS A= .0, B= 4.18, C= 4.93									
15	27	1	2	.0	6	0	0						
		2	19	.4	15	4	2						
		3	75	3.7	20	40	28						
		4	169	14.7	23	179	126						
		5	221	30.1	25	391	278	358	224	313	191	213	116
		6	147	28.9	26	387	276	364	249	334	227	269	178
		7	39	10.4	27	145	103	138	99	130	93	113	80
		8	3	1.0	28	15	11	14	11	14	10	13	9
			675	89.3		1161	824	874	583	791	521	608	383
				MEAN DIA. 4.9 INCHES, WEIBULL PARAMETERS A= .1, B= 5.12, C= 4.59									
20	34	2	7	.2	17	2	1						
		3	38	1.9	23	23	16						
		4	105	9.2	27	130	92						
		5	181	24.7	30	383	273	355	233	319	206	237	145
		6	190	37.3	32	615	439	582	411	545	383	461	320
		7	106	28.3	33	460	343	460	336	440	322	394	287
		8	26	9.1	34	158	113	153	113	148	111	137	102
		9	2	.9	35	16	11	15	11	15	11	14	11
			655	111.5		1807	1288	1565	1104	1467	1033	1243	865
				MEAN DIA. 5.6 INCHES, WEIBULL PARAMETERS A= .6, B= 5.27, C= 4.33									
25	40	2	3	.1	18	1	0						
		3	20	1.0	26	14	10						
		4	67	5.6	30	92	65						
		5	135	18.4	34	323	230	302	203	275	183	214	138
		6	178	35.0	36	647	462	616	441	582	415	503	356
		7	142	37.9	38	739	529	711	526	686	507	624	461
		8	62	21.6	39	432	310	418	310	408	308	382	287
		9	13	5.7	41	120	86	117	86	115	86	110	84
		10	1	.5	41	11	8	11	8	11	8	11	8
			621	126.1		2379	1700	2175	1574	2077	1507	1844	1334
				MEAN DIA. 6.1 INCHES, WEIBULL PARAMETERS A= 1.1, B= 5.36, C= 4.12									

Age	Av. dom. ht.	D.b.h.	Stems per acre	Basal area	Av. ht.	All trees, total stem		5-inch class and greater to o.b. tops of—					
								2 inches		3 inches		4 inches	
						O. b.	I. b.	O. b.	I. b.	O. b.	I. b.	O. b.	I. b.
Yrs	Ft	In	No.	Ft 2	Ft	— — — — — — — — — — — — — Ft 3 — — — — — — — — — — — — —							
30	45	2	1	.0	19	0	0						
		3	11	.5	28	8	6						
		4	43	3.8	34	66	47						
		5	97	13.2	37	253	180	237	162	218	147	174	115
		6	147	28.9	40	593	424	566	410	539	389	474	341
		7	149	39.8	42	857	614	826	614	800	596	736	548
		8	92	32.1	44	723	518	701	518	687	518	648	491
		9	32	14.1	45	325	233	317	233	312	233	299	229
		10	5	2.7	47	65	47	64	47	63	47	61	47
			577	135.2		2890	2069	2711	1984	2619	1930	2392	1771

MEAN DIA. 6.6 INCHES, WEIBULL PARAMETERS A= 1.5, B= 5.47, C= 3.95

Age	Av. dom. ht.	D.b.h.	Stems per acre	Basal area	Av. ht.	All trees, total stem		2 in O. b.	2 in I. b.	3 in O. b.	3 in I. b.	4 in O. b.	4 in I. b.
35	49	3	6	.3	29	5	3						
		4	27	2.4	35	43	30						
		5	67	9.1	40	188	134	177	123	164	113	134	90
		6	114	22.4	43	494	354	473	345	452	329	402	291
		7	135	36.1	45	831	596	803	596	780	584	722	540
		8	107	37.4	47	898	644	872	644	855	644	811	616
		9	53	23.4	49	586	421	571	421	564	421	543	417
		10	15	8.2	50	209	150	204	150	202	150	197	150
		11	2	1.3	51	34	25	34	25	33	25	33	25
			526	140.5		3288	2357	3134	2304	3050	2266	2842	2129

MEAN DIA. 7.0 INCHES, WEIBULL PARAMETERS A= 1.8, B= 5.60, C= 3.82

Age	Av. dom. ht.	D.b.h.	Stems per acre	Basal area	Av. ht.	All trees, total stem		2 in O. b.	2 in I. b.	3 in O. b.	3 in I. b.	4 in O. b.	4 in I. b.
40	53	3	3	.1	29	2	2						
		4	16	1.4	36	26	19						
		5	45	6.1	41	130	93	122	85	113	78	93	63
		6	83	16.3	45	376	269	361	264	345	253	309	225
		7	112	29.9	47	720	516	696	516	677	508	630	472
		8	106	37.0	50	946	679	920	679	903	679	860	655
		9	70	30.9	51	806	578	786	578	776	578	749	576
		10	29	15.8	53	428	307	418	307	415	307	404	307
		11	7	4.6	54	127	91	125	91	124	91	122	91
		12	1	.8	55	22	16	22	16	22	16	21	16
			472	143.1		3583	2570	3450	2536	3375	2510	3188	2405

MEAN DIA. 7.5 INCHES, WEIBULL PARAMETERS A= 2.1, B= 5.80, C= 3.70

23

SI 40

1,000 stems per acre

Age	Av. dom. ht.	D.b.h.	Stems per acre	Basal area	Av. ht.	All trees, total stem		5-inch class and greater to o.b. tops of—					
								2 inches		3 inches		4 inches	
						O. b.	I. b.	O. b.	I. b.	O. b.	I. b.	O. b.	I. b.
Yrs	Ft	In	No.	Ft 2	Ft	‑‑‑‑‑‑‑‑‑‑‑‑‑ Ft 3 ‑‑‑‑‑‑‑‑‑‑‑‑‑							
10	18	1	11	.1	6	1	0						
		2	111	2.4	12	19	12						
		3	343	16.8	15	141	96						
		4	343	29.9	17	271	189						
		5	70	9.5	18	90	64	80	44	65	34	33	10
		6	1	.2	19	2	1	2	1	2	1	1	1
			879	59.0		524	362	82	45	67	35	34	11
			MEAN DIA. 3.5 INCHES, WEIBULL PARAMETERS A= .0, B= 3.71, C= 4.81										
15	27	1	4	.0	7	0	0						
		2	38	.8	16	8	5						
		3	140	6.9	21	78	54						
		4	274	23.9	24	302	213						
		5	275	37.5	26	506	360	465	294	408	252	284	160
		6	116	22.8	27	317	226	298	205	275	188	223	150
		7	15	4.0	28	58	41	55	40	52	37	46	33
			862	95.9		1269	899	818	539	735	477	553	343
			MEAN DIA. 4.5 INCHES, WEIBULL PARAMETERS A= .1, B= 4.68, C= 4.48										
20	34	1	1	.0	8	0	0						
		2	14	.3	18	3	2						
		3	70	3.4	25	46	32						
		4	179	15.6	28	229	162						
		5	263	35.9	31	575	410	534	352	481	313	363	225
		6	211	41.4	33	704	503	667	473	627	442	533	372
		7	77	20.6	34	359	257	344	253	330	242	297	217
		8	10	3.5	35	63	45	60	45	59	44	54	41
			825	120.7		1979	1411	1605	1123	1497	1041	1247	855
			MEAN DIA. 5.2 INCHES, WEIBULL PARAMETERS A= .6, B= 4.84, C= 4.22										
25	40	2	5	.1	20	1	1						
		3	37	1.8	27	26	18						
		4	114	9.9	32	166	118						
		5	207	28.2	35	510	364	477	322	436	292	343	223
		6	228	44.8	37	851	609	811	582	768	550	667	474
		7	137	36.6	39	732	524	705	523	680	504	621	459
		8	39	13.6	40	279	200	270	200	264	199	247	186
		9	4	1.8	41	37	27	36	27	35	27	34	26
			771	136.9		2602	1861	2299	1654	2183	1572	1912	1368
			MEAN DIA. 5.7 INCHES, WEIBULL PARAMETERS A= 1.1, B= 4.94, C= 4.03										

Age	Av. dom. ht.	D.b.h.	Stems per acre	Basal area	Av. ht.	All trees, total stem		5-inch class and greater to o.b. tops of—					
								2 inches		3 inches		4 inches	
						O. b.	I. b.	O. b.	I. b.	O. b.	I. b.	O. b.	I. b.
Yrs	Ft	In	No.	Ft 2	Ft	— — — — — — — — — — — — Ft 3 — — — — — — — — — — — —							
30	45	2	2	.0	21	1	0						
		3	19	.9	29	14	10						
		4	72	6.3	34	111	79						
		5	151	20.6	38	404	288	379	260	349	238	282	187
		6	201	39.5	41	831	594	794	576	757	548	669	482
		7	166	44.4	43	977	700	943	700	914	682	843	628
		8	76	26.5	44	597	428	579	428	567	428	536	406
		9	17	7.5	46	177	127	172	127	170	127	163	125
		10	2	1.1	47	26	19	26	19	25	19	25	19
			706	146.8		3138	2245	2893	2110	2782	2042	2518	1847

MEAN DIA. 6.2 INCHES, WEIBULL PARAMETERS A= 1.5, B= 5.05, C= 3.88

Age	Av. dom. ht.	D.b.h.	Stems per acre	Basal area	Av. ht.	All trees, total stem		2 in O.b.	2 in I.b.	3 in O.b.	3 in I.b.	4 in O.b.	4 in I.b.
35	49	3	10	.5	30	8	5						
		4	44	3.8	36	72	51						
		5	105	14.3	40	295	211	278	192	257	177	210	142
		6	162	31.8	43	702	502	672	490	642	467	571	414
		7	164	43.8	46	1032	740	998	740	970	727	900	673
		8	104	36.3	47	872	626	848	626	831	626	788	599
		9	37	16.3	49	409	294	399	294	394	294	379	291
		10	7	3.8	50	97	70	95	70	94	70	92	70
		11	1	.7	51	17	12	17	12	17	12	16	12
			634	151.4		3504	2511	3307	2424	3205	2373	2956	2201

MEAN DIA. 6.6 INCHES, WEIBULL PARAMETERS A= 1.8, B= 5.18, C= 3.75

Age	Av. dom. ht.	D.b.h.	Stems per acre	Basal area	Av. ht.	All trees, total stem		2 in O.b.	2 in I.b.	3 in O.b.	3 in I.b.	4 in O.b.	4 in I.b.
40	53	3	5	.2	31	4	3						
		4	26	2.3	38	45	32						
		5	70	9.5	43	211	151	200	140	186	129	155	106
		6	121	23.8	46	561	401	538	394	516	378	463	338
		7	143	38.2	49	959	687	928	687	903	680	843	633
		8	116	40.5	51	1056	757	1027	757	1009	757	962	734
		9	59	26.1	52	692	497	675	497	668	497	644	496
		10	18	9.8	54	271	194	265	194	263	194	256	194
		11	3	2.0	55	56	40	54	40	54	40	53	40
			561	152.4		3855	2762	3687	2709	3599	2675	3376	2541

MEAN DIA. 7.1 INCHES, WEIBULL PARAMETERS A= 2.1, B= 5.37, C= 3.65

SI 40

1,250 stems per acre

Age	Av. dom. ht.	D.b.h.	Stems per acre	Basal area	Av. ht.	All trees, total stem		5-inch class and greater to o.b. tops of—					
								2 inches		3 inches		4 inches	
						O. b.	I. b.	O. b.	I. b.	O. b.	I. b.	O. b.	I. b.
Yrs	Ft	In	No.	Ft 2	Ft	‒ ‒ ‒ ‒ ‒ ‒ ‒ ‒ ‒ ‒ ‒ Ft 3 ‒ ‒ ‒ ‒ ‒ ‒ ‒ ‒ ‒ ‒ ‒							
10	18	1	23	.1	7	1	1						
		2	202	4.4	13	36	23						
		3	504	24.7	16	219	149						
		4	318	27.8	18	266	186						
		5	27	3.7	19	37	26	33	19	27	14	15	5
			‒‒‒‒	‒‒‒‒‒		‒‒‒‒	‒‒‒‒	‒‒‒‒	‒‒‒‒	‒‒‒‒	‒‒‒‒	‒‒‒‒	‒‒‒‒
			1074	60.7		559	385	33	19	27	14	15	5
					MEAN DIA. 3.2 INCHES, WEIBULL PARAMETERS A= .0, B= 3.41, C= 4.69								
15	27	1	7	.0	8	0	0						
		2	64	1.4	17	14	9						
		3	219	10.8	21	123	85						
		4	372	32.5	24	410	289						
		5	292	39.8	26	537	382	493	312	434	268	301	170
		6	83	16.3	27	227	162	213	147	197	135	160	107
		7	6	1.6	28	23	16	22	16	21	15	18	13
			‒‒‒‒	‒‒‒‒‒		‒‒‒‒	‒‒‒‒	‒‒‒‒	‒‒‒‒	‒‒‒‒	‒‒‒‒	‒‒‒‒	‒‒‒‒
			1043	102.4		1334	943	728	475	652	418	479	290
					MEAN DIA. 4.2 INCHES, WEIBULL PARAMETERS A= .1, B= 4.39, C= 4.37								
20	34	1	1	.0	9	0	0						
		2	23	.5	19	6	4						
		3	110	5.4	25	73	50						
		4	257	22.4	29	340	240						
		5	329	44.9	31	720	512	668	441	602	392	454	282
		6	209	41.0	33	697	498	661	468	621	438	528	369
		7	54	14.4	34	252	180	242	177	232	170	208	152
		8	4	1.4	35	25	18	24	18	23	18	22	16
			‒‒‒‒	‒‒‒‒‒		‒‒‒‒	‒‒‒‒	‒‒‒‒	‒‒‒‒	‒‒‒‒	‒‒‒‒	‒‒‒‒	‒‒‒‒
			987	130.1		2113	1502	1595	1104	1478	1018	1212	819
					MEAN DIA. 4.9 INCHES, WEIBULL PARAMETERS A= .6, B= 4.56, C= 4.13								
25	40	2	6	.2	21	2	1						
		3	58	2.8	28	43	30						
		4	167	14.6	33	251	178						
		5	273	37.2	36	692	493	648	440	594	400	471	309
		6	257	50.5	38	985	705	940	676	891	640	778	555
		7	122	32.6	39	652	467	628	465	606	449	553	409
		8	25	8.7	40	179	128	173	128	169	128	158	119
		9	2	.9	41	19	13	18	13	18	13	17	13
			‒‒‒‒	‒‒‒‒‒		‒‒‒‒	‒‒‒‒	‒‒‒‒	‒‒‒‒	‒‒‒‒	‒‒‒‒	‒‒‒‒	‒‒‒‒
			912	147.5		2823	2015	2407	1722	2278	1630	1977	1405
					MEAN DIA. 5.4 INCHES, WEIBULL PARAMETERS A= 1.1, B= 4.66, C= 3.94								

Age	Av. dom. ht.	D.b.h.	Stems per acre	Basal area	Av. ht.	All trees, total stem		5-inch class and greater to o.b. tops of—					
								2 inches		3 inches		4 inches	
						O. b.	I. b.	O. b.	I. b.	O. b.	I. b.	O. b.	I. b.
Yrs	Ft	In	No.	Ft²	Ft	– – – – – – – – – – – – – Ft³ – – – – – – – – – – – – –							
30	45	2	3	.1	22	1	1						
		3	30	1.5	30	24	16						
		4	105	9.2	35	167	118						
		5	205	28.0	39	562	401	529	364	488	334	397	266
		6	244	47.9	41	1009	722	964	700	919	666	812	585
		7	168	44.9	43	989	709	954	709	925	691	853	635
		8	60	20.9	44	471	338	457	338	448	338	423	320
		9	9	4.0	46	93	67	91	67	90	67	86	66
		10	1	.5	47	13	9	13	9	13	9	12	9
			825	156.9		3329	2381	3008	2187	2883	2105	2583	1881

MEAN DIA. 5.9 INCHES, WEIBULL PARAMETERS A= 1.5, B= 4.77, C= 3.80

Age	Av. dom. ht.	D.b.h.	Stems per acre	Basal area	Av. ht.	All trees, total stem O.b.	I.b.	2 in O.b.	I.b.	3 in O.b.	I.b.	4 in O.b.	I.b.
35	49	2	1	.0	22	0	0						
		3	15	.7	32	13	9						
		4	64	5.6	38	110	78						
		5	144	19.6	42	425	303	401	279	373	258	309	210
		6	205	40.3	45	930	665	891	652	853	624	764	556
		7	182	48.6	47	1170	839	1132	839	1101	826	1023	767
		8	94	32.8	49	822	590	799	590	785	590	746	568
		9	26	11.5	50	293	211	286	211	283	211	272	209
		10	3	1.6	51	43	31	42	31	41	31	40	31
			734	160.8		3806	2726	3551	2602	3436	2540	3154	2341

MEAN DIA. 6.3 INCHES, WEIBULL PARAMETERS A= 1.8, B= 4.90, C= 3.69

Age	Av. dom. ht.	D.b.h.	Stems per acre	Basal area	Av. ht.	All trees, total stem O.b.	I.b.	2 in O.b.	I.b.	3 in O.b.	I.b.	4 in O.b.	I.b.
40	53	3	7	.3	32	6	4						
		4	38	3.3	39	67	48						
		5	96	13.1	43	290	207	274	191	255	177	212	145
		6	156	30.6	46	723	517	693	508	665	487	597	436
		7	168	44.9	49	1126	807	1090	807	1061	799	990	744
		8	117	40.8	51	1065	764	1036	764	1018	764	970	740
		9	48	21.2	53	574	412	560	412	554	412	535	412
		10	11	6.0	54	165	119	162	119	161	119	156	119
		11	1	.7	55	19	13	18	13	18	13	18	13
			642	161.0		4035	2891	3833	2814	3732	2771	3478	2609

MEAN DIA. 6.8 INCHES, WEIBULL PARAMETERS A= 2.1, B= 5.07, C= 3.60

SI 40

1,500 stems per acre

Age	Av. dom. ht.	D.b.h.	Stems per acre	Basal area	Av. ht.	All trees, total stem		5-inch class and greater to o.b. tops of—					
								2 inches		3 inches		4 inches	
						O.b.	I.b.	O.b.	I.b.	O.b.	I.b.	O.b.	I.b.
Yrs	Ft	In	No.	Ft²	Ft	— — — — — — — — — — — Ft³ — — — — — — — — — — — —							
10	18	1	40	.2	8	2	1						
		2	318	6.9	13	57	36						
		3	640	31.4	16	279	190						
		4	259	22.6	18	216	151						
		5	9	1.2	19	12	9	11	6	9	5	5	2
			1266	62.4		566	387	11	6	9	5	5	2
		MEAN DIA. 3.0 INCHES, WEIBULL PARAMETERS A= .0, B= 3.18, C= 4.58											
15	27	1	11	.1	9	1	0						
		2	98	2.1	17	22	14						
		3	305	15.0	22	176	123						
		4	455	39.7	24	501	353						
		5	288	39.3	26	530	377	487	307	428	264	297	168
		6	58	11.4	27	159	113	149	103	138	94	112	75
		7	2	.5	28	8	5	7	5	7	5	6	4
			1217	108.1		1399	985	643	415	573	363	415	247
		MEAN DIA. 4.0 INCHES, WEIBULL PARAMETERS A= .1, B= 4.17, C= 4.26											
20	34	1	2	.0	10	0	0						
		2	35	.8	20	9	6						
		3	156	7.7	26	107	74						
		4	335	29.2	29	444	313						
		5	377	51.4	32	851	606	792	526	716	470	546	344
		6	197	38.7	33	657	469	623	441	585	413	498	347
		7	38	10.2	34	177	127	170	125	163	119	146	107
		8	2	.7	35	13	9	12	9	12	9	11	8
			1142	138.6		2258	1604	1597	1101	1476	1011	1201	806
		MEAN DIA. 4.7 INCHES, WEIBULL PARAMETERS A= .6, B= 4.35, C= 4.03											
25	40	2	12	.3	22	3	2						
		3	82	4.0	29	62	43						
		4	222	19.4	33	333	236						
		5	334	45.5	36	847	603	793	538	727	489	577	378
		6	272	53.4	38	1043	746	995	716	943	678	823	587
		7	107	28.6	39	572	409	550	408	531	394	485	359
		8	16	5.6	41	117	84	114	84	111	84	104	79
		9	1	.4	41	9	7	9	7	9	7	8	6
			1046	157.2		2986	2130	2461	1753	2321	1652	1997	1409
		MEAN DIA. 5.2 INCHES, WEIBULL PARAMETERS A= 1.1, B= 4.46, C= 3.86											

Age	Av. dom. ht.	D.b.h.	Stems per acre	Basal area	Av. ht.	All trees, total stem		5-inch class and greater to o.b. tops of—					
								2 inches		3 inches		4 inches	
						O. b.	I. b.	O. b.	I. b.	O. b.	I. b.	O. b.	I. b.
Yrs	Ft	In	No.	Ft 2	Ft	---	---	---	---	Ft 3	---	---	---
30	45	2	4	.1	23	1	1						
		3	42	2.1	31	34	24						
		4	141	12.3	36	231	163						
		5	257	35.0	40	723	516	681	470	630	433	515	347
		6	277	54.4	42	1173	839	1122	816	1070	777	949	686
		7	164	43.8	44	988	708	954	708	925	692	855	638
		8	47	16.4	45	378	271	367	271	359	271	340	258
		9	6	2.7	47	64	46	62	46	61	46	59	45
			938	166.8		3592	2568	3186	2311	3045	2219	2718	1974
			MEAN DIA. 5.7 INCHES, WEIBULL PARAMETERS A= 1.5, B= 4.56, C= 3.74										
35	49	2	1	.0	23	0	0						
		3	21	1.0	32	18	12						
		4	86	7.5	38	148	105						
		5	183	25.0	42	540	385	510	355	474	328	392	267
		6	240	47.1	45	1088	779	1043	763	999	730	894	651
		7	191	51.0	47	1228	880	1188	880	1155	867	1074	805
		8	84	29.3	49	735	527	714	527	701	527	667	508
		9	18	8.0	50	203	146	198	146	196	146	189	145
		10	2	1.1	51	28	20	28	20	28	20	27	20
			826	170.0		3988	2854	3681	2691	3553	2618	3243	2396
			MEAN DIA. 6.1 INCHES, WEIBULL PARAMETERS A= 1.8, B= 4.69, C= 3.64										
40	53	3	9	.4	33	8	5						
		4	50	4.4	39	88	63						
		5	123	16.8	44	380	271	359	252	335	234	281	193
		6	188	36.9	47	890	637	854	627	820	602	738	540
		7	186	49.7	49	1247	894	1206	894	1175	884	1096	824
		8	113	39.4	51	1028	738	1000	738	983	738	937	715
		9	39	17.2	53	466	335	455	335	450	335	435	335
		10	7	3.8	54	105	76	103	76	102	76	100	76
		11	1	.7	55	19	13	18	13	18	13	18	13
			716	169.4		4231	3032	3995	2935	3883	2882	3605	2696
			MEAN DIA. 6.6 INCHES, WEIBULL PARAMETERS A= 2.1, B= 4.86, C= 3.56										

SI 40

1,750 stems per acre

Age	Av. dom. ht.	D.b.h.	Stems per acre	Basal area	Av. ht.	All trees, total stem		5-inch class and greater to o.b. tops of—					
								2 inches		3 inches		4 inches	
						O.b.	I.b.	O.b.	I.b.	O.b.	I.b.	O.b.	I.b.
Yrs	Ft	In	No.	Ft²	Ft	-- Ft³ --							
10	18	1	64	.3	8	4	2						
		2	455	9.9	14	87	55						
		3	735	36.1	16	320	218						
		4	197	17.2	18	165	115						
		5	3	.4	19	4	3	4	2	3	2	2	1
			1454	64.0		580	393	4	2	3	2	2	1
			MEAN DIA. 2.8 INCHES, WEIBULL PARAMETERS A= .0, B= 3.00, C= 4.46										
15	27	1	17	.1	10	1	0						
		2	138	3.0	18	32	21						
		3	396	19.4	22	232	160						
		4	521	45.5	24	574	404						
		5	273	37.2	26	502	357	461	291	406	250	282	159
		6	42	8.2	27	115	82	108	74	100	68	81	54
		7	1	.3	28	4	3	4	3	3	2	3	2
			1388	113.7		1460	1027	573	368	509	320	366	215
			MEAN DIA. 3.9 INCHES, WEIBULL PARAMETERS A= .1, B= 4.00, C= 4.16										
20	34	1	3	.0	10	0	0						
		2	50	1.1	21	13	9						
		3	208	10.2	26	143	99						
		4	410	35.8	30	561	397						
		5	411	56.0	32	928	660	863	573	781	512	596	375
		6	181	35.5	33	604	431	573	405	538	379	457	319
		7	28	7.5	34	131	93	125	92	120	88	108	− 79
		8	1	.3	35	6	4	6	4	6	4	5	4
			1292	146.5		2386	1693	1567	1074	1445	983	1166	777
			MEAN DIA. 4.6 INCHES, WEIBULL PARAMETERS A= .6, B= 4.18, C= 3.94										
25	40	2	18	.4	23	5	3						
		3	109	5.4	30	86	60						
		4	279	24.3	34	431	306						
		5	386	52.6	37	1005	717	943	643	866	587	693	458
		6	279	54.8	39·	1098	785	1048	756	995	717	872	624
		7	92	24.6	40	504	361	486	361	469	349	430	316
		8	11	3.8	41	81	58	78	58	76	58	72	54
			1174	165.9		3210	2290	2555	1818	2406	1711	2067	1454
			MEAN DIA. 5.1 INCHES, WEIBULL PARAMETERS A= 1.1, B= 4.29, C= 3.79										

Age	Av. dom. ht.	D.b.h.	Stems per acre	Basal area	Av. ht.	All trees, total stem		5-inch class and greater to o.b. tops of—					
								2 inches		3 inches		4 inches	
						O.b.	I.b.	O.b.	I.b.	O.b.	I.b.	O.b.	I.b.
Yrs	Ft	In	No.	Ft 2	Ft	– – – – – – – – – – – – – Ft 3 – – – – – – – – – – – – –							
30	45	2	5	.1	24	2	1						
		3	56	2.7	32	47	33						
		4	178	15.5	37	299	212						
		5	306	41.7	40	861	614	810	560	750	515	613	413
		6	301	59.1	42	1275	912	1219	887	1163	845	1031	745
		7	157	42.0	44	946	678	913	678	886	662	819	611
		8	38	13.3	46	312	224	303	224	297	224	281	214
		9	4	1.8	47	42	30	41	30	41	30	39	30
			----	-----		----	----	----	----	----	----	----	----
			1045	176.2		3784	2704	3286	2379	3137	2276	2783	2013
					MEAN DIA. 5.6 INCHES, WEIBULL PARAMETERS A= 1.5, B= 4.40, C= 3.67								
35	49	2	1	.0	24	0	0						
		3	27	1.3	33	23	16						
		4	109	9.5	39	193	137						
		5	222	30.3	42	655	468	618	430	575	398	476	324
		6	272	53.4	45	1233	883	1182	865	1132	828	1013	738
		7	194	51.8	47	1248	894	1206	894	1173	881	1091	817
		8	74	25.8	49	647	464	629	464	618	464	587	447
		9	13	5.7	50	147	105	143	105	141	105	136	105
		10	1	.5	51	14	10	14	10	14	10	13	10
			----	-----		----	----	----	----	----	----	----	----
			913	178.5		4160	2977	3792	2768	3653	2686	3316	2441
					MEAN DIA. 6.0 INCHES, WEIBULL PARAMETERS A= 1.8, B= 4.52, C= 3.59								
40	53	3	12	.6	34	11	7						
		4	63	5.5	40	114	81						
		5	150	20.5	45	474	338	449	315	419	294	353	244
		6	217	42.6	48	1049	751	1007	741	968	712	874	640
		7	198	52.9	50	1354	971	1311	971	1278	963	1194	898
		8	108	37.7	52	1002	719	975	719	959	719	915	698
		9	32	14.1	54	390	280	381	280	376	280	364	280
		10	5	2.7	55	77	55	75	55	74	55	72	55
			----	-----		----	----	----	----	----	----	----	----
			785	176.6		4471	3202	4198	3081	4074	3023	3772	2815
					MEAN DIA. 6.4 INCHES, WEIBULL PARAMETERS A= 2.1, B= 4.69, C= 3.52								

SI 40

2,000 stems per acre

Age	Av. dom. ht.	D.b.h.	Stems per acre	Basal area	Av. ht.	All trees, total stem		5-inch class and greater to o.b. tops of—					
								2 inches		3 inches		4 inches	
						O.b.	I.b.	O.b.	I.b.	O.b.	I.b.	O.b.	I.b.
Yrs	Ft	In	No.	Ft 2	Ft	— — — — — — — — — — — Ft 3 — — — — — — — — — — — —							
10	18	1	97	.5	8	6	2						
		2	609	13.3	14	116	74						
		3	790	38.8	16	344	234						
		4	143	12.5	18	119	84						
		5	1	.1	19	1	1	1	1	1	1	1	0
			----	----		----	----	----	----	----	----	----	----
			1640	65.2		586	395	1	1	1	1	1	0
			MEAN DIA. 2.7 INCHES, WEIBULL PARAMETERS A= .0, B= 2.85, C= 4.35										
15	27	1	25	.1	10	2	1						
		2	184	4.0	18	43	28						
		3	469	24.0	22	286	198						
		4	573	50.0	25	657	463						
		5	253	34.5	26	466	331	427	270	376	232	261	147
		6	30	5.9	27	82	59	77	53	71	49	58	39
		7	1	.3	28	4	3	4	3	3	2	3	2
			----	----		----	----	----	----	----	----	----	----
			1555	118.8		1540	1083	508	326	450	283	322	188
			MEAN DIA. 3.7 INCHES, WEIBULL PARAMETERS A= .1, B= 3.86, C= 4.06										
20	34	1	4	.0	11	0	0						
		2	67	1.5	21	18	12						
		3	263	12.9	27	187	130						
		4	482	42.1	30	660	466						
		5	434	59.2	32	980	697	912	605	824	541	629	396
		6	166	32.6	33	554	395	525	372	493	348	419	293
		7	21	5.6	35	101	72	97	71	93	68	84	61
		8	1	.3	35	6	4	6	4	6	4	5	4
			----	----		----	----	----	----	----	----	----	----
			1438	154.2		2506	1776	1540	1052	1416	961	1137	754
			MEAN DIA. 4.4 INCHES, WEIBULL PARAMETERS A= .6, B= 4.04, C= 3.86										
25	40	2	24	.5	24	7	5						
		3	139	6.8	30	109	76						
		4	336	29.3	34	519	368						
		5	430	58.6	37	1120	798	1050	717	965	654	772	510
		6	280	55.0	39	1102	788	1051	759	999	720	875	627
		7	81	21.6	40	444	318	428	318	413	307	378	280
		8	8	2.8	41	59	42	57	42	55	42	52	39
			----	----		----	----	----	----	----	----	----	----
			1298	174.7		3360	2395	2586	1836	2432	1723	2077	1456
			MEAN DIA. 5.0 INCHES, WEIBULL PARAMETERS A= 1.1, B= 4.16, C= 3.71										

Age	Av. dom. ht.	D.b.h.	Stems per acre	Basal area	Av. ht.	All trees, total stem		5-inch class and greater to o.b. tops of—					
								2 inches		3 inches		4 inches	
						O. b.	I. b.	O. b.	I. b.	O. b.	I. b.	O. b.	I. b.
Yrs	Ft	In	No.	Ft 2	Ft	$Ft 3$							
30	45	2	7	.2	25	2	1						
		3	71	3.5	32	59	41						
		4	217	18.9	37	364	259						
		5	352	48.0	40	990	706	932	644	863	593	706	475
		6	319	62.6	43	1383	989	1324	965	1264	920	1125	815
		7	149	39.8	44	897	643	866	643	841	628	777	580
		8	31	10.8	46	255	183	247	183	242	183	229	174
		9	2	.9	47	21	15	21	15	20	15	20	15
			1148	184.7		3971	2837	3390	2450	3230	2339	2857	2059

MEAN DIA. 5.4 INCHES, WEIBULL PARAMETERS A= 1.5, B= 4.26, C= 3.61

Age	Av. dom. ht.	D.b.h.	Stems per acre	Basal area	Av. ht.	O. b.	I. b.	O. b.	I. b.	O. b.	I. b.	O. b.	I. b.
35	49	2	2	.0	25	1	0						
		3	34	1.7	34	30	21						
		4	133	11.6	39	235	167						
		5	259	35.3	43	782	558	739	516	689	479	573	393
		6	296	58.1	45	1342	961	1287	941	1232	901	1103	803
		7	195	52.1	47	1254	899	1213	899	1179	885	1096	822
		8	66	23.0	49	577	414	561	414	551	414	524	399
		9	10	.4.4	50	113	81	110	81	109	81	105	81
		10	1	.5	51	14	10	14	10	14	10	13	10
			996	186.9		4348	3111	3924	2861	3774	2770	3414	2508

MEAN DIA. 5.9 INCHES, WEIBULL PARAMETERS A= 1.8, B= 4.39, C= 3.54

Age	Av. dom. ht.	D.b.h.	Stems per acre	Basal area	Av. ht.	O. b.	I. b.	O. b.	I. b.	O. b.	I. b.	O. b.	I. b.
40	53	3	15	.7	35	14	10						
		4	77	6.7	41	143	102						
		5	176	24.0	45	556	397	526	370	492	345	414	286
		6	245	48.1	48	1184	848	1137	837	1093	804	986	723
		7	207	55.3	51	1444	1035	1398	1035	1364	1029	1276	961
		8	102	35.6	52	946	679	921	679	905	679	864	659
		9	26	11.5	54	317	227	309	227	306	227	296	227
		10	3	1.6	55	46	33	45	33	45	33	43	33
			851	183.6		4650	3331	4336	3181	4205	3117	3879	2889

MEAN DIA. 6.3 INCHES, WEIBULL PARAMETERS A= 2.1, B= 4.55, C= 3.48

SI 50

750 stems per acre

Age	Av. dom. ht.	D.b.h.	Stems per acre	Basal area	Av. ht.	All trees, total stem		5-inch class and greater to o.b. tops of—					
								2 inches		3 inches		4 inches	
						O.b.	I.b.	O.b.	I.b.	O.b.	I.b.	O.b.	I.b.
Yrs	Ft	In	No.	Ft 2	Ft	—	—	—	—	Ft 3	—	—	—
10	23	1	3	.0	5	0	0						
		2	30	.7	13	5	3						
		3	114	5.6	17	52	36						
		4	229	20.0	20	212	148						
		5	225	30.7	22	352	249	318	191	272	157	170	82
		6	83	16.3	23	194	138	181	121	164	109	127	81
		7	8	2.1	24	26	19	25	18	23	17	20	14
			692	75.4		841	593	524	330	459	283	317	177
				MEAN DIA. 4.5 INCHES, WEIBULL PARAMETERS A= .0, B= 4.74, C= 4.74									
15	34	2	8	.2	16	2	1						
		3	42	2.1	23	26	18						
		4	116	10.1	27	143	101						
		5	195	26.6	30	413	294	383	251	343	222	255	156
		6	193	37.9	32	624	446	591	417	554	389	468	325
		7	100	26.7	33	453	324	434	317	415	304	372	271
		8	22	7.7	34	134	96	129	96	125	94	116	87
		9	2	.9	35	16	11	15	11	15	11	14	11
			678	112.1		1811	1291	1552	1092	1452	1020	1225	850
				MEAN DIA. 5.5 INCHES, WEIBULL PARAMETERS A= .6, B= 5.20, C= 4.32									
20	43	2	2	.0	19	0	0						
		3	18	.9	27	13	9						
		4	62	5.4	32	90	64						
		5	131	17.9	36	332	237	311	211	285	192	226	148
		6	177	34.8	38	679	485	647	466	614	441	536	382
		7	153	40.9	40	838	600	808	600	780	580	714	529
		8	74	25.8	42	555	398	538	398	526	398	495	374
		9	18	8.0	43	175	125	170	125	167	125	160	122
		10	2	1.1	44	25	18	24	18	24	18	23	18
			637	134.7		2707	1936	2498	1818	2396	1754	2154	1573
				MEAN DIA. 6.2 INCHES, WEIBULL PARAMETERS A= 1.3, B= 5.28, C= 4.00									
25	50	3	7	.3	30	5	4						
		4	33	2.9	36	54	38						
		5	82	11.2	40	231	165	217	150	201	138	164	111
		6	135	26.5	44	599	428	574	419	549	400	489	356
		7	152	40.6	46	957	686	925	686	899	674	834	624
		8	109	38.0	48	934	670	907	670	891	670	846	643
		9	48	21.2	50	542	389	528	389	522	389	503	386
		10	11	6.0	51	156	112	153	112	151	112	147	112
		11	1	.7	52	18	13	17	13	17	13	17	13
			578	147.4		3496	2505	3321	2439	3230	2396	3000	2245
				MEAN DIA. 6.8 INCHES, WEIBULL PARAMETERS A= 1.8, B= 5.37, C= 3.75									

Age	Av. dom. ht.	D.b.h.	Stems per acre	Basal area	Av. ht.	All trees, total stem		5-inch class and greater to o.b. tops of—					
								2 inches		3 inches		4 inches	
						O. b.	I. b.	O. b.	I. b.	O. b.	I. b.	O. b.	I. b.
Yrs	Ft	In	No.	Ft²	Ft	-------------- Ft³ --------------							
30	56	3	2	.1	31	2	1						
		4	17	1.5	38	29	21						
		5	49	6.7	44	151	108	143	100	134	93	112	77
		6	92	18.1	47	436	312	418	307	401	294	361	264
		7	121	32.3	50	828	593	801	593	781	588	730	549
		8	114	39.8	53	1078	773	1049	773	1032	773	986	753
		9	73	32.3	54	889	639	868	639	859	639	830	639
		10	30	16.4	56	468	336	458	336	454	336	443	336
		11	9	5.3	57	154	110	150	110	150	110	147	110
		12	1	.8	58	23	17	23	17	23	17	22	17
			507	153.1		4058	2910	3910	2875	3834	2850	3631	2745
			MEAN DIA. 7.4 INCHES, WEIBULL PARAMETERS A= 2.3, B= 5.52, C= 3.56											
35	61	3	1	.0	32	1	1						
		4	8	.7	40	15	10						
		5	27	3.7	46	87	62	83	58	77	54	65	45
		6	57	11.2	50	287	205	276	204	266	196	241	177
		7	86	23.0	53	623	447	604	447	590	446	554	418
		8	97	33.9	56	969	695	944	695	930	695	890	682
		9	81	35.8	58	1060	761	1035	761	1025	761	994	761
		10	49	26.7	60	818	588	801	588	796	588	778	588
		11	21	13.9	62	438	315	430	315	428	315	421	315
		12	6	4.7	63	151	109	149	109	148	109	146	109
		13	1	.9	64	30	22	30	22	30	22	29	22
			434	154.5		4479	3215	4352	3199	4290	3186	4118	3117
			MEAN DIA. 8.1 INCHES, WEIBULL PARAMETERS A= 2.7, B= 5.78, C= 3.39											
40	66	4	3	.3	41	6	4						
		5	14	1.9	47	46	33	44	31	41	29	35	24
		6	33	6.5	52	173	124	166	123	160	119	146	108
		7	55	14.7	56	421	302	408	302	400	302	377	285
		8	70	24.4	59	737	529	718	529	708	529	680	521
		9	74	32.7	61	1018	731	995	731	986	731	958	731
		10	57	31.1	63	999	718	979	718	973	718	953	718
		11	35	23.1	65	766	550	751	550	749	550	737	550
		12	16	12.6	66	423	304	415	304	415	304	410	304
		13	5	4.6	68	160	115	157	115	157	115	156	115
		14	1	1.1	69	38	27	37	27	37	27	37	27
			363	152.9		4787	3437	4670	3430	4626	3424	4489	3383
			MEAN DIA. 8.8 INCHES, WEIBULL PARAMETERS A= 3.1, B= 6.19, C= 3.25											

SI 50

1,000 stems per acre

Age	Av. dom. ht.	D.b.h.	Stems per acre	Basal area	Av. ht.	All trees, total stem		5-inch class and greater to o.b. tops of—					
								2 inches		3 inches		4 inches	
						O. b.	I. b.	O. b.	I. b.	O. b.	I. b.	O. b.	I. b.
Yrs	Ft	In	No.	Ft 2	Ft	— — — — — — — — — — — — Ft 3 — — — — — — — — — — — —							
10	23	1	7	.0	6	0	0						
		2	65	1.4	14	12	8						
		3	221	10.8	18	107	73						
		4	350	30.5	20	323	227						
		5	217	29.6	22	339	240	307	184	262	152	164	79
		6	36	7.1	23	84	60	78	52	71	47	55	35
		7	1	.3	24	3	2	3	2	3	2	2	2
			897	79.8		868	610	388	238	336	201	221	116
			MEAN DIA. 4.0 INCHES, WEIBULL PARAMETERS A= .0, B= 4.28, C= 4.62										
15	34	1	1	.0	7	0	0						
		2	16	.3	18	4	2						
		3	81	4.0	24	51	36						
		4	200	17.5	28	256	181						
		5	283	38.6	31	619	441	575	379	518	337	390	242
		6	209	41.0	33	697	498	661	468	621	438	528	369
		7	68	18.2	34	317	227	304	223	292	214	262	191
		8	8	2.8	35	50	36	48	36	47	35	44	33
			866	122.4		1994	1421	1588	1106	1478	1024	1224	835
			MEAN DIA. 5.1 INCHES, WEIBULL PARAMETERS A= .6, B= 4.76, C= 4.21										
20	43	2	4	.1	21	1	1						
		3	33	1.6	29	25	17						
		4	110	9.6	34	170	121						
		5	207	28.2	37	539	384	506	345	465	315	372	245
		6	236	46.3	40	952	681	909	658	865	625	761	547
		7	155	41.4	42	891	639	860	639	833	620	766	570
		8	50	17.5	43	384	275	372	275	364	275	343	260
		9	7	3.1	44	70	50	68	50	67	50	64	49
			802	147.8		3032	2168	2715	1967	2594	1885	2306	1671
			MEAN DIA. 5.8 INCHES, WEIBULL PARAMETERS A= 1.3, B= 4.85, C= 3.90										
25	50	3	13	.6	31	11	7						
		4	58	5.1	38	100	71						
		5	134	18.3	42	395	282	373	260	347	240	287	196
		6	197	38.7	45	893	639	856	627	820	599	734	534
		7	181	48.4	47	1164	834	1125	834	1095	822	1018	763
		8	100	34.9	49	874	627	850	627	835	627	793	604
		9	30	13.3	51	345	248	337	248	333	248	321	247
		10	4	2.2	52	58	42	57	42	56	42	55	42
			717	161.4		3840	2750	3598	2638	3486	2578	3208	2386
			MEAN DIA. 6.4 INCHES, WEIBULL PARAMETERS A= 1.8, B= 4.92, C= 3.67										

Age	Av. dom. ht.	D.b.h.	Stems per acre	Basal area	Av. ht.	All trees, total stem		5-inch class and greater to o.b. tops of—					
								2 inches		3 inches		4 inches	
						O. b.	I. b.	O. b.	I. b.	O. b.	I. b.	O. b.	I. b.
Yrs	Ft	In	No.	Ft 2	Ft	- - - - - - - - - - - - - Ft 3 - - - - - - - - - - - - -							
30	56	3	4	.2	33	-3	2						
		4	29	2.5	40	53	37						
		5	81	11.0	45	256	183	242	170	227	159	190	132
		6	139	27.3	49	686	491	659	486	634	467	573	421
		7	162	43.3	52	1152	826	1116	826	1089	822	1021	770
		8	125	43.6	54	1204	864	1172	864	1154	864	1103	843
		9	61	26.9	56	771	553	752	553	745	553	721	553
		10	17	9.3	57	270	194	264	194	262	194	256	194
		11	3	2.0	58	59	42	57	42	57	42	56	42
			621	166.2		4454	3192	4262	3135	4168	3101	3920	2955

MEAN DIA. 7.0 INCHES, WEIBULL PARAMETERS A= 2.3, B= 5.06, C= 3.49

Age	Av. dom. ht.	D.b.h.	Stems per acre	Basal area	Av. ht.	All trees, total stem		2 inches		3 inches		4 inches	
35	61	3	1	.0	34	1	1						
		4	13	1.1	42	25	18						
		5	45	6.1	47	148	106	141	100	132	93	112	78
		6	89	17.5	52	466	334	448	332	432	320	394	291
		7	121	32.3	55	910	653	882	653	863	653	812	614
		8	120	41.9	58	1241	891	1209	891	1192	891	1144	877
		9	83	36.7	60	1123	807	1098	807	1088	807	1056	807
		10	38	20.7	61	645	463	632	463	628	463	614	463
		11	11	7.3	63	233	168	229	168	228	168	224	168
		12	2	1.6	64	51	37	50	37	50	37	50	37
			523	165.2		4843	3478	4689	3451	4613	3432	4406	3335

MEAN DIA. 7.6 INCHES, WEIBULL PARAMETERS A= 2.7, B= 5.29, C= 3.34

Age	Av. dom. ht.	D.b.h.	Stems per acre	Basal area	Av. ht.	All trees, total stem		2 inches		3 inches		4 inches	
40	66	4	5	.4	43	10	7						
		5	23	3.1	49	79	56	75	53	71	50	60	42
		6	51	10.0	54	277	199	267	198	258	191	236	175
		7	80	21.4	57	623	447	605	447	592	447	559	423
		8	94	32.8	60	1006	722	980	722	967	722	930	713
		9	84	37.1	63	1193	857	1167	857	1157	857	1125	857
		10	56	30.5	65	1013	728	992	728	987	728	967	728
		11	27	17.8	66	600	431	588	431	587	431	578	431
		12	9	7.1	68	245	176	241	176	240	176	238	176
		13	2	1.8	69	65	47	64	47	64	47	63	47
			431	162.2		5111	3670	4979	3659	4923	3649	4756	3592

MEAN DIA. 8.3 INCHES, WEIBULL PARAMETERS A= 3.1, B= 5.65, C= 3.21

SI 50

1,250 stems per acre

Age	Av. dom. ht.	D.b.h.	Stems per acre	Basal area	Av. ht.	All trees, total stem		5-inch class and greater to o.b. tops of—					
								2 inches		3 inches		4 inches	
						O.b.	I.b.	O.b.	I.b.	O.b.	I.b.	O.b.	I.b.
Yrs	Ft	In	No.	Ft²	Ft	----------- Ft³ -----------							
10	23	1	14	.1	7	1	0						
		2	114	2.5	15	23	15						
		3	344	16.9	19	176	121						
		4	433	37.8	21	419	294						
		5	177	24.1	23	289	205	263	160	226	133	146	74
		6	15	2.9	24	37	26	34	23	31	21	24	16
			1097	84.3		945	661	297	183	257	154	170	90
		MEAN DIA. 3.8 INCHES, WEIBULL PARAMETERS A= .0, B= 3.98, C= 4.50											
15	34	1	1	.0	8	0	0						
		2	28	.6	19	7	4						
		3	130	6.4	25	86	60						
		4	290	25.3	29	384	271						
		5	349	47.6	31	763	543	709	468	639	416	481	299
		6	201	39.5	33	670	479	636	450	597	421	508	354
		7	45	12.0	34	210	150	201	148	193	141	173	127
		8	3	1.0	35	19	13	18	13	18	13	16	12
			1047	132.4		2139	1520	1564	1079	1447	991	1178	792
		MEAN DIA. 4.8 INCHES, WEIBULL PARAMETERS A= .6, B= 4.47, C= 4.10											
20	43	2	6	.1	22	2	1						
		3	53	2.6	30	42	29						
		4	164	14.3	34	254	180						
		5	281	38.3	38	751	536	706	484	650	443	524	349
		6	276	54.2	40	1113	796	1064	770	1012	731	890	640
		7	145	38.2	42	822	589	793	589	768	572	707	525
		8	34	11.9	43	261	187	253	187	248	187	234	177
		9	3	1.3	44	30	21	29	21	29	21	27	21
			960	161.0		3275	2339	2845	2051	2707	1954	2382	1712
		MEAN DIA. 5.5 INCHES, WEIBULL PARAMETERS A= 1.3, B= 4.56, C= 3.81											
25	50	2	1	.0	23	0	0						
		3	20	1.0	32	17	12						
		4	88	7.7	38	152	108						
		5	189	25.8	42	558	398	526	366	489	339	405	276
		6	247	48.5	45	1120	802	1074	786	1028	752	920	670
		7	196	52.4	48	1287	923	1245	923	1212	911	1129	847
		8	86	30.0	49	752	540	731	540	718	540	682	520
		9	19	6.4	51	219	157	213	157	211	157	203	156
		10	2	1.1	52	29	21	28	21	28	21	27	21
			848	174.8		4134	2961	3617	2793	3686	2720	3366	2490
		MEAN DIA. 6.1 INCHES, WEIBULL PARAMETERS A= 1.8, B= 4.63, C= 3.59											

Age	Av. dom. ht.	D.b.h.	Stems per acre	Basal area	Av. ht.	All trees, total stem		5-inch class and greater to o.b. tops of—					
								2 inches		3 inches		4 inches	
						O. b.	I. b.	O. b.	I. b.	O. b.	I. b.	O. b.	I. b.
Yrs	Ft	In	No.	Ft 2	Ft	———————— Ft 3 ————————							
30	56	3	6	.3	34	5	4						
		4	44	3.8	41	82	58						
		5	115	15.7	46	371	265	352	248	330	232	278	193
		6	184	36.1	49	908	650	872	643	839	618	759	557
		7	192	51.3	52	1365	979	1323	979	1291	975	1210	912
		8	125	43.6	54	1204	864	1172	864	1154	864	1103	843
		9	49	21.6	56	619	444	604	444	598	444	579	444
		10	10	5.5	57	159	114	155	114	154	114	151	114
		11	1	.7	58	20	14	19	14	19	14	19	14
			726	176.7		4733	3392	4497	3306	4385	3261	4099	3077

MEAN DIA. 6.7 INCHES, WEIBULL PARAMETERS A= 2.3, B= 4.75, C= 3.42

Age	Av. dom. ht.	D.b.h.	Stems per acre	Basal area	Av. ht.	O. b.	I. b.	O. b.	I. b.	O. b.	I. b.	O. b.	I. b.
35	61	3	1	.0	35	1	1						
		4	20	1.7	43	39	28						
		5	65	8.9	48	219	156	208	147	195	138	166	116
		6	121	23.8	52	633	454	610	451	588	435	535	395
		7	152	40.6	55	1143	820	1108	820	1084	820	1020	771
		8	133	46.4	58	1376	987	1340	987	1322	987	1268	972
		9	77	34.0	60	1042	748	1018	748	1009	748	980	748
		10	29	15.8	61	492	354	482	354	479	354	469	354
		11	6	4.0	63	127	91	125	91	124	91	122	91
		12	1	.8	64	26	18	25	18	25	18	25	18
			605	176.0		5098	3657	4916	3616	4826	3591	4585	3465

MEAN DIA. 7.3 INCHES, WEIBULL PARAMETERS A= 2.7, B= 4.95, C= 3.29

Age	Av. dom. ht.	D.b.h.	Stems per acre	Basal area	Av. ht.	O. b.	I. b.	O. b.	I. b.	O. b.	I. b.	O. b.	I. b.
40	66	4	8	.7	43	16	11						
		5	33	4.5	50	116	83	110	78	104	74	89	63
		6	71	13.9	54	386	276	372	276	359	266	328	243
		7	105	28.1	58	832	597	808	597	791	597	747	567
		8	112	39.1	61	1218	874	1188	874	1172	874	1128	866
		9	89	39.3	63	1264	908	1236	908	1226	908	1192	908
		10	50	27.3	65	904	650	886	650	882	650	864	650
		11	19	12.5	66	422	303	414	303	413	303	407	303
		12	5	3.9	68	136	98	134	98	134	98	132	98
		13	1	.9	69	32	23	32	23	32	23	32	23
			493	170.3		5326	3823	5180	3807	.5113	3793	4919	3721

MEAN DIA. 8.0 INCHES, WEIBULL PARAMETERS A= 3.1, B= 5.28, C= 3.17

SI 50

1,500 stems per acre

Age	Av. dom. ht.	D.b.h.	Stems per acre	Basal area	Av. ht.	All trees, total stem		5-inch class and greater to o.b. tops of—					
								2 inches		3 inches		4 inches	
						O. b.	I. b.	O. b.	I. b.	O. b.	I. b.	O. b.	I. b.
Yrs	Ft	In	No.	Ft 2	Ft	------------- Ft 3 -------------							
10	23	1	23	.1	8	1	1						
		2	177	3.9	15	36	23						
		3	473	23.2	19	241	166						
		4	478	41.7	21	463	325						
		5	135	18.4	23	220	156	200	122	173	102	111	56
		6	6	1.2	24	15	10	14	9	12	8	10	6
			----	----		----	----	----	----	----	----	----	----
			1292	88.5		976	681	214	131	185	110	121	62
		MEAN DIA. 3.5 INCHES, WEIBULL PARAMETERS A= .0, B= 3.75, C= 4.38											
15	34	1	2	.0	9	0	0						
		2	44	1.0	20	11	7						
		3	187	9.2	26	128	89						
		4	379	33.1	29	502	355						
		5	396	54.0	31	866	617	805	531	725	472	546	339
		6	184	36.1	33	614	438	582	412	547	386	465	325
		7	30	8.0	34	140	100	134	98	129	94	116	84
		8	1	.3	35	6	4	6	4	6	4	5	4
			----	----		----	----	----	----	----	----	----	----
			1223	141.7		2267	1610	1527	1045	1407	956	1132	752
		MEAN DIA. 4.6 INCHES, WEIBULL PARAMETERS A= .6, B= 4.25, C= 4.00											
20	43	2	10	.2	23	3	2						
		3	78	3.6	30	61	43						
		4	223	19.5	35	355	251						
		5	346	47.2	38	925	660	869	595	800	545	645	429
		6	300	58.9	40	1210	866	1156	837	1100	795	968	695
		7	129	34.5	42	742	531	716	531	693	516	637	474
		8	23	8.0	43	177	127	171	127	168	127	158	120
		9	1	.4	44	10	7	10	7	10	7	9	7
			----	----		----	----	----	----	----	----	----	----
			1110	172.5		3483	2487	2922	2097	2771	1990	2417	1725
		MEAN DIA. 5.3 INCHES, WEIBULL PARAMETERS A= 1.3, B= 4.34, C= 3.72											
25	50	2	1	.0	24	0	0						
		3	30	1.5	33	26	18						
		4	121	10.6	39	214	152						
		5	243	33.1	43	734	524	693	484	646	449	538	368
		6	290	56.9	46	1344	962	1289	945	1236	905	1109	810
		7	200	53.5	48	1313	941	1270	941	1237	929	1152	864
		8	73	25.5	50	651	467	633	467	622	467	592	451
		9	13	5.7	51	150	107	146	107	144	107	139	107
		10	1	.5	52	14	10	14	10	14	10	14	10
			----	----		----	----	----	----	----	----	----	----
			972	187.4		4446	3181	4045	2954	3899	2867	3544	2610
		MEAN DIA. 5.9 INCHES, WEIBULL PARAMETERS A= 1.8, B= 4.41, C= 3.52											

Age	Av. dom. ht.	D.b.h.	Stems per acre	Basal area	Av. ht.	All trees, total stem		5-inch class and greater to o.b. tops of—					
								2 inches		3 inches		4 inches	
						O. b.	I. b.	O. b.	I. b.	O. b.	I. b.	O. b.	I. b.
Yrs	Ft	In	No.	Ft 2	Ft	———————————————— Ft 3 ————————————————							
30	56	3	9	.4	36	8	6						
		4	60	5.2	43	117	83						
		5	152	20.7	47	501	358	475	336	446	314	379	264
		6	226	44.4	51	1160	831	1116	825	1075	795	978	720
		7	212	56.7	53	1536	1102	1489	1102	1454	1099	1365	1030
		8	121	42.2	55	1187	852	1156	852	1138	852	1089	833
		9	39	17.2	57	501	360	490	360	485	360	470	360
		10	6	3.3	58	97	70	95	70	94	70	92	70
			825	190.2		5107	3662	4821	3545	4692	3490	4373	3277
		MEAN DIA. 6.5 INCHES, WEIBULL PARAMETERS A= 2.3, B= 4.52, C= 3.37											
35	61	3	2	.1	36	2	1						
		4	27	2.4	44	54	38						
		5	86	11.7	50	302	216	287	204	270	192	232	163
		6	152	29.8	54	826	592	796	591	769	570	703	520
		7	180	48.1	57	1403	1006	1361	1006	1332	1006	1257	952
		8	140	48.9	59	1473	1057	1435	1057	1416	1057	1360	1043
		9	70	30.9	61	963	692	941	692	933	692	906	692
		10	21	11.5	63	368	264	361	264	359	264	351	264
		11	4	2.6	64	86	62	84	62	84	62	83	62
			682	186.0		5477	3928	5265	3876	5163	3843	4892	3696
		MEAN DIA. 7.1 INCHES, WEIBULL PARAMETERS A= 2.7, B= 4.71, C= 3.24											
40	66	4	11	1.0	45	22	16						
		5	44	6.0	51	157	112	150	107	141	101	122	86
		6	91	17.9	56	513	367	494	367	478	356	439	326
		7	126	33.7	59	1016	729	986	729	966	729	914	694
		8	127	44.3	62	1404	1008	1369	1008	1352	1008	1301	1000
		9	90	39.8	64	1299	933	1270	933	1260	933	1226	933
		10	44	24.0	66	808	581	792	581	788	581	772	581
		11	14	9.2	68	320	230	314	230	314	230	309	230
		12	3	2.4	69	83	60	81	60	81	60	80	60
			550	178.0		5622	4036	5456	4015	5380	3998	5163	3910
		MEAN DIA. 7.7 INCHES, WEIBULL PARAMETERS A= 3.1, B= 5.01, C= 3.14											

SI 50

1,750 stems per acre

Age	Av. dom. ht.	D.b.h.	Stems per acre	Basal area	Av. ht.	All trees, total stem		5-inch class and greater to o.b. tops of—					
								2 inches		3 inches		4 inches	
						O. b.	I. b.	O. b.	I. b.	O. b.	I. b.	O. b.	I. b.
Yrs	Ft	In	No.	Ft 2	Ft	— — — — — — — — — — — — — Ft 3 — — — — — — — — — — — — —							
10	23	1	36	.2	8	2	1						
		2	254	5.5	16	54	35						
		3	598	29.4	19	305	210						
		4	493	43.0	21	477	335						
		5	101	13.8	23	165	117	150	91	129	76	83	42
		6	3	.6	24	7	5	7	5	6	4	5	3
			1485	92.5		1010	703	157	96	135	80	88	45
			MEAN DIA. 3.4 INCHES, WEIBULL PARAMETERS A= .0, B= 3.58, C= 4.26										
15	34	1	4	.0	10	0	0						
		2	63	1.4	20	16	11						
		3	250	12.3	26	171	119						
		4	465	40.6	29	616	435						
		5	425	58.0	31	930	662	864	569	778	506	586	364
		6	165	32.4	33	550	393	522	370	490	346	417	291
		7	21	5.6	34	98	70	94	69	90	66	81	59
		8	1	.3	35	6	4	6	4	6	4	5	4
			1394	150.6		2387	1694	1486	1012	1364	922	1089	718
			MEAN DIA. 4.4 INCHES, WEIBULL PARAMETERS A= .6, B= 4.08, C= 3.89										
20	43	2	14	.3	23	4	3						
		3	106	5.2	31	86	60						
		4	284	24.8	35	452	320						
		5	407	55.5	38	1088	776	1022	700	942	641	759	505
		6	312	61.3	40	1259	900	1202	870	1144	826	1006	723
		7	115	30.7	42	661	474	638	474	618	460	568	423
		8	17	5.9	43	131	94	127	94	124	94	117	88
		9	1	.4	44	10	7	10	7	10	7	9	7
			1256	184.2		3691	2634	2999	2145	2838	2028	2459	1746
			MEAN DIA. 5.2 INCHES, WEIBULL PARAMETERS A= 1.3, B= 4.17, C= 3.63										
25	50	2	2	.0	26	1	0						
		3	41	2.0	35	37	26						
		4	157	13.7	40	285	202						
		5	296	40.4	44	915	653	865	606	807	563	675	465
		6	325	63.8	47	1539	1101	1477	1085	1417	1040	1276	933
		7	199	53.2	49	1334	956	1291	956	1257	946	1173	881
		8	63	22.0	51	573	411	558	411	548	411	522	398
		9	9	4.0	52	108	76	103	76	102	76	98	76
			1092	199.1		4790	3425	4294	3134	4131	3036	3744	2753
			MEAN DIA. 5.6 INCHES, WEIBULL PARAMETERS A= 1.8, B= 4.24, C= 3.45										

Age	Av. dom. ht.	D.b.h.	Stems per acre	Basal area	Av. ht.	All trees, total stem		5-inch class and greater to o.b. tops of—					
								2 inches		3 inches		4 inches	
						O. b.	I. b.	O. b.	I. b.	O. b.	I. b.	O. b.	I. b.
Yrs	Ft	In	No.	Ft 2	Ft	— — — — — — — — — — — — Ft 3 — — — — — — — — — — — —							
30	56	3	13	.6	36	12	9						
		4	79	6.9	43	154	109						
		5	188	25.6	48	633	452	601	426	565	399	481	337
		6	264	51.8	51	1355	971	1304	964	1256	928	1142	841
		7	226	60.4	54	1669	1197	1618	1197	1581	1196	1486	1122
		8	114	39.8	56	1139	817	1109	817	1092	817	1046	801
		9	31	13.7	57	399	286	389	286	385	286	373	286
		10	4	2.2	58	65	46	63	46	63	46	61	46
			919	201.1		5426	3887	5084	3736	4942	3672	4589	3433
		MEAN DIA. 6.3 INCHES, WEIBULL PARAMETERS A= 2.3, B= 4.34, C= 3.31											
35	61	3	3	.1	37	3	2						
		4	35	3.1	45	71	51						
		5	108	14.7	50	379	271	360	256	339	241	291	205
		6	183	35.9	54	994	712	958	711	925	687	846	627
		7	201	53.7	57	1566	1123	1520	1123	1487	1123	1403	1063
		8	143	49.9	59	1505	1080	1466	1080	1446	1080	1389	1065
		9	63	27.8	61	867	622	847	622	840	622	816	622
		10	16	8.7	63	280	202	275	202	273	202	268	202
		11	2	1.3	64	43	31	42	31	42	31	41	31
			754	195.4		5708	4094	5468	4025	5352	3986	5054	3815
		MEAN DIA. 6.9 INCHES, WEIBULL PARAMETERS A= 2.7, B= 4.52, C= 3.20											
40	66	4	14	1.2	46	29	21						
		5	55	7.5	52	201	143	191	137	180	129	156	111
		6	111	21.8	56	625	448	603	448	583	434	536	398
		7	149	39.8	59	1202	862	1167	862	1143	862	1081	820
		8	137	47.8	62	1515	1087	1477	1087	1458	1087	1404	1078
		9	88	38.9	64	1270	912	1242	912	1232	912	1199	912
		10	38	20.7	66	698	501	684	501	680	501	667	501
		11	10	6.6	68	229	164	225	164	224	164	221	164
		12	2	1.6	69	55	40	54	40	54	40	54	40
			604	185.9		5824	4178	5643	4151	5554	4129	5318	4024
		MEAN DIA. 7.5 INCHES, WEIBULL PARAMETERS A= 3.1, B= 4.80, C= 3.11											

SI 50

2,000 stems per acre

Age	Av. dom. ht.	D.b.h.	Stems per acre	Basal area	Av. ht.	All trees, total stem		5-inch class and greater to o.b. tops of—					
								2 inches		3 inches		4 inches	
						O. b.	I. b.	O. b.	I. b.	O. b.	I. b.	O. b.	I. b.
Yrs	Ft	In	No.	Ft 2	Ft	‑ ‑ ‑ ‑ ‑ ‑ ‑ ‑ ‑ ‑ ‑ Ft 3 ‑ ‑ ‑ ‑ ‑ ‑ ‑ ‑ ‑ ‑ ‑							
10	23	1	53	.3	8	3	1						
		2	342	7.5	16	73	47						
		3	716	35.1	19	365	251						
		4	488	42.6	21	473	332						
		5	75	10.2	23	122	87	111	68	96	56	62	31
		6	1	.2	24	2	2	2	2	2	1	2	1
			1675	95.9		1038	720	113	70	98	57	64	32
			MEAN DIA. 3.2 INCHES, WEIBULL PARAMETERS A= .0, B= 3.43, C= 4.15										
15	34	1	5	.0	11	0	0						
		2	87	1.9	21	23	15						
		3	319	15.7	27	227	158						
		4	545	47.6	30	746	527						
		5	443	60.4	32	1000	712	930	618	841	552	642	404
		6	147	28.9	34	505	361	480	341	451	320	386	271
		7	16	4.3	35	77	55	74	54	71	52	64	47
			1562	158.7		2578	1828	1484	1013	1363	924	1092	722
			MEAN DIA. 4.3 INCHES, WEIBULL PARAMETERS A= .6, B= 3.94, C= 3.79										
20	43	2	20	.4	24	6	4						
		3	138	6.8	31	112	78						
		4	347	30.3	36	567	402						
		5	459	62.6	39	1259	898	1184	815	1094	748	888	595
		6	318	62.4	41	1315	940	1257	912	1197	867	1058	762
		7	102	27.3	42	587	420	566	420	548	406	504	375
		8	13	4.5	43	100	72	97	72	95	72	89	68
		9	1	.4	44	10	7	10	7	10	7	9	7
			1398	194.8		3956	2821	3114	2226	2944	2102	2548	1807
			MEAN DIA. 5.1 INCHES, WEIBULL PARAMETERS A= 1.3, B= 4.03, C= 3.55										
25	50	2	3	.1	26	1	1						
		3	54	2.7	35	49	34						
		4	195	17.0	41	362	257						
		5	347	47.3	45	1096	783	1038	729	970	680	816	564
		6	351	68.9	47	1662	1190	1595	1171	1531	1123	1378	1008
		7	196	52.4	49	1314	942	1271	942	1238	932	1155	868
		8	54	18.8	51	491	353	478	353	470	353	448	341
		9	7	3.1	52	82	59	80	59	79	59	76	59
			1207	210.3		5057	3619	4462	3254	4288	3147	3873	2840
			MEAN DIA. 5.7 INCHES, WEIBULL PARAMETERS A= 1.8, B= 4.10, C= 3.38										

Age	Av. dom. ht.	D.b.h.	Stems per acre	Basal area	Av. ht.	All trees, total stem		5-inch class and greater to o.b. tops of—					
								2 inches		3 inches		4 inches	
						O. b.	I. b.	O. b.	I. b.	O. b.	I. b.	O. b.	I. b.
Yrs	Ft	In	No.	Ft 2	Ft	– – – – – – – – – – – – Ft 3 – – – – – – – – – – – –							
30	56	3	17	.8	37	16	11						
		4	99	8.6	44	197	140						
		5	225	30.7	48	758	541	719	510	676	478	576	403
		6	297	58.3	51	1525	1092	1467	1084	1413	1044	1285	947
		7	235	62.8	54	1735	1244	1682	1244	1644	1243	1545	1167
		8	107	37.4	56	1069	767	1041	767	1025	767	982	752
		9	26	11.5	57	334	240	326	240	323	240	313	240
		10	3	1.6	58	48	35	47	35	47	35	46	35
			1009	211.7		5682	4070	5282	3880	5128	3807	4747	3544

MEAN DIA. 6.2 INCHES, WEIBULL PARAMETERS A= 2.3, B= 4.20, C= 3.25

Age	Av. dom. ht.	D.b.h.	Stems per acre	Basal area	Av. ht.	All trees, total stem		2 inches O.b.	2 inches I.b.	3 inches O.b.	3 inches I.b.	4 inches O.b.	4 inches I.b.
35	61	3	4	.2	38	4	3						
		4	44	3.8	45	90	64						
		5	130	17.7	50	456	326	433	309	408	290	351	247
		6	211	41.4	54	1147	821	1104	820	1067	792	976	723
		7	218	58.3	57	1699	1218	1648	1218	1613	1218	1522	1153
		8	143	49.9	59	1505	1080	1466	1080	1446	1080	1389	1065
		9	57	25.2	61	784	563	766	563	760	563	738	563
		10	13	7.1	63	228	164	223	164	222	164	217	164
		11	2	1.3	64	43	31	42	31	42	31	41	31
			822	205.0		5956	4270	5682	4185	5558	4138	5234	3946

MEAN DIA. 6.8 INCHES, WEIBULL PARAMETERS A= 2.7, B= 4.37, C= 3.16

Age	Av. dom. ht.	D.b.h.	Stems per acre	Basal area	Av. ht.	All trees, total stem		2 inches O.b.	2 inches I.b.	3 inches O.b.	3 inches I.b.	4 inches O.b.	4 inches I.b.
40	66	4	17	1.5	46	35	25						
		5	67	9.1	52	244	175	233	166	220	157	190	135
		6	131	25.7	56	738	529	712	529	688	512	632	469
		7	167	44.6	60	1369	982	1330	982	1304	982	1234	937
		8	145	50.6	62	1603	1151	1563	1151	1543	1151	1486	1141
		9	85	37.6	64	1227	881	1199	881	1190	881	1158	881
		10	33	18.0	66	606	435	594	435	591	435	579	435
		11	8	5.3	68	183	132	180	132	179	132	177	132
		12	1	.8	69	28	20	27	20	27	20	27	20
			654	193.2		6033	4330	5838	4296	5742	4270	5483	4150

MEAN DIA. 7.4 INCHES, WEIBULL PARAMETERS A= 3.1, B= 4.63, C= 3.08

SI 60
750 stems per acre

Age	Av. dom. ht.	D.b.h.	Stems per acre	Basal area	Av. ht.	All trees, total stem		5-inch class and greater to o.b. tops of—					
								2 inches		3 inches		4 inches	
						O. b.	I. b.	O. b.	I. b.	O. b.	I. b.	O. b.	I. b.
Yrs	Ft	In	No.	Ft²	Ft	— — — — — — — — — — — — Ft³ — — — — — — — — — — — — —							
10	28	1	2	.0	6	'0	0						
		2	19	.4	14	4	2						
		3	79	3.9	19	40	28						
		4	177	15.4	23	187	132						
		5	231	31.5	25	409	290	374	234	327	199	222	122
		6	150	29.5	26	395	282	371	254	341	232	274	182
		7	39	10.4	28	150	107	143	103	136	97	118	85
		8	3	1.0	28	15	11	14	11	14	10	13	9
			700	92.2		1200	852	902	602	818	538	627	398
					MEAN DIA. 4.9 INCHES, WEIBULL PARAMETERS A= .1, B= 5.06, C= 4.56								
15	41	2	3	.1	18	1	0						
		3	24	1.2	26	16	11						
		4	78	6.8	31	110	78						
		5	154	21.0	34	369	263	345	232	314	209	245	157
		6	195	38.3	37	728	521	694	498	657	470	571	406
		7	146	39.0	39	780	559	751	557	725	537	662	489
		8	58	20.2	40	414	297	401	297	392	296	367	277
		9	10	4.4	41	93	66	90	66	89	66	84	64
		10	1	.5	42	12	8	11	8	11	8	11	8
			669	131.6		2523	1803	2292	1658	2188	1586	1940	1401
					MEAN DIA. 6.0 INCHES, WEIBULL PARAMETERS A= 1.1, B= 5.21, C= 4.06								
20	51	3	7	.3	30	5	4						
		4	35	3.1	37	59	42						
		5	88	12.0	41	254	181	239	166	222	153	182	124
		6	143	28.1	45	648	464	622	455	595	435	533	388
		7	158	42.2	47	1016	728	982	728	956	717	888	666
		8	112	39.1	49	979	703	952	703	935	703	889	677
		9	48	21.2	51	552	397	539	397	532	397	513	395
		10	11	6.0	52	159	114	156	114	154	114	150	114
		11	1	.7	53	18	13	17	13	17	13	17	13
			603	152.7		3690	2646	3507	2576	3411	2532	3172	2377
					MEAN DIA. 6.8 INCHES, WEIBULL PARAMETERS A= 1.9, B= 5.24, C= 3.69								
25	60	3	1	.0	33	1	1						
		4	14	1.2	41	26	18						
		5	46	6.3	46	149	106	141	99	132	93	111	77
		6	89	17.5	50	448	321	431	318	415	306	376	277
		7	121	32.3	54	893	641	866	641	846	640	796	601
		8	118	41.2	56	1179	846	1148	846	1131	846	1083	829
		9	80	35.3	58	1047	752	1022	752	1013	752	982	752
		10	36	19.6	60	601	432	588	432	585	432	572	432
		11	10	6.6	61	205	148	201	148	201	148	197	148
		12	2	1.6	62	50	36	49	36	49	36	48	36
			517	161.7		4599	3301	4446	3272	4372	3253	4165	3152
					MEAN DIA. 7.6 INCHES, WEIBULL PARAMETERS A= 2.6, B= 5.36, C= 3.40								

46

Age	Av. dom. ht.	D.b.h.	Stems per acre	Basal area	Av. ht.	All trees, total stem		5-inch class and greater to o.b. tops of—					
								2 inches		3 inches		4 inches	
						O. b.	I. b.	O. b.	I. b.	O. b.	I. b.	O. b.	I. b.
Yrs	Ft	In	No.	Ft 2	Ft	— — — — — — — — — — — — Ft 3 — — — — — — — — — — — —							
30	67	4	4	.3	43	8	6						
		5	21	2.9	49	72	52	69	49	65	46	55	39
		6	49	9.6	54	266	191	256	190	248	184	227	168
		7	77	20.6	58	610	438	592	438	580	438	548	416
		8	92	32.1	61	1001	718	976	718	963	718	926	711
		9	83	36.7	64	1198	860	1171	860	1162	860	1131	860
		10	57	31.1	66	1047	752	1025	752	1021	752	1001	752
		11	29	19.1	67	654	470	641	470	640	470	630	470
		12	10	7.9	69	276	199	271	199	271	199	268	199
		13	3	2.8	70	99	71	97	71	97	71	96	71
			425	163.0		5231	3757	5098	3747	5047	3738	4882	3686

MEAN DIA. 8.4 INCHES, WEIBULL PARAMETERS A= 3.2, B= 5.64, C= 3.17

Age	Av. dom. ht.	D.b.h.	Stems per acre	Basal area	Av. ht.	All trees, total stem		2 inches O.b.	2 inches I.b.	3 inches O.b.	3 inches I.b.	4 inches O.b.	4 inches I.b.
35	74	4	1	.1	43	2	1						
		5	9	1.2	51	32	23	31	22	29	21	25	18
		6	24	4.7	56	135	97	130	97	126	94	116	86
		7	42	11.2	61	350	251	340	251	334	251	316	240
		8	57	19.9	64	650	467	634	467	627	467	604	465
		9	64	28.3	67	967	694	946	694	939	694	915	694
		10	57	31.1	69	1094	786	1072	786	1068	786	1048	786
		11	42	27.7	71	1004	721	985	721	983	721	969	721
		12	25	19.6	73	731	525	718	525	718	525	710	525
		13	12	11.1	74	417	300	410	300	410	300	407	300
		14	4	4.3	76	166	119	163	119	163	119	162	119
		15	1	1.2	77	48	35	47	35	47	35	47	35
			338	160.4		5596	4019	5476	4017	5444	4013	5319	3969

MEAN DIA. 9.3 INCHES, WEIBULL PARAMETERS A= 3.6, B= 6.16, C= 2.97

Age	Av. dom. ht.	D.b.h.	Stems per acre	Basal area	Av. ht.	All trees, total stem		2 inches O.b.	2 inches I.b.	3 inches O.b.	3 inches I.b.	4 inches O.b.	4 inches I.b.
40	79	5	3	.4	51	11	8	10	7	10	7	8	6
		6	10	2.0	57	57	41	55	41	54	40	49	37
		7	20	5.3	62	169	122	165	122	162	122	153	117
		8	30	10.5	66	353	253	344	253	340	253	329	253
		9	38	16.8	69	591	425	578	425	574	425	560	425
		10	43	23.5	72	861	619	844	619	841	619	826	619
		11	38	25.1	74	946	680	929	680	927	680	915	680
		12	31	24.3	76	943	678	927	678	927	678	918	678
		13	22	20.3	78	806	580	792	580	792	580	788	580
		14	14	15.0	79	603	433	593	433	593	433	591	433
		15	7	8.6	81	355	255	349	255	349	255	349	255
		16	3	4.2	82	175	126	172	126	172	126	172	126
		17	1	1.6	83	67	48	66	48	66	48	66	48
			260	157.5		5937	4268	5824	4267	5807	4266	5724	4257

MEAN DIA. 10.5 INCHES, WEIBULL PARAMETERS A= 4.1, B= 7.01, C= 2.80

SI 60
1,000 stems per acre

Age	Av. dom. ht.	D.b.h.	Stems per acre	Basal area	Av. ht.	All trees, total stem		5-inch class and greater to o.b. tops of—					
								2 inches		3 inches		4 inches	
						O.b.	I.b.	O.b.	I.b.	O.b.	I.b.	O.b.	I.b.
Yrs	Ft	In	No.	Ft²	Ft	Ft³							
10	28	1	4	.0	7	0	0						
		2	41	.9	16	9	6						
		3	152	7.5	21	85	59						
		4	293	25.6	24	323	227						
		5	287	39.1	26	528	375	485	306	426	263	296	167
		6	116	22.8	27	317	226	298	205	275	188	223	150
		7	14	3.7	28	54	38	51	37	49	35	43	30
			907	99.6		1316	931	834	548	750	486	562	347
						MEAN DIA. 4.5 INCHES, WEIBULL PARAMETERS A= .1, B= 4.61, C= 4.43							
15	41	2	6	.1	20	2	1						
		3	46	2.3	27	33	23						
		4	139	12.1	32	203	143						
		5	242	33.0	36	613	437	575	390	526	355	418	274
		6	250	49.1	38	959	685	914	658	867	623	757	540
		7	135	36.1	40	740	530	713	529	689	511	630	467
		8	33	11.5	41	242	173	234	173	229	173	215	162
		9	3	1.3	42	28	20	28	20	27	20	26	20
			854	145.5		2820	2012	2464	1770	2338	1682	2046	1463
						MEAN DIA. 5.6 INCHES, WEIBULL PARAMETERS A= 1.1, B= 4.76, C= 3.95							
20	51	3	13	.6	32	11	8						
		4	63	5.5	38	109	77						
		5	146	19.9	43	441	315	417	291	388	270	323	221
		6	212	41.6	46	982	703	942	691	904	662	811	592
		7	190	50.8	49	1274	913	1232	913	1200	903	1120	841
		8	102	35.6	50	910	653	885	653	869	653	827	631
		9	29	12.8	52	340	244	332	244	328	244	317	244
		10	4	2.2	53	59	42	58	42	57	42	56	42
			759	169.0		4126	2955	3866	2834	3746	2774	3454	2571
						MEAN DIA. 6.4 INCHES, WEIBULL PARAMETERS A= 1.9, B= 4.79, C= 3.60							
25	60	3	2	.1	35	2	1						
		4	25	2.2	43	49	35						
		5	78	10.6	48	263	188	249	177	234	166	200	140
		6	139	27.3	52	728	521	700	518	675	500	615	454
		7	168	44.9	55	1263	906	1225	906	1198	906	1127	852
		8	134	46.8	58	1386	995	1350	995	1331	995	1278	979
		9	70	30.9	59	932	669	910	669	902	669	875	669
		10	22	12.0	61	373	268	366	268	364	268	356	268
		11	4	2.6	62	83	60	82	60	82	60	80	60
			642	177.4		5079	3643	4882	3593	4786	3564	4531	3422
						MEAN DIA. 7.1 INCHES, WEIBULL PARAMETERS A= 2.6, B= 4.87, C= 3.32							

Age	Av. dom. ht.	D.b.h.	Stems per acre	Basal area	Av. ht.	All trees, total stem		5-inch class and greater to o.b. tops of—					
						O.b.	I.b.	2 inches		3 inches		4 inches	
								O.b.	I.b.	O.b.	I.b.	O.b.	I.b.
Yrs	Ft	In	No.	Ft 2	Ft	------------- Ft 3 -------------							
30	67	4	8	.7	44	16	11						
		5	36	4.9	51	129	92	122	87	116	82	100	70
		6	79	15.5	55	437	313	421	313	407	303	373	277
		7	114	30.5	59	919	659	892	659	874	659	827	628
		8	121	42.2	62	1338	960	1304	960	1288	960	1240	952
		9	91	40.2	64	1313	943	1284	943	1274	943	1240	943
		10	49	26.7	66	900	646	882	646	877	646	860	646
		11	18	11.9	68	412	296	404	296	403	296	397	296
		12	4	3.1	69	111	79	109	79	108	79	107	79
		13	1	.9	70	33	24	32	24	32	24	32	24
			521	176.7		5608	4023	5450	4007	5379	3992	5176	3915
		MEAN DIA. 7.9 INCHES, WEIBULL PARAMETERS A= 3.2, B= 5.10. C= 3.11											
35	74	4	2	.2	45	4	3						
		5	15	2.0	53	56	40	53	38	50	36	44	31
		6	39	7.7	58	228	163	220	163	213	159	196	146
		7	65	17.4	62	551	395	535	395	525	395	498	379
		8	83	29.0	66	977	701	953	701	942	701	909	700
		9	81	35.8	68	1242	892	1215	892	1206	892	1176	892
		10	62	33.8	71	1225	880	1200	880	1196	880	1174	880
		11	37	24.4	73	909	653	892	653	891	653	879	653
		12	17	13.4	74	504	362	495	362	495	362	490	362
		13	6	5.5	76	214	154	211	154	211	154	209	154
		14	1	1.1	77	42	30	41	30	41	30	41	30
			408	170.2		5952	4273	5815	4268	5770	4262	5616	4227
		MEAN DIA. 8.7 INCHES, WEIBULL PARAMETERS A= 3.6, B= 5.53, C= 2.93											
40	79	5	5	.7	53	19	13	18	13	17	12	15	10
		6	17	3.3	59	101	72	97	72	94	71	87	65
		7	32	8.6	64	280	201	272	201	267	201	254	193
		8	45	15.7	68	545	392	532	392	527	392	509	392
		9	53	23.4	71	848	609	830	609	825	609	805	609
		10	52	28.4	73	1056	759	1035	759	1032	759	1014	759
		11	43	28.4	75	1085	780	1065	780	1064	780	1050	780
		12	30	23.6	77	925	665	909	665	909	665	900	665
		13	18	16.6	79	668	480	657	480	657	480	653	480
		14	9	9.6	80	392	282	386	282	386	282	385	282
		15	4	4.9	82	205	147	202	147	202	147	202	147
		16	1	1.4	83	59	42	58	42	58	42	58	42
			309	164.5		6183	4442	6061	4442	6038	4440	5932	4424
		MEAN DIA. 9.9 INCHES, WEIBULL PARAMETERS A= 4.1, B= 6.26, C= 2.77											

SI 60

1,250 stems per acre

Age	Av. dom. ht.	D.b.h.	Stems per acre	Basal area	Av. ht.	All trees, total stem		5-inch class and greater to o.b. tops of—					
								2 inches		3 inches		4 inches	
						O. b.	I. b.	O. b.	I. b.	O. b.	I. b.	O. b.	I. b.
Yrs	Ft	In	No.	Ft 2	Ft	------------ Ft 3 ------------							
10	28	1	8	.0	7	0	0						
		2	72	1.6	16	15	10						
		3	240	11.6	21	134	93						
		4	399	34.8	24	440	310						
		5	303	41.3	26	558	396	512	323	450	278	313	176
		6	81	15.9	27	222	158	208	143	192	131	156	104
		7	5	1.3	28	19	14	18	13	17	12	15	11
			----	-----		----	----	----	----	----	----	----	----
			1108	106.8		1388	981	738	479	659	421	484	291
			MEAN DIA. 4.2 INCHES, WEIBULL PARAMETERS A= .1, B= 4.31, C= 4.31										
15	41	2	11	.2	21	3	2						
		3	75	3.7	28	55	38						
		4	210	18.3	33	315	223						
		5	323	44.0	36	819	584	767	521	703	473	558	365
		6	277	54.4	38	1062	759	1013	729	960	690	838	598
		7	115	30.7	40	630	451	607	451	587	436	537	398
		8	20	7.0	41	146	105	142	105	139	105	130	98
		9	1	.4	42	9	7	9	7	9	7	9	7
			----	-----		----	----	----	----	----	----	----	----
			1032	158.8		3039	2169	2538	1813	2398	1711	2072	1466
			MEAN DIA. 5.3 INCHES, WEIBULL PARAMETERS A= 1.1, B= 4.47, C= 3.85										
20	51	2	1	.0	24	0	0						
		3	22	1.1	33	19	13						
		4	97	8.5	39	172	122						
		5	208	28.4	43	628	449	594	415	553	385	460	315
		6	267	52.4	47	1264	905	1213	891	1164	855	1048	766
		7	205	54.8	49	1374	985	1330	985	1295	975	1208	908
		8	87	30.4	51	792	568	770	568	757	568	721	550
		9	18	8.0	52	211	152	206	152	204	152	197	151
		10	2	1.1	53	30	21	29	21	29	21	28	21
			----	-----		----	----	----	----	----	----	----	----
			907	184.6		4490	3215	4142	3032	4002	2956	3662	2711
			MEAN DIA. 6.1 INCHES, WEIBULL PARAMETERS A= 1.9, B= 4.49, C= 3.51										
25	60	3	4	.2	36	4	3						
		4	39	3.4	43	76	54						
		5	114	15.5	49	392	280	372	264	350	248	300	210
		6	188	36.9	53	1003	718	966	716	932	691	851	629
		7	202	54.0	55	1519	1089	1473	1089	1440	1089	1356	1025
		8	138	48.2	58	1427	1025	1391	1025	1371	1025	1316	1008
		9	58	25.6	60	785	564	767	564	760	564	738	564
		10	14	7.6	61	238	171	233	171	231	171	226	171
		11	2	1.3	62	42	30	41	30	41	30	40	30
			----	-----		----	----	----	----	----	----	----	----
			759	192.8		5486	3934	5243	3859	5125	3818	4827	3637
			MEAN DIA. 6.8 INCHES, WEIBULL PARAMETERS A= 2.6, B= 4.55, C= 3.25										

Age	Av. dom. ht.	D.b.h.	Stems per acre	Basal area	Av. ht.	All trees, total stem		5-inch class and greater to o.b. tops of—					
								2 inches		3 inches		4 inches	
						O. b.	I. b.	O. b.	I. b.	O. b.	I. b.	O. b.	I. b.
Yrs	Ft	In	No.	Ft 2	Ft	– – – – – – – – – – – – Ft 3 – – – – – – – – – – – –							
30	67	4	13	1.1	46	27	19						
		5	54	7.4	52	197	141	187	134	177	126	153	109
		6	111	21.8	57	636	456	614	456	594	443	547	406
		7	149	39.8	60	1222	876	1187	876	1163	876	1101	836
		8	139	48.5	63	1561	1121	1523	1121	1504	1121	1449	1113
		9	90	39.8	65	1319	947	1290	947	1280	947	1246	947
		10	40	21.8	67	746	536	731	536	727	536	713	536
		11	11	7.3	69	255	184	251	184	250	184	246	184
		12	2	1.6	70	56	40	55	40	55	40	54	40
			609	189.0		6019	4320	5838	4294	5750	4273	5509	4171
MEAN DIA. 7.5 INCHES, WEIBULL PARAMETERS A= 3.2, B= 4.74, C= 3.06													
35	74	4	3	.3	46	6	4						
		5	22	3.0	53	82	58	78	56	74	53	64	45
		6	56	11.0	59	332	238	321	238	311	232	287	214
		7	89	23.8	63	766	550	745	550	731	550	694	528
		8	102	35.6	66	1200	862	1171	862	1158	862	1117	860
		9	92	40.6	69	1431	1028	1400	1028	1391	1028	1357	1028
		10	61	33.3	71	1205	866	1181	866	1176	866	1155	866
		11	31	20.5	73	762	547	747	547	746	547	736	547
		12	11	8.6	74	326	234	320	234	320	234	317	234
		13	3	2.8	76	107	77	105	77	105	77	105	77
		14	1	1.1	77	42	30	41	30	41	30	41	30
			471	180.5		6259	4494	6109	4488	6053	4479	5873	4429
MEAN DIA. 8.4 INCHES, WEIBULL PARAMETERS A= 3.6, B= 5.11, C= 2.89													
40	79	5	8	1.1	55	31	22	29	21	28	20	24	17
		6	24	4.7	61	147	105	142	105	138	103	128	95
		7	44	11.8	65	391	280	380	280	373	280	355	271
		8	61	21.3	69	750	539	732	539	725	539	701	539
		9	67	29.6	72	1088	781	1064	781	1058	781	1033	781
		10	59	32.2	75	1231	884	1207	884	1203	884	1183	884
		11	44	29.0	77	1140	819	1119	819	1118	819	1104	819
		12	27	21.2	79	854	614	839	614	839	614	832	614
		13	13	12.0	80	489	351	480	351	480	351	478	351
		14	5	5.3	81	221	159	217	159	217	159	217	159
		15	2	2.5	83	104	75	102	75	102	75	102	75
			354	170.7		6446	4629	6311	4628	6281	4625	6157	4605
MEAN DIA. 9.4 INCHES, WEIBULL PARAMETERS A= 4.1, B= 5.77, C= 2.74													

SI 60

1,500 stems per acre

Age	Av. dom. ht.	D.b.h.	Stems per acre	Basal area	Av. ht.	All trees, total stem		5-inch class and greater to o.b. tops of—					
								2 inches		3 inches		4 inches	
						O. b.	I. b.	O. b.	I. b.	O. b.	I. b.	O. b.	I. b.
Yrs	Ft	In	No.	Ft 2	Ft				Ft 3				
10	28	1	13	.1	8	1	0						
		2	111	2.4	17	25	16						
		3	339	16.6	21	190	131						
		4	489	42.7	24	539	379						
		5	296	40.4	26	545	387	500	316	440	272	306	172
		6	56	11.0	27	153	109	144	99	133	91	108	72
		7	2	.5	28	8	5	7	5	7	5	6	4
			1306	113.7		1461	1027	651	420	580	368	420	248

MEAN DIA. 4.0 INCHES, WEIBULL PARAMETERS A= .1, B= 4.09, C= 4.19

Age	Av. dom. ht.	D.b.h.	Stems per acre	Basal area	Av. ht.	All trees, total stem O.b.	I.b.	2 in O.b.	I.b.	3 in O.b.	I.b.	4 in O.b.	I.b.
15	41	2	17	.4	22	5	3						
		3	111	5.4	30	87	61						
		4	285	24.9	34	441	312						
		5	396	54.0	37	1031	735	967	660	889	602	711	470
		6	288	56.5	39	1133	810	1081	780	1027	740	900	645
		7	97	25.9	41	545	390	525	390	508	378	466	346
		8	12	4.2	42	90	65	87	65	85	65	80	61
			1206	171.3		3332	2376	2660	1695	2509	1785	2157	1522

MEAN DIA. 5.1 INCHES, WEIBULL PARAMETERS A= 1.1, B= 4.25, C= 3.74

Age	Av. dom. ht.	D.b.h.	Stems per acre	Basal area	Av. ht.	All O.b.	I.b.	2 in O.b.	I.b.	3 in O.b.	I.b.	4 in O.b.	I.b.
20	51	2	1	.0	25	0	0						
		3	33	1.6	34	29	20						
		4	136	11.9	40	247	175						
		5	270	36.8	44	834	596	789	553	736	514	616	424
		6	314	61.7	47	1487	1064	1427	1048	1369	1005	1233	901
		7	209	55.9	49	1401	1004	1356	1004	1320	994	1232	925
		8	74	25.8	51	673	483	655	483	644	483	613	468
		9	12	5.3	52	141	101	137	101	136	101	131	101
		10	1	.5	53	15	11	14	11	14	11	14	11
			1050	199.5		4827	3454	4378	3200	4219	3108	3839	2830

MEAN DIA. 5.9 INCHES, WEIBULL PARAMETERS A= 1.9, B= 4.27, C= 3.42

Age	Av. dom. ht.	D.b.h.	Stems per acre	Basal area	Av. ht.	All O.b.	I.b.	2 in O.b.	I.b.	3 in O.b.	I.b.	4 in O.b.	I.b.
25	60	3	6	.3	37	6	4						
		4	56	4.9	44	111	79						
		5	153	20.9	49	526	376	500	355	470	333	402	282
		6	234	45.9	53	1248	894	1202	891	1160	860	1059	783
		7	227	60.7	56	1738	1246	1686	1246	1649	1246	1554	1176
		8	136	47.5	58	1407	1010	1371	1010	1351	1010	1297	994
		9	48	21.2	60	650	466	635	466	629	466	611	466
		10	9	4.9	61	153	110	150	110	149	110	145	110
		11	1	.7	62	21	15	20	15	20	15	20	15
			870	206.9		5860	4200	5564	4093	5428	4040	5088	3826

MEAN DIA. 6.6 INCHES, WEIBULL PARAMETERS A= 2.6, B= 4.32, C= 3.19

Age	Av. dom. ht.	D.b.h.	Stems per acre	Basal area	Av. ht.	All trees, total stem		5-inch class and greater to o.b. tops of—					
								2 inches		3 inches		4 inches	
						O. b.	I. b.	O. b.	I. b.	O. b.	I. b.	O. b.	I. b.
Yrs	Ft	In	No.	Ft²	Ft	— — — — — — — — — — — — Ft³ — — — — — — — — — — — —							
30	67	4	18	1.6	47	38	27						
		5	74	10.1	53	275	197	262	188	248	177	215	153
		6	143	28.1	57	820	587	791	587	766	571	704	524
		7	180	48.1	61	1501	1076	1458	1076	1429	1076	1355	1030
		8	151	52.7	63	1696	1218	1654	1218	1634	1218	1574	1210
		9	86	38.0	66	1280	919	1252	919	1242	919	1210	919
		10	32	17.5	67	597	429	584	429	582	429	571	429
		11	7	4.6	69	163	117	159	117	159	117	157	117
		12	1	.8	70	28	20	28	20	28	20	27	20
			692	201.4		6398	4590	6188	4554	6088	4527	5813	4402

MEAN DIA. 7.3 INCHES, WEIBULL PARAMETERS A= 3.2, B= 4.48, C= 3.01

Age	Av. dom. ht.	D.b.h.	Stems per acre	Basal area	Av. ht.	O. b.	I. b.	O. b.	I. b.	O. b.	I. b.	O. b.	I. b.
35	74	4	4	.3	48	9	6						
		5	30	4.1	55	116	83	110	79	105	75	91	65
		6	74	14.5	60	447	320	431	320	418	313	387	289
		7	112	29.9	64	979	703	952	703	935	703	889	677
		8	121	42.2	67	1445	1038	1410	1038	1395	1038	1347	1037
		9	98	43.3	70	1547	1111	1513	1111	1503	1111	1467	1111
		10	58	31.6	72	1162	835	1139	835	1135	835	1115	835
		11	25	16.5	74	623	447	611	447	610	447	602	447
		12	7	5.5	76	213	153	209	153	209	153	207	153
		13	2	1.8	77	72	52	71	52	71	52	71	52
			531	189.9		6613	4748	6446	4738	6381	4727	6176	4666

MEAN DIA. 8.1 INCHES, WEIBULL PARAMETERS A= 3.6, B= 4.81, C= 2.85

Age	Av. dom. ht.	D.b.h.	Stems per acre	Basal area	Av. ht.	O. b.	I. b.	O. b.	I. b.	O. b.	I. b.	O. b.	I. b.
40	79	5	10	1.4	55	39	28	37	26	35	25	30	22
		6	32	6.3	61	196	141	190	141	184	138	170	127
		7	57	15.2	66	514	369	500	369	491	369	468	357
		8	75	26.2	69	922	662	901	662	891	662	862	662
		9	78	34.5	72	1266	909	1239	909	1231	909	1203	909
		10	64	34.9	75	1335	959	1309	959	1305	959	1283	959
		11	42	27.7	77	1088	782	1068	782	1067	782	1054	782
		12	23	18.1	79	727	523	715	523	715	523	709	523
		13	10	9.2	80	376	270	369	270	369	270	368	270
		14	3	3.2	82	134	96	132	96	132	96	132	96
		15	1	1.2	83	52	37	51	37	51	37	51	37
			395	177.9		6649	4776	6511	4774	6471	4770	6330	4744

MEAN DIA. 9.1 INCHES, WEIBULL PARAMETERS A= 4.1, B= 5.40, C= 2.72

SI 60

1,750 stems per acre

Age	Av. dom. ht.	D.b.h.	Stems per acre	Basal area	Av. ht.	All trees, total stem		5-inch class and greater to o.b. tops of—					
								2 inches		3 inches		4 inches	
						O.b.	I.b.	O.b.	I.b.	O.b.	I.b.	O.b.	I.b.
Yrs	Ft	In	No.	Ft²	Ft	— — — — — — — — — — — Ft³ — — — — — — — — — — —							
10	28	1	20	.1	8	1	0						
		2	159	3.5	17	36	23						
		3	444	21.8	22	260	180						
		4	560	48.9	24	617	435						
		5	277	37.8	26	510	362	468	296	411	254	286	161
		6	40	7.9	28	113	81	107	74	99	68	81	55
		7	1	.3	28	4	3	4	3	3	2	3	2
			----	-----		----	----	----	----	----	----	----	----
			1501	120.1		1541	1084	579	373	513	324	370	218
		MEAN DIA. 3.8 INCHES, WEIBULL PARAMETERS A= .1, B= 3.91, C= 4.08											
15	41	2	25	.5	23	7	5						
		3	152	7.5	30	119	83						
		4	362	31.6	34	560	397						
		5	455	62.0	37	1185	845	1111	758	1021	692	817	540
		6	290	56.9	39	1141	816	1089	786	1034	745	907	649
		7	82	21.9	41	460	330	444	330	429	319	394	292
		8	8	2.8	42	60	43	58	43	57	43	54	40
			----	-----		----	----	----	----	----	----	----	----
			1374	183.3		3532	2519	2702	1917	2541	1799	2172	1521
		MEAN DIA. 4.9 INCHES, WEIBULL PARAMETERS A= 1.1, B= 4.08, C= 3.64											
20	51	2	2	.0	26	1	0						
		3	46	2.3	35	42	29						
		4	179	15.6	41	332	236						
		5	331	45.1	45	1046	747	990	696	926	648	778	538
		6	350	68.7	48	1692	1211	1625	1196	1561	1148	1409	1032
		7	208	55.6	50	1423	1020	1377	1020	1342	1011	1254	943
		8	63	22.0	52	584	419	569	419	559	419	533	407
		9	9	4.0	53	108	77	105	77	104	77	100	77
			----	-----		----	----	----	----	----	----	----	----
			1188	213.3		5228	3739	4666	3408	4492	3303	4074	2997
		MEAN DIA. 5.7 INCHES, WEIBULL PARAMETERS A= 1.9, B= 4.09, C= 3.34											
25	60	3	9	.4	38	9	6						
		4	75	6.5	46	156	111						
		5	193	26.3	51	690	493	657	469	620	441	534	377
		6	277	54.4	54	1505	1078	1450	1076	1401	1040	1281	949
		7	246	65.7	57	1917	1375	1860	1375	1820	1375	1718	1301
		8	131	45.7	59	1378	989	1343	989	1325	989	1272	976
		9	40	17.7	61	550	395	538	395	533	395	518	395
		10	6	3.3	62	104	74	101	74	101	74	99	74
			----	-----		----	----	----	----	----	----	----	----
			977	220.1		6309	4521	5949	4378	5800	4314	5422	4072
		MEAN DIA. 6.4 INCHES, WEIBULL PARAMETERS A= 2.6, B= 4.14, C= 3.12											

Age	Av. dom. ht.	D.b.h.	Stems per acre	Basal area	Av. ht.	All trees, total stem		5-inch class and greater to o.b. tops of—					
								2 inches		3 inches		4 inches	
						O. b.	I. b.	O. b.	I. b.	O. b.	I. b.	O. b.	I. b.
Yrs	Ft	In	No.	Ft²	Ft	– – – – – – – – – – – – – Ft³ – – – – – – – – – – – – – –							
30	67	4	24	2.1	48	52	37						
		5	95	13.0	53	353	252	336	241	318	228	276	196
		6	176	34.6	58	1027	736	991	736	960	716	884	658
		7	205	54.8	61	1709	1226	1660	1226	1628	1226	1543	1173
		8	159	55.5	64	1814	1302	1770	1302	1748	1302	1685	1296
		9	80	35.3	66	1191	855	1164	855	1156	855	1126	855
		10	26	14.2	68	492	353	482	353	480	353	471	353
		11	5	3.3	69	116	83	114	83	114	83	112	83
		12	1	.8	70	28	20	28	20	28	20	27	20
			771	213.5		6782	4864	6545	4816	6432	4783	6124	4634
			MEAN DIA. 7.1 INCHES, WEIBULL PARAMETERS A= 3.2, B= 4.28, C= 2.96										
35	74	4	5	.4	49	11	8						
		5	39	5.3	56	153	109	146	105	139	100	121	87
		6	92	18.1	61	564	404	545	404	529	396	490	366
		7	134	35.8	65	1190	854	1157	854	1137	854	1082	824
		8	137	47.8	68	1661	1192	1621	1192	1603	1192	1550	1192
		9	101	44.6	70	1594	1145	1560	1145	1549	1145	1512	1145
		10	53	28.9	72	1062	763	1041	763	1037	763	1019	763
		11	20	13.2	74	498	358	489	358	488	358	482	358
		12	5	3.9	76	152	109	149	109	149	109	148	109
		13	1	.9	77	36	26	36	26	36	26	35	26
			587	199.0		6921	4968	6744	4956	6667	4943	6439	4870
			MEAN DIA. 7.9 INCHES, WEIBULL PARAMETERS A= 3.6, B= 4.58, C= 2.82										
40	79	4	1	.1	49	2	2						
		5	13	1.8	57	52	37	50	36	47	34	41	30
		6	41	8.1	62	256	183	247	183	240	180	222	166
		7	71	19.0	67	650	466	632	466	621	466	592	452
		8	87	30.4	71	1101	791	1075	791	1064	791	1030	791
		9	86	38.0	74	1435	1031	1404	1031	1396	1031	1365	1031
		10	66	36.0	76	1395	1003	1368	1003	1364	1003	1342	1003
		11	40	26.4	78	1050	755	1031	755	1030	755	1017	755
		12	19	14.9	80	609	437	598	437	598	437	593	437
		13	7	6.5	81	266	191	262	191	262	191	261	191
		14	2	2.1	83	90	65	89	65	89	65	89	65
			433	183.2		6906	4961	6756	4958	6711	4953	6552	4921
			MEAN DIA. 8.8 INCHES, WEIBULL PARAMETERS A= 4.1, B= 5.12, C= 2.69										

SI 60
2,000 stems per acre

Age	Av. dom. ht.	D.b.h.	Stems per acre	Basal area	Av. ht.	All trees, total stem		5-inch class and greater to o.b. tops of—					
								2 inches		3 inches		4 inches	
						O. b.	I. b.	O. b.	I. b.	O. b.	I. b.	O. b.	I. b.
Yrs	Ft	In	No.	Ft²	Ft	— — — — — — — — — — — Ft³ — — — — — — — — — — —							
10	28	1	30	.2	8	2	1						
		2	216	4.7	17	48	31						
		3	550	27.0	22	322	222						
		4	611	53.3	25	700	494						
		5	255	34.8	26	469	333	431	272	379	234	263	148
		6	29	5.7	28	82	59	77	54	72	49	59	40
		7	1	.3	28	4	3	4	3	3	2	3	2
			----	-----		----	----	----	----	----	----	----	----
			1692	125.9		1627	1143	512	329	454	285	325	190
			MEAN DIA. 3.7 INCHES, WEIBULL PARAMETERS A= .1, B= 3.76, C= 3.96										
15	41	2	36	.8	23	10	7						
		3	199	9.8	30	156	109						
		4	440	38.4	35	700	496						
		5	503	68.6	38	1345	959	1263	866	1164	792	938	624
		6	286	56.2	40	1154	825	1102	797	1048	758	923	663
		7	70	18.7	41	393	282	379	282	367	273	336	250
		8	6	2.1	42	45	32	44	32	43	32	40	30
			-----	-----		----	----	----	----	----	----	----	----
			1540	194.5		3803	2710	2788	1977	2622	1855	2237	1567
			MEAN DIA. 4.8 INCHES, WEIBULL PARAMETERS A= 1.1, B= 3.93, C= 3.54										
20	51	2	2	.0	27	1	0						
		3	61	3.0	36	57	40						
		4	225	19.6	42	428	304						
		5	390	53.2	46	1259	899	1193	841	1118	785	944	655
		6	379	74.4	48	1832	1312	1760	1295	1691	1243	1526	1118
		7	203	54.3	50	1388	995	1344	995	1310	987	1224	921
		8	55	19.2	52	510	366	496	366	488	366	466	356
		9	7	3.1	53	84	60	82	60	81	60	78	60
			-----	-----		----	----	----	----	----	----	----	----
			1322	226.8		5559	3976	4875	3557	4688	3441	4236	3110
			MEAN DIA. 5.6 INCHES, WEIBULL PARAMETERS A= 1.9, B= 3.95, C= 3.26										
25	60	3	12	.6	39	12	9						
		4	96	8.4	46	200	142						
		5	234	31.9	51	837	598	796	568	751	535	648	457
		6	315	61.9	55	1743	1249	1680	1249	1624	1207	1488	1104
		7	259	69.2	57	2018	1447	1958	1447	1917	1447	1808	1370
		8	125	43.6	59	1315	944	1282	944	1264	944	1214	931
		9	34	15.0	61	468	336	457	336	453	336	440	336
		10	5	2.7	62	86	62	84	62	84	62	82	62
			----	-----		----	----	----	----	----	----	----	----
			1080	233.3		6679	4787	6257	4606	6093	4531	5680	4260
			MEAN DIA. 6.3 INCHES, WEIBULL PARAMETERS A= 2.6, B= 3.99, C= 3.06										

	Age	Av. dom. ht.	D.b.h.	Stems per acre	Basal area	Av. ht.	All trees, total stem		5-inch class and greater to o.b. tops of—					
									2 inches		3 inches		4 inches	
							O. b.	I. b.	O. b.	I. b.	O. b.	I. b.	O. b.	I. b.
	Yrs	Ft	In	No.	Ft 2	Ft	— — — — — — — — — — — — — Ft 3 — — — — — — — — — — — — —							
	30	67	3	1	.0	41	1	1						
			4	31	2.7	49	69	49						
4 243 148			5	118	16.1	55	455	325	434	312	411	296	359	257
			6	207	40.6	59	1228	880	1186	880	1150	859	1061	791
· 39 45			7	227	60.7	62	1923	1380	1869	1380	1833	1380	1739	1323
· 3 2			8	163	56.9	65	1889	1356	1843	1356	1821	1356	1757	1352
			9	75	33.1	67	1133	814	1108	814	1100	814	1072	814
15 325 291			10	21	11.5	69	403	290	395	290	393	290	386	290
: 3.96			11	3	2.0	70	71	51	69	51	69	51	68	51
				----	-----		----	----	----	----	----	----	----	----
				846	223.6		7172	5146	6904	5083	6777	5046	6442	4878
				MEAN DIA. 7.0 INCHES, WEIBULL PARAMETERS A= 3.2, B= 4.12, C= 2.91										
	35	74	4	7	.6	50	16	11						
32 938 624			5	49	6.7	57	196	140	187	135	178	128	156	112
58 923 663			6	111	21.8	62	692	496	669	496	650	487	602	450
			7	157	42.0	66	1416	1016	1377	1016	1353	1016	1288	982
75 534 250			8	148	51.7	69	1820	1307	1777	1307	1758	1307	1700	1307
32 45 38			9	101	44.6	72	1639	1178	1604	1178	1595	1178	1557	1178
95 2237 1367			10	48	26.2	74	988	710	969	710	966	710	949	710
: 3.54			11	16	10.6	75	404	290	396	290	396	290	391	290
			12	3	2.4	77	92	66	91	66	91	66	90	66
				----	-----		----	----	----	----	----	----	----	----
				640	206.4		7263	5214	7070	5198	6987	5182	6733	5095
				MEAN DIA. 7.7 INCHES, WEIBULL PARAMETERS A= 3.6, B= 4.39, C= 2.79										
795 944 655	40	79	4	1	.1	49	2	2						
243 1506 1118			5	17	2.3	57	68	49	65	47	62	44	54	39
			6	50	9.8	63	317	227	306	227	298	223	276	207
597 1024 921			7	84	22.4	67	769	552	748	552	735	552	701	535
366 466 366			8	100	34.9	71	1265	909	1236	909	1223	909	1184	909
65 78 60			9	93	41.1	74	1551	1114	1519	1114	1510	1114	1476	1114
			10	66	36.0	76	1395	1003	1368	1003	1364	1003	1342	1003
441 4237 3039			11	37	24.4	78	971	698	953	698	952	698	941	698
C: 3.26			12	16	12.6	80	512	368	504	368	504	368	499	368
			13	5	4.6	81	190	137	187	137	187	137	186	137
			14	1	1.1	83	45	33	44	33	44	33	44	33
535 648 457				----	-----		----	----	----	----	----	----	----	----
1488 1104				470	189.3		7085	5092	6930	5088	6879	5081	6703	5043
1487 1878 1379				MEAN DIA. 8.6 INCHES, WEIBULL PARAMETERS A= 4.1, B= 4.90, C= 2.67										

944 1274 931
536 440 336
62   82   62

4531 3685 4298
C: 3.58

57.

Smalley, Glendon W., and Bailey, Robert L.

    1974. Yield tables and stand structure for shortleaf
          pine plantations in Tennessee, Alabama, and
          Georgia highlands. South. For. Exp. Stn., New
          Orleans, La. 57 p. (USDA For. Serv. Res. Pap.
          SO-97)

Detailed schedules of trees per acre, basal area, mean tree height,
and cubic-foot yields in eight volume categories by 1-inch diameter
classes are presented for all combinations of four site indexes,
seven ages from seed, and six planting densities.

Additional keywords: *Pinus echinata*, diameter distributions,
Weibull function, survival, height-diameter relationships.

Smalley, Glendon W., and Bailey, Robert L.

    1974. Yield tables and stand structure for shortleaf
           pine plantations in Tennessee, Alabama, and
           Georgia highlands. South. For. Exp. Stn., New
           Orleans, La. 57 p. (USDA For. Serv. Res. Pap.
           SO-97)

Detailed schedules of trees per acre, basal area, mean tree height,
and cubic-foot yields in eight volume categories by 1-inch diameter
classes are presented for all combinations of four site indexes,
seven ages from seed, and six planting densities.

Additional keywords: *Pinus echinata*, diameter distributions,
Weibull function, survival, height-diameter relationships.

U.S. Department of Agriculture
Forest Service Research Paper SO-98

Development

of the

Shaping-Lathe Headrig

Peter Koch

OCT 20 1975

Southern Forest Experiment Station
Forest Service
U.S. Department of Agriculture

1974

Contents

Development

of the

Shaping-Lathe Headrig

Peter Koch

Southern Forest Experiment Station
Forest Service
U.S. Department of Agriculture

1974

Development

of the

Shaping-Lathe Headrig

Peter Koch

Southern Forest Experiment Station
Forest Service
U.S. Department of Agriculture

1974

KOCH SHAPING-LATHE HEADRIG

The drawing on the opposite page depicts the commercial version of the Koch shaping-lathe headrig. This first production model, design of which is based on data derived from prototype trials described in the present paper, will be commercially making southern hardwood flakes and pallet cants by early spring of 1975. It carries a 54-inch-long, six-knife cutterhead with 12-inch cutting circle. The cutterhead is turned at 3,600 r/min by a 300-hp motor designed to momentarily carry a 200-percent overload without pullout from synchronous speed. The workpiece is driven from one end with a 5-hp, variable-speed motor that provides rotational speeds from 9 to 27 r/min. The headrig will accept bolts 3.5 to 12 inches in diameter and 40 to 53 inches in length. Feed rate is estimated at six bolts per minute. This initial production model was built under a Southern Forest Experiment Station contract with Stetson-Ross, Seattle, Wn.; funds were provided by the Branch of State and Private Forestry, USDA Forest Service.

Summary of Results From Prototype Trials

The headrig, cutting in the 0-90 mode, requires 3 to 6 seconds to machine end-chucked short hardwood or softwood logs. Output consists of cants having any desired polygonal shape, or round posts, plus flakes or pulp chips.

Removal of flakes leaves cants with excellent surface quality and dimensional accuracy. Up-milling is more practical than down-milling. Water-soaked wood cut at 160°F required 5.5 percent less net cutterhead power than green wood cut at 72°F. Net specific cutting energy showed positive linear correlation with wood specific gravity and negative correlation with flake thickness; for up-milling of hot and cold loblolly pine, sweetgum, hickory, and southern red oak, it averaged 9.92 hp minutes per cubic foot of wood removed as flakes 0.015, 0.025, and 0.035 inch thick.

To manufacture 0.015-inch-thick flakes with a six-knife head rotating at 3,600 r/min, average cutterhead demand will be about 267 hp when machining green, unheated hardwoods of 0.75 density, 12-inch diameter, and 53-inch length into 8- by 8-inch cants. Peak demand, assuming a maximum depth of cut of 3 inches, will be 510 hp.

If flakes 0.030 inch thick are cut with a three-knife head, average cutterhead demand will be about 206 hp when machining 8-inch squares from green, dense 12-inch bolts 53 inches long. Peak demand during 3-inch-deep cuts will be 356 hp.

One-and-one-half horsepower (net delivered to the workpiece spindle at all spindle speeds), should be sufficient to turn against peak forces exerted by the cutterhead. Bolt deflections in bending and torsion should not be severe when cants having diameters above 6 inches are being machined. If bolts are highly eccentric (for example, if 3-inch-deep cuts are needed to make 4-inch-round posts), however, bending deflections may be as much as 1/16- to 1/8-inch, and torsional deflection may be 2 degrees.

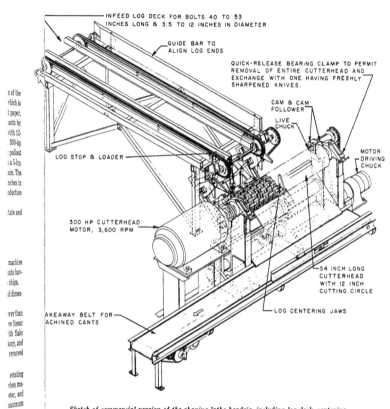

INFEED LOG DECK FOR BOLTS 40 TO 53
INCHES LONG & 3.5 TO 12 INCHES IN DIAMETER

GUIDE BAR TO
ALIGN LOG ENDS

QUICK-RELEASE BEARING CLAMP TO PERMIT
REMOVAL OF ENTIRE CUTTERHEAD AND
EXCHANGE WITH ONE HAVING FRESHLY
SHARPENED KNIVES.

CAM & CAM
FOLLOWER

LIVE
CHUCK

MOTOR
DRIVING
CHUCK

LOG STOP & LOADER

300 HP CUTTERHEAD
MOTOR, 3,600 RPM

54 INCH LONG
CUTTERHEAD
WITH 12 INCH
CUTTING CIRCLE

AKEAWAY BELT FOR
ACHINED CANTS

LOG CENTERING JAWS

Sketch of commercial version of the shaping-lathe headrig, including log deck, centering device, charger, and takeaway conveyor for machined cants. Flakes are blown from the cutterhead hood (removed for purposes of this illustration) for conveying to flakeboard plant. Design feed rate is six logs per minute. Smoothly machined cants will have the shape and dimensions of replaceable cams mounted on the workpiece spindle. (Drawing from Stetson-Ross, Seattle.)

iii

Koch, Peter

    1974. Development of the shaping-lathe headrig. South. For. Exp. Stn., New Orleans, La. 20 p. (USDA For. Serv. Res. Pap. SO-98)

A prototype yielded data for designing a commercial headrig capable of machining end-chucked hardwood or softwood bolts into cants for pallet boards or posts of any desired round or polygonal shape. Wood removed is in the form of flakes or pulp chips. Net specific cutting energy showed positive linear correlation with wood specific gravity and negative correlation with flake thickness. A 54-inch cutterhead carrying six knives and rotating at 3,600 r/min will average 270 horsepower when machining green, unheated hardwoods of 0.75 density, 12-inch diameter, and 53-inch length into cants 8 inches square and flakes 0.015 inch thick; peak power requirements for cuts 3 inches deep will be about 510 horsepower.

Additional keywords: Southern pine, *Pinus taeda, Liquidambar styraciflua, Quercus falcata, Carya* spp., power requirements, specific cutting energy, chipping headrigs and edgers, pallets, short logs, bolts, machining, milling, cants, lumber, flakers, flaking, chippers, chipping, pulp chips, surface quality, rake angle, clearance angle, sharpness angle, chip quality, flakeboard, particleboard.

Koch

adapted for use in structural flakeboard. Such flakes are commonly 2 to 3 inches long, 0.015 to 0.045 inch thick, and perhaps 3/8-inch wide.

Three current trends in forest resources and markets favor the development and application of a shaping-lathe headrig. First, supplies of high-quality logs have diminished, while small hardwoods of low quality remain relatively abundant throughout much of Eastern United States. Second, demand for pallet lumber has risen rapidly in recent years, and the headrig is particularly well suited to convert small hardwoods into pallet cants. Finally, structural exterior flakeboard is being test-marketed and appears likely to take over part of the market for sheathing grades of plywood. The product is currently manufactured from softwoods and aspen (*Populus* sp.), but research indicates that it can be made from mixed eastern hardwoods.

This paper describes a second prototype of sufficient size to provide information for designing and building a commercial model suited to hardwood logs 40 to 53 inches long and 3.5 to 12 inches in diameter.

The new prototype, built under contract with Stetson-Ross of Seattle, Washington, is capable of chucking a 12-inch-diameter bolt 6.5 inches long (fig. 2). Bolts to be machined are clamped in the chucks of the workpiece spindle, which turns at about 15 r/min. Attached to the spindle is a replaceable cam having the shape and dimensions of the desired cant. The cam rotates and moves with the workpiece until it strikes a follower aligned with the cutterhead. As the workpiece makes a single revolution, the center distance between cutterhead and workpiece changes in response to the cam, and the workpiece (log) is machined to the shape and dimensions of the cam. Since the log makes only a single revolution while being sized, machining time is brief—approximately 4 seconds.

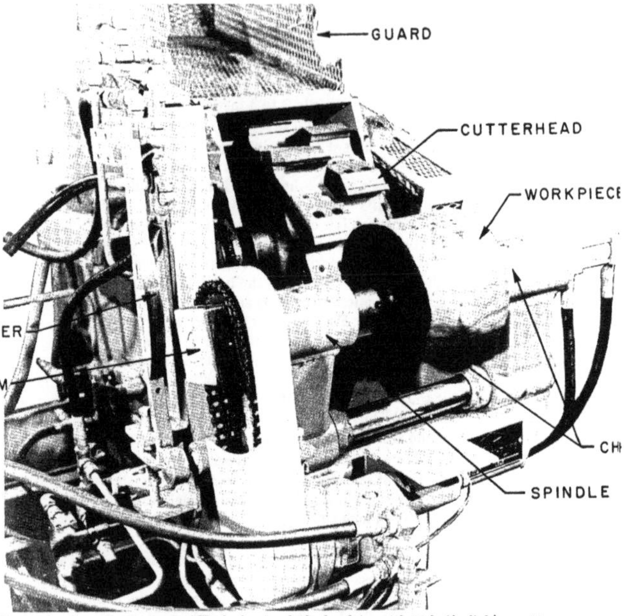

Figure 2.—*Second prototype of Koch shaping-lathe headrig, with 10-knife flaking cutter in place. Cam is 4 inches square. The bolt, when machined, will also measure 4 inches square.*

The cutterhead is 18 inches in diameter and is turned at 1,800 r/min by a 30-hp motor. Configuration of knives is shown in figure 3. To mimimize power requirement and enhance flake quality, rake angle is large—43°. Clearance angle, at 5°, is considered the minimum necessary to avoid undue interference with the workpiece. The resulting sharpness angle of 42° yields a cutting edge moderately resistant to nicking.

The cutterhead is in two segments, each 3-1/2 inches long and slotted for 10 knives; if desired, half the knives can be removed, so that only five are cutting. The two segments are indexed 18° from each other, to cause flake severance at their junction.

A test program was executed in the Se factory of Stetson-Ross to evaluate cutter action when reducing 6-inch rounds to 4 squares. Factors were:

Species
  Sweetgum (*Liquidambar styraciflua*)
  Southern red oak (*Quercus falcata* vai cata)
  Hickory (*Carya* sp.)
Wood temperature
  Stored in water at room temperature
  Heated in water at 180° F
Cutting direction
  Up-milling
  Down-milling

2

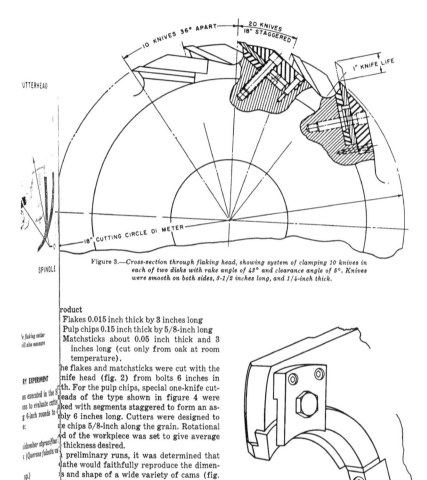

Figure 3.—*Cross-section through flaking head, showing system of clamping 10 knives in each of two disks with rake angle of 43° and clearance angle of 5°. Knives were smooth on both sides, 3-1/2 inches long, and 1/4-inch thick.*

UTTERHEAD

10 KNIVES 36° APART

20 KNIVES
18" STAGGERED

1" KNIFE LIFE

18" CUTTING CIRCLE DIAMETER

SPINDLE

'e flaking cutter
sill also measure

RY EXPERIMENT

as executed in the S
oss to evaluate cutte
g 6-inch rounds to (
e:

ridambar styraciflua
c (Quercus falcata va

sp.)
:
st room temperature
at 180° F

roduct
Flakes 0.015 inch thick by 3 inches long
Pulp chips 0.15 inch thick by 5/8-inch long
Matchsticks about 0.05 inch thick and 3 inches long (cut only from oak at room temperature).

he flakes and matchsticks were cut with the nife head (fig. 2) from bolts 6 inches in th. For the pulp chips, special one-knife cut- eads of the type shown in figure 4 were ked with segments staggered to form an as bly 6 inches long. Cutters were designed to e chips 5/8-inch along the grain. Rotational d of the workpiece was set to give average thickness desired.

preliminary runs, it was determined that athe would faithfully reproduce the dimen- s and shape of a wide variety of cams (fig. The flaking head yielded very smooth sur- s on all three species tested; figure 6 shows lts on sweetgum and red oak. Surfaces made he pulp chip cutterhead were comparatively rh (fig. 7).

Figure 4.—*Detail of the pulp chip knife shown in figure 11. Cutting-circle diameter was 18 inches.*

3

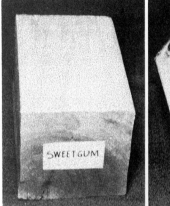

Figure 6.—*Cants smoothly machined by removal of flakes 0.015 inch thick.*

Flake appearance differed greatly with spe-es. In general, sweetgum (and southern pine) akes were wide and flat; hickory flakes were ide but tended to roll up into cigarette shape. ak flakes tended to be splinter-like. Flakes cut t averaged wider than those cut cold, since they lintered less. Up-milled flakes were less splint-er-like than down-milled flakes (figs. 8 and 9).

To make matchstick-like particles for the structural matrix in foamed urethane products[1], oak was revolved at a speed calculated to yield

[1] Marra, A. A. Analysis of the utility of southern hardwoods as a furnish for low-density composite products. USDA For. Serv. South. For. Exp. Stn., Alexandria, La., Final Report FS-SO-3201-2.42. March 1974.

Figure 8.—*Up-milled flakes 0.015 inch thick and 3 inches long cut from sweetgum (top), southern red oak (center), and hickory (bottom) soaked in water at 72°F (left) and 160°F (right).*

Figure 9.—*Down-milled flakes 0.015 inch thick and 3 inches long cut rm sweetgum (top), southern red oak (center), and hickory (bottom) soa i in water at 72°F (left) and 160°F (right).*

splinters 0.05 inch thick plus a hexagon measuring 4 inches across the flats. Cant surface remained good with no tearout around knots. The matchsticks tended not to be severed into 3-inch-long particles; many were 6 inches long, i.e., the length of the bolt (fig. 10).

The stacked disks, v tch had gullets somewhat obstructed, yielded p p chips that contained high proportions of pi : hips, particularly in oak. In an effort to impro chip flow, a new single-knife disk was const cted in which the gullet was smooth and ca cious. To eliminate the omplication of severing the fibers of each chip length, the workpiece was reduced to 5/8-inch long the grain (fig. 11). Resulting chips are depicted in figures 12 and 13. Motion pictures taken at 6,000 frames per second indicated that chambering action was unimpeded. It is likely, therefore, that refinement of head design will

Figure 12.—*Up-milled pulp chips 0.15 inch thick cut from sweetgum (top), southern red oak (center), and hickory (bottom) soaked in water at 72°F (left) and 160°F (right). Bruised fiber ends were not caused by chipping knife but resulted from prior machining of bolt to establish 5/8-inch chip length.*

PROCEDURE IN DETERMINING POWER REQUIREMENT

At the conclusion of the Seattle trials, the machine was displayed under power in Atlanta, Georgia, where it stimulated great interest at the June 1973 machinery show of the Southern Forest Products Association (Mason 1973). From Atlanta, it was shipped to the Southern Forest Experiment Station's laboratory at Pineville, Louisiana.

Figure 11.—*One-knife head for making pulp chips 5/8-inch long.*

7

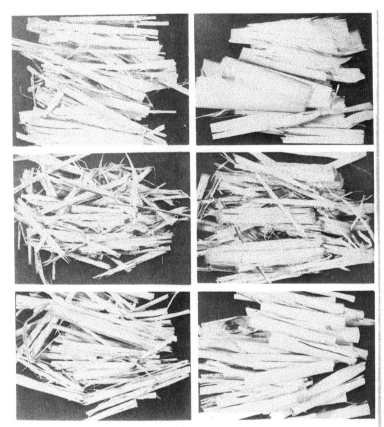

Figure 9.—*Down-milled flakes 0.015 inch thick and 3 inches long cut from sweetgum (top), southern red oak (center), and hickory (bottom) soaked in water at 72°F (left) and 160°F (right).*

splinters 0.05 inch thick plus a hexagon measuring 4 inches across the flats. Cant surface remained good with no tearout around knots. The matchsticks tended not to be severed into 3-inchlong particles; many were 6 inches long, i.e., the length of the bolt (fig. 10).

The stacked disks, which had gullets somewhat obstructed, yielded pulp chips that contained high proportions of pin chips, particularly in oak. In an effort to improve chip flow, a new singleknife disk was constructed in which the gullet was smooth and capacious. To eliminate the

not improve chip quality greatly. While chips of the type illustrated are less than desirable for chemical pulp, they should make excellent furnish for mechanical pulp produced in disk refiners.

Figure 12.—*Up-milled pulp chips 0.15 inch thick cut from sweetgum (top), southern red oak (center), and hickory (bottom) soaked in water at 72°F (left) and 160°F (right). Bruised fiber ends were not caused by chipping knife but resulted from prior machining of bolt to establish 5/8-inch chip length.*

PROCEDURE IN DETERMINING POWER REQUIREMENT

At the conclusion of the Seattle trials, the machine was displayed under power in Atlanta, Georgia, where it stimulated great interest at the June 1973 machinery show of the Southern Forest Products Association (Mason 1973). From Atlanta, it was shipped to the Southern Forest Experiment Station's laboratory at Pineville, Louisiana.

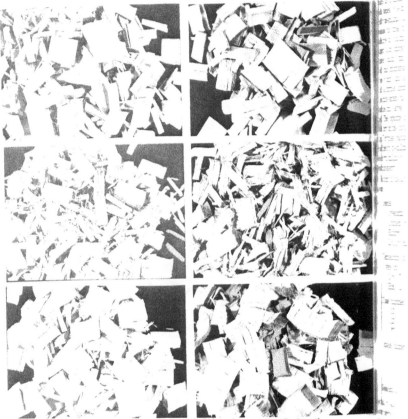

Figure 13.—*Down-milled pulp chips 0.15 inch thick cut from sweetgum (top), southern red oak (center), and hickory (bottom) soaked in water at 72°F (left) and 160°F (right). Bruised fiber ends were not caused by chipping knife, but resulted from prior machining of bolt to establish 5/8-inch chip length.*

The purpose of the Louisiana phase of the study was determination of specific cutting energy and maximum cutterhead power demand when making flakes from several species under a variety of cutting conditions. The flaking head was the same one used in Seattle (figs. 2 and 3).

It had been determined that power to turn the 6-inch-long workpieces past the flaking head was very small; the net was about 0.05 hp. Further observations of workpiece spindle power were not made.

At the outset, it was observed that flakes cut

from the wood outside the 4-inch central core of the mated slice.

Guidelines for running the experiment were:

- The replications were blocks in time.

- Room-temperature wood was cut in the mornings; bolts to be flaked hot were heated 4 hours in 160° F water during the morning and cut in the afternoon. While awaiting evaluation, bolts (with accompanying slices) were stored outside under water.

- In each day of operation, flakes of a single thickness were cut.

- Flakes 0.015 inch thick were cut with 10 knives in the cutterhead; for 0.025- and 0.035-inch flakes, alternate knives were removed to leave only five cutting. Workpiece r/min are shown in table 1; they ranged from 11.8 to 20.2 and were computed to yield the desired flake thickness at the bolt radius midway between 2 inches and the outer radius. It would have been desirable to do all cutting with 10 knives, but the machine design is such that workpiece speeds of 10 to 20 r/min are most practical; a workpiece speed of near 40 r/min would have been required to cut 0.035-inch flakes with a 10-knife head. The cutterhead (18-inch cutting circle diameter) turned at 1,800 r/min under no load, and had negligible drop in r/min under the average load imposed by the tests.

Table 1.—*Schedule of actual workpiece rotational speeds with target and achieved average flake thicknesses*[1]

Target flake thickness[2]	Achieved flake thickness[2]	Bolt outside diameter		
		6	7	8
— — —Inch— — —		— — —R/min — — —		
[3]0.015	0.0149	17.0	15.6	14.3
[4] .025	.0248	14.3	12.9	11.8
[4] .035	.0350	20.2	18.0	16.6

[1] The cutterhead rotated at 1,800 r/min.
[2] Figured at average depth of cut from outside diameter to the 4-inch cylindrical core.
[3] The cutterhead carried 10 knives.
[4] The cutterhead carried 5 knives.

Data recorded for each bolt (in addition to moisture content and specific gravity of the outer portion of the mated slice) were as follows: bolt length, workpiece r/min, watt-seconds consumed by the 30-horsepower cutterhead motor, and maximum watts demanded by the cutterhead motor. The workpiece r/min and instantaneous watt requirement of the cutterhead were recorded on strip charts moving at 25 mm per second to permit later computation of average flake thickness, average cutterhead power demand, and watt-seconds of energy to make each entire cut. From these data, plus a knowledge of cutterhead power demand when idling, it was possible to compute the energy input per cubic foot of wood removed.

Finally, an analysis was made of brake horsepower output related to watts input to the cutterhead motor (fig. 14). These data permitted computation of net specific cutting energy expressed as net horsepower minutes per cubic foot of wood removed. This computation was the primary objective of the experiment, as the information was needed to size the cutterhead motor on the commercial Koch lathe being designed at the time. The cutterhead on the commercial model is 54 inches long. For convenience in analysis and application of average and maximum net horsepower information, the data are presented for bolts 50 inches long.

At the beginning of the experiment, knives were freshly sharpened. By the time all 432 bolts had been run (main experiment) some knives were beginning to develop small nicks and slight dulling.

The central residual pieces were saved from the last two replications, and their surface quality was visually evaluated.

Cutting began in September 1973, proceeded at the rate of 24 bolts per day, and was completed October 19, 1973.

Limited down-milling data were taken a few days later on green sweetgum bolts only, cut at 72° F to 4-inch diameter, in a factorial arrangement:

Cutting direction
Up-milling
Down-milling

Replication	Net specific cutting energy	Maximum net cutting power per 50 inches of bolt length per knife
	$Hp\ minutes/ft^{3}$	$Horsepower$
1	9.40	18.05
2	9.34	17.60
3	9.95	18.15
4	10.26	19.18
5	10.20	19.24
6	10.39	19.58

While machining the 432 bolts, the cutterhead was actually cutting wood for about 51 minutes. At a charging rate of six bolts per minute for the commercial machine, about 1,200 bolts may be machined during a morning's work, with actual machining time of 150 minutes or less. Extension of the data in the foregoing tabulation to 1,200 bolts indicates a likely increase in net cutterhead horespower of near 20 percent from morning installation of freshly sharpened knives to mid-shift replacement.

Specific Gravity

Specific gravity of the flaked portion of bolts varied significantly (0.05 level) according to species and bolt diameter in the following interaction:

Species	Bolt diameter, inches			
	6	7	8	Average
Sweetgum	0.48	0.46	0.47	0.47
Loblolly pine	.46	.50	.52	.49
Southern red oak	.61	.60	.58	.58
Hickory	.70	.69	.68	.69

Specific gravity of hickory and red oak varied inversely with bolt diameter, that of loblolly was positively correlated with bolt diameter. No diameter-related trend was evident for sweetgum. Average specific gravity of all bolts up-milled was 0.56.

Bolts assigned to the three flake-thickness and two soaking treatments did not vary in specific gravity.

Moisture Content

With data from all bolts pooled, moisture content of portions flaked was 110.6 percent. Bolts heated in hot water before machining had 4.3

11

percentage points less moisture content (108.4 percent) than those stored in cold water (112.7 percent); the difference was significant at the 0.05 level.

Partly in consequence of the specific gravity relationships just discussed, hickory and red oak had lower moisture contents than loblolly pine and sweetgum:

Species	Bolt diameter, inches			
	6	7	8	Average
	— Percent of dry weight —			
Hickory	66.2	68.8	68.9	68.0
Southern red oak	91.2	95.3	100.8	95.8
Loblolly pine	145.1	127.1	123.3	131.8
Sweetgum	145.5	149.1	145.3	146.7

In hickory and oak the highest moisture contents occurred in the 8-inch bolts; in loblolly pine, however, 6-inch bolts had the highest.

Idle Horsepower

The relationship between input wattage to the cutterhead motor and output mechanical horsepower was proven linear by a prony brake test (fig. 14).

When the 7-inch-long, 18-inch-diameter cutterhead was vee-belted to this motor, 2.9 horsepower were required to rotate the head at 1,800 r/min. This idling horsepower was the same whether the cutterhead was fitted with 10 knives or 5.

Horsepower to idle is primarily comprised of bearing friction and windage. The first of these components should be proportional to the weight of the cutterhead (and thereby the length); the second component should be directly related to cutterhead length.

From the test data, it can be predicted that idle power required to drive a 54-inch-long cutterhead will be about 22 horsepower, i.e., (54/7) (2.9). This value is for a cutterhead speed of 1,800 r/min with cutting circle 18 inches in diameter. It is likely that a head 12 inches in diameter rotating at 3,600 r/min would not differ substantially in idling demand.

Net Specific Cutting Energy

The statistic of foremost importance in determining average power requirements of a wood machining operation is net specific cutting energy, that is, the energy (over and above idling energy) required to remove a unit volume of wood in a unit time.

For the entire up-milling experiment, specific cutting energy averaged 9.92 horsepower minutes per cubic foot. Uncomplicated by interactions with other factors, wood soaked in 160° water required 5.5 percent less specific cutting energy than wood held in water at 72° F (9.6 vs. 10.21 horsepower minutes per cubic foot).

Specific gravity of wood was positively correlated with specific cutting energy (fig. 15); flake thickness had a negative correlation (fig.

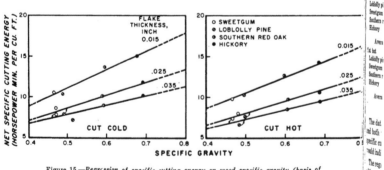

Figure 15.—*Regression of specific cutting energy on wood specific gravity (basis of ovendry weight and green volume) for wood cut at 72°F (left) and 160°F (right). Plotted points are averages for the species named.*

horsepower minutes per cubic foot), are as follows:

Wood cut at 72° F
0.015-inch flakes ($R^2 = 0.72$; $S_e = 1.29$)
$$\hat{Y} = -0.1937 + 22.6177 \, X$$
0.025-inch flakes ($R^2 = 0.68$; $S_e = 1.01$)
$$\hat{Y} = 0.9244 + 15.5586 \, X$$
0.035-inch flakes ($R^2 = 0.61$; $S_e = 0.84$)
$$\hat{Y} = 2.2545 + 11.2798 \, X$$

Wood cut at 160° F
0.015-inch flakes ($R^2 = 0.66$; $S_e = 1.28$)
$$\hat{Y} = 0.9034 + 19.4060$$
0.025-inch flakes ($R^2 = 0.67$; $S_e = 0.95$)
$$\hat{Y} = 0.5955 + 15.0228 \, X$$
0.035-inch flakes ($R^2 = 0.72$; $S_e = 0.70$)
$$\hat{Y} = 1.6473 + 11.4714 \, X$$

From figure 15 it is evident that specific cutting energy is greatest when dense wood is cut cold into thin flakes. If it is assumed that 0.8 is near the upper limit of density for mixed eastern hardwoods, then (by the first equation listed above) the specific cutting energy when making 0.015-inch flakes will be maximum at 17.9 horsepower minutes per cubic foot of wood removed.

Because of contrasting species patterns of variation in bolt specific gravity with bolt diameter, net specific cutting energy also varied significantly with bolt diameter. With all flake thicknesses and both temperatures pooled, averages were as follows:

Species	Bolt diameter, inches			
	6	7	8	Average
	———$Hp \, min/ft^3$———			
Loblolly pine	8.46	8.06	8.46	8.33
Sweetgum	8.95	8.65	8.63	8.75
Southern red oak	11.34	10.52	10.03	10.63
Hickory	12.56	11.83	11.57	11.99
Average	10.33	9.77	9.67	

Loblolly and sweetgum diameters showed little relationship with specific cutting energy. For southern red oak and sweetgum, however, specific cutting energy decreased as bolt diameter increased. While the differences related to bolt diameter are statistically significant, they are probably not large enough to cause problems for users of the commercial machine.

13

Average Net Cutterhead Horsepower Per 50 Inches of Bolt Length

Power requirement during time in cut averaged highest when machining 0.015-inch-thick flakes from cold hickory bolts 8 inches in diameter. For the 10-knife cutterhead rotating at 1,800 r/min, net average power demand was 120.0 horsepower (per 50 inches of bolt length) for the six hickory bolts so machined. One of the six bolts required an average of 142.3 horsepower (per 50 inches of bolt length). For the 54-inch commercial machine, addition of 22 idling horsepower would raise average demand to 164 horsepower for a 10-knife head (1,800 r/min) that is removing 0.015-inch flakes while machining a cold, dense hickory bolt from 8-inch to 4-inch diameter (bolt length 50 inches).

Thicker flakes were cut with only five knives mounted in the head, and therefore had less average net cutting power demand when machining cold hickory bolts 8 inches in diameter:

Flake thickness	Number of knives cutting	Average net cutterhead power per 50 inches of bolt length
Inch		*Horsepower*
0.015	10	120.0
.025	5	81.1
.035	5	83.2

As may be deduced from table 2 (specific cutting energy), loblolly pine required only about two-thirds the average net cutterhead power needed for hickory.

Average Net Cutterhead Horsepower Per 50 Inches of Bolt Length Per Knife

Horsepower per knife is a useful statistic in designing cutterheads. At 1,800 r/min the overall average was 9.94 net per knife per 50 inches of bolt length. (Doubling the cutterhead speed would double this value.) Highest average readings occurred when machining 8-inch hickory to 4-inch rounds at 72° F:

Flake thickness	Average cutterhead power per 50 inches of bolt length per knife
Inch	*Horsepower*
0.015	12.00
.025	16.23
.035	16.63

The highest individual value observed when cutting 0.015-inch cold hickory flakes was 14.2. This value is important because it is a determinant of motor size on the production machine. Thus a 10-knife, 1,800 r/min head making a 2-inch-deep cut in a 50-inch-long, cold hickory bolt would have a net average requirement of 142 horsepower when cutting 0.015-inch-thick flakes.

For the commercial machine, a cutterhead speed of 3,600 r/min would permit the use of fewer knives than one turning at 1,800, and a head with six knives cutting would allow more flexibility than one with 10 knives. For a six-knife head turning at 3,600 r/min the average horsepower needed would be 170, i.e., (14.2) (6) (3,600/1,800). To this value must be added idling horsepower. To cut flakes thicker than 0.015 inch, the number of knives could be reduced by removing every other knife, thereby keeping about the same workpiece rotational speed but reducing the average power requirement.

Average Net Cutterhead Horsepower Per 50 Inches of Bolt Length Per Knife Per Inch Depth of Cut

The foregoing paragraphs discussed the average horsepower required to cut an 8-inch bolt to a 4-inch cylinder, i.e., to take a cut 2 inches deep. The data permit analyses of cuts 1, 1.5, and 2 inches deep. Figure 17 shows that the relationship between cutting depth and horsepower is linear in this range. It will be noted, however, that the plots do not pass through the origin, but have a Y intercept (about 15 horsepower). That is, a doubling of depth of cut does not quite double the net average horsepower required.

For all 432 bolts, the average for the 1,800 r/min cutterhead was 6.75 net horsepower per 50 inches of bolt length per knife per inch depth of cut. Highest averages occurred with cold hickory:

Flake thickness	Average cutterhead power per 50 inches of bolt length per knife per inch depth of cut
Inch	*Horsepower*
0.015	6.89
.025	10.35
.035	9.94

14

It is hard to visualize an operator calling for such a deep cut, but he may be pressed to fill orders when he has only large wood on hand.

Maximum Net Cutterhead Horsepower Per 50 Inches of Bolt Length

Motor size in the lathe headrig is determined not only by average power drawn during the few seconds the head is in the cut, but also by the maximum power drawn during each cut. These maximums, occurring about six times per minute, must not cause motor overheating nor exert sufficient torque to slow the motor from synchronous speed, i.e., "pull out."

The power curves displayed no sharp peaks when 6-, 7-, and 8-inch bolts were cut to 4-inch cylinders, but maximum values were easily noted where wood density patterns were eccentric. Power maximums were greatest when 0.015-inch flakes were cut from cold hickory bolts 8 inches in diameter. Then maximum power demand by a 10-knife head at 1,800 r/min was 218.0 horsepower per 50 inches of bolt length for the six hickory bolts so machined. One of the six bolts peaked at 232.9 net horsepower.

For all 432 bolts the mean value of maximum net horsepower per 50 inches of bolt length was 116.5. Bolts at 160° F needed 5.5 percent less (113.2) than those cut at 72° F (119.8).

The trend evident in figure 17 also appears in the linear relationship of maximum power to depth of cut, as follows:

Bolt diameter	Depth of cut	Maximum net cutterhead power per 50 inches of bolt length
Inches	*Inches*	*Horsepower*
6	1.0	83.2
7	1.5	117.0
8	2.0	149.3

Loblolly pine drew only 70 percent of the maximum required by hickory, and the 10-knife head cutting 0.015-inch-thick flakes drew maximums substantially greater than the five-knife head cutting 0.025- and 0.035-inch flakes:

| Species | Flake thickness, inch | | | |
	0.015	0.025	0.035	Average
	Maximum net horsepower per 50 inches of bolt length			
Loblolly pine	121.3	79.1	94.0	98.1
Sweetgum	120.6	88.9	106.1	105.2
Southern red oak	150.5	100.8	119.0	123.4
Hickory	167.7	118.9	131.2	139.2
Average	140.0	96.9	112.6	

Maximum Net Cutterhead Horsepower Per 50 Inches of Bolt Length Per Knife

Overall mean maximum horsepower per knife per 50 inches of bolt length was 18.6. The largest maximums occurred when machining cold 8-inch hickory to 4-inch rounds:

Flake thickness	Maximum net cutterhead power per 50 inches of bolt length per knife
Inch	*Horsepower*
0.015	21.80
.025	28.30
.035	32.94

The highest individual net maximum obse when cutting 0.015-inch cold hickory flake inch depth of cut) was 23.3 per 50 inches o length per knife.

Thus a 10-knife, 1,800-r/min head cutt inches deep in a 50-inch-long bolt would dr net maximum of 233 horsepower when m: 0.015-inch flakes from cold, dense hickory. six knives at 3,600 r/min, the maximum v likely be 280 horsepower, i.e., (23.3) (6) (3 1,800) —or about 302 with idling horsep added.

Maximum Net Cutterhead Horsepower Per 50 Inches of Bolt Length Per Knife Per Inch Depth of Cut

Since the relationship between maximui horsepower and depth of cut is linear, it is u to express maximum net horsepower in t of depth of cut. For all 432 bolts, the averag the 1,800 r/min cutterhead was 12.56 p inches of bolt length per knife per inch der cut. Wood soaked in hot (160° F) water maximums 5.3 percent less than wood macl at 72° F (12.22 vs. 12.91 horsepower).

The maximums were positively corre with specific gravity (fig. 18) and with thickness (table 3 and fig. 19).

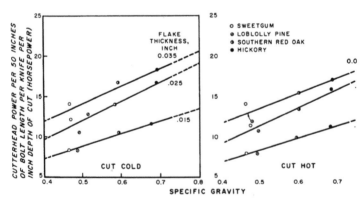

Figure 18.—*Regression of maximum observed (per cutting cycle) net cutting horsepower per 50 inches of bolt length per knife per inch depth of cut, on wood specific gravity (basis of ovendry weight and green volume). Plotted points are 18-bolt averages for the species named and are based on cuts 1.0, 1.5, and 2.0 inches deep. Cutterhead speed was 1,800 r/min.*

0.035-inch flakes ($R^2 = 0.46$; $S_e = 2.22$)

$$\hat{Y} = 3.2659 + 21.7839\ X$$

Wood cut at 160° F

0.015-inch flakes ($R^2 = 0.51$; $S_e = 1.33$)

$$\hat{Y} = 0.9257 + 14.8163\ X$$

0.025-inch flakes ($R^2 = 0.54$; $S_e = 1.87$)

$$\hat{Y} = 0.1725 + 22.2934\ X$$

0.035-inch flakes ($R^2 = 0.58$; $S_e = 1.57$)

$$\hat{Y} = 4.2001 + 18.5873\ X$$

From figure 18 it is evident that the maximum power demand occurs when the most dense wood is machined cold. If 0.8 specific gravity is the upper limit for mixed eastern hardwoods, then (by the first equation listed above) the maximum net cutterhead power for 0.015-inch-thick flakes will be about 13.6 horsepower per 50 inches of bolt length per knife per inch depth of cut. Under these conditions, a 10-knife head rotating at 1,800 r/min would draw 272 maximum net horsepower in machining an 8-inch square from a cold bolt 12 inches in diameter and 50 inches long. Net maximum for a six-knife head rotating at 3,600 r/min would be 326.

Should the depth of cut be increased from the 2 inches assumed above to 3 inches, maximum net horsepower would be 408 for the 10-knife head (1,800 r/min) and 490 for the six-knife head (3,600 r/min).

These maximums perhaps overstate the requirement somewhat because the relationship between depth of cut and horsepower—while linear—is not quite directly proportional; i.e., doubling the depth of cut in oak and hickory does not quite double the maximum net horsepower required. In the following tabulation, all factors are pooled except depth of cut and species:

Species	Depth of cut, inches			
	1.0	1.5	2.0	Average
	Maximum net horsepower per 50 inches of bolt length per knife per inch depth of cut			
Loblolly pine	10.52	9.97	10.64	10.38
Sweetgum	11.50	11.42	11.21	11.38
Southern red oak	14.48	13.36	12.25	13.36
Hickory	16.41	15.25	13.74	15.13
Average	13.23	12.50	11.96	

Because the 6-inch hickory bolts were most dense, the highest net maximum occurred with

17

Species	Flake thickness, inch			
	0.015	0.025	0.035	Average
	Maximum net horsepower per 50 inches of bolt length			
Loblolly pine	121.3	79.1	94.0	98.1
Sweetgum	120.6	88.9	106.1	105.2
Southern red oak	150.5	100.8	119.0	123.4
Hickory	167.7	118.9	131.2	139.2
Average	140.0	96.9	112.6	

Maximum Net Cutterhead Horsepower Per 50 Inches of Bolt Length Per Knife

Overall mean maximum horsepower per knife per 50 inches of bolt length was 18.6. The largest maximums occurred when machining cold 8-inch hickory to 4-inch rounds:

Flake thickness	Maximum net cutterhead power per 50 inches of bolt length per knife
Inch	*Horsepower*
0.015	21.80
.025	28.30
.035	32.94

The highest individual net maximum observed when cutting 0.015-inch cold hickory flakes (2. inch depth of cut) was 23.3 per 0 inches of bolt length per knife.

Thus a 10-knife, 1,800-r/min head cutting 2 inches deep in a 50-inch-long bolt would draw a net maximum of 233 horsepower when making 0.015-inch flakes from cold, dense hickory. With six knives at 3,600 r/min, the maximum would likely be 280 horsepower, i.e., (3.3) (6) (3,600/1,800)—or about 302 with idling horsepower added.

Maximum Net Cutterhead Horsepower Per 50 Inches of Bolt Length Per Knife Per Inch Depth of Cut

Since the relationship between maximum net horsepower and depth of cut is linear, it is useful to express maximum net horsepower in terms of depth of cut. For all 432 bolt, the average for the 1,800 r/min cutterhead was 12.56 per 50 inches of bolt length per knife per inch depth of cut. Wood soaked in hot (16° F) water had maximums 5.3 percent less than wood machined at 72° F (12.22 vs. 12.91 horsepower).

The maximums were positively correlated with specific gravity (fig. 1) and with flake thickness (table 3 and fig. 19.

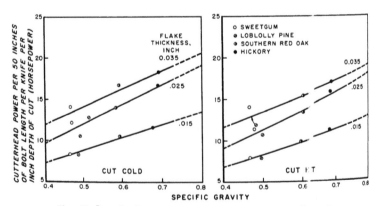

Figure 18.—*Regression of maximum observed (per cutting cycle) net cutting horsepower per 50 inches of bolt length per knife per inch depth of cut, on wood specific gravity (basis of ovendry weight and green volume). Plotted p ints are 18-bolt averages for the species named and are based on cuts 1.0 .5, and 2.0 inches deep. Cutterhead speed was 1,800 r/min.*

Table 3.—*Maximum net cutterhead horsepower per 50 inches of bolt length per knife per inch depth of cut*

Temperature and species	Flake thickness, inch			
	0.015	0.025	0.035	Average
	— — — — Horsepower — — — —			
Cut cold				
Loblolly pine	8.40	10.63	12.94	10.66
Sweetgum	8.40	12.19	14.21	11.60
Southern red oak	10.64	14.04	16.86	13.85
Hickory	11.56	16.70	18.36	15.54
Average	9.75	13.39	15.59	
Cut hot				
Loblolly pine	7.92	10.52	11.85	10.10
Sweetgum	7.93	11.47	14.07	11.16
Southern red oak	9.84	13.36	15.43	12.88
Hickory	11.20	15.91	17.08	14.73
Average	9.22	12.81	14.61	

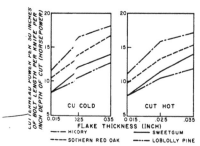

Figure 19.—*Maximum net cutterhead horsepower observed in each machining cycle per knife per 50 inches of bolt length per inch depth of cut, related to average flake thickness, species, and wood temperature. Data from 18 bolts were averaged to locate each of the 24 points plotted. Cutterhead speed was 1,800 r/min.*

The regression equations graphed in figure 18, with squared correlation coefficients and square roots of error mean squares (values of S_e are in horsepower minutes per cubic foot), are as follows:

Wood cut at 72° F
0.015-inch flakes ($R^2 = 0.61$; $S_e = 1.16$)
$$\hat{Y} = 1.027 + 15.6451 \, X$$
0.025-inch flakes ($R^2 = 0.52$; $S_e = 2.15$)
$$\hat{Y} = 0.151 + 23.6785 \, X$$

0.035-inch flakes ($R^2 = 0.46$; $S_e = 2.22$)
$$\hat{Y} = 3.2659 + 21.7839 \, X$$

Wood cut at 160° F
0.015-inch flakes ($R^2 = 0.51$; $S_e = 1.33$)
$$\hat{Y} = 0.9257 + 14.8163 \, X$$
0.025-inch flakes ($R^2 = 0.54$; $S_e = 1.87$)
$$\hat{Y} = 0.1725 + 22.2934 \, X$$
0.035-inch flakes ($R^2 = 0.58$; $S_e = 1.57$)
$$\hat{Y} = 4.2001 + 18.5873 \, X$$

From figure 18 it is evident that the maximum power demand occurs when the most dense wood is machined cold. If 0.8 specific gravity is the upper limit for mixed eastern hardwoods, then (by the first equation listed above) the maximum net cutterhead power for 0.015-inch-thick flakes will be about 13.6 horsepower per 50 inches of bolt length per knife per inch depth of cut. Under these conditions, a 10-knife head rotating at 1,800 r/min would draw 272 maximum net horsepower in machining an 8-inch square from a cold bolt 12 inches in diameter and 50 inches long. Net maximum for a six-knife head rotating at 3,600 r/min would be 326.

Should the depth of cut be increased from the 2 inches assumed above to 3 inches, maximum net horsepower would be 408 for the 10-knife head (1,800 r/min) and 490 for the six-knife head (3,600 r/min).

These maximums perhaps overstate the requirement somewhat because the relationship between depth of cut and horsepower—while linear—is not quite directly proportional; i.e., doubling the depth of cut in oak and hickory does not quite double the maximum net horsepower required. In the following tabulation, all factors are pooled except depth of cut and species:

Species	Depth of cut, inches			
	1.0	1.5	2.0	Average
	Maximum net horsepower per 50 inches of bolt length per knife per inch depth of cut			
Loblolly pine	10.52	9.97	10.64	10.38
Sweetgum	11.50	11.42	11.21	11.38
Southern red oak	14.48	13.36	12.25	13.36
Hickory	16.41	15.25	13.74	15.13
Average	13.23	12.50	11.96	

Because the 6-inch hickory bolts were most dense, the highest net maximum occurred with

17

Species	Flake thickness, inch			
	0.015	0.025	0.035	Average
	Maximum net horsepower per 50 inches of bolt length			
Loblolly pine	121.3	79.1	94.0	98.1
Sweetgum	120.6	88.9	106.1	105.2
Southern red oak	150.5	100.8	119.0	123.4
Hickory	167.7	118.9	131.2	139.2
Average	140.0	96.9	112.6	

**Maximum Net Cutterhead Horsepower
Per 50 Inches of Bolt Length
Per Knife**

Overall mean maximum horsepower per knife per 50 inches of bolt length was 18.6. The largest maximums occurred when machining cold 8-inch hickory to 4-inch rounds:

Flake thickness	Maximum net cutterhead power per 50 inches of bolt length per knife
Inch	*Horsepower*
0.015	21.80
.025	28.30
.035	32.94

The highest individual net maximum obs
when cutting 0.015-inch cold hickory flake
inch depth of cut) was 23.3 per 50 inches o
length per knife.

Thus a 10-knife, 1,800-r/min head cutt
inches deep in a 50-inch-long bolt would d
net maximum of 233 horsepower when m
0.015-inch flakes from cold, dense hickory.
six knives at 3,600 r/min, the maximum v
likely be 280 horsepower, i.e., (23.3) (6) (3
1,800)—or about 302 with idling horsep
added.

**Maximum Net Cutterhead Horsepower
Per 50 Inches of Bolt Length Per
Knife Per Inch Depth of Cut**

Since the relationship between maximu
horsepower and depth of cut is linear, it is u
to express maximum net horsepower in t
of depth of cut. For all 432 bolts, the averaį
the 1,800 r/min cutterhead was 12.56 p
inches of bolt length per knife per inch dep
cut. Wood soaked in hot (160° F) wate
maximums 5.3 percent less than wood mac
at 72° F (12.22 vs. 12.91 horsepower).

The maximums were positively corre
with specific gravity (fig. 18) and with
thickness (table 3 and fig. 19).

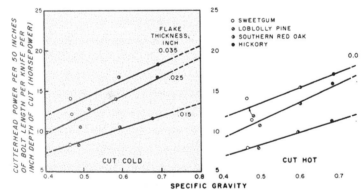

Figure 18.—*Regression of maximum observed (per cutting cycle) net cutting horsepower per 50 inches of bolt length per knife per inch depth of cut, on wood specific gravity (basis of ovendry weight and green volume). Plotted points are 18-bolt averages for the species named and are based on cuts 1.0, 1.5, and 2.0 inches deep. Cutterhead speed was 1,800 r/min.*

0.035-inch flakes ($R^2 = 0.46$; $S_e = 2.22$)

$$\hat{Y} = 3.2659 + 21.7839\ X$$

Wood cut at 160° F
0.015-inch flakes ($R^2 = 0.51$; $S_e = 1.33$)

$$\hat{Y} = 0.9257 + 14.8163\ X$$

0.025-inch flakes ($R^2 = 0.54$; $S_e = 1.87$)

$$\hat{Y} = 0.1725 + 22.2934\ X$$

0.035-inch flakes ($R^2 = 0.58$; $S_e = 1.57$)

$$\hat{Y} = 4.2001 + 18.5873\ X$$

From figure 18 it is evident that the maximum power demand occurs when the most dense wood is machined cold. If 0.8 specific gravity is the upper limit for mixed eastern hardwoods, then (by the first equation listed above) the maximum net cutterhead power for 0.015-inch-thick flakes will be about 13.6 horsepower per 50 inches of bolt length per knife per inch depth of cut. Under these conditions, a 10-knife head rotating at 1,800 r/min would draw 272 maximum net horsepower in machining an 8-inch square from a cold bolt 12 inches in diameter and 50 inches long. Net maximum for a six-knife head rotating at 3,600 r/min would be 326.

Should the depth of cut be increased from the 2 inches assumed above to 3 inches, maximum net horsepower would be 408 for the 10-knife head (1,800 r/min) and 490 for the six-knife head (3,600 r/min).

These maximums perhaps overstate the requirement somewhat because the relationship between depth of cut and horsepower—while linear—is not quite directly proportional; i.e., doubling the depth of cut in oak and hickory does not quite double the maximum net horsepower required. In the following tabulation, all factors are pooled except depth of cut and species:

Species	Depth of cut, inches			
	1.0	1.5	2.0	Average
	Maximum net horsepower per 50 inches of bolt length per knife per inch depth of cut			
Loblolly pine	10.52	9.97	10.64	10.38
Sweetgum	11.50	11.42	11.21	11.38
Southern red oak	14.48	13.36	12.25	13.36
Hickory	16.41	15.25	13.74	15.13
Average	13.23	12.50	11.96	

Because the 6-inch hickory bolts were most dense, the highest net maximum occurred with these bolts when cut cold:

Flake thickness	Maximum net cutterhead power per 50 inches of bolt length per knife per inch depth of cut
	Horsepower
0.015	12.61
.025	19.57
.035	19.35

Of the six 6-inch, cold-cut hickory bolts machined to yield 0.015-inch flakes, one had 13.96 maximum requirement. This is not much different from the value of 13.6 previously postulated from figure 18. These are the values which will size the motor, because flakes thicker than 0.015 inch can be cut by removing every other knife from the cutterhead, thus reducing peak power demands.

CONCLUSIONS

The purpose of this experiment was to determine the minimum cutterhead motor that will permit the commercial shaping-lathe headrig to machine six bolts per minute. In reaching conclusions, some design parameters of the cutterhead and feedworks are assumed, as follows:

Cutterhead

Direction of workpiece rotation	upmilling
Cutterhead speed	3,600 r/min
Cutterhead length	54 inches
Diameter of cutting circle	12 inches
Number of knives cutting when machining flakes 0.015-inch thick	6
Alternate numbers of knives (by removal)	3
Cutterhead idling power	22 horsepower

Workpiece

Maximum workpiece diameter	12 inches
Maximum density of wood to be machined (basis of ovendry weight and green volume)	0.75
Condition of workpiece	green at 72° F
Maximum length of workpiece	53 inches

Feedworks and Sets

Maximum depth of cut (at knots, crooks, eccentricities) on which to base estimate of momentary maximum horsepower demand .... 3 inches

Maximum removal of cross section on which to base average net cutterhead power demand.... 12-inch round to 8-inch square

Speed ratios on variable-speed gearmotor driving the workpiece 3:

Average Horsepower During Cutting Cycle

To convert specific cutting energy data to average horsepower demand, the time in cut must be known. In the discussion that follows, it will be assumed that 0.015-inch flakes are being cut with a six-knife head turning at 3,600 r/min. The workpiece—not rotating—approaches the cutterhead at 5.4 inches per second until the follower strikes the cam. Thereupon, the workpiece rotates through 360 at 10.31 r/min.

Time in cut is therefore comprised of two components:

Plunge time in cut (e.g., 2 inches when machining an 8-inch square from a 12-inch round)

Time to rotate 360 degrees

Plunge time is 2/5.4 seconds or 0.37 second, and time to rotate 360 degrees is 5.82 seconds. Therefore total time in cut is 0.37 + 5.82 = 6.19 seconds or 0.103 minute.

Volume of wood removed from each 53-inch-long, 12-inch-diameter bolt machined to an 8- by 8-inch-square cant amounts to 1.506 cubic feet, i.e., $(0.5^2 \pi - .67^2)$ (53/12).

At a wood specific gravity of 0.75, specific cutting energy for 0.015-inch flakes is 16.7 horsepower minutes per cubic foot of cold wood removed (fig. 15, left). Average net horsepower demand to reduce a 12-inch bolt to an 8-inch square is therefore 245 horsepower, i.e., (16.7) (1.506)/0.103. To this must be added the 22 horsepower required to turn the cutterhead while idling. Total average cutterhead horsepower expended over the 6.19-second machining cycle will therefore be 267.

If the operator reduced the number of knives cutting from six to three, so as to increase average flake thickness from 0.015 to 0.030 inch, the average cutterhead power would be reduced

18

available by operating with six or three knives cutting:

Number of knives cutting (and average flake thickness, inch)	Workpiece diameter at which flake thickness is measured, inches		
	5	7.5	10
	— — *Workpiece r/min* — —		
6 knives cutting			
0.015	20.62	15.47	10.31
.020	27.49	20.63	13.75
3 knives cutting			
.25	17.18	12.89	8.59
.30	20.62	15.47	10.31

Power Required to Rotate Workpiece, and Workpiece Deflection in Torsion

In order for a 12-inch-diameter cutterhead rotating at 3,600 r/min to have average demand of 245 horsepower while cutting, it must exert a force of 715 pounds tangential to its cutting circle.

If 0.015-inch flakes are to be cut at an average diameter of 10 inches, with workpiece rotation speed of 10.31 r/min, only 0.58 hp is required to rotate the workpiece while an 8-inch square is cut from a 12-inch round, i.e., (π) (10/12) (10.31) (715)/33,000.

Torque exerted by the driving chuck must not unduly twist the workpiece, or machined squares will tend to assume a propeller shape. The following calculations are intended to show that the maximum torsional deflections anticipated will not be of substantial practical importance.

Assume that the 53-inch-long workpiece is driven from one end and that the 715-pound force noted in the previous paragraph is applied as a concentrated load to the idle chuck end in a manner to deliver (715) (5 inches) or 3,575 inch-pounds of torque. Further assume that the machined workpiece is an 8-inch cylinder. Both assumptions are conservative, i.e., the workpiece is actually 8 inches square and the load of 715 pounds is uniformly distributed along the bolt.

Under these assumptions, the workpiece would deflect (at the idle-chuck end from the driven-chuck end) through an angle of 0.270 degree, as follows:

Angle of deflection $\theta = \dfrac{TL}{JG}$; where θ is in radians, T is torque (inch pounds), L is bolt

19

length (inches), J is polar moment of inertia $(\frac{\pi D^4}{32}$, inches$^4)$, and G is modulus of rigidity (assumed to be 100,000 lb/in^2)

Therefore:

$$\theta = \frac{(3,575)\ (53)}{(402.1)\ (100,000)} = 0.00471 \text{ radians}$$
$$= 0.270 \text{ degree}$$

This would deflect a point on the periphery of the workpiece 0.019 inch, e.g. $(\frac{0.270}{360})$ (8π), from the unstressed position.

If this same 715-pound force was applied at 2.5-inch radius to twist a 4-inch post, the torque would be halved and the polar moment of inertia decreased 16 times; therefore the post would twist through an angle θ of (0.270) $(16)/2 =$ 2.16 degrees, and a point on the periphery at one end would deflect 0.075 inch, e.g., (2.16) (π) $(4)/$ 360.

It is likely that these torsional deflections are within acceptable limits. Should they prove to be excessive, however, application of power to both workpiece chucks—rather than only one—would diminish them by a factor of 4.

Lateral Deflection of the Workpiece

Assuming a tangentially and uniformly applied bending load—due to parallel cutting forces —of 715 pounds over the 53-inch length of an 8-inch machined cylinder (when cutting a dense wood such as *Quercus bicolor* Willd.), the cylinder would deflect 0.0043 inch, e.g., 5wL^4/ 384EI. This assumes an E of 1,590,000 lb/in^2.

From tables 19-8 and 19-10 of Agriculture Handbook 420 (Koch 1972), it is seen that the ratio of knife normal force to parallel cutting force is $\frac{-0.4}{1.7}$ when 0.015-inch-thick chips are cut in the 0-90 direction from saturated wood of high density (loblolly latewood) with a knife having a 45-degree rake angle. Therefore, the workpiece

will tend to be drawn into the cutterhead by a uniformly distributed force of —168 pounds, i.e., (—0.4) (715)/1.7. Resulting deflection (toward the cutterhead) at midpoint of an 8-inch cylinder will be 0.0010 inch.

These same loads applied to a 4-inch round post will cause deflections 16 times larger, i.e., 0.069 inch in a downward direction, and 0.016 inch toward the cutterhead.

It is probable that these workpiece bending deflections are within acceptable limits for manufacture of pallet parts and other industrial lumber products.

LITERATURE CITED

Koch, P.
  1964. Square cants from round bolts without slabs or sawdust. For. Prod. J. 14: 332-336.

Koch, P.
  1967a. Development of the chipping headrig. Rocky Mountain For. Ind. Conf. Proc. 1967: 135-155. Fort Collins, Colo.

Koch, P.
  1967b. Too many crooked small logs? Consider a shaping lathe headrig. South. Lumberman 215 (2680): 147-149.

Koch, P.
  1968. Converting southern pine with chipping headrigs. South. Lumberman 217 (2704): 131-138.

Koch, P.
  1972. Utilization of the southern pines. USDA Agric. Handb. 420, 1663 p.

Mason, R.
  1973. Lathe creates hardwood flakes for manufacture of "super strong" flake board. Wood and Wood Prod. 78 (10): 32-34.

20

U. S. Department of Agriculture
Forest Service Research Paper SO-100

Development of
Pine-Hardwood Stands in North Alabama
Following Improvement Cuttings

Glendon W. Smalley

Southern Forest Experiment Station
Forest Service
U. S. Department of Agriculture
1974

Acknowledgments

The author acknowledges with appreciation the work of K. A. Brinkman in designing and establishing this study, and of Frank Freese and H. O. Sontag in recording and summarizing early results.

U.S. Department of Agriculture
Forest Service Research Paper SO-99

of Slash Pine to
AND PHOSPHORUS APPLICATION
In Southeastern Flatwoods

W. F. Mann, Jr.
and
I. M. McGilvray

Southern Forest Experiment Station
Forest Service
U.S. Department of Agriculture

1974

Mann, W. F., Jr., and McGilvray, J. M.

1974. Response of slash pine to bedding and phos-
phorus application in southeastern flatwoods.
South. For. Exp. Stn., New Orleans, La. 9 p.
(USDA For. Serv. Res. Pap. SO-99)

Phosphorus fertilization and mechanical seedbed preparation influenced growth of slash pine on three Plummer sand sites that
varied in wetness. On dry and intermediate sites, tree heights at
age 8 years were about equal on flat-disked strips, beds elevated
about 5 inches, and beds elevated about 10 inches. On the wet site,
high beds excelled. Top dressing with phosphorus at age 2 years
stimulated growth by 2 to 5 feet on all seedbeds of the wet and
intermediate sites.

Additional keywords: *Pinus elliottii*, phosphorus fertilization,
seedbed preparation, bedding, growth.

Response of Slash Pine to Bedding and Phosphorus Application in Southeastern Flatwoods

W. F. Mann, Jr., and J. M. McGilvray[1]

Phosphorus fertilization and mechanical seedbed preparation influenced growth of slash pine on three Plummer sand sites that varied in wetness. On dry and intermediate sites, tree heights at age 8 years were about equal on flat-disked strips, beds elevated about 5 inches, and beds elevated about 10 inches. On the wet site, high beds excelled. Top dressing with phosphorus at age 2 years stimulated growth by 2 to 5 feet on all seedbeds of the wet and intermediate sites.

Additional keywords: *Pinus elliottii,* phosphorus fertilization, seedbed preparation, bedding, growth.

Slash pine (*Pinus elliottii* var. *elliottii* Engelm.) grows slowly on many poorly drained, sandy soils in the Southeast. During much of the year the water table is so high that growth is retarded. Some stands never reach merchantable size. Drainage, bedding, and fertilization—alone and in combination—are often used to boost the productivity of these sites (*2,3,4,5,6*).

Bedding, in which large disks are used to make ridges 5 to 6 inches high, is the most common practice. Bed height is governed by available equipment, and not by information about effects on tree growth. In the study reported here, a major purpose was to compare standard beds with ones about twice as high. Fertilization with phosphorus was also tested, since other investigators have reported substantial growth responses to this nutrient on similar soils (*3*). All treatments were installed on one prevalent soil type, but sites varied in wetness.

METHODS

The study area is in northwest Florida, about 15 miles northwest of Port St. Joe, in the low, flat area of the Lower Coastal Plain. The soil is a Plummer fine sand—relatively deep, and with about 2 percent of silt

[1] The authors are Chief Silviculturist and Biological Laboratory Technician, Southern Forest Experiment Station, USDA Forest Service, Pineville, La. 71360. They thank Dr. James B. Baker, Soil Scientist, Southern Forest Experiment Station, Stoneville, Miss., for assistance in prescribing and applying phosphorus.

and clay in the upper 4 feet (1). Phosphorus and nitrogen contents are low. Total phosphorus is about 30 ppm and total nitrogen varies from 20 to 430 ppm at depths to 42 inches. The carbon-nitrogen ratio is approximately 18:1 (1).

Elevation varies 1 or 2 feet with a gradual gradient to a shallow drain. On the lower portion, water stands on or near the surface for 8 to 10 months of the year. The drier part, on higher ground, is free of surface water during all but 2 or 3 months in spring. Vegetation was primarily native grasses intermixed with small stems of gallberry (*Ilex glabra* (L.) Gray) that sometimes occurred in clumps.

Three areas were selected for study: a wet, a dry, and an intermediate site that are within 1,000 feet of each other. Twelve plots, each measuring 50 by 200 feet and 20 feet apart, were established on each area. All were burned in January 1965, and the mechanical site treatments were installed soon thereafter. Treatments, replicated three times on each site, included:

Burn only.

Flat-disk strips twice to measure effects of competition control. Four strips, each 7.5 feet wide, were prepared on a plot (fig. 1).

With a disk, elevate beds 5 to 6 inches high and 7.5 feet wide, edge-to-edge. Four such beds were made on each plot.

Prepare beds about 10 to 12 inches high and 7.5 feet wide—4 beds per plot (fig. 2). A road grader was used on the dry site and a small dragline on the wet and intermediate sites.

Figure 1.—*Flat-disked strips gave excellent temporary control of competing grasses.*

itrogen contents;
iitrogen varies f
nitrogen ratio is

radient to a shal
r the surface for!
ound, is free of
Vegetation was
s of gallberry (
aps.
iry, and an inter
. Twelve plots, e
established on a
echanical site tr
ts, replicated th

:ompetition cont
ared on a plot (f

7.5 feet wide, e
.ch plot.
5 feet wide—1 b
n the dry site an
: sites.

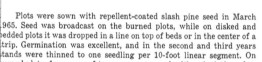

Figure 2.—*High beds on intermediate site were elevated about 10 to 12 inches with a small dragline. Natural pines in background are stunted, and few will reach merchantable size.*

Plots were sown with repellent-coated slash pine seed in March 1965. Seed was broadcast on the burned plots, while on disked and bedded plots it was dropped in a line on top of beds or in the center of a strip. Germination was excellent, and in the second and third years stands were thinned to one seedling per 10-foot linear segment. On burned plots, four rows of trees were reserved with a deviation of ± 2 feet from the centerline. Subsequent mortality resulted in some un-stocked segments, especially on unbedded plots of the wet site.

In March 1967, end 50-foot segments of beds were selected at ran-dom and broadcast with 100 pounds of elemental phosphorus per acre in the form of triple superphosphate.

Five plots on the wet area—three burned and two with standard beds—were destroyed by wildfire in the spring of 1968.

Total height has been measured annually on the two interior rows of each plot. Sample trees were flagged initially and used throughout the study. Where substitution was necessary, it was usually possible to find a replacement of about the same vigor and size as the original pine.

l of competing pine

3

RESPONSE TO SEEDBED PREPARATIONS

Dry Site

Data in figure 3 were collected on 150-foot, unfertilized segments of plots. Seedbed preparations had a significant (0.05 level) effect on heights every year from age 2 to 8 on all sites, but rankings of treatments were not the same on all sites.

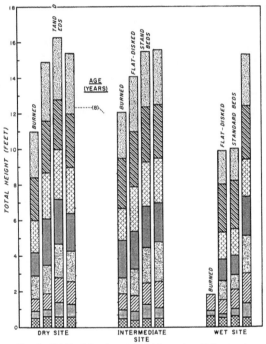

Figure 3.—*Heights of pines at ages 1 through 8 years, by seedbed treatments.*

4

Dry Site

After eight growing seasons on the dry site, heights on flat-disked, standard-bedded, and high-bedded plots did not differ significantly. Heights averaged 15.5 feet, and the maximum variation was 1.4 feet. On burned seedbeds, by contrast, heights averaged only 11.0 feet.

Height differences became evident in the early years, and the ranking at 8 years was the same as 2. In the last 4 years total growth in the three best treatments has been almost identical. Growth on the burned plots has been less in every year than on the other seedbeds, so that differences are still increasing.

Intermediate Site

Growth on standard and high beds on the intermediate site has been nearly the same throughout the study, with heights totaling 15.5 at 8 years. The greatest difference at any time was 0.3 foot, and growth in a single year never varied by more than this amount. At 8 years, pines on flat-disked plots averaged about 1.5 feet less than on both bedded plots. Most of this difference occurred in the first 4 years. Figure 4 shows 2-year-old trees on standard beds.

Figure 4.—*Two-year-old slash pines on standard beds on intermediate sites. Native grasses have reinvaded the area.*

Growth on burned and flat-disked seedbeds was nearly the same for the first 3 years, but in each year thereafter the trees on disked plots excelled. Disking is usually considered an aid to early growth. Thus, the delayed response was surprising, especially as the reduction of grass competition lasted no more than 2 years. At age 8, heights on burned plots averaged 2.0 to 3.5 feet less than on disked and bedded plots—differences that were significant and important.

Wet Site

The loss of five plots by fire after the third growing season complicates comparisons of seedbeds on the wet site.

As on the other sites, differences in height developed early. At age 3, pines were significantly taller on high beds than in all other treatments, exceeding heights on standard beds, the second-ranked treatment, by 43 percent. Heights on burned and flat-disked plots ranked third. Here as on the intermediate sites, disking was not better than burning alone.

Beginning with the fourth year, after two plots had been destroyed by fire, data for standard beds are based on one replication. It is noteworthy that at age 3 trees on the undamaged plot were shorter than in the other two replications—1.7 versus 2.1 and 2.4 feet. If all replications had remained intact, 8-year heights for the standard beds might have averaged between 12 and 14 feet instead of 10 feet. Foresters in this region know that bedding is essential on wet areas and that flat disking is not enough. What cannot be resolved, however, is whether high beds are superior to standard beds and, if so, what is the magnitude of the difference.

EFFECTS OF FERTILIZATION

The information on response to seedbed preparation was from 150-foot, unfertilized segments of plots. As data on the effects of fertilization were collected on 50-foot segments, subplots for comparisons are unequal in size.

On the dry site, fertilization with phosphorus had no significant effect on heights at 8 years on any of the seedbeds.

By the end of the fourth year fertilization had significantly increased heights on all seedbeds on the intermediate site (fig. 5). While differences were in the range from 0.3 to 1.0 foot, it is important to note that pines responded quickly to the phosphorus applied as a top dressing.

The fertilized trees continued to gain each year after age 4. By age 8 they were from 2.5 to 5.4 feet taller than the unfertilized, with no indications that gains were diminishing. It seems likely that phosphorus will promote growth for at least a few more years.

6

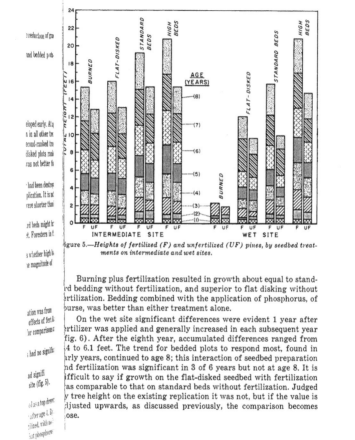

Figure 5.—*Heights of fertilized (F) and unfertilized (UF) pines, by seedbed treatments on intermediate and wet sites.*

Burning plus fertilization resulted in growth about equal to standard bedding without fertilization, and superior to flat disking without fertilization. Bedding combined with the application of phosphorus, of course, was better than either treatment alone.

On the wet site significant differences were evident 1 year after fertilizer was applied and generally increased in each subsequent year (fig. 6). After the eighth year, accumulated differences ranged from 4 to 6.1 feet. The trend for bedded plots to respond most, found in early years, continued to age 8; this interaction of seedbed preparation and fertilization was significant in 3 of 6 years but not at age 8. It is difficult to say if growth on the flat-disked seedbed with fertilization was comparable to that on standard beds without fertilization. Judged by tree height on the existing replication it was not, but if the value is adjusted upwards, as discussed previously, the comparison becomes close.

Figure 6.—*These 4-year-old pines are growing on high beds on the wet site. Those on the man's left were fertilized with phosphorus at age 2, and are obviously taller than unfertilized pines on his right.*

DISCUSSION

Results from small plots such as this are often conservative, especially as to effects of seedbed preparation. Many of the treatments interrupt natural drainage and tend to retard runoff or impound water. On large tracts, for example, beds would be oriented so as to improve surface drainage by channeling water into ditches and natural courses.

High beds, one of the main features of the study, were not more effective than standard beds on dry and intermediate sites. They were probably superior on the wet site, even if 8-year heights on standard beds are adjusted upward to compensate for the loss of the two best replications. But practical and economical equipment to build high beds is not available. Even if an efficient machine were developed, it is doubtful if the increased growth on the wet site was enough to offset the cost of moving so much soil.

Flat disking was as effective as standard bedding on dry and intermediate sites. Early growth was slightly better on the beds during the first 3 or 4 years, but almost identical after that. It is doubtful, then, if differences will increase in the future. Two major factors should enter into the decision on how to treat such sites: (1) difference in cost between flat disking and bedding, and (2) expected problems in managing and cutting timber on bedded tracts.

8

The in
ᵤᵣᵣoborate
growth on ɛ

Fertiliː
the dry site
ᵤₗₑₗ growtₕ
ₜₕrough the
these are at

Phosph
ₐₙd interm
without ferₜ
plications o
ₑₑₐₜₑ site

1. Gammon
1953. P
F

2. Maki, T
1969. D
T

3. Pritche
1964. I
ₑ

4. Pritchel
1969. I
o

5. Schultz
1968.

6. White,
1970. ₁

The inadequacy of flat disking on the wet site has been amply
orroborated in commercial operations. Burning alone gave the poorest
rowth on all sites, and some form of mechanical seedbed preparation
s justified even in the driest situation.

Fertilization has not improved growth on any of the seedbeds on
he dry site. On the intermediate and wet sites, however, it has stimu-
ated growth appreciably. Responses have become larger each year
hrough the eighth, and some differences are now as great as 6 feet.
These are strong indications that phosphorus applications will pay their
vay.

Phosphorus applied to burned and flat-disked seedbeds on the wet
nd intermediate sites gave comparable growth to standard bedding
vithout fertilization. For landowners who cannot finance bedding, ap-
blications of phosphorus may be a substitute. Of course, bedding and
ertilization were superior to either treatment alone on wet and intermediate sites.

LITERATURE CITED

1. Gammon, N., Jr., Henderson, J. R., et al.
   1953. Physical, spectrographic and chemical analyses of some virgin
   Florida soils. Fla. Agric. Exp. Stn. Bull. 524, 130 p.

2. Maki, T. E.
   1969. Drainage and soil moisture control in forest production. *In*
   The Ecology of Southern Forests, p. 139-157. Proc., La. State
   Univ. 17th Annu. For. Symp.

3. Pritchett, W. L.
   1964. Fertilizing slash pine on the sandy soils of Florida. Proc.,
   23rd Annu. Meet. Soil & Crop Sci. Soc. Fla., p. 104-108.

4. Pritchett, W. L.
   1969. Improvement in soil fertility as a means of increasing growth
   of forest trees. *In* The Ecology of Southern Forests, p. 183-198.
   Proc., La. State Univ. 17th Annu. For. Symp.

5. Schultz, R. P.
   1968. Soil or foliar fertilization of well-drained and flooded slash
   pine seedlings. USDA For. Serv. Res. Pap. SE-32, 8 p. Southeast. For. Exp. Stn., Asheville, N.C.

6. White, E. H., and Pritchett, W. L.
   1970. Water table control and fertilization for pine reproduction
   in the flatwoods. Fla. Agric. Exp. Stn. Bull. 743, 41 p.

The
in design
Sontag in

Contents

Smalley, Glendon W.

    1974. Development of pine-hardwood stands in north Alabama following improvement cuttings. South. For. Exp. Stn., New Orleans, La. 9 p. (USDA For. Serv. Res. Pap. SO-100)

Depleted pine-hardwood stands with adequate seed sources of desirable species were rehabilitated by stand-improvement measures. Nineteen-year gains in stand volume averaged 778 cubic feet per acre for a commercial operation which removed all merchantable low-value trees. Where the commercial cut was supplemented by removal of all culls and undesirable trees to a 2-inch diameter limit, gains were 1,498 cubic feet. Unmanaged check stands grew 427 cubic feet.

Additional keywords: Intermediate cutting, *Pinus taeda*, *Pinus echinata*, natural regeneration.

DEVELOPMENT OF PINE-HARDWOOD STANDS IN NORTH ALABAMA FOLLOWING IMPROVEMENT CUTTINGS

Glendon W. Smalley [1]

Depleted pine-hardwood stands with adequate seed sources of desirable species were rehabilitated by stand-improvement measures. Nineteen-year gains in stand volume averaged 778 cubic feet per acre for a commercial operation which removed all merchantable low-value trees. Where the commercial cut was supplemented by removal of all culls and undesirable trees to a 2-inch diameter limit, gains were 1,498 cubic feet. Unmanaged check stands grew 427 cubic feet.

Additional keywords: Intermediate cutting, *Pinus taeda, Pinus echinata*, natural regeneration.

In 1947 a study was begun in northern Alabama to appraise the effects of stand improvement in pine-hardwood forests that had been dly depleted by cutting and fire but still ntained seed trees of desirable species. The inds were typical of those found today on ousands of acres of the Cumberland Plateau. The Plateau area is predominantly pine untry. While the improvement cuttings vad in intensity, all had the object of increas-; pine stocking and reducing the amount of ace occupied by hardwoods of undesirable 'm or species. On the better sites, however, :h hardwoods as white oak and yellow-pop-· were favored along with pine.

Measurements made periodically for 19 irs showed that all intensities of stand imvement produced substantial gains in stock-;, basal area, and cubic volumes. These gains re obtained with minimum disturbance of

incipal Soil Scientist at the Silviculture Laboratory, untained at Sewanee, Tennessee, by the Southern Fores/ periment Station in cooperation with the University of › Sou/h.

the sites and at low capital investment. Treatments of the kind described should therefore prove of interest to owners of small as well as large tracts.

This paper describes stand development and discusses implications for forest management in the area today. The experimental plots were on the Flat Top Experimental Forest, maintained near Birmingham in cooperation with the U. S. Pipe and Foundry Co. The study ended in 1966, when the experimental forest was closed. Ten-year results were published earlier (Smalley 1958), and the present report is final.

STUDY AREA

The stands are in Jefferson County, Alabama, on rolling terrain ranging from 260 to 620 feet above sea level. Precipitation averages 53 inches annually with 25 inches falling during April through September. Average annual temperature is 64° F with an April-through-September average of 76°. The growing season is 240 days.

Physiographically, the area is at the southern extremity of the Cumberland Plateau (Fenneman 1938). The interbedded sandstone and shale give rise to moderately deep Enders and shallow Townley soils on the ridgetops. Soils on slopes are commonly shallow and belong to the Montevallo series. Stream-bottom soils, derived from colluvium and alluvium of the uplands, are members of the Pope-Philo-Stendal-Atkins drainage catena. Surface texture ranges from sandy loam to shaly silt loam.

Shortleaf pine *(Pinus echinata* Mill.) was the most common conifer, but many of the larger trees were dying of littleleaf disease

(Phytophthora cinnamoni Rands). Loblolly *(P. taeda* L.), Virginia *(P. virginiana* Mill.), and longleaf pines *(P. palustris* Mill.) of seed-bearing age were present, but fires had killed most of their reproduction. The predominant young pine growth was shortleaf. which had sprouted after a fire 5 years earlier and was mostly under 6 feet tall.

Hardwoods were chiefly oaks *(Quercus* spp.), hickories *(Carya* spp.), and blackgum *(Nyssa sylvatica* Marsh.) under 12 inches d.b.h., of low quality, and suitable only for mine props or crossties. White oak *(Q. alba* L.) and yellow-poplar *(Liriodendron tulipifera* L.) reproduction was scarce.

On the average, the original stands contained 168 trees (> 3.5 inches d.b.h.) per acre with a basal area of 47 square feet and a volume of 501 cubic feet. Pines accounted for 23 percent of the trees, 28 percent of the basal area, and 41 percent of the volume.

Eighty-six percent of the trees were less than 10 inches d.b.h., with about equal numbers occurring in the 4- to 5- and the 6- to 9-inch classes. Diameter distribution within species groups varied considerably. Nearly one-half of the pines were 6 to 9 inches, and 20 percent were 10 inches and larger. Of the sound hardwoods, 49 percent were in the 4- to 5-inch class and 41 percent in the 6- to 9-inch class.

METHODS

Treatments

Initially the study tested three intensities of stand improvement. It was reasoned that a landowner's choice of treatment would depend on the amount of capital he wished to invest and how soon he desired a well-stocked stand.

Treatment 1 (COM) was a commercial improvement cut that was expected to provide a small net profit. All undesirable trees (that is, other than pines, white oak, and yellow-poplar) were removed, provided that any tree cut would at least pay for itself under current logging and market conditions. When left, undesirable trees were of relatively good quality and were not overtopping desirable trees. Some desirable trees were cut where thinning was needed.

Thus the study originally consisted of eight replications of four treatments randomized in four blocks. The null hypothesis that volume and growth variables of the treated plots did not differ from those of the check was tested by analyses of variance at the 0.05 probability level. Differences between treatment means were detected by Duncan's multiple-range test.

After the treatments were repeated in 1958, the design consisted of eight treatments in each of four blocks. The total number of plots remained the same, but there were no replications within blocks.

Inventories

All stems > 3.5 inches d.b.h. were inventoried in 1947, 1953, 1957, 1962, and 1966. A reinventory was made after the cut of 1948, but after the treatments of 1958 and 1962 the residual stands were determined by subtracting the cut tally from the pre-cut inventory. Volumes of merchantable trees were calculated from a local table in terms of gross cubic feet and gross board feet (International ¼-inch rule), inside bark. During the treatments, a tally was kept of the number of trees removed in classes below 3.5 inches.

IMMEDIATE STAND CHANGES

Effects of the initial treatments are reported as averages of the combined single and repeat plots, since there was no distinction between them in 1948.

One-half or more of the trees in each of the three largest diameter classes were harvested from the COM & COMREP stands (table 1). The result was a reduction of 33 percent in stocking, 48 percent in basal area, and nearly 50 percent in volume. The ratio of pine to total stocking increased from 28 to 42 percent.

In the 6IN and 6INREP treatments, 61 percent or more of the trees in the three largest diameter classes were harvested or deadened. The result was a reduction of 38 percent in stocking, 52 percent in basal area, and 44 percent in volume. Pine composition increased from 30 to 45 percent.

In the 2IN and 2INREP stands about 75 percent of the trees in the three largest diameter classes and 94 percent of those in the 4- and 5-inch classes were harvested or deadened. In addition 99 nonmerchantable (< 3.5 inches d.b.h.) hardwoods were deadened to release seedlings and saplings of pine or desirable hardwoods (table 2). Basal area was reduced 78 percent and volume 66 percent. Pine constituted 14 percent of the stands before and 85 percent after cutting.

Following the initial treatments the stands were two-storied, having an overstory of residual pines and desirable hardwoods and a patchy understory of pine seedlings and saplings.

RESPONSE AFTER 19 YEARS

Long-term stand response depended on the intensity of release, number of treatments, amount of established reproduction, and species composition.

Table 1.—*Stand structure per acre, before and after initial treatments*

Stand characteristic	COM & COMREP		6IN & 6INREP		2IN & 2INREP	
	Before	After	Before	After	Before	After
	– – – – – – – – – – Number – – – – – – – – – –					
Diameter class (inches)						
4-5	87	87	70	70	67	4
6-9	84	30	75	25	61	13
10-15	18	8	19	8	24	6
16+	2	1	2	1	2	1
	– – – – – – – – – – – Ft² – – – – – – – – – – –					
Basal area	48	25	46	22	44	10
	– – – – – – – – – – – Ft³ – – – – – – – – – – –					
Volume	513	261	499	279	476	160

3

Table 2.—*Per acre summary of unmerchantable hardwoods removed by treatment, diameter, and date*

Treatment and year	Trees per acre						Basal area
	Diameter breast height (inches)					Total	
	1	2	3	4	5		
	– – – – – – – – Number – – – – – – – –						Ft2
2IN							
1948	0	48	37	31	26	142	9
2INREP							
1948	0	62	51	41	28	182	11
1958	256	170	26	9	3	464	8
1963	77	144	14	5	4	244	5
Total	333	376	91	55	35	890	24
2INDEF							
1958	112	58	18	28	20	236	8
1963	339	141	17	7	2	506	7
Total	451	199	35	35	22	742	15

Stocking, Diameter Distribution, and Species Composition

Stocking, mostly pine, was increased substantially by all treatments (fig. 1). Increases ranged from 2.2-fold on the 2INDEF to 16-fold on the 2IN and 2INREP. The 6INREP had more trees than the 6IN, while the COM had more than the COMREP. There were about equal numbers on the 2IN and 2INREP plots.

Following initial treatment the 2INDEF was better stocked than the 2IN and 2INREP and it responded more quickly to release. In 5 years stocking nearly quadrupled. More than 500 unmerchantable trees were deadened when the treatment was repeated in 1963 (table 2).

For repeat treatments, but not for single treatments, pine stocking increased with intensity of release. Pine composition averaged 74 percent on the moderate treatments (COM, 6IN, COMREP, and 6INREP) and 93 percent on the intensive treatments (2INDEF, 2IN, 2INREP). Almost 900 unmerchantable trees were deadened in the 2INREP treatments.

Diameter distributions of the moderate treatments were similar throughout the experiment even though total stocking differed (fig. 2). Changes in the proportion of saw log size trees (10 inches d b.h. and above) were slight. On intensively managed plots, trees 10-15 inches in d.b.h. increased rapidly in number, and by 1966 constituted a larger proportion of the stands than on all other

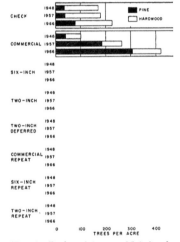

Figure 1.—*Number of trees > 3.5 inches d. by treatment, date, and species gr*

treatments except the CHK. Figure 2 s the development of this diameter class on REP plots. Numbers of saw log size t should continue to increase substantially ing the next decade.

4

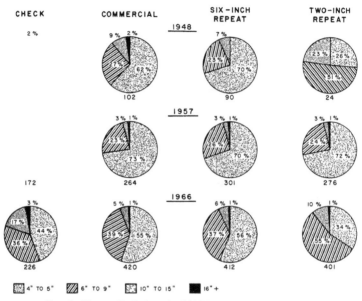

Figure 2.—*Diameter distributions after initial treatment (1948), and 10 (1957) and 19 (1966) years later. Number of trees per acre > 3.5 inches d.b.h. is shown below each chart.*

Repetition of the intensive treatment resulted in faster growth of residual trees and continued ingrowth. In 1966, the 2INREP had more 6- to 9-inch trees than any other treatment.

Stocking on the CHK increased about 50 trees. Most of these were pines which had struggled to measureable size. As a result, hardwood composition decreased from 79 to 64 percent.

There was a small but continuing loss of shortleaf pine from littleleaf disease; the average for all plots was 4.4 cubic feet per acre per year. Pine mortality from other causes was slight. Hardwood mortality was obviously related to treatment. On the 6IN, 2IN, 6IN-REP, and 2INREP the loss of hardwoods was negligible because those most apt to die were harvested or deadened. On the CHK hardwood losses averaged (by coincidence) 4.4 cubic feet per acre per year—mostly red oaks and hickories.

Basal Area

In 1966 the basal area of all stands exceeded 70 square feet (fig. 3). Pine made up 73 to 98 percent of the basal area on the treated plots, 34 percent on the CHK.

On the repeat, but not on the single treatments, both total and pine basal area increased with intensity of release. Total and pine basal area on the 2INREP and 6INREP were greater than on the corresponding single treatments. Total basal area on the COM exceeded that on the COMREP; pine basal area was equal.

5

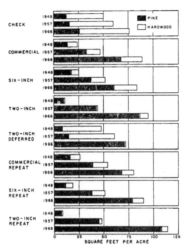

Figure 3.—*Basal area of trees* > *3.5 inches d.b.h.*

Basal area growth was computed by subtracting 1948 values from those present in 1966 and then adding the amount removed in the 1958 and 1963 cuts. No allowance was made for trees that died. As figure 4 shows, pine and total basal area growth on the repeat treatments increased with intensity of release. Ac-

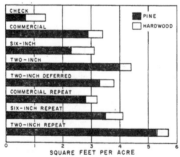

Figure 4.—*Mean annual basal area growth of trees* > *3.5 inches d.b.h., 1948-1966.*

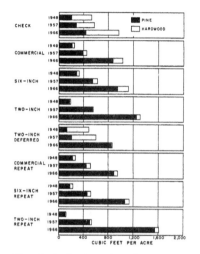

Figure 6.—*Stand volume of sound trees > 3.5 inches d.b.h.*

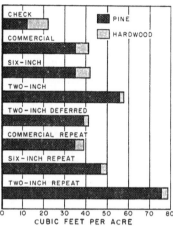

Figure 7.—*Mean annual volume growth of sound trees > 3.5 inches d.b.h., 1948-1966.*

creasing intensity of release for both single and repeat cuts. Growth on the 6INREP and 2INREP exceeded that on the corresponding single treatments.

The rate of volume growth on the 2INREP was significantly greater and that on the CHK was significantly smaller than the rates on all other treatments (table 3). Mean annual increments of the moderate treatments did not differ significantly. Except on the CHK, pine accounted for most of the growth.

Table 3.—*Summary of analyses of variance of cubic foot volume and growth* [1]

				STAND VOLUME—1966				
Treatment	2INDEF	CHK	COMREP	COM	6IN	6INREP	2IN	2INREP
Mean [2]	432.5	482.1	510.8	514.8	563.4	569.7	653.6	803.6

			STAND MEAN ANNUAL INCREMENT—1948-66					
Treatment	CHK	2INDEF	COM	6IN	COMREP	6INREP	2IN	2INREP
Mean	11.2	20.4	20.4	21.0	21.4	25.0	29.2	39.4

				PINE VOLUME—1966				
Treatment	CHK	2INDEF	COM	COMREP	6IN	6INREP	2IN	2INREP
Mean	218.8	427.9	438.1	473.2	474.0	538.8	627.1	775.0

			PINE MEAN ANNUAL INCREMENT—1948-66					
Treatment	CHK	6IN	COM	2INDEF	COMREP	6INREP	2IN	2INREP
Mean	6.9	17.4	17.8	18.3	19.1	23.7	28.0	37.9

[1] Means underscored are not significantly different at the 0.05 level of probability.
[2] Means are the average of four 0.5-acre plots.

Pine Volume

Deadening culls and small competing hardwoods (2IN) resulted in a 6.6-fold increase in pine volume, and where this intensive treatment was repeated (2INREP) pine volume was 14.2 times greater (fig. 6). The moderate treatments registered gains of 3.3 to 5.9-fold. Pine volume increased with intensity of release for both single and repeat treatments. Volumes on repeat treatments exceeded those on the corresponding single treatments, but in the moderate treatments the differences were small.

Pine volume was significantly smallest on CHK plots (table 3). Other treatments ranked as they did for stand volume.

Pine growth rates increased with intensity of release for repeat cuts (fig. 7). Rates on repeat treatments exceeded those on single treatments, but the COM and COMREP differed by only 0.7 cubic foot. Results of the statistical analysis (table 3) were the same as for stand volume growth.

For most treatments, 1963-1966 periodic annual increment is a better indicator of future growth than are averages for the entire study. During this period, 2INREP plots had an annual increment of 124 cubic feet of pine (fig. 8). In stands with good pine stocking after initial treatment, as in 2INDEF (58 pines per acre), volume growth in excess of 100 cubic feet per acre can probably be achieved in about a decade. Estimated average growth potential of Alabama forests is 90 cubic feet per acre annually (Murphy 1973).

Saw Log Volume

Insufficient time elapsed for the stands to develop much volume in trees > 9.5 inches d.b.h. In 1966 saw log volume was greatest on the CHK—3,900 fbm, of which 52 percent was hardwood. On the other treatments sawtimber was mostly pine with a small amount of white oak and yellow-poplar. The 1966 diameter distributions (fig. 2) suggest an early and substantial gain in board-foot volume.

MANAGEMENT AND SILVICULTURAL IMPLICATIONS

A combination of practices—cleaning, improvement, liberation, salvage, and some thin-

north slopes are favored for yellow-poplar (Russell et al. 1970, Smalley 1969, Smalley and Pierce 1972), and breaking up large areas of pine types with quality hardwoods is advantageous to wildlife habitat as well as aesthetically pleasing.

LITERATURE CITED

Campbell, W. A., and Copeland, O. L.
    1954. Littleleaf disease of shortleaf and loblolly pine. USDA Circ. 940, 41 p.

Fenneman, N. M.
    1938. Physiography of Eastern United States. 714 p. McGraw-Hill Book Co., N. Y.

Jemison, G. M., and Hepting, G. H.
    1949. Timber stand improvement in the southern Appalachian region. USDA Misc. Pub. 693, 80 p.

Murphy, P. A.
    1973. Alabama forests: trends and prospects. USDA For. Serv. Resour. Bull. SO-42, 36 p. South. For. Exp. Stn., New Orleans, La.

Russell, T. E., Loftus, N. S., Mignery, A. L., and Smalley, G. W.
    1970. Planting yellow-poplar in central Tennessee and northern Alabama. USDA For. Serv. Res. Pap. SO-63, 17 p. South. For. Exp. Stn., New Orleans, La.

Smalley, G. W.
    1958. Stand improvement pays off. South. Lumberman 197(2465): 100-102.

Smalley, G. W.
    1969. Ten-year growth of yellow-poplar planted in north Alabama. J. For. 67: 567-568.

Smalley, G. W., and Pierce, K.
    1972. Yellow-poplar, loblolly pine, and Virginia pine compared in Cumberland Plateau plantations. USDA For. Serv. Res. Note SO-141, 6 p. South. For. Exp. Stn., New Orleans, La.

U.S. Department of Agriculture
Forest Service Research Paper SO-101

Wood Machining Abstracts, 1972 and 1973

Charles W. McMillin

Southern Forest Experiment Station
Forest Service
U.S. Department of Agriculture
1974

Contents

WOOD MACHINING ABSTRACTS, 1972 AND 1973

Charles W. McMillin

The purpose of this paper is to abstract significant od-machining literature that has not been viously digested in English-language texts and liographies.[1] The entries mainly represent lications of a research nature, but trade journal ar-es have been included when judged to be of in-est. While most of the literature is from 1972 and '3, an attempt was made to include important ers that had been missed in the earlier searches.

'rincipal sources of information were archival rnals in wood science and technology, but *Forestry stracts*, published by the Commonwealth ricultural Bureaux, Farnham Royal, England, and Bureaux' Card Title Service were invaluable. The rch also included the *Abstract Bulletin of the titute of Paper Chemistry* and publications lists l annual and periodic reports issued by the regional DA forest experiment stations and the Forest ducts Laboratory, Madison, Wisconsin. Patents e largely noted from the listing carried in the 1972 l 1973 issues of the *Forest Products Journal*.

Many abstracts of English-language publications e written by the author after a reading of the ginal. Abstracts of foreign-language publications e drawn largely from *Forestry Abstracts* and the stract *Bulletin,* though with minor editorial nges, deletions, or amplifications. The author ognizes that limits of time and linguistic ability ve probably caused omissions and interpretations.

'or continuity and convenience to the reader, the hor has attempted to duplicate the style and nat of the previous reviews in this series.[1] Each er is abstracted under its major subject-matter l; the large categories are subdivided. To enhance ty, papers are cross-referenced in all pertinent fields. Cross-references take the form:

Doe. 1972. (Sawing; Bandsawing)

Such a reference indicates that the full citation appears in the Sawing category, under the field Bandsawing. The subject-matter heads are summarized in the table of contents. An index of authors appears on page 44.

This review indicated that the total research effort is continuing at a high level, if not accelerating. About 66 publications per year appeared in archival journals for the years 1963 through 1965. The reviews of 1966 through the present have references as follows:

Review and year	References
Koch (1968)	
1966	138
1967	103
McMillin (1970)	
1967	36
1968	187
1969	89
Koch (1973)	
Prior to 1969 or undated	35
1969	71
1970	153
1971	156
Present review	
Prior to 1971 or undated	16
1971	64
1972	199
1973	148

hief previous compilations
ittee on Recent Wood Machining Literature **Wood machining abstracts, 1957-1958.** For Prod Res Soc., Madison, Wis 20 p 1959
ittee on Recent Wood Machining Literature **Wood machining abstracts, 1958-1959.** For Prod Res Soc., Madison, Wis 19 p [1960]
ittee on Recent Wood Machining Literature **Wood machining abstracts, 1959-1961.** For Prod Res Soc., Madison, Wis 18 p 1961
P **Wood Machining Processes.** 530 p N Y.: Ronald Press Co 1964
P., and McMillin, C. W **Wood machining review, 1963 through 1965.** For Prod J Part I, 16 (9) 76-82, 107-115, Part II, 16 (10) 43-48
966
P **Wood machining abstracts, 1966 and 1967.** USDA For Serv. Res Pap SO-34, 38 p South. For Exp Stn., New Orleans, La 1968
P **Wood machining abstracts, 1970 and 1971.** USDA For Serv Res Pap SO-83, 46 p South For Exp Stn., New Orleans, La 1973
llin, C W **Wood machining abstracts, 1968 and 1969.** USDA For Serv Res Pap SO-58, 35 p South For Exp Stn., New Orleans, La
970

Table 1. — Number of references on wood machining research for 1972 and 1973, classified into 43 fields of activity

Subject	Primary listings	Cross references	Total
History and general texts	9	4	13
Properties of wood	13	9	22
Orthogonal cutting	15	4	19
Peripheral milling parallel to grain	7	1	8
Barking	28	1	29
Sawing			
General	20	9	29
Bandsawing	8	9	17
Sash gangsawing	11	4	15
Circular sawing	8	20	28
Chain sawing	12	10	22
Machining with high-energy jets	5	4	9
Machining with light	3	1	4
Shearing	9	3	12
Total sawing	76	60	136
Jointing, planing, molding, and shaping	10	12	22
Turning	1	1	2
Boring, routing, and carving	6	0	6
Mortising and tenoning	0	1	1
Machining with coated abrasives	13	6	19
Veneer cutting	28	21	49
Chipping, flaking, and grinding			
Chipping	12	9	21
Chipping headrigs and edgers	7	4	11
Mobile chippers	3	3	6
Flaking	1	0	1
Grinding	0	1	1
Total chipping, etc.	23	17	40
Defibrating	56	9	65
Properties of the cutting edge and cutter			
Tool material	2	2	4
Dulling	6	0	6
Fitting and sharpening	7	1	8
Stability	21	4	25
Temperature	1	4	5
Friction	3	1	4
Total cutting edge and cutter	40	12	52
Computer, tape, and card control of machines	7	1	8
Research instrumentation and techniques	14	11	25
Nomenclature	1	2	3
Machinability of species and reconstituted wood			
Hardwoods	5	13	18
Softwoods	9	18	27
Plywood	0	0	0
Particleboards and sandwich boards	3	2	5
Fiberboards, hard and soft	1	0	1
Total machinability	18	33	51
Bibliographies	2	1	3
Reports of research programs and industrial developments	10	2	12
Patents	19	0	19
Safety, including noise	31	1	32
	427	209	636

literature pertinent to the Society's 32 subject-matter divisions. These divisions range from harvesting timber to manufacture of finished products — including mechanical, but not chemical pulp. Users can order abstracts directly from the Society or can obtain reference data in the form of computer printouts. Also available to subscribers is a simple desk-top microfiche system. Rapid manual retrieval of citations and abstracts is made possible through comprehensive keyword and species indexes.

It is intended that members of the Southern Forest Experiment Station staff will continue to search the literature and add citations and abstracts to the AIDS system as pertinent material is uncovered. For this reason, the Station will discontinue publication of "Wood Machining Abstracts" with the present paper.

GENERAL TEXTS

Koch, P. 1972. UTILIZATION OF THE SOUTHERN PINES. VOL. I: THE RAW MATERIAL. VOL. II: PROCESSING. U.S. Department of Agriculture, Agriculture Handbook 420, 1663 p.
Of this 1,663 page textbook, 193 pages (including 116 illustrations) are devoted to wood machining. While the subject is not treated to the depth of a previous textbook by the author (Koch, P 1964. WOOD MACHINING PROCESSES. 530 p. The Ronald Press Co. New York.) much recent research and new data specific to the machining of southern pine is included. The chapter cites 148 references. The book is available from the Superintendent of Documents, U. S. Government Printing Office, Washington, D.C. Subject matter areas covered are: Historical background, Orthogonal cutting, Shearing and cleaving, Peripheral milling parrallel to grain, Barking, Converting with chipping headrigs, Sawing, Planing and moulding, Machining with coated abrasives, Veneer cutting, Chipping, Boring, Machining with high-energy jets, Machining with lasers and Literature cited.

Koch. 1973. (Reports of research programs and industrial developments)

Larionov, A. I., Kuritsyn, V. N., and Lukashin, M. M. 1972. FEATURES ON THE CUTTING OF FROZEN WOOD 57 p. Izadatel'stvo Lesnaya Promyshlennost'. Moscow.
A compilation of data on effects of low temperature on the physical and mechanical properties of wood; fundamental aspects of cutting frozen wood, and practical use of circular saws and chain saws. Barking of frozen wood is but briefly discussed

Martyr. 1973. (Sawing; General)

Metz, J. 1972. MACHINING AND SANDING — DATELINE **1980.** Wood and Wood Products 77(5):42-43.
A brief, speculative look at woodworking machinery of the future

Soine, H. 1973. MODERN FURNITURE MANUFACTURE. PART 2. Sizing and trimming of boards, working of edges, folding,

boring, wrapping, auxiliary equipment. Holz als Roh-und
Werkstoff 31:181-197. Munich.
*A survey of processing steps in the manufacture of furniture.
Environmental aspects are given some consideration.*

PROPERTIES

Anonymous. 1972. THE COMMERCIAL TIMBERS OF
SOUTHEAST ASIA. Revue du Bois et de ses Applications
27(12):43-51. Paris.
*A tabular summary of useful information on the more important
Southeast Asian woods, arranged alphabetically by commercial
names. Coverage includes distribution and general characteristics of
the tree, physical and mechanical properties, working
characteristics, durability, and uses of the wood.*

Borgin, K. 1971. THE COHESIVE FAILURE OF WOOD
STUDIED WITH THE SCANNING ELECTRON
MICROSCOPE. Journal of Microscopy 94, Part 1:1-11. Oxford.
*Wood samples, mainly pines, were stressed to failure in tension
parallel to the grain and shear parallel or perpendicular to the grain
and examined in the scanning electron microscope. Rate of loading
and temperature were varied between wide limits. In tension,
fissures developed between microfibrils causing rupture
perpendicular to the longitudinal axis. At both high and low
temperatures, the matrix in which the microfibrils are embedded,
and to a much lesser extent the middle lamella between fibers, failed
by plastic flow when the load was applied slowly and strands or
bundles of microfibrils were pulled out and fractured in an irregular
pattern. Cohesive failure in tension caused by sudden impact was
characterized by limited slippage of microfibrils and a near brittle
failure of the embedding matrix. A clean cleavage approaching a
true brittle failure was observed when samples rapidly stressed in a
direction perpendicular to the grain. When stressed in shear along
the grain and tension across the grain, cohesive failures showed that
the weakest part of the ultrastructure were the bonds between
parallel strands of microfibrils. However, the layer between the
middle lamella and the cell wall (or the middle lamella itself) failed
when the wood was subjected to cleavage parallel to the grain. If
cleavaged rapidly, as in machining, rheological properties of the
middle lamella did not permit plastic deformation with subsequent
failure, and the bulk structure of the tissue behaved almost as a
homogeneous material*

Chow, Walters, and Guther. 1971. (Defibrating)

DeBaise, G. R. 1972. MORPHOLOGY OF WOOD SHEAR
FRACTURE ASTM Journal of Testing and Evaluation 7(4):568-
572.
*The morphologies of parallel and perpendicular to the grain shear
fractures were found to depend upon the testing temperature and
moisture content of the wood. At -50°C, parallel shear failures
produced smooth fracture surfaces dominated by intracellular
failures At room temperature and above, parallel shear failures
produced a fuzzy surface composed of pulled out fibers; their
number was related to moisture content. As the moisture content
approached the fiber saturation point, the number of pulled out
fibers and the transition from intracellular to intercellular fracture
increased. In perpendicular to the grain shear, fracture morphology
was always smooth. It was predominantly intracellular at low
moisture content and intercellular at high moisture content. The
velocity of crack propogation had a direct effect on the type of
fracture generated Rapid crack extension was characterized by*

angle accounted for 60 to 90 percent of the variation in ultimate load. For deflection, the corresponding range was from 40 to 95 percent. It is concluded that for an accurate analysis of small cantilever beams in both wood cutting and structural problems, the orthotropic theory of elasticity must be modified as a function of beam depth.

Saiki, H., Furukawa, I., and Harada, H. 1972. AN OBSERVATION ON TENSILE FRACTURE OF WOOD BY SCANNING ELECTRON MICROSCOPE. Kyoto University Forestry Bulletin 43, p. 309-319. Kyoto, Japan.
A detailed examination of mainly softwood samples designed to reveal the structural features of ruptured cell wall surfaces from microtensile specimens or from the tensile surface of beams Contains 14 micrographs.

Scurfield, G., Silva, S. R., and Wold, M. B. 1972. FAILURE OF WOOD UNDER LOAD APPLIED PARALLEL TO GRAIN· A STUDY USING SCANNING ELECTRON MICROSCOPY. Micron 3:160-184. London.
A scanning electron microscope was used to examine the morphology and anatomy of compression failure zones in blocks loaded parallel to the grain or in specimens removed from the compression side of beams subjected to ¼-point loading. The specimens were prepared from (a) 17 species of hardwood; (b) Eucalyptus regnans before and after treatment with liquid ammonia or neutral sulphite pulping liquor; (c) 8 species of hardwood impregnated with polymethyl methacrylate polymerized from the monomer by gamma-irradiation; and (d) the reaction of wood of E. goniocalyx. The origin and mode of propagation of compression failure are discussed in relation to the observations. Quantitative estimates indicated that in untreated samples, strain at failure increased and the ratio of compressive strength to specific gravity decreased with increasing fiber length. Values for the compressive strength and strain at failure of the polymer impregnated specimens differed from those predicted by an additive model suggesting an interaction between cell walls and the polymer.

Simonov and Maiorova. 1971. (Barking)

Tomin, M. 1972. INFLUENCE OF ANISOTROPY ON FRACTURE TOUGHNESS OF WOOD. Wood Science 5:118-121.
It is shown that the anisotropic characteristics of wood strongly influence the rate of dissipation of strain (energy release rate) with crack extension. For a body in a state of generalized plane stress two basic crack extension modes are affected. The energy release rate for the opening mode is governed by elastic properties perpendicular to the crack plane. In the second forward sliding mode, the influence of the different elastic properties in the principal directions is more pronounced. It is concluded that the model of linear fracture mechanics for orthotropic bodies provides a better description of fracture behavior in wood than constitutive equations derived for isotropic bodies.

AL CUTTING

Splitter, mounted on a 2-ton lorry, and can quarter logs 96 inches long and 28 inches in diameter The log is placed on the splitter by a knuckle-boom loader installed behind the cab of the lorry. Both loader and splitter are operated hydraulically by power supplied from an auxiliary 60-hp engine fitted at the back of the cab. The quartered pieces fall on either side of the lorry. The splitter can handle about 60 oversize logs per hour.

Altman, J. A. 1973. MOVEABLE KNIFE LOG SPLITTER. Technical Release 73-R-11, 2 p. American Pulpwood Association. New York.
A brief description of a hydraulically operated log splitter having a 4-way knife to quarter hardwood and softwood logs as large as 28 inches in diameter and 96 inches long.

Anonymous. [n.d.] RUPTURE PATTERN IN CUTTING WOOD. CSIRO Division of Building Research. Forest Products Laboratory. Melbourne.
A 16 mm, black and white sound film showing rupture patterns that occur when wood is cut in each of the three cardinal directions (0/90, 90/0, and 90/90) and in the pulp chipping orientation (90/45)

Azarenok, V. A. 1972. INVESTIGATION OF THE PROCESS OF CUTTING WITH AND WITHOUT THE FORMATION OF CHIPS Lesnoi Zhurnal 15(1).78-82. Arkhangelsk.

Hamamoro, K., and Mori, M. 1972. FUNDAMENTAL STUDIES ON LOW FREQUENCY CUTTING OF WOOD. II. MECHANICS OF LATERAL VIBRATORY CUTTING. Journal of the Japan Wood Research Society 18:337-342. Tokyo.
The effects of lateral vibratory amplitude (0.03-4.4 mm), frequency (10-70 Hz), and feed speed (0.003-8.62 mm/sec) on chip formation and wood surface quality were studied using a reciprocating slide block apparatus having a 3-component dynamometer on the feed carriage Tests were made for the 90°-90°, 90°-0°, and 0°-90° cutting directions using samples of Formosan cypress, red lauan, and Formosan cypress, respectively. In the 90°-90° cutting direction, the lateral motion of the knife (rake angle 50° reduced ·fiber deformation ahead of the cutting edge. Since chips were of the shear type, surface quality was improved For 90°-0° vibratory cutting with a rake angle of 30°, surfaces were smooth as long as the grain was parallel to the cutting plane. For veneer cutting (0°-90° direction) with a rake angle of 70°, the frequency of lathe checks was greater with vibratory cutting than when vibration was not used The radius of chip curvature decreased with increasing frequency and decreasing feed speed Vibratory amplitude had little effect on roughness for all three cutting directions after an initial small amplitude was established. Figures and tables have English captions

Hamamoro, K., and Mori, M. 1972. FUNDAMENTAL STUDIES ON LOW FREQUENCY CUTTING OF WOOD. III. EFFECTS OF VIBRATORY FACTORS ON CUTTING FORCE IN LATERAL VIBRATORY CUTTING. Journal of the Japan Wood Research Society 18:343-348. Tokyo.
The cutting apparatus and dynamometer used in previous work were used to study the effect of vibratory frequency (10-70 Hz), amplitude (0.03-4 4 mm), and feed speed (0.003-8 62 mm/sec) on cutting forces for each of the 3 principal cutting directions. Results are presented in graphs with English captions. The ratios of parallel and normal cutting-force components in lateral vibratory cutting to those in conventional cutting increased as the ratio of feed speed to the maximum velocity of the vibrating knife edge increased.

Hamamoro, K., and Mori, M. 1972. FUNDAMENTAL STUDIES ON LOW FREQUENCY CUTTING OF WOOD. IV. RELATION BETWEEN EFFECTIVE RAKE ANGLE AND CUTTING FORCE IN LATERAL VIBRATORY CUTTING. Journal of the Japan Wood Research Society 18:387-392. Tokyo.

of inclination 10°, clearance angle 6-16°, and rake angle 39-49°

Sukhanov 1973. (Sawing; Sash gangsawing)

Tomin. 1972. (Properties of wood)

Varbanov, I 1972. THE POSSIBILITY OF REPLACING THE SANDING OF WOOD BY SCRAPING WITH A FIXED KNIFE. Drevo 27(3).68-69, 76. Prague.
Oak, beech, and beech-veneered panels at 8 and 10 percent moisture content were forced past an electrically heated (20-400°C) fixed knife having negative rake angles of -10 to -14 degrees At a cutting speed of 12-30 m/min, a surface layer 0 1 to 0.5 mm was removed by a scraping action Surface smoothness values are tabulated as functions of feed speed, workpiece moisture content, and knife temperature (optimum was 300°C) Surface quality was as good as that obtained by abrasive sanding

Vries, J. de. 1973. THE SPLITTING OF PULPWOOD LOGS INTO BILLETS. Leaflet 117, 32 p Forestry and Timber Bureau Australia.
Describes, with illustrations, various stationary and mobile machines and manual methods used in Australia for splitting pulpwood logs into billets. Species processed, production rates, and costs are tabulated

G PARALLEL TO GRAIN

CHIPS. Bumazhaya Promyshlennost' (4):22. Moscow.
Experiments were made to determine the effect of sharpness angle of the chip breaker (30 to 90°) and the distance between the breaker and the cutter (6 to 19 mm) on chip thickness and quality Chip thickness increased when the sharpness angle was greater than 52° and decreased with increasing distance. Chips of the highest quality were obtained with sharpness angles in the range of 30-50° and distances less than 19 mm

Sima, P., Mihut, I., and Lazar, M. V. 1972. A NEW METHOD OF IMPROVING THE QUALITY OF MECHANICALLY PROCESSED SURFACES. Industria Lemnului 23(10) 421-426. Bucharest.
Describes a system for imparting axial oscillations to a rotary cutterhead (perpendicular to the feed direction and parallel to the cutting plane) and the kinematics of the knife edge Trials of planing softwood with a knife having a deliberately notched edge showed that such oscillatory action greatly improved surface quality, especially at low feed speeds

Taranu, N. 1971. STUDY OF THE VARIATION IN THE DYNAMIC ELEMENTS OF PERIPHERAL MILLING AS A FUNCTION OF THE REGIME DURING THE CUTTING PROCESS. Industria Lemnului 22(12) 479-485. Bucharest
Reports, in a series of graphs and polynomial regression equations, the influence of the cutting speed, feed speed, depth and width of cut, and rake angle on cutting force and power, and on feed force in the milling of Carpinus betulus, Fagus sylvatica and Quercus robur

Tikhonov, V. A. 1971. CALCULATION OF THE GEOMETRY AND FORCE CHARACTERISTICS OF A CUTTING INSTRUMENT FOR LONGITUDINAL CUTTING OF WOOD. Lesnoi Zhurnal 14(5).79-86. Arkhangelsk.
A theoretical analysis of forces in planing

quartered pieces fall on either side of the lorry. The splitter can handle about 60 oversize logs per hour.

Altman, J. A. 1973. MOVEABLE KNIFE LOG SPLITTER. Technical Release 73-R-11, 2 p. American Pulpwood Association. New York.
A brief description of a hydraulically operated log splitter having a 4-way knife to quarter hardwood and softwood logs as large as 28 inches in diameter and 96 inches long.

Anonymous [n.d.] RUPTURE PATTERN IN CUTTING WOOD. CSIRO Division of Building Research. Forest Products Laboratory. Melbourne.
A 16 mm. black and white sound film showing rupture patterns that occur when wood is cut in each of the three cardinal directions (0/90, 90/0, and 90/90) and in the pulp chipping orientation (90/45).

Azarenok, V. A. 1972. INVESTIGATION OF THE PROCESS OF CUTTING WITH AND WITHOUT THE FORMATION OF CHIPS. Lesnoi Zhurnal 15(1):78-82. Arkhangelsk.

Hamamoto, K., and Mori, M. 1972. FUNDAMENTAL STUDIES ON LOW FREQUENCY CUTTING OF WOOD. II. MECHANICS OF LATERAL VIBRATORY CUTTING. Journal of the Japan Wood Research Society 18:337-342. Tokyo.
The effects of lateral vibratory amplitude (0 03-4.4 mm), frequency (10-70 Hz), and feed speed (0.003-8.62 mm/sec) on chip formation and wood surface quality were studied using a reciprocating slide block apparatus having a 3-component dynamometer on the feed carriage. Tests were made for the 90°-90°, 90°-0°, and 0°-90° cutting directions using samples of Formosan cypress, red lauan, and Formosan cypress, respectively. In the 90°-90° cutting direction, the lateral motion of the knife (rake angle 50° reduced fiber deformation ahead of the cutting edge. Since chips were of the shear type, surface quality was improved. For 90°-0° vibratory cutting with a rake angle of 30°, surfaces were smooth as long as the grain was parallel to the cutting plane. For veneer cutting (0°-90° direction) with a rake angle of 70°, the frequency of lathe checks was greater with vibratory cutting than when vibration was not used. The radius of chip curvature decreased with increasing frequency and decreasing feed speed. Vibratory amplitude had little effect on roughness for all three cutting directions after an initial small amplitude was established Figures and tables have English captions

Hamamoto, K., and Mori, M. 1972. FUNDAMENTAL STUDIES ON LOW FREQUENCY CUTTING OF WOOD. III. EFFECTS OF VIBRATORY FACTORS ON CUTTING FORCE IN LATERAL VIBRATORY CUTTING. Journal of the Japan Wood Research Society 18:343-348. Tokyo.
The cutting apparatus and dynamometer used in previous work were used to study the effect of vibratory frequency (10-70 Hz), amplitude (0 03-4 4 mm), and feed speed (0.003-8.62 mm/sec) on cutting forces for each of the 3 principal cutting directions. Results are presented in graphs with English captions. The ratios of parallel and normal cutting-force components in lateral vibratory cutting to those in conventional cutting increased as the ratio of feed speed to the maximum velocity of the vibrating knife edge increased

Hamamoto, K., and Mori, M. 1972. FUNDAMENTAL STUDIES ON LOW FREQUENCY CUTTING OF WOOD. IV. RELATION BETWEEN EFFECTIVE RAKE ANGLE AND CUTTING FORCE IN LATERAL VIBRATORY CUTTING. Journal of the Japan Wood Research Society 18:387-392. Tokyo.

6

of inclination 10°, clearance angle 6-16°, and rake angle 39-49°

Sukhanov. 1973 (Sawing; Sash gangsawing)

Tomin. 1972. (Properties of wood)

Varbanov, I. 1972. THE POSSIBILITY OF REPLACING THE SANDING OF WOOD BY SCRAPING WITH A FIXED KNIFE Drevo 27(3):68-69, 76. Prague.
Oak, beech, and beech-veneered panels at 8 and 10 percent moisture content were forced past an electrically heated (20-400°C)·fixed knife having negative rake angles of -10 to -14 degrees At a cutting speed of 12-30 m/min, a surface layer 0 1 to 0.5 mm was removed by a scraping action Surface smoothness values are tabulated as functions of feed speed, workpiece moisture content, and knife temperature (optimum was 300°C). Surface quality was as good as that obtained by abrasive sanding

Vries, J. de. 1973. THE SPLITTING OF PULPWOOD LOGS INTO BILLETS. Leaflet 117, 32 p Forestry and Timber Bureau. Australia.
Describes, with illustrations, various stationary and mobile machines and manual methods used in Australia for splitting pulpwood logs into billets. Species processed, production rates, and costs are tabulated

G PARALLEL TO GRAIN

CHIPS. Bumazhaya Promyshlennost' (4):22. Moscow.
Experiments were made to determine the effect of sharpness angle of the chip breaker (30 to 90°) and the distance between the breaker and the cutter (6 to 19 mm) on chip thickness and quality Chip thickness increased when the sharpness angle was greater than 52° and decreased with increasing distance Chips of the highest quality were obtained with sharpness angles in the range of 30-50° and distances less than 19 mm.

Sima, P , Mihut, I., and Lazar, M. V. 1972. A NEW METHOD OF IMPROVING THE QUALITY OF MECHANICALLY PROCESSED SURFACES. Industria Lemnului 23(10).421-426. Bucharest.
Describes a system for imparting axial oscillations to a rotary cutterhead (perpendicular to the feed direction and parallel to the cutting plane) and the kinematics of the knife edge Trials of planing softwood with a knife having a deliberately notched edge showed that such oscillatory action greatly improved surface quality, especially at low feed speeds

Taranu, N. 1971. STUDY OF THE VARIATION IN THE DYNAMIC ELEMENTS OF PERIPHERAL MILLING AS A FUNCTION OF THE REGIME DURING THE CUTTING PROCESS. Industria Lemnului 22(12)·479-485. Bucharest.
Reports, in a series of graphs and polynomial regression equations, the influence of the cutting speed, feed speed, depth and width of cut, and rake angle on cutting force and power, and on feed force in the milling of Carpinus betulus, Fagus sylvatica and Quercus robur

Tikhonov, V. A. 1971 CALCULATION OF THE GEOMETRY AND FORCE CHARACTERISTICS OF A CUTTING INSTRUMENT FOR LONGITUDINAL CUTTING OF WOOD. Lesnoi Zhurnal 14(5) 79-86. Arkhangelsk
A theoretical analysis of forces in planing

BARK

Anonymous 1972. NEW BARKING SYSTEM. Pulp and Paper Magazine of Canada 73(7):106.

Very briefly describes and illustrates a small-log barking machine that can be adapted to suit the requirements of individual mills by varying the number of barker arms (2 to 6) and other components. The unit reportedly can bark softwoods, hardwoods, and frozen logs

Anonymous. 1973. LOW-TEMPERATURE PRE-TREATMENT FOR BARKING SOFTWOOD LOGS. Department of the Environment. Canadian Forestry Service Research News 161(1):2-3. Ottawa.

When using a ring type debarker, bark was more easily removed from frozen logs when a radially directed rolling load was applied to the log surface prior to the barking operation. A prototype device, currently being field tested, consists of 16 air cylinders mounted radially on four structural rings. A roller bearing attached to each piston transmits the load to the bark surface as the log is fed through the device in an axial direction.

Arola, R. A., and Erickson, J. R. 1973. COMPRESSION DEBARKING OF WOOD CHIPS. U. S. Department of Agriculture Forest Service Research Paper NC-85, 11 p. North Central Forest Experiment Station, St. Paul, Minn.

Presents the results from 2 years of testing a single-pass compression debarking process for chips of aspen, hard maple, jack pine, and loblolly pine While nip pressure, nip setting, method of chip storage, and season of cut were all found to influence the efficiency of bark removal, the most significant variable was season of cut About 70 percent of the bark was removed from wood cut in the growing season while approximately 45 percent was removed from wood cut in the dormant season The percent of bark removal and wood recovery varied little between topwood and bolewood More variation in bark removal was noted for material stored 9 months in roundwood form than for material chipped fresh. A nip pressure of 1,400 pounds per lineal inch and a nip spacing of 0.020-inch yielded reasonable bark removal with limited wood loss. Presizing or grading of chips prior to compression debarking did not improve bark removal

Arola, R. A., and Hillstrom, W. A. 1972. COMPRESSION DEBARKING OF BRANCHWOOD CHIPS FROM FINLAND. U. S. Department of Agriculture Forest Service Research Note NC-143, 4 p North Central Forest Experiment Station, St. Paul, Minn.

Tests were conducted to remove bark and needles from spruce and pine branchwood chips that had been previously subjected to microorganic action It was found that bark removal was better when chips were aged and moist than when they were dry. Of the treatment combinations tested, a three-pass compression debarking process yielded least residual bark in moist spruce, moist pine, and dry pine The aged, moist spruce, which had a high input needle content, still contained 14.8 percent needles after the third pass. With moist and dry pine, the residual needle content was about 1 percent. With the three-pass system, wood loss was greater for pine than for spruce. A single-pass compression system used in conjunction with conditioning treatments (steaming and drubbing) removed more bark and needles than a single compression pass. Further reductions of bark and needle content were obtained when air flotation was used prior to steaming or drubbing Analysis indicated that 1,000 pounds of aged, moist spruce branchwood chips could yield about 200 pounds of wood fiber containing only 4 percent residual bark and needles

Holz-Zentralblatt 98(100):1409-1410.
Describes work studies mainly in spruce.

Ohira, Y. 1969. FUNDAMENTAL STUDIES ON
DEBARKING BY MEANS OF SMALL SIZE DRUM ROTOR
(III). Bulletin of the Government Forest Experiment Station
(221):21-37. Tokyo.
*Reports test results using three types of barkers: (1) progressive type
with two drums, (2) single drum with barking bars of various
heights, and (3) single drum with barking bars and flange walls.*

Pokryshkin, O. V. 1972. EFFECT OF BARKING SPEED ON
THE QUALITY OF BARKING. Derevoobrabatyvayushchaya
Promyshlennost' (8):13-14. Moscow.
*Describes experiments on the effect of peripheral barking speed (1-
3.8 m/s, equivalent to rotor speeds of 78-300 rev./min) on the
quality of barking various categories of Spruce wood in a rotor-type
barker. Quality of barking decreased with increasing barking speed.
The following barking speeds are recommended (for OK-66M, OK-
63, Cambio-66, and VK-26 barkers): 120-130 rev./min for floated
frozen wood; 150-160 rev /min for frozen freshly felled wood, and
180-190 rev /min for freshly felled and floated summer-felled wood
In further experiments, logs of 20, 30, 40, and 50 cm in diameter
were barked at peripheral speeds of 1 8, 2.7, 3.6, and 4.5 m/s.
The specific resistance to barking decreased, and the resistance to
cutter penetration increased with increasing log diameter, i.e., with
increasing peripheral barking speed*

Polozov, M. I. 1971. THEORETICAL INVESTIGATIONS ON
THE PROCESS OF BARKING WOOD WITH AN
ULTRASONIC JET OF AIR WITH A SOLID FILLER. Lesnoi
Zhurnal 14(6):56-61. Arkhangelsk.

Polozov, M. 1972. A PROMISING METHOD OF BARKING
WOOD. Lesnaya Promyshlennost' (2):21-22. Moscow.
*Gives results of investigations on barking spruce, pine, aspen, and
birch with jets of compressed air containing a sawdust filler. Data
are given on the degree of bark removal for various categories of
logs (recently felled, floated, dried, frozen), and for various air
pressures, jet angles, and distances between the nozzle and the
wood. The best filler was sawdust (particle size 1.5 to 4 mm)
produced by ripsawing dry wood. The optimum mixture was about
99 percent air and 0.4 percent sawdust (by volume), or 65 percent air
and 35 percent sawdust (by weight). Nozzles should be arranged
transversely for barking freshly felled, floated, and partly dried
wood, and longitudinally for dry and frozen wood, slabs, and logs
with sweep Calculated throughput of the pneumatic barking unit is
480 m³ per 7-hour shift.*

Simonov, M. N., and Maiorova, A. G. 1971. THE STRENGTH
OF WOOD AND BARK AT LOW TEMPERATURES. Lesnoi
Zhurnal 14(2):81-86. Arkhangelsk.
*In connection with mechanical debarking studies, data are tabulated
on shear and compressive strengths of wood and of bark for spruce,
pine, aspen, and birch at temperatures from 0° to -30°C.*

Sokol'skii, G. K., and Singalevich, M. S. 1972. BATCH
BARKING OF FLOATED TIMBER. Lesnaya Promyshlennost'
(8) 10-11. Moscow.
*Describes the design of a floating debarker developed in Russia In
the installation, floated bundles of logs are opened, transferred by
conveyor to the barking hopper, and then to a bunching unit before
being returned to the water.*

Rushnov, N. P., Pryakhin, E. A., and Maslovskii, V. I. 1973. THE

9

BARKING OF CHIPS IN CHIPPERS. Lesnaya Promyshlennost' (7):25 Moscow.

Shiryaev, O. 1973. BARKING WOOD WITH STEAM. Bumazhnaya Promyshlennost' (5):17-18 Moscow.
A note on a method of drum barking birch and aspen logs using steam at 150-160°C

Zimmermann, G. 1971. NEW BARKING MACHINES AT THE HANOVER FAIR. Holz-Zentralblatt 97(72/73):1063. Stuttgart.

SAW

GENERAL

Abegg, B. 1972 CAPACITIES AND COSTS OF STATIONARY LOG PROCESSING PLANTS. Holz als Roh-und Werkstoff 30 215-219 Munich.

Airth and Calvert. 1973. (Computer, tape, and card control of machines)

Anonymous. 1972. (Properties of the cutting edge and cutter; Fitting and sharpening)

Anonymous 1972. (Computer, tape, and card control of machines)

Burridge, M. S. 1973. TIMBER CUTTING METHODS. Journal of the Institute of Wood Science 6(3):13-16. London.
A summary, with 33 references, of more recent research to improve the efficiency of wood machining operations and methods of reducing waste Included are both conventional cutting methods (bandsaws, circular saws, and chipping headrigs) and non-conventional methods (vibration, lasers, plasma, and water jets).

Buzinov, O A., Sysoev, N. N., and Chikuchin, A. E. 1972. SEGMENTAL TURNING DEVICES FOR SLEEPER LENGTHS. Lesnaya Promyshlennost' (8):16. Moscow.
Describes design details of a device for turning logs and cants in a machine designed to saw railroad ties Output per shift is claimed double that of machines equipped with conventional turning devices.

Conway. 1973 (History and general texts)

Cummins, L. K., and Culbertson, D. D. 1972. SAWMILL SIMU-LATION MODEL FOR MAXIMIZING LOG YIELD VALUES Forest Products Journal 22(10).34-40.
A mathematical model for computing the log breakdown sequence that results in maximum value yield is presented (restricted to nominal 12-inch diameter logs of Douglas-fir). Numerical examples show the sensitivity of value yield to log diameter, saw kerf, and lumber dimensions.

Deutsch, I., and Serbu, A. 1972. THE DIMENSIONS OF SOFTWOOD CANT SLABS HAVING THE SAME STRENGTH AS SAWN TIMBER OF RECTANGULAR CROSS-SECTION. Industria Lemnului 23(1):11-13. Bucharest.

Dobie, J. 1972 GUIDELINES FOR STUDY OF SAWMILL PERFORMANCE. Department of the Environment. Canadian Forestry Service Information Report VP-X-93, 75 p Western Forest Products Laboratory, Vancouver.

yields in sawmilling, and gives data on utilization of oak logs in a Yugoslavian mill

Thunell, B. 1972. MACHINERY AND MILL LAYOUTS — PRESENT TRENDS IN SCANDINAVIA. Svenska Traforskningsinstitutet. Series B. Number 110. Stockholm.
Flow diagrams and photographs are used to illustrate a discussion on equipment trends in modern Scandinavian sawmills.

Vetsheva, V. F. 1972. AN EFFECTIVE METHOD OF BREAKING DOWN LARGE LOGS. Derevoobrabatyvayushchaya Promyshlennost' (3):4-6. Moscow.
A theoretical study of the overrun obtained by more than 20 different methods of log breakdown. General trends and recommendations are given.

Warren, W. G. 1973. HOW TO CALCULATE TARGET THICKNESS FOR GREEN LUMBER. Department of the Environment. Canadian Forestry Service Information Report VP-X-112, 11 p. Western Forest Products Laboratory. Vancouver.
Details a statistical procedure for determining lumber target thickness whereby the proportion of undersized material may be kept below a prescribed minimum. The estimation of the separate components of variation may be used to indicate which parts of the production process are in need of tightened quality control.

White, V. S. 1973. MODERN SAWMILL TECHNIQUES. Volume 1. Proceedings of the First Sawmill Clinic. 312 p. Miller Freeman Publications, Inc. San Francisco.
A compilation of papers and discussions related to technological developments in the sawmill industry. The 19 chapters provide the reader with practical information on such subjects as log scaling, handling and sorting, log breakdown, best-opening-face, drying, maintenance, and electronic control.

BANDSAWING

Breznjak and Moen. 1972. (Properties of the cutting edge and cutter; Stability)

Cremisio and Brabec. 1973. (Patents)

Dobie, J., and Sturgeon, W. J. 1972. AN ASSESSMENT OF THE ECONOMIC BENEFITS OF DOUBLE-CUT HEADRIGS. Forest Products Journal 22(2):22-24.
In a study of three British Columbia mills, the productivity of double-cutting band headrig saws was 14 to 23 percent greater than that of single-cutting rigs While sawing accuracy was slightly reduced, investment in double-cutting rigs may prove economic at present.

Curtu, I. 1972. EXPERIMENTAL STUDIES ON THE STATIC RIGIDITY OF A BANDSAW FRAME. Industria Lemnului 23(6):229-232. Bucharest.
Describes the geometry of deformations occurring in the frame of a Raiman B-3 ripping bandsaw caused by tensioning. The deformations, although relatively small, had a significant effect on sawing accuracy They were considerably reduced by fixing the lower free end of the saw frame to the base with an upright metal bar

Curtu, I. 1972. STATIC AND DYNAMIC TENSION STRESSES INSIDE A BANDSAW FRAME. Industria Lemnului 23(7):284-290. Bucharest.

Describes the experimental method and concludes that operating stresses increase with increasing feed speed and decrease with increasing cutting speed

Fahey and Hunt. 1972. (Chipping, flaking, and grinding; Chipping headrigs and edgers)

Hallock, H., and Lewis, D W. 1971. INCREASING SOFTWOOD DIMENSION YIELD FROM SMALL LOGS — BEST OPENING FACE. U S. Department of Agriculture Forest Service Research Paper FPL-166, 12 p. Forest Products Laboratory, Madison, Wis.
Presents a system for precisely locating where to make the opening cut in a softwood log so as to achieve maximum yield. An example is given for five common log breakdown methods using logs 4 6 to 20 5 inches in diameter Of these, the variable-opening-face live sawing method consistently gave the highest yield. Automated mill systems may be designed employing presently available diameter-measuring equipment, numerically controlled log positioners, a mini-computer, and the opening face data from the BOF computer program.

Kersavage, P. C. 1972 SAWING METHOD EFFECT ON THE EFFECT OF PRODUCTION OF CHERRY LUMBER. Forest Products Journal 22(8).33-40.
Cherry lumber was manufactured in northern Pennsylvania mill equipped with a rosser-head debarker, a single-cut, 6-foot band headrig with automatic, air-operated carriage and setworks, a 3-saw edger, and an air-lift, multi-saw trimmer. Two sawing methods were used — conventional or grade sawing and live sawing. Live sawing resulted in a considerable increase in lumber output per man-hour with only a slight decrease in National Hardwood Lumber Association lumber grade recovery Sawmill productivity, expressed on the basis of lumber value per hour of operation, increased significantly with live sawing. Live-sawn lumber was wider, contained more edge grain, and had less all-heart or all-sap. Cost analysis revealed that live sawing could have resulted in a 45 percent increase of gross profit per day assuming changes in quality characteristics of live-sawn lumber were acceptable to the consumer Results indicate it might be advantageous to sort logs by grade, live saw grade 2 logs, and conventionally saw others

Korgushov, A S. 1972. CALCULATION OF THE WORKING VOLUME OF THE GULLET IN A BANDSAW. Lesnoi Zhurnal 15(6) 130-133. Arkhangelsk

Montague 1971. (Sawing; General)

Pahlitzsch and Puttkammer. 1972. (Properties of the cutting edge and cutter, Stability)

Pahlitzsch and Puttkammer. 1973. (Properties of the cutting edge and cutter; Stability)

Rawat, B. S., and Bharnagar, R. C. 1973. A NOTE ON TIME-MOTION STUDY IN SAWMILLING Indian Forester 99(4) 218-234. Debra Dun.
Tabulates the time required to convert logs of Shorea robusta into lumber using a horizontal bandsaw.

St-Laurent 1971. (Orthogonal cutting)

Thunell 1972. (Properties of the cutting edge and cutter; Stability)

Yur'ev and Veselkov. 1972. (Properties of the cutting edge and cutter, Stability)

SASH GANGSAWING. 104 p. Izdatel'stvo Lesnaya Promyshlennost'. Moscow.

Describes instruments for measuring the cross-sectional dimensions and surface roughness of lumber, outlines the principles of quality control, and presents a quality control procedure applicable to sash gangsawing

Sukhanov, V. G. 1973 RESULTS OF EXPERIMENTS ON THE VIBRATION SAWING OF WOOD. Derevoobrabatyva-yushchaya Promyshlennost' (2):15-16. Moscow.

Briefly reviews research at the Moscow Forestry & Wood Technology Institute on vibration cutting of wood with toothed and toothless blades of various kinds, and describes results of experiments on vibration sawing. The process differs from sash gangsawing in the considerably reduced stroke (less than the height of the kerf), and increased speed (1500-3000 strokes/min) Optimum parameters were symmetrical teeth without set, rake angle 90 to 120°, pitch 5 to 7 mm, height 3 mm and balde thickness 0.8 mm (for a depth of cut of about 100 mm). Sawn-surface quality and accuracy of cut were best with a stroke of 15 to 20 mm. Workpieces as thick as 50 mm thick could be sawn at feed speeds up to 17.5 m/min, and it was possible to saw with several blades simultaneously.

Thunell. 1972. (Properties of the cutting edge and cutter; Stability)

Tukherm, Kh. 1971. KINEMATIC CALCULATION OF THE CUTTING MECHANISMS OF SASH GANGSAWS. Latvijas Lauksaimniecibas Akademijas Raksti (36)'78-88. Russia.

Proposes an analytical method for describing the kinematics of sash gangsawing.

Yur'ev, Yu. I., and Potyarkin, L. P. 1971. CALCULATION OF THE FRICTION FORCES IN THE GUIDES OF A SASH GANGSAW. Lesnoi Zhurnal 14(5):86-89. Arkhangelsk.

CIRCULAR SAWING

Anonymous. [n.d.] (Properties of the cutting edge and cutter; Fitting and sharpening)

Anonymous. 1971. CIRCULAR SAWS GIVE RELIABLE RESULTS. Sagverken Travaruindustrien (8)'493, 495, 499. Stockholm.

A study at two Swedish mills using the ARI circular saw system concluded that acceptable accuracy can be achieved

Cheremnykh and Chizhevskii. 1971. (Safety, including noise)

Chomcharn, A. 1972. CUTTING RESISTANCE OF WOOD-POLYMER COMPOSITES IN RIPSAWING. Master of Science Thesis. State University College of New York. Syracuse.

Dogaru, V. 1971. THE INFLUENCE OF TOOTH GULLET SIZE ON FEED SPEED IN WOOD SAWING. Industria Lemnului 22(6)'215-220. Bucharest.

Dushkov. 1971. (Properties of the cutting edge and cutter; Fitting and sharpening)

Koga and Nanasawa. 1973. (Properties of the cutting edge and cutter; Dulling)

Lunstrum, S. J. 1972. CIRCULAR SAWMILLS AND THEIR EFFICIENT OPERATION. 86 p. U S. Department of Agriculture Forest Service. State and Private Forestry, Atlanta, Ga.

A practical manual written for circular sawmill operators The text considers only primary log breakdown but coverage includes sawing accuracy, equipment selection, set-up, operation, and maintenance. A useful troubleshooting guide is provided to help locate causes of more common problems found in sawmill operations

Malcolm, F. B., and Hallock, H. 1972. EFFECTS OF THREE SAWING METHODS ON WARP OF HARD MAPLE DIMENSION CUTTINGS. Forest Products Journal 22(4).57-60.

Hard maple logs were circular sawn into standard 4/4 factory lumber by three methods' (1) Grade sawn with faces orientated so knots were generally at edges, (2) grade sawn with faces orientated so knots were generally in the middle; and (3) live sawn with all saw cuts in the same plane. The lumber was then kiln-dried, and planed to final thickness Freedom of planing skips on the surfaced cutting was the basis for warp evaluation. It was found that both grade sawing methods produced greater areas of skip-free cuttings than did the live-sawing method Waste loss due to warp only (original dry lumber basis) amounted to as much as 36 percent for the grade sawing methods and 50 percent for the live-sawing method

Mazurkin, P. M. 1971. CHIPLESS CUTTING OF WOOD. Lesnoi Zhurnal 14(4): 54-56. Arkhangelsk.

Gives a brief account of experiments on cutting wood with rotating steel disks. Early studies used flat disks having symmetrical wedge-like edges but were discontinued because of disk overheating In later work, flat disks with teeth (but without set) were used and feed speeds up to 0 3 m/sec were achieved When cutting 25 mm thick air-dry spruce, the specific cutting energy ranged from 1.5 to 3 5 kg.m/cm' as peripheral cutting speed increased from 10 to 100 m/sec. Lateral deflections or overheating of the blade were not observed It was possible to cut at various angles to the grain and surface quality was superior to that of planed wood. No surface discoloration was noted.

McKenzie. 1973. (Properties of the cutting edge and cutter; Stability)

McLauchlan. 1972. (Properties of the cutting edge and cutter; Stability)

Montague. 1971. (Sawing; General)

More and Nieh. 1973. (Properties of the cutting edge and cutter; Stability)

Pearson. 1972. (Patents)

Petrov. 1972. (Properties of the cutting edge and cutter; Stability)

Priest, D T. 1972. THE CHOICE OF STEEL PLATE STELLITE-TIPPED OR TUNGSTEN-TIPPED CIRCULAR SAWS Timber Trades Journal. Annual Special Issue. S/11, S/13, S/15. London.

Compares capital and maintenance costs, and service life of three types of circular saws and tabulates their advantages and disadvantages

Prokes. 1973. (Properties of the cutting edge and cutter; Stability)

Saplin. 1971. (Properties of the cutting edge and cutter; Stability)

Shkutko. 1973. (Properties of the cutting edge and cutter; Dulling)

Stakhiev and Lyzhin. 1972. (Properties of the cutting edge and cutter, Stability)

Stephenson and Plank. 1972. (Properties of the cutting edge and cutter, Fitting and sharpening)

St-Laurent, A 1973. IMPROVING THE SURFACE QUALITY OF RIPSAWN DRY LUMBER. Forest Products Journal 23(12):17-24.
White spruce, white and red pine, sugar maple, and yellow birch lumber at 12 percent moisture content was ripsawn with a 12-inch diameter, 36-tooth saw swage-set for hardwoods and spring-set for softwoods As expected, power demand increased with increasing lumber thickness, saw and feed speed, and wood specific gravity; a monogram is provided to estimate the power required at the sawblade in terms of these variables. Roughness of the sawn surface (measured by a stylus method) increased with increasing feed speed and protrusion of the blade above the surface of the workpiece At minimum protrusion, smoothness approached that of a planed surface and long ribbon-like chips (suitable for kraft pulp) were contained in the waste material Surface roughness and power requirements increased as wear progressed By using a succession of saws, each cutting only with the top of its blade and having less teeth, may be one way to improve whole-log utilization. Chipping headrigs produce cants that would be suitable for subsequent breakdown into lumber and pulpable kerf material by this method.

Thrasher. 1972. (Patents)

Thunell. 1972 (Properties of the cutting edge and cutter; Stability)

Turikov and Shevchenko. 1973. (Properties of the cutting edge and cutter; Stability)

Vinogradov. 1973. (Safety, including noise)

Ward. 1972. (Computer, card, and tape control of machines)

CHAIN SAWING

Aho 1971. (Safety, including noise)

Bryan, R. W 1972. INEXPENSIVE WAY TO CUT SHORTWOOD AT THE LOADER. Forest Industries 99(1):70-71.
Describes a grapple that bucks tree length stems into bolts. The grapple houses a hydraulically powered Omark chain saw with a 39½-inch bar and is attached to a Copeland Model 500 Hydro-Crane Vertical movement of the saw is provided by a 1-inch hydraulic cylinder In operation, the grapple moves several stems to a position along the loader, the ends are evened, and the bundle bucked to the desired lengths while held in the grapple

Firth. 1972. (Safety, including noise)

Fushimi. 1972. (Safety, including noise)

Fushimi, Waranabe, Shigaki, and Ezaki 1972. (Safety, including noise)

Heding, N. 1972. CROSS-CUTTING AND SORTING HARDWOOD WITH AN OSA-770 GRAPPLE-SAW. AN NSR (NORDISKA SKOGSARBETSSTUDIERNAS RAD) REPORT. Dansk Skovforenings Tidsskrift 57(2):175-187. Denmark

increases as the horsepower of the engine increases. Cutting time is slightly increased and the average cutting force is greatly reduced when the number of teeth on the sprocket are increased.

Sakurai and Tanaka. 1972. (Safety, including noise)

Srolarık, Srutek, and Bouse. 1970. (Safety, including noise)

Swan, D. A. 1971. THE REVOLUTIONARY RING SAW. Technical Release 71-R-60, 3 p. American Pulpwood Association. New York.
A portable, gas-engine driven saw similar in function to a conventional chain saw but having a circular, ring-shaped cutting blade and no bar. The tendency to "kick back" is minimal and most of the blade is guarded. A splitter blade located in the lower half of the ring is thicker than the inside of the blade, yet thinner than the kerf, so much binding is eliminated. A friction drive allows the blade to slip at a predetermined load. The saw is manufactured by Ring Saws, Inc., Waltham, Massachusetts.

Thompson. 1973. (Safety, including noise)

Uhorskai, O. 1972. THE CUTTING CAPACITY OF ONE-MAN POWER SAWS IN CROSS-CUTTING ROUND TIMBER. Lesnicky Casopis 18(4):341-362. Bratislava, Czechoslovakia.
Provides comparative performance data for the Stihl 050 AVL, Stihl 041 AV, Homelite XL 903, and Homelite XL 913 chain saws.

Uspenskii, V. A., Ievlev, A. I., and Maksimov, V. M. 1973. INVESTIGATION OF THE LONGITUDINAL STABILITY OF SAW CHAINS. Lesnoi Zhurnal 16(2):106-109. Arkhangelsk.

Welch. 1971. (Safety, including noise)

MACHINING WITH HIGH ENERGY JETS

Anonymous. 1972. NEW FLUID JET CUTTING SYSTEM. McCartney Manufacturing Co., Inc. Baxter Springs, Kansas.
A manufacturer's brochure, with illustrations, describing a commercially available fluid-jet cutting system. The unit operates at 30,000 to 60,000 psi using small, especially designed nozzles with orifice diameters ranging from 0.002 to 0.015 inch. A water-polymer fluid is used at a rate of 3 to 50 gallons per hour.

Behr, E. A. 1971. IMPREGNATION OF WOOD BY HIGH ENERGY JETS. Journal of the Institute of Wood Science 5(5):24-27. London.
Roundwood sections of southern pine, red pine, and western red cedar were subjected to liquid streams containing pentachlorophenol at pressures up to 45,000 psi. The volume of liquid injected ranged from 40 to 250 ml. About 6 seconds were required to inject 125 ml but 85 ml could be injected in about 3 seconds. Depth of penetration increased with increasing jet pressure and was about 5 cm for red pine. There was more separation of wood around the entrance hole with red cedar than with the pines and less with wet wood than with dry wood.

Burridge. 1973. (Sawing; General)

Chamberlin, F. B. 1973. PAPERBOARD APPLICATIONS FOR THE HIGH-ENERGY FLUID JET CUTTER. TAPPI 56(8):78-80.
Describes the commercial application of a high-energy, fluid jet cutting system used to cut furniture components from laminated convolutely and spirally wound paper tubes. An intensifier pump,

driven by a hydraulic power unit, brings water (containing a long chain polymer) to a pressure of about 40,000 p.s i The fluid then passes through a small diameter nozzle and penetrates the workpiece. The potential use of the equipment as a replacement for trim knives on a paper corrugator is briefly examined.

Eberle, J. F. 1973. FLUID JET CUTTING TAPPI 56(10) 84-87
A general discussion of a newly commercialized fluid jet cutting system. Water, containing a low concentration of a long chain polymer, is pressurized to 30,000-50,000 psi and released through small orifices of 0 002-0.015 inch diameter The energy is concentrated in a needle-like stream having a velocity of 2,000-3,000 feet per second The 30-40 gallons per hour of fluid used is biodegradable Noise levels during operation are low (about 80 decibels) and the system is regarded as less hazardous than many other woodworking tools. A complete basic system including a single pump/intensifier unit, accumulator, polymer mixing module, valves, filter, nozzles, gauges, and controls is estimated to cost (1973 basis) between $25,000 and $30,000.

Johnson. 1973. (Patents)

Polozov. 1972 (Barking)

Szymani, R. 1972. A STUDY OF CORRUGATED BOARD CUTTING BY HIGH VELOCITY LIQUID JETS Forest Products Journal 22(8):17-25.
Eleven types of corrugated board having a wide range of properties were cut with a commercially available liquid jet apparatus (See Anonymous. 1972. (Sawing, Machining with high energy jets)) Specimens were cut perpendicular to the corrugations using combinations of the following·
Cutting fluid — filtered tap water or water with 0 3 % polymer additive.
Nozzle diameter — 0 0082-inch and 0 102-inch
Feed rate — 300, 500, and 700 fpm
Pressure — 20,000, 30,000, and 40,000 psi.
It was found that acceptable quality cuts could be made with liquid jets at speeds in excess of current production methods A substantial increase in efficiency was achieved when the jet fluid contained a low concentration of polyethylene oxide Wetting of edges was minor and can be neglected Corrugated board cut by the jet had an edgewise compressive strength nearly twice that obtained with a conventional slitter. Some specific recommendations for additional research are provided.

MACHINING WITH LIGHT

Burridge. 1973. (Sawing; General)

Klante, D. 1972. LASER ENTRY IN PAPER CONVERTING: STILL A PIPE DREAM? Allgemeine Papier-Rundschau (9):346-348.
Concludes that lasers should soon find application in paper converting, especially for cutting the running web and in fabricating steel rule die blocks.

McMillin, C. W. 1972. LASER MACHINING — A STATUS REPORT. Southern Lumberman 225(2795) 19-20.
When cutting with an air-jet-assisted carbon-dioxide laser of 240 watts' output power, maximum feed speed at the point of full penetration of the beam decreased with increasing workpiece thickness. For samples at 12 percent moisture, feed speeds averaged 111 and 17 inches per minute for samples 0 25 and 1 00 inch thick.

Somewhat lower speeds were required for wet wood. In wet wood, maximum speed was unrelated to specific gravity. In dry wood, slightly slower speeds were required when wood density was high than when it was low. The laser cut along and across the grain with equal speed and kerf width was about 0.012 inch. Micrographs showed that the laser-cut surfaces, while blackened, were far smoother than saw surfaces. Char could be easily removed with compressed air or by light brushing.

Raroff, P. 1973. LASER APPLICATIONS IN THE PAPER INDUSTRY. Pulp and Paper 47(3):128-129.
Briefly reviews the development of laser cutting and suggests a number of applications in the paper industry (cutting, perforating, slitting, and scoring).

SHEARING

Arola, R. A. 1972. ESTIMATING FORCE AND POWER REQUIREMENTS FOR CROSSCUT SHEARING OF ROUNDWOOD. U.S. Department of Agriculture Forest Service Research Paper NC-73, 8 p. North Central Forest Experiment Station, St. Paul, Minn.
Presents a practical procedure for estimating (with nomographs) the force required to crosscut shear frozen and unfrozen roundwood of various specific gravities and diameters with shear blades ranging in thickness from ¼- to ⅞-inch. Also provided are nomographs to evaluate hydraulic cylinder sizes, pump capacities, and motor horsepower requirements to effect the cut.

Azarenok. 1972. (Orthogonal cutting)

Bryan, R. W. 1972. NEW HARVESTER PROCESSES 50 TO 60 STEMS PER HOUR. Forest Industries 99(12):60.
Caterpillar's 950 tree-length harvester is being successfully used to log slash and longleaf pine pulpwood in natural stands.
See also: Pickard. 1972. (Sawing; Shearing)

Johnston, St-Laurent, and L'Ecuyer. 1973. (Reports of research programs and industrial developments)

Liebold, E., Kirmer, H., and Schilke, A. 1972. CHIPLESS CROSS-CUTTING OF LOGS WITH SEVERAL KNIVES. Holztechnologie 13(3):156-161. Leipzig.

McNeil and Deline. 1972. (Patents)

Pickard, D. 1972. CAT TREE-LENGTH HARVESTER PRODUCING 4½ CORDS AN HOUR. Canadian Forest Industries 92(7):27-29.
The new Caterpillar 950 tree-length harvester is felling, limbing, topping, and bunching one tree every 50-55 seconds in a 25-cord-per-acre, 14-trees-per-cord Ontario stand. The operating cycle begins as the machine approaches a tree with the felling and processing head in the vertical position and grapple arms open. The grapple grasps the tree and the felling head shears it at the butt. The tree is lowered to the horizontal position as the machine backs up to the pile while a drive mechanism propels it through delimbing arms and shears top it to the desired length. As the log drops on the pile, the harvester moves forward and the cycle is repeated.

Powell, L. H. 1970. EVALUATION OF LOGGING-MACHINE PROTOTYPES: DROTT FELLER-BUNCHER. Woodlands Report Number WR/29, 14 p. Pulp and Paper Research Institute of Canada. Pointe Claire.
For logging operations, the usual attachments of the Drott 35YC

MOLDING, AND SHAPING

TEMPLATE. Derevoobrabatyvayushchaya Promyshlennost'
(1):9-10. Moscow.
A theoretical study of design principles.

Rushnov, Pryakhin, and Maslovskıı 1973. (Peripheral mıllıng
parallel to grain)

Sima, Mıhut, and Lazar. 1972. (Peripheral milling parallel to graın)

Srewart, H. 1972. ABRASIVE VS. KNIFE PLANING. Wood and
Wood Products 77(8) 73-76.
*Comparisons are drawn between abrasive and knife planing for
effects that machine variables, such as width of cut, depth of cut,
feed speed, and grain orientation, have on power requirements and
surface quality.*

Srewart, H A. 1973. CROSS-GRAIN KNIFE PLANING MAY
IMPROVE COTTONWOOD-ASPEN UTILIZATION.
Southern Lumberman 227(2818):15-18.
*When cottonwood and aspen samples were planed across the grain
(in contrast to the conventional parallel to the grain planing) surfaces were smooth and had minimal tear out and chipped and fuzzy
grain. Flakes produced by cross-grain planing appear suitable for
high strength particleboard A major disadvantage of the system is
the limited length of boards that can be processed*

Srock, H. 1969. MODERN MOLDING MACHINES — THEIR
ADVANTAGES IN WOODWORKING INDUSTRY. Timber
6(11) 13, 15. Johannesburg.
*Reviews developments in multi-spindle molding machines and
tabulates recommended cutting angles for kiln-dried and green
timber of 22 common species.*

Stovpyuk. 1971. (Orthogonal *cutting*)

Tanaka, C., Sakurai, T., and Horie, T. 1972. VIBRATION AND
NOISE OF HAND FEED PLANER. III. VIBRATION
BEHAVIOR ON THE TABLE. Journal of the Japan Wood
Research Society 18 435-441. Tokyo.
*Vibrations on the table of a hand feed planer were measured in three
mutually perpendicular directions during no-load operation and
during planing. In idle operation, acceleration forces in a direction
perpendicular to the table were least (about 0.15 g) near the cutter
and increased to about 0 45 g at the outer erds of both the infeed
and outfeed tables. Acceleration forces parallel to the surface were
essentially uniform over the entire surface (0 15 g for both directions) Accelerations were greater in all three directions during planing than during idle operation. Forces (especially in the perpendicular direction) increased with increasing board width, depth of
cut, and feed speed As the linear length of wood planed increased,
acceleration forces perpendicular to the table increased However,
forces remained constant for the two parallel directions The
relationship between table vibration and power consumption was
good*

Tikhonov. 1971. (Peripheral milling parallel to graın)

17

Somewhat lower speeds were required for wet wood. In wet wood, maximum speed was unrelated to specific gravity. In dry wood, slightly slower speeds were required when wood density was high than when it was low. The laser cut along and across the grain with equal speed and kerf width was about 0.012 inch. Micrographs showed that the laser-cut surfaces, while blackened, were far smoother than saw surfaces. Char could be easily removed with compressed air or by light brushing.

Raroff, P. 1973. LASER APPLICATIONS IN THE PAPER INDUSTRY. Pulp and Paper 47(3):128-129.
Briefly reviews the development of laser cutting and suggests a number of applications in the paper industry (cutting, perforating, slitting, and scoring).

SHEARING

Arola, R. A. 1972. ESTIMATING FORCE AND POWER REQUIREMENTS FOR CROSSCUT SHEARING OF ROUNDWOOD. U.S. Department of Agriculture Forest Service Research Paper NC-73, 8 p. North Central Forest Experiment Station, St. Paul, Minn.
Presents a practical procedure for estimating (with nomographs) the force required to crosscut shear frozen and unfrozen roundwood of various specific gravities and diameters with shear blades ranging in thickness from ¼- to ⅞-inch. Also provided are nomographs to evaluate hydraulic cylinder sizes, pump capacities, and motor horsepower requirements to effect the cut.

Azarenok. 1972. (Orthogonal cutting)

Bryan, R. W. 1972. NEW HARVESTER PROCESSES 50 TO 60 STEMS PER HOUR. Forest Industries 99(12):60.
Caterpillar's 950 tree-length harvester is being successfully used to log slash and longleaf pine pulpwood in natural stands.
See also: Pickard. 1972. (Sawing; Shearing)

Johnston, St-Laurent, and L'Ecuyer. 1973. (Reports of research programs and industrial developments)

Liebold, E., Kirtner, H., and Schilke, A. 1972. CHIPLESS CROSS-CUTTING OF LOGS WITH SEVERAL KNIVES. Holztechnologie 13(3):156-161. Leipzig.

McNeil and Deline. 1972. (Patents)

Pickard, D. 1972. CAT TREE-LENGTH HARVESTER PRODUCING 4½ CORDS AN HOUR. Canadian Forest Industries 92(7):27-29.
The new Caterpillar 950 tree-length harvester is felling, limbing, topping, and bunching one tree every 50-55 seconds in a 25-cord-per-acre, 14-trees-per-cord Ontario stand. The operating cycle begins as the machine approaches a tree with the felling and processing head in the vertical position and grapple arms open. The grapple grasps the tree and the felling head shears it at the butt. The tree is lowered to the horizontal position as the machine backs up to the pile while a drive mechanism propels it through delimbing arms and shears top it to the desired length. As the log drops on the pile, the harvester moves forward and the cycle is repeated.

Powell, L. H. 1970. EVALUATION OF LOGGING-MACHINE PROTOTYPES: DROTT FELLER-BUNCHER. Woodlands Report Number WR/29, 14 p. Pulp and Paper Research Institute of Canada. Pointe Claire.
For logging operations, the usual attachments of the Drott 35YC

JOINTING, PLANING, MOLDING, AND SHAPING

Anonymous. 1973. (Machining with coated abrasives)

Bunemovich, E. E., Vekshin, A. M., and Kleba, N. P. 1971. USE OF HELICAL KNIVES IN PLANING MACHINES. Derevoobrabatyvayushchaya Promyshlennost' (9):6-9. Moscow.
Discusses the advantages of helical knives for planers and thicknessers, and describes some crescent-shaped, flexible knives developed in the USSR. Research and commercial experience with these knives indicate that higher feed speeds may be used without reducing surface quality

Cheremnykh and Chizhevskii. 1973. (Safety, including noise)

Deaner. 1972. (Safety, including noise)

Hart. 1972. (Safety, including noise)

Hojan and Ozimek. 1971. (Safety, including noise)

Hoshi, T., and Sato, S. 1972. MILLING AND DRYING OF 13 WOOD SPECIES FROM NEW GUINEA. Wood Industry 27(7):17-19. Tokyo.

Konishi, C. 1972. CUTTING ABILITY OF KNIFE WITH PLANER. I. RELATIONSHIP BETWEEN CONDITION OF WOOD AND CUTTING FORCE. Journal of the Japan Wood Research Society 18:223-229. Tokyo.
Cutting force, measured with a workpiece dynamometer, was least when planing parallel to the grain and increased with increasing wood specific gravity. The effect of knife wear on cutting force is graphically presented for several species from East Asia.

Murga and Ivanovskii. 1972. (Peripheral milling parallel to grain)

Pahlitzsch and Sandvoss. 1972. (Properties of the cutting edge and cutter; Dulling)

Pease. 1972. (Safety, including noise)

Perry, J. D. 1973. ESTIMATING PLANING MILL RESIDUES. Southern Lumberman 227(2824):185.
Thirty-five bundles of southern pine lumber ranging in size from 1 x 4's to 2 x 12's were weighed and measured before and after surfacing. Shavings averaged 767 pounds (ovendry) per MBF; trim, 88 pounds; and end-trim sawdust, 5.8 pounds.

Rawat, B. S., Rajput, S. S., and Pant, B. C. 1973. A STUDY ON POWER REQUIREMENT IN THICKNESS PLANING. Indian Forester 99(1):23-32. Dehra Dun.
Sixteen Indian timber species were planed with a commercial thicknesser. Tables and graphs show the variation in power demand (Kw/cm width) in relation to feed rate (10, 14, and 22 cm/sec), depth of cut (0 4, 0.8, 1 2, and 1.6 mm), wood specific gravity, and shear strength.

Rukin, V. V., and Barin, I. V. 1972. A MACHINE FOR MANUFACTURING TRIANGULAR ITEMS WITH RADIALLY CURVED ANGLES WITHOUT THE USE OF A TEMPLATE. Derevoobrabatyvayushchaya Promyshlennost' (1):9-10. Moscow.
A theoretical study of design principles.

Rushnov, Pryakhin, and Maslovskii. 1973. (Peripheral milling parallel to grain)

Sima, Mihut, and Lazar. 1972. (Peripheral milling parallel to grain)

Stewart, H. 1972. ABRASIVE VS. KNIFE PLANING. Wood and Wood Products 77(8):73-76.
Comparisons are drawn between abrasive and knife planing for effects that machine variables, such as width of cut, depth of cut, feed speed, and grain orientation, have on power requirements and surface quality

Stewart, H. A. 1973. CROSS-GRAIN KNIFE PLANING MAY IMPROVE COTTONWOOD-ASPEN UTILIZATION. Southern Lumberman 227(2818):15-18.
When cottonwood and aspen samples were planed across the grain (in contrast to the conventional parallel to the grain planing) surfaces were smooth and had minimal tear out and chipped and fuzzy grain. Flakes produced by cross-grain planing appear suitable for high strength particleboard. A major disadvantage of the system is the limited length of boards that can be processed

Stock, H. 1969. MODERN MOLDING MACHINES — THEIR ADVANTAGES IN WOODWORKING INDUSTRY. Timber 6(11):13, 15. Johannesburg.
Reviews developments in multi-spindle molding machines and tabulates recommended cutting angles for kiln-dried and green timber of 22 common species.

Stovpyuk. 1971. (Orthogonal cutting)

Tanaka, C., Sakurai, T., and Horie, T. 1972. VIBRATION AND NOISE OF HAND FEED PLANER. III. VIBRATION BEHAVIOR ON THE TABLE. Journal of the Japan Wood Research Society 18:435-441. Tokyo.
Vibrations on the table of a hand feed planer were measured in three mutually perpendicular directions during no-load operation and during planing. In idle operation, acceleration forces in a direction perpendicular to the table were least (about 0.15 g) near the cutter and increased to about 0 45 g at the outer ends of both the infeed and outfeed tables. Acceleration forces parallel to the surface were essentially uniform over the entire surface (0 15 g for both directions). Accelerations were greater in all three directions during planing than during idle operation Forces (especially in the perpendicular direction) increased with increasing board width, depth of cut, and feed speed. As the linear length of wood planed increased, acceleration forces perpendicular to the table increased. However, forces remained constant for the two parallel directions. The relationship between table vibration and power consumption was good.

Tikhonov. 1971 (Peripheral milling parallel to grain)

17

TURNING

Loof 1973 (Patents)

Werner, H 1971 THE MANUFACTURE OF ROUND RODS FROM NATURAL WOOD. Holzindustrie 24(10):303-305. Leipzig.

Gives the main conclusions of work to develop a cutter for manufacture of broom handles in which the workpiece moves axially through a rotating annular head with inwardly directed knives.

BORING, ROUTING, AND CARVING

Noguchi, M , Fujiwara, K., and Sugihara, H. 1973. USE OF ULTRASONIC VIBRATION IN TURNING WOOD. Wood Science 5 211-222.
Studies the effects of amplitude, direction of vibration and cutting velocity on cutting force, surface quality, chip length, and tool wear when turning samples of clear Fagus crenata at about 9 5 percent moisture content In general, the parallel tool force decreased with increasing vibration amplitude when vibration was in the tangential and radial directions Lateral vibration had little effect on cutting force With tangentially directed vibration, but not with radially directed vibration, the normal cutting force decreased with increasing amplitude. Vibration had a negligible effect on surface roughness With radially directed vibration, but not with lateral vibration, chip length decreased with increasing amplitude The rate of tool wear was reduced when cutting with the tangentially directed system; with radially directed vibrations wear rates increased. Cutting velocity had no effect on forces, surface roughness, chip length, or tool wear

Nowak, K., Biniek, P., and Staszewski, A. 1972. EFFECT OF THE DIAMETER OF PRE-DRILLED HOLES ON NAIL-HOLDING STRENGTH OF WOOD AND PARTICLEBOARDS. Przemysl Drzewny 23(9) 16-18. Warsaw
Holes 0 to 100 percent of the diameter of nails measuring 2.2 to 2 8 mm were bored in samples of Pinus sylvestris, Betula verrucosa, and a three layer particleboard. Optimum hole diameter was about 70 percent of the nail diameter for pine and birch wood, and 55 to 70 percent of nail diameter for particleboard

March, B. W. 1973. PIERCING OF HOLES IN WOOD AND WOOD PRODUCTS. FIRA 13(43):20. Srevenage, England.
Small diameter, blind holes were made using a solid punch that produces no chips but forms the hole by mechanical displacement of adjacent material In tests with solid wood, hardboard, plywood, and plastic overlays, the method proved suitable for holes up to 3.2 mm in diameter and 9 6 mm deep. While such holes are too small for dowels, the process could be used to machine starting holes for screws

Mater, M. H. 1972 BORING 40-FOOT LONG UTILITY POLES FOR CONDUIT PASSAGE. Forest Products Journal 22(6) 28-31.
Describes the design of a machine for boring 2½-inch diameter holes lengthwise through 40-foot long electrical transmission poles at a rate of 12 ft/min Three air actuated centering clamps hold and straighten the pole during boring A movable boring tool-alignment tube assembly is axially positioned at both ends of the pole An internal tube carries the boring head and rotates at 1800 r/min while the outer tube maintains accurate alignment; it is rotated at 10 r/min to prevent jamming The boring bits (resembling end-milling heads) meet at the pole center with an axial alignment tolerance of ¼ inch. Scavenging air is introduced through a rotary seal in the assembly, flows through the system between the tubes and returns carrying sawdust through the inner tube One-inch electrical conduit was readily pushed through the resulting hole In tests with southern pine and Douglas-fir, bored poles were found to dry faster and absorb more preservative Strength reduction on an 8-inch diameter, 30-foot long pole was calculated to be about 1.0 percent in bending and 9 8 percent in shear Pole weight was reduced about 1 1 pounds per foot of length.

McMillin, C. W., and Woodson, G. E. 1972. MOISTURE CONTENT OF SOUTHERN PINE AS RELATED TO THRUST, TORQUE, AND CHIP FORMATION IN BORING. Forest Products Journal 22(11): 55-59.
Holes 3½ inches deep were bored with a 1-inch spur machine bit in southern pine having specific gravity of 0.53 (ovendry weight and volume at 10 4 percent moisture) The bit was rotated at 2,400 rpm and removed chips 0 020 inch thick. For wood moisture contents ranging from ovendry to saturation, thrust was lower when boring along the grain (average 98 pounds) than across the grain (average 138 pounds), while torque was higher when boring along the grain (average 42 inch-pounds) than across the grain (average 33 inch-pounds). For both boring directions, torque and thrust increased with increasing moisture content to a maximum at about 5 to 10 percent, then decreased to a constant value at about the fiber-saturation point. Chip types resembled those obtained in orthogonal cutting. The type of bit tested, net power at the spindle required to cut 0 020-inch-thick chips at speeds of 2,400 rpm or less should not exceed 2 hp., regardless of boring direction or moisture content, thrust should not exceed 200 pounds. Chip types resembled those obtained in orthogonal cutting.

Woodson, G. E., and McMillin, C. W. 1972. BORING DEEP HOLES IN SOUTHERN PINE. Forest Products Journal 22(4):49-53.
When holes 10½ inches deep and 1 inch in diameter were made with either a ship auger or a double-spur, double-twist machine bit, clogging occurred at a shallower depth (avg. 6.5 inches) when boring across the grain than when boring along the grain (avg 10.1 inches) In both boring directions, thrust force and torque demand for unclogged bits were less for the ship auger than for the machine bit Generally, torque and thrust were positively correlated with chip thickness and specific gravity; they were unrelated to spindle speed when the thickness of chips was held constant. For the machine bit, thrust was less in wet than in dry wood Although the ship auger required less horsepower than the machine bit, it was slightly less efficient; i e., more energy was required to remove a unit volume of wood In boring along the grain, the ship auger made better holes than the machine bit when the wood was dry; in wet wood hole quality did not differ between bit types. When boring across the grain, the machine bit made better holes in both wet and dry wood.

18

MORTISING AND TENONING

Hojan and Ozimek. 1971. (Safety, including noise)

MACHINING WITH COATED ABRASIVES

Anonymous. 1972. SANDING. Wood and Wood Products 77(3): 41, 75.
A short discussion of reader response to a questionnaire regarding sanding problems and how they solve them.

Anonymous. 1973. (Reports of research programs and industrial developments)

Anonymous. 1973. ABRASIVE PLANING. Wood and Wood Products 78(9):24-27.
Describes the abrasive planing operation at a California mill that produces about 450,000 board feet of Douglas-fir, ponderosa pine, and sugar pine 2 x 4s and 2 x 6s per double shift. The 52-inch wide sander is designed to surface 13 2 x 4s or 8 2 x 6s simultaneously using two opposed heads. The·first head uses 40 grit belts while the second uses 24 grit. Dimensions are held to about ±0.010 inch. With motors totaling 800 horsepower, the planer is capable of processing 36,000 board feet per hour.

Anonymous. 1973. MORE HELP IN SANDING COMPLEX SHAPES: BONDED WHEELS. Wood and Wood Products 78(9):35, 82.
Bonded abrasive wheels, individually shaped to·fit the contour of the workpiece, are ·finding increasing application in sanding moldings and other complex shapes. The·finish obtained in one or two passes of the wheel is claimed to be superior to that obtained by successive grit sizes on conventional equipment and down time is less. Depending on species, a 6-inch diameter wheel is rotated at about 800 to 1800 r.p.m. using a constant pressure of 5 to 10 pounds per inch of contact area. Feed speeds of 15 fpm can be expected.

Anonymous. 1973. ABRASIVE SLEEVES. Wood and Wood Products 78(9):36.
Describes a shock-resistant steel sleeve with tungsten carbide grits. The sleeve slips over a specially designed arbor that expands with centrifugal force to rigidly hold the sleeve. While the standard sleeve length is 16 inches, the ends are beveled and they may be ganged

Anonymous. 1973. (Nomenclature)

Anonymous. 1973. SANDING AND ABRASIVES. Wood and Wood Products 78(9):22-23.
A general discussion from industrial sources on sanding and abrasive problems, machinery purchases, operational changes, and suggestions for new equipment.

Efremova and Mitusov. 1972 (Safety, including noise)

Efremova and Mitusov. 1973. (Safety, including noise)

Hamamoto, K., and Mori, M. 1973. FUNDAMENTAL STUDIES ON LOW FREQUENCY CUTTING OF WOOD. V. EFFECTS OF VIBRATORY FACTORS ON SERVICE LIFE OF COATED ABRASIVES IN VIBRATORY SANDING. Journal of the Japan Wood Research Society 19:305-310. Tokyo.
Samples of a three layer particleboard (specific gravity 0.75) were

machined with a 100 mm diameter drum-type sander using AA No. 60 grit paper The workpiece was vibrated in a direction perpendicular to the feed at frequencies from 10 to 30 Hz and double amplitudes ranging up to 12 mm. The end faces of samples were sanded for 300 minutes using a constant drum speed of 150 r/min and a sanding pressure of 200 g/cm². It was found that the rate of stock removal (and total stock removed) was greater for vibratory sanding than for conventional sanding. The average rate of stock removal increased with increasing frequency and amplitude. Loading of the abrasive surface was less with vibratory sanding than with conventional sanding. The percentage of loaded area decreased with increasing frequency. It also decreased with increasing double amplitude but above 0.5 mm the change was small.*

Hamamoto, K., and Mori, M. 1973. FUNDAMENTAL STUDIES ON LOW FREQUENCY CUTTING OF WOOD. VI. RATE OF MATERIAL REMOVAL AND SURFACE ROUGHNESS OF WORKPIECE IN VIBRATORY SANDING. Journal of the Japan Wood Research Society 19:379-384. Tokyo.
Tests were conducted on clear air dry samples of Formosan cypress (Chamaecyparis taiwanensis) using a vibrating drum sander having a frequency range of 0.1 to 60 Hz, double amplitude of 0.15 to 7.0 mm, and rotational speed of 12 to 450 r/min. It was found that the rate of material removal was greater for vibratory sanding at all amplitudes and frequencies than for conventional sanding, providing the contact pressure of the drum did not exceed a critical value. Surface·finish was better for samples sanded with vibration. Roughness values are closely correlated with the sinusoidal curve form followed by the grit edge; they decreased rapidly with increasing inclination angle of the wave form up to 20 degrees, after which the rate of change became less. The improved efficiency of vibratory sanding was attributed to the greater grit velocity and an apparent self-sharpening effect. A maximum sanding pressure exists (all other factors held constant) above which higher efficiencies cannot be obtained.

Kato, C., Fukui, H., and Ono, M. 1972. THE CUTTING PERFORMANCE OF COATED ABRASIVE BELT SANDING CROSS SECTION SURFACE BY SMALL-SIZED BELT SANDER. Journal of the Japan Wood Research Society 18:123-130. Tokyo.
Samples of Fagus crenata were sanded on the radial and transverse surfaces using eight types of abrasive belts and the rate of stock removal measured. It was found that the rate of removal decreased with increasing sanding time and was greatest when sanding the radial surface parallel to the grain. Sanding defects such as checks and burning were more severe on transverse surfaces than on radial surfaces. The rate of stock removal increased with increasing belt velocity and with sanding pressure for both surfaces Belts that performed well in terms of pressure required to remove 2 5 μm of material per 1 m of belt passage yielded rough surfaces

Lyubimov. 1972. (Machinability of species and reconstituted wood; Particleboards and sandwich boards)

19

Sweum, L., and Furrer, W. 1973. SANDING PARTICLEBOARD ON WIDE BELT MACHINERY. Forest Industries 100(4):71.
A short troubleshooting guide for wide belt sanding of particleboards

Varbanov. 1972 (Orthogonal *cutting*)

Venkararaman, J., Raghavendra, B. G., and Nagaraju, S. 1972. THICKNESS CONTROL OF FINISHED PLYWOOD. Journal of the Indian Academy of Wood Science 2(1):38-46. Bangalore.
The variation in thickness of plywood and hardboard samples was measured in an Indian plywood factory before and after sanding. For the operation, the amount of material removed (average 0.3 mm) was much greater than necessary Better control of veneer thickness during peeling was recommended.

Yatsyuk, A. I., and Zayars, I. M. 1972. EFFECT OF THE GRAIN SIZE OF THE ABRASIVE DISK ON PROCESS

VENEER C

Anonymous 1973 VENEER SPECIES OF THE WORLD. Interim Report. IUFRO. 150 p. Forest Products Laboratory, Madison.
Data is presented in a highly condensed tabular format that provides information on locality of growth, estimated volume, specific gravity, green moisture content, ease of cutting, recommended cutting temperature, dry veneer characteristics, widthwise shrinkage, color and·figure, and relative suitability for various end uses Data tabulated for about 100 conifers and several hundred hardwoods In one table common names of the species are tabulated alphabetically opposite the scientific names. In another table, the scientific names are arranged alphabetically opposite the common names

Anonymous. 1973. (Research instrumentation and techniques)

Achterberg, Bucher, and Matschey. 1972. (Machinability of species and reconstituted wood, Hardwoods)

Bohlen, J. C. 1972. LVL. LAMINATED-VENEER LUMBER — DEVELOPMENT AND ECONOMICS. Forest Products Journal 22(1):18-26.
By rotary peeling sawlogs and laminating thick veneer into long, wide planks, which are subsequently sawn into structural dimension lumber, substantial increases in yield (up to 47 percent), are obtained. Although processing costs of laminating are higher, these costs are offset by the improved yield. The laminated-veneer lumber (LVL) processes make possible the production of structural lumber of any required length Long beams will enable a "two span" beam design to be applied to conventional dwelling-unit joist requirements The cost of joists per dwelling-unit indicates a saving of 8 5 percent at 1971 average logging costs Although ¼ less sawdust production and increased yield from logs improve the environmental and conservation aspects of lumber manufacture, these advantages may be somewhat offset unless precautions are taken for glue waste disposal methods and air-pollution control of veneer dryer exhaust emissions

Boulloud, J C 1972. RESEARCH ON VENEER PEELING AT THE CENTRE TECHNIQUE FORESTIER TROPICAL. Bois et Forets des Tropiques (142):35-51. Nogent-sur Marne.
Gives an illustrated account of the methods used to study veneer peeling involving the use of high speed motion pictures and a tool

Feihl, O., and Carroll, M. N. 1973. THICKNESS VARIATION IN VENEER PEELED WITH A FLOATING BAR. Department of the Environment. Canadian Forestry Service Information Report OP-X-62, 14 p. Eastern Forest Products Laboratory, Ottawa.
In tests with a variety of Canadian hardwood and softwood species, it was found that thickness variation in veneer peeled with a floating bar can be held within the same limits as in veneer peeled with fixed bars. With the floating bar, forward movement must be limited — a minimum horizontal gap of 75 to 85 percent of the feed proved adequate. Proper thickness was reached with fewer revolutions of the bolt when veneer was peeled with a floating bar than when peeled with a fixed bar.

Feihl, O., and Carroll, M. N. 1973. PEELING VENEER WITH A FLOATING BAR: EFFECT OF BAR PRESSURE ON VENEER QUALITY. Forest Products Journal 23(12):28-31.
Veneer cutting trials with poplar, white elm, and red pine were conducted on a 4-foot lathe equipped with a pressure controlled floating flat bar (face inclined 14-degrees to the vertical), and later with a floating ¾-inch diameter, motor driven roller bar. It was found that surface roughness and depth of lathe checks decreased with increasing pressure in the range of 20 to 90 pounds per linear inch (pli) of log. The effect of pressure was greater for veneer peeled with the flat bar than for veneer peeled with the roller bar and was more pronounced on thick than on thin veneer. Pressure had little effect on veneer thickness. Bars set with a low vertical gap generally yielded tighter and smoother veneer, but quality was sensitive to variations in pressure. A pressure range of 20 to 60 pli for the flat bar, and 20 to 80 pli for the roller bar proved satisfactory. The quality of veneer peeled with either bar was equal to that peeled with a standard nonfloating bar

Feihl, O., and Godin, V. 1970. SETTING VENEER LATHES WITH AID OF INSTRUMENTS. Department of Fisheries and Forestry. Canadian Forestry Service Publication 1206, 42 p Ottawa.
Describes, in practical terms, various instruments and procedures needed to set up veneer lathes A discussion of knife grinding and some of the problems encountered in controlling settings during lathe operation is also given. An appendix contains a list of publications on veneer cutting available from the Ottawa Forest Products Laboratory, addresses of manufacturers of lathe setting instruments, tables of recommended settings for various Canadian species, and heating schedules for veneer logs.

Feihl and Godin. 1973. (Machinability of species and reconstituted wood; Hardwoods)

Feihl and Godin. 1973. (Machinability of species and reconstituted wood, Softwoods)

Feihl, O., and Godin, V. 1971. WEAR, PLAY, AND HEAT DISTORTION IN VENEER LATHES. Department of Fisheries and Forestry. Canadian Forestry Service Publication 1188, 12 p. Ottawa.
A lathe, while properly adjusted when a new knife or pressure bar is installed, may not remain so after the peeling operation is started as a result of play or heat distortions These two sources of difficulty are described, including their effect on veneer quality, together with practical corrective procedures.

George and Miller. 1970. (Research instrumentation and techniques)

Hailey and Hancock. 1973. (Research instrumentation and techniques)

Hailey, Hancock, and Walser. 1973. (Machinability of species and reconstituted wood, Softwoods)

Hirz. 1971. (Patents)

Hse. 1972. (Research instrumentation and techniques)

Huang, Y. S. 1971. A NEW ANALYSIS OF KINEMATICS IN ROTARY VENEER CUTTING. Experimental Forest of National Taiwan University Technical Bulletin 86, 9 p Taiwan.
Develops equations for the change in clearance angle with the position of the knife edge; the relation between the actual and theoretical veneer thickness; cutting velocity; and total cutting power.

Jain, Gupta, and Bagga. 1973. (Machinability of species and reconstituted wood; Hardwoods)

Koch, P 1973. STRUCTURAL LUMBER LAMINATED FROM ¼-INCH ROTARY-PELLED SOUTHERN PINE VENEER. Forest Products Journal 23(7): 17-25.
By the lamination process evaluated, 60 percent of total log volume ended as kiln-dry, end-trimmed, sized, salable 2 by 4's — approximately 50 percent more than that achieved by conventional bandsawing of matched logs. Morever, modulus of elasticity of the laminated 2 by 4's (adjusted to 12 percent moisture content) averaged 1,950,000 psi compared to 1,790,000 for the sawn 2 by 4's Allowable fiber stress in edgewise bending was 2,660 psi for the laminated 2 by 4's, more than twice the value obtained (1,270 psi) for the sawn 2 by 4's A manufacturing procedure for 6-ply lumber is described that calls for no major innovation in equipment When the price per MBF of structural lumber is 50 percent greater than the price per M ft.² of ½-inch sheathing plywood, it should be more profitable to manufacture laminated lumber than plywood

Kolyashev, D. 1971. THE CHANGE IN THE CLEARANCE ANGLE OF THE KNIFE DURING PEELING OF LOGS. Nauchni Trudove. Vissh Lesotekhnicheski Institut (18) 41-45 Sofiya
A mathematical analysis

Lutz, J. F. 1971. WOOD AND LOG CHARACTERISTICS AFFECTING VENEER PRODUCTION. U. S. Department of Agriculture Forest Service Research Paper FPL-150, 31 p Forest Products Laboratory, Madison, Wis.
Describes the physical and mechanical properties of wood important to veneer production and use A bibliography contains 41 references

Lutz, J. F. 1972. VENEER SPECIES THAT GROW IN THE UNITED STATES. U. S. Department of Agriculture Forest Service Research Paper FPL 167, 129 p. Forest Products Laboratory, Madison, Wis.
Describes properties of 156 U S. tree species that affect their manufacture and use as veneer or products made from veneer For ready availability, much information is condensed into eight summary tables listed early in the report. Factors covered in the tables include volume and log characteristics, physical and mechanical properties, cutting and drying of veneer, and quality and uses of veneer Information that does not readily fit into the tables is presented in narrative form by individual species

Meriluoto, J. 1973 TEMPERATURE IN THE PEELING OF VENEER. Paperi ja Puu 55(10):763,771. Helsinki.

Myronuk. 1972. (Research instrumentation and techniques)

Newall, Hudson. and White. 1969. (Machinability of species and reconstituted wood; Softwoods)

Nikolaev, A. F. 1969. THE STABILITY OF A VENEER BOLT DURING PEELING. Lesnoi Zhurnal 12(4).117-124. Arkhangelsk.
A theoretical study, supported by experimental data, on the stability of small-diameter veneer bolts when axial compressive forces exerted on the bolt are (1) constant, (2) are changed periodically, or (3) changed periodically at high frequency

Nikolaev, A. F. 1971. LOSSES OF WOOD IN VENEER BOLT CORES IN RELATION TO PEELING PARAMETERS. Lesnoi Zhurnal 14(2):114-117. Arkhangelsk.
A theoretical analysis of losses resulting from premature removal of the core.

Nilberg. 1973. (Parents)

Olischlager, K. 1970. SPRUCE VENEER FROM PRUNED TREES. Allgemeine Forstzeitschrift 25(12):251-254. Munich.
Results of small-scale factory peeling and slicing trials suggest that veneer could be profitably produced from pruned spruce if logs are no less than 30 m long with a mid-diameter greater than 30 cm The knotty core should be less than 25 percent of the cross-section.

Orlov and Strizhev. 1972. (Research instrumentation and techniques)

Osterfeld, H 1972. VENEER PRODUCTION BY PEELING. Holz-und Kunststoffverarbeitung 7(2).108-111. Stuttgart.
Reviews the application of technology to veneer peeling, with special reference to steaming, checks, and lathe settings and operation

Palka, L. C., and Holmes, B. 1973. EFFECT OF LOG DIAMETER AND CLEARANCE ANGLE ON THE PEEL QUALITY OF 0 125-INCH-THICK DOUGLAS-FIR VENEER. Forest Products Journal 23(7):33-41.
Two-inch wide disks cut from five green Douglas-fir logs were converted to 0 125-inch-thick veneer on an experimental mini-lathe. A constant roller-bar compression of 13 percent was maintained with a freely rotating 0 625-inch diameter bar. Mean cutting speed was 14 5 feet per minute and the knife had a 23° main bevel angle without microbevelling Average veneer thickness and lathe-check frequency appeared positively correlated with specific gravity The number of growth rings per inch was negatively correlated with lathe-check depth and veneer roughness. A clearance angle of 0°30' is recommended to optimizing thickness control for logs 13-inch to 5-inch in diameter For best control of roughness, the clearance angle should increase from 0°00' at the periphery to 0°30' in the central heartwood, then decrease to -0°30' near the core

Pearson. 1972. (Parents)

Peltonen 1973. (Parents)

Peters. C. C., Mergen, A. F., and Panzer, H R. 1972. THICK SLICING OF WOOD EFFECTS OF WOOD AND KNIFE INCLINATION ANGLE. Forest Products Journal 22(9):84-91.

One-half-inch-thick slices of clear, flat-grained red oak and southern pine were cut with the knife and workpiece inclined, relative to the cutting motion, at angles of 45°, 67.5°, and 90° The wood, heated to 190°F., was cut at 50 feet per minute using a special experimental machine. A fixed pressure bar applied compression and restraint. While the results are considered tentative. it was found that depth of fracture was least when using inclination-angle differences of 45°. Depths ranged from about 0.10 to 0.40-inch in both species. However, at the 45° setting, uniformity of thickness was much poorer, the surface was rougher and there was considerable twist. The degree of cupping decreased with increasing angular inclination. Cutting energy (the product of the combined cutting forces and the distance through which they act) was unaffected by angular difference. The best overall slice quality (shallowest fractures, smoothest surface, and most uniform thickness) occurred when the knife was not inclined relative to the work.

Porter. 1972. (Parents)

Rauch, P., and Boiciuc, M. 1972. THERMAL TREATMENT IN AN AUTOCLAVE OF CANTS FOR DECORATIVE VENEER. Industria Lemnului 23(3):101-105. Bucharest.
Veneer cants were loaded on wagons and steamed in an autoclave at 116-118°C and 2 atm pressure. Treatment times for five native and imported species were reduced by 40-50 percent compared with treatment at atmospheric pressure. Other advantages were improved veneer quality, largely owing to more uniform heating; and a decrease in the power consumed in slicing, the decrease varying with species and being particularly large in Entandrophragma cylindricum.

Reed. 1972. (Parents)

Sugiyama, S., and Mori, M. 1973. FUNDAMENTAL STUDIES ON MECHANISM OF VENEER CUTTING. III. NUMERICAL ANALYSIS OF STRESS DISTRIBUTION IN WORKPIECE UNDER PRESSURE BAR COMPRESSION. Journal of the Japan Wood Research Society 19:385-391. Tokyo.
Water saturated samples of clear red lauan (Shorea negrosensis) were end coated with pressure sensitive tape imprinted with a grid pattern for test with various types of nosebars. The depth of nosebar indentation ranged from 0.2 to 3.0 mm; feed speed was held constant at 144.9 mm/min. Vertical and horizontal nosebar forces increased linearly with increasing indentation for both roller and fixed nosebars. For fixed nosebars, the coefficient of friction increased with increasing indentation (range 0.05 to 0.20) but remained constant at about 0 40 for roller bars The workpiece area in contact with the nosebar surface increased with increasing indentation; for a given indentation friction was greater for a roller bar than for a double or single faced fixed nosebar The amount of permanent workpiece deformation increased with increasing indentation but did not exceed 0.5 mm. Equations are provided to estimate the stress distribution within the nosebar.

Tochigi, T., and Hayashi, D. 1972. VENEER CUTTING WITH JET AIR PRESSURE. III. Journal of the Japan Wood Research Society 18:273-281. Tokyo.
Reports further results using a high-pressure air jet in place of a nosebar. In the present study, the velocity of the jet after leaving the nozzle, and its velocity after impinging on the knife and veneer are examined The force of the jet is resolved into horizontal and vertical components, depending on the angle of inclination, that prevent curling of the veneer and compress the workpiece. The effect of inclination angle on cutting force is shown graphically for

22

Venkataraman, Raghavendra, and Nagaraju. 1972. (Machining with coated abrasives)

Walser, D. C. 1972. AN INSTRUMENT FOR MEASURING KNIFE HEIGHT ON VENEER LATHES. Forest Products Journal 22(3):59-60.
Describes a simple, accurate tool to assist the lathe operator with knife height adjustment. The cost of the device is low (about $40 00) and accuracies within $\pm$ 0 001-inch can be maintained.

Woodfin and Lane. 1971 (Machinability of species and reconstituted wood, Softwoods)

Yerkes and Woodfin. 1972. (Machinability of species and reconstituted wood; Softwoods)

NG, AND GRINDING

experimental data. The pseudo-viscosity appears to be a function of concentration only. Error analyses for both modes of flow showed that all experimental data points were within 5 percent of values predicted by the proposed correlation models

Hakkila, P , and Saikku, O. 1972. MEASUREMENT OF BARK PERCENTAGE IN SAWMILL CHIPS. Folia Forestalia 135, 12 p. Institutum Forestale Fenniae, Helsinki.
In an experiment designed to develop a simple method of estimating the percentage of bark in sawmill chips used for pulp, it was found that a 3-liter sample per truckload was sufficient provided chips were taken at a distance greater than 30 cm from the surface. Samples from various loads should be mixed and the percent bark calculated on an ovendry weight basis

Harron, J. V., and Keays, J L. 1972. STATISTICAL ANALYSIS OF CHIP GEOMETRY AND MOISTURE ON KRAFT PULP YIELD. Department of the Environment. Canadian Forestry Service Information Report VP-X-88, 20 p. Western Forest Products Laboratory, Vancouver.
It was found that western hemlock chips of 2-mm. thickness pulp most uniformly and give maximum screened yield of permanganate number 20. Ovendry chips (as compared to green chips) generally gave lower screened yield, higher screen rejects, and cooked to higher Kappa numbers, consuming more alkali in the process

Harron, J. V , and Keays, J. K. 1973. EFFECT OF CHIP GEOMETRY AND MOISTURE ON YIELD AND QUALITY OF KRAFT PULPS FROM WESTERN HEMLOCK AND BLACK SPRUCE. Pulp and Paper Magazine of Canada 74(1):79-87.
Kraft cooks were made of laboratory cut western hemlock chips ranging in thickness from 2 to 10 mm and in lengths of 3/16 to 1-5/8 inch. Analysis showed that 2 mm green chips of all lengths gave maximum screened yield at KMnO, number 20, gave minimum screen rejects, and pulp most uniformly A chip thickness of 4 mm proved only slightly less favorable With increasing chip thickness above 4 mm and with increasing chip length, screen yield was reduced, screen rejects increased, and pulping became progressively more heterogeneous. When chips were ovendried, screen yield was lower, rejects higher, KMnO, and Kappa numbers increased, and more effective alkali was used Pulp quality of unbleached hemlock

kraft pulps at 300-ml CSF was apparently not impaired by decreasing chip thickness and grain length, except for some decrease in burst for thin chips of short length

Hudson and Williams 1973. (Research instrumentation and techniques)

Keating, J., and Albertson, G. 1972. IMPROVED FLOW REDUCES CHIP DAMAGE IN CHIPPER. British Columbia Lumberman 56(12) 24.
Describes a potential wood waste chipper having an improved method of releasing the chips out the rear housing into a discharge tunnel thereby reducing chip damage and breakup. Tests showed chips impacted the discharge tunnel at an angle of 15-degrees and were discharged at a velocity of about 90 feet per second in contrast to velocities of about 150 feet per second in some conventional chippers

Krysinki and Michalski. 1972. (Properties of the cutting edge and cutter. Dulling)

Neubaumer, H. G. 1971. THE HORIZONTAL WOOD HOG. Proceedings of the 26th Annual Northwest Wood Products Clinic, p 75-89 Engineering Extension Service, Washington State University, Pullman.
An illustrated presentation of different types of horizontal hogs with emphasis on units of recent manufacture

Radulescu, V. M. 1972. SOME PROBLEMS IN THE CONSTRUCTION AND USE OF WOOD CHIPPERS. Industria Lemnului 23(5) 187-196. Bucharest.
Compares the advantages and limitations of disk and drum type chippers with tablulated data on commercial models In general, drum chippers are preferred because they can process waste wood as well as roundwood and yield a uniform product Details are given of the construction and operation of two Rumanian drum chippers: a large pressure-fed machine, and a small gravity-fed machine for waste wood only

Rushnov, Pyrakhin, and Maslovskii. 1973. (Barking)

Schmutzler, W. 1972. THE FURTHER DEVELOPMENT OF THE CUTTER-BLOCK FOR THE MZM-10 CYLINDER-TYPE CHIPPER. Holzindustrie 25(1):10-12. Leipzeig.

Sturos, J A. 1973 PREDICTING SEGREGATION OF WOOD AND BARK CHIPS BY DIFFERENCES IN TERMINAL VELOCITIES. U.S Department of Agriculture Forest Service Research paper NC-90, 10 p. North Central Forest Experiment Station, St. Paul, Minn.
Relates terminal velocities of wood and bark chips for eight pulpwood species (aspen, sugar maple, jack pine, red pine, white spruce, balsam fir, slash pine, and loblolly pine) to their length, width, thickness, moisture content, and specific gravity Differences in the terminal velocities of the wood and bark chips caused by differences in these properties were used to predict the degree of segregation possible by air flotation.

Tanaka, Takahashi, Sakurai, Hirasa, and Nakamura. 1972. (Safety, including noise)

Tereshchenko, V I., and D'yachkov, V. P 1972. DETERMINING THE SURFACE AREA OF WOOD PARTICLES. Derevoobrabatyvayushchaya Promyshlennost' (8) 12-13. Moscow.
Tabulates values of the coefficient K (total chip surface area/chip

Discusses the theory of impact crushing of branchwood, and describes experiments made in a special chamber where the weight and speed of the impacting hammer head could be varied. Experiments were made on specimens of varying moisture content, 50 to 100 mm in diameter and 4 cm long, taken from branches of poplar, birch, spruce, pine and oak Results showed that it was best to comminute branches green, at temperatures above freezing Specific impact energy should be 3 to 6 kWh/m³, the impact surface of the hammer should be flat, and the walls of the chamber should not prevent the free redistribution of the wood within the area. Single-layer particleboards made from impact-crushed material had comparatively high strength.

Sellaeg, Kucera, and Mengkrog. 1972. (Machinability of species and reconstituted wood; Particleboards and sandwich boards)

Sellaeg and Gislerud 1972. (Machinability of species and reconstituted wood; Fiberboards, hard and soft)

Steiner, K. 1972. CHIPPERS FOR LONG LOGS FOR THE PARTICLEBOARD INDUSTRY. DEVELOPMENT AND STATE OF TECHNOLOGY. Holz als Roh-und Werkstoff 30. 207-214. Munich.
Describes and illustrates various types of chippers for processing tree length logs into chips for particleboards.

FLAKING

Radulescu, V. M. 1972. MODERN COMMINUTION PROCESSES AND EQUIPMENT FOR THE PRODUCTION OF PARTICLES REQUIRED BY PARTICLEBOARD FACTORIES. Industria Lemnului 23(11).477-482. Bucharest
Describes the cutting action of different types of machines used to convert chips and wood waste into small particles of various sizes.

GRINDING

Efremova and Mitusov 1972. (Safety, including noise)

ATING

Sawdust is also used as a secondary source of raw material for paperboard pulp. Some specific data on properties of thermomechanical pulp from different species refined under varying conditions are tabulated

Alexander. 1972. (Patents)

Anonymous. 1972. REFINER GROUNDWOOD MILL PULPS ASPEN AT ESCANABA. Pulp and Paper 46(4) 85-87.
Details are given of Mead Corporation's new highly automated 80-ton-per-day refiner groundwood mill. The refiner pulp at present is combined with pulp from the stone groundwood mill and used for production of publication-grade papers. It is planned to shut down the stone groundwood mill and double the capacity of the refiner groundwood operation.

Asplund, A. 1973. DEVELOPMENT OF THE THERMO-MECHANICAL PULPING METHOD. *In* International Mechanical Pulping Conference Proceedings, 20 p. Session III. Sponsored by The Swedish Association of Pulp and Paper Engineers and European Liaison Committee for Pulp and Paper. Stockholm.

*Tensile and tear strengths of newsprint produced by the thermo-
mechanical pulping process are 50 to 100 percent higher than those
obtained by grinding or disk-refining under atmospheric conditions.
Optical characteristics are similar to those of non-pressurized
refiner mechanical pulp but the shive content is less and the energy
consumption is lower The influence of process variables upon the
energy consumption and pulp characteristics are discussed and
equipment needs are detailed with flow diagrams*

Atack. 1972. (Reports of research programs and industrial
developments)

Atack, D. 1972. CHARACTERIZATION OF PRESSURIZED
REFINER MECHANICAL PULPS. Svensk Papperstidning
75(3) 89-94. Stockholm.
*Structural differences between pressurized and nonpressurized
refiner groundwood pulps are described and the importance of
relating basic ·fiber characteristics to pulp testing methods is
discussed The character of pulp is shown to markedly change when
chips are refined at pressures in the range of 10 to 100 psig (115° to
170°C) These changes appear related to the temperature at which
lignin undergoes a major reversible structural transition from an
elastic glassy state to an elastic rubbery one Some implications of
this transition on the interaction of refining variables, pulp quality,
and characterization are discussed. Article in English.*

Atack, D., and Heffell, L. R. 1972. HIGH SPEED GRINDING
OF EASTERN BLACK SPRUCE. PART I. Pulp and Paper
Magazine of Canada 73(9) 78-82.
*Samples of Picea mariana were ground for combinations of stone
speed (4000, 6000, 8000, and 10,000 ft/min) and grinding pressure
(20, 30, 40, and 50 lb/in²). A fully conditioned stone having 900
active grits/sq. in at the surface was used at a pit consistency of 1
and 3 percent. It was found that production rate increased with both
speed and pressure Specific energy consumption maxima occurred
at about 8000 and 6000 ft/min for low and high consistency
operations, respectively The Forgacs S factor increased with increasing speed up to 8000 ft/min then leveled off while the L factor
decreased as speed increased from 4000 to 8000 ft/min then leveled
off or increased slightly Freeness increased with increasing speed
and pressure at 1 percent consistency. At 3-percent consistency, it
rose to a maximum and fell to a minimum at a pressure of 50 lb/in².
At constant pressure, sheet strength properties decreased with
increasing speed, with some leveling off at speeds approaching
10,000 ft/min As speed increased brightness decreased as much as
4 to 10 points, the decrease being larger at lower grinding pressures.*

Atack, D., and Wood, P. N. 1973. ON THE HIGH
CONSISTENCY OPERATION OF LARGE DOUBLE
ROTATING DISK REFINERS. In International Mechanical
Pulping Conference Proceedings, 25 p. Session III. Sponsored by
The Swedish Association of Pulp and Paper Engineers and
European Liaison Committee for Pulp and Paper. Stockholm.
*Operating instabilities occur in large double disk refiners when chips
are refined at high consistencies. These instabilities can limit the
operating refining consistencies in the primary stage of many
industrial multistage systems to values¨ below the optimum
discharge consistency range of 20 to 35 percent. It was found that
when dilution water was fed directly into the eye of a refiner in the
form of high speed jets from conduits symmetrically located around
the feed spout the refiner could be operated at discharge
consistencies of between 30 to 35 percent. Increased refining
consistency has resulted in newsgrade mechanical pulps of
equivalent or higher quality at reduced levels of specific energy.*

26

A general review of developments in mechanical pulping of southern pine from 1940 to the present.

Chow, P., Walters, C. S., and Guiher, J. K. 1971. pH MEASUREMENTS FOR PRESSURE-REFINED PLANT-FIBER RESIDUES. Forest Products Journal 21(12):50-51.
The pH of shavings, chips, sawdust, and bark of various hardwoods and non-woody residues differed after processing in a Bauer 418 steam pressurized refiner. Thus, when such fibers are to be used in boards, adhesive selection should be based on an after processing pH value Apart from refined maple shavings, unrefined cottonwood shavings, and cornstalks, all materials tested had a pH less than 7. The pH of the cold water extract was only slightly higher than that of hot water extracts.

Corson, S. R. 1971. APPLICATION OF SIZE REDUCTION THEORIES TO DISK REFINER PULP PRODUCTION. New Zealand Journal of Forestry Science 1(1):125-127. Rotorua
The disk refining process for mechanical pulp production is a well established operation. Present trends indicate that it will play a major role in the production of sanitary tissue, newsprint, magazine paper, and higher grades of paper It is shown that general comminution theories may be used to describe the breakdown of wood chips in the disk refiner This will lead to a better understanding of the disk refining process and enable the production of more uniform quality pulp at lower cost

Corson, S. R. 1973. DYNAMIC BEHAVIOR OF A DISK REFINER. *In* International Mechanical Pulping Conference Proceedings, 16 p. Session I. Sponsored by The Swedish Association of Pulp and Paper Engineers and European Liaison Committee for Pulp and Paper. Stockholm.
Proposes a mathematical model describing the breakdown of wood chips to pulp fibers in terms of particle size. The model relates the cumulative weight percentage of pulp fibers of size equal to or less than (y) to the input chip size (x) in terms of a simple linear function. The parameters of this function are themselves related to the reduction ratio (z) by second-order polynomial functions. The model is based on a single probabilistic function which assumes that wood chips of a specific size breakdown in a consistent manner to form mechanical pulp and that they are not influenced by chips of other sizes in the refining zone. In its present form, the model provides a basis for a theory of the disk refining process similar to those already developed for other unit operations.

Dobrovol'skii, P. P. 1972. DEFIBRATION OF WOOD. 231 p. Izdatel'stvo Lesnaya Promyshlennost' Moscow.
A monograph dealing with the defibration process, grinding stones, the mechanics of defibration, the material and thermal balance of the defibration process, and the design of defibrators and control systems

Dobrovolsky, P. P., and Gorelov, V. V. 1973 IMPROVEMENT OF CHAIN GRINDER FEEDING MECHANISM. *In* International Mechanical Pulping Conference Proceedings, 11 p. Session VI. Sponsored by The Swedish Association of Pulp and Paper Engineers and European Liaison Committee for Pulp and Paper. Stockholm.
Develops a mathematical expression for the maximum charge pressure without causing structure damage to the wood.

Esterer, A. K 1972. HIGH-YIELD DRY PULPING STILL A DREAM? Pulp and Paper 46(10).98-100.
Maximum wood fiber utilization can be achieved when all

lignocellulose is retained in the product It is proposed that this can be accomplished by mechanical defiberization, followed by blocking of reactive groups in lignocellulose with a suitable chemical which will stabilize the fibers against discoloration and disintegration by ultraviolet light and atmospheric oxygen

Freeman, C. W., and Smart, D. C. 1973. GREAT NORTHERN INCREASES CAPACITY WITH REFINER MECHANICAL PULPING. Pulp and Paper 47(9):96-97.
Using aspen chips, a 150 ton per day chip groundwood system meets the requirements of a new twin-wire paper machine at Millinocket, Maine, where six Sprout-Waldron 42-1B refiners (in two lines of three refiners each) and all auxiliaries are operated by one man per shift from a central control room. A second man assists by doing sampling, testing, cleaning, etc Mill engineers estimate that the same two-man crew could operate a similar plant of three times the present capacity

Furukawa, Saiki, and Harada. 1973. (Properties of wood)

Gros, G., and Johnsson, B 1973. A COMPUTER SYSTEM AS A TOOL FOR OBTAINING UNIFORM GROUNDWOOD PULP QUALITY. *In* International Mechanical Pulping Conference Proceedings, 20 p. Session VI. Sponsored by The Swedish Association of Pulp and Paper Engineers and European Liaison Committee for Pulp and Paper. Stockholm.
Describes the computerized quality control system used at the Holmens Bruk AB groundwood mill, Norrkoping, Sweden.

Harvey. 1972. (Patents)

Hill, J., and Eriksson, L. 1973. MECHANICAL PULPING PROCESSES EVALUATED BY AN OPTICAL DEVICE. *In* International Mechanical Pulping Conference Proceedings, 13 p. Session II. Sponsored by The Swedish Association of Pulp and Paper Engineers and European Liaison Committee for Pulp and Paper. Stockholm
Describes an optical device capable of characterizing pulp properties

Hoglund, H., and Tistad, G. 1973. ENERGY UPTAKE BY WOOD IN THE MECHANICAL PULPING PROCESS. *In* International Mechanical Pulping Conference Proceedings, 26 p. Session I. Sponsored by The Swedish Association of Pulp and Paper Engineers and European Liaison Committee for Pulp and Paper. Stockholm.
In the grinding and chip refining processes, wood is broken down into pulp by periodical, high frequency forces. A comparison between the feed rate of the wood and the peripheral speed and coarseness of the stone shows that only a few of the loading cycles actually induce failure in the wood structure. A similar comparison between the rate of input of chips and the rotational speed of the disk also shows that only a minority of the loading cycles, cause rupture of the chips. In the majority of the loading cycles no new surface is created and the mechanical energy is converted into heat. Plastic models of wood structures were loaded in a manner similar to the mechanical action occurring in the grinding and chip refining processes and the distribution of stress studied utilizing polarized light It was shown that the stress distribution depends on stiffness. In grinding, stress appears to be maximum at some distance from the actual grinding zone.

Jensen 1973. (Reports of research programs and industrial developments)

Since pulp properties varied with refining conditions, it was concluded that no temperature sensitive mechanisms were involved The variation in pulp quality was attributed to the hydrodynamic effect of changes in the refining zone consistency. In second stage pulping, a temperature gradient extended across the refining zone which differed with changes in operating parameters. These gradients were not associated with changes in pulp properties, and it was concluded that no temperature sensitive mechanisms were operating. In two stage refining, pulp quality appeared to be determined only by the specific energy used in either or both stages, and the refining zone consistency in the first stage There was no evidence of self-pressurization of the refiner under any of the conditions used.

McCarty, E. F. 1973. TECHNOLOGY DEVELOPED FOR HIGH HARDWOOD GROUNDWOOD NEWSPRINT. Paper Trade Journal 157(24).38-42.
A summary of 5 years work to develop an 85 percent hardwood groundwood newsprint and ultimately, newsprint of 100 percent hardwood.

McMillin, C. W. 1973. DYNAMIC TORSIONAL UNWINDING OF SOUTHERN PINE TRACHEIDS AS OBSERVED IN THE SCANNING ELECTRON MICROSCOPE. *In* International Mechanical Pulping Conference Proceedings, 21 p. Session I. Sponsored by The Swedish Association of Pulp and Paper Engineers and European Liaison Committee for Pulp and Paper. Stockholm.
In previous research on the process for making groundwood in a double-disk refiner, a theoretical stress analysis indicated that tracheids of Pinus taeda *L. can fail while under torsional stress and unwind into ribbonlike elements. Such elements have been shown to provide the coherence necessary for strength development in these pulps. Depending upon their physical state, tracheids may also buckle or shear and yield no ribbons In this study, macerated earlywood and latewood tracheids of loblolly pine were stressed in torsion with a specially designed fixture and observed at high magnification in a scanning electron microscope. Three previously postulated failure types were identified. Type I was characterized by the formation of a crack, generally parallel to the zone of weakness delineated by the fibril helix, followed by unwinding into a ribbon with further twisting. Type II failures were by shear perpendicular to the fiber axis, while Type III were by continuous diagonal buckling. For earlywood fibers rotated in a clockwise direction, the proportion of Type I failures decreased with increasing axial separation (the distance between the points of applied torque). The proportion was unaffected when tracheids were heated to about 100°C. In latewood, the proportion of Type I failures increased slightly with increasing axial separation and was greater for hot than for cold fibers. For the test conditions, total length of the ribbonlike portion was generally shorter for latewood than for earlywood. For both earlywood and latewood, the proportion of Type II failures was greatest when the axial separation was small and the fibers were unheated Type III failures were principally associated with wide axial separation. Counterclockwise rotation of the fiber proved undesirable, since no Type I failures were generated Scanning electron micrographs of pulp produced by double-disk refining confirmed the presence of these failure types*

Meinecke, A. 1972. PRODUCTION OF BLEACHED REFINER GROUNDWOOD FROM HARDWOODS. Pulp and Paper Magazine of Canada 73 (1).80-85.
Gentle alkaline pretreatment of hardwoods permits the production of pulp whose properties are similar to those made from softwoods. Extensive pilot plant tests were conducted on poplar and birch using

9

various alkaline pretreatments. Impregnated chips were defibrated in two stages at about 15 percent consistency in a 5E Voith disk refiner It was found that specific energy requirements were about the same as for conventional groundwood but refiner groundwood pulp properties were superior. Fines content was low and the percentage of long ·fibers was high. Sheet properties were comparable to softwood refiner groundwood except for tear which was adversely affected by the shorter hardwood ·fiber. Pulp brightness was higher for poplar than for birch. Large quantities of poplar refiner groundwood were used in test runs on paper machines in mixtures with sulphite chemical pulps and stone groundwood without adverse effects on machine operation or paper properties. For the process to be economically feasible, hardwood must be 15 percent cheaper than softwood.

Mihelich, W. G., Wild, D. J., Beaulieu, S. B., and Beath, L. R. 1972. SINGLE-STAGE CHIP REFINING — SOME MAJOR OPERATING PARAMETERS AND EFFECTS ON PULP QUALITY. Pulp and Paper Magazine of Canada 73(5):78-82.
A collation of various studies on common operating variables encountered in single-stage refining of softwood chips, and how they affect pulp quality.

Mukherjea, V. N., Sharma, Y. K., and Guha, S. R. D. 1971. NEWSPRINT GRADE REFINER GROUNDWOOD PULP FROM JUTE STICKS. Indian Forester 97(2):101-104. Dehra Dun.
Green·jute chips 1 to 2 cm long were multi-pass defibrated in a 30.5-cm single-disk refiner (1440 r/min) using plate clearances of 0.38 to 0 25 mm and consistencies of 4 to 10 percent. After analysis of the jute pulp and paper properties, samples were mixed with bleached beaten sulphite pulp (40 percent) and sheet properties evaluated It was found that the quality of the jute pulp improved as the plate clearance was reduced and that refining should be regulated so as to produce a pulp with about 150 ml CSF and with about 64 percent of the pulp passing through a 100-mesh screen classification. Handsheets of 60.40 mix had sufficient strength and brightness to warrant consideration of jute sticks as a potential papermaking material. The economics of collecting the sticks and the possibility of deterioration during storage should be given further consideration.

Nurminen, H., Holm, H., and Rahkonen, R. 1973. NEW REFINER INSTALLATION AT RAUMA. In International Mechanical Pulping Conference Proceedings, 22 p. Session IV. Sponsored by The Swedish Association of Pulp and Paper Engineers and European Liaison Committee for Pulp and Paper. Stockholm
The groundwood mill at Rauma, Finland, was enlarged at the beginning of 1972 when a refiner groundwood line was introduced. At the same time the capacity of the reject-refiner line was also enlarged Capacity of the sawdust and refiner groundwood lines was increased to 100 tons a day, and the reject-refiner line is capable of handling 175 tons a day. Each line consists of two Enso-Bauer 488 disk refiners The third stage of the sawdust and refiner groundwood lines utilize Enso-Bauer 411 and 412 pump-through refiners Operational experience and quality control procedures are detailed.

Orchard, W. M., and Charters, M. T. 1973. STEAM PRESSURIZED DOUBLE-DISK REFINING — A NEW TECHNIQUE IN HIGH-YIELD SULFITE PULPING. TAPPI 56(5):142-145.
In a series of trials it was found that a premium quality bisulfite pulp in the 75 to 80 percent yield range could be made in a single

pass through a steam pressurized double-disk refining system at a power application of one-third that required in conventional atmospheric disk refiners. The shive content of the pulp was exceptionally low and bonding properties were excellent in mixed furnishes with groundwood.

Ota, M., Maraki, Y., and Kawabe, J. 1973. ON THE RELATION BETWEEN MECHANICAL DEFIBRATE CONDITION AND THE PROPERTIES OF SIS HARDBOARD, MANUFACTURED FROM VARIOUS HARDWOODS OF THE KYUSHU DISTRICT. Bulletin of the Kyushu University Forests (47):311-317. Fukuoka, Japan.
Six hardwood species having a broad range of densities were pulped by the Asplund process and manufactured into SIS hardboards. Board properties proved negatively related to length of refining time; the effect was greater with species of low density.

Page, El-Hasseiny, Winkler, and Bain. 1972. (Properties of wood)

Pehu-Lehronen, J. 1973. STONE GROUNDWOOD. Pulp and Paper International 15(6):38, 40, 42. Brussels.
Stone groundwood, recently considered obsolete, is the subject of renewed interest High-load grinding, improved stone cleaning techniques, and application of automatic controls, have brought about significant improvements in efficiency and pulp quality. Stone groundwood and refiner mechanical pulping are not seen as rivals, but as complementing each other. It is predicted that further development of both processes will make it possible to produce a better newsprint using only refiner and stone groundwood pulp than can be made using either one with chemical pulp.

Perez, A. R. 1972. ROD MILL REFINING OF HARDWOOD AMMONIA PLASTICIZED CHIPS. Master of Science Thesis. State University College of New York. Syracuse.

Peterson, H. E., and Nelson, D. G. 1972. PRODUCTION OF REFINER MECHANICAL PULP FROM ASPEN FOR PUBLICATION PRINTING PAPERS. TAPPI 55:396-401.
Essentially shive-free refiner groundwood pulp was produced from aspen for use in publication grades of printing papers. Only small differences were noted between refiner groundwood pulp and pulp produced on stones

Peterson, V., and Dahlquist, G. 1973. SINGLE DISK AND DOUBLE DISK REFINERS IN DIFFERENT ARRANGEMENTS. THE EFFECT ON PULP PROPERTIES AND POWER CONSUMPTION. In International Mechanical Pulping Conference Proceedings, 20 p. Session III. Sponsored by The Swedish Association of Pulp and Paper Engineers and European Liaison Committee for Pulp and Paper. Stockholm.
The effect on pulp quality of single and two stage processing with both single disk and double disk refiners was studied. Single stage refining was found preferable with both types of refiners, except when mechanical pulp with a very low shive content is required. For such pulps, two stage refining with one double-disk refiner in each stage yielded best results. Both types of refiners produced pulps of equivalent quality at the same freeness level, except that the double-disk refiner yielded pulps with significantly lower shives and consumed less energy than did the single-disk refiner.

Reiner, P. L., and Jackson, M. 1973. AN APPROACH TO GROUNDWOOD CHARACTERIZATION USING ULTRASONICS. In International Mechanical Pulping Conference Proceedings, 35 p. Session II. Sponsored by The

Swedish Association of Pulp and Paper Engineers and European Liaison Committee for Pulp and Paper. Stockholm.

A test cell was constructed with two ultrasonic transducers (a transmitter and receiver) positioned on opposite walls and low consistency pulp suspension pumped through the cell under turbulent flow conditions. After amplification the signal is then resolved, according to its energy content, as a function of frequency. By comparing the energy spectra obtained for pulp and water, an attenuation coefficient for fibers in the suspension can be expressed as a function of wavelength. In tests with stone and chip refiner groundwood, chemimechanical and chemical pulps, it was found that the attenuation coefficient was influenced by fiber characteristics, temperature, air content, and concentration of the suspension. The influence of stock temperature and air content was eliminated by thermo-controls and deaerating and the attenuation data normalized to a reference concentration. The attenuation coefficient at certain discrete wavelengths provided directly related to the characteristic and papermaking potential of the pulps.

Reinhall. 1972. (Patents)

Rudstrom, L., Samuelsson, L. G., and Uhlin, K. 1973. MECHANICAL PULP IN GRAPHIC PAPERS. *In* International Mechanical Pulping Conference Proceedings, 14 p. Session II. Sponsored by The Swedish Association of Pulp and Paper Engineers and European Liaison Committee for Pulp and Paper. Stockholm.

The generally held opinion is that mechanical pulps have properties inferior to those of chemical pulps when used in graphic papers. High shives content, low brightness and poor brightness stability are considered as serious drawbacks In an investigation made in the Swedish market an analysis was made of the need for brightness and brightness stability in graphic papers. The needs of end users were discussed with printers, publishers, advertising agencies, etc. It was found that for a dominative part of the market, the papermakers and the printers request for quality is exaggerated and was considerably higher than the demand of the end users. It is expected that the need for brightness and brightness stability of pulps used in graphic paper production will decrease in the future. Thus, mechanical pulps represent an important potential raw material for high quality graphic papers.

Snyder, J. W., and Forstrom, S. W. 1973. PRESSURIZED REFINING OF CHIPS AND SAWDUST FOR NEWSPRINT. *In* International Mechanical Pulping Conference Proceedings, 13 p. Session IV. Sponsored by The Swedish Association of Pulp and Paper Engineers and European Liaison Committee for Pulp and Paper. Stockholm.

A production scale, three-stage chip/sawdust refiner line utilizing a Bauer 418 pressurized refiner preceding two pre-existing Bauer 412 atmospheric discharge refiners has been in use at Publishers Paper Co., Newberg, Oregon, pulp and newsprint mill since December 1972. Trial news machine runs utilizing this pulp have enabled reductions in chemical furnish, and/or the use of increased quantities of sawdust in the pressurized refiner furnish. News machine trials are continuing with the objective of producing an offset printing grade of newsprint.

Steinmetz, P. E. 1973. PRODUCING HARDBOARDS FROM RED OAK. U. S. Department of Agriculture Forest Service Research Paper FPL 219, 13 p. Forest Products Laboratory, Madison, Wis

To improve hardboards produced from Asplund pulped red oak, extractives were removed, pulp yield reduced, resin content increased, and fiber pH adjusted. The methods were applied to both wet-formed and dry-formed, phenolic-bonded, 1/4-inch-thick boards having specific gravities of 1 0. Although a large portion of the extractives were removed by washing the pulp with hot water, board quality was unaffected. A loss in all properties occurred when the pulp was washed with a 1-percent sodium hydroxide solution that removed a greater amount of extractives than did the water. Lowering the pulp yield from 88 to 80 percent improved board strength but adversely affected linear stability. Increasing resin content from 2 to 4 percent improved all properties for both types of boards. Dry-formed boards were generally improved by adjusting the pH of the fiber to 4.5 prior to resin application.

Stothert, W. D., and Crotogino, H. F., 1972. THE "SELECTIVE-REFINING" MANUFACTURE OF REFINER GROUNDWOOD FOR FINE GROUNDWOOD PAPERS. Pulp and Paper Magazine of Canada 73(3):45-49

Chips and sawdust of mixed lodgepole pine and white spruce were refined separately in two passes at about 20 percent consistency in an Asplund defibrator until the CSF of the chip pulp was 250 ml and the sawdust pulp was 175 ml. Energy requirements were approximately 60 and 50 hpd per odt. The pulps were then screened over a 0.060-inch round-hole screen The reject fraction of the chip pulp (40 percent) was further refined in one pass at 15-percent consistency to 120 ml CSF The resulting chip pulps were combined and cleaned in two passes through a Berg Centricleaner. For sawdust pulping, the screen reject rate was less than 2 percent and could be neglected Sheets from the combined chip and sawdust pulps had low bulk, high printing quality, low show-through and strike-through, and were virtually free of shives and linting debris.

Takashai, H., Endoh, K., and Suzuki, H. 1972. STUDIES ON HANDLING PROCESS OF WOOD FIBER FOR DRY PROCESSING OF FIBERBOARD. III. EFFECT OF GRINDING POWER REQUIRED ON SIZE OF FIBER. Journal of the Japan Wood Research Society 18:57-61. Tokyo.

Tests were made to determine the effect of energy consumption on fiber size when seven species were defibrated in a Bauer single and double disk refiner. Before refining, chips were steam cooked at pressures of 4, 6, and 8 kg/sq. cm.

Vaarasalo, J., and Mannstrom, B. 1973. SCANDINAVIAN PINE AS A RAW MATERIAL FOR MECHANICAL PULP. *In* International Mechanical Pulping Conference Proceedings, 19 p. Session I. Sponsored by The Swedish Association of Pulp and Paper Engineers and European Liaison Committee for Pulp and Paper. Stockholm.

The influence of wood quality was examined by grinding fine-textured, fast-grown Pinus sylvestris samples of different ages in a laboratory grinder. Both had similar densities, heartwood percentages, and acetone extractives. It was found that older wood generally produced a somewhat stronger groundwood, and handsheets were slightly more porous and coarser than young wood. Wood from the top of the stem yielded weaker pulps than did wood from other parts of the tree. The initial wet web strength of the pine pulp was 45 percent, breaking length 33 percent, and tear factor 24 percent inferior to spruce pulp beaten to the same freeness. The porosity of sheets made from pine was more than 70 percent higher than that obtained from spruce Strength properties of pine pulps were considerably improved by soaking the logs for at least 16 hours at a temperature of about 90°C before grinding However, the treatment increased energy consumption, pulp brightness diminished, and smoothness was impaired Adding sodium sulphite to the soaking water reduced the darkening during treatment, and further improved strength properties However, properties remained less than those obtained with spruce. In a preliminary test using an 8-inch Bauer laboratory refiner, it was found that a two-hour impregnation in a 2 percent solution of Na SO₄ at 90°C improved the strength of pine refiner groundwood above the level of spruce refiner groundwood although papermaking properties such as porosity and smoothness were impaired. It was concluded that of the two pulping processes, disk refining is preferable.

Voss, K. 1972. REFINER GROUNDWOOD MAKES ITS WAY. Allgemeine Papier-Rundschau (29):1066-1067.
The manufacture of groundwood using disk refiners followed by high-consistency post refining and cleaning-screening systems has found commercial application in various countries for production of

PROPERTIES OF THE CUT

TOOL MATERIAL

Dem'yanovskii, K. I. 1972. THERMAL TREATMENT OF WOODWORKING TOOLS. 104 p. Izdatel'stvo Lesnaya Promyshlennost'. Moscow.
A booklet giving information on the properties of the main steels used for woodworking tools and on thermal treatment of tools for moulding, planing and drilling Special attention is paid to the thermal and thermomechanical treatment of sash gangsaws.

Dem'yanokii, K. I., and Vasil'eve, Z. K. 1972. AN IMPACT TESTER FOR DETERMINING THE IMPACT STRENGTH OF THE WORKING PART OF WEDGE-SHAPED CUTTERS. Lesnoi Zhurnal 15(3):84-87. Arkhangelsk.
Describes the design and operation of a pendulum-type apparatus for testing the impact strength of steel cutters used in woodworking.

Koga and Nanasawa. 1973. (Properties of the cutting edge and cutter; Dulling)

Priest. 1972. (Sawing, Circular sawing)

DULLING

Koga, T., and Nanasawa, N. 1973. LIFE CHARACTERISTICS OF TUNGSTEN CARBIDE TIPPED CIRCULAR SAW. I. EFFECTS OF MACHINING FACTORS AND TIP MATERIAL. Journal of the Japan Wood Research Society 19:311-316 Tokyo.
Reports the wear characteristics of circular saws tipped with five types of tungsten carbide in relation to number of teeth, feed speed, and blade protrusion. Saw life increased with increasing Vickers' hardness of the carbide tip

Koga, T, and Nanasawa, N. 1973. LIFE CHARACTERISTICS OF TUNGSTEN CARBIDE TIPPED CIRCULAR SAW. II. EFFECT OF CUTTING SPEED. Journal of the Japan Wood Research Society 19:317-322. Tokyo.
Develops tool life equations for carbide tipped and chromium-vanadium circular saws when cutting wood and particleboard at various peripheral speeds.

Kryazhev, N. A. 1971. THE GEOMETRY AND INTERVAL BETWEEN SHARPENING OF ROTARY CUTTING TOOLS. Derevoobrabatyvayushchaya Promyshlennost' (8):10-12 Moscow.
The sharpness angle of knives should be about 35 degrees when

very sensitive to a steady deflecting force) increased from 59°C before slotting to 170°C when the ratio of slot length to peripheral radius was 0 39. At the same speed and with a rim temperature of 100°C, resonant frequencies in vibration modes having two and three modal diameters reached maximum values at a slot-length ratio of about 0 3, at higher ratios the frequencies decreased rapidly.

McLauchlan, T. A. 1972. RECENT DEVELOPMENTS IN CIRCULAR RIP SAWING. Forest Products Journal 22(6):42-48.
In 1967, total kerf losses in producing a 2 by 4 from a 4-inch-deep cant amounted to about 19/32-inch per cut With recently developed techniques these losses have been reduced about 70 percent to 1/6-inch per cut. Basic to the new technology is the concept of placing flat almost friction free pads on either side of the saw near its leading edge so that vibration and lateral movement is greatly restricted. Four basic types are currently used In one type, parallel plates are mounted several thousandths of an inch from the saw surface, lubrication is provided by water entering at the leading edge. In another version, water is inserted at horizontal slots near the center of parallel guide faces A third type consists of cup-like structures through which fluid is forced to provide a friction-free bearing surface The fourth type uses an integral part of the guide to wipe a lubricating film onto the surface of the saw plate Preliminary reports on British Columbia mills using guided saws having kerfs of 1/8- to 3/16-inch are promising

More, C. D., and Nieh, L. T. 1973. ON THE FOUNDATION OF CIRCULAR-SAW STABILITY THEORY. Wood and Fiber 5:160-169.
Buckling and critical speed saw stability theories for predicting the flatness of a circular saw are presented The concepts were tested in experiments involving accurate measurement of disk natural frequencies, frequency spectrums, and thermal effects. It was concluded that buckling and critical speeds are significant factors and are the instability mechanism for symmetrical circular saws

Pahlitzsch, G. and Ekkehard, F. 1973. THE TENSIONING OF CIRCULAR SAWBLADES, PART II. CALCULATION AND EXPERIMENTAL DETERMINATION OF TENSIONING STRESSES IN SAWBLADES. Holz als Roh- und Werkstoff 31:457-463. Munich.

Pahlitzsch, G., and Puttkammer, K. 1972. THE LOADING OF BANDSAW BLADES. Holz als Roh-und Werkstoff 30:165-174. Munich.
The stresses in bandsaws at rest, idling, and cutting are analyzed and their importance illustrated by examples Stresses in tensioned blades were measured with strain gauges to verify the theoretical relations. The ratio between minimum and maximum longitudinal stress are due mainly to pre-loading, bending, and tensioning but transverse stresses also occur. Large differences between maximum and minimum stresses may induce cracking and failure of the blade.

Pahlitzsch, G., and Puttkammer, K. 1973. DETERMINATION OF THE STIFFNESS OF BANDSAW BLADES. Holz als Roh-und Werkstoff 31.161-167. Munich.
Analyses of forces in the plane of a bandsaw blade and transversely to it suggest that the stiffness of the machine is limited by the low stiffness of the blade A theoretical approach is given for calculating blade stiffness under bending, tension and torsional stress that shows the blade is more sensitive to bending than to torsion Measurements of the lateral blade stiffness suggest that stiffness depends on tension in the blade and its length. Stiffness was relatively unaffected by blade width and thickness.

Pearson 1972 (Parents)

Petrov, V. I. 1972. EFFECT OF AEROSTATIC (COMPRESSED AIR) GUIDES ON THE LONGITUDINAL STABILITY OF CIRCULAR SAWS. Derevoobrabatyvayushchaya Promyshlennost' (10):11-12. Moscow.
Describes experiments with a special test rig in which blades 400 mm in diameter (2, 2.2 and 2 5 mm thick) were operated with compressed-air guides having jet nozzle diameters of 1 mm, a jet chamber pressure of 1 kg/cm³, and a gap of 0.1-0 5 mm between nozzle and blade surface. Stability was best when the gap was 0.2 mm

Prokes, S. 1973. STATIC RESONANT FREQUENCIES OF SAW DISKS. Drevo 28(7):203-207. Bratislava, Czechoslovakia.

Prokof'ev, G. F. 1971. EXPERIMENTAL INVESTIGATIONS ON THE STABILITY OF SASH GANGSAW BLADES. Lesnoi Zhurnal 14(6):161-165. Arkhangelsk.
Experimental results are given that confirm the validity of a previous theoretical study.

Radu, A., and Curtu, I. 1972. ASPECTS OF THE DYNAMICS OF MACHINE TOOLS FOR CUTTING WOOD. Industria Lemnului 23(8):336-342. Bucharest.
Presents theoretical concepts of machine tools as dynamic and elastic (vibrational) systems. The rigidity and stability characteristics of a bandsaw and a planer are provided as examples.

Saplin, V. S. 1971. AERODYNAMIC RESISTANCE IN THE ROTATION OF CIRCULAR SAWS. Lesnoi Zhurnal 14(4):140-144. Arkhangelsk.
Presents a theoretical approach to calculating the aerodynamic resistance of circular saws. In experiments with saws of various sizes and types, it was found that blades lose their transverse stability at 700 to 900 revolutions per minute but stability is regained at speeds greater than 900 revolutions per minute.

Stakhiev, Yu., and Lyzhin, F. V. 1972. STABILITY OF THE BLADES IN PLATE CIRCULAR SAWS. Lesnoi Zhurnal 15(1) 163-168. Arkhangelsk.
Results indicate that blades half the normal thickness can be successfully used.

Thunell, B. 1972. ON DIMENSIONAL ACCURACY IN RESAWING WITH BANDSAWS. Holztechnologie 13(1):28-33. Leipzig
Discusses factors affecting sawing accuracy, loss of material in cutting, and the forces acting on the blade. Stability is determined primarily by tension and by the distance between guide wheels. In practice, blade tension should be higher than theoretically necessary. Stresses in various parts of the sawblade are analyzed and it is noted that accuracy in sawing is also dependent of the accuracy of the feedworks.

Thunell, B 1972 THE STRESSES IN A BAND SAW BLADE. Paperi ja Puu 54(11):759-764. Helsinki.
A review of previous work on stresses in bandsaw blades and their relation to sawing accuracy It is shown that blades are subjected to unneeded stress (from the point of stability) and that the margin of safety before blade failure is small.

Thunell, B. 1972. DIMENSIONAL ACCURACY IN SAWING. Publication Series B-NR 109, 17 p. Swedish Forest Products Laboratory. Stockholm.

Ivanovskiĵ, E. G., Goronok, B. M., and Vasılevskaya, P. V. 1971.
EFFECT OF CERTAIN PHYSICAL AND MECHANICAL
PROPERTIES OF WOOD ON THE COEFFICIENT OF
FRICTION DURING THE SLIDING OF A CHIP ON THE
CUTTER. Derevoobrabatyvayushchaya Promyshlennost' (10):7.
Moscow.

*A note giving data on the friction coefficient of pine, larch, oak, and
birch (natural and compressed) of various densities when
orthogonally cut at 0.1 mm/sec. The coefficient decreased linearly
with increasing density (in the range of 0.39 to 1.0 g/cc), regardless
of species. With a further increase in density, the friction coefficient
approached a limiting value of about 0.19.*

Knudson and Schniewind. 1972. (Properties of wood)

RD CONTROL OF MACHINES

*constraints and defect location in the board) and will predict the
maximum yield of cuttings of a specified grade, for a softwood cut-
up and edge- and end-gluing operation by locating ripping saw
kerfs.*

Mundler, H. V. 1971. ELECTRONIC CONTROLLING IN THE
WOOD WORKING INDUSTRY — PART II: ELECTRONIC
MEASUREMENT OF LENGTH, DIAMETER, AND
VOLUME OF LOGS. Holz als Roh-und Werkstoff 29:456-461.
Munich.
*An illustrated account of equipment for sorting, grading, and
scaling logs.*

Porter, A. W. 1972. INNOVATIONS IN SOFTWOOD
LUMBER PRODUCTION METHODS. Canadian Forest
Industries 92(9):37, 39-41.
*A review of developments in Europe and North America dealing
with electronic log scaling systems, minicomputers with programs
for yield optimization; numerically-controlled setworks, one-man
chipping headrigs, self-feeding barkers and edgers, accurate thin
kerf saws, and automatic trim saws and lumber sorters.*

Ward, D. 1972. OLIVER'S COMPUTER-CONTROLLED
CUTOFF MACHINE. Woodworking and Furniture Digest
74(8):36-40.
*Describes a computer controlled cutoff saw that can be
programmed to simultaneously produce as many as 10 different
lengths within a tolerance of ⅛-inch. The computer determines the
location of cuts (without manual handling or preset stops) that will
yield the optimum combination of lengths as a function of board
length, where end trim is required, and where (as indicated by the
operator) defects are to be removed. In field tests of a prototype
machine, through-put was 34 feet per minute or about 10,000 board
feet in 8 hours.*

White. 1973. (Sawing; General)

RESEARCH INSTRUMENT/

Anonymous. 1973. (Safety, including noise)

Anonymous. 1973. REMOTE TEMPERATURE SENSING OF BOLT TEMPERATURE WHILE PEELING VENEER. Department of the Environment. Canadian Forestry Service Research News 16(1):6-7. Ottawa.
A brief description of an infrared sensor designed to continuously monitor the surface temperature of hardwood veneer bolts during peeling

Bagley, J. M., and Jacobsen, C. F. 1972. A PRECISION SPLITTER FOR LABORATORY CHIP PRODUCTION. Pulp and Paper Magazine of Canada 73(8):72-77.
Specifications and design details are provided. The unit is particularly useful when precision chips are needed for pulping studies

Borgin. 1971. (Properties of wood)

Boulloud. 1972. (Veneer cutting)

Dem'yanokii and Vasil'eve. 1972. (Properties of the cutting edge and cutter; Tool material)

Furukawa, Saiki, and Harada. 1973. (Properties of wood)

George, P., and Miller, D. G. 1970. DETECTION OF ROUGHNESS IN MOVING DOUGLAS-FIR VENEER. Forest Products Journal 20(7):53-59.
Describes a device for detecting excessive roughness in moving veneer. It employs a photoelectric system that senses the depth of depressions passing under it and can be adjusted to respond to either a single depression or to a numerical concentration of depressions greater than a preset depth. The accuracy of the detector is affected by anatomical characteristics, veneer speed variation, and moisture content; cumulative error is about 16 percent. The approximate cost of the detector as described is about $1,500.

Hailey, J. R. T., and Hancock, W. V. 1973. METHODS AND TECHNIQUES FOR VENEER PEELING RESEARCH. Department of the Environment. Canadian Forestry Service Information Report VP-X-107, 23 p. Western Forest Products Laboratory, Vancouver.
Defines the following attributes of veneer quality and describes routine procedures for measuring them: veneer thickness, veneer roughness, lathe-check depth, and moisture content. Statistical procedures for determining sample size and the method of variance analysis is discussed. A veneer-sampling plan and peeler-bolt sampling plan are presented to ensure statistically sound procedures and to assist in maintaining uniformity of method between studies

Hill and Eriksson. 1973. (Defibrating)

Hse, C. 1972. METHOD FOR COMPUTING A ROUGHNESS FACTOR FOR VENEER SURFACES. Wood Science 4:230-233.
Equations for determining the roughness factor (ratio of true surface to apparent area) of rotary-cut veneer were derived from an assumed tracheid model. With data measured on southern pine veneers, the equations indicated that the roughness factor of latewood was near unity, whereas that of earlywood was about 2. The results were expected since earlywood has larger lumens and thinner cell walls than latewood and therefore is more likely to be cut across cell walls.

Reiner and Jackson. 1973. (Defibrating)

Scurfield, Silva, and Wold. 1972. (Properties of wood)

Sobolev. 1971. (Sawing; Sash gangsawing)

Treiber, E. 1971. THE USE OF SCANNING ELECTRON MICROSCOPES IN WOOD RESEARCH. Svensk Papperstidning 74(17):509-514. Stockholm.

CLATURE

belt sanders and a troubleshooting guide.

Anonymous. 1973. (Safety, including noise)

Horten. 1971. (History and general texts)

AND RECONSTITUTED WOOD

Hoshi. 1972. (Machinability of species and reconstituted wood; Softwoods)

Jain, N C, Gupta, R. C, and Bagga, J. K. 1973. PEELING CHARACTERISTICS OF INDIAN TIMBERS — PART 10 DALBERGIA LATIFOLIA (ROSEWOOD). Holzforschung und Holzverwertung 25(1):18-21. Vienna.
In tests with rosewood, optimum knife angle to produce strong, smooth veneer was 91.2 degrees. Temperature and cutting velocity did not have an appreciable effect.

Kato, Fukui, and Ono. 1972. (Machining with coated abrasives)

Lutz. 1971. (Veneer cutting)

Lutz. 1972. (Veneer cutting)

McCarty. 1973. (Defibrating)

Meinecke. 1972. (Defibrating)

Muhammad, S. 1970. USE OF HARDWOODS FOR MAKING MATCHSTICKS. Van Vigyan 8(¾):207-209. Dehra Dun.
Lists hardwood species found suitable for making matchsticks and matchboxes and discusses the merits of Salai (Boswellia serrata) as a raw material for matchsticks Article in English.

Ota, Maraki, and Kawabe 1973. (Defibrating)

Péters, Mergen, and Panzer. 1972. (Veneer cutting)

Rawat, B. S., Rajput, S. S., and Pant, B. C. 1972. A NOTE ON THE WORKING QUALITIES OF SOME INDIAN TIMBERS. Indian Forester 98(11).669-676. Dehra Dun.
Describes test conditions for four woodworking operations (planing, boring, mortising and turning), and tabulates machinability in terms of incidence of defects (roughness, fuzzy, or torn grain, chip marks etc) for Tectona grandis, Adina cordifolia, Mangifera indica, and Terminalia myriocarpa.

Steinmetz. 1973. (Defibrating)

St-Laurent. 1973. (Sawing; Circular sawing)

SOFTWOODS

Anonymous. 1973 (Veneer cutting)

Atack and Heffell. 1972 (Defibrating)

Baldwin. R. F 1972. WAYS TO INCREASE SAWMILL OVER-
RUN. Wood and Wood Products 77(8):110B-110D.
*A five-step sequence has proved useful in establishing an effective
size control program (1) survey present performance, (2) evaluate
potential savings, (3) rank the attainable savings opportunities, (4)
establish goals, and (5) implement a management plan.*

Blechschmidt. 1970. (Defibrating)

Blechschmidt and Unger. 1973. (Defibrating)

Carpenter. 1973. (Defibrating)

Feihl and Carroll 1973. (Veneer cutting)

Feihl, O , and Godin, V. 1973. ROTARY CUTTING OF BLACK
SPRUCE FROM CHIBOUGAMAU, QUEBEC. Department of
the Environment. Canadian Forestry Service Information Report
OP-X-64, 15 p. Eastern Forest Products Laboratory, Ottawa.
*Unheated black spruce logs were cut without difficulty on a 4-foot
lathe equipped with either a flat pressure bar or a ⅜-inch roller bar
A 30-degree microbevel is recommended for industrial operations to
help the knife resist damage from hard knots Bolts heated to 150°F
were excessively soft and did not peel as smoothly as unheated bolts
During winter months when logs are thawed in steam or hot water, a
cool-down period to about 110°F may be desirable. In those spruce
logs of reasonably good quality, veneer was comparable to that of
any other softwood species peeled in Canada*

Hailey, J. R. T., Hancock, W. V., and Walser, D. C. 1973.
PEELING WESTERN RED CEDAR FOR SANDED PANELS.
Department of the Environment. Canadian Forestry Service
Information Report VP-X-109, 14 p. Western Forest Products
Laboratory, Vancouver.
*Concludes that with tight control on lathe setting it is feasible to
produce cedar veneer with reasonably uniform thickness*

Hallock, H., and Malcolm, F. B. 1972. SAWING TO REDUCE
WARP IN PLANTATION RED PINE STUDS. U. S.
Department of Agriculture Forest Service Research Paper FPL-
164, 26 p. Forest Products Laboratory, Madison, Wis.
*The study examines the relationship of two industrial (conventional
and Scragg) and one experimental FPL sawing method (modified
Scragg) on subsequent warp in studs of red pine (Pinus resinosa).
Results generally indicated that red pine is a suitable species for
studs The experimental modified Scragg method proved superior to
the normal Scragg method in terms of the percentage of studs
meeting "stud" grade warp requirements Both the experimental
and normal Scragg methods were superior to the conventional
method Similar results were indicated when mean warp was used as
a criterion The modified Scragg method proved especially effective
in reducing crook and twist but had little effect on bow. More studs
from upper log met crook and bow limits of "stud" grade than from
butts, the reverse was true for twist. The same effect was obtained
when mean warp was evaluated Studs sawn from the outside of logs
were always lower in warp than those from the inner area and is the
main reason for the superiority of the experimental method When
all aspects of warp were considered, the best industry method
yielded 73 percent from butt logs and 84 percent from upper logs*

Ferkes, V. P., and Woodfin, R. O. 1972. VENEER RECOVERY FROM BLACK HILLS PONDEROSA PINE. U. S. Department of Agriculture Forest Service Research Paper RM-82, 23 p. Rocky Mountain Forest Experiment Station, Fort Collins, Colo.

Veneer recovered from a selected sample of 144 Black Hills ponderosa pine sawtimber trees was sufficient in both volume and grades to allow production of at least ⅓ inch C-D plywood. Proportions of C and better grades of veneer increased with veneer block diameter but decreased with tree d.b.h. and block heights in the tree. This apparently conflicting trend results from the greater number of large knots in the middle-stem blocks of larger trees. Recovery ratios increased with both tree diameter and block diameter and were higher for defective blocks than sound blocks. Due to the smaller net scale for defective blocks. Nearly 45 percent of the cubic-foot volume was utilized as veneer.

PARTICLEBOARDS AND SANDWICH BOARDS

Glebov, I. T, and Kucherov, I. K. 1972. CUTTING FORCE IN ROTARY CUTTING OF HOT-PRESSED PARTICLEBOARDS MADE WITHOUT BINDERS. Lesnoi Zhurnal 15(4):102-107. Arkhangelsk.

Boards made by hot pressing larch chips without the use of binders had an edge compression strength of 210 kg/cm² and Janka hardness of 1542 kg/cm². The forces recorded in edge milling the boards were analyzed in relation to cutting speed and tool parameters, and compared with values for conventional particleboard and with birch wood. It was found that the specific cutting energy in machining the hot-pressed boards was less than that required for conventional particleboard but greater than that for solid birch.

Lyubimov, V. G. 1972. INVESTIGATION OF SURFACE ROUGHNESS IN SANDING COMPREG LAMINATE. Lesnoi Zhurnal 15(5):95-100. Arkhangelsk.

In tests on sanding the end and side surfaces of compreg laminates with a glass abrasive disk, disk performance and durability was satisfactory and surface quality was good.

Radulescu. 1972. (Chipping, flaking, and grinding; Flaking)

Sellaeg, M., Kucera, B., and Mengkrog, O. 1972. CHIPPING OF BIRCH FOR PARTICLEBOARD PRODUCTION. Tidsskrift for Skogbruk 80(4):433-444. Oslo.

Green, whole-tree birch chips were produced with a mobile chipper and manufactured into particleboard. Board properties were unaffected and the yield of usable chips was 20 to 50 percent greater.

Sweum and Furrer. 1973. (Machining with coated abrasives)

FIBERBOARDS, HARD AND SOFT

Sellaeg, M., and Gislerud, O. 1972. GREEN CHIPS FROM THINNINGS FOR USE IN FIBERBOARD PRODUCTION. Tidsskrift for Skogbruk 80(4):423-432. Oslo.

Small Norway spruce logs were field chipped with a tractor powered mobile chipper and blown into a truck for transport to a fiberboard plant. It was found that yield of fiber was 30 to 60 percent greater when whole trees were chipped than when the delimbed merchantable volume was chipped. In preliminary tests with mixtures containing up to 20 percent green chips there was no appreciable change in the properties of insulation board.

BIBLIOGRAPHIES

Keays and Leask. 1973. (Reports of research programs and industrial developments)

Koch, P. 1973. WOOD MACHINING ABSTRACTS, 1970 and 1971. U. S. Department of Agriculture Forest Service Research Paper SO-83, 46 p. Southern Forest Experiment Station, New Orleans, La.

Lists 415 references, most with abstracts. An introduction mentions

notable papers in various fields of endeavor and surveys trends in mechanical processing of wood.

Weiner, J., and Pollock, V. 1972. BARKERS AND BARKING OF PULPWOOD. Bibliographic Series 190, 56 p. Supplement II. Institute of Paper Chemistry, Appleton.

Two-hundred-forty-five references, most with abstracts, covering the period 1967-1971.

REPORTS OF RESEARCH PROGRAMS AND INDUSTRIAL DEVELOPMENTS

Anonymous. 1973. THE CHIPPING HEADRIG: A MAJOR INVENTION OF THE 20TH CENTURY. Forests and People 3(2):18-20, 36-37.

A nontechnical article dealing with the development of chipping headrigs. Some projected trends in pine utilization (procurement practices, harvesting taproots, and laminated lumber) are also given.

Anonymous. 1973. 3M OPENS NEW COATED ABRASIVE RESEARCH CENTER. Wood and Wood Products 78(9):34.

Short description of 3M's new 32,000 square feet research facility.

Arola and Erickson. 1973. (Barking)

Atack, D. 1972. MECHANICAL PULPING AT THE INSTITUTE. PART V. REFINER MECHANICAL PULPING. Trend — The Activities of the Pulp and Paper Research Institute of

Canada (20) 6-12.

An account, with 20 references, of the refiner groundwood pulping program at the Pulp and Paper Institute of Canada, from its inception in 1954 to 1972.

Bryant, B. S. 1972. THE ROLE OF RESEARCH IN THE FOREST PRODUCTS INDUSTRY — OBSERVATIONS OF AN INVOLVED OBSERVER. Forest Products Journal 22(5):9-12.

A discussion of some factors related to the apparent retreat of corporate level research within the forest products industry.

Erickson. 1972. (Barking)

Ivanovskij, E. G., Vasilevskaya, P. V., and Laurner, E. M. 1972. RECENT INVESTIGATIONS ON THE CUTTING OF WOOD. 129 p. Izatel'stvo Lesnaya Promyshlennost', Moscow.

39

A monograph of previously published research at the Forest and Wood Technology Academy at Leningrad.

Jensen, W. 1973. FUTURE TECHNICAL NEEDS AND TRENDS OF THE PAPER INDUSTRY — CONVERSION TO PULP. TAPPI 56(5) 64-69.
Includes a discussion of trends in debarking wood and chips, and manufacture of mechanical pulp Dry barking (or a two-stage wet/dry process) is likely to be used by 1980 and bark will find extended use other than as fuel. Harvesting of crowns, branches, and stumps will necessitate removal of unwanted debris from the wood Pressurized disk refining will be more fully evaluated and will find greatest application in printing papers and boards. The stone groundwood process is likely to be made more efficient and its level of automation increased. A number of specific research problems are listed for each area.

Johnston, J. S., St-Laurent, A., and L'Ecuyer, A. 1973. DAMAGE TO SAW TIMBER FROM TREE-FELLING AND BUCKING SHEARS. Department of the Environment. Canadian Forestry Service Information Report OP-X-77, 20 p. Eastern Forest Products Laboratory, Ottawa.
A report of the discussion between delegates to a seminar on shear induced damage to timber during felling.

Keays, J. L., and Leask, R. A. 1973. REFINER MECHANICAL PULP: PAST, PRESENT, AND POTENTIAL. Paper Trade Journal 157(35):20-29.
The history and present status of refiner groundwood pulping is reviewed with consideration to developments which may affect its future growth. Factors affecting pulp quality are discussed,

including equipment, processing, and raw materials. The long-term potential of refiner groundwood is considered in terms of the development of a single-furnish newsprint which would serve to lower costs, conserve wood, and reduce steam pollution One-hundred-twenty-two references are provided

Koch, P. 1973. WHOLE-TREE UTILIZATION OF SOUTHERN PINE ADVANCED BY DEVELOPMENTS IN MECHANICAL CONVERSION. Forest Products Journal 23(10) 30-33.
In 1963 approximately 30 percent of the dry weight of above- and below-ground parts of southern pine trees ended as dry-surfaced lumber or paper; the remaining 70 percent was largely unused. By 1980, computer-controlled chipping headrigs, thin-kerf saws, lamination of lumber from rotary-cut veneer, high-yield pulping processes, and more intensive use of roots, bark, and tops will likely double the yield of lumber, paper, bark products, and reconstituted board to 60 percent of the dry weight of above- and below-ground tree parts

Koenigshof, G. A. 1973. DOUBLE THE TREE? Southern Lumberman 227(2824):125-126.
Describes, in general terms, the COM-PLY conversion process whereby veneer is manufactured from the outer portion of the log and the interior is chipped or flaked. The two materials are then reconstituted into 2 x 4 studs having two or three 1/10- or 1/8-inch thick veneers on each wide face of a particleboard core. While no specific data are provided, strength, stiffness, nail holding, and dimensional stability are deemed adequate. It is claimed that COM-PLY studs can be produced at a cost equal to conventional methods and yield twice as many studs per log.

PATENTS

Alexander, D. K. 1972. WOOD GRINDER. (U. S. Patent No. 3,690,568) U.S. Patent Office, Washington, D. C.
Describes a stone grinder designed to automatically maintain specific grinding energy at a substantially constant value.

Bodline, A. G. 1972. SONIC TIMBER CUTTING METHOD AND APPARATUS. (U. S. Patent No. 3,688,824) U. S. Patent Office, Washington, D. C.
A cutting device employing unidirectional pulses of energy.

Cremisio, R. S., and Brabec, C. J. 1973. BI-METAL SAW BLADE STOCK AND METHOD OF MAKING THE SAME. (U S. Patent No. 3,766,808) U. S. Patent Office, Washington, D. C
A bi-metal bandsaw having high speed steel teeth formed in the powered state by hot isostatic pressing.

Daniels, D H. 1973. PROCESS AND APPARATUS FOR MAKING SHAKES. (U. S. Patent No. 3,754,586) U. S. Patent Office, Washington, D. C.
A method and apparatus for making shakes especially adapted for wood blocks having angular deviations of the grain.

Harvey, H C. 1972. GREEN-WOOD FIBRATING MEANS AND METHOD. (U. S. Patent No. 3,674,219) U. S. Patent Office, Washington, D. C.
A process and apparatus for converting solid pieces of timber into slender, splinter-like strands of wood fiber of more or less uniform thickness Timber first passes through a series of rools revolving at uniform or variable controlled speeds. The rools have varying

thread-like configurations and are vertically activated under controlled pressure of such magnitude as to cause the solid wood to be separated into splinter-like strands along the grain. The spongy mass of loosely matted fiber strands are then passed through a scrubbing device to be pulled apart.

Hirz, J. R. 1971. APPARATUS FOR SLICING LUMBER. (U. S. Patent No. 3,614,968) U. S. Patent Office, Washington, D. C.
A device for slicing slabs from a cant using two opposing knives each adapted to move half way through the cant from opposite ends.

Johnson, D. L. 1972. METHOD OF JET BREAKING VENEER INTO NARROW FLAKES. (U. S. Patent No. 3,670,791) U. S. Patent Office, Washington, D. C.
A fluid jet is used to break a continuous sheet of veneer into controlled narrow widths as the sheet is formed from the bolt.

Kennemore, P. H. 1971. CHIP EJECTOR CHAIN TOOTH. (U. S. Patent No. 3,625,266) U. S. Patent Office, Washington, D C.
A chain saw tooth designed to more effectively eject chips from the kerf.

Loof, N.O.T. 1973. WOOD-TURNING MACHINE. (U. S Patent No. 3,739,822) U. S. Patent Office, Washington, D. C.
Describes a tool holder in the shape of a spherical segment that carries an easily replaceable cutting tool. The tool holder is capable of linear movement in three directions and is also rotatable about an axis perpendicular to the rotation axis of the workpiece.

Porter, A. W. 1972. VENEER MACHINE ROLLER PRESSURE BAR ASSEMBLY. (U. S. Patent No. 3,670,790) U. S. Patent Office, Washington, D. C.
A housing that supports a rollerbar by the hydrostatic lubrication of oil, water, or air.

Reed, G. O. 1972. LOG PRE-CENTERING APPARATUS FOR VENEER LATHES. (U. S. Patent No. 3,664,395) U. S. Patent Office, Washington, D. C.
A device for aligning the axial centerline of a veneer log with the rotational axis of the lathe chucks.

Reinhall, R. B. 1972. PULP FIBERIZING GRINDING PLATE. (U. S. Patent No. 3,674,217) U. S. Patent Office, Washington, D C.
A refiner groundwood disk having ridges and ribs designed to minimize fiber breakage.

Thrasher, E. W. 1972. METHOD AND APPARATUS FOR OPERATING A ROTARY SAW. (U. S. Patent No. 3,645,304) U. S. Patent Office, Washington, D. C.
An apparatus and method for operating circular saws with guides that apply lateral pressure to the saw surface.

Welborn, R. E. 1971. STUMP REMOVAL MACHINE. (U. S. Patent No. 3,625,267) U. S. Patent Office, Washington, D. C.
A vehicle-mounted stump removal machine employing a cutter-carrying boom for lateral movement.

UDING NOISE

Anonymous. 1973. TEST CODE FOR EVALUATING THE NOISE EMISSION OF WOODWORKING MACHINERY. 16 p. Woodworking Machinery Manufacturers of America. Philadelphia.
A preliminary noise level code for woodworking machines specifying test conditions, machine operating modes, measuring instruments and procedures, method of reporting test results, and nomenclature

Bryant, L. H. 1972. POLLUTION CONTROL IN THE SAWMILLING INDUSTRY — PRESENT AND FUTURE PROBLEMS. Australian Forest Industries Journal 38(10) 66-71. Sydney.
A general presentation of ways to dispose of sawmill residues, including burning in incinerators, land-fills, and return to the forest Also considered are noise pollution and environmental aspects of wood preservatives and extractives in streams.

Cheremnykh, N. N., and Chizhevskii, M. P. 1971. DESIGN FEATURES FOR DAMPING CIRCULAR SAWS TO REDUCE NOISE Lesnoi Zhurnal 14(6):149-152. Arkhangelsk.
Describes, with engineering drawings, new collars and spring dampeners reportedly successful in reducing circular sawblade noise

Cheremnykh, N. N., and Chizhevskii, M. P 1973. Reducing the noise of planers. Lesnoi Zhurnal 16(1).88-95. Arkhangelsk.
A summary account of various ways noise can be reduced in planers, thicknessers, and molders by inserting material in guards

and by insulating screens, hoods, and chambers.

Deaner, F. R. 1972. AN INVESTIGATION OF NOISE REDUCTION IN A SINGLE SURFACE PLANER BY MEANS OF A DIRECTIONAL AIR FLOW CONTROL DUST COLLECTION HOOD. Master of Science Thesis. State University College of New York, Syracuse.

Efremova, T. K., and Mitusov, V. A. 1972. CALCULATION OF THE LOWER CONCENTRATION THRESHOLD FOR THE EXPLOSION OF WOOD DUST. Lesnoi Zhurnal 15(2):80-86. Arkhangelsk.
A theoretical study, supported by experiments on dust of spruce, pine and larch wood and of pine and spruce bark from a wood-flour factory. It was found that the lower threshold for the explosion of wood dust is a concentration of about 30-35 g/m^3. Dust fractions of particle size smaller than 120 µm were the most dangerous, and resinous and volatile substances increased the danger of explosion. The explosiveness of the dust decreased with increasing moisture content and with the admixture of bark dust

Efremova, T. K., and Mitusov, V. A. 1973. PREVENTION OF WOOD-DUST EXPLOSIONS. Derevoobrabatyvayushchaya Promyshlennost' (2):8-10. Moscow.

Firth, R. D. 1972. CHAIN SAW TESTING BY NEW ZEALAND FOREST SERVICE EVALUATES NOISE, VIBRATION, POWER CHARACTERISTICS. Forest Industries Review 3(5):2-9. Auckland.
Gives test results of the noise, vibration, and power characteristics of 15 chain saws ranging from ultra-light to heavy.

Fushimi, T. 1972. THE MEASURED VALUES OF A CHAINSAW HANDLE'S VIBRATION AND THEIR CHANGES BY MEASURING CONDITIONS. Memoirs of the College of Agriculture, Ehime University 16(2):235-248. Japan.

Fushimi, T., Wotanabe, K., Shigaki, H., and Ezaki, T. 1972. THE RELATION BETWEEN VIBRATION VALUES OF CHAIN SAWS AND THEIR PERMISSIBLE LIMIT. Bulletin of the Ehime University Forests (9):51-56. Ehime, Japan.

Glorig, A. 1972. MEDICAL ASPECTS OF NOISE CONTROL. TAPPI 699-704.
A general discussion of the nature of hearing loss in individuals is given It is noted that noise levels above 90 dB can cause a significant reduction of ability to hear speech if continued over several years (100 dB is the approximate noise level of a subway). While hearing protection can be easily provided to prevent such loss, control at the source is the eventual solution. The nonauditory health effects of noise (fatigue, vertigo, etc.) have not been established It is concluded that enough information is currently known to institute hearing conservation programs in industry and prevent hearing loss in nearly all exposed individuals.

Hart, F. D 1972. A THEORETICAL AND EXPERIMENTAL STUDY OF WOOD PLANER NOISE AND ITS CONTROL. Ph.D. Dissertation. North Carolina State University at Raleigh.
The dominant source of planer noise was identified as the board being surfaced Noise increased 6 decibels for each doubling of board width. A theoretical model was developed to define the vibrational field set up in the board and as a guide for improved cutterhead design. Experimental studies identified the effect of various parameters on noise levels and suggested some techniques for reduction.

4

chipping than when idle running; there was little difference for drum type chippers. With a disk chipper, noise was greater when processing wood of high than of low density; noise was unrelated to density with drum chippers (avg. 98 decibels). When chippers were installed below ground level, overall noise levels were reduced about 10 decibels during idle running and about 5 decibels during chipping.

Tanaka, Sakurai, and Horie. 1972. (Joinring, planing, molding, and shaping)

Thompson, S. E. 1973. CHAIN SAW SAFETY, VIBRATION AND NOISE. Publication Number 730702, 12 p. Society of Automotive Engineers, New York.
Reviews, in terms of severity and solution, the three principal hazards to chain saw operators — cuts and bruises, vibration injury, and hearing impairment. Some progress is indicated. Chain brakes and special chains have been introduced in an attempt to solve the kickback problem and some units meet most existing European vibration standards. Noise levels remain higher than desirable, but newer mufflers are beginning to reduce exhaust noise toward mechanical noise levels.

Varava, V. I., Pomogaev, S. A., and Fil'Kevich, I. V. 1973. DETERMINING THE PARAMETERS OF LONGITUDINAL SHOCK ABSORBERS FOR LOGGING MACHINES. Lesnoi Zhurnal 16(2):42-47. Arkhangelsk.
A theoretical study.

Vinogradov, E. G. 1973. HOW TO MAKE WORK SAFE ON CIRCULAR SAWS. Lesnaya Promyshlennost' (1):12-14. Moscow.

Welch, R. 1971. VIBRATION IN CHAIN SAWS. Australian Forestry 35(4):215-225. Canberra.
The vibration characteristics of 12 chainsaws, six having anti-vibration handles, are described. The adequacy of current Australian vibration safety limits based on displacement and frequency are questioned.

Index of Authors

45

McMillin, Charles W.

1974. Wood machining abstracts, 1972 and 1973.
    South. For. Exp. Stn., New Orleans, La.
    46 p. (USDA For. Serv. Res. Pap. SO-101)

Lists 427 references, most with abstracts.

U.S. Department of Agriculture
Forest Service Research Paper SO-102

Forest Tax Trends in
Nine Louisiana Parishes

William C. Siegel

Southern Forest Experiment Station
Forest Service
U.S. Department of Agriculture

1974

U.S. Department of Agriculture
Forest Service Research Paper SO-102

Forest Tax Trends in
Nine Louisiana Parishes

William C. Siegel

Southern Forest Experiment Station
Forest Service
U.S. Department of Agriculture

1974

Siegel, William C.

    1974. Forest tax trends in nine Louisiana parishes.
       South. For. Exp. Stn., New Orleans, La. 12 p.
       (USDA For. Serv. Res. Pap. SO-102)

Two principal taxes presently affect woodland owners in Louisiana—the property or ad valorem tax on forest land, and the severance tax on harvested timber. Updated time-series trends for nine sample parishes indicate that average assessments, millage, and taxes are rising. The paper also discusses certain other fiscal measures affecting Louisiana timber growers and dealers.

Additional keywords: Forest property tax, timber severance tax, millage, assessment.

U.S Department of Agriculture
Forest Service Research Paper SO-103

$\backslash$

Forest Tax Trends in Nine Louisiana Parishes

William C. Siegel

Two principal taxes presently affect woodland owners in Louisiana—the property or ad valorem tax on forest land, and the severance tax on harvested timber. Current provisions of both levies are part of the Forest Tax Law of 1954, which took effect on January 1, 1955, as an amendment to the Louisiana Constitution. This paper discusses updated, 5-year interval, time series trends in both taxes for nine parishes. Data on assessments, millage, and taxes paid are included in the same format as in previous papers (*2, 11*). Certain other fiscal measures affecting Louisiana timber growers and dealers are also discussed.

BACKGROUND

Louisiana's tax structure has changed considerably during the last 50 years. The number of major State taxes has grown from 4 to more than 40, and the emphasis on various taxes has shifted. The State government now collects more than 7 of every 10 nonfederal tax dollars levied in Louisiana. The national average is about 50 percent. As a result the State—as compared to the parishes (counties) and other local governments—presently provides the highest percentage of total public services of all 50 States. And, again in contrast to the situation elsewhere in the U.S., taxes are still the biggest source of State revenue in Louisiana, currently providing more than half the annual budget.

This situation has arisen largely because taxes are the source of State revenue most susceptible to legislative control. Thus, tax provisions have tended to become embedded in the State's financial structure and have not been reevaluated as frequently as economic conditions have warranted. During the last few years, however, there have been indications that this situation may be slowly changing.

Real Property Taxation in Louisiana

In the spring of 1974, Louisiana's voters adopted a new constitution that incorporated a number of important property tax changes. These will take effect on January 1, 1978, although the remainder of the new

The author is Principal Economist at the Southern Forest Experiment Station, USDA Forest Service, New Orleans, Louisiana.

constitution becomes official on January 1, 1975. The present instrument specifies that property be taxed in direct proportion to its actual cash value, and that real estate be assessed at actual market value and so listed on the tax rolls. The legal mandate of market value, however, has been largely ignored by parish assessors over the years. Thus most real property assessments—including those of woodland—are considerably lower. The tax statutes permit parishes to levy on as little as 25 percent of assessment. To offset the common under-valuations, however, most apply millage to the entire amount.

The new constitution departs drastically from this procedure. It specifies three classes of property as subject to ad valorem taxation: land, improvements for residential purposes, and other property. These are to be assessed at 10, 10, and 15 percent of fair market value, respectively. The only exception is that agricultural, horticultural, marsh, and timberlands are to be assessed at 10 percent of use value rather than fair market value. All property subject to taxation is to be reappraised and revalued at intervals of not more than 4 years.

Assessments under both constitutions are subject to annual review by parish governing authorities and then by the Louisiana Tax Commission. The old constitution requires that all taxes shall be uniform upon the same class of property throughout the territorial limits of the tax levying authority. Thus at present, upon receipt of parish assessment abstracts, the Tax Commission can adjust group valuations within and among parishes—ostensibly to achieve equalization, as provided for by law. At certain times in the past, however, assessments have been raised to help meet State budget requirements since the ad valorem tax rate was fixed in the constitution and could not be increased.

The State Homestead Exemption Act of 1934, which will prevail until superseded by the new constitution, provides that any homestead—upon the owner's application—may be exempted from taxes on the first $2,000 of assessed valuation; for war veterans, the ceiling is $5,000. In rural areas, parcels of land separate from the home tract may be included, either to an aggregate of 160 acres or to the amount of the exemption. Exemptions apply to most local property taxes, but there are some exceptions. City taxes are exempt only in New Orleans; and water, sewerage, and lighting district taxes only if they are parishwide.

The new constitution raises the general homestead exemption to $3,000, and authorizes the legislature to make further increases to a maximum of $5,000 by a two-thirds vote of both houses. The exemption for veterans continues at $5,000 and a like exemption has been granted to nonveterans who are 65 years of age and older.

Until 1973 the State also levied an ad valorem tax on property, in addition to the parish taxes. This levy, set by law at 5¾ mills, accounted for $34.6 million in 1972. A constitutional amendment in that year re-

2

pealed the tax, but the new constitution gives the legislature the right to restore it at a rate not to exceed 5¾ mills.

The parishes, unlike the State, have wide latitude in setting tax rates. Thus, as local revenue needs have increased, the general practice has been to raise millage. Even so, however, local levies are often below constitutional limits. Property taxes as a percentage of total State and local tax revenues are about half the national average. Except in Jackson and Orleans Parishes, the general parish tax may not rise above 4 mills without voter approval. General municipal taxes have a 7 mill limit. The new constitution retains these restrictions. Unlike the old document, however, it sets no limit on school maintenance millage and public improvement levies.

Property tax revenue from timberland is important to local governments. Commercial woodland is found in 63 of the 64 parishes and constitutes a significant portion of the property tax assessment base. In 1959 property taxes on Louisiana forest acreage totaled $4 million (11). The 1969 yield was approximately $5.2 million, a 30-percent increase. In 1974 it will probably be close to $6 million. Woodland assessments in 14 parishes account for more than 10 percent of the total parish ad valorem assessment, including buildings and improvements. Forest land is also becoming a more significant portion of the State's rural land assessment base. In the late 1950's it accounted for about 28 percent of total rural land valuations (4). By 1962 the proportion had risen to 33 percent (5), by 1969 to 34½ percent (6), and by 1973 it was 36 percent (7).

Forest Tax Legislation

Louisiana's current forest property tax and forest severance tax statutes were enacted in 1954 in response to widespread demand for forest tax reform. As part of the Forest Tax Law of 1954, they took effect January 1, 1955. Under this legislation, only the land is subject to ad valorem taxation. Taxes on timber are postponed until time of harvest under the yield tax principle, although the term "severance tax" is still officially used. Woodlands had previously been on a severance tax basis, with land and growing timber assessed together for annual property taxes and the timber taxed again when cut.

The levy under the pre-1955 severance tax had ranged from 25 cents to $1.50 per M for sawtimber, depending on species, and had been 15 cents per cord for pulpwood. The entire severance revenue had been allocated to the State Forestry Commission. The present law taxes sawtimber at 2¼ percent and pulpwood at 5 percent of current average stumpage value by species. Values are determined annually by joint action of the Louisiana Forestry Commission and the Louisiana Tax Commission, with three-fourths of the severance revenue being re-

3

turned to the parish of origin. The remainder goes into the State's general fund.

Virgin timber, for practical purposes nonexistent, was not made subject to the 1954 law. Until 1972 it was still theoretically subject, upon harvest, to the old severance tax rates until that legislation was repealed in 1972. As a matter of practice, however, the new rates had been applied all along to the little virgin timber that was cut. Such timber is still subject to property taxes.

Also not supplanted by the 1954 law was a reforestation contract law passed in 1922 and a reforestation contract severance tax enacted in 1926. These acts and subsequent amendments permit landowners who reforest denuded land to enter into a contract with the State and parish for a fixed assessment on both the land and the planted timber. A 6-percent severance tax is paid on products cut. After a maximum of 50 years from the date of contract, the land and any timber remaining thereon becomes subject to regular ad valorem taxation. Three-fourths of this severance revenue is returned to the parishes where the timber was cut and the remainder goes into the State's General Fund. A 1926 amendment to the basic act provides that contracts entered into can be made retroactive.

The first reforestation contracts were let in 1933, but in recent years contract acreage has been declining. The peak was in 1953, when 670,-000 acres were reported under agreement (3). By 1973 only 190,000 acres remained, distributed among eight parishes (7).

The 1954 law requires all woodlands except reforestation contract acreage to be assessed in one of four categories: tidewater cypress, hardwood, longleaf pine, and other pine. Prior to the law's enactment, forest land had been classified among a multitude of categories according to the whims of individual assessors. The new requirement was largely met by 1963 (5, 9). In most parishes, however, no more than two of these types are found. The vast majority of timberland today is classified as either other pine or hardwood. The 1973 tax rolls list 7.9 million acres of other pine land, 3.4 million of hardwood, 0.4 million of cypress, and 81,000 of longleaf pine (7). The ranges in average parish assessments for these categories are shown in table 1.

The only discrepancy in meeting the requirements of the 1954 law continues to be the large amount of forest acreage classed as "miscellaneous" land, which is defined by the State Tax Commission as "land not permitting of more specific classification" (8). Such acreage has steadily increased on the tax rolls since 1955, particularly in recent years. Between 1963 and 1973, the number of "miscellaneous" forest acres nearly tripled from about 800,000 to 2.3 million (5, 7). Gains in the classification have been the primary reason for a steady decline during the last decade in land classified as forest by the parish assessors—from

4

TABLE 1.—*Range in parish assessment averages for woodland categories, 1973*

Assessment category	Parishes	Lowest average assessment	Highest average assessment	Average assessment, all parishes
	No.	— — — — — — Dollars — — — — — —		
Hardwood	53	2.82 (Lafourche)	119.59 (Plaquemines)	6.45
Longleaf pine	8	4.64 (West Feliciana)	8.33 (Winn)	6.21
Other pine	33	3.06 (Assumption)	18.40 (E. Baton Rouge)	6.80
Tidewater cypress	4	1.71 (Iberia)	4.76 (Lafourche)	3.65
Reforestation	8	3.00 (Allen)	18.05 (St. Tammany)	5.61

a high of 14.7 million acres to a low of 11.9 million in 1973. In most parishes with large acreages of woodland classified as "miscellaneous," such lands are assessed at about the same value as woodland that is classed as forest.

Specific legislative proposals for implementing the present use provision of the new constitution are presently being developed by forestry interests. It is entirely possible that the same categories of forest land outlined in the 1954 legislation will continue in effect.

STUDY METHOD

The nine parishes studied are Beauregard, Concordia, Iberville, Jackson, La Salle, Richland, Sabine, Washington, and Webster. Time-series data on woodland assessments, millage, and taxes at 5-year intervals from 1939 through 1959 have previously been published (*2, 11*). In order to continue the established pattern, the same information is presented in this paper for 1964 and 1969. Complete 1974 data will not be available from State records until 1976.

The nine parishes constitute a good geographic distribution, typify ownership conditions, and afford proportional representation of forest types throughout the State. Each has a predominant timber type.

The data were obtained essentially as before, i.e., by sampling the 1964 and 1969 tax rolls for each parish. The State constitution requires that duplicate copies of the tax rolls be submitted annually to the State auditor before November 15. After audit, the records—arranged by ward within parishes—are filed by the Tax Commission and made available for public inspection. Aggregate holdings within wards are usually listed alphabetically by ownership. Within holdings, assessments by type are usually given separately but with only the total tax recorded.

Listings with forest property were sampled on a systematic basis by taking a random start from among the first 30 on a parish roll; data were taken from this entry and every thirtieth thereafter. Limitations of time and expense precluded a complete random sampling procedure. The data included forest type, forest acreage, forest assessment, total assessment (all assets combined), forest protection tax, and total tax. Average assessment per acre, average millage per dollar of assessment, and average tax per acre were then determined for each parish from this information.

ASSESSMENT LEVELS AND TRENDS

The assessor's approach to valuation is the foundation of the property tax system. Even for those who are well trained and experienced, assessing the worth of property—particularly woodland—is more art than science. But even if all similar forest properties within a parish are assessed consistently, their values may still be set above or below the same types of properties in other parishes. Should such differences occur, however, they may or may not have a substantial effect on taxes. Taxes actually paid are determined by applying current millage rates to assessments, and millage also varies widely among parishes.

Table 2 shows average forest assessments for the nine sample parishes at 5-year intervals from 1949 through 1969. The previously reported years of 1949, 1954, and 1959 (2, 11) have been included to facilitate comparison.

TABLE 2.—*Average assessment per forest acre for nine sample parishes*

Parish	1949	1954	1959	1964	1969
	— — — — — — Dollars — — — — — —				
Beauregard	4.03	4.57	5.70	5.93	6.40
Concordia	1.98	2.29	3.59	6.11	5.03
Iberville	1.78	1.68	1.72	5.05	4.24
Jackson	3.84	4.19	5.11	5.33	5.65
La Salle	4.66	5.28	6.85	6.29	6.44
Richland	7.26	5.98	7.29	7.64	7.62
Sabine	3.73	3.61	4.92	5.00	5.00
Washington	4.11	4.29	4.33	8.68	8.57
Webster	5.82	5.96	6.22	6.50	6.41

The assessment level between 1959 and 1964 continued its earlier upward trend for eight of the nine parishes. Increases ranged from a low of 1.6 percent in Sabine Parish to a high of 194 percent in Iberville. Only in La Salle Parish was the average assessment lower (by 8 percent) in 1964 than in 1959.

6

But from 1964 to 1969 the trend reversed somewhat. The average dropped in five parishes and remained the same in another. In no case did it revert to the 1959 level. The biggest decline was 17 percent in Concordia Parish and the smallest was less than 1 percent in Richland. The 1969 valuations in three parishes, despite decreases in average assessment during the preceding 5 years, were substantially above those of a decade earlier. The 10-year increase in Concordia was 40 percent, in Iberville 247 percent, and in Washington 98 percent.

Iberville Parish, despite its large increase in assessments since 1959, has consistently had the lowest valuations of the nine parishes. Its woodlands are mostly swamp hardwood. Washington Parish's assessments now average the highest, having overtaken Richland's in 1964. Predominantly a pine parish, Washington's large increase can probably be attributed to its proximity to the large New Orleans urban area. In consequence, its woodlands have had steadily increasing use for recreation and other nontimber purposes and have been purchased for speculation and second-home sites.

The average woodland assessment for the entire State in 1969 was $5.85 per acre (6). Averages for four of the sample parishes were below this figure and those for five were above. The ranges from which the averages were determined for 1964 and 1969 are shown in table 3. In six of the nine parishes, most of the high valuations had been placed on small tracts. If forest holdings of less than 20 acres are eliminated from the sample data for these parishes, average assessment levels decrease —sometimes by as much as 30 percent. Smaller holdings are obviously being valued for uses other than timber growing. In Sabine and Webster Parishes, however, size of tract makes no appreciable difference in assessment level. And in La Salle Parish, the larger holdings are generally assessed higher than the smaller tracts.

TRENDS

TABLE 3.—*Range of sample assessments used to determine average assessments per forest acre*

Parish	1964		1969	
	High	Low	High	Low
	— — — — — — — *Dollars* — — — — — — —			
Beauregard	25.00	4.32	50.00	4.51
Concordia	30.00	1.74	12.50	2.00
Iberville	16.67	1.42	13.33	1.01
Jackson	10.00	4.62	13.14	4.71
La Salle	10.00	4.97	12.25	5.00
Richland	10.00	6.00	10.00	5.00
Sabine	5.00	5.00	5.00	5.00
Washington	100.00	3.07	25.00	3.50
Webster	6.64	6.33	7.00	4.17

TAX AND MILLAGE LEVELS AND TRENDS

The general upward trend in millage for the nine sample parishes reported earlier (11) for the years prior to 1960 continued during the subsequent decade (table 4). Although average millage dropped in two parishes between 1959 and 1964, it rose in these parishes during the next 5 years. Thus the average in all nine parishes was higher in 1969 than in 1959. Jackson Parish registered the largest increase, one of 22 mills, a rise of 50 percent. Iberville's 2-mill increase was the smallest. Between 1959 and 1964, for the nine parishes combined, millage increased an average of 12.5 percent. The corresponding rise during the subsequent 5 years was 10 percent.

TABLE 4.—*Average millage per dollar of assessment for nine sample parishes*

Parish	1949	1954	1959	1964	1969
			— Mills —		
Beauregard	43	40	47	50	62
Concordia	48	46	57	71	66
Iberville	53	58	49	46	51
Jackson	37	38	44	58	66
La Salle	38	55	55	52	68
Richland	41	45	40	43	49
Sabine	40	50	56	63	63
Washington	28	35	47	58	58
Webster	30	34	40	47	49

The tax pattern (table 5) has generally continued to be one of substantial increases since 1959, which was the last year reported in earlier papers. Average taxes rose in all but one of the nine parishes between 1959 and 1964 and declined in only two between 1964 and 1969. For the decade as a whole, however, all nine experienced heavy increases. The largest spurts were in Iberville Parish (175 percent) and in Washing-

TABLE 5.—*Average taxes per forest acre for nine sample parishes*

Parish			Dollars Cents		
Beauregard	17	18	27	30	39
Concordia	9	11	21	43	33
Iberville	9	10	8	23	22
Jackson	14	16	22	31	37
La Salle	18	26	37	33	45
Richland	30	27	29	33	38
Sabine	15	18	28	31	31
Washington	12	15	21	50	50
Webster	18	20	25	31	32

8

ton Parish (136 percent). Two others—Concordia and Jackson—jumped more than 50 percent. The increases in Iberville and Washington are due largely to the tremendous rise in forest assessments whereas the increases in Concordia and Jackson represent spurts in both assessments and millage.

Reflecting large post-war increases, the average tax per forest acre in 1969 was more than 30 cents in eight of the nine parishes. In seven parishes, woodland taxes increased more than 100 percent between 1949 and 1969. The rise in Concordia was more than 200 percent and in Washington it was more than 300 percent. Despite these averages, however, the ranges on the sample properties vary widely as shown in table 6. Most of the woodland tracts represented at the high ends of the ranges are small. Although the actual tax per acre on such holdings is high, the total tax bill is not enough to concern most owners—particularly if their primary purpose for owning the property is for other than timber growing.

TABLE 6.—*Range in woodland taxes per acre for sample parishes*

Parish	1964		1969	
	High	Low	High	Low
	— — — — — — *Dollars* — — — — — —			
Beauregard	1.12	0.17	3.49	0.26
Concordia	2.05	.12	.85	.12
Iberville	.77	.03	.72	.05
Jackson	.56	.24	.84	.30
La Salle	.52	.25	.81	.33
Richland	.47	.21	.53	.24
Sabine	.40	.23	.39	.22
Washington	1.69	.17	1.41	.19
Webster	.41	.20	.37	.22

OTHER REVENUE MEASURES AFFECTING WOODLANDS

Severance Taxes

Severance taxes on cut timber, of which 75 percent are returned to the parishes, continue to be an ever-expanding source of revenue. Collections have been rising sharply in recent years with steadily increasing harvests of both sawtimber and pulpwood. The annual yield exceeded $1 million for the first time in 1967, and in 1972 it rose above $2 million. The severance income in 1973 was $2.36 million, nearly three times the $800,000 of only a decade earlier.

Petroleum products account for the bulk of the severance taxes collected in Louisiana. Eight other resources—including timber—provide

9

the remainder. As a source of severance income, timber has gradually increased in importance. Whereas a decade ago it ranked second to sulphur in the nonpetroleum group (10), it now ranks first. In this group the proportion attributable to timber has recently been about 35 percent as compared to about 20 percent 10 years ago. Of the nine parishes, those in the pine areas—as may be expected—have contributed the most to the increased severance revenue.

Timber severance collections seem certain to continue their upward trend as Louisiana's forest industries keep expanding. Increasing severance revenue, however, has not kept woodland property taxes from rising in the sample parishes. But as demands for public services have mushroomed, it may well have prevented even greater increases in forest property taxes than have occurred.

Reforestation Contract Taxes

Receipts from the reforestation contract severance tax have been declining as contract acreage has diminished. The average annual collection during the 5-year period ending with fiscal year 1972-73 was $41,000 as compared to average annual receipts for the preceding 5 fiscal years of $96,000. Of the nine parishes, only Beauregard and Washington contain contract acreage. The reforestation contract law will probably be of even more limited importance in future years, as its most advantageous features were incorporated in the Forest Tax Law and will undoubtedly also be reflected in the new legislation that will implement use assessment in 1978.

Forest Acreage Tax

A forest acreage or protection tax has been authorized in Louisiana since 1944. The tax is paid by woodland owners and collected by the parishes. Revenues are placed in a forestry fund administered by the State for protecting woodlands from fire and other damage. The State and Federal governments contribute at least twice the amount deposited each year by the participating parishes. Allocations from the fund are then prorated to each of these parishes, of which there were 37 in 1973—an increase of six since 1965.

The maximum tax of 2 cents per acre is collected in 33 parishes. It is 0.5 cent in Concordia, Tensas, Franklin, and West Carroll. Most of the parishes that have chosen not to assess the levy are in southern Louisiana; a few are in the northeast part of the State. Of the nine sample parishes, only Iberville and Richland do not participate.

WHAT OF THE FUTURE?

Three factors commonly cause taxes to affect timber growing adversely. They are the increasingly higher costs of local government,

10

faulty administration of the property tax, and the substantial acreages of forest land that offer little prospect of early income from which to meet annual taxes (*12*). All three conditions exist in Louisiana.

The forest owner, as such, has little control over the first of these. Local government expenditures in Louisiana probably will continue to spiral upward. In recent years, however, several factors have lightened the load for those who pay property taxes—including the forest taxpayer. Consolidation in the number of local government units has somewhat slowed the rate of growth of local public expenditures. And secondly, the property tax's portion of such outlays is slowly declining with the gradual substitution of other forms of revenue.

The second cause of difficulty—faulty property tax administration—results in uneven distribution of the tax load through failure to assess like properties at uniform percentages of value. Although the 1954 Forest Tax Law improved the forest tax climate, inequities still exist. The wide range in assessments and taxes—both within many parishes, and between others—has been pointed out. In all fairness, however, it must be acknowledged that officials have had a difficult task in fully implementing the law and also providing increased revenue. Most assessors have no knowledge of valuation techniques applicable to forest land.

Two courses of action are basic if the quality and uniformity of forest assessment practices in Louisiana are to be improved: (1) standardization of assessment procedures among parishes, and (2) the instruction of parish assessors in woodland valuation techniques or else making available to them trained forest appraisers. Neither course, however, has appeared to be urgent under the present law—especially when comparisons are made with previous tax practices. But with the advent of use assessment in 1978, as mandated by the State's new constitution, assessment practices and standards will have to be improved if valuations are to be fair and equitable. Local and State officials can hardly be expected to develop a workable system of taxing woodlands on the basis of actual use, a method more complex than that used now, without technical forestry assistance. At the same time, assessment procedures must be kept relatively simple to meet the needs of local assessors and to gain their cooperation. The inherent disadvantage of the property tax in respect to deferred yield from forests is likely to cause trouble if assessment procedures become too highly developed. Complex modifications appear unnecessary and could well prove disastrous. Preferential property taxes appear to operate without cost to State and local governments. Nevertheless they do, in fact, impose a forced expense on taxpayers to whom the burden has been shifted, complicate administration, and progressively weaken the entire property tax system (*1*).

Because of its broad constitutional powers, the State Tax Commission is in an ideal position to provide leadership for accomplishing the task that lies ahead.

11

An ideal solution would be the employment of professional foresters by the Commission. Their primary functions would be to prepare forest assessment guides, recommend values for various classifications, and assist local assessors in implementing the guides.

LITERATURE CITED

1. Advisory Commission on Intergovernmental Relations.
   1963. The role of the States in strengthening the property tax. Vol. 1 (A–17): 1–187.
2. Craig, R. B.
   1942. Taxes on forest property in nine selected parishes of Louisiana, 1937–40, in nine counties of Alabama, 1937–41, and in seven counties of Mississippi, 1936–41. USDA For. Serv. South. For. Exp. Stn. Occas. Pap. 101, 23 p.
3. Louisiana Tax Commission.
   1954. Sixth biennial report of the Louisiana Tax Commission for the years 1952–53. 316 p.
4. Louisiana Tax Commission.
   1960. Ninth biennial report of the Louisiana Tax Commission for the years 1958–59. 355 p.
5. Louisiana Tax Commission.
   1964. Eleventh biennial report of the Louisiana Tax Commission for the years 1962–63. 302 p.
6. Louisiana Tax Commission.
   1970. Fourteenth biennial report of the Louisiana Tax Commission for the years 1968–69. 208 p.
7. Louisiana Tax Commission.
   1974. Sixteenth biennial report of the Louisiana Tax Commission for the years 1972–73. 195 p.
8. Siegel, W. C.
   1962. Forest land assessments rising in Louisiana. USDA For. Serv. South. For. Exp. Stn. South. For. Notes 142.
9. Siegel, W. C.
   1964. Forest tax law at work. For. Farmer 24(3): 14–16.
10. Siegel, W. C.
    1966. Forest tax trends in Louisiana. USDA For. Serv. Res. Note SO–48, 6 p. South. For. Exp. Stn., New Orleans, La.
11. Siegel, W. C., and Perry, J. D.
    1961. Forest taxation in Louisiana. USDA For. Serv. South. For. Exp. Stn. Occas. Pap. 187, 14 p.
12. Williams, E. T.
    1961. Trends in forest taxation. Natl. Tax J. 14(2): 113–144.

U S. Department of Agriculture
Forest Service Research Paper SO-103

S

2. Craig, R. B.
   1942. Taxes on forest property in nine selected parishes of Louisiana, 1937–40, in nine counties of Alabama, 1937–41, and in seven counties of Mississippi, 1936–41. USDA For. Serv. South. For. Exp. Stn. Occas. Pap. 101, 23 p.

3. Louisiana Tax Commission.
   1954. Sixth biennial report of the Louisiana Tax Commission for the years 1952–53. 316 p.

4. Louisiana Tax Commission.
   1960. Ninth biennial report of the Louisiana Tax Commission for the years 1958–59. 355 p.

5. Louisiana Tax Commission.
   1964. Eleventh biennial report of the Louisiana Tax Commission for the years 1962–63. 302 p.

6. Louisiana Tax Commission.
   1970. Fourteenth biennial report of the Louisiana Tax Commission for the years 1968–69. 208 p.

7. Louisiana Tax Commission.
   1974. Sixteenth biennial report of the Louisiana Tax Commission for the years 1972–73. 195 p.

8. Siegel, W. C.
   1962. Forest land assessments rising in Louisiana. USDA For. Serv. South. For. Exp. Stn. South. For. Notes 142.

9. Siegel, W. C.
   1964. Forest tax law at work. For. Farmer 24 (3): 14–16.

10. Siegel, W. C.
    1966. Forest tax trends in Louisiana. USDA For. Serv. Res. Note SO–48, 6 p. South. For. Exp. Stn., New Orleans, La.

11. Siegel, W. C., and Perry, J. D.
    1961. Forest taxation in Louisiana. USDA For. Serv. South. For. Exp. Stn. Occas. Pap. 187, 14 p.

12. Williams, E. T.
    1961. Trends in forest taxation. Natl. Tax J. 14 (2): 113–144.

U.S. Department of Agriculture
Forest Service Research Paper SO-103

Mechanized Row-Thinning Systems In Slash Pine Plantations

Walter C. Anderson
and
James E. Granskog

Southern Forest Experiment Station
Forest Service
U S. Department of Agriculture

1974

ACKNOWLEDGMENT

Cooperation in the form of timber, manpower, and equipment was received from International Paper Company, St. Regis Paper Company, and Owens-Illinois, Inc.

Mechanized Row-Thinning Systems In Slash Pine Plantations

Walter C. Anderson
and
James E. Granskog

Southern Forest Experiment Station
Forest Service
U.S. Department of Agriculture

1974

MECHANIZED ROW-THINNING SYSTEMS
IN SLASH PINE PLANTATIONS

Walter C. Anderson and James E. Granskog

Over the next decade or two, most of the 15 to million acres of pine plantations in the South l become ready for a first commercial thinng. The magnitude and nature of the job is strated by the situation in slash pine—the st extensively planted of the southern pines. Slash pine plantations are heavily concented in a belt extending on either side of the orgia-Florida State line. During the peak nting years of 1957 through 1960, 200,000 to),000 acres were planted each year on private nindustrial ownerships under the impetus of Conservation Reserve of the Soil Bank proam. In addition, the forest industry planted out 100,000 acres of its lands annually during s period. Although planting on nonindustrial operties has declined sharply since the Soil nk program expired, it has increased substanlly on forest industry lands.

Each year plantations on 200,000 or more es are reaching the preferred age for a first nmercial thinning. The task is enormous and 'ficient labor is not available to do the job by rent labor-intensive thinning methods. Also, nning is becoming more expensive because of constantly rising wage rates.

Foresters, therefore, are looking for faster ys of accomplishing the job with limited, inasingly costly manpower. Mechanized sys-1s for row thinning might be the answer. itching from labor-intensive to capital-inten-e technology would reduce labor requirements acre. Because of the greater productivity of chanized crews, workers could be paid higher ges. In addition, loggers would acquire higher tus as machine operators, and the otherwise rous job of thinning young stands would be-1e less difficult and more attractive. Row nning is specifically designed for patterned ests. Since entire rows are removed, there no costs for marking trees to be cut, and the red rows serve as convenient roads for skid-g.

This study measured the productivity of the principal types of mechanized logging systems that could be used in row thinning slash pine plantations, identified the factors that affected their performance, and evaluated their effect on output in various plantation environments. Information of this kind is needed to estimate thinning costs, determine economical thinning opportunities, and obtain an optimum balance among the machines combined into a system.

The study was limited to mechanical row thinning of slash pine plantations growing in sandy soils on flat terrain. Results are applicable to like operations in plantations on similar sites throughout the slash pine belt. However, they cannot be applied directly to stands of other species on different topographies.

The efficiencies of alternative systems were not compared since the operations observed were at different levels of development. One was a fully balanced, commercial operation; another was a pilot test of a prototype machine; and a third was an experimental run conducted solely for this study. Neither did the various systems carry processing to the same level of end product. For two systems, further processing took place at the mill. Comparisons were also inappropriate because with a single example of each system, operator performance could strongly affect results. In addition, the manager selecting a thinning system must consider more than machine efficiency. Railroad transportation, for example, may necessitate selection of a shortwood system. Availability of parts and service can also be a determining factor in the choice of a particular machine.

MACHINES

Three types of harvesting systems were appraised in the production of pulpwood from thinnings. These were the shortwood, long-log, and whole-tree systems, as classified by the output from the harvester. Individual machines were selected on the basis of their availability for study. However, the cycle times for the harvesters might be considered representative of other harvesters in similar systems. The other principal machine in each system was a skidder.

Shortwood System

The shortwood system included a Timberline TH-100[1] thinner-harvester—a machine designed for thinning southern pine plantations (fig. 1). It can cut trees with stump diameters of up to 12 inches. Mechanical arms hold the tree while the shear closes to fell it. The bole is then fed butt first into the processing mechanism, delimbed by a set of fixed knives, and cut by a buckshear into bolts of a predetermined length (from 5 to 7 feet) down to a 2-inch top. Each bolt drops into a cradle as it is bucked. When a third of a cord has accumulated in the cradle, the bundle of bolts is deposited on the ground alongside the row being cut. The harvester proceeds down one row and returns up the next one to be removed.

A Franklin 133 prehauler gathered the bolts left by the TH-100 and removed them from the plantation. The grapple mounted on the machine loads the woodrack, which has a capacity of approximately 2 cords. The load is transported to

[1] Mention of trade names is solely to identify equipment used and does not imply endorsement by the U.S. Department of Agriculture.

the landing, where the grapple unloads the bolts onto a set-out trailer.

Long-Log System

The long-log system featured a Caterpillar 950 tree harvester, which felled, limbed, and bunched (fig. 2). It can sever trees with a stump diameter of up to 18 inches and can delimb a bole down to a top diameter of 2½ inches. The tree is gripped by two grapple arms and sheared. The bole is then tilted forward so that it is fed horizontally through the delimbing knives while the machine backs up to the selected piling area. There the butt is placed even with those of the other piled stems. The machine then moves forward as the delimbing process continues. The stem is dropped as the shear tops the tree, completing the cycle. The harvester operates only when traveling away from the landing so that all stems will be piled with their butts toward the landing. When a row is completed, it returns to the landing area to begin removal of the next row.

The bunched stems left by the harvester were skidded to the landing by a Caterpillar 518 skidder equipped with a 94-inch Fleco grapple. Because of the large capacity of this machine, the operator would proceed to the most distant bunch of logs in a row and then pick up successive bunches to complete a load as he headed back toward the landing.

FIGURE 1.—*Timberline TH 100 shortwood harvester*

Whole-Tree System

A Soderhamn Go-Go harvester did the felling and bunching in a whole-tree system (fig. 3). This machine was developed for thinning southern pine plantations by modifying a Go-Go skidder. The shear can sever a tree with a stump diameter of up to 12-inches. The severed tree falls into an accumulator mounted on the left side of the machine. As many as seven trees may be collected in the accumulator before the dumping arm drops the bundle off to the side. To bunch stems with their butts toward the landing,

the Soderhamn harvester fells trees only as it proceeds toward the landing. A workman with a chainsaw limbs and tops the trees in each pile before they are skidded from the plantation.

A Dunham Log Hog skidder, which is essentially a modified farm tractor equipped with 40-inch grapple, did the skidding for the whole-tree system. It can pick up one full-size pile of logs dropped by the harvester or two small piles of only two or three stems each. The operator would begin skidding logs from a row by picking up the bunch nearest the landing and then proceed to pick up successive bunches down the row.

FIGURE 2.—*Caterpillar 950 long-log harvester.*

FIGURE 3.—*Soderhamn whole-tree harvester.*

STUDY AREA

All field data were collected in north Florida, where mechanized row thinning operations were being carried on in the area's extensive plantations. One reason for this activity is that there are large pulpwood markets nearby for the small diameter trees removed in thinnings. Also, logging conditions here are favorable for machines. The plantations in which mechanized thinning was studied are fairly typical. They are on dry, sandy sites that will support heavy machinery. The terrain is level to slightly sloping. Ground cover consists largely of gallberry, palmetto, and blackberry bushes.

The three plantations were on land with a site index of 70 at age 25, and in all three survival was generally good. The shortwood and long-log operations were in 17-year-old stands; the whole-tree operation was in a 14-year-old stand. Initial spacing in the plantation thinned by the shortwood harvesting system was 6 feet between rows and 8 feet within rows. In the plantation thinned by the long-log method, trees had been planted at 10 feet between and 6 feet within rows. In that thinned by the whole-tree method spacing was at 8 and 6 feet, respectively. In all thinning operations, every third row was taken out.

DATA COLLECTION

Time and production data were collected on the tree harvester and the prehauler or skidder in each operation. The harvester was the key piece of equipment; skidding data were needed only for the purpose of balancing the system. In each operation, data on harvesting and skidding were taken on the same machine and same machine and same operator. All operators were experienced men whose performance was considered above average.

Felling and Processing

Sample plots were line segments in rows being cut. Segments were selected to cover a range of tree diameters and spacing intervals. Segment length was limited by the extent of uniformity of the tree diameters and intervals between trees. Consequently, segment length ranged from 12 to 60 feet, and number of trees varied from two to six.

On each operation, a total of 30 sample plots were installed in rows to be cut. For each sample plot the following information was recorded before cutting:

(1) Length of segment from the first tree to the first tree beyond the segment, measured to the nearest tenth of a foot;

(2) D.b.h. of each tree, measured to the nearest tenth of an inch;

(3) Total height of each tree, measured to the nearest foot.

Time required to cut and process the trees on a plot was recorded to the nearest tenth of a second. Timing began when the first tree in the segment was sheared and ended when the first tree beyond the segment was sheared.

While the TH-100 was working within a sample plot, the number of bolts processed from each tree was counted, and the number of times the cradle was dumped was also recorded. After the Cat 950 had completed processing trees in a sample plot, log lengths and top diameters were measured. The same measurements were taken on the plot after the trees cut in a sample by the Soderhamn had been limbed and topped.

Skidding

For the Franklin 133 prehauler, data were collected on 13 roundtrips. Time spent in travel empty, loading, travel loaded, and unloading was recorded for each trip. Measurements for each trip were distance travelled empty, distance traveled loaded, distance traveled while loading, number of stops to pick up piles of bolts, and cords carried per load.

Total time was kept for each roundtrip by the Cat 518 and the Dunham Log Hog skidders. Timing began the moment the previous load of logs was dropped and continued until the load being timed was dropped. Before skidding began, the distance from the landing to each bunch of logs was measured, and log measurements were taken. During each skidder trip, the following were recorded:

(1) Distance from the landing to the first bunch of logs picked up,

(2) Number of bunches picked up for the load,

(3) Number of logs in the load.

Volumes per trip were calculated from the log measurements. Time and measurement data were obtained for 14 Cat 518 skidder trips and for 36 Dunham Log Hog skidder trips.

In all operations, times were taken to the near-

4

t tenth of a second, distances were measured to
e nearest foot, and volumes were calculated to
e nearest tenth of a cord.

ANALYSIS OF HARVESTING TIMES

Data collected for each harvester were sub-
cted to regression analysis. A stepwise regres-
on procedure was used to select the best predic-
on equation.

The dependent variable was time, expressed
minutes per hundred feet of plantation row.
ll data were converted to a hundred-foot basis.
he factors tested for their statistical signifi-
nce were:

(1) Number of trees

(2) Total length of stems cut

(3) Average d.b.h., i.e., sum of d.b.h. ÷ num-
:r of trees

(4) Average d.b.h. squared, i.e., sum of d.b.h.^{2}
- number of trees

(5) Total length of stems cut ÷ average d.b.h.
juared

(6) Sum of d.b.h. squared ÷ sum of d.b.h.

umber of 7-foot bolts and sum of log lengths in
:et were used as two alternative measures of
ital length of stems cut, depending on the form
f output by the harvesting system.

Variables 5 and 6 were found to be significant
: the 5 percent level. The first is a measure of
ital log lengths per unit of basal area. The sec-
1d is a measure of the dispersion of tree size
round the mean. Together these two measures
imbine all the variables tested.

Regression coefficients, multiple correlation
efficients, and standard errors of estimate for
e estimating equations for the three harvesters
'e shown in table 1.

Another equation can be substituted for the
andard form in the case of the whole-tree harv-

ester, which unlike the other two machines fells
but does not further process trees. This equation
is simpler since it has only one variable—total
length of stems cut. The constant and regression
coefficient are:

$$Y = -1.013 + 0.012X$$

where:

Y = Time per 100 feet of row, in minutes

X = Sum of log lengths, in feet.

The alternative equation for the whole-tree harvester gives a better fit than the standard form,
especially for estimates at points distant from
the mean. The proportion of variance explained
is improved from 77.7 to 84.5 percent; and the
standard error of estimate is lowered from 0.87
to 0.71 minute.

The equations apply to productive time only.
Therefore, allowance must be made for machine
downtime, idle time, and turn-around time depending on the method of operation.

HARVESTER PRODUCTIVITY AND COST

To determine how rates and costs of production will vary in different plantation situations,
use of each of the three types of harvesters was
simulated.

The Plantations

Detailed descriptions of stand structures were
obtained from Bennett and Clutter's *Multiple
Product Estimates of Unthinned Slash Pine
Plantations*,[2] which gives stand and height tables
by age, site index, and density. Tables were chos-
en for plantation ages 15 and 20 years, a range
which covers the preferred age for thinning.

[2] Bennett, F. A., and Clutter, J. L. *Multiple-product yield estimates
of unthinned slash pine plantations—pulpwood, sawtimber, gum.*
USDA For. Serv. Res. Pap. SE-35, table 1, page 9, and table 3, p. 12.
Southeast. For. Exp. Stn., Asheville, N.C. 1968.

iBLE 1.—*Estimating equations for time to harvest 100 feet of plantation row*

irvesters	Equations	R²	SE
ortwood	$Y = -13.142 + 4.462(\Sigma B \div \frac{\Sigma D^2}{N}) + 2.377(\Sigma D^2 \div \Sigma D)$	79.6	1.47
ng-log	$Y = -6.390 + 0.743(\Sigma L \div \frac{\Sigma D^2}{N}) + 0.949(\Sigma D^2 \div \Sigma D)$	83.7	0.57
iole-tree	$Y = -8.551 + 0.395(\Sigma L \div \frac{\Sigma D^2}{N}) + 1.264(\Sigma D^2 \div \Sigma D)$	77.7	0.87

Y = Time per 100 feet in minutes; D = D.b.h. in inches; L = Log length in feet; B = Bolts; N = Number of trees.

5

Also, site indices of 60 and 70 feet at 25 years were selected because they represent the preponderance of sites where slash pine is planted. Merchantable trees were limited to those in the 4-inch diameter class and larger. These sizes were typically utilized in the thinning operations observed; smaller trees were simply pushed down.

For each density class, intervals between trees within rows were calculated for plantations spaced 5, 6, 8, and 10 feet between rows. Tree lengths were computed to a 2-inch top diameter for bolts and a 3-inch one for logs with formulas developed by Bennett et al.[3] Tree volumes were figured to the same top diameters. Stand characteristics were then used with the estimating equations to obtain time per hundred feet of row.

Estimating Productivity

In simulating hourly production for each harvester, we assumed that a 40-acre tract—20 chains on a side—was being thinned and that each row would be 1,320 feet long.

The shortwood harvester could travel up one row, turn, and return down the next row to be felled since the prehauler used with this system could enter a row from either end. Grapple skidders, however, travel with a load of logs in the direction the butts are pointing. Therefore, to limit skidding distance, the long-log and whole-tree harvesters would cut half of the row with butts facing one way and half facing the other. They would do this by entering a row from one end, cutting 660 feet, returning to the end of the row, and then entering the next row to be removed. Because of these differences in cutting patterns, nonproductive operating time for the long-log and whole-tree harvesters is considerably greater than that for the shortwood harvester, which loses time only while turning at the end of a row. Return time for the tree-length harvester is about 2 minutes. The whole-tree harvester takes twice as long because it is driven over the bunched trees on the return trip to breakdown tops to ease the job of limbing and topping.

Hourly output in cords for the TH-100, Cat 950, and Soderhamn is shown in figures 4, 5, and

6. For the shortwood harvester, productivity is positively correlated with spacing between rows. This pattern persists throughout the four combinations of age and site index assessed. The difference between the curves is least where the trees are smallest (site index 60, age 15 years) and greatest where the trees are largest (site index 70, age 20 years). Each curve peaks, indicating the density for which the production rate is maximum for that particular spacing between rows.

For the long-log harvester, only 8- and 10-foot spacing between rows was considered because the width of the harvester make it infeasible to operate where rows are closer. The curves for the two spacings are nearly identical over the range of densities. The production function approximates a negative sloping curve extending from about 7 cords per hour at 300 trees per acre to 4¼ cords at 800 trees, for site index 70 and age 15. Other combinations of site index and age show similar patterns, with the height of the curve rising with each increase in site index and age.

The sheaves of curves for the full-tree harvester, which were computed with the alternative equation, display negative slopes for all combinations of plantation ages and site indices. Thus, productivity is higher in stands with fewer trees per acre. The spread of curves was about the same for all combinations of site index and age except site index 60 at age 15. For all stands considered, curves increase in height with each reduction in spacing between rows.

Harvester Costs

Cost per cord of mechanical thinning in a particular plantation depends on the rate of production and on the cost of owning and operating the machines, including wages.

The hourly cost of the harvesters (table 2) is the sum of the fixed and variable machine costs plus the cost of the operator. In calculating these costs, we assumed a machine life of 5 years and a salvage value of 20 percent of the delivered price. Operating time, figured on the basis of one 8-hour shift per day for 250 days per year, was 2,000 hours per year, or 10,000 hours for the life of the machine.

The fixed costs of ownership include depreciation, interest charges, and insurance payments.

[3] Bennett, F. A., Swindel, B. F., and Schroeder, J. G. Estimating veneer and residual pulpwood volumes for planted slash pine trees. USDA For. Serv. Res. Pap. SE-112, p 2. Southeast. For. Exp. Stn., Asheville, N C 1974

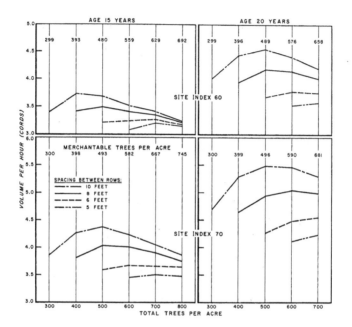

FIGURE 4.—*Potential hourly output of the shortwood harvester.*

TABLE 2.—*Hourly cost of owning and operating three harvesters*

Expenditure	TH-100	Cat 950	Soderhamn
	— — — — — — — — — *Dollars* — — — — — — — — —		
Fixed costs:			
Depreciation[1]	2.80	5.20	1.84
Interest and insurance	1.19	2.21	.78
TOTAL FIXED COSTS	3.99	7.41	2.62
Variable costs:			
Repair and maintenance	3.50	6.50	2.30
Fuel	.50	.93	.29
Lubrication and engine oil	.05	.09	.03
TOTAL VARIABLE COSTS	4.05	7.52	2.62
Operator	3.50	3.50	3.50
TOTAL MACHINE COSTS	11.54	18.43	8.74

[1] Approximate delivered price is $35,000 for the TH-100, $65,000 for the Cat 950, and $23,000 for the Soderhamn.

7

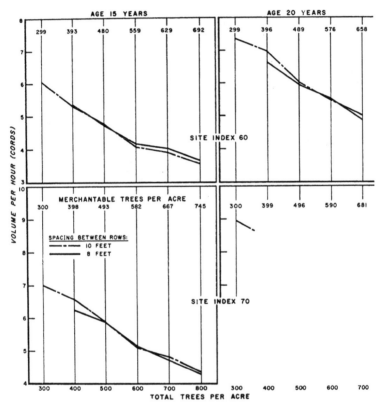

FIGURE 5.—*Potential hourly output of the long-log harvester.*

8

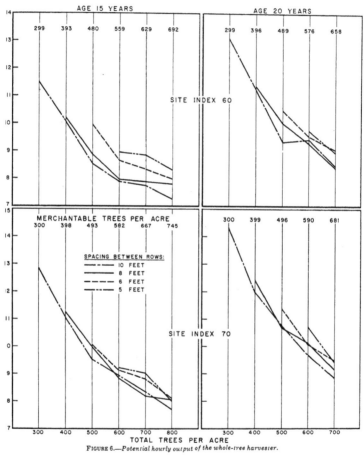

FIGURE 6.—*Potential hourly output of the whole-tree harvester.*

Straight line depreciation was calculated as follows:

$$\frac{\text{delivered price—salvage value}}{\text{hours of machine life}} = $$

depreciation per hour.

Interest and insurance were figured as 10 percent of the average investment. The standard formula for average investment is:

$$AI = \frac{(I-R)\ (N+1)}{2N} + R$$

where:

AI—Average investment in dollars
I—Delivered price in dollars
R—Salvage value in dollars
N—Life in years.

The variable costs, which are associated with the time the machine operates, include repairs and maintenance, fuel, and lubrication and oil. The charge for repair and maintenance over the life of the harvester was assumed to be equal to its delivered price. Fuel costs were figured on the basis of 1 gallon per hour per 70 bhp at a price of $0.50 per gallon. Lubrication and oil changes were priced at 10 percent of the cost of fuel. These components of variable costs were also put on an hourly basis.

The cost of the machine operator was set at $3.50 per hour, which included wages plus Social Security taxes, State unemployment compensation, and workmen's insurance paid by the employer. No allowance was made for supervision.

To estimate harvester costs per cord, the hourly machine cost may be divided by the hourly output of the machine for the selected stand situation. For example, cost per cord for a Cat 950 was figured for a 15-year-old stand planted in rows 8 feet apart on site index 70 land. There were 500 trees per acre of which 480 met utilization standards. In such a plantation, this particular harvester would turn out 5.9 cords per operating hour at a cost of slightly more than $3 per cord. At a density of 300 trees per acre, cost per cord would be more than 15 percent less; and at a density of 800 trees, more than 35 percent greater. In younger stands growing on poorer sites cost per cord would be higher; in older stands on better sites, it would be lower.

The preceding example illustrates the procedure for estimating costs per cord of a machine for various stand conditions. Those making their own analyses, however, should use these data with care. Hourly production data for each harvester represent potential hourly output (i.e., at 100 percent machine availability). It would be unrealistic to expand hourly production, to a daily basis by simply multiplying by the number of hours per shift. Likewise, in estimating costs per cord, the hourly production rates can be reduced by whatever figure one wants to assume for machine availability. In the example, if 80 percent availability were assumed, output would be 20 percent less and the cost per cord raised from $3 to $4. Still lower availability would reduce output and raise the cost per cord proportionately. Also, costs per cord for each harvester are not directly comparable since the output from each is in a different stage of production.

BALANCING THE SYSTEM

The three tree harvesters, of course, are only a part of the respective logging systems, although the most important part. To complete each system, supporting equipment and men must be added. The major piece of supporting equipment is the skidder. The number of skidders and harvesters needed to make a balanced system depends on the relative capacities of the two machines.

Skidder Production

For the shortwood system, the total time involved per trip in running bolts from the woods with the prehauler is the sum of the travel times empty and loaded plus the times spent loading and unloading the prehauler. The equation for travel time empty is:

$$Y = 0.2605 + 0.0012X$$

and that for travel time loaded is:

$$Y = 0.1004 + 0.00188X$$

where:

Y—Time (in minutes)
X—Distance (in feet).

The proportion of variation in travel time explained by distance was 74.2 percent for the first equation and 80.4 percent for the second. The standard error of estimate was 0.15 and 0.20 minute, respectively, for the two equations. The average time required to load the prehauler was 8.25±0.90 minute, and to unload it was 3.85 ±0.90 minute at the 67-percent confidence level. Average volume per load was 2.28 cords.

10

For example, a 2:3 ratio on the shortwood operation where timing was done reflects the usage of the prehauler for other purposes such as straightening the loads on the set-out trailers and the machine availability that had been experienced. Machine availability can vary widely depending on operators and maintenance practices. Thus, the formulas for the various skidders should be used only as a rough guide to balance the number of machines needed with the harvesters.

DISCUSSION

This study has related stand characteristics of slash pine plantations to the performance of selected tree harvesters used for row thinning. Additional factors must be considered, however, in relation to mechanized thinning and the selection of machines.

Tract size and volume to be removed are important factors in the success of any mechanized system. Small tracts could not be profitably thinned by this method unless several could be combined to provide a sufficient volume to be harvested from a given area.

Productivity of the harvesters will vary for other species. Those with larger crowns and heavier limbs than slash pine will require longer processing times.

Limitations in machine design will affect performance and selection of harvesters. Nonmerchantable trees can be a problem if the machine is not designed so small trees can merely be ridden down. If they must be removed with the shear, productivity is lowered. Of course, shear size will preclude the use of some machines in older plantations.

The operator too plays a large role in controlling productivity. The marginal worker cannot be entrusted with the operation and care of the complex and expensive equipment. Skilled labor is needed. Ideally, individuals possessing the dexterity, judgment, and disposition to qualify as operators would be selected by screening tests and then given formalized training and supervised experience on the job.

This investigation of mechanized row thinning in slash pine plantations can serve as a guide for other studies of this type. Those studying a particular machine can follow the steps specified in this report. It illustrates the data collection procedure, analytical method, and application of results. With the movement toward capital-intensive thinning systems, the need for such studies increases because the consequence of poor planning and administration become more and more costly.

Anderson, Walter C., and Granskog, James E.

1974. Mechanized row-thinning systems in slash pine plantations. South. For. Exp. Stn., New Orleans, La. 12 p. (USDA For. Serv. Res. Pap. SO-102)

From machine times and tree measurements taken on row-thinning operations in slash pine plantations, equations were developed to estimate productivity of harvesters in shortwood, long-log, and whole-tree systems. Output and costs were calculated for specific stand conditions.

Additional keywords: Harvesters, mechanization, productivity, and logging costs.

U. S. Department of Agriculture

Forest Service Research Paper SO-104

Natural Regeneration
and Development
of Nuttall Oak
and
Associated Species

Robert L. Johnson

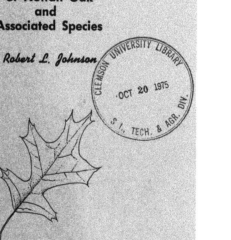

Southern Forest Experiment Station
Forest Service
U. S. Department of Agriculture

1975

Johnson, Robert L.

1975. Natural regeneration and development of Nuttall oak and associated species. South. For. Exp. Stn., New Orleans, La. 12 p. (USDA For. Serv. Res. Pap. SO-104)

Nuttall oaks from a dense seedling catch survived for 5 to 10 years in complete shade and for 15 years with 1 or 2 hours of daily sunlight. They made little growth, but they responded well when released after periods of 1 to 9 years. Green ash, water hickory, and sugarberry survived equally well and outgrew the oaks after the overstory was removed.

Additional keywords: *Quercus nuttallii*, green ash, *Fraxinus pennsylvanica*, water hickory, *Carya aquatica*, sugarberry, *Celtis laevigata*, stand development.

Johnson, Robert L.

1975. Natural regeneration and development of Nuttall oak and associated species. South. For. Exp. Stn., New Orleans, La. 12 p. (USDA For. Serv. Res. Pap. SO-104)

Nuttall oaks from a dense seedling catch survived for 5 to 10 years in complete shade and for 15 years with 1 or 2 hours of daily sunlight. They made little growth, but they responded well when released after periods of 1 to 9 years. Green ash, water hickory, and sugarberry survived equally well and outgrew the oaks after the overstory was removed.

Additional keywords: *Quercus nuttallii*, green ash, *Fraxinus pennsylvanica*, water hickory, *Carya aquatica*, sugarberry, *Celtis laevigata*, stand development.

Natural Regeneration and Development
Of Nuttall Oak and Associated Species

Robert L. Johnson [1]

Nuttall oaks from a dense seedling catch survived for 5 to 10 years in complete shade and for 15 years with 1 or 2 hours of daily sunlight. They made little growth, but they responded well when released after periods of 1 to 9 years. Time of release did not affect subsequent survival or growth. Green ash, water hickory, and sugarberry survived equally well and outgrew the oaks after the overstory was removed. After 15 years, the largest ashes were 5 to 10 feet taller and 1.5 to 2 times larger in diameter than most Nuttalls. On the better sites and where openings were one-fifth acre or larger, trees emerged from the vines in 5 to 10 years.

Additional keywords: *Quercus nuttallii*, green ash, *Fraxinus pennsylvanica*, water hickory, *Carya aquatica*, sugarberry, *Celtis laevigata*, stand development.

Most bottom-land hardwood forests in the Midsouth are mixtures of many species and age or size classes. Conditions are the results of single-tree selection cutting in stands under some management and of diameter-limit cutting in unmanaged stands. In many forests, frequent harvesting has greatly decreased the proportion of oaks. This decline is significant, since oaks are the species most used by southern hardwood industries.

The study reported here offers strong evidence that silviculture must become more intensive than at present, if oaks are to continue to predominate in bottom-land hardwood forests. The data apply specifically to Nuttall oak *(Quercus nuttallii* Palmer), a red oak that is common to the Mississippi Delta and adjacent river bottoms. But ongoing research indicates that similar results can be expected with other oaks on other sites.

THE STUDY

The study was in a mixed stand of uneven-aged hardwoods on the Delta Experimental Forest near Stoneville, Mississippi. Soils were deep clay, classified as either Sharkey or Alligator. There

[1] The author is Principal Silviculturist at the Southern Hardwoods Laboratory, which is maintained at Stoneville, Mississippi, by the Southern Forest Experiment Station, Forest Service—USDA, in cooperation with the Mississippi Agricultural and Forestry Experiment Station and the Southern Hardwood Forest Research Group.

were three major sites, differing mainly in elevation: ridges, wit[
the sweetgum-water oak combination; flats, with the elm-ash-hack[
berry-Nuttall oak association; and sloughs, with the overcup oak[
bitter pecan type. On such sites, difference in elevation betwee[
ridges and sloughs may be no more than 6 feet, while ridges an[
flats are almost indistinguishable to the uninitiated.

Between November 1957 and May 1958, unusual circumstance[
resulted in the establishment of thousands of Nuttall oak seedling[
directly under large seed trees. A very good crop of acorns mature[
in the fall of 1957, and rainfall from November through May wa[
heavy. Water standing on the study area protected and stratifie[
the acorns and provided a moist seedbed. High June temperature[
triggered rapid germination before the surface soil dried and becam[
unsuitable for seedling establishment (fig. 1).

Fifty stands of Nuttall reproduction were selected over a 40[
acre area. All were directly under seed trees that averaged 60 t[
80 years old, 100 feet tall, and 16 to 24 inches in d.b.h. A full[
shaded 4-milacre plot, 13.2 by 13.2 feet, was located in each stand[

Figure 1.—*Typical dense stocking of 1-year-old Nuttall oak seedlings o[
study plots. Inset: Seedbed conditions were so favorable tha[
some seedlings originated from acorns lying on the surface
normally acorns must germinate underground if seedlings ar[
to become established.*

2

lots had from 211 to 1,521 Nuttalls; the average was 552, or the equivalent of 138,000 trees per acre.

Initially, five treatments were assigned at random: release after 1, 2, 3, and 4 years, and no release (check plots). Each treatment was applied to 10 plots. As the study progressed, however, three of the checks were found to have consistently heavier seedling stands than the other 47 plots. When it had become clear that 4-year-old Nuttalls would respond to release, these three plots were released after 9 years. In the rest of this paper, they are considered as a sixth treatment, and information about the check treatment is from seven plots only.

Annually for 12 years, and then after 15 years, trees on each plot were counted and tallied by species. Heights of the 12 tallest oaks per plot were measured and tagged. As other oaks became dominant, they too were tagged and measured. After the first year, plastic expansion rings were placed around the base of each oak in a plot to maintain identity of original trees.

After the fifth year, heights of the tallest tree of each competing species were tallied on each plot. Starting a year later, conditions in and around each plot were described. Noted were size of opening, interspecies competition, weed and vine competition, growth characteristics by species, depth of standing water (relative soil moisture conditions), and rodent damage. In 2 separate years, size of opening was estimated by using a range finder to measure distance to the surrounding overstory.

Plots were released by cutting all competing trees 1 inch d.b.h. and larger during the dormant season. Diameter of the opening varied but usually was equal to at least half the height of the surrounding overstory.

RESULTS

Released Plots

Except for one or two cottonwoods in each of three plots, practically no new seedlings became established after release. But it became apparent immediately that the Nuttalls would have to compete with advance reproduction and sprouts of green ash *(Fraxinus pennsylvanica* Marsh.), water hickory *(Carya aquatica* (Michx. f.) Nutt.), sugarberry *(Celtis laevigata* Willd.), and a scattering of other species. Usually the advance reproduction—of which ash was the most common—was lopped back to ground level when the overstory was removed, but some trees 6 to 7 feet tall were left intact. While sprouts were mainly from cut reproduction, some were from saplings and poles (fig. 2).

3

Figure 2.—*Much of the green ash reproduction originated as sprouts from stumps of cut trees. All sprouts in these photos are 6 years old. Note that the largest emerge from the largest stump.*

Average reduction in stocking followed a similar trend over time for stands released from 1 through 4 years (fig. 3). Starting with 125,000 to 150,000 trees per acre, stocking declined about 95 percent in 15 years—to 9,000 trees per acre, of which 6,000 were Nuttall oaks. Approximately three-fourths of the oak mortality occurred during the first 7 years. Trees of other species accounted for less than 1 percent of stocking in 1958 but for about one-third in 1972. Half of all trees greater than 10 feet tall in 1972 were of species other than Nuttall.

All plots followed a similar pattern of development. Within a year or two after release, they were invaded by vines, mostly peppervine *(Ampelopsis arborea* (L.) Koehne) and trumpetcreeper *(Campsis radicans* (L.) Seem.). In 3 or 4 years, tree reproduction was completely covered by a dense mat of vines. The vines persisted until stands averaged 15 to 20 feet tall, or on the better sites for about 8 to 10 years after release. Plots with less than 10,000 trees per acre, those in openings of less than 1/5-acre, or those with poor moisture conditions were slow to emerge from the vines. Almost one plot in three was still in the vine stage 11 to 14 years after release.

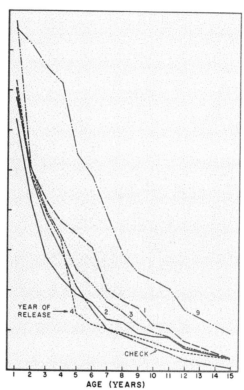

3.—*Stand density (all species) by year of release.*

Species varied in reaction to the vine competition. Nuttall oak and water hickory reproduction tended to be stout and slow to add height growth. Trees of these two species developed en masse, pushing the vines up until eventually the whole stand was clear. Both sprouts and advance reproduction of ash grew rapidly above the oaks and water hickory. Vines climbed individual ash trees and pulled the terminals over to within 3 or 4 feet of the ground. At this stage, it appeared that the ash would not survive or at best would develop very poor form. But within a year or two, a lateral became dominant at the point where the old terminal was bent over (fig. 4). The old terminal died and dropped off, and the bole had only a slight bend. Sugarberry reacted to vines about as ash did, but was not so consistent.

Figure 4.—*This ash terminal (indicated by the machete point) was pulled over by vines. A new leader took over at point of bend. The old terminal will soon drop off, leaving the main stem with a curve so slight that it will be of little consequence when the tree is harvested.*

Among the four most common species, green ash was the fastest grower and Nuttall oak the slowest. Growth varied considerably both within and between treatments (table 1). Eleven to 14 years after release, dominant ash were 30 to 35 feet tall and 3.0 to 3.5 inches d.b.h. or about 5 to 10 feet taller and 1½ to 2 times larger

6

Table 1.—*Average size and range,*[1] *after 15 years, of the tallest Nuttall oak per plot and tallest tree among competing species*

Age at release (years)	Plots	Height		Diameter at breast height	
		Oak	Competitor	Oak	Competitor
	No.	– – – – – – Feet – – – – – –		– – – – Inches – – – –	
1	10	7.1 **16.9** 34.0	8.9 **20.9** 33.0	0.1 **1.2** 2.6	1.0 **1.9** 3.7
2	10	7.0 **15.2** 25.0	16.2 **22.8** 36.0	.2 **1.0** 1.9	1.0 **2.0** 3.4
3	10	7.1 **13.5** 19.3	8.8 **20.5** 31.0	.3 **.8** 1.3	.7 **1.9** 2.9
4	10	2.5 **15.3** 25.0	11.6 **22.6** 38.0	.0 **1.0** 1.9	.7 **2.4** 5.1
9	3	14.8 **15.2** 15.5	19.5 **22.7** 25.5	.7 **.8** 1.0	1.6 **2.4** 3.6
Not released	7	2.2 **2.9** 3.3	2.6 **10.1** 13.5	...	...

[1] Values in bold face are averages, others are ranges.

in d.b.h. than most Nuttalls (fig. 5). Ash was dominant on 44 percent of the plots, sugarberry on 15 percent, Nuttall oak on 13 percent, water hickory on 5 percent, and other species on 23 percent. When trees outside as well as on the plots were considered, ash was either dominant or codominant with one or more other species on 71 percent of the plots. Sugarberry was dominant or codominant on 24 percent, Nuttall oak on 10 percent, and water hickory on 7 percent.

Annual growth was not uniform by species, years, or plots. Oak seedlings changed crown positions almost annually, generally because of vine development or a sudden spurt of growth. Few oaks retained dominance throughout the study.

Growth was best on low plots where water usually stood from December to mid-May or early June and where openings were at least 100 feet in diameter (fig. 6). Standing water on these plots probably had the same effect as in impoundments where, according to Broadfoot, diameter growth of Nuttall oak was increased by 38 percent when water was held until July 1. Broadfoot ascribed this effect to infiltration that increased the supply of soil moisture available after surface water was gone.[2]

Nine-year-old Nuttall oaks responded as well to release as 1- to 4-year-olds did. The most notable difference was that some trees died back to 4 to 6 inches above ground level and resprouted. Trees released late underwent the same kind and degree of competition but grew as well as or better than trees released earlier. The three plots released after 9 years were low in elevation, a circumstance that may account for the better survival (fig. 3). Also, release of the 9-year-old trees was more complete than the average release

[2] Broadfoot, W. M. **Shallow-water impoundment increases soil moisture and growth of hardwoods.** Soil Sci. Soc. Am. Proc. 31: 562-564. 1967.

7

Figure 5.—*Typical stand 13 years after the overstory was cut. The large, clearly dominant tree is green ash; Nuttall oaks are no more than half the diameter of the ash and 5 to 10 feet shorter.*

8

Figure 6.—*Opening in top photo, of approximately 1/10-acre, is too small for good development of reproduction. Opening in bottom photo is about ½-acre, minimum recommended size. Both openings are 6 years old. Larger trees in bottom photo are green ash.*

9

given to 1- to 4-year-old trees; at least ½-acre was cleared around all three plots.

Check Plots

Stocking on the seven check plots averaged 150,000 trees per acre during the fall of 1958. Average survival was similar to that on released plots. Four of the seven plots still had living Nuttalls after 10 years, and two after 15 years. Oaks survived beyond 8 to 10 years only where they received an hour or two of direct sunlight from an opening somewhere in the canopy. Soil moisture supplies also appeared important, for plots where seedlings survived for 15 years usually had standing water into the growing season.

The trees grew from ½ inch to 2 inches in height annually but did not die back and resprout. The largest living Nuttalls after 15 years were 3 feet tall. Diameters at the rootcollar were less than 0.5 inch, and growth rings could not be recognized. Root systems were small and in balance with tops. Except for the dark, rough appearance of the stems, 15-year-old Nuttalls resembled 1- to 2-year-old trees (fig. 7).

New Nuttall oak reproduction appeared in just two measurements, 1965 and 1966, and on only a few plots. None of the plots had more than 10 new trees.

Green ash, water hickory, and sugarberry demonstrated even more shade tolerance than the oaks by surviving in the understory for 15 years. In fact, most were older than 15 years, since they had been in the understory when the study began. Few new nonoak seedlings became established.

In complete shade, or with very little sunlight, green ash was the fastest-growing species, followed by water hickory and red maple. But even the ash did not exceed 6 inches of height growth in a year. Sugarberry survived but did not grow well in the understory. On some plots sugarberry was nipped back to the ground almost annually by rabbits. It was the only species to be so attacked.

There appeared to be a limit beyond which understory trees would not grow. Green ash and water hickory, for example, would reach 10 to 15 feet tall, then die back sometimes to the rootcollar, resprout, and regrow to about the same height.

DISCUSSION AND CONCLUSIONS

It is important, silviculturally, that seedlings of several hardwood species are highly tolerant of shade and will ultimately respond well to release. Advance reproduction of green ash, water hickory,

10

Figure 7.—*Nuttall oak on the left is 25 inches tall and has been in the understory on a check plot 15 years; Nuttall on the right is 21 inches tall and 1 year old from the nursery. The two trees differ considerably in color and texture of the bark.*

and sugarberry can be stored in the understory for more than 15 years; Nuttall oaks will endure for 5 to 10 years. All four species can be kept alive even longer by selective cutting of the overstory to allow 1 to 2 hours of daily sunlight to reach the understory. Green ash may grow up to 6 inches in height annually in the understory, whereas Nuttall may not exceed an inch or two. Root systems develop correspondingly. A larger root system supports the taller ash and results in more rapid shoot growth following release.

11

Harvesting of the overstory can be postponed when there is advance reproduction of tolerant species. Conversely, the new stand is already set, so that the type of harvesting or silvicultural system adopted will have little effect on the species composition that follows. New seedlings will be unimportant, since the few that become established grow too slowly to compete. The main consideration in selecting a harvesting system will be the size of opening created. An opening less than 1/5-acre is too small for good growth and development of reproduction. Moreover, small openings offer no advantage in reducing vines, since any opening large enough for tree growth will support vine growth. If tree development beyond 15 years is considered, openings of ½-acre or more—as recommended for upland hardwood forests in the Central States[1]—would appear suitable.

If Nuttall oak is to be a favored species, reproduction must be established before the overstory is removed. Because a good stand occurred only once in 15 years during the present study, there appears to be ample reason to manage Nuttall properly when it does occur in the forest. On sites where the species does well the forest manager has at least 5 years to release the reproduction. A single release may not suffice, since the seedlings grow more slowly than most common associates. After 11 to 14 years of release, sprouts of green ash, the most important competitor, may be two to three times as large in d.b.h. and nearly half again as tall as dominant oaks. The second release might best take place when dominant oaks average about 2 inches d.b.h. or, to judge from results in this study, in about 10 to 15 years on a moist site.

Microsites are important to species development but are difficult to recognize. Nuttall oaks will grow best in depressional areas where in a normal year water stands into the early part of the growing season. Condition of the larger Nuttalls is a good indicator of site potential. Data gathered in this study show that Nuttall oaks will sometimes start growing where they are not well suited. But when this happens, the young trees do not survive long with or without overstory release.

Vines are part of the ecological complex in bottom-land forests on Sharkey clay soil. They hinder, but do not halt, the development of reproduction. Species vary in their reaction to vines. Green ash and, to a lesser extent, sugarberry respond by sending up new leaders after the terminal has been pulled over. Nuttall oak and water hickory, which tend to grow more slowly, overcome the vines by sheer numbers. With good stocking, stands are usually out of the vine stage when dominants are 15 to 20 feet tall.

[1] Sander, I. L., and F. B. Clark. Reproduction of upland hardwood forests in the Central States. USDA Agric. Handb. 405, 25 p. 1971.

12